AF528374

Organic Chemistry

Transition from High School to College

Organic Chemistry

Transition from High School to College

Dipak K. Mandal
Former Professor of Chemistry
Presidency College/University, Kolkata

ALLIED PUBLISHERS PVT. LTD.

Noida • Mumbai • Kolkata • Chennai • Bangalore

ALLIED PUBLISHERS PRIVATE LIMITED

D-5, Sector-2, **Noida**–201 301
Ph. Nos.: 0120-4320295/2542557/4352866 • E-mail: delhi.books@alliedpublishers.com

17 Chittaranjan Avenue, **Kolkata**–700072
Ph.: 033-22129618 • E-mail: cal.books@alliedpublishers.com

15 J.N. Heredia Marg, Ballard Estate, **Mumbai**–400001
Ph.: 022-42126969 • E-mail: mumbai.books@alliedpublishers.com

No. 25/10, Commander-in-Chief Road, Ethiraj Lane (Next to Post Office)
Egmore, **Chennai**–600008
Ph.: 044-28223938 • E-mail: chennai.books@alliedpublishers.com

P.B. No. 9932, No. 15, 3rd Floor (Next to Vijaya Bank), 5th Cross, Gandhinagar,
Karnataka, **Bangalore**–560009, Ph.: 080-41530285 / 22386239
E-mail: bngl.journals@alliedpublishers.com / apsabng@airtelmail.in

Website: www.alliedpublishers.com

ISBN: 978-93-89934-94-6

Published by Sunil Sachdev and printed by Ravi Sachdev at Allied Publishers Pvt. Ltd.,
D-5, Sector-2, Noida–201 301

Preface

The study of organic chemistry is not simply learning a list of facts, but a way of thinking and understanding. During my teaching first year BSc (Honours) students for many years, I had gathered an impression that students did not learn organic chemistry in a way as it should be in their high school years. There seemed to be a disconnect between what high schools teach and what colleges expect. I had thought that I would address the issue at some point of time in future.

Now I have written this textbook on foundational organic chemistry addressed principally to advanced high school students, which will provide a seamless transition from high school to college. The book is organized into four parts, and each part is factored into several chapters.

The book starts with – what is organic chemistry? This chapter (Chapter 1) is an introduction and a guide to the subject as to what organic chemistry is, where it plays a role, its origin and development as a science, and what we are going to learn in our study.

Part I deals with the structure of organic compounds, and consists of five chapters (Chapters 2–6). Chapter 2 makes us familiar with different classes of organic compounds based on functional groups. In Chapter 3, we come to the theory of bonding in organic molecules. Then in Chapter 4, we begin to think about structures of organic molecules in three dimensions, that is, stereochemistry. In Chapter 5, we see how structure of organic molecules relates to their physical properties. The last chapter of Part I is spectroscopy, which introduces mass spectrometry (MS), UV, IR and NMR spectroscopy to determine the structure of organic molecules. Inclusion of a chapter on spectroscopy at this level might be debatable. However, the chapter focuses more on the application (which is not difficult to understand) rather than its theory. I believe this inclusion will give the high school students a 'head-start' in organic chemistry.

The knowledge of structures leads logically to the reactivity of organic molecules in Part II, which covers the reactions of organic compounds. Part II consists of ten chapters (Chapters 7–16). At this level, I have organized chapter topics mainly with reference to classes of organic compounds that undergo addition, elimination, substitution or rearrangement reactions. I have also included some basic mechanisms in each chapter. Chapter 7 gives a general overview of classification and mechanisms of organic reactions. Addition reactions are described in Chapters 8 and 9. Chapter 8 presents addition reactions of alkenes and alkynes, while Chapter 9 describes addition reactions of aldehydes and ketones. Chapter 10 deals with elimination reactions of alkyl halides and alcohols. Substitution reactions are described in three chapters (Chapters 11–13). Chapter 11 introduces substitution reactions of alkanes and alkyl halides at a saturated carbon atom. Chapter 12 describes substitution reactions of carboxylic acids and their derivatives at carbonyl carbon atom, while substitution reactions of aromatic compounds are described in Chapter 13. Chapter 14 presents reactions of amines, diazonium compounds and phenols. Chapter 15 focuses on the chemistry of enolates, including the aldol and Claisen reactions. Finally, rearrangement reactions are described in Chapter 16.

Acquiring the knowledge of the structure and reactions, we proceed logically to Part III (Chapters 17 and 18), the synthesis of organic compounds. Organic synthesis is a creative chemistry, which needs planning and designing. A general strategy to design the synthesis of organic compounds is known as the retrosynthetic analysis. The retrosynthetic strategy is included briefly in Chapter 17. This inclusion, I believe, would be a huge plus for the students when they enrol in an advanced undergraduate course. Using the retrosynthetic approach, the synthesis of different classes of organic compounds including some simple and common drugs is presented in Chapter 18.

Finally, an overview of large molecules called macromolecules is included as special topics in Part IV (Chapters 19 and 20). Chapter 19 deals with mainly polymers of synthetic origin, commonly known as "plastics". Chapter 20 presents a brief introduction of four principal classes of biomolecules, namely, carbohydrates, lipids, proteins and nucleic acids.

As an element of learning and pedagogy, I have provided many in-chapter problems with detailed answers, reinforcing the main themes in the text. Besides, several end-of-chapter supplementary problems with a range of difficulty levels are provided for practice. A section on 'Further reading' is also included to enable the students to go further into the subject.

The approach presented in this book is distinct and class tested. Further, the chapters on various topics have been written in a way that will provide the teachers considerable latitude in when they cover them in their courses.

I hope advanced high school students and beginning undergraduates will find this book invaluable for their academic progression and also for competitive entrance examinations. Students in undergraduate medicine, pharmaceutics, polymer science and medicinal chemistry will also find this book very useful for a solid understanding in organic chemistry as it relates to their disciplines.

I would like to thank the editor and his colleagues at Allied Publishers for their excellent support and cooperation. Finally, I thank all members of my family for their interest and encouragement during the preparation of this book.

Dipak K. Mandal

Contents

PART–II
Reactions of Organic Compounds

PART–III
Synthesis of Organic Compounds

PART–IV
Special Topics

CHAPTER 1

What is Organic Chemistry?

"Organic chemistry nowadays almost drives me mad. To me it appears like a primeval tropical forest full of the most remarkable things, a dreadful endless jungle into which one does not dare enter for there seems to be no way out".

—Friedrich Wöhler 1835

Chemistry is the science of atoms and molecules. For its study, the subject has been traditionally divided into several branches. Organic chemistry, inorganic chemistry and physical chemistry are the three major divisions. Yet, it should be appreciated that these are just artificial divisions of an otherwise unified discipline of chemistry.

1.1 ORGANIC CHEMISTRY AND CARBON COMPOUNDS

The word 'organic' seems to attach some notion of life or naturalness to a substance. Thus the term **organic chemistry** was applied during the early nineteenth century to the study of substances related to living or once-living organisms. In contrast, the term **inorganic chemistry** was used to the study of substances such as rocks and minerals that have no life.

As times passed, chemists realized that compounds containing carbon atoms actually share many of the same properties. After isolating and characterizing the vast number of different substances, **organic chemistry has been defined as the chemistry of carbon-containing compounds**. Such a classification, however, does break down in certain cases. Materials that are salts of carbonates such as sodium carbonate (Na_2CO_3) and salts of cyanides such as potassium cyanide (KCN) and gases such as carbon monoxide (CO) and carbon dioxide (CO_2) are considered as inorganic compounds, but almost all other compounds containing carbon atoms fall into the classification of organic compounds. The list of organic compounds is endless; millions of organic compounds are known, either as a result of discovery in nature or preparation in the laboratory.

Carbon is not the only atoms present in organic molecules; hydrogen atoms are almost always present. Besides, organic molecules often contain oxygen and nitrogen atoms, some contains phosphorus or sulphur atoms as well, and some also have halogen atoms (F, Cl, Br, and I). These are the *main elements* of organic compounds. However, many more elements in the periodic table, particularly lithium, boron, silicon, magnesium, and transition metals like zinc, copper, tin and palladium have also been incorporated into organic compounds.

At the molecular level, life runs on organic chemistry. Molecules of life constitute very large organic molecules that may contain hundreds or thousands of atoms, called biomolecules. Deoxyribonucleic acid (DNA) is the giant biomolecule that contains the genetic information for all living organisms. Proteins are biomolecules that are found in all living cells, and make up our blood, muscle and skin. Enzymes are also proteins that catalyse the reactions in our bodies and run the chemistry of life.

1.2 ORGANIC CHEMISTRY IS ALL AROUND US

Organic chemistry is everywhere in our everyday life. They are present in our food, domestic materials and commercial products.

The clothes we wear are made up of organic compounds that are polymers, either natural such as cotton and wool, or synthetic such as polyester and nylon. Different types of plastics that we use are synthetic polymers. Soaps and detergents that are used as cleansing agents are also organic compounds.

Petrochemicals are important organic chemicals that are derived from crude oil. Fractional distillation separates the crude oil into different fractions; one such fraction is gasoline or petrol that propels our automobiles. Simple hydrocarbons are also available as fuels, such as methane (natural gas), and propane and butane (liquefied petroleum gas, LPG).

We experience a variety of colours, tastes and smell in our daily life. The colourful dyes and coatings are organic. The common sugar that has a sweet taste is sucrose, an organic compound isolated from sugarcane. The familiar flavouring agents or perfumes are organic substances. For example, menthol is a flavouring compound obtained from the essential oil of spearmint, and *cis*-jasmone is a perfume distilled from jasmine flowers.

Most of the drugs that we use to get relief of our suffering from various diseases are organic. These include the most commonly used analgesics such as aspirin and paracetamol, and antibiotics such as penicillins that are our defence against bacteria.

Thus organic chemistry is associated with almost every aspect of our lives. As you will learn organic chemistry, you will appreciate how remarkable it is.

1.3 THE ORIGIN OF ORGANIC CHEMISTRY

As a science, organic chemistry is about 200 years old. Its origin dates back to the early part of the nineteenth century. Since all organic compounds studied during that period came from living sources, either plant or animal, there was a strong belief that such compounds possessed some special **"vital force"**. It was proposed that the intervention of a vital force was necessary for the formation of organic compounds, and this vitalism was a key difference between organic and inorganic substances. Most chemists at the time also believed that synthesis of compounds possessing vital force could take place only in living organisms and could not take place in the laboratory from inorganic materials.

The vital force theory was found to be erroneous afterwards and therefore rejected. In 1828, the German chemist Friedrich Wöhler accomplished the first laboratory synthesis of **urea** (an organic compound isolated from urine) from **ammonium cyanate** (an inorganic compound) which is the ammonium salt of cyanic acid (HOCN). This experiment marked the beginning of the *end of the vital force theory*.

$$\underset{\textbf{Ammonium cyanate}}{NH_4^{+}\ {}^{-}OCN} \xrightarrow{\text{Heat}} \underset{\textbf{Urea}}{NH_2CONH_2}$$

Subsequently, a number of compounds from organic sources were prepared from inorganic substances during 1828–1850. And with the demise of the vital force theory, the science of organic chemistry began to flourish since 1850.

1.4 ORGANIC CHEMISTRY AND THE STRUCTURAL THEORY

An important observation in the Wöhler experiment was that ammonium cyanate and urea have the same molecular formula (CH_4N_2O). It was generally believed by most chemists at the time that a molecular formula is *unique* for a particular compound. But that idea failed to account for the Wöhler's observation, and different compounds that have the same molecular formula are since known to be **isomers**.

Between 1858 and 1861, a new idea called the **structural theory** was developed by August Kekulé, Archibald Couper and Alexander Butlerov as one of the most fundamental theories in chemistry. The theory allowed the chemists to solve the **problem of isomerism** in organic molecules.

The structural theory addressed two fundamental questions:

1. **What is the combining ability or valence of the carbon atom?**

A carbon atom can make four bonds, that is, **carbon is tetravalent**. Similarly, oxygen is divalent, and hydrogen is monovalent. A bond is represented by a line.

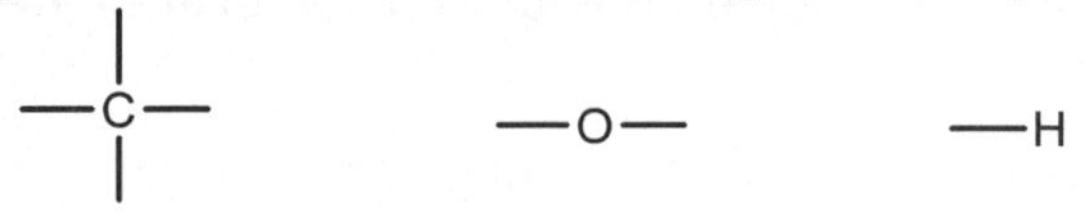

—C— —O— —H

Tetravalent **Divalent** **Monovalent**

2. **How is a carbon atom connected to other carbon atoms?**

A carbon atom can use one or more of its valence to form bonds to other carbon atoms. Thus there can be a **single bond**, a **double bond** or a **triple bond** between two carbon atoms.

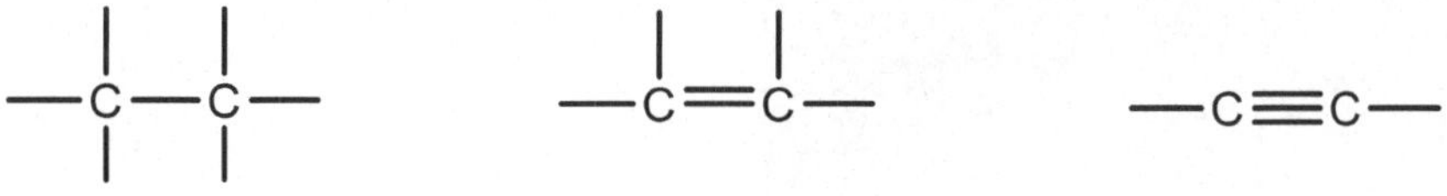

—C—C— —C=C— —C≡C—

Single bond **Double bond** **Triple bond**

It was Kekulé and Couper who independently pointed out that the high percentage of carbon atoms and the structural complexity of organic compounds could be explained if carbon atoms are bonded to other carbon atoms.

Problem of Isomerism

Ethyl alcohol and **dimethyl ether** are two different compounds that have the same molecular formula (C_2H_6O) but different physical properties. For example, ethyl alcohol is a liquid at room temperature and has a boiling point of 78.5°C, whereas dimethyl ether is a gas at room temperature with a boiling point of –24.9°C. As these compounds have the same molecular formula, they are **isomers**.

The **structural theory** can explain the isomerism in terms of the connectivity between atoms. The crucial difference is: **ethyl alcohol has a C–C–O linkage, while dimethyl ether has a C–O–C linkage**. The two carbon atoms are tetravalent, the oxygen atom is divalent, and all hydrogen atoms are monovalent in each compound.

```
     H   H                    H         H
     |   |                    |         |
  H—C—C—O—H              H—C—O—C—H
     |   |                    |         |
     H   H                    H         H
```

(C_2H_6O) **Ethyl alcohol**

(C_2H_6O) **Dimethyl ether**

Three-Dimensional Model of Molecular Structure

The structural theory gives molecular structures that represent the connection of atoms in a two-dimensional space (e.g. in the plane of the paper). The extension of the ideas of the structural theory to **three-dimensional space** was made independently by J. H. van't Hoff and J. A. Le Bel in 1874. They proposed a **tetrahedral model** for the four single bonds to a carbon atom.

Consider, for example, **methane** (CH_4). According to this model, the four C–H bonds are so arranged in the three-dimensional space that they point toward the corners of a regular tetrahedron, with the carbon atom at its centre, as shown in Figure 1.1.

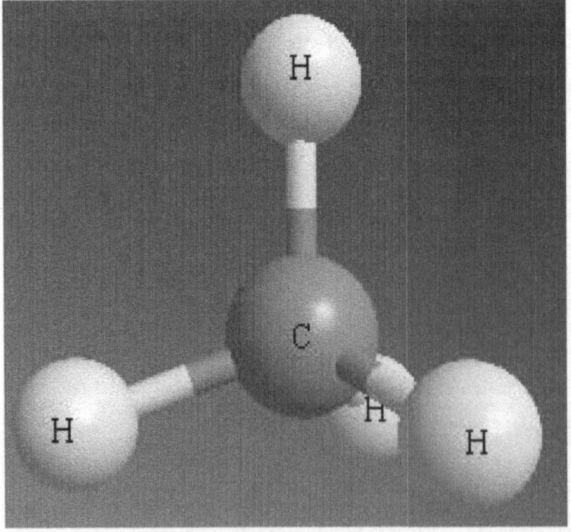

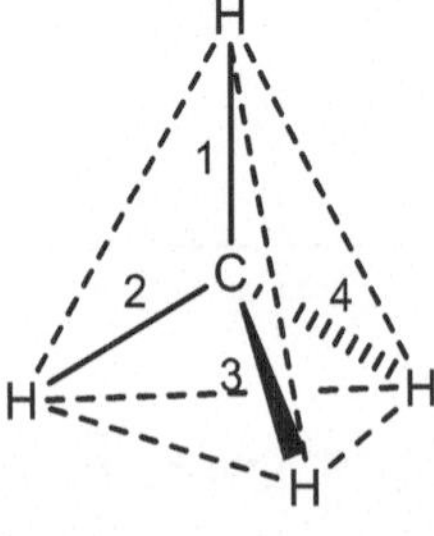

Figure 1.1: The tetrahedral structure for methane: (left) a ball and stick model; (right) a three-dimensional or perspective structure, in which four hydrogen atoms occupy the four corners of a regular tetrahedron. The four C–H bonds are labelled 1, 2, 3, and 4. Bonds 1 and 2 (line bonds) lie on the plane of the paper, bond 3 (wedged bond) is above the plane, and bond 4 (hashed bond) is below the plane.

A ball and stick model of the tetrahedral structure for methane is shown on the left, while a three-dimensional or perspective structure resembling the model is shown on the right.

In the perspective structure, two C–H bonds are in the plane of the paper when one C–H bond must be above the plane and the remaining C–H bond lies below the plane.

Cyclic or Ring Structure

Benzene is an organic compound which has been known since 1825. In 1834, the German chemist Eilhardt Mitscherlich showed that benzene has molecular formula C_6H_6. The molecular formula itself looked very surprising (the number of hydrogen atoms being the same as the number of carbon atoms), since most compounds that were known then had a far greater proportion of hydrogen atoms. For a long time, the structure of benzene had eluded chemists, although many possible structures for the compound had been proposed.

In 1865, August Kekulé proposed that carbon atoms can join to one another to form **chains**, but sometimes the **carbon chains can be closed to form rings**. With that idea, Kekulé offered an answer to the problem of the structure for benzene.

The Kekulé's structure of benzene was a **cyclic or ring structure**, which is represented as,

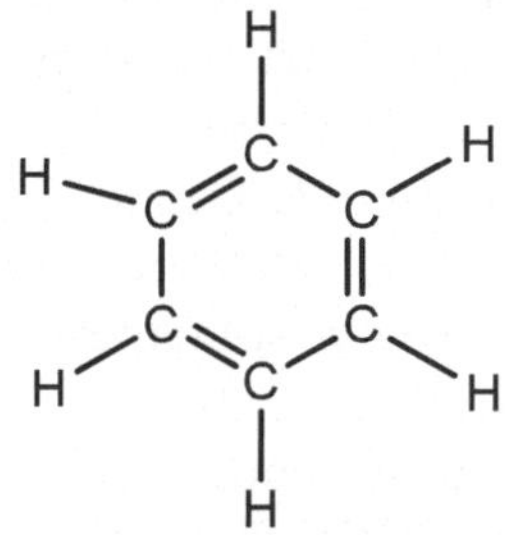

(C_6H_6)

Kekulé structure for benzene

According to a legend, the structure of benzene appeared to Kekulé in a dream. In his dream, Kekulé saw a snake devouring its own tail, forming a ring. When he awoke, he thought that benzene must be a ring, and worked out the structure of the compound. As Kekulé described it many years later (1890):

> *"...the atoms were gambling before my eyes...all twisting and turning in snake-like motion. But look! What was that? One of the snakes had seized hold of its own tail, and the form whirled mockingly before my eyes. As if by a flash of lightning I woke;... I spent the rest of the night working out the consequences of the hypothesis. Let us learn to dream, gentlemen, and then perhaps we shall learn the truth."*

1.5 ORGANIC CHEMISTRY AND THIS BOOK

Organic chemistry as a subject for study is best divided into three major themes: **structure**, **reactions** and **synthesis**. Along this line of division, this book is organized into three parts (Parts I–III). In addition, the book includes special topics which are covered in Part IV.

The book starts with an introductory chapter (Chapter 1):

1. What is organic chemistry?

   This chapter is an introduction and a guide to the subject as to what organic chemistry is, where it plays a role, its origin and development as a science, and what we are going to learn in our study.

   Part I of this book refers to the **structure of organic compounds** and it consists of five chapters (Chapters 2–6).
2. Organic compounds: structural formulas, functional groups, classification and nomenclature
3. Bonding in organic molecules
4. Stereochemistry
5. Physical properties, acidity and basicity
6. Spectroscopy

Organic chemistry is a network of interconnecting ideas and knowledge. Thus a particular chapter cannot be read in isolation; it requires some knowledge of organic compounds presented in other chapters. To this end, **Chapter 2** makes us familiar with different classes of organic compounds based on functional groups, their structural formulas and nomenclature. Then in **Chapter 3**, we come to the theory of bonding, resonance and aromaticity. Thereafter, in **Chapter 4**, we begin to think about structures of organic molecules in three dimensions, and understand the concept of chirality, configuration and conformation. In **Chapter 5**, we see how structure of organic molecules relates to their physical properties, acidity and basicity. Finally, in **Chapter 6** (the last chapter of Part I), we introduce spectroscopy as the most important experimental technique for determining the structure of organic molecules. This chapter deals with mass spectrometry (MS), ultraviolet-visible (UV-Vis) spectroscopy, infrared (IR) spectroscopy and nuclear magnetic resonance (NMR) spectroscopy.

With the knowledge of the structure, we then proceed logically to the reactivity of organic molecules in **Part II**, which describes the **reactions of organic compounds**. This part contains ten chapters (Chapters 7–16).

7. Introduction to organic reactions: classification and mechanisms
8. Alkenes, alkynes and conjugated dienes: addition reactions
9. Aldehydes, ketones and conjugated carbonyl compounds: addition reactions
10. Alkyl halides and alcohols: elimination reactions
11. Alkanes, alkyl halides, alcohols and ethers: substitution reactions
12. Carboxylic acids and their derivatives: substitution reactions
13. Aromatic compounds: substitution reactions
14. Amines, diazonium salts and phenols: substitution and elimination reactions
15. Reactions of enolates: the aldol and Claisen reactions
16. Rearrangement reactions: 1, 2-shifts and aromatic rearrangements.

Organic reactions can be classified as addition, elimination, substitution and rearrangement. On the molecular level, a particular class of reactions (say, addition) may proceed by different mechanisms depending on the type of functional group present in organic compounds. For example, both alkenes and ketones undergo addition reactions; however, alkenes react mainly by *electrophilic addition* mechanism, whereas ketones react by *nucleophilic addition* mechanism. However, at this introductory level, we have desired to organize chapter topics mainly with reference to different classes of organic compounds that undergo addition, elimination, substitution or rearrangement reactions. We have introduced some basic mechanisms in each chapter; however, detailed mechanistic discussion is beyond the scope of this text.

We begin with a brief overview of classification and mechanisms of organic reactions in **Chapter 7**. **Chapter 8** introduces addition reactions of alkenes, alkynes and conjugated dienes, while **Chapter 9** presents addition reactions of aldehydes, ketones and conjugated carbonyl compounds. **Chapter 10** deals with elimination reactions of alkyl halides and alcohols. Substitution reactions are described in three chapters (Chapters 11–13). **Chapter 11** introduces substitution reactions of alkanes, alkyl halides, alcohols and ethers that take place at a *saturated carbon atom*. **Chapter 12** describes substitution reactions of carboxylic acids and their derivatives that take place at an *unsaturated carbon atom* (*carbonyl carbon*). Substitution reactions of aromatic compounds involving substitution at the *unsaturated carbon atom of the benzene ring* are described in **Chapter 13**. **Chapter 14** presents reactions of amines, diazonium salts and phenols which involve various substitution and elimination reactions. Importantly, a carbonyl group can impart a special reactivity pattern with reference to the acidity of α-hydrogen atoms, which leads to the formation of *enolate ions*. **Chapter 15** focuses on the reactions of enolates, particularly the aldol reaction and the Claisen reaction, which provide important methods for making carbon-carbon bonds. The chapter also includes reactions of enolates derived from acetoacetic ester and malonic ester, which are highly useful in organic synthesis. Finally, rearrangement reactions are included in **Chapter 16**. The most common types of rearrangement reactions involve migration of a group from one atom to the next atom within the molecule (called 1, 2-shifts) and rearrangements within an aromatic ring (called aromatic rearrangements).

Acquiring the knowledge of the structure and reactions, we then proceed logically to **Part III**, the **synthesis of organic compounds**. Part III consists of two chapters (Chapters 17 and 18).

17. Synthetic strategy: retrosynthetic analysis.
18. Synthesis of organic compounds: the retrosynthetic approach.

Organic synthesis is a creative chemistry, that is, to create or make molecules. A general *strategy* to designing the synthesis of organic compounds is known as the *retrosynthetic analysis*. The most important question when faced with the problem of preparing a new compound is: what will be the starting materials or the compounds to start with? Retrosynthetic analysis can provide that answer. In retrosynthetic analysis, we *work backward* from the product towards the starting materials using our knowledge of *known and reliable reactions*. **Chapter 17** presents a brief overview of retrosynthetic analysis as the synthetic strategy. Using the retrosynthetic approach, **Chapter 18** describes the synthesis of different classes of organic compounds including difunctionalized compounds and cyclic molecules. The chapter also includes the synthesis of some simple and important drugs.

So far, we have talked about rather small molecules in our study of organic chemistry. However, there are also very large molecules containing hundreds and thousands of atoms. These are called macromolecules, which are usually polymers consisting of many repeating units or monomers. An overview of these macromolecules, both natural and synthetic, is included as **special topics** in **Part IV**. Two chapters (Chapters 19 and 20) are devoted to this part.

19. Polymers and polymerization reactions
20. Biomolecules: carbohydrates, lipids, proteins and nucleic acids

Chapter 19 deals with mainly addition and condensation polymers of synthetic origin, commonly known as "plastics". The development of the processes by which these polymers are made has been responsible for the remarkable growth of the chemical industry. **Chapter 20** presents molecules of life, which are biopolymers or biomolecules. A brief introduction of four principal classes of biomolecules, namely, carbohydrates, lipids, proteins and nucleic acids is provided in this chapter.

So **this book** is about all these things. This is the landscape through which we are going to travel. The language of organic chemistry is all pictorial, and the journey ahead seems to be exciting, rewarding and sometimes challenging too.

PART-I

Structure of Organic Compounds

CHAPTER 2

Organic Compounds: Structural Formulas, Functional Groups, Classification and Nomenclature

IN THIS CHAPTER

- Conventions for drawing organic structures: Dash, condensed and bond-line formulas
- Lewis structures for organic molecules; formal charges and arrow formalism
- Functional groups and classification of organic compounds: Alkanes and cycloalkanes, alkenes and cycloalkenes, alkynes, arenes, alkyl and aryl halides, alcohols, phenols, ethers, amines, nitro compounds, aldehydes, ketones, carboxylic acids, acyl halides, acid anhydrides, esters, amides and nitriles
- Classification of carbon and nitrogen atoms: Primary, secondary, tertiary and quaternary
- Constitutional isomers
- Alkyl, alkenyl, alkynyl, aryl and acyl groups
- Common names and systematic IUPAC nomenclature for different classes of organic compounds

Organic compounds are called carbon compounds. Carbon has the unique ability to combine with other carbon atoms to form **carbon chains** or **carbon rings**.

Compounds containing only carbon and hydrogen are called **hydrocarbons**. Besides carbon and hydrogen, most organic compounds also contain oxygen, nitrogen or other atoms, which are linked to a carbon chain or carbon ring called a **hydrocarbon framework**. The carbon chains can be *linear* (straight-chain) or *branched.*

The open chain or acyclic compounds containing single, double or triple bonds but no rings are called **aliphatic** compounds. The closed chain or cyclic compounds containing usually a benzene ring are called **aromatic** compounds. If cyclic compounds are not aromatic, they are called aliphatic cyclic or **alicyclic** compounds. When a ring of the cyclic compounds contains one or more heteroatoms (such as oxygen, nitrogen or sulphur), they are called **heterocyclic** compounds.

2.1 CONVENTIONS OF DRAWING ORGANIC MOLECULES

Organic molecules can be drawn in several ways as: dash structures, condensed structures or bond-line structures.

- In **dash structures**, all C atoms and all C–H bonds are drawn.
- In **condensed structures**, the C–H bonds and often the C–C bonds are not shown.

- In **bond-line structures**, a C–C bond is indicated by a line and the chain of carbon atoms is drawn as zig-zag. The H atoms attached to carbon atoms are not shown; however, the H atoms attached to heteroatoms are always shown.

These three conventions of drawing organic structures are illustrated with examples in Figure 2.1.

Aliphatic Compound: Pentane

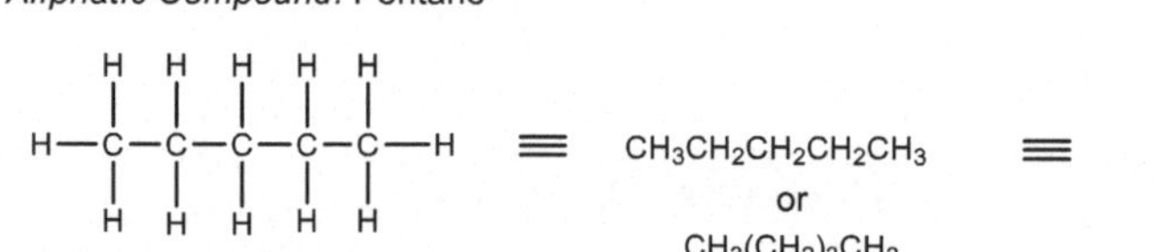

Dash formula **Condensed formula** **Bond-line formula**

Alicyclic compound: Cyclohexane

Dash formula **Condensed formula** **Bond-line formula**

Aromatic compound: Benzene

Dash formula **Condensed formula** **Bond-line formula**

Heterocyclic compounds: Furan and tetrahydrofuran

Dash formula **Bond-line formula** **Condensed formula** **Bond-line formula**

Furan *Tetrahydrofuran*

Figure 2.1: Structural representations of some organic compounds by dash formulas, condensed formulas and bond-line formulas.

Note: The above methods for drawing organic structures are **conventions**, *not* rules. So it is not unusual to see structures that involve a combination of two or three different drawing methods. Chemists often have good reasons for doing this, to emphasize or highlight certain important parts of a molecule.

For example, a structure of **ethylbenzene** is often drawn as a combination of bond-line formula for the ring and a condensed formula for the ethyl group, as shown here.

CH_2CH_3 ← **Condensed formula**

← **Bond-line formula**

Ethylbenzene

PROBLEM 2.1: Determine the number of hydrogens on the following bond-line structure of 1-butene and 2-butene:

1-Butene **2-Butene**

Answer

A carbon has four bonds. So, if a carbon on the bond-line structure has only one bond to other atom, it must have 3 hydrogens. Similarly, if a carbon has two, three or four bonds to other atoms, it must have 2, 1 and 0 hydrogens respectively.

2.2 LEWIS STRUCTURES

Lewis structures of organic molecules show *which* **atoms** are connected to each other, and *where* the **electrons** in the molecule reside. A Lewis structure is thus represented by **covalent bonds** between the atoms and **lone pairs (nonbonding pairs)** on the atoms on which they reside. These structures are named after the American chemist Gilbert N. Lewis, who proposed his model of chemical bonds in 1916.

In Lewis structures, the bonding electron pairs, together with lone pairs, give *each* **carbon, nitrogen, oxygen** or **halogen** atom an **octet of electrons** (eight electrons) in the **valence shell**, which correspond to the stable electronic configuration of the noble gas. This is known as the **octet rule**.

Notably, the octet rule applies to carbon, nitrogen, oxygen and halogens (F, Cl, Br, I). But a **hydrogen atom** can get a **duet** (two electrons) from one covalent bond or bond pair in its valence shell, which also corresponds to the electronic configuration of a noble gas, helium.

Lewis structures are routinely used to describe organic reactions and their mechanisms. In that sense, they constitute the vocabulary to study organic chemistry. Drawing correct Lewis structures for organic molecules is thus of extreme importance.

2.2.1 Drawing Lewis Structures of Simple Molecules

To draw Lewis structures, **one shared electron pair** between two atoms is indicated by a **single bond** (–), **two shared pairs** by a **double bond** (=) and **three shared pairs** by a **triple bond** (≡). Heteroatoms (N, O or halogen atom) usually possess lone pairs of electrons. **Each lone pair** is shown as a pair of **two dots.**

Figure 2.2 shows the normal valence (or the number of bonds) and the number of lone pairs associated with hydrogen, carbon, nitrogen, oxygen and halogen atoms in organic molecules.

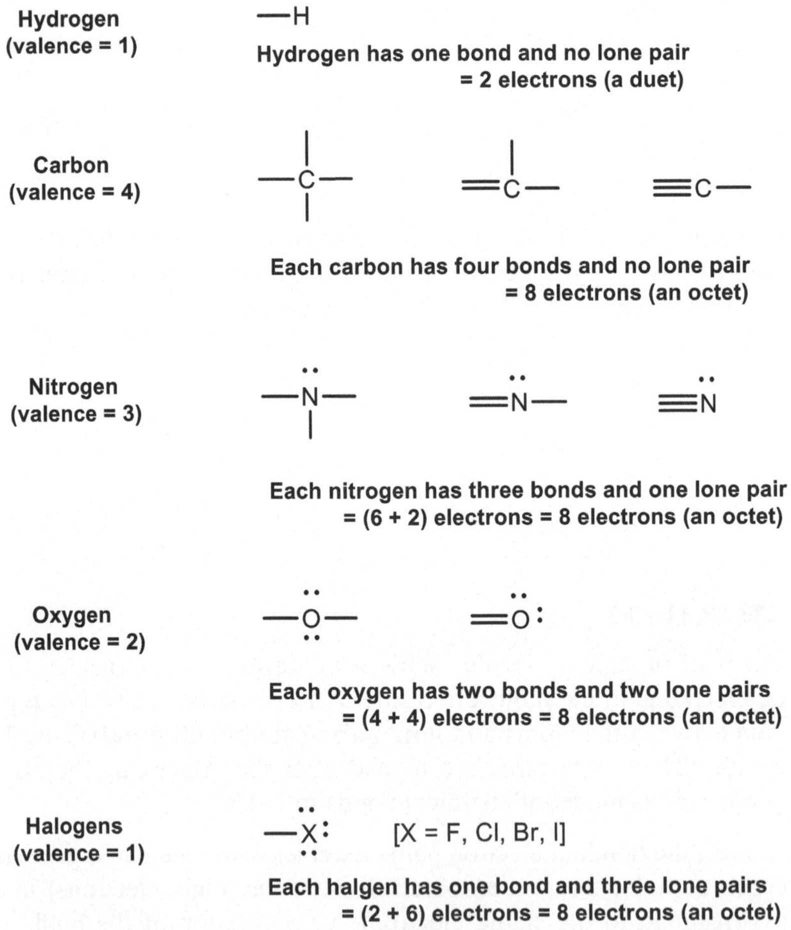

Hydrogen (valence = 1)

—H

Hydrogen has one bond and no lone pair = 2 electrons (a duet)

Carbon (valence = 4)

Each carbon has four bonds and no lone pair = 8 electrons (an octet)

Nitrogen (valence = 3)

Each nitrogen has three bonds and one lone pair = (6 + 2) electrons = 8 electrons (an octet)

Oxygen (valence = 2)

Each oxygen has two bonds and two lone pairs = (4 + 4) electrons = 8 electrons (an octet)

Halogens (valence = 1)

[X = F, Cl, Br, I]

Each halgen has one bond and three lone pairs = (2 + 6) electrons = 8 electrons (an octet)

Figure 2.2: Normal valence (number of bonds) and lone pairs for hydrogen, carbon, nitrogen, oxygen and halogen atoms. A lone pair is indicated by a pair of two electron dots.

To illustrate, the Lewis structures of some simple molecules in which the atoms have their normal valence are shown in Figure 2.3.

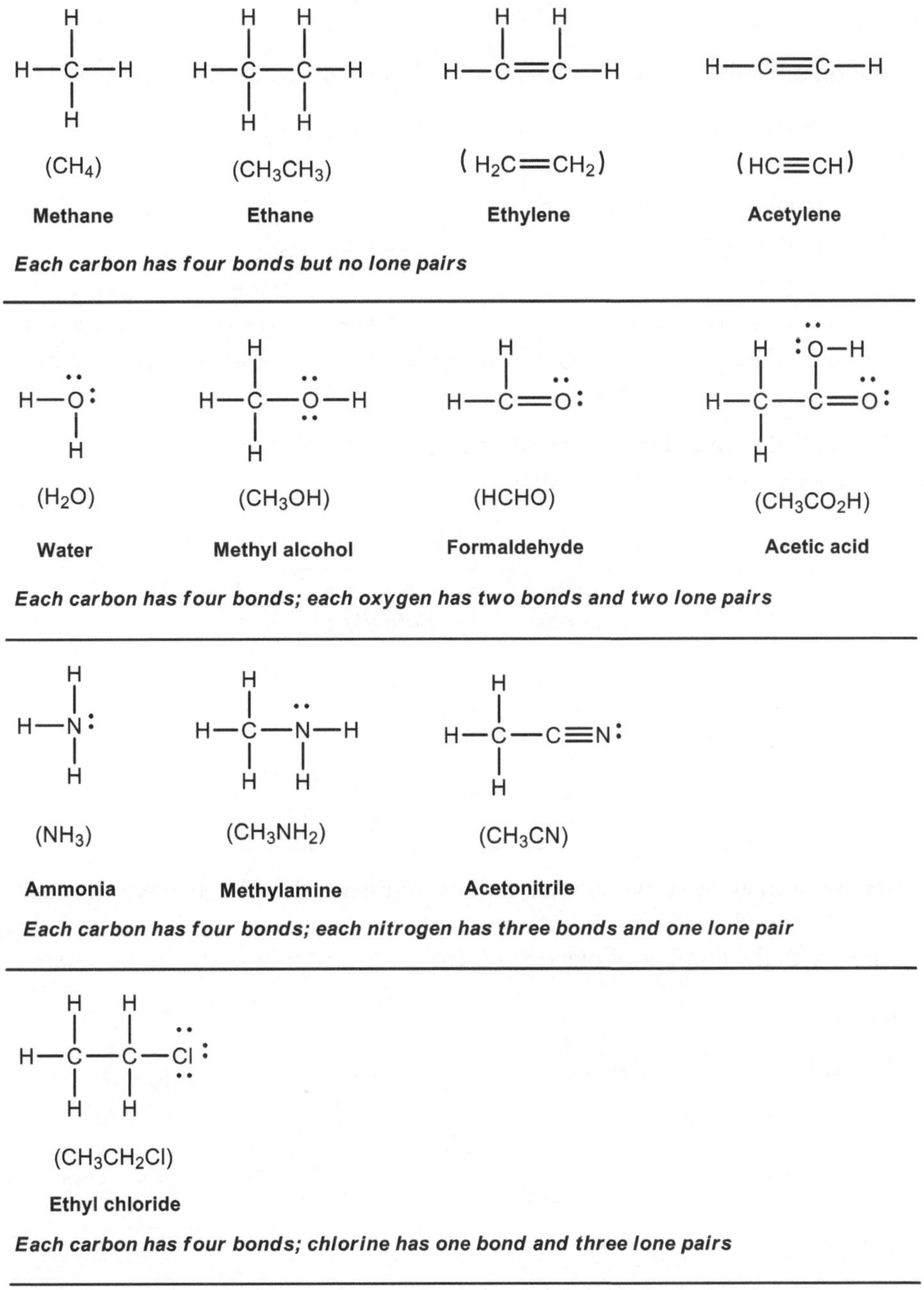

(CH_4) **Methane** (CH_3CH_3) **Ethane** ($H_2C{=}CH_2$) **Ethylene** ($HC{\equiv}CH$) **Acetylene**

Each carbon has four bonds but no lone pairs

(H_2O) **Water** (CH_3OH) **Methyl alcohol** ($HCHO$) **Formaldehyde** (CH_3CO_2H) **Acetic acid**

Each carbon has four bonds; each oxygen has two bonds and two lone pairs

(NH_3) **Ammonia** (CH_3NH_2) **Methylamine** (CH_3CN) **Acetonitrile**

Each carbon has four bonds; each nitrogen has three bonds and one lone pair

(CH_3CH_2Cl) **Ethyl chloride**

Each carbon has four bonds; chlorine has one bond and three lone pairs

Figure 2.3: Lewis structures of some simple compounds having normal valence of atoms.

2.2.2 Formal Charge in Lewis Structures

If an atom has an *apparent* **abnormal number of bonds** in a Lewis structure, it will carry a **formal charge**. A formal charge on an atom is calculated as:

Formal charge = (number of valence electrons) – (number of bonds) – (2 × number of lone pairs)

The **sum** of all formal charges is equal to the **net charge** on the molecule or ion.

The calculation of formal charge is illustrated with the following cases for N and O atoms:

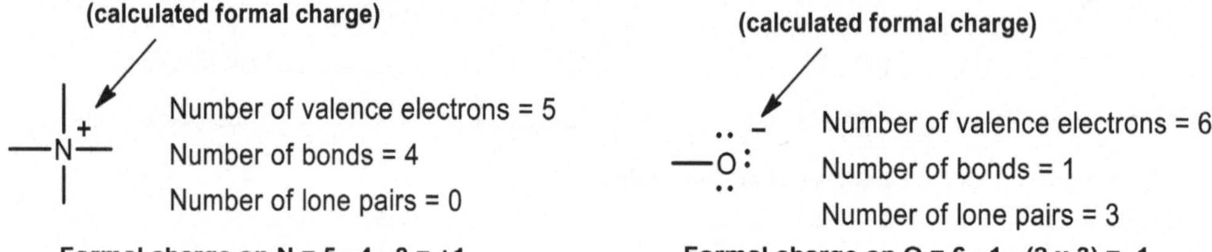

Formal charge on N = 5 - 4 - 0 = +1 **Formal charge on O = 6 - 1 - (2 x 3) = -1**

Remember that although nitrogen has 5 valence electrons, the **maximum valence of N is 4**, which gives an octet of electrons.

Table 2.1 shows the **calculated formal charges** on C, N and O atoms associated with different number of bonds and lone pairs.

Table 2.1: Formal charges on different atoms in Lewis structures

Atom	*Number of Bonds*	*Number of Lone Pairs*	*Formal Charge*
C	3	0	+1
	3	1	−1
N	4	0	+1
	2	2	−1
O	3	1	+1
	1	3	−1

The **Lewis structures** of some cations, anions and neutral molecules with proper formal charges are illustrated in Figure 2.4. Atoms with a positive or negative formal charge are also specified with the number of electrons (zero, duet, sextet or octet) in the valence shell.

Cations:

H^+	Methyl cation	Hydronium ion	Ammonium ion
(H has no electron)	**(C has 6 electrons or sextet)**	**(O has 8 electrons or octet)**	**(N has 8 electrons or octet)**
Hydrogen ion or proton	**Methyl cation**	**Hydronium ion**	**Ammonium ion**

Anions:

$H:^-$	$H—C≡C:^-$	$H—O:^-$	$H—N:^-$ (with H)
(H has 2 electrons or duet)	**(C has 8 electrons or octet)**	**(O has 8 electrons or octet)**	**(N has 8 electrons or octet)**
Hydride ion	**Acetylide ion**	**Hydroxide ion**	**Amide ion**

Molecules (net charge = 0):

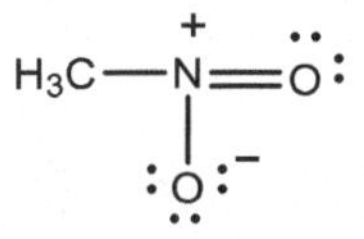

$H_3C{-}\overset{+}{N}(=\ddot{O}{:})({-}{:}\ddot{O}{:}^{-})$

(CH_3NO_2)

(C, N and each O having octet)

Nitromethane

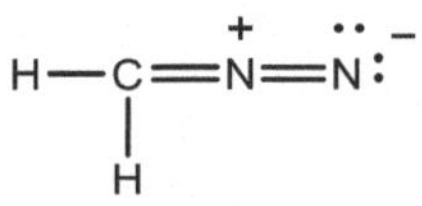

$H{-}C(H){=}\overset{+}{N}{=}\ddot{N}{:}^{-}$

(CH_2N_2)

(C and each N having octet)

Diazomethane

Figure 2.4: Lewis structures of some cations, anions, and neutral molecules with formal charges.

Note: In a Lewis structure, the formal charges on atoms must be shown, but the lone pairs on atoms may or may not be shown explicitly. If the lone pairs are not shown, it is assumed that they are there on the appropriate atoms. In some cases, only one of the lone pairs is shown when there is a reason to include this.

It may be pointed out that the *number of lone pairs on an atom* (when its formal charge and number of bonds are specified) can be easily determined using the above equation for formal charge or using Table 2.1.

2.2.3 Arrow Formalism and Multiple Lewis Structures

An organic molecule is often represented by *two or more* Lewis structures. The **arrow formalism** can be used to draw other possible structures from a given Lewis structure.

In the arrow formalism, a **curved or curly arrow** denotes the **movement of an electron pair** (bond pair or lone pair) from one position to another within the molecule. In the process, **the atom from which the electron pair is moving away gets a positive charge** (or no charge if it has already a negative charge), while **the atom to which the electron pair moves gets a negative charge** (or no charge if it has already a positive charge). The **net charge** in *each* of the Lewis structures must be the **same**.

The **double-headed arrows (↔)** are used to draw two or more Lewis structures of the same molecule. For example, two Lewis structures for **formaldehyde** are written as:

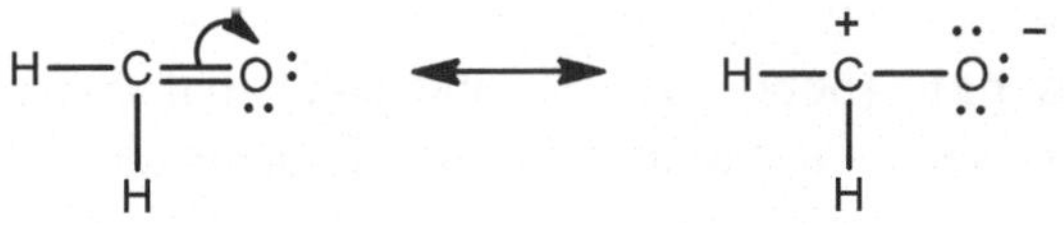

$H{-}C(H){=}\ddot{O}{:} \longleftrightarrow H{-}\overset{+}{C}(H){-}\ddot{O}{:}^{-}$

(net charge = 0) **(net charge = +1 -1 = 0)**

Two Lewis structures of formaldehyde

A **sequence of curved arrows** that involve the movement of bond pairs and lone pairs can be used to draw **multiple Lewis structures**, as shown in Figure 2.5.

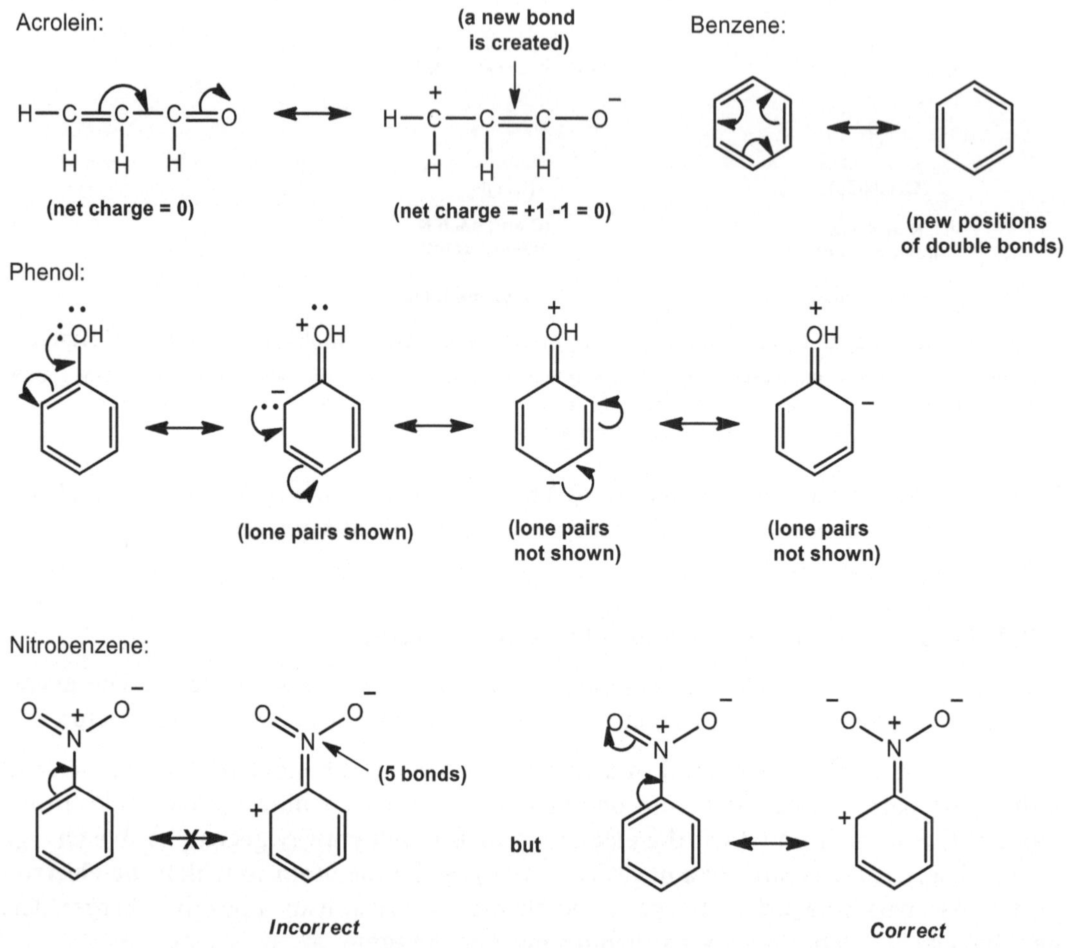

Acrolein:

(a new bond is created)

Benzene:

(net charge = 0)

(net charge = +1 -1 = 0)

(new positions of double bonds)

Phenol:

(lone pairs shown)

(lone pairs not shown)

(lone pairs not shown)

Nitrobenzene:

(5 bonds)

but

Incorrect

Correct

Figure 2.5: Drawing multiple Lewis structures for organic compounds (acrolein, benzene, phenol and nitrobenzene) using curved arrows.

As shown, an **electron pair** (bond pair or lone pair) that comes between two atoms **creates a bond** between them, and the **movement of bonds** along the chain or ring **shifts their positions**.

As lone pairs on atoms are not always explicitly shown, **a curved arrow may be drawn from a '–' charge on the atom,** which implies that the arrow is drawn from a lone pair residing on the same atom.

All atoms in a Lewis structure must have proper number of bonds. Thus, the Lewis structure for nitrobenzene in which **N has five bonds** is **incorrect** because nitrogen cannot be pentavalent.

PROBLEM 2.2: Draw the proper Lewis structure for each of the following:

(a) 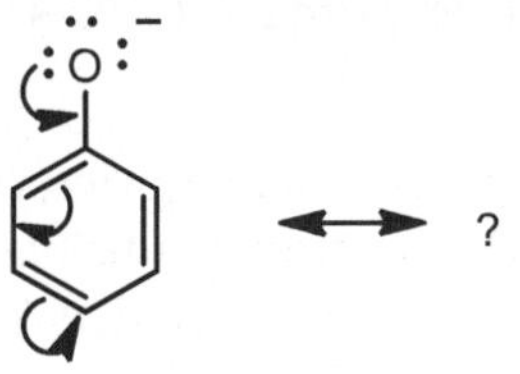 ⟷ ?

(b) 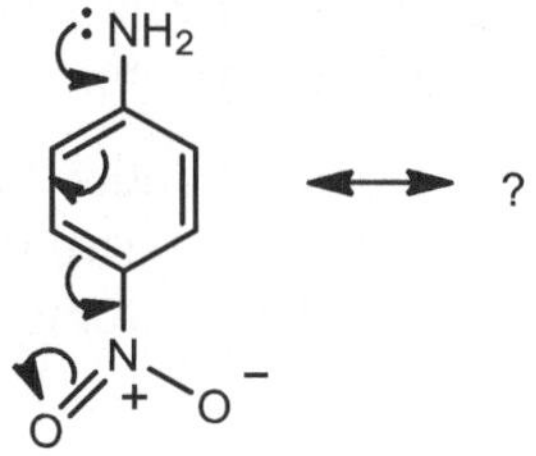 ⟷ ?

Answer

(a)

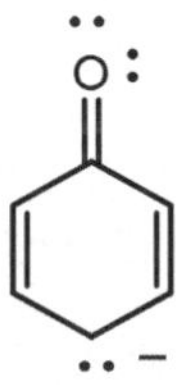

(b)

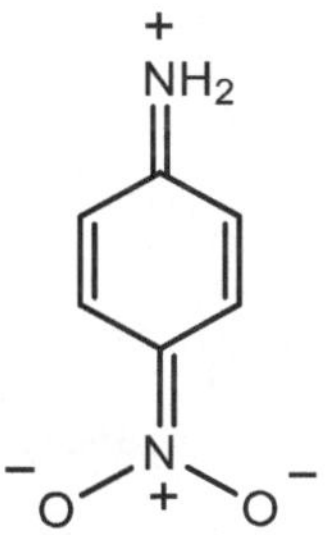

2.3 FUNCTIONAL GROUPS

The **reactivity** of an organic compound is usually determined by a **part** of the molecule comprising a **small group of atoms**. Since this small group dictates how a molecule will react or function, it is called a **functional group**. If we study the general reactions of a particular functional group, we can essentially learn the reactions of thousands of specific molecules which contain that functional group.

Based on functional groups, the vast majority of organic compounds can be divided into a relatively small number of **classes**. These are described in the following sections.

2.3.1 Alkanes: Classification of Carbons, Constitutional Isomers and Alkyl Groups

Alkanes are hydrocarbons in which each carbon atom is connected to a maximum of four atoms, i.e. each carbon is **saturated**. They contain **only C–C and C–H single bonds**, and do *not* possess a functional group.

In general, alkanes are much less reactive, and their reactions are extremely limited. Because of their lack of reactivity, alkanes are often used as **solvents**. However, alkanes can easily burn (react with oxygen) and hence are useful as **fuels**.

Alkanes have the general formula C_nH_{2n+2}. The simplest member (C_1 alkane) is **methane** (CH_4). The successive C_2, C_3, C_4, C_5 and C_6 members are called **ethane** (C_2H_6), **propane** (C_3H_8), **butane** (C_4H_{10}), **pentane** (C_5H_{12}) and **hexane** (C_6H_{14}).

A succession of alkanes differing by one methylene (CH_2) group is known as a **homologous series**. Thus the higher homologue of methane (CH_4) is ethane (C_2H_6) and the next higher homologue is propane (C_3H_8), and so on.

Classification of Carbon Atoms

A carbon atom in an alkane chain can be classified as primary (1°), secondary (2°), tertiary (3°) or quaternary (4°).

A carbon atom which is bonded to **only one** other C atom is called a primary (1°) carbon. Similarly, a secondary (2°) carbon is bonded to **two** other C atoms, a tertiary (3°) carbon is bonded to **three** other C atoms, and a quaternary (4°) carbon is bonded to **four** other C atoms.

It is of note that the single carbon in methane (CH_4) is unique; it is *not* a primary (1°) carbon.

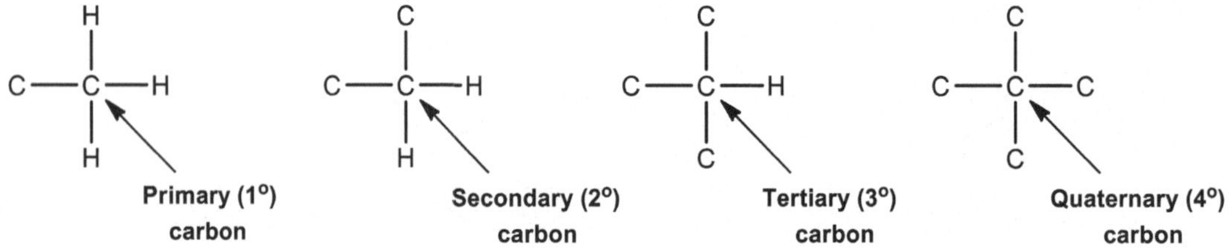

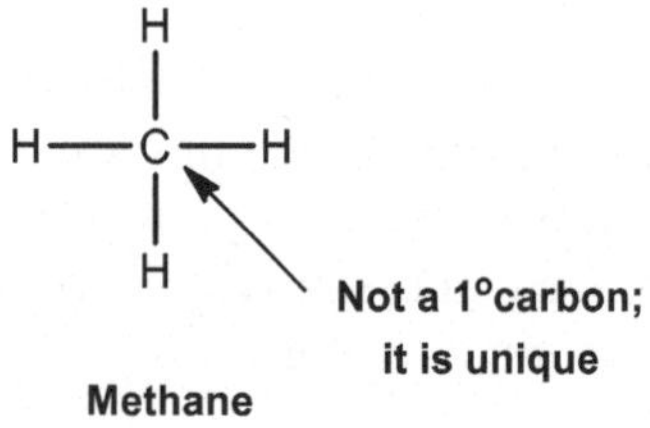

Constitutional isomers

A **carbon chain** in an alkane can be a **straight chain** or a **branched chain**. A straight chain is a single continuous chain that allows for going from one end to the other end through all carbon atoms; otherwise, the chain is branched.

A straight-chain alkane is used to be called a normal alkane or *n*-alkane. However, the prefix *n* is now generally omitted. For example, a **straight-chain C_4 alkane** is simply called **butane**.

A chain of three carbon atoms is necessarily a straight chain; the **branching** of carbon chain occurs from **C_4 alkane onwards**, which gives rise to the existence of **isomers**. Isomers are different compounds that have the same molecular formula.

A particular class of isomers is called constitutional isomers. Constitutional isomers are isomers that differ in the bonding connectivity of atoms.

Figure 2.6 shows the possible constitutional isomers and their common or trivial names for C_4 and C_5 alkanes. The C_3 alkane (which must be a straight-chain) has no constitutional isomers.

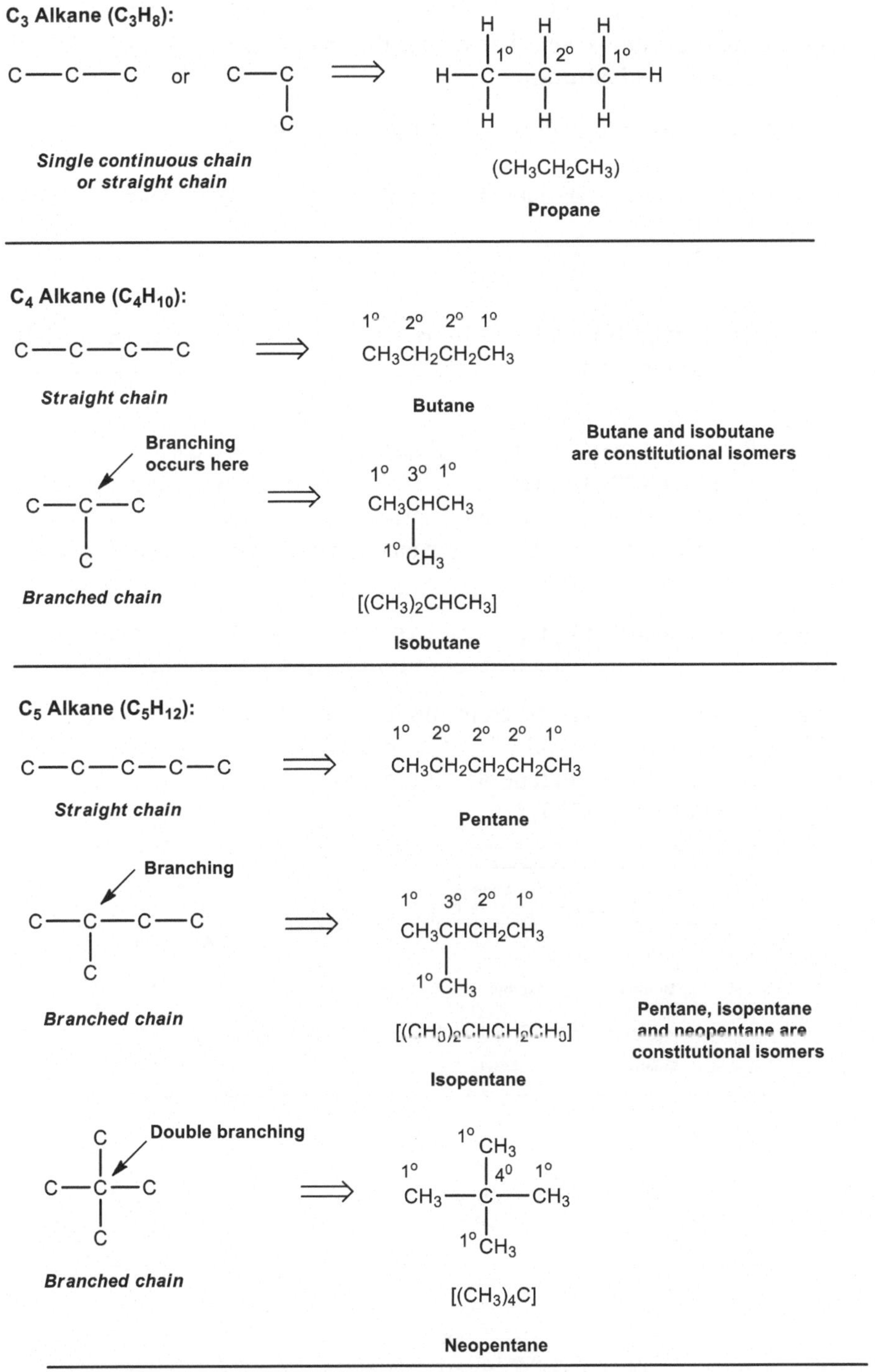

C_3 Alkane (C_3H_8):

C—C—C or C—C(—C) ⟹ H_3C—CH_2—CH_3 (1°, 2°, 1°)

Single continuous chain or straight chain

($CH_3CH_2CH_3$)

Propane

C_4 Alkane (C_4H_{10}):

C—C—C—C ⟹ 1° 2° 2° 1° $CH_3CH_2CH_2CH_3$

Straight chain

Butane

Branching occurs here

C—C(—C)—C ⟹ 1° 3° 1° CH_3CHCH_3 with 1° CH_3 branch

Branched chain

[$(CH_3)_2CHCH_3$]

Isobutane

Butane and isobutane are constitutional isomers

C_5 Alkane (C_5H_{12}):

C—C—C—C—C ⟹ 1° 2° 2° 2° 1° $CH_3CH_2CH_2CH_2CH_3$

Straight chain

Pentane

Branching

C—C(—C)—C—C ⟹ 1° 3° 2° 1° $CH_3CHCH_2CH_3$ with 1° CH_3 branch

Branched chain

[$(CH_3)_2CHCH_2CH_3$]

Isopentane

Pentane, isopentane and neopentane are constitutional isomers

Double branching

C—C(—C)(—C)—C ⟹ 1° CH_3—C(4°)(1° CH_3)(1° CH_3)—1° CH_3

Branched chain

[$(CH_3)_4C$]

Neopentane

Figure 2.6: Structures and common names for constitutional isomers of C_4 and C_5 alkanes. The C_3 alkane has no constitutional isomer. The carbon atoms are designated as primary (1°), secondary (2°), tertiary (3°) and quaternary (4°), as applicable.

As shown, there are two constitutional isomers for the **C_4 alkane**. These isomers have different connectivity of atoms: one has a straight chain called **butane**, while the other has a branched chain called **isobutane**.

The **C_5 alkane** has three constitutional isomers, namely, **pentane**, **isopentane** and **neopentane.**

Note: If an alkane contains a $(CH_3)_2CH$ group (called an isopropyl group), it is called an isoalkane. Both isobutane and isopentane possess an isopropyl group.

Alkyl groups

The groups that are derived from alkanes by *removal of a hydrogen atom* are called **alkyl groups**. Therefore, alkyl groups have one residual valence for combining with other atom or group.

The H atoms attached to a primary, secondary and tertiary carbon atom are called **primary (1°), secondary (2°) and tertiary (3°) hydrogen atoms**, respectively. Since no hydrogen atom is attached to a quaternary carbon, quaternary hydrogen atom does not exist. Removal of a primary, secondary or tertiary H atom from alkanes leads to different alkyl groups.

Figure 2.7 shows the **possible alkyl groups** derived from **methane, ethane, propane** and **butanes** (butane and isobutane), and their **common names** and **abbreviations.**

Only *one* alkyl group can be derived from methane or ethane. But *two* alkyl groups can be derived from propane by removal of a 1° or a 2° H, whereas *four* alkyl groups can be derived from butanes (*two* from butane by removal of a 1° or 2° H, and another *two* from isobutane by removal of a 1° or 3° H).

Alkane − (minus) H = Alkyl

Alkane	− H	= Alkyl	Abbreviation
CH_4 (**Methane**)	− H	= CH_3— (**Methyl**)	**Me**
CH_3CH_3 (**Ethane**)	− H	= CH_3CH_2— (**Ethyl**)	**Et**
$CH_3CH_2CH_3$ (2°, 1°) (**Propane**)	− H (1°)	= $CH_3CH_2CH_2$— (**Propyl**)	**Pr**
	− H (2°)	= CH_3CHCH_3 = $(CH_3)_2CH$— (**Isopropyl**)	***i*-Pr**
$CH_3CH_2CH_2CH_3$ (2°, 1°) (**Butane**)	− H (1°)	= $CH_3CH_2CH_2CH_2$— (**Butyl**)	**Bu**
	− H (2°)	= $CH_3CH_2CHCH_3$ (***sec*-Butyl**)	***s*-Bu**

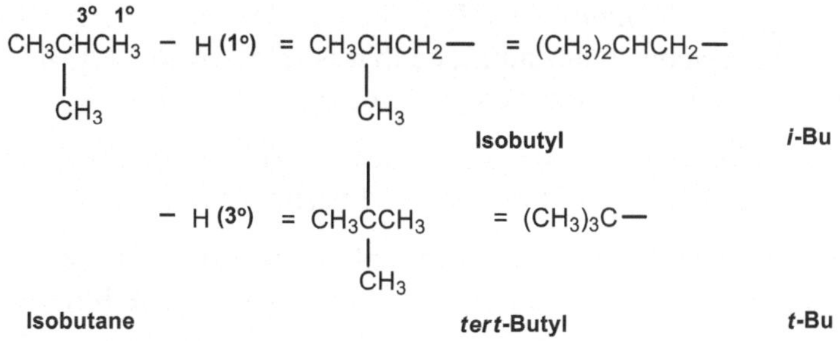

$$\underset{\displaystyle |\atop CH_3}{\overset{3^\circ\;\;1^\circ}{CH_3CHCH_3}} - H(1^\circ) = \underset{\displaystyle |\atop CH_3}{CH_3CHCH_2—} = (CH_3)_2CHCH_2—$$

Isobutyl *i*-Bu

$$- H(3^\circ) = \underset{\displaystyle |\atop CH_3}{\overset{|}{CH_3CCH_3}} = (CH_3)_3C—$$

Isobutane ***tert*-Butyl** ***t*-Bu**

Figure 2.7: Alkyl groups derived from methane, ethane, propane and butanes (butane and isobutene), and their abbreviations. *sec* and *tert* denote *secondary* and *tertiary* respectively. In abbreviations, *i*, *s* and *t* stand for iso, *sec* and *tert* respectively.

Note: For higher alkanes, all hydrogen atoms of a particular kind (1°, 2° or 3°) are not always equivalent. Thus, if there are two sets of equivalent primary hydrogen atoms in an alkane, then removal of a primary hydrogen atom from each of the two sets will give two different alkyl groups (see Problem 2.15).

PROBLEM 2.3: Suggest the structure of a hydrocarbon with the molecular formula C_8H_{18} that has only primary hydrogen atoms.

Answer

The molecular formula (C_8H_{18}) corresponds to the general formula C_nH_{2n+2}, where $n = 8$. Therefore, the hydrocarbon is an alkane. Since all 18 hydrogen atoms are primary, there are six CH_3 groups. The structure of the hydrocarbon is

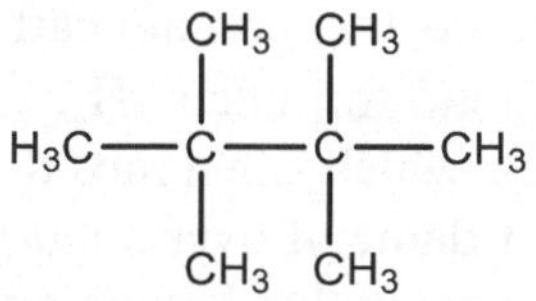

$$H_3C—\underset{CH_3}{\overset{CH_3}{C}}—\underset{CH_3}{\overset{CH_3}{C}}—CH_3$$

2.3.2 Alkenes, Alkynes and Alkadienes

Alkenes

Alkenes (also called **olefins**) are unsaturated hydrocarbons. They are called **unsaturated** because they have two carbons, each of which is attached to *three* atoms while the maximum number of atoms each carbon is able to accommodate is *four*.

Alkenes contain carbon-carbon double bond (C=C) as a functional group. The general form of alkenes is:

General form of alkenes

$$\begin{matrix} R & & R \\ & C{=}C & \\ R & & R \end{matrix}$$

$(R_2C{=}CR_2)$

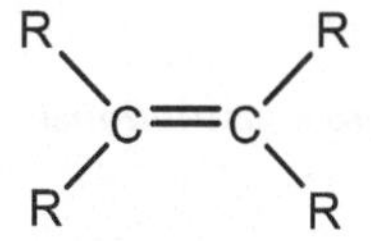

R = H, alkyl or other hydrocarbon group

Alkenes have the general formula C_nH_{2n}. The simplest member is **ethylene** or **ethene** (C_2H_4). Ethylene is a gaseous plant hormone that is released to start ripening of the fruit. Farmers often spray ethylene on their crops to force the fruit to ripen.

Alkenes also make a **homologous series**; the next two higher homologues of ethylene are **propylene or propene** (C_3H_6) and **butylene or butene** (C_4H_8).

Like alkanes, alkenes also exhibit **constitutional isomerism**. For example, the C_4 alkene (C_4H_8) can exist as three constitutional isomers, namely, **1-butene, 2-butene** and **isobutene**.

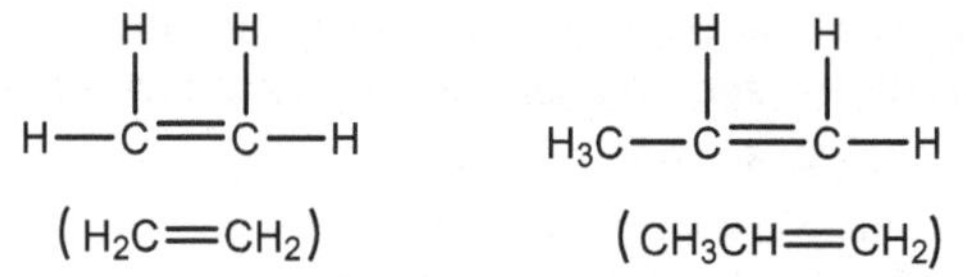

$H-C(H)=C(H)-H$ ($H_2C=CH_2$)

Ethylene (ethene)

$H_3C-C(H)=C(H)-H$ ($CH_3CH=CH_2$)

Propylene (propene)

C_4 alkene (C_4H_8):

$CH_3CH_2CH=CH_2$ (C-1 numbered) — **1-Butene**

$CH_3CH=CHCH_3$ (C-2, C-1 numbered) — **2-Butene**

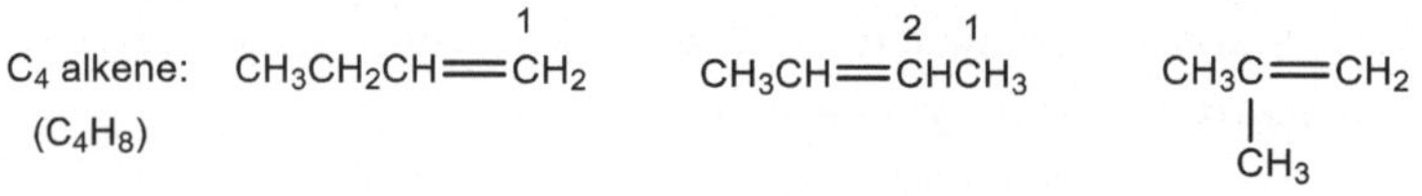

$CH_3C(CH_3)=CH_2$ — **Isobutene**

Three constitutional isomers

Alkynes

Alkynes are unsaturated hydrocarbons that contain carbon-carbon triple bond (C≡C) as a functional group, and have the general formula C_nH_{2n-2}. The simplest alkyne is acetylene or ethyne (C_2H_2). Acetylene is a gas, which when mixed with oxygen in proper proportions, burns cleanly to produce a very hot flame of over 3000°C. The oxyacetylene flame is used to weld metals and melt glass. The next higher homologue of acetylene is propyne (C_3H_4).

The **constitutional isomers** of the C_4 alkyne (C_4H_6) are **1-butyne** and **2-butyne.**

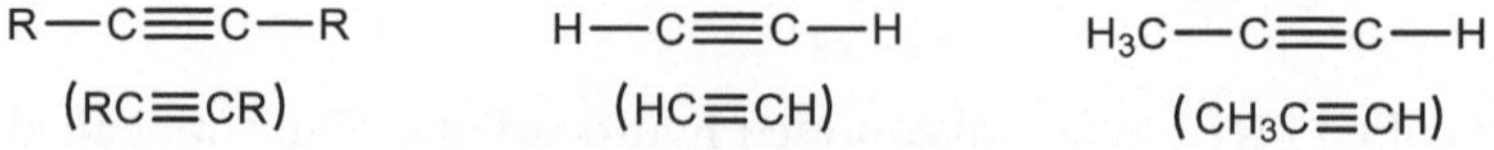

$R-C\equiv C-R$ ($RC\equiv CR$) — ***General form of alkynes***

$H-C\equiv C-H$ ($HC\equiv CH$) — **Acetylene (ethyne)**

$H_3C-C\equiv C-H$ ($CH_3C\equiv CH$) — **Propyne**

C_4 alkyne (C_4H_6):

$CH_3CH_2C\equiv CH$ (C-1 numbered) — **1-Butyne**

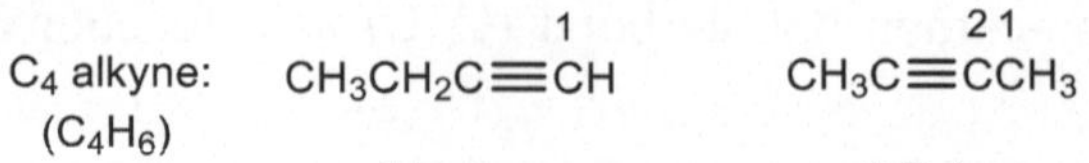

$CH_3C\equiv CCH_3$ (C-2, C-1 numbered) — **2-Butyne**

Two constitutional isomers

Alkadienes

Alkadienes (or simply, dienes) are *doubly* unsaturated hydrocarbons that contain two C=C double bonds. They have the same general formula (C_nH_{2n-2}) as for alkynes.

Dienes are of three types:

(a) **Conjugated dienes**, in which two double bonds are separated by *only one* C–C single bond. The simplest conjugated diene is **1, 3-butadiene** (C_4H_6).

(b) **Nonconjugated or isolated dienes**, which have two double bonds separated by *more than one* C–C single bond. An example is **1, 4-pentadiene** (C_5H_8).

(c) **Cumulated dienes**, where a *single* carbon has two double bonds. The simplest cumulated diene is **allene** (C_3H_4).

Conjugated diene:

$$H{-}\underset{|}{\overset{H}{C}}{=}\overset{H}{C}{-}\overset{H}{C}{=}\overset{H}{C}{-}H \equiv \overset{4}{H_2C}{=}\overset{3}{CH}{-}\overset{2}{CH}{=}\overset{1}{CH_2}$$

(two double bonds are separated by only one single bond)

1,3-Butadiene

Nonconjugated diene:

$$H{-}\overset{H}{C}{=}\overset{H}{C}{-}\overset{H}{\underset{H}{C}}{-}\overset{H}{C}{=}\overset{H}{C}{-}H \equiv \overset{5}{H_2C}{=}\overset{4}{C}H\overset{3}{C}H_2\overset{2}{C}H{=}\overset{1}{CH_2}$$

(two double bonds are separated by more than one single bond)

1,4-Pentadiene

Cumulated diene:

$$H{-}\overset{H}{C}{=}C{=}\overset{H}{C}{-}H \equiv H_2C{=}C{=}CH_2$$

(a single carbon has two double bonds)

Allene

Note: Unsaturated hydrocarbons with three and four double bonds are known as trienes and tetraenes respectively. In general, hydrocarbons with multiple double bonds are referred to as polyenes.

2.3.3 Cycloalkanes and Cycloalkenes

Cycloalkanes

Cycloalkanes are cyclic saturated hydrocarbons and have the **same general formula (C_nH_{2n}) as for alkenes**. Thus, in terms of atomic composition, **a ring is a *double bond equivalent (DBE)*** (see Section 6.1.4).

The simplest cycloalkane is **cyclopropane** (C_3H_6) since at least three carbon atoms are required to make a ring.

The following are the structures of C_3–C_6 cycloalkanes (**cyclopropane**, **cyclobutane**, **cyclopentane** and **cyclohexane**), a substituted cyclopentane (**methylcyclopentane**) and a disubstituted cyclohexane (**1, 2-dimethylcyclohexane**).

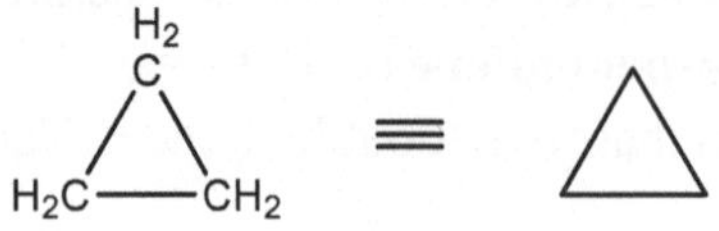

Cyclopropane (C_3H_6)

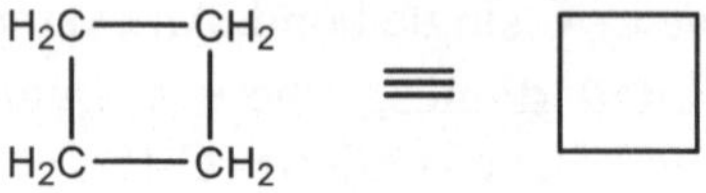

Cyclobutane (C_4H_8)

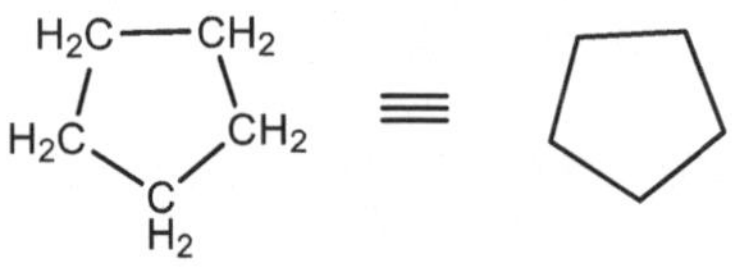

Cyclopentane (C_5H_{10})

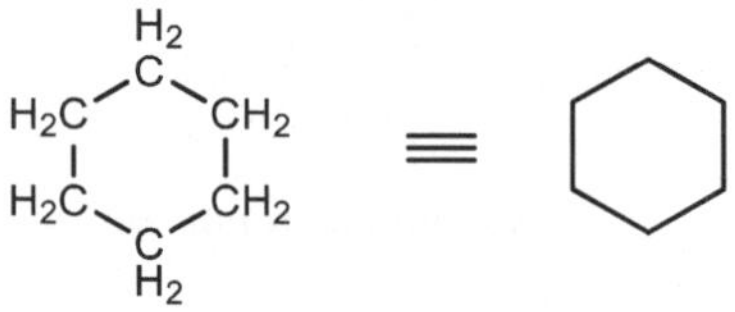

Cyclohexane (C_6H_{12})

Substituted cycloalkanes

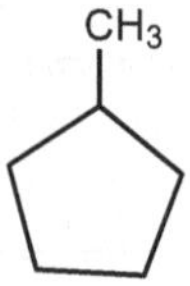

Methylcyclopentane

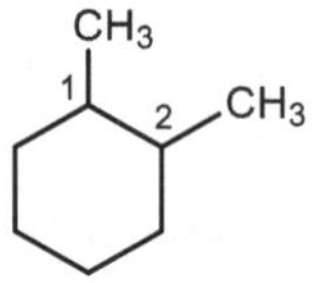

1,2-Dimethylcyclohexane

Cycloalkenes

Cycloalkenes are cyclic unsaturated hydrocarbons containing a **double bond**. Since they possess *one ring* (a double bond equivalent) and *one double bond*, they have the same **general formula (C_nH_{2n-2}) as for dienes.**

Two examples are **cyclopentene** and **cyclohexene**.

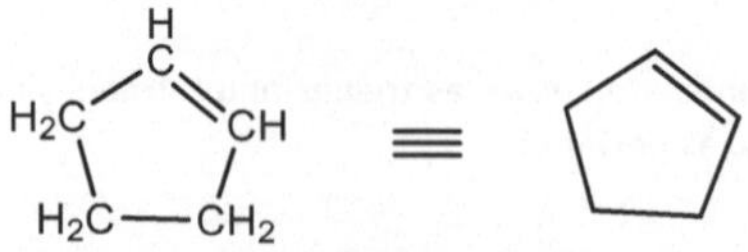

Cyclopentene (C_5H_8)

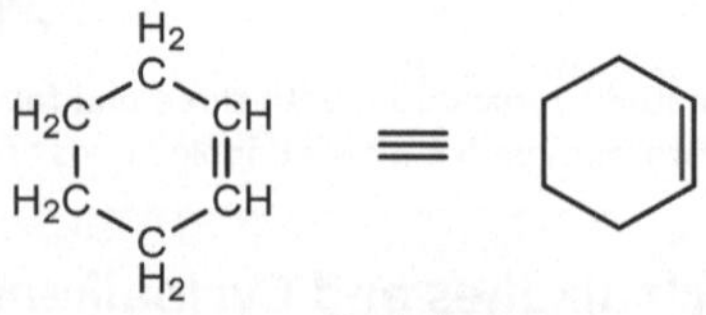

Cyclohexene (C_6H_{10})

2.3.4 Aromatic Hydrocarbons or Arenes

Aromatic hydrocarbons are a class of cyclic compounds that contain conjugated double bonds. They are also called **arenes.**

The name *aromatic* comes from the fact that many of the simple aromatic compounds isolated initially have a good smell or fragrance, for example, vanillin (an aromatic compound isolated from vanilla) has a lovely odour. But some aromatic compounds have also unpleasant smells, or are odourless.

Many simple aromatic compounds are obtained commercially from coal tar. The lord of aromatic compounds is **benzene** (C_6H_6), a six-carbon ring compound containing three conjugated or alternating double bonds.

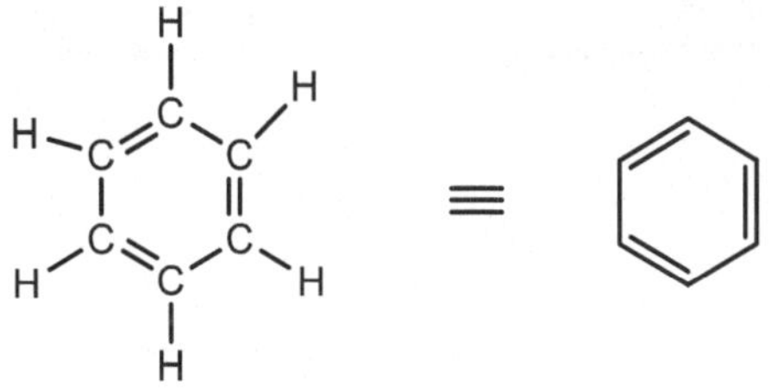

Benzene
(C_6H_6)

A **benzene ring** comprising one ring and three double bonds gives the **double bond equivalent (DBE) = 4**.

The cyclic conjugation in benzene is unique which is distinctly different from acyclic conjugation, and makes benzene more stable and less reactive than its acyclic counterpart. Aromatic compounds with a **benzene ring as a functional group** are known as **benzenoid aromatic compounds**.

Besides benzenoid aromatics, there are also *nonbenzenoid* aromatic compounds and *heteroaromatic* compounds.

Most aromatic compounds are derivatives of benzene with one or more substituents on the benzene ring. For example, **toluene** (methylbenzene) is a monosubstituted benzene with a methyl group as the substituent.

Disubstituted benzenes are commonly indicated by the prefixes ***ortho*- (*o*), *meta*- (*m*) or *para* (*p*)**. The prefix *ortho*- implies that substituents are in a relative 1, 2-position, *meta*- indicates a 1,3-position, and *para*- indicates a 1,4-position. **Xylenes** are disubstituted benzenes in which two substituents are methyl groups. Three xylenes are possible: ***o*-xylene**, ***m*-xylene** and ***p*-xylene**.

It should be noted that with respect to a substituent (X), there are two *ortho* positions, two *meta* positions and one *para* position.

Monosubstituted:

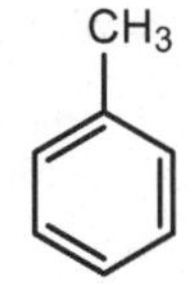

CH_3

Toluene
(methylbenzene)

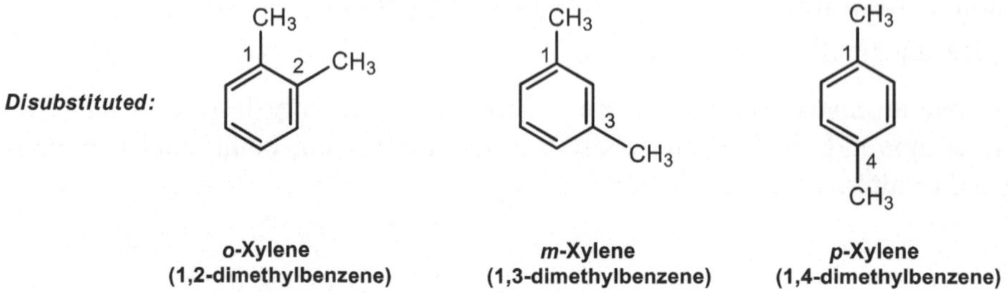
Disubstituted:
CH3
1
2
CH3
CH3
1
3
CH3
CH3
1
4
CH3
o-Xylene
(1,2-dimethylbenzene)
m-Xylene
(1,3-dimethylbenzene)
p-Xylene
(1,4-dimethylbenzene)

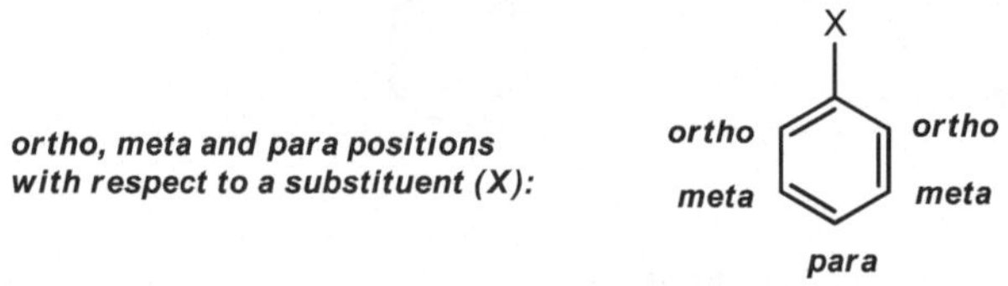
ortho, meta and para positions
with respect to a substituent (X):
X
ortho
ortho
meta
meta
para

2.3.5 Alkyl, Alkenyl, Alkynyl and Aryl Groups from Unsaturated Hydrocarbons

Alkyl groups

We have already seen that the removal of a hydrogen atom from a saturated carbon in alkanes gives an alkyl group (see Figure 2.7). Similarly, the removal of a hydrogen atom from a **saturated carbon** in alkenes, alkynes or arenes gives another type of **alkyl groups**.

The most important of these alkyl groups are derived from the saturated carbon (alkyl carbon) adjacent to the unsaturated carbon.

For example, the removal of a H atom from CH_3 group of **propene** gives an **allyl** group. Similarly, **propargyl** and **benzyl** groups are derived from the removal of a H atom from CH_3 group of **propyne** and **toluene** respectively. The abbreviation for the **benzyl** group is **Bn**.

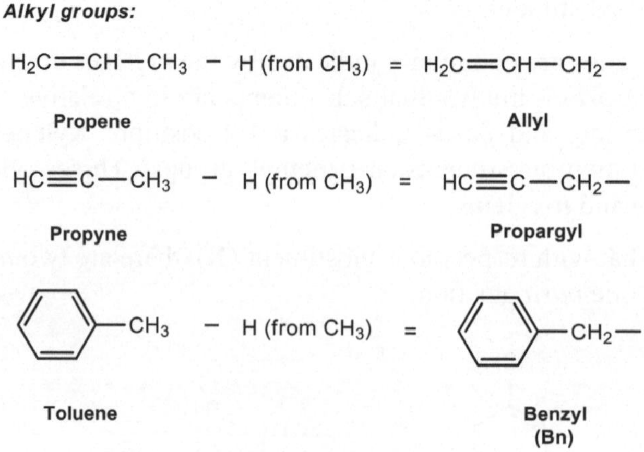
Alkyl groups:
H2C=CH—CH3 − H (from CH3) = H2C=CH—CH2—
Propene
Allyl
HC≡C—CH3 − H (from CH3) = HC≡C—CH2—
Propyne
Propargyl
CH3 − H (from CH3) = CH2—
Toluene
Benzyl
(Bn)

Alkenyl and alkynyl groups

An alkenyl or alkynyl group is derived from the removal of a hydrogen atom from an unsaturated carbon (alkene or alkyne carbon).

Only one **alkenyl** group can be derived from **ethylene**, which is called a **vinyl** group. However, two alkenyl groups (**propenyl** and **isopropenyl**) can be derived from **propene**.

An alkynyl group derived from acetylene is called an ethynyl or acetylenic group.

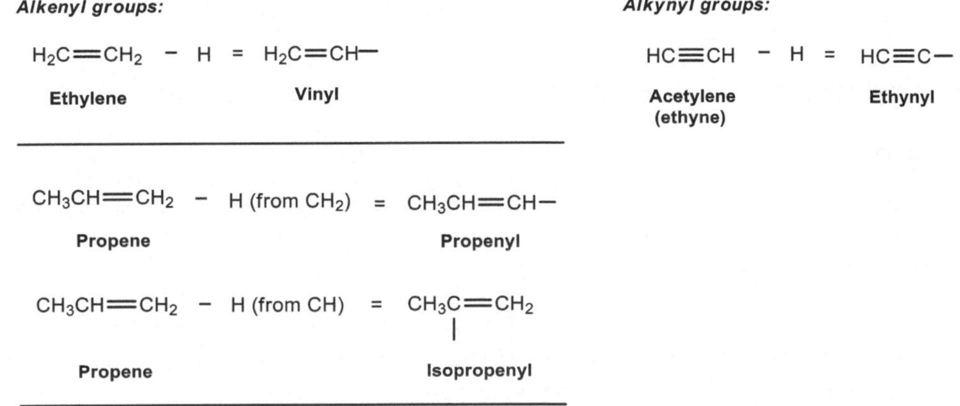

Aryl groups

The removal of hydrogen atom from an unsaturated carbon of **benzene** gives the simplest aryl group called a **phenyl** group. Phenyl can be denoted by **C_6H_5–** or **Ph–** or **ϕ–**. Note carefully that C_6H_5– is called phenyl, *not* benzyl.

In general, a group derived from **aromatic compounds** by removal of a *ring H atom* is called an **aryl group (Ar)**. For example, removal of an *o*-hydrogen atom from **methylbenzene (toluene)** gives ***o*-methylphenyl (or *o*-tolyl)** group. Similarly, removal of the *p*-hydrogen atom from **methoxybenzene (anisole)** gives ***p*-methoxyphenyl (or *p*-anisyl)** group.

–H − H =

(C_6H_5-)

Benzene

Phenyl
(Ph or ϕ)

X –H − H = X

(X = substituent at
***o*-, *m*- or *p*-position)**

Aromatic compound

Aryl
(Ar)

Examples of aryl groups:

CH_3 –H − H (*ortho*) = CH_3 (*o*-$CH_3C_6H_4$-)

Methylbenzene
(toluene)

***o*-Methylphenyl**
or *o*-tolyl

$H_3CO-C_6H_4-H$ − H (*para*) = $H_3CO-C_6H_4-$ (p-$H_3COC_6H_4$-)

Methoxybenzene (anisole) **p-Methoxyphenyl or p-anisyl**

2.3.6 Alkyl Halides and Aryl Halides

Alkyl halides

Alkyl halides or haloalkanes contain fluoro (–F), chloro (–Cl), bromo (–Br) or iodo (–I) functional group. They are derived from hydrogen halides (HX; X = halogen) by replacement of the H atom by an alkyl group.

Alkyl halides are classified as **primary (1°)**, **secondary (2°)** and **tertiary (3°)** based on the attachment of the halogen to a primary, secondary or tertiary carbon atom. **Methyl** halides are *unique*; they are *not* classifiable as primary halides.

H—X (X = F, Cl, Br or I) **Hydrogen halide**

R—X (R = alkyl) **Alkyl halides**

1° carbon: (RCH_2X) **Primary alkyl halides**

2° carbon: (R_2CHX) **Secondary alkyl halides**

3° carbon: (R_3CX) **Tertiary alkyl halides**

Not 1° carbon: (CH_3X) **Methyl halides**

The following are some examples of **primary (1°)**, **secondary (2°)** and **tertiary (3°)** alkyl halides:

Primary alkyl halides:

1°: $CH_3CH_2—Cl$ **Ethyl chloride (Chloroethane)**

1°: $CH_3CH(CH_3)CH_2—Cl$ **Isobutyl chloride (*i*-BuCl)**

1°: $CH_3C(CH_3)_2CH_2—Br$ **Neopentyl bromide**

1°, allylic: $H_2C{=}CHCH_2Br$ **Allyl bromide**

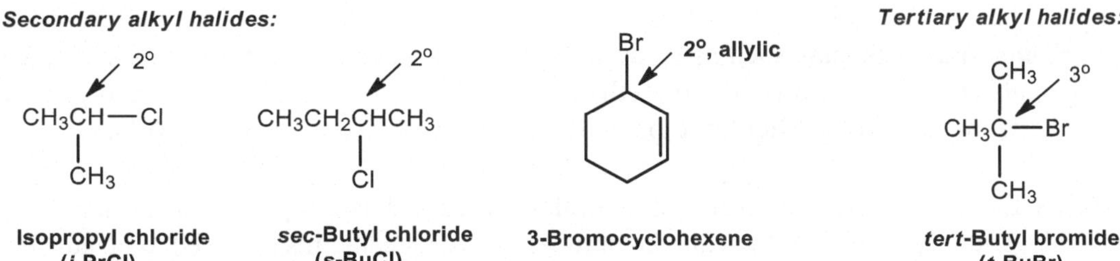

Isopropyl chloride (*i*-PrCl) **sec-Butyl chloride (*s*-BuCl)** **3-Bromocyclohexene** ***tert*-Butyl bromide (*t*-BuBr)**

Note that **3-bromocyclohexene** is a **secondary** as well as an **allylic** bromide (Br is attached to an allylic carbon adjacent to the double bond).

PROBLEM 2.4: How many alkyl chlorides are possible with the molecular formula C_4H_9Cl? Give their structural formulas and common names, and classify them as primary, secondary or tertiary alkyl chloride.

Answer

Four butyl chlorides are possible, of which two are primary, one secondary and the other tertiary.

1°	1°	2°	3°
$CH_3CH_2CH_2CH_2Cl$	CH_3CHCH_2Cl (with CH_3 on C-2)	$CH_3CHCH_2CH_3$ (with Cl on C-2)	$CH_3C(CH_3)_2$—Cl
Butyl chloride (primary)	**Isobutyl chloride (primary)**	***sec*-Butyl chloride (secondary)**	***tert*-Butyl chloride (tertiary)**

Aryl halides

Aryl halides or haloarenes are organohalogen compounds in which the **halogen atom is directly attached to the aromatic (benzene) ring**. They are derived from hydrogen halides (HX) by replacement of the H atom by an **aryl group**.

The following are some examples of aryl halides:

Ar—X (Ar = aryl)

Aryl halides

Cl on benzene ring (C_6H_5Cl)	Cl and CH_3 on adjacent ring carbons	Br and CH_3 in meta positions	I and CH_3 in para positions
Chlorobenzene (Ar = phenyl)	***o*-Chlorotoluene (Ar = *o*-tolyl)**	***m*-Bromotoluene (Ar = *m*-tolyl)**	***p*-Iodotoluene (Ar = *p*-tolyl)**

Organic halides are seldom found as natural products. They are commercially used as propellants in aerosol cans such as spray paint, solvents such as chloroform ($CHCl_3$), and refrigerants such as dichlorodifluoromethane (CF_2Cl_2). Organic halides are also used as insecticides such as DDT (dichlorodiphenyltrichloroethane) (see Section 18.6.1).

PROBLEM 2.5: Draw bond-line structural formulas of isopropyl chloride, *tert*-butyl chloride, and allyl chloride. Indicate the number of hydrogen atoms assumed to be attached to each of the carbon atoms.

Answer

The normal valence of carbon is four. Therefore, the number of H atoms attached to each carbon = 4 – (number of bonds drawn to each carbon).

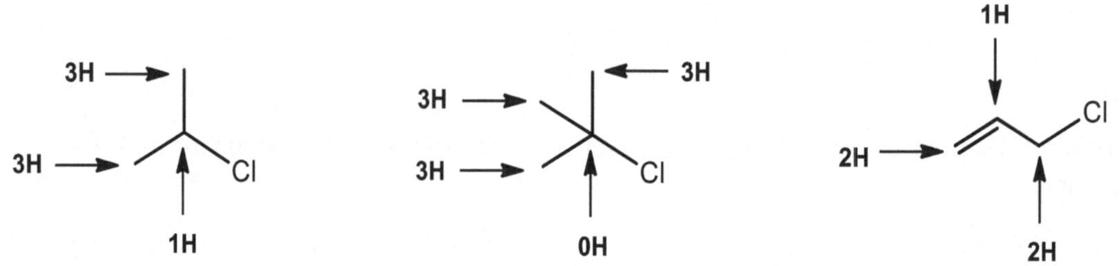

3H → 3H → 1H Cl

Isopropyl chloride

3H → 3H → 3H → 0H Cl

***tert*-Butyl chloride**

1H 2H → 2H Cl

Allyl chloride

2.3.7 Alcohols, Phenols and Ethers

Alcohols

Alcohols contain the hydroxyl (–OH) functional group. They are derived from **water** by replacement of one of the H atoms by an **alkyl group**.

Similar to alkyl halides, **alcohols** are classified as being **primary (1°)**, **secondary (2°)** and **tertiary (3°)** based on the type of carbon (primary, secondary or tertiary) to which the OH group is directly attached.

H—OH

Water

R—OH (R = alkyl)

Alcohols

1° carbon: (RCH_2OH) **Primary alcohols**

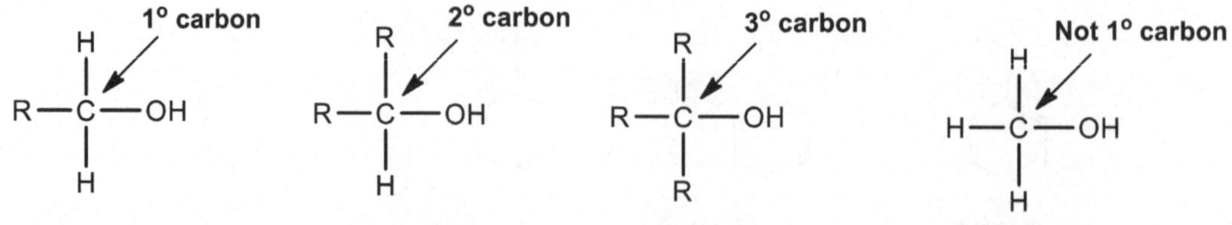

2° carbon: (R_2CHOH) **Secondary alcohols**

3° carbon: (R_3COH) **Tertiary alcohols**

Not 1° carbon: (CH_3OH) **Methyl alcohol**

The simplest alcohol is **methyl alcohol** (CH_3OH), which is *not* a primary alcohol.

The simplest *primary* alcohol is **ethyl alcohol** (CH_3CH_2OH). Ethyl alcohol (or ethanol) is the most familiar as a drinking alcohol, which is found in beer, wine and liquor. However, ethanol is toxic in very large doses.

The simplest *secondary* alcohol is **isopropyl alcohol** (or isopropanol), which is a rubbing alcohol used in cleaning solutions and as disinfectants.

The following are some examples of **primary**, **secondary** and **tertiary** alcohols.

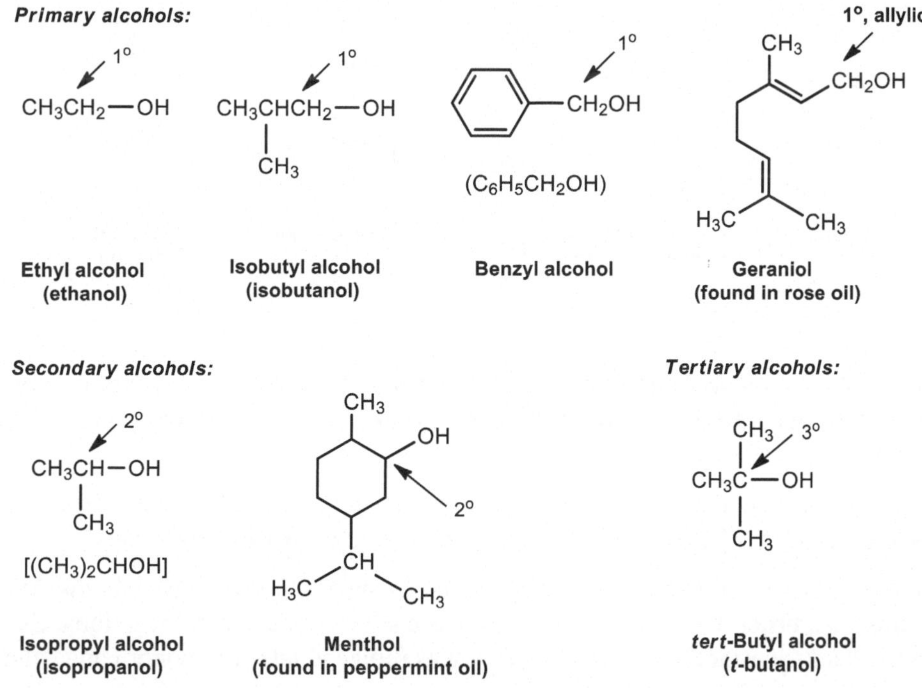

Ethyl alcohol (ethanol) **Isobutyl alcohol (isobutanol)** **Benzyl alcohol** **Geraniol (found in rose oil)**

Isopropyl alcohol (isopropanol) **Menthol (found in peppermint oil)** ***tert*-Butyl alcohol (*t*-butanol)**

Alcohols containing two, three or more OH groups are called **dihydric**, **trihydric** or **polyhydric** alcohols. **Ethylene glycol** is a dihydric alcohol, used in antifreeze. **Glycerol** (commonly known as glycerine) is a trihydric alcohol.

CH_2OH
|
CH_2OH

Ethylene glycol (a dihydric alcohol)

CH_2OH
|
$CHOH$
|
CH_2OH

Glycerol (a trihydric alcohol)

Phenols

Phenols contain the –OH functional group which is directly attached to a carbon atom of benzene ring. They are derived from water by replacement of one of the H atoms by an aryl group.

The chemical properties of phenols are different from those of alcohols. **Phenol** itself is hydroxybenzene, also known as **carbolic acid**, which is used as a disinfectant.

The following are some examples of **phenols** (one OH group) and **dihydric phenols** (two OH groups).

H—OH Ar—OH (Ar = aryl)

Water **Phenols**

Phenols:

OH

(C_6H_5OH)

Phenol

OH CH_3

***o*-Cresol (*o*-methylphenol)**

Dihydric phenols:

OH OH

Resorcinol

OH OH

Quinol (hydroquinone)

Ethers

Ethers differ from alcohols and phenols in that **the oxygen atom of ethers is bonded to two carbon atoms, and they contain the group (–O–)**. The two carbon groups attached to the oxygen atom may be **alkyl** or **aryl** groups.

If the two groups attached to the oxygen atom are the same, the ethers are said to be **symmetrical**, and if different, the ethers are called **mixed or unsymmetrical**.

The simplest ether is **dimethyl ether**. However, **the most common ether is diethyl ether (also known as 'ether')**, which is widely used as a solvent in chemical reactions. Ether was once used as an anaesthetic for surgery but it was replaced later by modern anaesthetics.

The following are some examples of **symmetrical** and **mixed** ethers:

R—O—R' (R, R' = alkyl or aryl)

Ethers

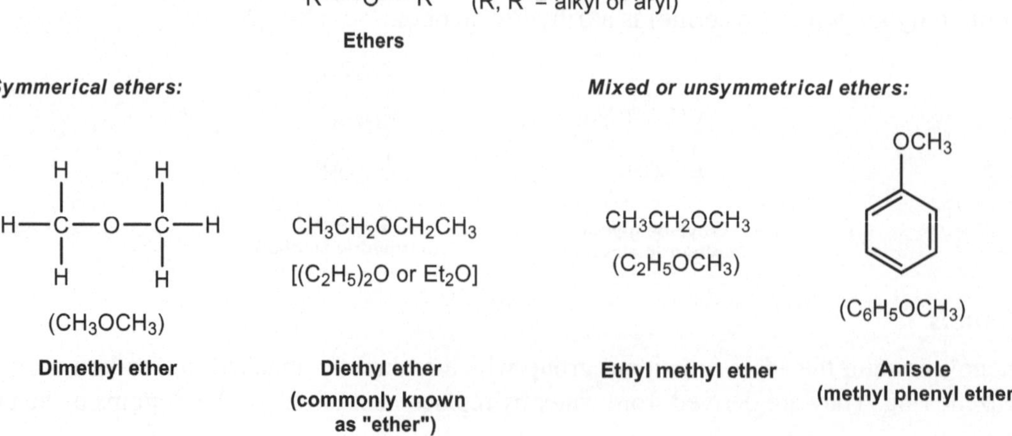

Cyclic ethers as three-membered rings are called **epoxides or oxiranes**. The simplest epoxide is **ethylene oxide.**

Tetrahydrofuran (THF) is five-membered ring ether, which is an efficient solvent in organic reactions.

Ethylene oxide
(an epoxide or oxirane)

Tetrahydrofuran
(THF)

2.3.8 Amines

Amines contain the amino (–NH_2, –NHR or –NR_2) functional group. They are derived from ammonia (NH_3) by replacement of one, two or three H atoms by **alkyl or aryl groups**.

Amines are classified as **primary (1°)**, **secondary (2°)** and **tertiary (3°)** based on the type of **nitrogen atom as being primary** (nitrogen bonded to only one carbon atom), **secondary** (nitrogen bonded to two carbon atoms) or **tertiary** (nitrogen bonded to three carbon atoms). Note that this classification of amines is based on the *type of nitrogen*, whereas the classification of alkyl halides or alcohols is based on the *type of carbon* attached to the functional group.

Like carbon, nitrogen can also form four bonds as in ammonium ion (NH_4^+). Replacement of all four hydrogen atoms of ammonium ion by alkyl groups produces a **quaternary ammonium ion** because the **nitrogen is quaternary (4°)**, being bonded to four carbon atoms. Organic compounds, known as **quaternary ammonium salts**, are well known.

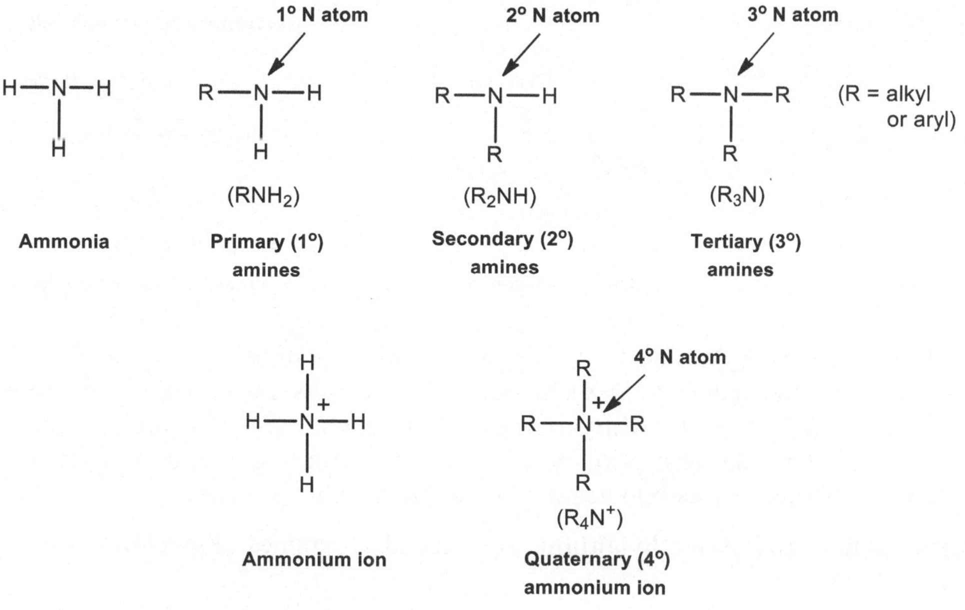

Amines are very useful as nitrogenous *organic bases*. Many of the amines have been isolated from plants, animals and other organisms. Nicotine, cocaine, morphine and amphetamine are some familiar compounds containing amino groups.

The following are some examples of primary, secondary and tertiary amines, and a quaternary ammonium salt.

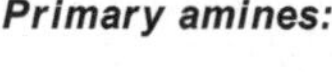

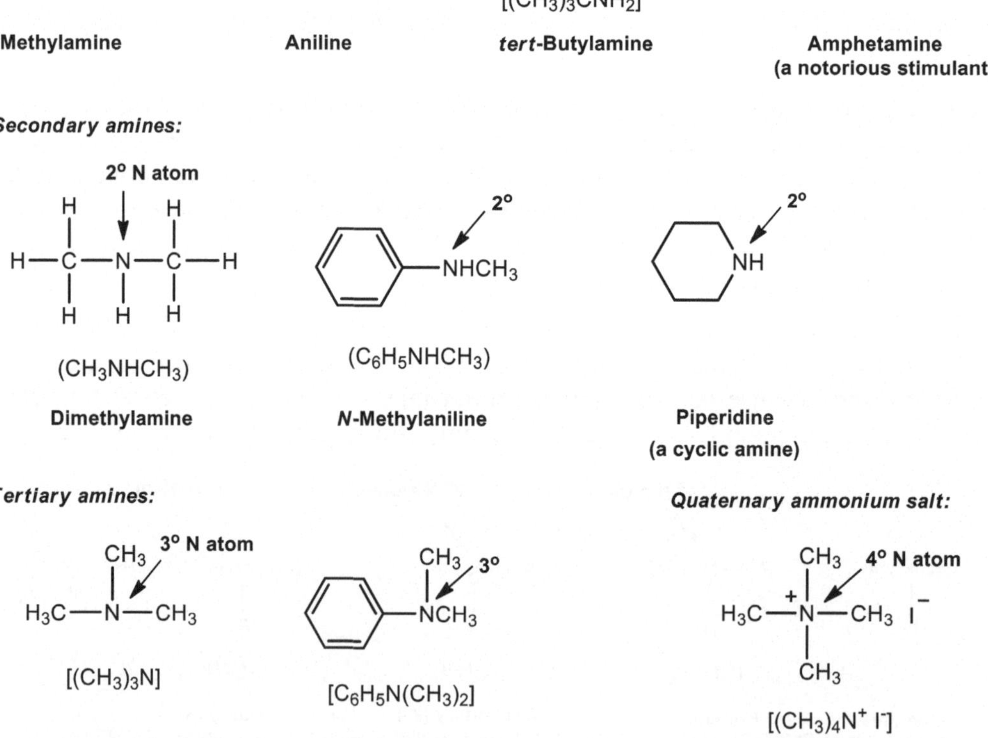

It is of note that **methylamine** (CH_3NH_2) is a **primary amine**, whereas methyl alcohol (CH_3OH) or methyl halide (CH_3X) is *not* a primary alcohol or halide. **Aniline** is a primary *aromatic* amine. ***tert*-Butylamine** and **amphetamine** (a neurologically active dangerous stimulant) are also primary amines since the nitrogen is primary (bonded to one carbon), although the carbon bearing amino group is not primary.

Dimethylamine and *N*-methylaniline are secondary amines. **Piperidine** is a *cyclic* secondary amine.

Trimethylamine and *N,N*-dimethylaniline are tertiary amines.

Tetramethylammonium iodide is a quaternary ammonium salt.

Diamines

Amines containing *two* amino groups are called **diamines**. Two examples are **ethylenediamine** and **putrescine.**

The foul smell of decaying animals comes from putrescine which is formed as meat decays.

$H_2N—CH_2CH_2—NH_2$	$H_2NCH_2CH_2CH_2CH_2NH_2$
$[H_2N(CH_2)_2NH_2]$	$[H_2N(CH_2)_4NH_2]$
Ethylenediamine	**Putrescine**

PROBLEM 2.6: Answer the following:

(a) How many primary amines are possible with molecular formula C_3H_9N? Give their structures and names.

(b) How many secondary amines are possible with molecular formula $C_4H_{11}N$? Give their structures and names.

(c) How many tertiary amines are possible with molecular formula $C_4H_{11}N$? Give their structures and names.

Answer

(a) Two primary amines are possible. The following are their structures:

$CH_3CH_2CH_2NH_2$ — **Propylamine**

CH_3CHNH_2 with CH_3 on the central carbon ($CH_3CH(CH_3)NH_2$) — **Isopropylamine**

(b) Three secondary amines are possible with the following structures:

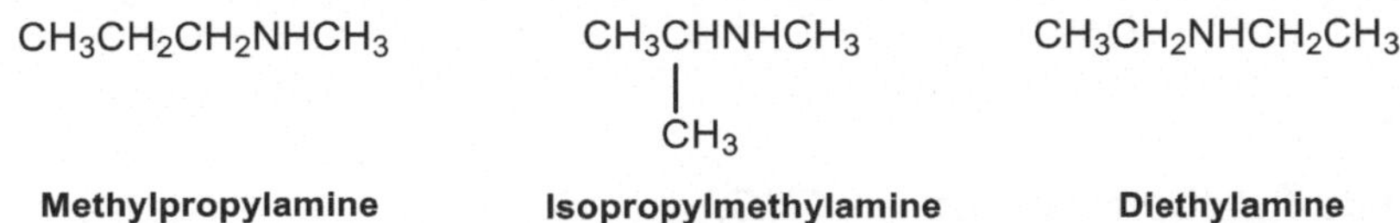

Methylpropylamine **Isopropylmethylamine** **Diethylamine**

(c) Only one tertiary amine is possible. The structure of the tertiary amine is:

$CH_3CH_2NCH_3$ with CH_3 on N ($CH_3CH_2N(CH_3)CH_3$)

Ethyl(dimethyl)amine

2.3.9 Nitro Compounds

Nitro compounds contain the nitro (–NO_2) functional group. In the NO_2 group, N has four bonds with a positive formal charge (+) on N and a negative formal charge (–) on one of the O atoms.

Nitro compounds are derived from **nitric acid** by the replacement of OH group by an **alkyl or aryl group**.

The following are some examples of nitro compounds:

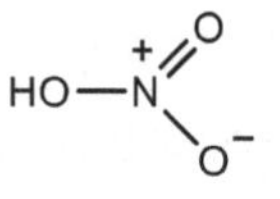

($HONO_2$ or HNO_3)
Nitric acid

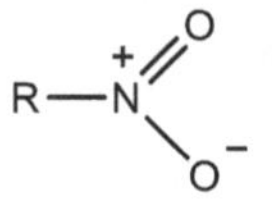

(RNO_2) (R = alkyl, aryl)
Nitro compounds

Examples

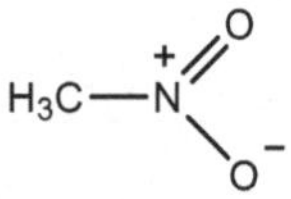

(CH_3NO_2)
Nitromethane

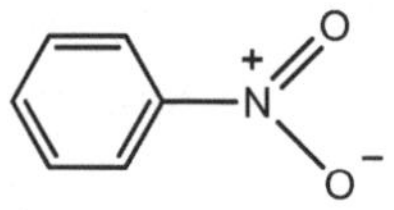

($C_6H_5NO_2$)
Nitrobenzene

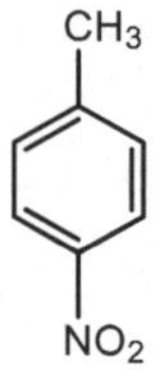

***p*-Nitrotoluene**

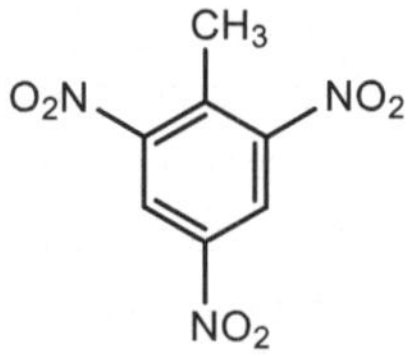

2,4,6-Trinitrotoluene (TNT)

Nitromethane is an *aliphatic* nitro compound, while **nitrobenzene, *p*-nitrotoluene** and **2,4,6-trinitrotoluene** are *aromatic* nitro compounds. Note that 2,4,6-trinitrotoluene (**TNT**) is a highly explosive compound.

2.3.10 Aldehydes and ketones

Both aldehydes and ketones contain a carbonyl group, which is a C=O group. In **aldehydes**, the carbonyl group is bonded to *at least one H atom*, whereas in **ketones**, the carbonyl group is bonded to *two carbon atoms*.

Aldehydes

The aldehyde functional group is –CHO.

The simplest aldehyde is **formaldehyde** (HCHO) in which the carbonyl group (C=O) is bonded to two H atoms. Formaldehyde has many uses, including as a preservative. The smell that often comes out of a biology laboratory is due to formaldehyde used to preserve biological specimens.

The simplest *aromatic* aldehyde is **benzaldehyde** (C_6H_5CHO) that gives almonds its odour.

The **general formula** and some examples of aldehydes including **vanillin** (a common flavouring agent used in foods and beverages) and **cinnamaldehyde** (a key component of the flavour of cinnamon) are given below.

General formula: R—C(=O)—H or R—CHO (R = H, alkyl or aryl)

Aldehydes

Examples

H—C(=O)—H ($H_2C{=}O$ or HCHO)

Formaldehyde

H_3C—C(=O)—H (CH_3CHO)

Acetaldehyde

C_6H_5—C(=O)—H (C_6H_5CHO)

Benzaldehyde

MeO, HO—C_6H_3—CHO

Vanillin

Present in vanilla beans

C_6H_5—CH=CH—CHO

Cinnamaldehyde

A key component of flavour of cinnamon

Ketones

The simplest ketone is **dimethyl ketone or acetone** (CH_3COCH_3). Acetone is a common laboratory solvent. It is the smelly substance found in nail polish remover.

The **general formula** and some examples of ketones are shown here.

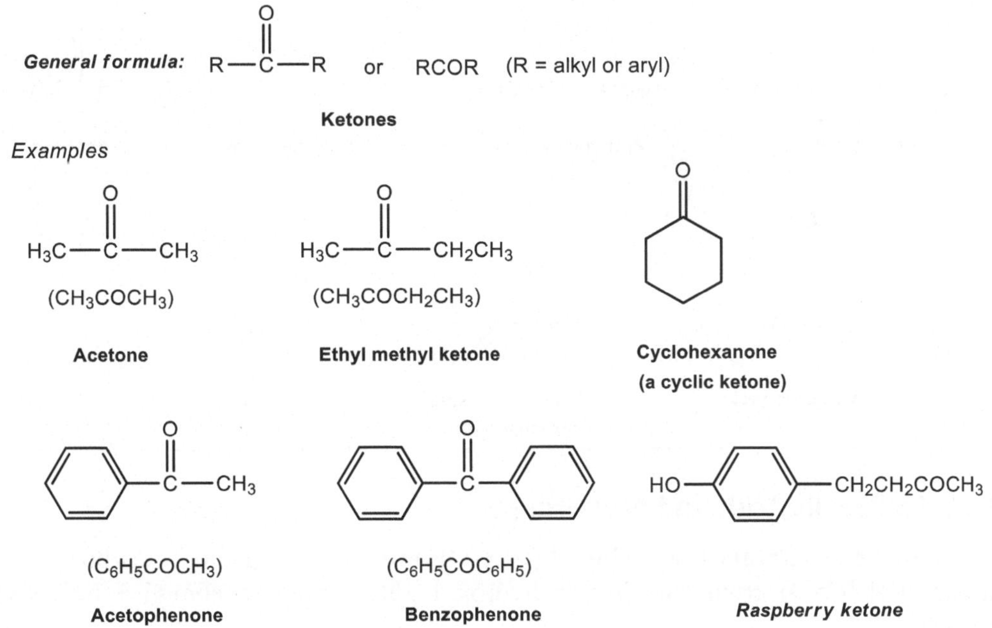

Ethyl methyl ketone is the next higher homologue of acetone. **Cyclohexanone** is a cyclic ketone. **Acetophenone** and **benzophenone** are two *aromatic* ketones. The key component of the flavour of raspberries is attributed to a ketone called ***Raspberry ketone.***

PROBLEM 2.7: Draw condensed and bond-line structural formulas for butyraldehyde and isopropyl methyl ketone.

Answer

$CH_3CH_2CH_2CHO$ ≡ CHO or O

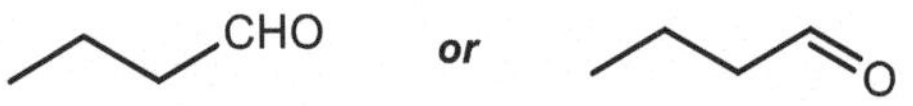

Butyraldehyde

$(CH_3)_2CHCOCH_3$ ≡ O

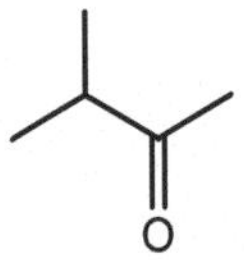

Isopropyl methyl ketone

PROBLEM 2.8: Write structural formulas for all possible constitutional isomers with the molecular formula C_3H_6O.

Answer

The isomers contain either a double bond (C=C or C=O) or a ring. There are seven possible constitutional isomers comprising two alcohols (acyclic and cyclic), three ethers (one acyclic and two cyclic), an aldehyde and a ketone.

$H_2C{=}CH{-}CH_2OH$ **Allyl alcohol**

$H_2C{=}CH{-}OCH_3$ **Methyl vinyl ether**

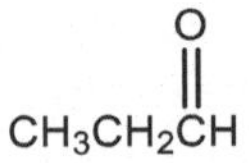

$CH_3CH_2CH{=}O$ **Propionaldehyde**

$CH_3C({=}O)CH_3$ **Acetone**

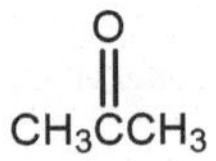

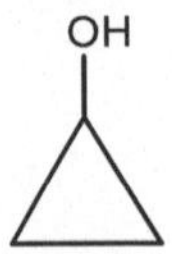

OH **Cyclopropanol**

O, CH_3 **Propylene oxide (2-methyloxirane)**

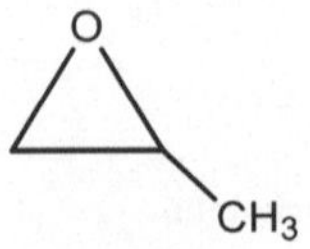

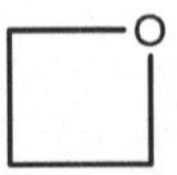

O **Oxetane**

2.3.11 Carboxylic Acids and Acyl Groups

Carboxylic acids contain the carboxyl (–COOH) functional group. A carboxyl group is a carbonyl (C=O) group attached to hydroxyl (OH) group [**carbo**nyl + hydr**oxyl** = **carboxyl**].

The simplest carboxylic acid is **formic acid** ($HCOOH$ or HCO_2H). The next higher homologue is **acetic acid** (CH_3COOH or CH_3CO_2H), which is responsible for the flavour and bitterness of vinegar.

The simplest *aromatic* carboxylic acid is **benzoic acid** (C_6H_5COOH or $C_6H_5CO_2H$), which is widely used as a food preservative and in the manufacture of various chemicals.

The **general formula** and some examples of carboxylic acids are shown below.

General formula: $R-C(=O)-OH$ or $R-COOH$ or $R-CO_2H$

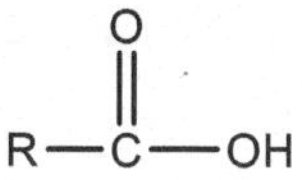

(R = H, alkyl or aryl)

Carboxylic acids

Examples

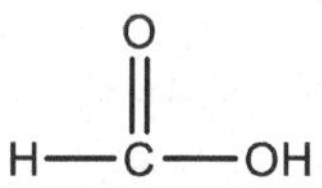

$H-C(=O)-OH$ (HCO_2H)	$H_3C-C(=O)-OH$ (CH_3CO_2H)	$CH_3CH_2CO_2H$	$CH_3CH_2CH_2CO_2H$
Formic acid	**Acetic acid** ***Present in vinegar***	**Propionic acid**	**Butyric acid**

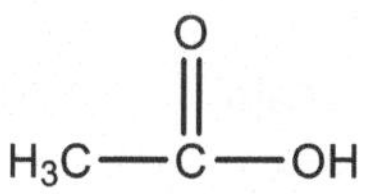

$CH_3CH_2CH_2CH_2CO_2H$	$CH_3CH_2CH_2CH_2CH_2CO_2H$
Valeric acid	**Caproic acid**

First six members of the homologous series and their common names

$H_2C{=}CHCO_2H$	$C_6H_5-C(=O)-OH$ ($C_6H_5CO_2H$)	o-$HOC_6H_4CO_2H$
Acrylic acid ***Unsaturated carboxylic acid***	**Benzoic acid**	**Salicylic acid (o-hydroxybenzoic acid)**

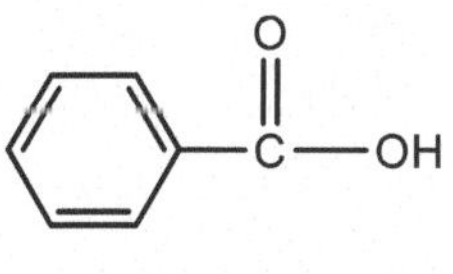

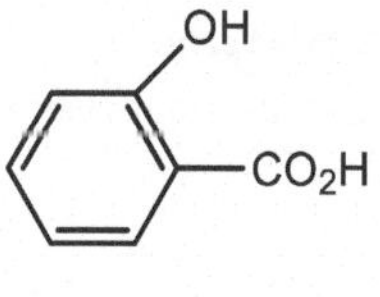

Aromatic carboxylic acids

Note that **formic acid** (C_1 acid), **acetic acid** (C_2 acid), **propionic acid** (C_3 acid), **butyric acid** (C_4 acid), **valeric acid** (C_5 acid) and **caproic acid** (C_6 acid) are the *common names* of *first six members of the homologous series* of carboxylic acids.

Acrylic acid is an *unsaturated* carboxylic acid containing a C=C bond. **Benzoic acid** and **salicylic acid** (*o*-hydroxybenzoic acid) are two examples of *aromatic* carboxylic acids.

Dicarboxylic acids

Carboxylic acids containing *two* carboxyl (CO_2H) groups are called **dicarboxylic ands.**

The simplest *aliphatic* dicarboxylic acid is **oxalic acid**. Some higher homologues of oxalic acid are **malonic acid, succinic acid, glutaric acid** and **adipic acid.**

An *aromatic* dicarboxylic acid is **phthalic acid.**

The structural formulas of the above dicarboxylic acids and some dicarboxylic acids found in fruits such as **malic acid** (in apples) **tartaric acid** (in grapes) and **citric acid** (in lemons) are shown below.

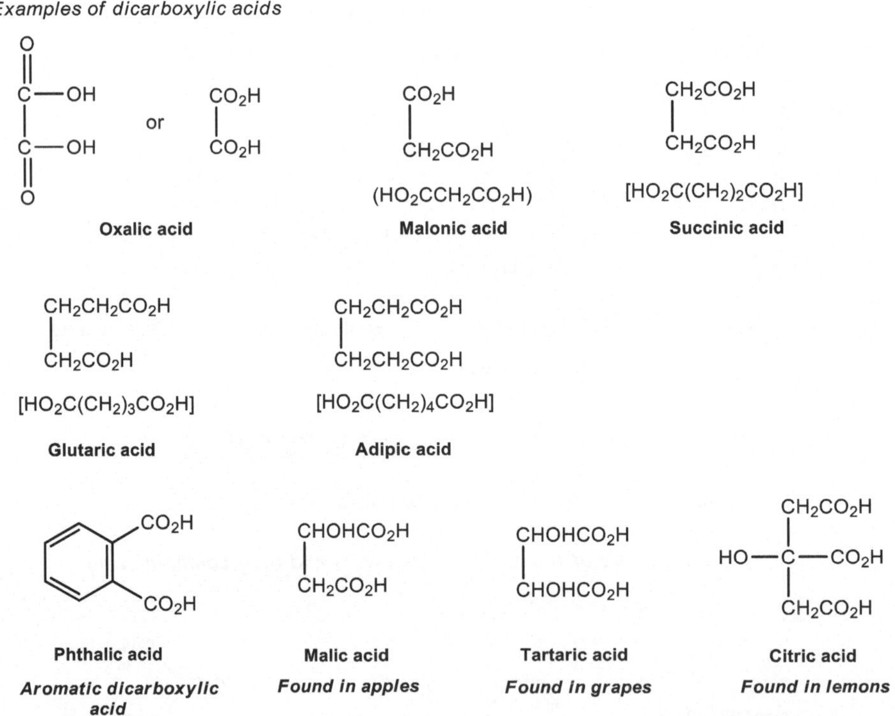

PROBLEM 2.9: Draw condensed and bond-line structural formulas for butyric acid and tartaric acid.

Answer

$CH_3CH_2CH_2CO_2H$ ≡ [bond-line: CO_2H] or [bond-line: O, OH]

Butyric acid

$HO_2CCHOHCHOHCO_2H$ ≡ [bond-line: OH, CO_2H, HO_2C, OH]

Tartaric acid

Acyl groups

The removal of hydroxyl group (OH) from a **carboxylic acid** (RCOOH) gives an **acyl group**.

The general formula of the acyl group is RCO– (or –COR).

Two common acyl groups are **acetyl (–$COCH_3$)** and **benzoyl (–COC_6H_5)**. The **abbreviations** for acetyl and benzoyl are **Ac** and **Bz** respectively.

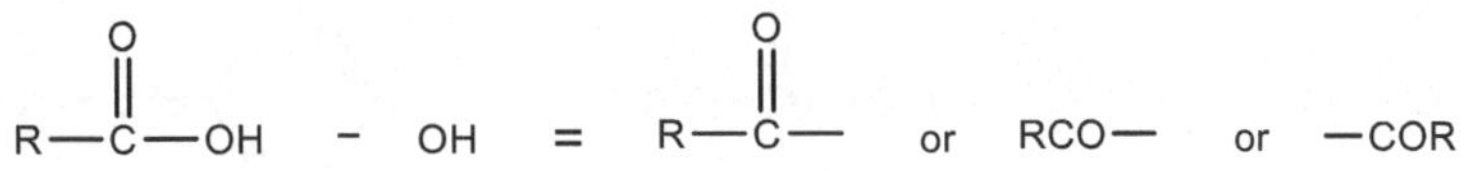

R—C(=O)—OH − OH = R—C(=O)— or RCO— or —COR

Carboxylic acid **Acyl group**

Examples

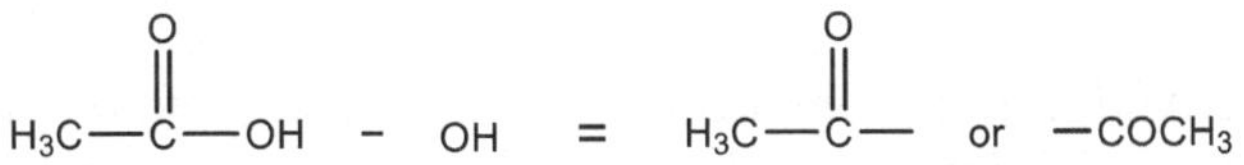

H_3C—C(=O)—OH − OH = H_3C—C(=O)— or —$COCH_3$

Acetic acid **Acetyl (Ac)**

C_6H_5—C(=O)—OH − OH = C_6H_5—C(=O)— or —COC_6H_5

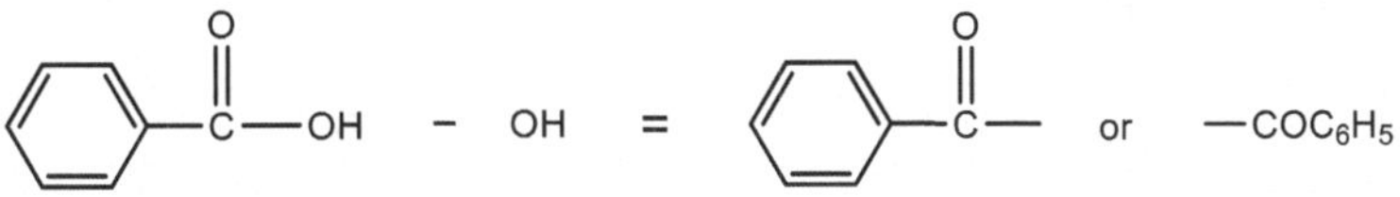

Benzoic acid **Benzoyl (Bz)**

2.3.12 Carboxylic acid Derivatives: Acyl Halides, Acid Anhydrides, Esters and Amides

The replacement of the **OH group** of a carboxylic acid (RCOOH) by another **heteroatom group (Y)** gives a **carboxylic acid derivative** (RCOY).

When the group (Y) is **halogen** (X) **carboxylate** (OCOR), **alkoxy** (OR) and **amino** (NH_2), the derivative is called an **acyl halide**, **acid anhydride**, **ester** and **amide** respectively.

R—C(=O)—OH **Carboxylic acid**

R—C(=O)—Y **Carboxylic acid derivatives**

R—C(=O)—X (Y = X = halogen) **Acyl halide**

R—C(=O)—OCOR (Y = OCOR) **Acid anhydride**

R—C(=O)—OR (Y = OR) **Ester**

R—C(=O)—NH_2 (Y = NH_2) **Amide**

Acyl chlorides

Acyl chlorides (also called acid chlorides) contain the –COCl functional group. They are useful reagents in organic synthesis.

Two common acyl chlorides are **acetyl chloride** (CH_3COCl or AcCl) and **benzoyl chloride** (C_6H_5COCl or BzCl).

Examples

$R-C(=O)-Cl$ (RCOCl) ***Acyl chloride***

$H_3C-C(=O)-Cl$ (CH_3COCl or AcCl) **Acetyl chloride**

$C_6H_5-C(=O)-Cl$ (C_6H_5COCl or BzCl) **Benzoyl chloride**

Acid anhydrides

Acid anhydrides are derived from *two* molecules of a carboxylic acid by the *loss of water*. They behave similarly as acyl chlorides, and are widely used in chemical synthesis.

Acetic anhydride [$(CH_3CO)_2O$ or Ac_2O] and **phthalic anhydride** are two common reagents.

$$R-C(=O)-OH + HO-C(=O)-R \longrightarrow R-C(=O)-O-C(=O)-R + H_2O$$

(RCOOCOR) ***Acid anhydride***

Examples

$H_3C-C(=O)-O-C(=O)-CH_3$ [$(CH_3CO)_2O$ or Ac_2O] **Acetic anhydride**

Phthalic anhydride

Esters

Esters contain the –COOR (or $–CO_2R$) functional group. A common ester is ethyl acetate.

Esters have often pleasant smells and flavours, and many of the fruit smells are due to esters. For example, smells of apples, pineapples and bananas are due to **ethyl butyrate**, **propyl valerate** and **isoamyl acetate** respectively.

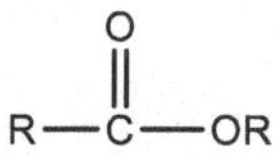

$$R-\overset{O}{\overset{\|}{C}}-OR$$

(RCO_2R)

Ester

Examples

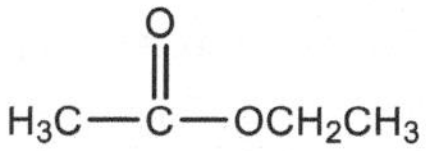

$$H_3C-\overset{O}{\overset{\|}{C}}-OCH_2CH_3$$

($CH_3CO_2CH_2CH_3$)

Ethyl acetate

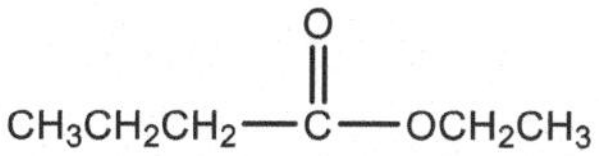

$$CH_3CH_2CH_2-\overset{O}{\overset{\|}{C}}-OCH_2CH_3$$

($CH_3CH_2CH_2CO_2CH_2CH_3$)

Ethyl butyrate
(apples)

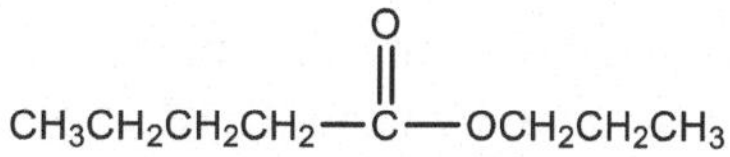

$$CH_3CH_2CH_2CH_2-\overset{O}{\overset{\|}{C}}-OCH_2CH_2CH_3$$

($CH_3CH_2CH_2CH_2CO_2CH_2CH_2CH_3$)

Propyl valerate
(pineapples)

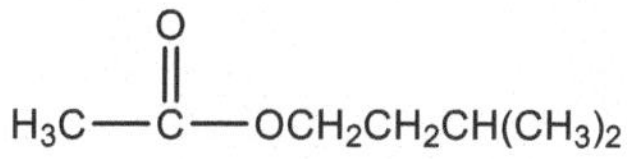

$$H_3C-\overset{O}{\overset{\|}{C}}-OCH_2CH_2CH(CH_3)_2$$

[$CH_3CO_2CH_2CH_2CH(CH_3)_2$]

Isoamyl acetate
(bananas)

Amides and anilides

Amides contain the $-CONH_2$ functional group. Two common examples are acetamide and benzamide.

Amides are quite abundant in nature. Proteins are polyamides. The common antibiotic penicillin contains amide bonds.

Amides include *N*-substituted amides (RCONHR) and *N,N*-disubstituted amides ($RCONR_2$). *N,N*-Dimethylformamide (DMF) is a common organic solvent.

An *N*-aryl substituted amide is called an **anilide** (RCONHAr). Thus **anilides contain the –CONHAr functional group**. Two examples are **acetanilide** and **benzanilide**.

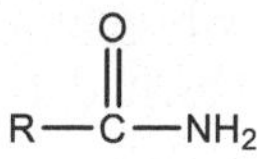

$$R-\overset{O}{\overset{\|}{C}}-NH_2$$

($RCONH_2$)

Amide

Examploo

$$H_3C-\overset{O}{\overset{\|}{C}}-NH_2$$

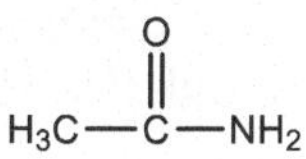

(CH_3CONH_2)

Acetamide

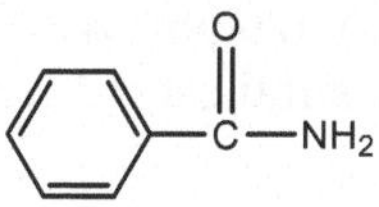

$$C_6H_5-\overset{O}{\overset{\|}{C}}-NH_2$$

($C_6H_5CONH_2$)

Benzamide

$$H-\overset{O}{\overset{\|}{C}}-N(CH_3)_2$$

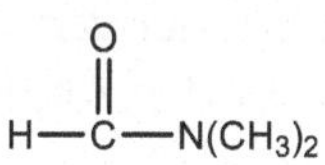

[$HCON(CH_3)_2$]

***N,N*-Dimethylformamide**
(DMF)

$$R-\overset{O}{\overset{\|}{C}}-NHAr$$

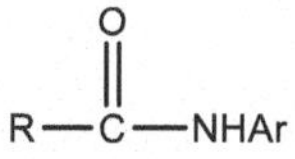

Anilide
(N-aryl substituted amide)

Examples

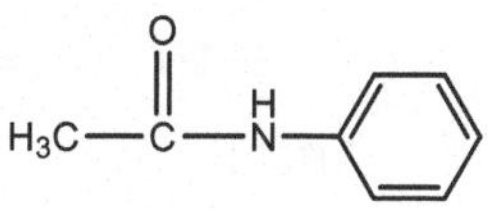

$$H_3C-\overset{O}{\overset{\|}{C}}-\overset{H}{N}-C_6H_5$$

($CH_3CONHC_6H_5$)

Acetanilide

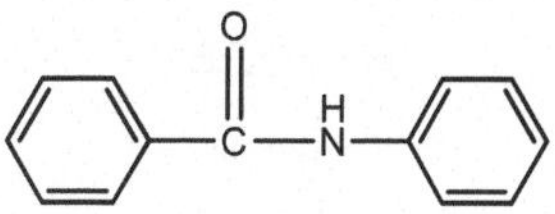

$$C_6H_5-\overset{O}{\overset{\|}{C}}-\overset{H}{N}-C_6H_5$$

($C_6H_5CONHC_6H_5$)

Benzanilide

PROBLEM 2.10: Answer the following:

(a) Draw bond-line structural formulas for ethyl acetate and acetic anhydride.

(b) Indicate the anilide group in the structures of acetanilide and benzanilide.

Answer

(a) The following are the bond-line formulas for ethyl acetate ($CH_3COOCH_2CH_3$) and acetic anhydride ($CH_3COOCOCH_3$):

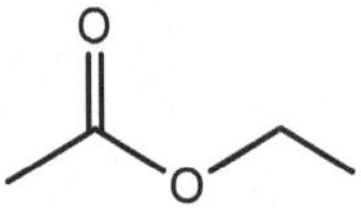

Ethyl acetate

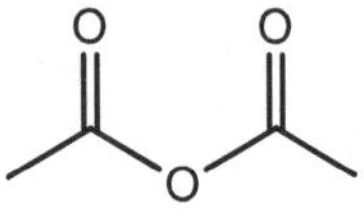

Acetic anhydride

(b) The anilide group in both acetanilide and benzanilide is an *N*-phenyl substituted amide group ($-CONHC_6H_5$).

Anilide group

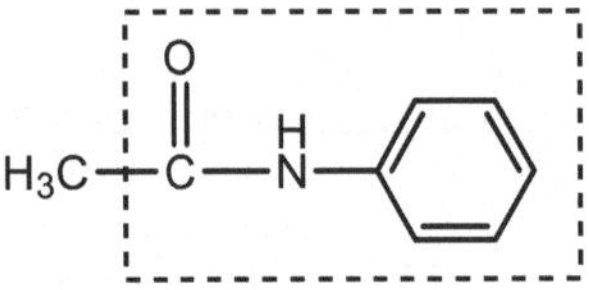

$H_3C-C(=O)-NH-C_6H_5$

$-CONHC_6H_5$

Acetanilide

Anilide group

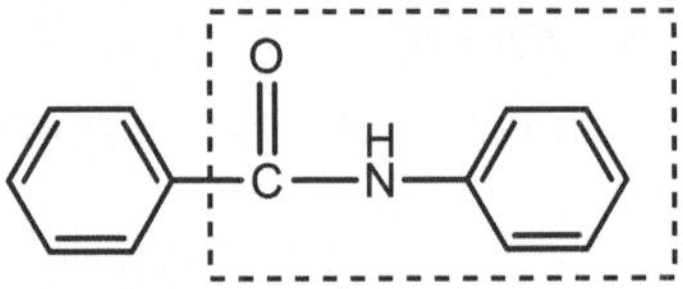

$C_6H_5-C(=O)-NH-C_6H_5$

$-CONHC_6H_5$

Benzanilide

2.3.13 Nitriles

Nitriles contain the cyano (–CN) functional group. They are also called **cyanides**. The cyano group has a carbon-nitrogen triple bond (C≡N) and is linear.

Nitriles are derived from **hydrogen cyanide** by replacement of the hydrogen atom with **alkyl or aryl groups**. The simplest nitrile is **acetonitrile** (CH_3CN), which is a common organic solvent.

Nitriles are easily converted into carboxylic acids and *their nomenclature has a relation to carboxylic acids*. For example, the nitrile CH_3CN is related to CH_3CO_2H called acet*ic* acid. **Replacing *ic* of acetic with *onitrile* gives the name acetonitrile** for CH_3CN.

The following are some examples of nitriles:

H—C≡N

(HCN)

Hydrogen cyanide

R—C≡N (R = alkyl, aryl)

(RCN)

Nitriles

Examples

$H_3C—C≡N$ (CH_3CN)	$(CH_3)_2CHCN$	$H_2C═CH—CN$	$C_6H_5—C≡N$ (C_6H_5CN)
Acetonitrile (methyl cyanide)	**Isobutyronitrile (isopropyl cyanide)**	**Acrylonitrile (vinyl cyanide)**	**Benzonitrile (phenyl cyanide)**

Note that the names **isobutyronitrile** is derived from isobuty*ric* acid and **acrylonitrile** from acryl*ic* acid. The name **benzonitrile** is derived from benz*oic* acid by replacing *oic* with *onitrile.*

2.4 SYSTEMATIC IUPAC NOMENCLATURE

So far, the names of most organic compounds are given according to their **common or trivial names**, many of which are still used widely.

However, to name a vast pool of organic compounds requires a systematic nomenclature which has been proposed by the International Union of Pure and Applied Chemistry (IUPAC). The **IUPAC nomenclature**, first put forward in 1982, has been revised several times to keep it up to date.

Here we will give a brief outline of the IUPAC nomenclature for different classes of organic compounds.

2.4.1 General features

The IUPAC name of an organic compound is composed of **four parts**:

- **Parent** chain or the *longest carbon chain*
- **Suffix** for the *principal functional group*
- **Prefix(es)** for *substituents* and *minor functional groups*
- **Positions** of *substituents* and *functional groups*

A. *Parent Chain*

The **parent chain** is the **longest continuous carbon chain** to which a functional group, if present, is directly attached. **The parent chain is named as an alkane**, although the compound may not be an alkane.

The names of alkanes in terms of the number of carbon atoms in a chain (up to ten carbons) are:

Number of C atoms	1	2	3	4	5	6	7	8	9	10
Alkane	methane	ethane	propane	butane	pentane	hexane	heptane	octane	nonane	decane

In the IUPAC nomenclature, the parent **alkane** is shortened as **alk** for hydrocarbons, and **alkan** for other compounds containing heteroatom functional groups.

To illustrate, consider the structures of the following compounds: a hydrocarbon (**X**) and an alcohol (**Y**) that contains a heteroatom functional group OH. The longest carbon chain for each of these compounds is a chain of five carbon atoms; therefore, the parent alkane is **pentane**.

For the **hydrocarbon** (**X**), the shortened form is **pent**, but for the **alcohol** (**Y**), the shortened form is **pentan**.

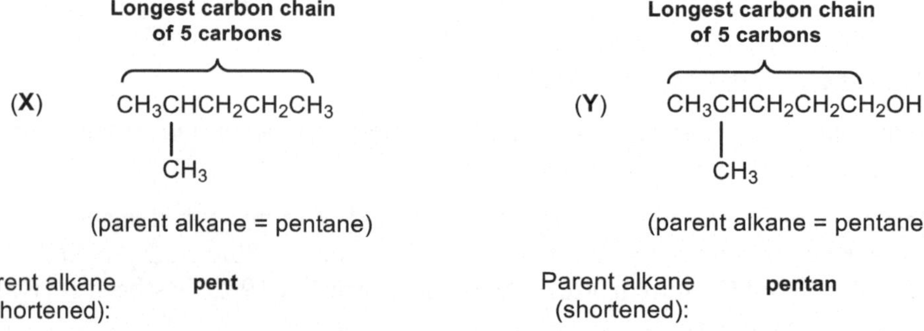

B. *Suffix*

The **suffix** is used for a **functional group** or the **principal functional group** when more than one functional group is present.

The principal or major functional group is decided by **functional group priorities**. The priorities for some common functional groups are the following:

$$CO_2H > CN > CHO > C{=}O \text{ (keto)} > OH > NH_2 > C{=}C > C{\equiv}C > X \text{ (halo)} > NO_2$$

The **suffix** is added to the parent (**alk** or **alkan**) to give the name of the **class** of the compound.

Table 2.2 shows the suffixes for some common functional groups and names of corresponding classes.

Table 2.2: Functional groups, their suffixes and names of classes

Functional Group	*Parent*	*Suffix*	*Class*
None	alk	-ane	alkane
C=C	alk	-ene	alkene
C≡C	alk	-yne	alkyne
OH	alkan	-ol	alkanol
NH_2	alkan	-amine	alkanamine
CHO	alkan	-al	alkanal

Functional Group	*Parent*	*Suffix*	*Class*
C=O (ketone)	alkan	-one	alkanone
CO_2H	alkan	-oic acid	alkanoic acid
COCl	alkan	-oyl chloride	alkanoyl chloride
CO_2R	alkan	alkyl -oate	alkyl alkanoate
COOCOR	alkan	-oic anhydride	alkanoic anhydride
$CONH_2$	alkan	-amide	alkanamide
CN	alkan	-enitrile	alkanenitrile

Notes:

1. When two or more *identical* functional groups are present, the suffix is labelled with a multiplying prefix: **di**- for two groups, **tri**- for three, and so on.
   In these cases, the parent alkane chain is denoted by **alka** (instead of alk) for hydrocarbons, and **alkane** (instead of alkan) for others.
   For example, hydrocarbons containing two C=C double bonds have the suffix -**diene** and the class is **alkadiene**. Similarly, for alcohols with two OH groups, the suffix is **-diol** and the class is **alkanediol.**
2. If a compound has a C=C double bond and other functional groups, the suffix for the compound will be obtained by combining the suffix –ene with the suffix for the other functional group as follows:
   ene + yne = **enyne**; ene + ol = **enol**; ene + one = **enone**; ene + oic acid = **enoic acid**, etc.

C. *Prefix*

The **alkyl (or aryl) substituents** and the **minor functional groups** (based on the functional group priorities) on the parent chain are denoted by their **prefixes.**

Table 2.3 shows the prefixes of some substituents and minor functional groups.

Table 2.3: Prefixes for substituents and minor functional groups

Substituent or Minor Functional Group	*Prefix*
R or Ar	alkyl- or aryl-
C=C	alkenyl-
C≡C	alkynyl-
X (F, Cl, Br, I)	halo- (fluoro-, chloro-, bromo-, iodo-)
OH	hydroxy-
OR	alkoxy-
NH_2	amino-
CHO	formyl-
C=O (ketone)	oxo-
CO_2H	carboxy-
CO_2R	alkoxycarbonyl-
CN	cyano-

Note: If the number of *identical* substituents or functional groups is two or more, the multiplying prefix **di**- for two, **tri**-for three, **tetra**- for four, and so on, is added to the prefix for the relevant substituent or functional group.

For example, a compound with two methyl groups gets the prefix **di**methyl-.

D. *Position of Substituent or Functional Group*

The **position** of attachment of *each* **substituent** or **functional group** to the parent chain is indicated by a **number**. The number is separated from words by a **hyphen**.

A **word** consisting of parent, suffix and prefix is written **without a space** between the parts.

The **carbon atoms of the parent chain** for *alkanes* and *compounds containing functional groups* are **numbered** as follows:

Alkanes

(a) Number the parent chain from the end which gives the **lower number to the substituent**.
(b) If two or more substituents are present, select that numbering which gives the **lower sum of numbers for the substituents**.
(c) In the case of **identical sum**, number the chain according to the **alphabetical order of substituents**.
(d) If two longest chains have **equal length**, select the chain with the **greater number of substituents as the parent chain**.

Compounds containing functional groups

(a) Number the parent chain from the end which is **nearer to the functional group** or **nearer to the principal functional group** when two or more functional groups are present.
(b) If **substituents** are present, select the numbering which gives them **lower numbers**.

Note: When two or more different substituent groups are present, an **alphabetical arrangement of the groups** (irrespective of their numbers) is necessary to name the organic compound.

E. *IUPAC name: Combination of parent, suffix, prefix and position*

Consider the hydrocarbon **(X)** and the alcohol **(Y)** shown above (Section 2.4.1A). The **IUPAC names** of these compounds are assigned by combining the four parts: parent, suffix, prefix (or prefixes) and position (or positions).

As shown below, the IUPAC name of the hydrocarbon **(X)** is **2-methylpentane**, while the IUPAC name of the alcohol **(Y)** is **4-methyl-1-pentanol**.

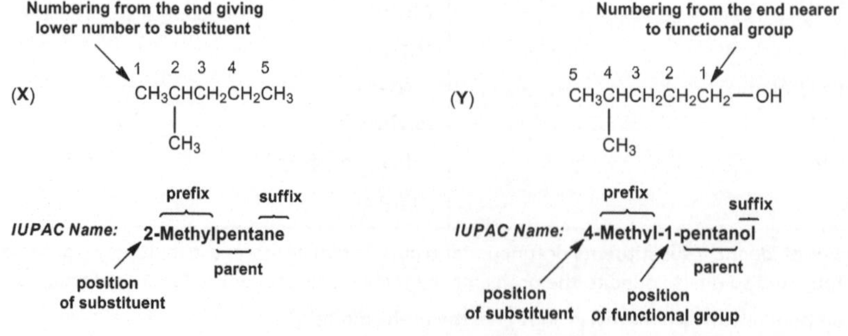

2.4.2 Illustrative Examples

In this section, we will illustrate the **IUPAC nomenclature** of organic compounds from different classes. An IUPAC name is determined by the combination of four parts: **parent**, **suffix**, **prefix** and **positions**, as described in Section 2.4.1.

For some organic compounds, **common or trivial names** are still widely used. So IUPAC names along with common names are indicated in certain cases.

Alkanes

The following examples show how the IUPAC names for **alkanes** are assigned.

Note that the **correct numbering** of the parent chain [see Section 2.4.1D] is indicated with a **tick (√)**, and the **incorrect numbering** is indicated with a **cross (×)**.

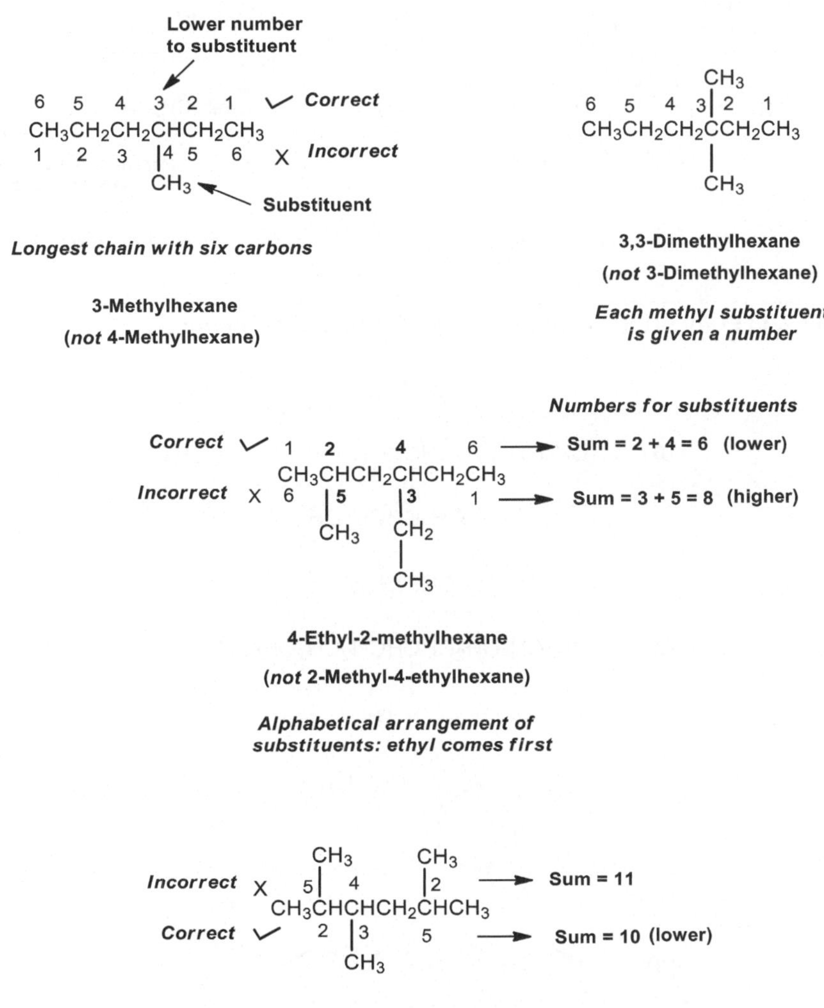

Numbering based on alphabetical priority: ethyl > methyl

✓ 7 5 3 1 → **Sum = 8**

$CH_3CH_2CHCH_2CHCH_2CH_3$

X 1 3 5 7 → **Sum = 8**

CH_3 (Methyl) CH_2CH_3 (Ethyl) ***Same value***

3-Ethyl-5-methylheptane

(*not* 5-Ethyl-3-methylheptane)

1 2 3 $CH_3CH_2CH_2$ — C4 (H_3C—C—CH_3) — $CH_3CH_2CHCH_2CH_2CH_2CH_3$ 5 6 7 8 9

Longest chain with nine carbons

5-Ethyl-4,4-dimethylnonane

(*not* 4,4-Dimethyl-5-ethylnonane)

Multiplying prefix di- is not considered in alphabetical choice: ethyl comes first

Parent chain of seven carbons has four substituents

Correct ✓ 1 2 3 4 5 6 7

$CH_3CHCHCHCHCH_2CH_3$ with CH_3, CH_3, CH_3, $CH_2CH_2CH_3$ **(propyl)**

2,3,5-Trimethyl-4-propylheptane

OR

1 2 3 4 $CH_3CHCHCHCHCH_2CH_3$ with CH_3, CH_3 **(*sec*-butyl)**, CH_3, $CH_2CH_2CH_3$ 5 6 7 X ***Incorrect***

Parent chain of seven carbons has three substituents

(*not* 4-*sec*-Butyl-2,3-dimethylheptane)

PROBLEM 2.11: Give IUPAC names of the following compounds:

(a) $CH_3CH_2CH_2CHCHCH_2CH_2CH_2CH_2CH_3$ with substituents $CH(CH_3)_2$ and $CH_3CHCH_2CH_3$

(b) $CH_3CH_2CH_2CHCHCH_2CH_2CH_2CH_3$ with substituents $CH(CH_3)_2$ and CH_3

(c) cyclohexane with H_3C and $C(CH_3)_3$

(d)

Answer

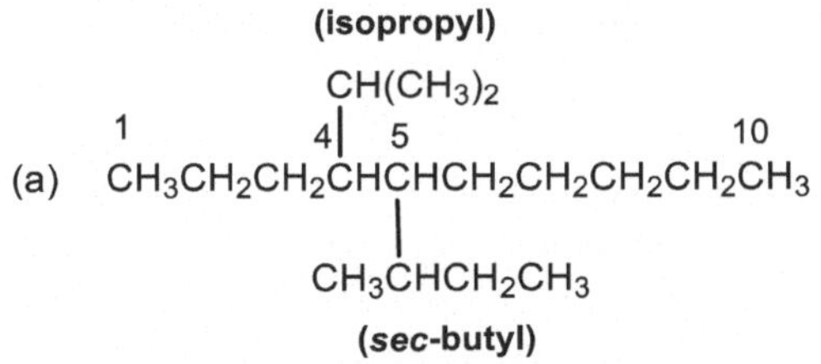

(a)

```
                 (isopropyl)
                  CH(CH3)2
      1          4|  5                 10
(a)  CH3CH2CH2CHCHCH2CH2CH2CH2CH3
                      |
                 CH3CHCH2CH3
                  (sec-butyl)
```

5-*sec*-Butyl-4-isopropyldecane

sec is not considered in alphabetical choice: sec-butyl comes before isopropyl

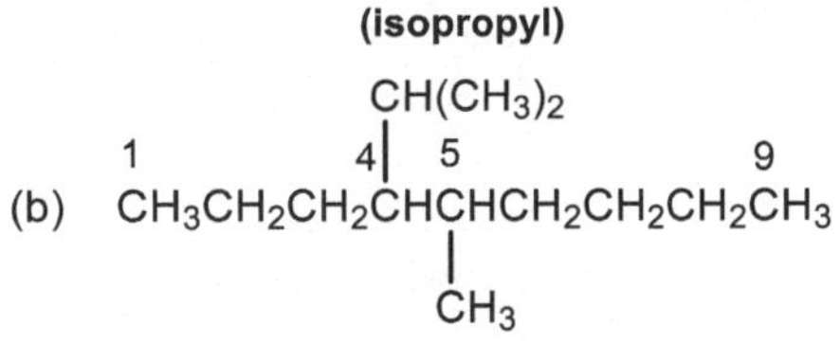

(b)

```
                 (isopropyl)
                  CH(CH3)2
      1          4|  5            9
(b)  CH3CH2CH2CHCHCH2CH2CH2CH3
                      |
                     CH3
```

4-Isopropyl-5-methylnonane

Alphabetical choice: isopropyl comes before methyl

(c) H_3C 3, 1 $C(CH_3)_3$ (cyclohexane ring)

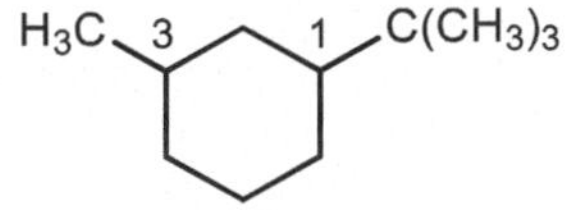

1-*tert*-Butyl-3-methylcyclohexane

tert is not considered in alphabetical priority in numbering and in alphabetical arrangement

(d) 1, 5, 10; (1), (2), (3)

(neopentyl or 2,2-dimethylpropyl)

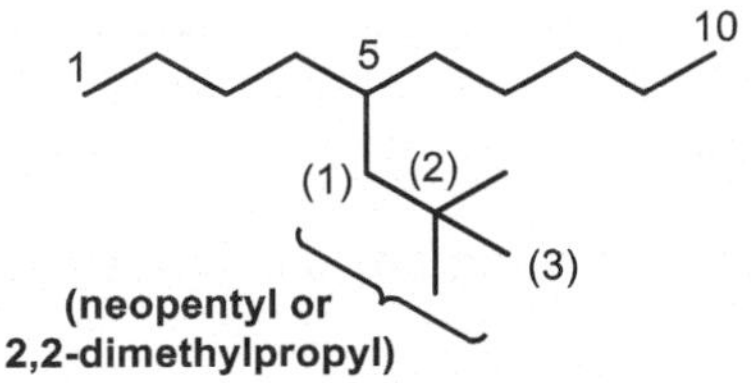

5-Neopentyldecane

or

5-(2,2-Dimethylpropyl)decane

Alkenes

The following examples illustrate the IUPAC nomenclature for **alkenes** and **cycloalkenes** containing the C=C functional group.

If the position of C=C in the structure is unique, then the *number* indicating the position of the functional group (C=C) is *not* specified. For instance, $CH_3CH{=}CH_2$ is named **propene** (*not* 1-propene).

The IUPAC nomenclature for **alkadienes** (hydrocarbons containing two C=C double bonds) is also illustrated with an example. Note that the parent diene is named **pentadiene** (*not* pentdiene) [see Section 2.4.1B].

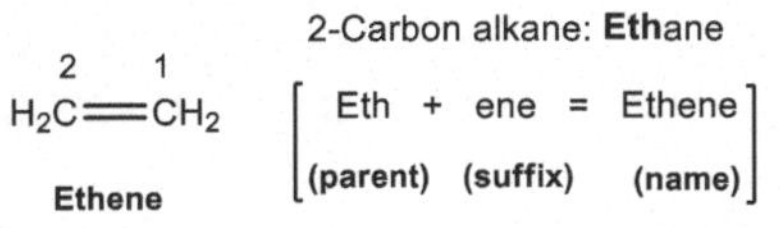

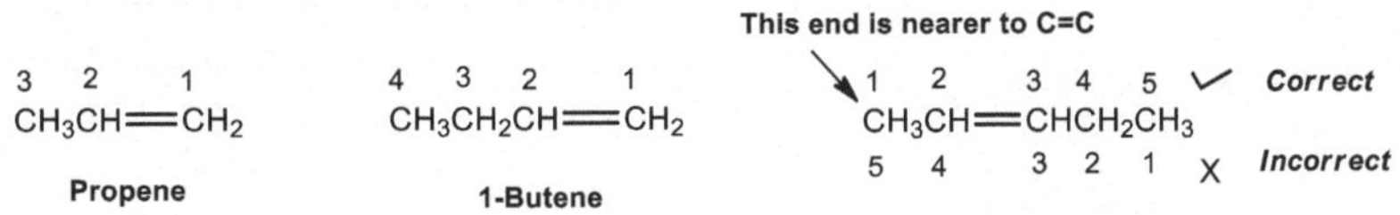

2-Carbon alkane: **Eth**ane

```
 2    1
H2C==CH2      [ Eth     +  ene     =  Ethene ]
              [(parent)   (suffix)    (name) ]
 Ethene
```

This end is nearer to C=C

```
3   2    1          4  3   2    1           1  2    3  4  5  ✓  Correct
CH3CH==CH2          CH3CH2CH==CH2           CH3CH==CHCH2CH3
                                            5  4    3  2  1  X  Incorrect
  Propene              1-Butene
                                              2-Pentene
                                           (not 3-Pentene)
```

Substituent has lower number

1 2 3 4 ✓
$CH_3C{=}CHCH_3$
4 3 2 1 X
CH_3

2-Methyl-2-butene
(*not* 3-Methyl-2-butene)

Lower number to C=C

5 4 3 2 1 ✓
$CH_3C{=}CHCH{=}CH_2$
1 2 3 4 5 X
CH_3

4-Methyl-1,3-pentadiene
(*not* 2-Methyl-2,4-pentdiene)

(propyl)
$CH_2CH_2CH_3$

1-Propylcyclohexene

Numbering from this end of the double bond to give lower numbers to substituents

CH_3CH_2 CH_3

4-Ethyl-3-methylcyclohexene

Alkynes

The following are the IUPAC names of some **alkynes** containing the C≡C functional group.

If the position of the functional group is unique, then the *number* indicating its position is *not* specified. For instance, $CH_3C{\equiv}CH$ is named **propyne** (*not* 1-propyne).

The IUPAC names of hydrocarbons containing both C≡C and C=C are also illustrated.

2 1
$HC{\equiv}CH$
Ethyne

2-Carbon alkane: **Eth**ane

[Eth + yne = Ethyne]
[**(parent)** **(suffix)** **(name)**]

---

3 2 1
$CH_3C{\equiv}CH$
Propyne

4 3 2 1
$CH_3CH_2C{\equiv}CH$
1-Butyne

5 4 3 2 1
$CH_3CHCH_2C{\equiv}CH$
CH_3
4-Methyl-1-pentyne

3 1
$CH_3CH{=}CHC{\equiv}CH$
(ene + yne = enyne)

3-Penten-1-yne

1 2 3 4 5 ✓ ***Correct***
$CH_2{=}CHCH_2C{\equiv}CH$
5 4 3 2 1 X ***Incorrect***

(functional group priority: C=C > C≡C ; between them, C=C gets lower number 1)

1-Penten-4-yne
(*not* 4-Penten-1-yne

Alkyl halides

The following examples illustrate the IUPAC nomenclature for **alkyl halides**. The **halo** functional group is treated as a **substituent** group.

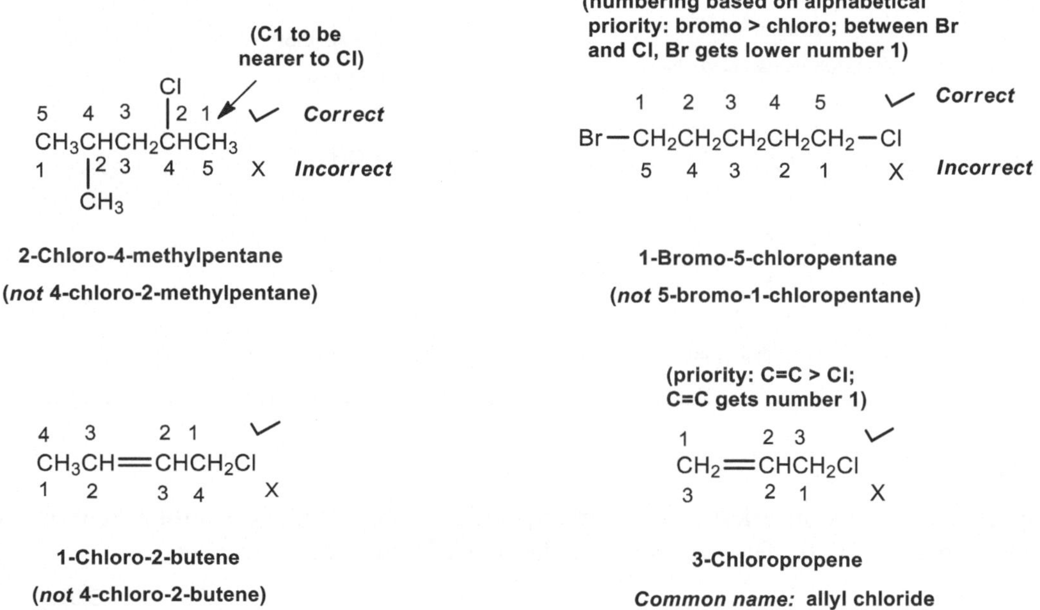

Alcohols

The alcohol functional group is –OH, and the class is **alkanol** (one OH group). When two OH groups are present, the class is **alkanediol**.

The IUPAC nomenclature for alcohols and dihydric alcohols is illustrated with the following examples:

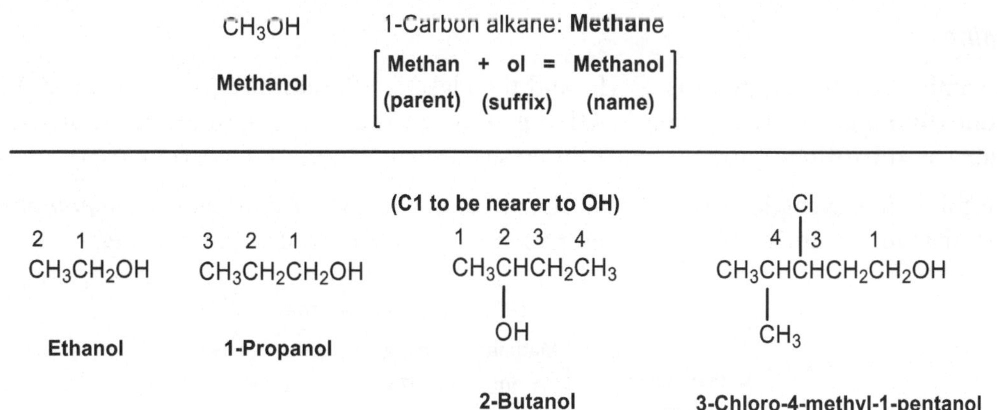

Major functional group

$H_2C{=}CHCH_2OH$ (2 1)

(priority: OH > C=C)

2-Propen-1-ol

***Common name:* allyl alcohol**

CH_2OH (1) – CH_2OH (2)

1,2-Ethanediol

(*not* 1,2-Ethandiol)

2,5-Dimethylcyclopentanol

2-Cyclohexen-1-ol

Ethers

An ether contains an **alkoxy** (OR) group, which is regarded as a **substituent** in the IUPAC nomenclature. The following are the IUPAC names of some ethers:

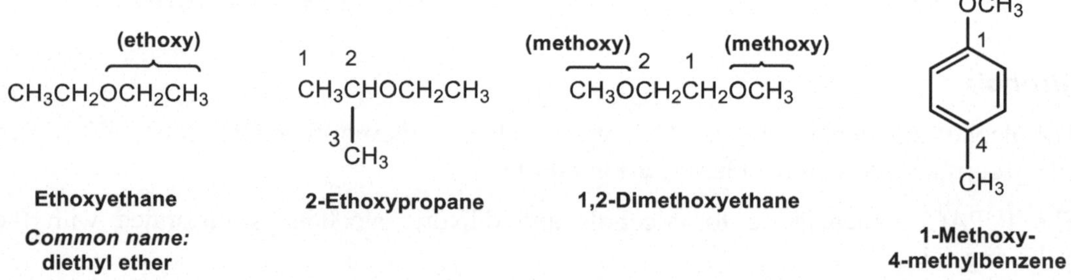

Amines

The amine functional group is $-NH_2$, and the class is **alkanamine** (one NH_2 group) and **alkanediamine** (two NH_2 groups). **Alkyl groups on the nitrogen atom** of the amine are treated as **substituents** and the name of the substituent is prefaced by ***N*- or *N,N*-**.

The following examples illustrate the IUPAC nomenclature for amines. *Common names* of methanamine, ethanamine and 2-propanamine are also given within *brackets*.

CH_3NH_2 1-Carbon alkane: **Methane**

Methanamine

(methylamine)

Methan	**+**	**amine**	**=**	**Methanamine**
(parent)		**(suffix)**		**(name)**

---

2 1
$CH_3CH_2NH_2$

Ethanamine
(ethylamine)

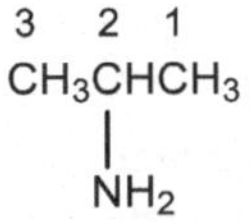

3 2 1
CH_3CHCH_3
|
NH_2

2-Propanamine
(isopropylamine)

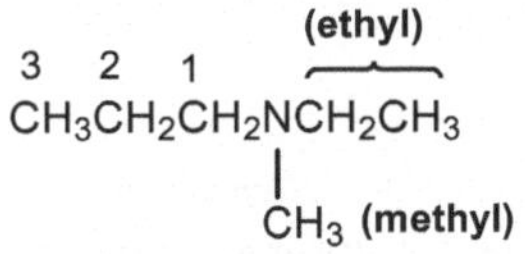

3 2 1 (ethyl)
$CH_3CH_2CH_2NCH_2CH_3$
|
CH_3 (methyl)

***N*-Ethyl-*N*-methyl-1-propanamine**

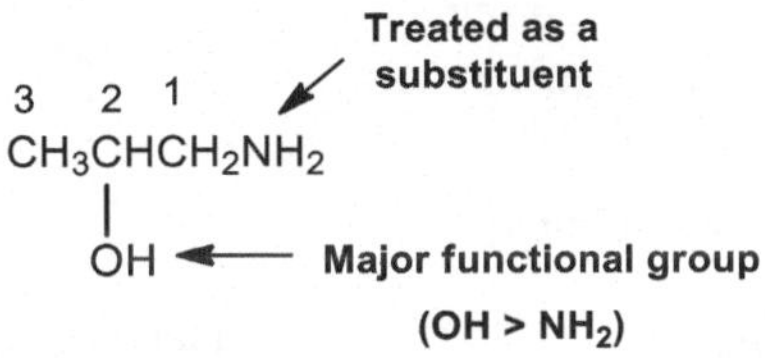

3 2 1
$CH_3CHCH_2NH_2$ ← **Treated as a substituent**
|
OH ← **Major functional group ($OH > NH_2$)**

1-Amino-2-propanol

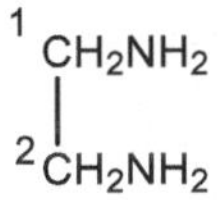

1 CH_2NH_2
|
2 CH_2NH_2

1,2-Ethanediamine
(*not* 1,2-ethandiamine)

Nitro compounds

Nitro (NO_2) group is the lowest priority functional group, and is treated as a **substituent** in the IUPAC system. The following are the IUPAC names of some nitro compounds:

CH_3NO_2

Nitromethane

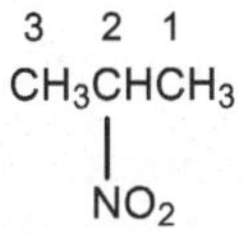

3 2 1
CH_3CHCH_3
|
NO_2

2-Nitropropane

OCH_3 (1), NO_2 (2)

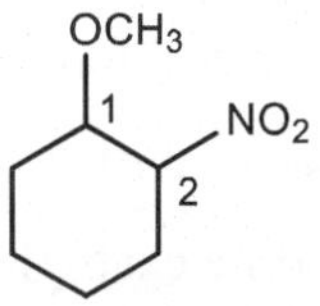

1-Methoxy-2-nitrocyclohexane

Aldehydes

The aldehyde functional group is –CHO, and the class is **alkanal**. Since the aldehyde carbon is normally C1, the position 1 of the aldehyde group is not indicated.

If the CHO group is directly attached to a **cycloalkane ring**, the suffix is carbaldehyde, and the compound is called **cycloalkanecarbaldehyde**.

The following examples illustrate the IUPAC nomenclature for aldehydes. *Common names* of methanal, ethanal, propanal, butanal, 2-methylpropanal and 2-propenal are also shown within *brackets*.

HCHO — 1-Carbon alkane: **Methane**

Methanal
(formaldehyde)

[**Methan** + **al** = **Methanal**]
[**(parent)** **(suffix)** **(name)**]

CH_3CHO	CH_3CH_2CHO	$CH_3CH_2CH_2CHO$	$CH_3CH(CH_3)CHO$
Ethanal (acetaldehyde)	**Propanal (propionaldehyde)**	**Butanal (butyraldehyde)**	**2-Methylpropanal (isobutyraldehyde)**

$CH_2{=}CHCHO$	$CH_3CH(OH)CH_2CHO$	$C_6H_{11}CHO$
2-Propenal (acrolein or acraldehede)	**3-Hydroxybutanal** (Major functional group (CHO > OH); OH ← Substituent)	**Cyclohexanecarbaldehyde**

Ketones

The suffix for the keto (C=O) functional group is –one, and the class is **alkanone**. When two keto groups are present, the class is **alkanedione**.

If the position of the keto group in the structure is unique, then the number indicating its position is not specified. For instance, $CH_3COCH_2CH_3$ is named **butanone** (not 2-butanone) since the position of the keto group must be C2.

When a **keto group** is present as a **substituent**, it is called **oxo** (as a prefix).

The following are some examples of the IUPAC nomenclature for ketones. *Common names* of propanone, 4-methyl-3-penten-2-one and 2,4-pentanedione are also given within *brackets*.

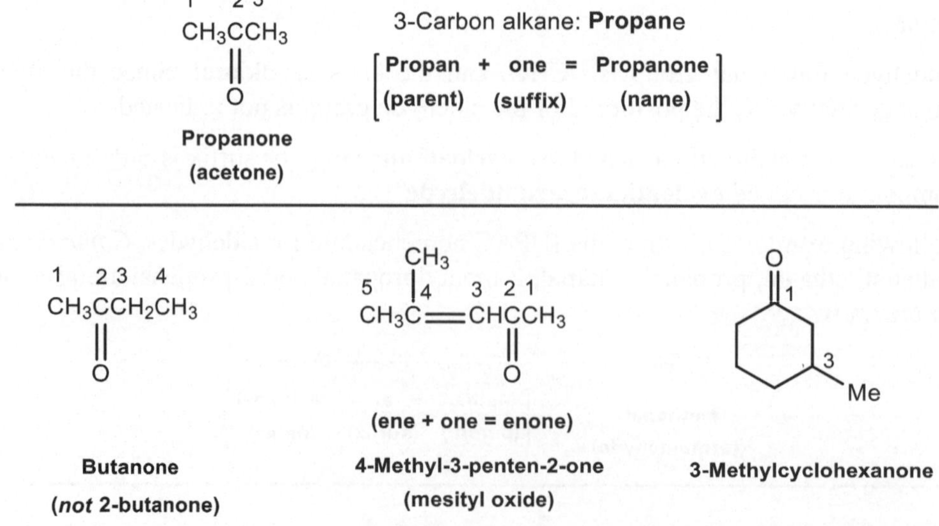

4 2
$CH_3CCH_2CCH_3$ (each numbered C is C=O)

2,4-Pentanedione
(acetylacetone)

5 4 3 2 1
$CH_3CCH_2CH_2CHO$ (C-4 is C=O)
← **Substituent**
Major functional group **(CHO > C=O)**

4-Oxopentanal

Carboxylic acids

The carboxylic acid functional group is $-CO_2H$, and the class is **alkanoic acid**. When two CO_2H groups are present, the class is **alkanedioic acid**.

If the CO_2H group is directly attached to a **cycloalkane or benzene ring**, the **suffix carboxylic acid** is added to the name of the **ring**. For example, if CO_2H is attached to a cyclohexane ring, the compound is called **cyclohexanecarboxylic acid.**

The simplest aromatic carboxylic acid is **benzoic acid**, and this name is retained as the preferred IUPAC name. As a systematic IUPAC name, it is called **benzenecarboxylic acid.**

The following are the IUPAC names of some carboxylic acids. *Common names* of first six members of alkanoic acids are also shown within *brackets.*

For α-amino acids (in which NH_2 is at 2-position), common names are generally used. For example, **alanine** is the common name of **2-aminopropanoic acid.**

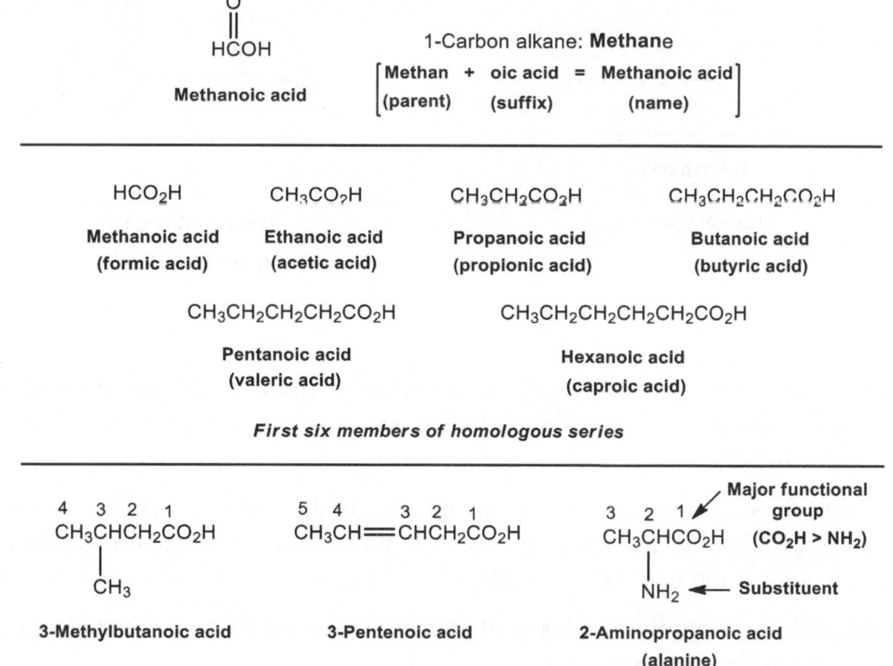

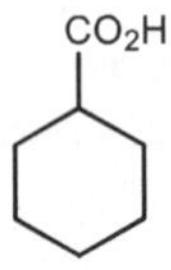

CO_2H on cyclohexane ring — **Cyclohexanecarboxylic acid**

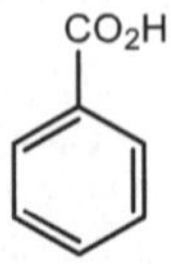

CO_2H on benzene ring — **Benzoic acid or benzenecarboxylic acid**

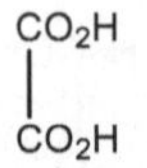

CO_2H–CO_2H — **Ethanedioic acid (oxalic acid)**

Esters

The ester functional group is $-CO_2R$, and the class is **alkyl alkanoate** (R = alkyl). R may also be an **alkenyl** or **aryl** group. The carboxylic part of esters may also be derived from aromatic carboxylic acids such as benzoic acid.

The following examples illustrate the IUPAC nomenclature for esters. *Common names* of some esters are mentioned within *brackets*.

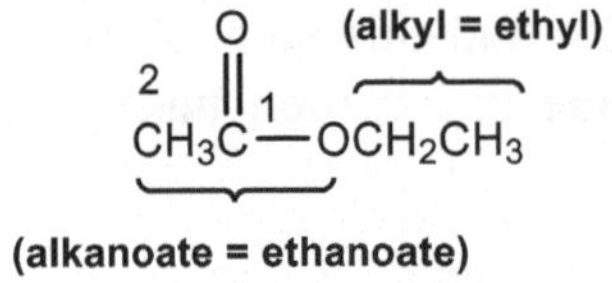

$CH_3C(=O)-OCH_2CH_3$ (alkyl = ethyl); (alkanoate = ethanoate)

Ethyl ethanoate
(ethyl acetate)

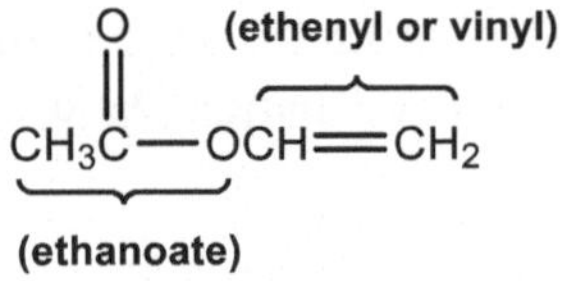

$CH_3C(=O)-OCH=CH_2$ (ethenyl or vinyl); (ethanoate)

Ethenyl ethanoate
(vinyl acetate)

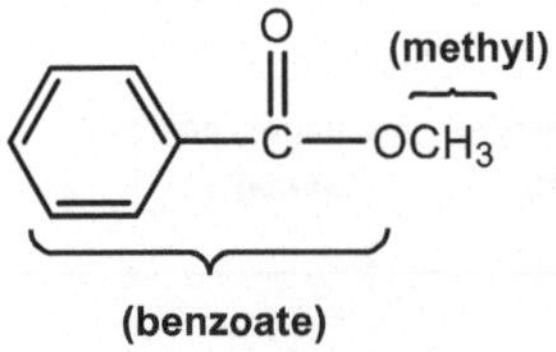

$C_6H_5-C(=O)-OCH_3$ (methyl); (benzoate)

Methyl benzoate

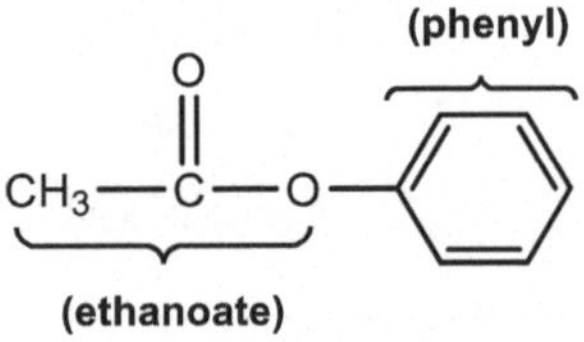

$CH_3-C(=O)-O-C_6H_5$ (phenyl); (ethanoate)

Phenyl ethanoate
(phenyl acetate)

Amides

The amide functional group is $-CONH_2$, and the class is **alkanamide. Alkyl groups on the amide nitrogen atom** are treated as **substituents** and the name of the substituent is prefaced by ***N*- or *N,N*-**.

Note that the nomenclature for amides is related to the corresponding alkan*oic* acids. Replacing *oic* of alkanoic by *amide* gives alkanamide. Similarly, **aromatic amides** are named. For example, the name **benzamide** arises by replacing ***oic*** of benzoic with ***amide***.

The IUPAC nomenclature for amides is illustrated with the following examples. *Common names* of some amides are given within *brackets*.

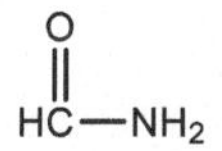

$HC(=O)-NH_2$

Methanamide
(formamide)

1-Carbon alkane: **Methane**

[**Methan** (parent) + **amide** (suffix) = **Methanamide** (name)]

$\overset{2}{C}H_3\overset{1}{C}ONH_2$

Ethanamide
(acetamide)

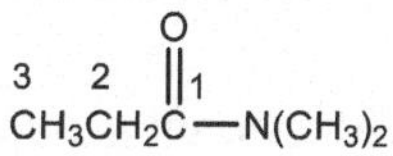

$CH_3CONHCH_3$

***N*-Methylethanamide**
(*N*-Methylacetamide)

$\overset{3}{C}H_3\overset{2}{C}H_2\overset{1}{C}(=O)-N(CH_3)_2$

***N.N*-Dimethylpropanamide**

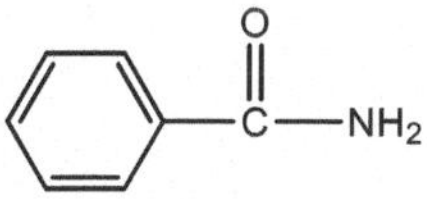

$C_6H_5-C(=O)-NH_2$

Benzamide

Nitriles

The nitrile functional group is –CN, and the class is **alkanenitrile**. If a C=C is present, the class is **alkenenitrile**.

If the CN group is directly attached to a **benzene or cycloalkane ring**, the **suffix carbonitrile** is added to the name of the **ring**.

When a nitrile group is present as a **substituent**, it is called **cyano** (as a prefix).

The following are the IUPAC names of some nitriles. *Common names* of some nitriles are shown within *brackets*.

$\overset{2}{C}H_3\overset{1}{C}\equiv N$

Ethanenitrile
(acetonitrile)

2-Carbon alkane: **Ethane**

[**Ethan** (parent) + **enitrile** (suffix) = **Ethanenitrile** (name)]

$\overset{3}{H_2C}=\overset{2}{C}H-\overset{1}{C}\equiv N$

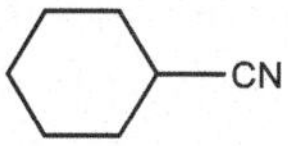

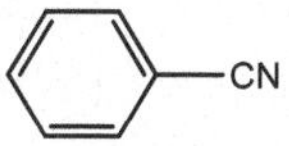

Propenenitrile
(acrylonitrile)

$C_6H_{11}-CN$

Cyclohexanecarbonitrile

C_6H_5-CN

Benzenecarbonitrile
(benzonitrile)

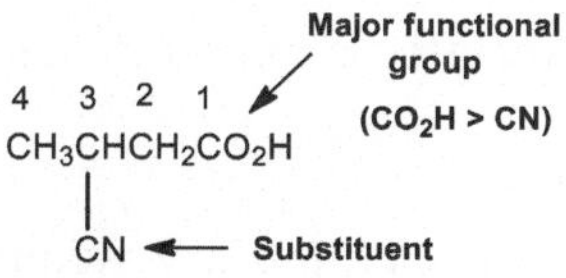

$\overset{4}{C}H_3\overset{3}{C}H(CN)\overset{2}{C}H_2\overset{1}{C}O_2H$

Major functional group (CO_2H > CN)

CN ← **Substituent**

3-Cyanobutanoic acid

SUPPLEMENTARY PROBLEMS

2.12

(a) Write Lewis structures for nitronium ion (NO_2^+) and nitrite ion (NO_2^-) indicating the formal charge and lone pairs on the proper atoms.

(b) With the help of curved arrows, draw five Lewis structures for *p*-nitrophenol.

2.13 How many functional groups can you identify in the following compound?

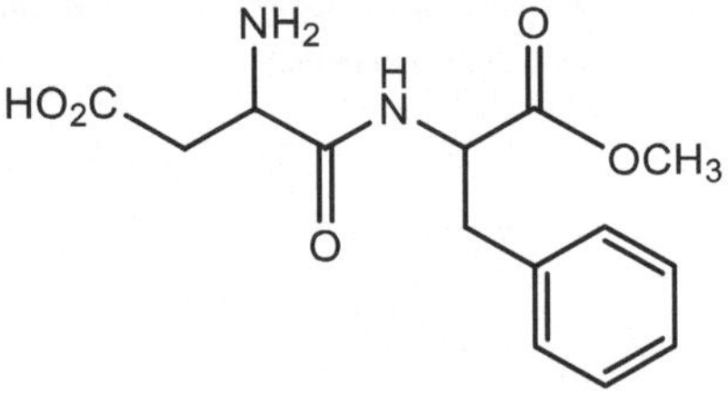

Aspartame
(an artificial sweetener)

2.14 Suggest the structure of a hydrocarbon of molecular formula C_5H_8 that has only secondary hydrogen atoms.

[*Hint:* four CH_2 groups and two rings with a common carbon atom (known as a spiro compound)]

2.15 Write the possible alkyl groups that can be derived from the following alkane:

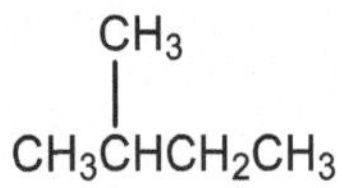

$$CH_3CH(CH_3)CH_2CH_3$$

Isopentane

[*Ans*: two sets of equivalent primary hydrogens, one set of equivalent secondary hydrogens and a tertiary hydrogen give four alkyl groups: $(CH_3)_2CHCH_2CH_2$– (isopentyl or isoamyl or 3-methylbutyl), $CH_3CH_2CH(CH_3)CH_2$–(2-methylbutyl), $CH_3CH(CH_3)CH(CH_3)$–(3-methyl-2-butyl) and $(CH_3)_2C(CH_2CH_3)$–(*t*-pentyl or 2-methyl-2-butyl)]

2.16 Justify or criticize the following:

Sec-Butyl bromide defines a specific compound whereas *sec*-pentyl bromide denotes more than one compound.

2.17 How many primary, secondary and tertiary alcohols are possible with the molecular formula $C_5H_{11}O$? Write their structural formulas and IUPAC names.

[*Hint:* eight isomeric alcohols: four primary, three secondary and one tertiary]

2.18 Write structural formulas and give IUPAC names for the following compounds:

(a) Aldehydes with molecular formula C_4H_8O.

(b) Ketones with molecular formula $C_5H_{10}O$.

[*Hints:* (a) two aldehydes; (b) three ketones]

2.19 Draw a structural formula for each of the following:

(a) 1-(2-propenyl)cyclohexene

(b) 4-Bromo-5-methyl-1-hexyne

(c) 3-Ethyl-1,3-hexanediol

(d) 2-Methyl-2-butenal

(e) 5-Methyl-4-hexen-3-one

(f) 4-Methyl-2-cyclohexenone

(g) 4-Hydroxy-2-pentenoic acid

(h) 2-Butyl 4-methylpentanoate

(i) 3-Phenylbutanamide

(j) Tetrabutylammonium fluoride

2.20 Provide the IUPAC name for each of the following:

(a) $HC\equiv CCH_2OH$

(b) $CH_3CH_2OCH_2CHCH_3$ (OH on CH)

(c) $CH_3CCH_2CHCH_2OH$ (=O on C; Cl on CH)

(d) $CH_3CHCH_2C(CH_3)CH_3$ (NO_2 on CH; CH_3 and OH on C)

(e) $H_2C{=}C(CH_3)CHCH_2CHO$ (OH on CH)

(f) $CH_3CHCH_2CHCH_3$ (CN on first CH; Br on second CH)

(g) NH_2 O OH

(h) O O O

(i) O O

(j) H_2N OH

CHAPTER 3

Bonding in Organic Molecules

IN THIS CHAPTER

- Atomic orbitals: s and p; energy diagram
- Molecular orbitals: σ and π; energy diagram
- Hybridization of carbon, nitrogen and oxygen; formation of C–H, C–C, C–O, C–N, C=C, C=O and C≡C bonds; molecular shapes; predicting hybridization of atoms in molecules and ions
- Bond lengths and bond angles; hybridization and VSEPR theories
- Bond energies; estimation of heat of reaction (ΔH^0)
- Conjugation and delocalization (π–π, π–p); no delocalization for orthogonal π bonds
- Aromaticity, Frost circle and Hückel rules
- Resonance, rules for resonance and resonance energy
- Hyperconjugation (σ–p, σ–π)
- Inductive effects (+*I*, –*I*), mesomeric effects (+*M*, –*M*) and steric effects

In terms of simple **Lewis theory**, a covalent bond is formed by sharing an electron pair between two atoms.

However, modern understanding of chemical bonding is based on the **quantum theory**, proposed independently by Erwin Schrödinger (Nobel prize in physics, 1933), Werner Heisenberg (Nobel prize in physics, 1932) and Paul Dirac (Nobel prize in physics, 1933).

The idea that **electron** can behave both as a **wave** and a **particle** (wave-particle duality) was proposed by Louis de Broglie (Nobel prize in physics, 1929), and it is an experimentally observable fact that electrons can have the character of a particle (cathode rays) as well as a wave (electron diffraction).

The Schrödinger's formulation of quantum theory based on the wave nature of the electron, known as **wave mechanics**, is most often used by chemists.

3.1 ATOMIC ORBITALS

Question: Where the electrons are in an atom?

Answer: The electrons are in atomic orbitals.

Electrons in an atom do not rotate round the nucleus in defined orbits like planets around the sun. Their behaviour is not governed by Newton's laws of motion (classical mechanics). The world of electrons in atoms is different and complex, not like our everyday perception; it is an abstract, quantum world.

An important quantum-mechanical principle, called the Heisenberg's uncertainty principle, states that it is impossible to determine simultaneously the exact position and momentum of an electron. Thus electrons can never be precisely located in an atom. We can only think of electrons in atoms as having a probability of being in a region of three-dimensional space.

The region of space where the probability of finding the electron is a maximum is called an atomic orbital. We can give an *analogy* here. The atomic orbital is the apartment in which the electron resides, but the doors and windows of the apartment are closed so that we can never peek in and know exactly where in the apartment the electron is at a given moment. However, we know for sure that the electron is in the apartment, which is its orbital.

The shapes of **atomic orbitals** are determined by mathematical functions, known as **wavefunctions**. Schrödinger developed a wave equation as a mathematical model for an electron associated with the atomic nucleus. The **Schrödinger equation** is a differential equation, and solving the Schrödinger equation for a hydrogen atom gives a **series of solutions as simple equations**. These solutions to the wave equation, called the wavefunctions, can be graphed in three-dimensions. Such **graphs** represent the **atomic orbitals** in a pictorial fashion.

The solutions for a one-electron hydrogen atom can also be used for the **orbitals of multi-electron atoms** with necessary modifications.

Wavefunction, Phase Sign and Probability

An **atomic orbital** is described by a **wavefunction**, which is usually designated by the symbol ϕ. However, it is the **square of the wavefunction (ϕ^2)** which denotes the **probability** of finding the electron at any point in the three-dimensional space available to it.

Thus if ϕ^2 is *large* in a unit volume of space, the probability of finding the electron in that volume, i.e. the **electron probability density** is *high*. Similarly, if ϕ^2 is *small* in some other unit volume of space, the electron density is *low*. The **sum of all these probabilities** gives a smeared out picture of the electron's habitat, called an **electron cloud**. We can imagine the electron cloud by blinking our eyes repeatedly for a large number of times and plotting the points at which the electron is at each blink.

For a hydrogen atom, the **atomic orbital (1s)** has a spherical region of electron probability about the nucleus and the probability of finding the electron within a sphere of radius 1.7 Å is greater than 95%.

In general, the value of the **wavefunction** ϕ for a particular point in space may be **positive or negative**, which indicates the **phase** of the orbital (see below). The value of ϕ may also be **zero** and the point where $\phi = 0$ is called the **node** of the orbital. **At the node, there is zero probability ($\phi^2 = 0$) of finding the electron**, and on **opposite sides** of a node, **ϕ has opposite signs.**

It should be noted that the sign of ϕ indicates an **algebraic sign** and must not be confused with positive or negative charge. **An atomic orbital as a graphical description of the wave function ϕ always shows a phase sign.**

An atomic orbital can also be depicted in terms of the electron probability distribution or electron cloud. The probability (ϕ^2) is always **positive** irrespective of the sign of ϕ (+ or –). Thus **an atomic orbital in terms of probability (ϕ^2) distribution has no phase sign**.

Electron occupancy

Atomic orbitals provide *a series of possible energy values* for the electron. Thus each orbital is associated with a particular **energy level or energy state**.

An orbital may contain zero, one or two electrons. If the orbital contains *no* electrons, it is said to be **empty or vacant**, and if the orbital contains a maximum of *two* electrons, it is called a **filled** orbital. But when the orbital contains a *single* electron, it is called a **half-vacant or half-filled** orbital.

3.1.1 *s* and *p* atomic orbitals

A hydrogen atom has only 1s orbital, which is the simplest atomic orbital. The atoms with which we are most concerned in organic chemistry are C, N and O, which have 2s and 2p atomic orbitals besides 1s orbital.

1s orbital

The **1s orbital** is **spherically symmetrical** about the nucleus. It is *conventionally* drawn as a **circle** which is a section through a spherical contour. Importantly, the 1s orbital has a **single phase sign** (say, +).

A simple *analogy* will help us understand the nature of the **phase sign**. Consider the vibration of a string fixed at both ends. We know that such a string has a fundamental frequency. In this vibration, every point on the string moves up or moves down at the same time, although to a different extent. Since the *direction* of movement of every point is the *same* [upward (say, +) or downward (–)], the **vibration of the string is in the same phase**.

Figure 3.1 shows **1s orbital** in spherical and conventional representations with a single phase sign in relation to the fundamental vibration of a string, and as a probability distribution or electron cloud.

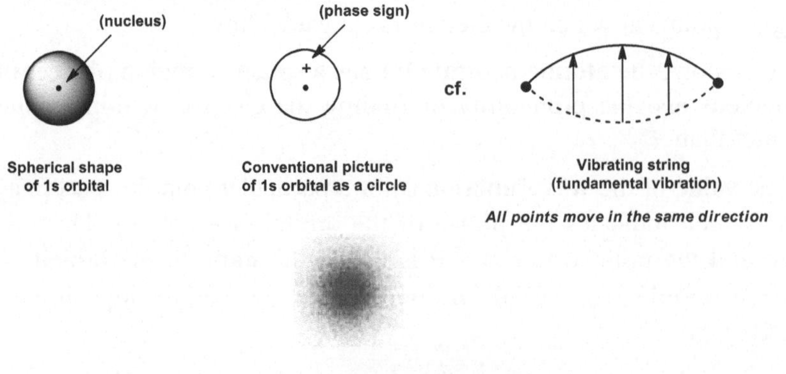

Figure 3.1: Representations of 1s orbital: spherical shape, conventional picture with a single phase sign (say, +) vis-à-vis the phase of a vibrating string, and electron cloud.

2s orbital

Like 1s orbital, the **2s orbital** is also spherically symmetrical but possesses a **spherical node** which divides the orbital into two parts with **opposite phase signs** (+ and –). The node is close to the nucleus and hence the *inner sphere is not important for bonding.* Conventionally, the 2s orbital is drawn as a single circle with a single sign, omitting the inner sphere.

2s orbital with a spherical node and opposite phase signs

Conventional picture of 2s orbital omitting the inner sphere

2p orbitals

Unlike an s orbital, a **p orbital** is dumbbell-shaped and has **two lobes** with **opposite phase signs** (+ and –), which are separated by a single **nodal plane**. The opposite phases of two lobes can also be indicated by **shading**: one lobe is shaded and the other lobe is not shaded.

An *analogy* of a **vibrating string** (first harmonic) or a **transverse wave** moving across a lake might be helpful to understand the different phases in a p orbital.

There are two halves of the string of which the left half moves upwards and the right half moves downwards; these two halves have opposite phases with a change of phase at the node.

For the transverse wave, the wave in the crests above the average level is positive, whereas the wave in the troughs below the average level is negative. There is a change of phase at the node where the wave is at the average level.

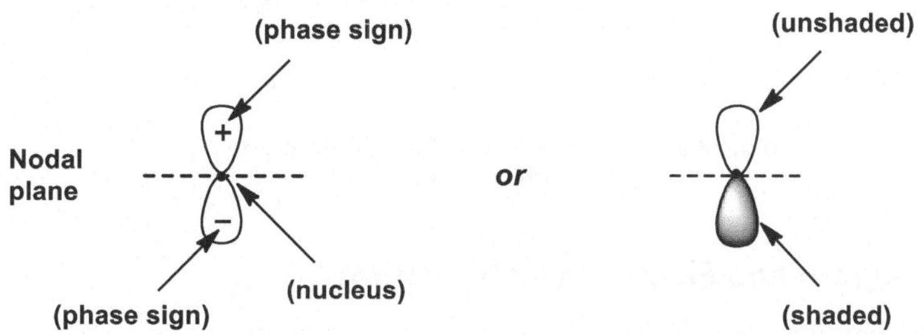

p orbital with two lobes of opposite phase signs

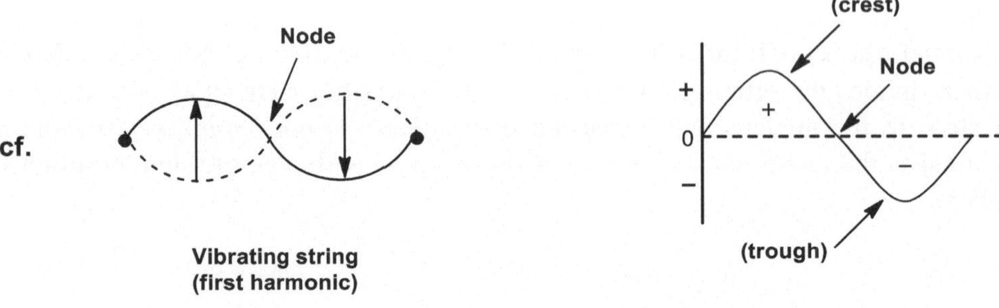

Transverse wave

Left part moves upwards and right part movesdownwards; two parts separated by a node

Opposite phases of the wave in crest and trough; change of phase at node

Unlike the 1s or 2s orbital which is spherical and non-directional, the **2p orbital is directional and oriented along an axis in three-dimensional space**. There are three possible orientations along x-, y- and z-axes, which give **2p$_x$**, **2p$_y$** and **2p$_z$** orbitals.

In terms of the *probability* distribution, a p orbital has *no* phase sign.

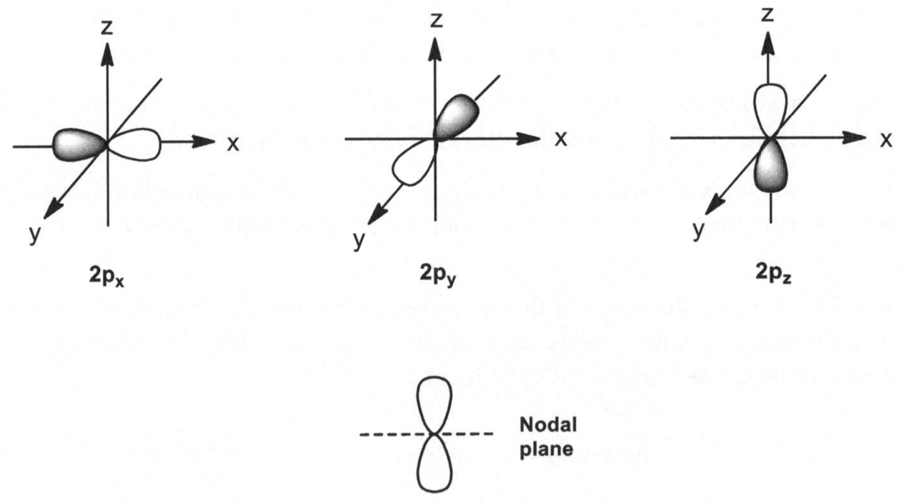

p orbital in terms of probability distribution
(no phase sign)

3.1.2 Energy Diagram and Electron Configuration

A hydrogen atom has only 1s orbital and it has only 1 electron. Therefore, the **electron configuration of hydrogen (H) is 1s^1**.

However, carbon (C), nitrogen (N) and oxygen (O) atoms have 1s, 2s and three 2p (2p$_x$, 2p$_y$, 2p$_z$) orbitals. A **carbon** atom has **6 electrons**, a **nitrogen** atom has **7 electrons** and an **oxygen** atom has **8 electrons**.

Relative energies of atomic orbitals

For any atom, **an electron in a 1s orbital has the lowest energy** since it is closest to the positively charged nucleus and under the influence of greatest electrostatic attractive force. In a hydrogen atom, there is only one electron which occupies the 1s orbital. The **next higher energy orbital is the 2s orbital** which has a spherical node.

The three 2p orbitals have the same energy. Orbitals of equal energy are called **degenerate orbitals**; thus **$2p_x$, $2p_y$ and $2p_z$ orbitals are degenerate**. The **2p orbital** has a planar node and it has **higher energy** than the 2s orbital.

The relative energies of the orbitals are $1s < 2s < 2p$.

Electron spin

An electron has the property of **spin**. There are **two possible spin orientations**, usually indicated as **up** and **down** by arrows. If the spins of two electrons are **both up** or **both down**, they are said to be **parallel** but if one spin is **up** and the other spin is **down**, they are called **opposite or antiparallel.**

An atomic orbital can contain a maximum of two electrons, when their spins must be opposite, and they form an electron pair. This is a consequence of the Pauli exclusion principle.

With degenerate orbitals such as the **three p orbitals**, electrons occupy each p orbital **singly** with **parallel spins** until they are forced to pair up. This is known as the **Hund's rule**. This arrangement allows the electrons to be farther apart and thereby to minimize the electron-electron repulsion.

Ground state electron configuration

The **energy diagram** and **electron configuration** of carbon, nitrogen and oxygen atoms are shown in Figure 3.2. The lowest energy or most stable electron configuration is called the **ground state electron configuration.**

For each atom (C, N and O), electrons are filled up in atomic orbitals in order of increasing energy starting from the lowest energy 1s orbital, obeying the Pauli exclusion principle and the Hund's rule.

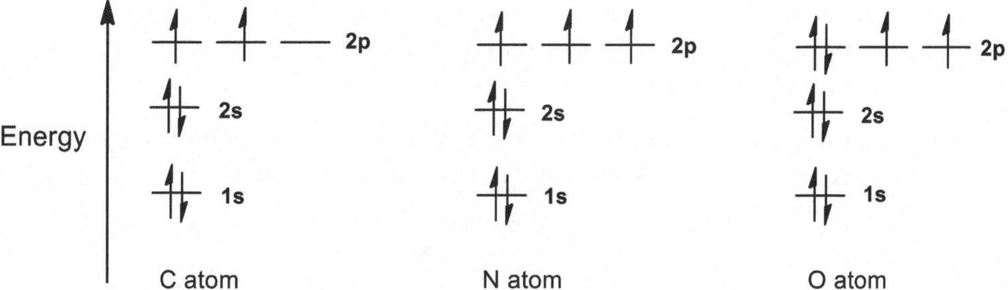

Figure 3.2: The ground state electron configuration of carbon, nitrogen and oxygen atoms: C $1s^2 2s^2 2p_x^1 2p_y^1$; N $1s^2 2s^2 2p_x^1 2p_y^1 2p_z^1$; O $1s^2 2s^2 2p_x^2 2p_y^1 2p_z^1$.

A carbon atom has 6 electrons. Therefore, in the ground state or the lowest energy state, the electron configuration of **carbon** (C) is $\mathbf{1s^22s^22p_x^12p_y^1}$. Similarly, the ground state electron configuration of **nitrogen** (N) atom having 7 electrons and **oxygen** (O) atom having 8 electrons is $\mathbf{1s^22s^22p_x^12p_y^12p_z^1}$ and $\mathbf{1s^22s^22p_x^22p_y^12p_z^1}$, respectively.

3.1.3 Remarks

In an atomic orbital, electrons may be found anywhere except in a node. The **1s orbital has no node**; thus the electron in this orbital may be found in any place in the sphere *including* the nucleus! But the size of the nucleus is very small, so the probability of finding the electron in it is also extremely small.

In a **p orbital**, an electron may be found in any lobe; the **two lobes together form the orbital or electron cloud**. When a p orbital contains *two* electrons, it *does not mean* that each lobe has one electron; it *means* that both electrons could be anywhere in the two lobes.

All orbitals of an atom are **superimposed** on each other. 1s and 2s orbitals are separate orbitals, although the *2s orbital occupies some of the space of the 1s orbital.* Three p orbitals are mutually perpendicular but if they are added together, a **spherical distribution** of electron density results.

3.2 MOLECULAR ORBITALS

Question: Where the electrons are in a molecule?

Answer: The electrons are in molecular orbitals.

Molecules are formed by combination of atoms. So, molecular orbitals are formed by combination of atomic orbitals. Mathematically, this combination of atomic orbitals is taken to be a simple addition or subtraction, called the **linear combination of atomic orbitals (LCAOs).**

Two atomic orbitals (wavefunctions ϕ_1 and ϕ_2) combine to give two **new wavefunctions** (ψ_1 and ψ_2) that describe the **two molecular orbitals**:

$$\psi_1 = \phi_1 + \phi_2 \quad \ldots (3.1)$$

$$\psi_2 = \phi_1 - \phi_2 \quad \ldots (3.2)$$

Note that the linear combination uses only the **first power** of wavefunctions, i.e. ϕ_1, ϕ_2, **not** ϕ_1^2, ϕ_2^2, etc. (cf. equation of a straight line). In general, ϕ_1 and ϕ_2 each in the above equations bears a *coefficient* (a fractional number). But to keep the equations simple, the coefficients are not included here.

The **combination of atomic orbitals** can be understood with the help of a simple *analogy* of combining transverse waves as **constructive (in-phase)** or **destructive (out-of-phase)**:

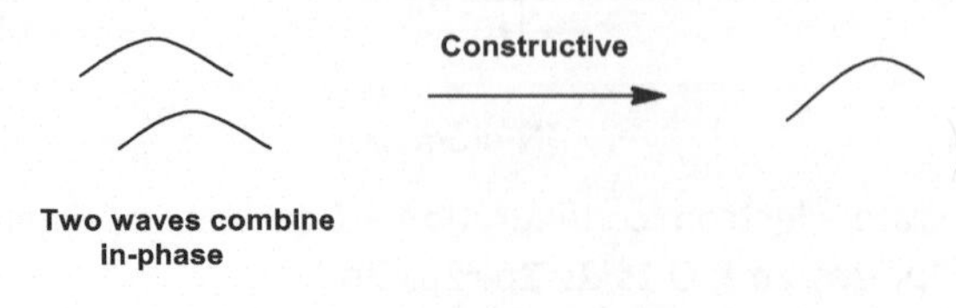

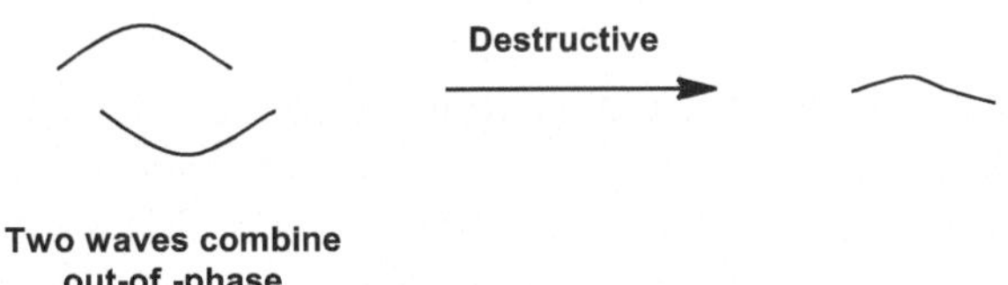

Eq. (3.1) corresponds to **in-phase** combination since both ϕ_1 and ϕ_2 have the **same sign** (+), while **Eq. (3.2)** corresponds to **out-of-phase** combination as ϕ_1 and ϕ_2 have **opposite signs** (+ and –).

The combination of atomic orbitals can give two kinds of molecular orbitals, called **sigma (σ) molecular orbitals** and **pi (π) molecular orbitals**.

3.2.1 Sigma (σ) Molecular Orbitals: s orbital–s orbital Combination

Consider the combination of **two s orbitals** (ϕ_1 = s, ϕ_2 = s) on two atoms as they approach each other. Since s orbitals are spherical and non-directional, the direction in which they approach each other does not matter.

Two molecular orbitals ψ_1 and ψ_2 are formed by additive (in-phase) and subtractive (out-of-phase) combinations, as shown below.

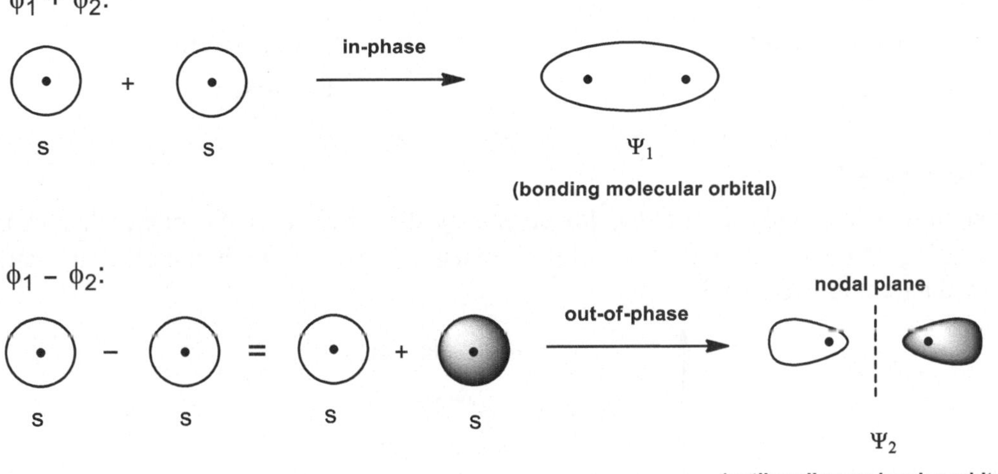

The molecular orbitals encompass *both* nuclei. In the **in-phase** combination, the electron density is quite large in the region of space between the two positively charged nuclei. Thus the electrons would exert a strong attractive force on both the nuclei and hold them together. The result is a bond between the two atoms. The **in-phase molecular orbital** (ψ_1) is therefore called a **bonding molecular orbital**.

On the other hand, the **out-of-phase** combination leads to the destruction of electron density in between the two nuclei and creates a **node** (nodal plane). If electrons are put into this molecular orbital (ψ_2), they could be found anywhere *except between the nuclei* where a node is present. There would be a very small electron density in the region of space between the two nuclei, and the repulsive force between the two positively charged nuclei would be much greater than the small attractive force of the electrons for the nuclei. This is a situation which works against bonding. For this reason, the **out-of-phase molecular orbital (ψ_2)** is called an **antibonding molecular orbital**.

Symmetry

The bonding and antibonding molecular orbitals formed from the s–s overlap have an important feature of **symmetry**: they are **cylindrically symmetrical**. This implies that if either molecular orbital is rotated around an axis joining the two nuclei and viewed from end-on, it remains indistinguishable from the original one before rotation.

The molecular orbitals with cylindrical symmetry are called **σ (sigma) molecular orbitals**. The **bonding molecular orbital** is denoted by **σ** and the **antibonding molecular orbital** is denoted by σ* (* indicates antibonding character).

Thus ψ_1 is σ (or σ_s) and ψ_2 is σ^* (or σ_s^*). The subscript s is often used to indicate that the molecular orbital results from combination of two s orbitals.

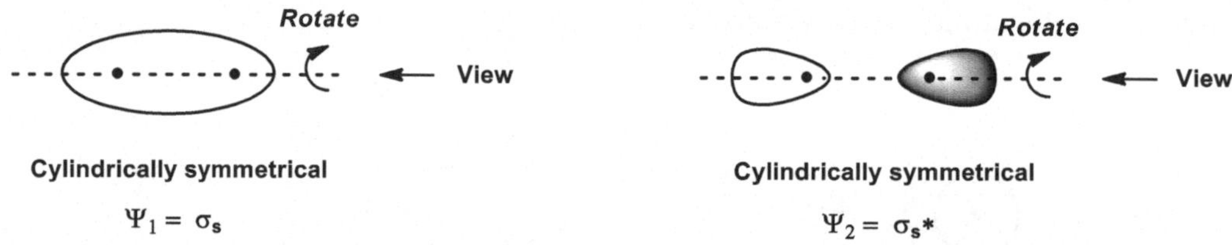

Energy diagram

The bonding σ molecular orbital has **lower energy** than the s atomic orbital, whereas the antibonding σ* molecular orbital has **higher energy**. This is shown in a molecular orbital energy diagram in Figure 3.3.

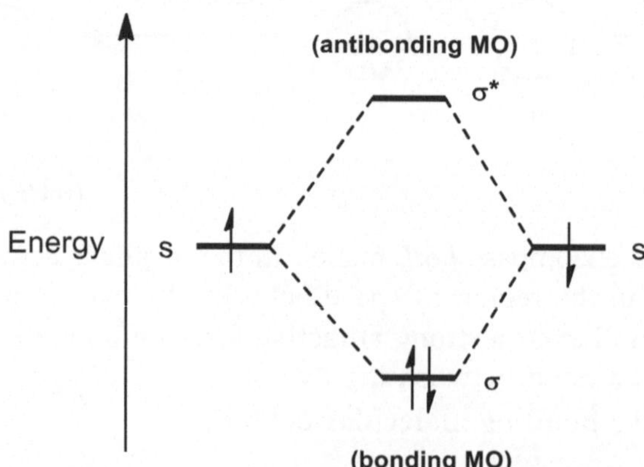

Figure 3.3: Molecular orbital (MO) energy diagram for the interaction between two s atomic orbitals.

Like an atomic orbital (AO), a **molecular orbital (MO)** can also accommodate a *maximum* of **two electrons** with **opposite spins**. When two AOs, each with one electron, overlap, both electrons go into the lower energy σ MO. The antibonding σ* MO remains vacant.

The total energy of the two electrons will thus be *lower* than it was in the separated atoms, and *a bond is formed*. Thus **bonding is a stabilizing phenomenon**.

Note: The above energy diagram represents a **general pattern** for combination of *any two* atomic orbitals.

Example

The simplest case of s–s overlap is between **two 1s orbitals of H atoms**. Two electrons of hydrogen molecule would occupy the bonding σ orbital. Thus the electrons are most likely to be found **in between** the hydrogen nuclei. The result is the **formation of a σ bond** between two hydrogen atoms.

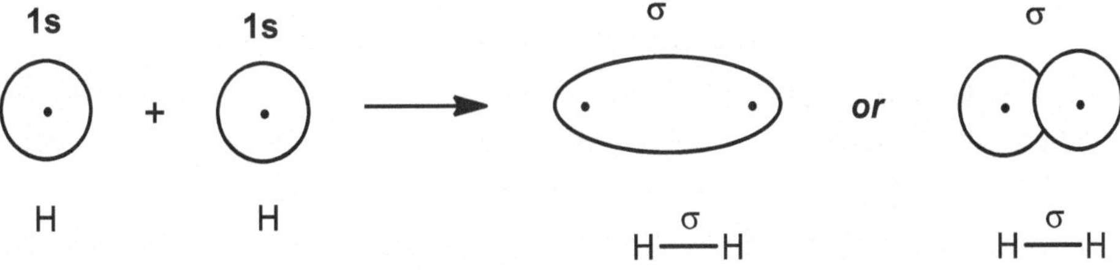

Note that the σ molecular orbital can also be sketched as overlapping s atomic orbitals.

3.2.2 Sigma (σ) Molecular Orbitals: p orbital–p orbital combination

Unlike an s orbital, a p atomic orbital is directional (p_x, p_y or p_z). Therefore, *the direction in which two p orbitals approach each other influences their overlap*.

The **end-on overlap** of two p orbitals generates two molecular orbitals that are **cylindrically symmetrical** around an axis joining the two nuclei. These molecular orbitals are therefore of **σ type**. It is of note that **each p orbital has a node that remains in each molecular orbital**.

The **in-phase** overlap does *not* create any extra node; there is large electron density between the two nuclei which would exert an attractive force on the nuclei to bind them together. The **in-phase molecular orbital is therefore a bonding molecular orbital (σ_p)** with **no node**. (Remember that only extra nodes are counted; the nodes of p orbitals that must remain in the molecular orbital cannot be counted).

In contrast, the **out-of-phase** overlap creates a **node** between the two nuclei. Thus the probability of finding the electron in the region of space between the nuclei is very small in this molecular orbital. The **out-of-phase molecular orbital is therefore an antibonding molecular orbital (σ_p*)**.

Note that the subscript p in σ_p and σ_p* indicates that the molecular orbital results from combination of two p orbitals.

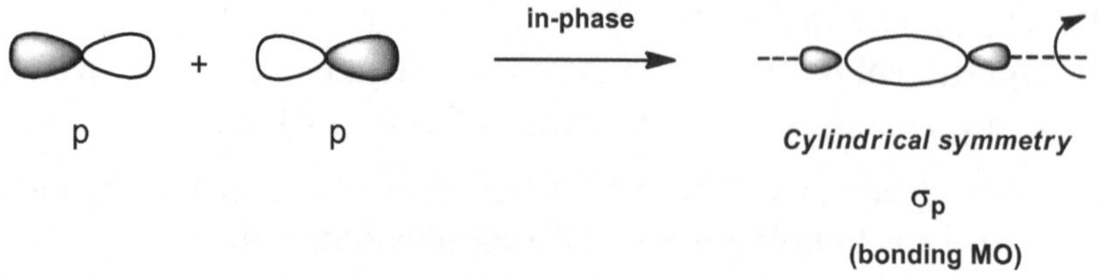

in-phase

p p

Cylindrical symmetry

σ_p

(bonding MO)

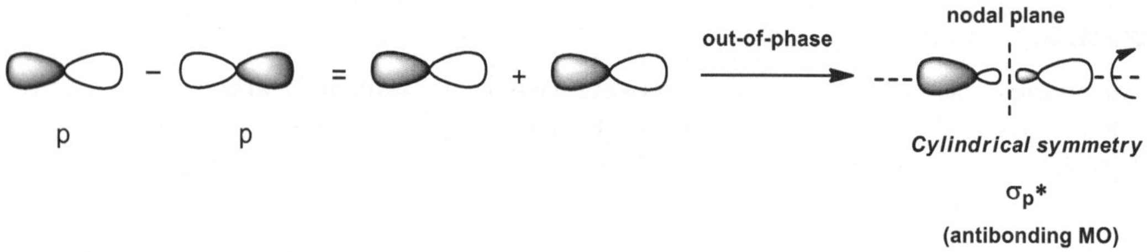

nodal plane

out-of-phase

p p

Cylindrical symmetry

σ_p*

(antibonding MO)

3.2.3 Pi (π) Molecular Orbitals: p orbital–p orbital combination

The **side-on or lateral overlap of two p orbitals** can take place when they are **parallel** to each other, which creates **two molecular orbitals**. The p orbital overlap occurs above and below the two nuclei, and not directly between them.

These molecular orbitals are **not cylindrically symmetrical** because if we rotate the orbital about the axis joining the two nuclei, we do not get the orbital indistinguishable from the original one. They are called **π molecular orbitals**.

The **in-phase** overlap gives a **bonding π molecular orbital** with *no* extra node, whereas the **out-of-phase** overlap leads to an **antibonding π* molecular orbital** with an extra **node** (nodal plane). An extra node in a molecular orbital implies the node that results from the p orbital overlap, *not* the node present originally in the p orbital itself.

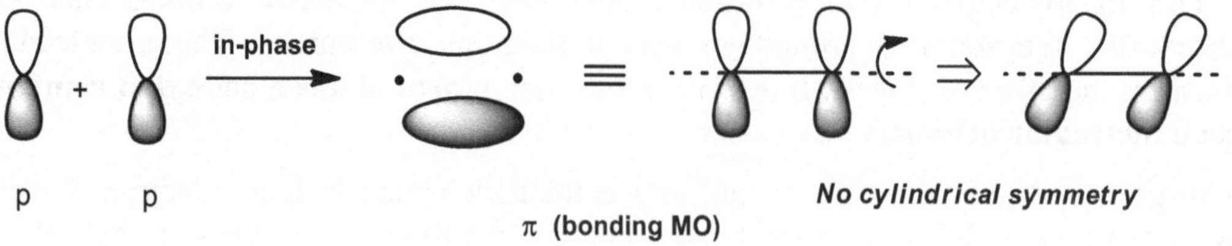

in-phase

p p

No cylindrical symmetry

π (bonding MO)

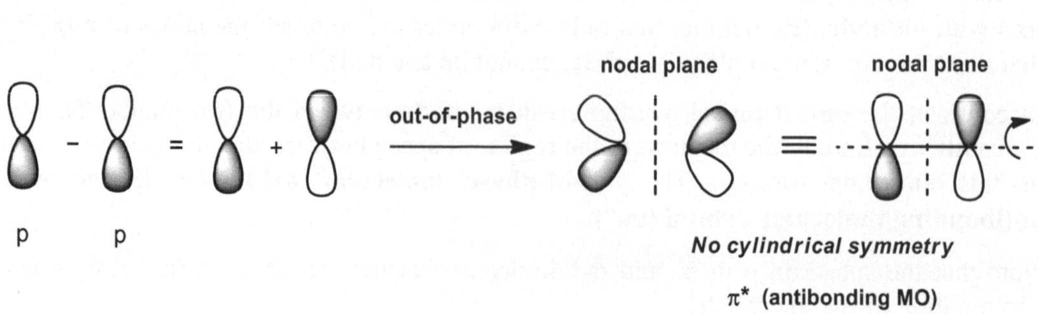

nodal plane nodal plane

out-of-phase

p p

No cylindrical symmetry

π* (antibonding MO)

Note that the π and π* orbitals can also be drawn simply as 'uncombined' p orbitals with proper phase signs of the lobes.

Example

An example where *both* σ and π bonds result from p–p overlaps is **nitrogen molecule (N_2)**. The molecule has a nitrogen-nitrogen triple bond (N≡N) which consists of **a σ bond** and **two π bonds**.

Each nitrogen atom has three degenerate p orbitals ($2p_x^1 2p_y^1 2p_z^1$). The σ bond is formed by an **end-on** 2p–2p overlap (say, $2p_x$–$2p_x$), whereas each of the two π bonds is formed by a **side-on** 2p–2p overlap ($2p_y$–$2p_y$ or $2p_z$–$2p_z$).

3.2.4 Energy Diagram for σ and π Molecular Orbitals

The **end-on** overlap between two atomic orbitals is **more effective** than when they overlap **side-on**. As a result, **a σ bond becomes stronger than a π bond**.

Thus the energy of bonding σ MO will be *lower* than that of bonding π MO, while the energy of antibonding σ* MO will be *higher* than that of antibonding π* MO.

Figure 3.4 shows a schematic **energy diagram** for σ and π molecular orbitals, both resulting from the overlap of p orbitals. As shown, two p orbitals combine to form two σ orbitals (bonding σ and antibonding σ*), while another two p orbitals combine to form two π orbitals (bonding π and antibonding π*).

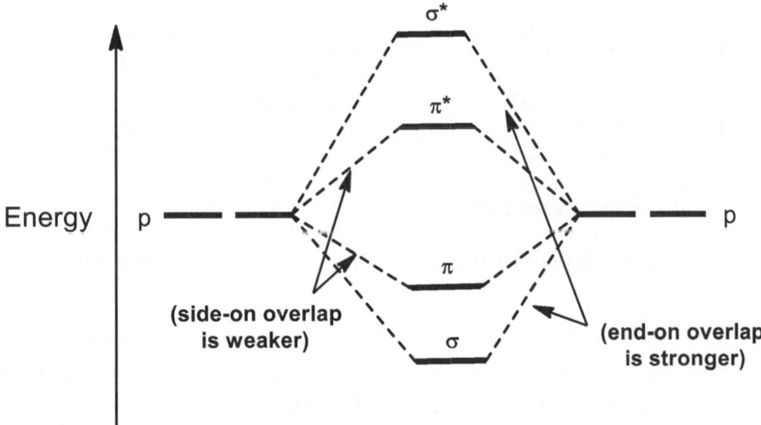

Figure 3.4: Relative energies of σ and π molecular orbitals from the overlap of p orbitals.

3.2.5 Remarks

The **out-of-phase** combination of two atomic orbitals does not destroy the orbitals themselves as for the classical waves; rather it leads to the depletion of electron density in the region of space between the two nuclei where a small area (nodal plane) has no electron density at all.

The out-of-phase molecular orbital is a full orbital like the in-phase molecular orbital but here the electron would spend most of its time not between the nuclei but outside this region.

When *two filled atomic orbitals* with a total of *four electrons* combine, the *new molecular orbitals* must also describe the distribution of four electrons in space, thereby requiring two molecular orbitals (not just one!), since a molecular orbital, like an atomic orbital orbital, can contain a maximum of two electrons.

In general, the number of atomic orbitals combined is equal to the number of molecular orbitals formed.

3.3 HYBRIDIZATION AND MOLECULAR SHAPES

Carbon has the ground state electron configuration $1s^2 2s^2 2p_x^1 2p_y^1$, which shows that the 2s orbital in the valence shell is completely filled. So, only *two* electrons in the valence orbitals $2p_x$ and $2p_y$ are available to be shared in the formation of two covalent bonds. Yet, carbon has a valence of 4 and it can form **four covalent bonds**.

This is possible if one of the 2s electrons is **promoted** to the vacant $2p_z$ orbital to give the **excited configuration** $\mathbf{1s^2 2s^1 2p_x^1 2p_y^1 2p_z^1}$. This leaves the carbon with four valence orbitals, each containing one electron, which can make four covalent bonds.

Promoting a 2s electron into the higher-energy 2p orbital of course requires energy, but this promotion also allows the formation of two extra bonds, which more than compensates the energetic cost in electron promotion.

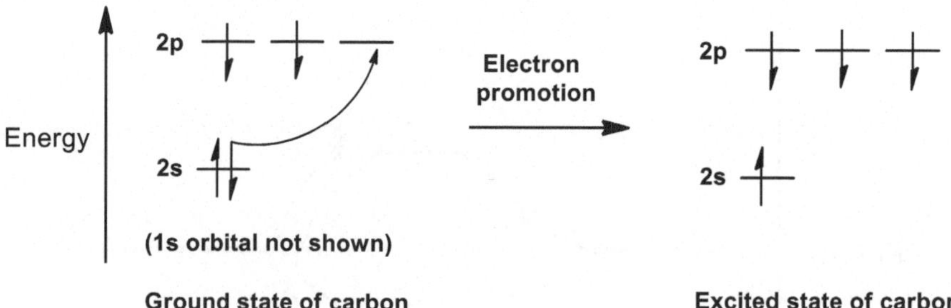

However, this simple picture is inadequate to account for many of the structural characteristics of organic molecules, and an alternative model of bonding has been provided by introducing the concept of **hybridization**.

3.3.1 Alkanes: sp^3 Hybridization of Carbon

Consider the simplest alkane, **methane** (CH_4). If four C–H bonds of methane are formed by involving an s orbital (2s) and three p orbitals ($2p_x$, $2p_y$, $2p_z$) of carbon, *one* of the C–H bonds would be necessarily *different* from the other three. But the molecular properties of CH_4 suggest that all four C–H bonds are **equivalent**. This is possible if we can start with four **equivalent valence orbitals** on carbon, which we don't have. To do this, we use the concept of **hybridization**.

Hybridization involves mixing of 2s and 2p atomic orbitals in various proportions to produce a *new set* of atomic orbitals, called **hybrid orbitals**.

For example, the combination of 2s and three 2p orbitals can be used to generate *four* new, equivalent hybrid orbitals, indicated as **2sp³** or simply **sp³**.

$$2s + 2p_x + 2p_y + 2p_z \xrightarrow[\text{(mix and divide into four equivalent hybrid orbitals)}]{\text{Hybridize}} 2sp^3 + 2sp^3 + 2sp^3 + 2sp^3 = 4 \times (sp^3)$$

Four equivalent sp³ hybrid orbitals

Each sp³ hybrid orbital has one part s and three parts p (i.e. **s = 25%, p = 75%**). Therefore, *each* sp³ hybrid orbital results from mixing of one-quarter of the 2s orbital with three-quarters of the p orbital. The resulting **sp³ hybrid orbital has two lobes with opposite phase signs**, but unlike a p orbital, the two lobes have **different sizes.**

The unequal size of two lobes arises from in-phase (same sign) combination on one side of the carbon nucleus and out-of-phase combination (opposite signs) on the other side.

(1/4 s) + (3/4 p) —Combine→ in-phase ¦ out-of-phase → An sp³ hybrid

(• denotes carbon nucleus)

The **in-phase** mixing gives a **large lobe** on the left side, whereas a **small lobe** on the right side is generated from the **out-of-phase** mixing.

Like p orbitals, **sp³ hybrid orbitals are directional**, and four sp³ orbitals point to the four corners of a regular **tetrahedron** and the angle between any two sp³ orbitals is **109°28′**.

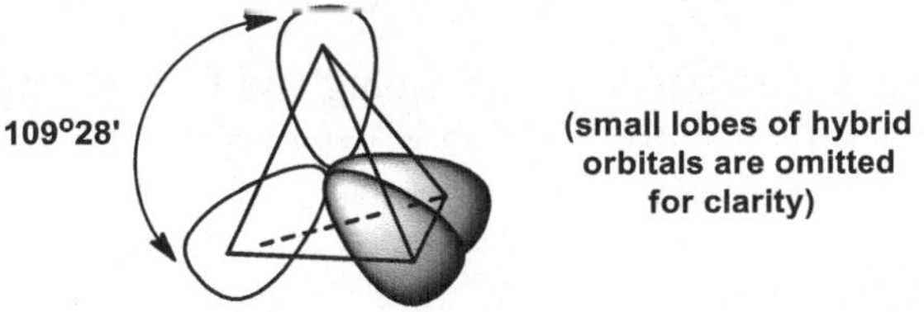

Tetrahedral arrangement of sp³ hybrid orbitals

Bonding in methane

In **methane**, carbon is sp³-hybridized, and each sp³ hybrid orbital contains one electron. Each hydrogen atom has 1s orbital containing one electron.

C: $(sp^3)^1(sp^3)^1(sp^3)^1(sp^3)^1$ H: $1s^1$

Each **C–H bond** in CH_4 results from the overlap of an **sp^3 hybrid orbital on C** with the **1s orbital on H**. It is the **large lobe of sp^3 orbital** with greater electron density that is used in bonding.

The C–H bond is a **σ bond** because it is cylindrically symmetrical about the line joining the two nuclei, as shown below. Note that the σ molecular orbital is sketched as 'uncombined' orbitals with proper phase signs. In terms of molecular orbital theory, there will be also an antibonding σ* orbital but that orbital is empty.

C(sp^3) + **H(1s)** ⟶ **C—H (σ)**

C—H ⟹ (sp^3-s) σ

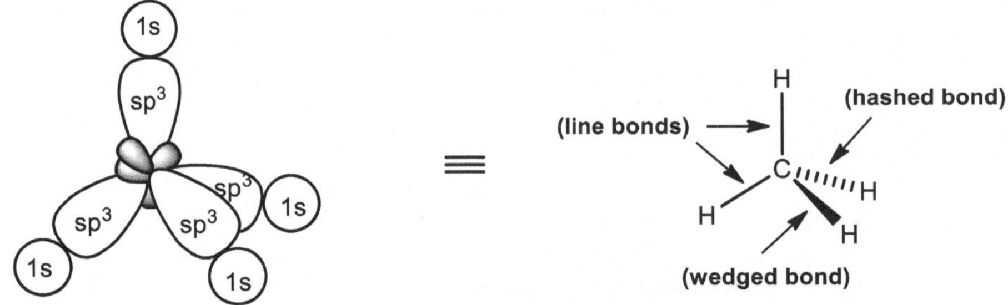

Tetrahedral representation of methane

Since each C–H bond is *directed along the axis of sp^3 hybrid orbital*, the structure of CH_4 is **tetrahedral** (see also Figure 1.1). In the tetrahedral representation, two **line bonds** lie on the plane of the paper, the **wedged bond** is above the plane, and the **hashed bond** is below the plane.

Because hybrid orbitals use a large lobe in bonding and form stronger bonds, the concept of hybridization is generally used for bonding in organic molecules.

Bonding in ethane

When a carbon atom is bonded to four atoms, irrespective of whether they are the same or different, carbon needs four hybrid orbitals, which can be provided by **sp^3** hybridization.

In **ethane** (C_2H_6), each carbon is bonded to four atoms (one carbon and three hydrogens); so **each carbon is sp^3-hybridized**. The **C–C** single bond (**σ bond**) is formed by the end-on overlap of an **sp^3** hybrid orbital on one carbon atom with an **sp^3** hybrid orbital on the second carbon atom. Each **C–H σ bond** in ethane also results from the overlap of an **sp^3** orbital on carbon with **1s** orbital on hydrogen, as for methane.

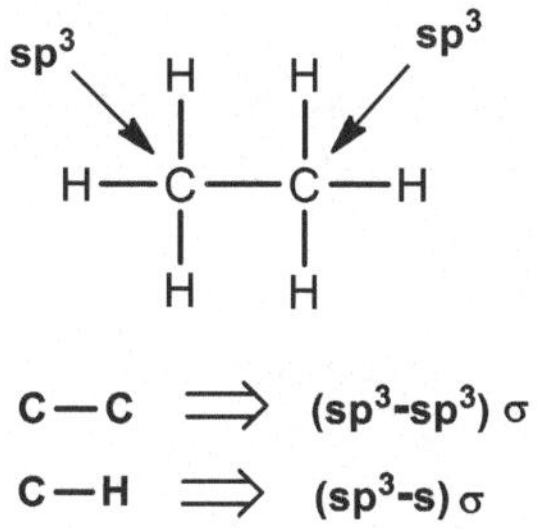

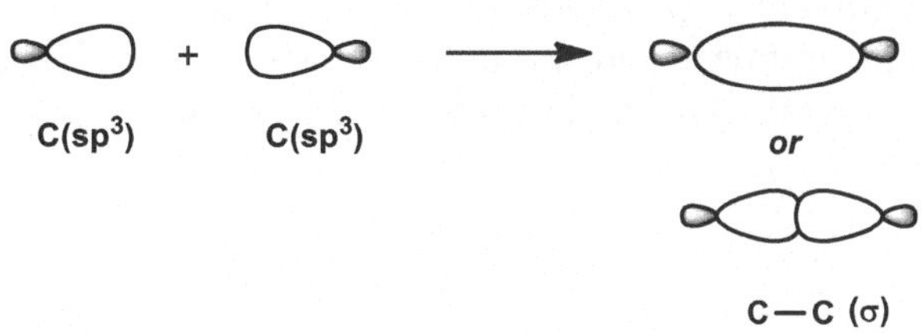

sp³ H H sp³
H—C—C—H
H H

C—C ⇒ $(sp^3\text{-}sp^3)\ \sigma$
C—H ⇒ $(sp^3\text{-}s)\ \sigma$

C(sp³) + C(sp³) ⟶ or C—C (σ)

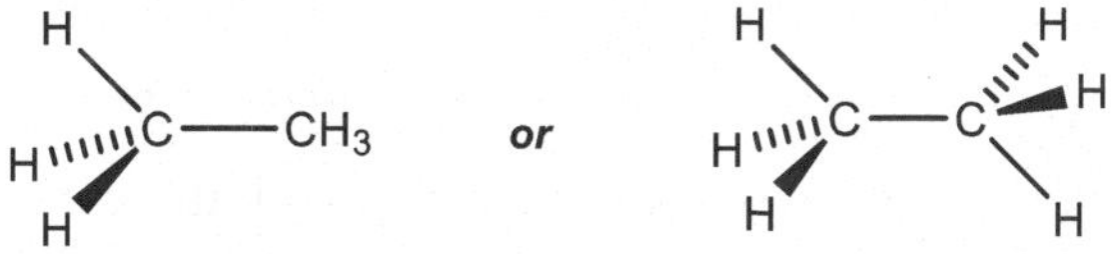

Tetrahedral representation of ethane

Ethane is **tetrahedral** and can be represented by substitution of one H atom of methane with a CH_3 group. It can be fully represented in terms of the tetrahedral structure for both carbons when each carbon is placed at one corner of the tetrahedron for other carbon, and hydrogen atoms occupy the other three corners of each tetrahedron.

Remarks: **Hybrid orbitals are atomic orbitals**, *not* molecular orbitals. Hybrid orbitals are generated by the linear combination of atomic orbitals on the *same atom*, while molecular orbitals are formed by the linear combination of atomic orbitals on *different atoms*. As in the case of molecular orbitals, the **number of hybrid orbitals produced is equal to the number of atomic orbitals combined or mixed**. Notably, hybridization applies only to the orbitals, *not* to the electrons.

3.3.2 Alkenes: sp² Hybridization of Carbon

Consider the simplest alkene, **ethylene** ($H_2C{=}CH_2$) containing a carbon-carbon double bond. However, **hybrid orbitals are used to make single or σ bonds**. In ethylene, each carbon is bound to three atoms (one carbon and two hydrogens). Therefore, a set of **three hybrid orbitals** is needed for *each* carbon to make three σ bonds.

The three hybrid orbitals can be obtained by **sp² hybridization** which involves mixing of 2s orbital with two 2p orbitals (say, $2p_x$ and $2p_y$), leaving behind one **unhybridized $2p_z$ orbital** on each carbon.

$$\underbrace{2s + 2p_x + 2p_y}_{\text{Hybridize}} + 2p_z \longrightarrow \underbrace{2sp^2 + 2sp^2 + 2sp^2}_{\text{Three equivalent sp}^2\text{ hybrid orbitals}} + \underbrace{2p_z}_{\text{Unhybridized p orbital}} = 3 \times (sp^2) + p_z$$

Thus the valence shell electronic configuration of carbon is: $(2sp^2)^1(2sp^2)^1(2sp^2)^1(2p_z)^1$.

Each **sp^2 hybrid orbital** has one part s and two parts p (i.e. **s = 33.3%, p = 66.7%**). The three equivalent sp^2 hybrid orbitals point to the three corners of an **equilateral triangle** and are **coplanar**, and the unhybridized p orbital is **perpendicular** to the plane of the hybrid orbitals. The angle between any two sp^2 orbitals is **120°**.

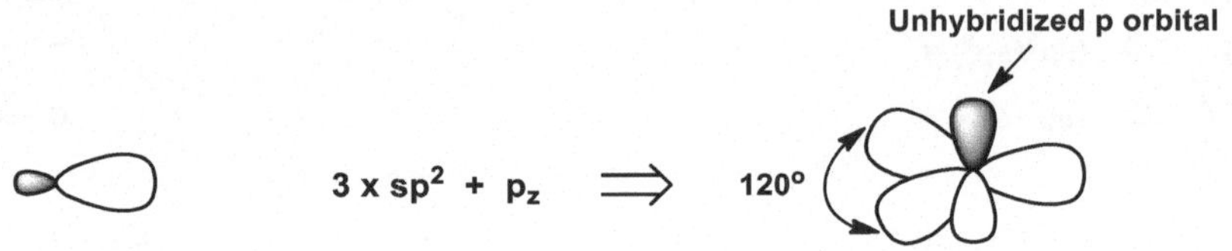

Unhybridized p orbital

3 x sp^2 + p_z ⇒ 120°

An sp^2 hybrid orbital

Three sp^2 orbitals are trigonal planar

Small lobes are omitted for clarity

The three sp^2 orbitals on each carbon can overlap with three other orbitals (one sp^2 orbital from the other carbon and two hydrogen 1s orbitals) to form three **σ bonds** (C–C, C–H, C–H). The side-on or parallel overlap of two $2p_z$ orbitals, one on each carbon, forms the **π bond** between the carbons.

Therefore, C=C ⇒ (sp^2–sp^2) σ + (p–p) π; each C–H ⇒ (sp^2–s) σ.

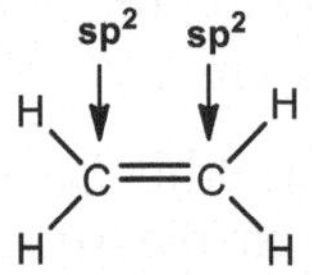

Ethylene

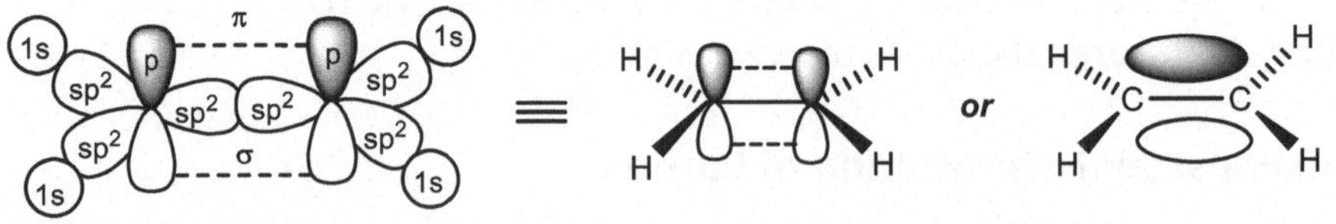

Trigonal planar

C=C ⇒ (sp^2-sp^2) σ + (p-p) π

C—H ⇒ (sp^2-s) σ

The shape of ethylene molecule is **trigonal planar** with bond angles close to 120°. The molecular plane contains two carbons and four hydrogens. The π bond is **perpendicular** to the molecular plane. Note that in the above diagram, the molecular plane is perpendicular to the plane of the paper and the π bond lies in the plane of the paper.

3.3.3 Alkynes: sp Hybridization of Carbon

Consider the simplest alkyne, **acetylene** (HC≡CH) that contains a carbon-carbon triple bond. In acetylene, *each* carbon is bound to two atoms (one carbon and one hydrogen). Thus each carbon requires a set of **two hybrid orbitals** for making two σ bonds.

The two hybrid orbitals are generated by **sp hybridization** which involves mixing of 2s orbital with one 2p orbital (say, $2p_x$), leaving behind **two unhybridized p orbitals** ($2p_y$ and $2p_z$) on each carbon.

$$\underbrace{2s + 2p_x}_{\text{Hybridize}} + 2p_y + 2p_z \longrightarrow \underbrace{2sp + 2sp}_{\text{Two equivalent sp hybrid orbitals}} + \underbrace{2p_y + 2p_z}_{\text{Unhybridized p orbitals}} = \mathbf{2 \times (sp) + p_y + p_z}$$

Thus the valence shell electronic configuration of carbon is: $(2sp)^1(2sp)^1(2p_y)^1(2p_z)^1$.

Each **sp hybrid orbital** has one part s and one part p (i.e. **s = 50%, p = 50%**). The two equivalent sp hybrid orbitals are oriented at an angle of **180°** with respect to each other, and the two unhybridized $2p_y$ and $2p_z$ orbitals are **perpendicular** to each other and to the line containing the sp orbitals.

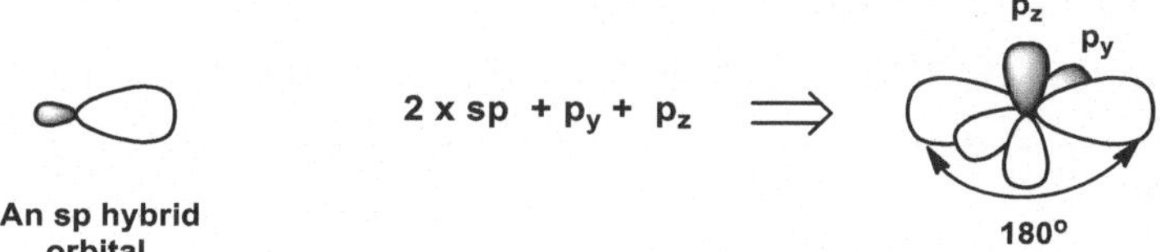

Two sp orbitals are oriented at 180°

Small lobes are omitted for clarity

The two sp orbitals on each carbon overlap with one sp orbital from the other carbon and one hydrogen 1s orbital to form **two σ bonds** (C–C and C–H). The parallel or side-on p_y–p_y and p_z–p_z overlaps form **two π bonds** between the carbons. A carbon-carbon triple bond thus consists of one σ bond and two π bonds.

Therefore, C≡C ⇒ (sp–sp) σ + 2 × (p–p) π; each C–H ⇒ (sp–s) σ.

sp sp

H—C≡C—H

Acetylene

C≡C ⇒ (sp-sp) σ + 2 x (p-p) π

C—H ⇒ (sp-s) σ

A cylinder of π electron cloud as combination of two π bonds

Acetylene molecule is **linear with bond angles of 180°**, and the two π bonds are **mutually perpendicular** which create a **cylinder of π electrons** around the central C–C σ bond.

3.3.4 Hybridization of Nitrogen and Oxygen

Nitrogen and oxygen atoms in organic molecules often possess one or more lone pairs of electrons. **Hybrid orbitals are used not only for σ bonding but also to accommodate lone pairs.**

The **hybridization of nitrogen or oxygen** is determined by the **sum of the number of atoms** bonded to that particular atom and the **number of lone pairs**. If the sum is **four**, the atom is **sp^3**-hybridized, if **three**, the atom is **sp^2**-hybridized, and if **two**, the atom is **sp**-hybridized.

Ammonia, amines and quaternary ammonium ion

In **ammonia or amines (1°, 2° or 3°)**, N is bonded to **3 atoms** and it has **1 lone pair**, making a total of 4. The N atom is therefore **sp^3-hybridized** and the lone pair resides in an sp^3 hybrid orbital.

The N–H σ bond is formed by the overlap of an sp^3 orbital of N with 1s orbital of H, while the N–C σ bond is formed by the overlap of an sp^3 orbital of N with an sp^3 orbital of carbon.

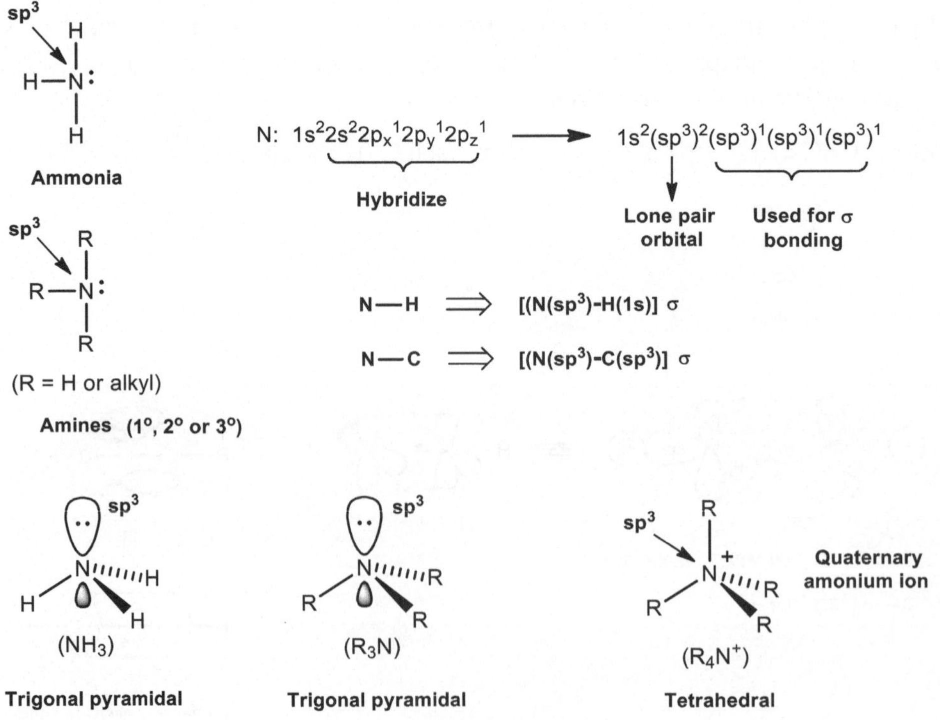

The shape of **ammonia or amine** is said to be **trigonal pyramidal** since one corner of the tetrahedron is occupied by a lone pair of electrons, instead of an atom.

In the case of **quaternary ammonium ion** (R_4N^+), the N atom is bonded to **4 atoms**, and hence its shape is **tetrahedral.**

Water, alcohols and ethers

In **water, alcohols or ethers, O** is bonded to **2 atoms** and it has **2 lone pairs**, giving a total of 4. The O atom is therefore **sp³-hybridized**, and two lone pairs are accommodated in two sp³ hybrid orbitals.

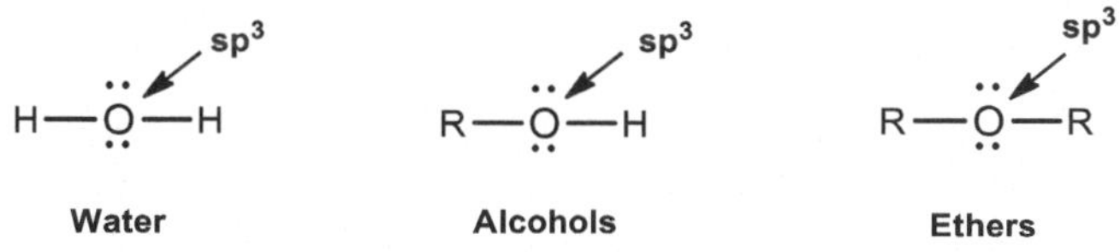

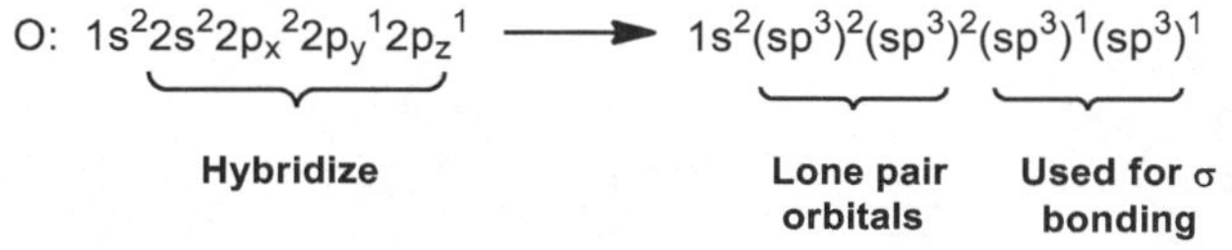

sp³ H—Ö—H **Water**

sp³ R—Ö—H **Alcohols**

sp³ R—Ö—R **Ethers**

O: $1s^22s^22p_x^22p_y^12p_z^1$ ⟶ $1s^2(sp^3)^2(sp^3)^2(sp^3)^1(sp^3)^1$

Hybridize **Lone pair orbitals** **Used for σ bonding**

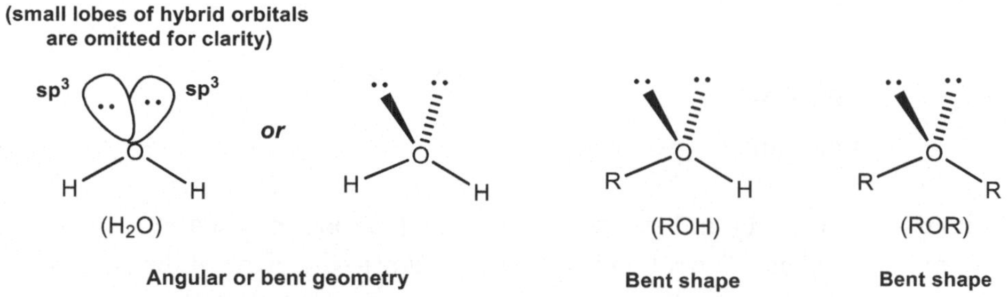

(small lobes of hybrid orbitals are omitted for clarity)

sp³ sp³ (H_2O) *or* (ROH) (ROR)

Angular or bent geometry **Bent shape** **Bent shape**

Since two corners of the tetrahedron are occupied by two lone pairs instead of two atoms, the molecule has an **angular or bent** geometry. It is *not* linear.

Aldehydes and ketones

Aldehydes or ketones contain a **carbonyl group (C=O)**.

Consider a **ketone** in which the carbonyl **C** is bonded to **3 atoms** (one O atom and two C atoms), and it has **no lone pair**. Therefore, the C atom in the carbonyl group is **sp²-hybridized.**

The three sp² hybrid orbitals of the **carbonyl carbon** are used to form **one C–O σ bond** and **two C–C σ bonds** to alkyl groups (R).

On the other hand, the **O atom** in the carbonyl group is bonded to **one C atom** only, but it has **two lone pairs**, giving a sum of 3. Thus the O atom is also **sp²-hybridized**.

The **C–O σ bond** is formed by the end-on overlap of an sp^2 orbital on carbonyl carbon with an sp^2 orbital on oxygen atom. The side-on overlap between the unhybridized p orbital on carbonyl C and the unhybridized p orbital on O gives the **π bond** of the carbonyl group.

Therefore, C=O ⇒ [C(sp^2)–O(sp^2)] σ + [C(p)–O(p)] π

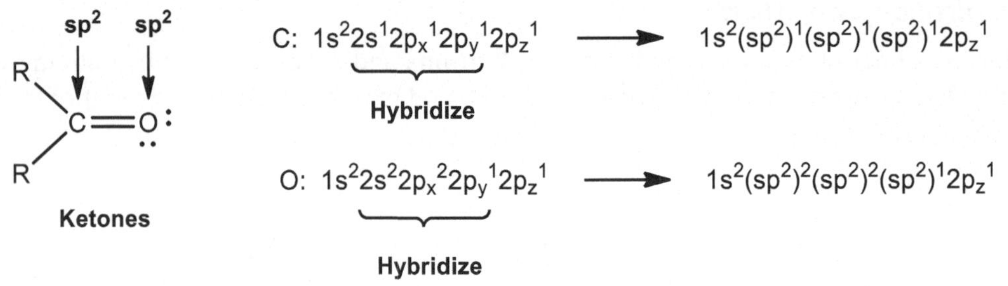
sp2 sp2
R
C=O:
R
Ketones
C: $1s^2 2s^1 2p_x^1 2p_y^1 2p_z^1$ ⟶ $1s^2(sp^2)^1(sp^2)^1(sp^2)^1 2p_z^1$
Hybridize
O: $1s^2 2s^2 2p_x^2 2p_y^1 2p_z^1$ ⟶ $1s^2(sp^2)^2(sp^2)^2(sp^2)^1 2p_z^1$
Hybridize

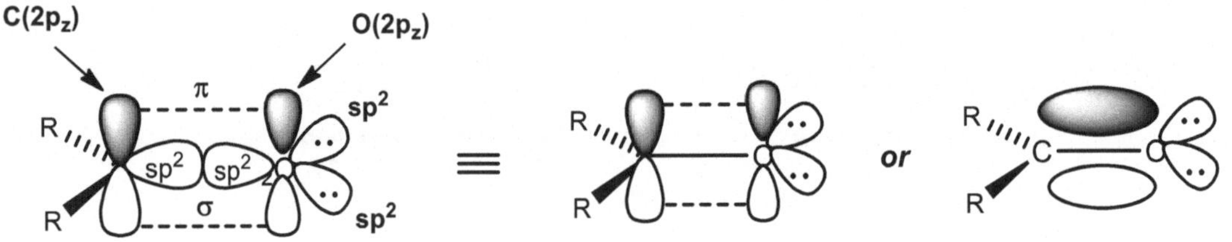
C(2p$_z$)
O(2p$_z$)
π
sp2
sp2
σ
R
or

C=O ⇒ [C(sp^2)-O(sp^2)] σ + [C(p)-O(p)] π

Trigonal planar

The carbonyl group is **trigonal planar**; the carbonyl carbon attached to two carbons and one oxygen lie in a plane. The π bond is *perpendicular* to the plane of the carbonyl group.

Note: Spectroscopic evidence indicates that *one* of the two lone pairs on O atom is in a p orbital, which suggests that the O atom is sp-hybridized.

3.3.5 Determining Hybridization for Atoms

We have learned that hybrid orbitals are used for making single or σ bonds between atoms and also to accommodate lone pairs of electrons.

The number of σ bonds to an atom is equal to the number of atoms bound to that atom. Thus the **hybridization** of an atom can be predicted by counting the **number of atoms** bonded to that particular atom (the **number of σ bonds**) and the **number of lone pairs** of electrons around that atom.

The **sum of the numbers** [number of atoms bound (i.e. number of σ bonds) + number of lone pairs] gives the hybridization for the atom (Table 3.1). If the sum is four, the hybridization of the atom is sp^3, if the sum is three, the hybridization of the atom is sp^2, and if the sum is two, the hybridization of the atom is sp.

Table 3.1: Predicting Hybridization for C, N and O Atoms

Number of Atoms Bound (i.e. number of σ bonds) + Number of lone pairs	*Hybridization*	*Approximate Bond Angle*	*Geometry (or shape)*
2	sp	180°	Linear
3	sp^2	120°	Trigonal planar
4	sp^3	109.5°	Tetrahedral

The following examples illustrate how to predict the hybridization for atoms.

Example 1

3 2 1
$CH_3CH{=}CH_2$ ≡ H–C3(H)(H)–C2(H)=C1(H)–H ⇒ C1: sp^2; C2: sp^2; C3: sp^3

Propene

C3: (bound to 4 atoms or 4 σ); C2: (bound to 3 atoms or 3 σ); C1: (bound to 3 atoms or 3 σ)

Example 2

5 43 2 1
$HC{\equiv}CCH_2CH{=}CH_2$ ≡ H–C5≡C4–C3(H)(H)–C2(H)=C1(H)–H ⇒ C1: sp^2; C2: sp^2; C3: sp^3; C4: sp; C5: sp

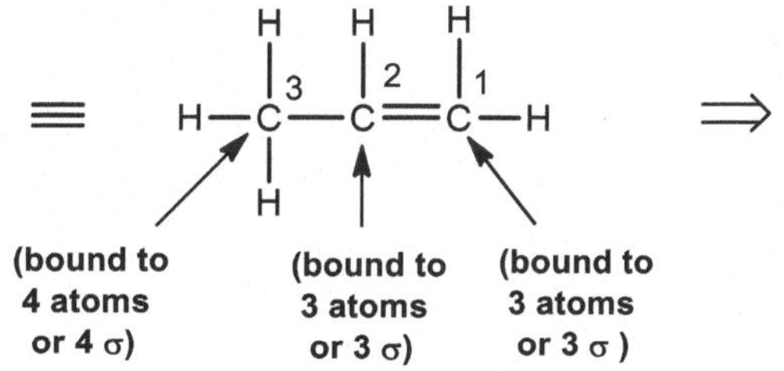

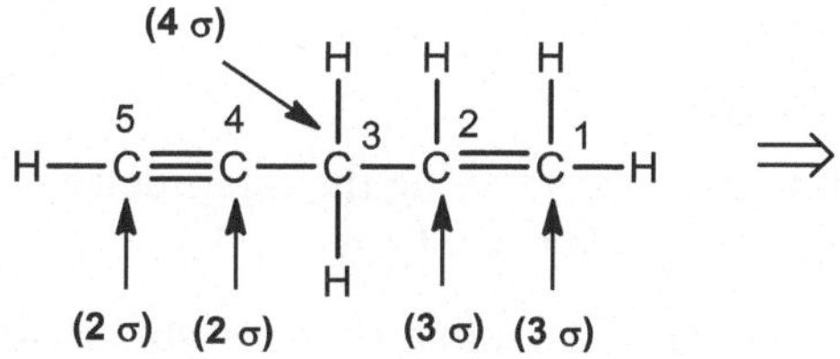

1-Penten-4-yne

C5: (2 σ); C4: (2 σ); C3: (4 σ); C2: (3 σ); C1: (3 σ)

Example 3

CH_3CN ≡ H–C(H)(H)–C≡N:

Acetonitrile

CH₃ carbon: (4 σ: sp^3); nitrile carbon: (2 σ: sp); N: (1 σ + 1 lone pair; sum = 2) N: sp

---

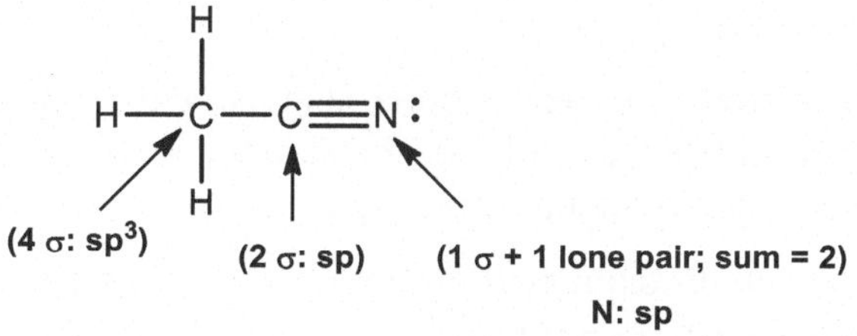

PROBLEM 3.1: Predict the hybridization of the indicated atoms in the following species:

(a) CH_3^+ (c) $H_2C{=}\overset{+}{O}H$

(b) CH_3^- (d) BH_4^-

Answer

The hybridization of an atom is determined by the *sum* of the number of σ bonds to that atom (i.e. the number of atoms bound to that particular atom) and the number of lone pairs on that atom. If the sum is four, the atom is sp^3, if three, the atom is sp^2, and if two, the atom is sp. Thus the hybridization of the indicated atom in each case is assigned. Note that lp denotes lone pair.

(a) CH_3^+ (H–C⁺ with two further H on C) C: 3 σ, **sp^2**

(b) CH_3^- (H–C:⁻ with two further H on C) C: (3 σ + 1 lp), **sp^3**

(c) $H_2C{=}\overset{+}{\ddot{O}}{-}H$ O: (2 σ + 1 lp), **sp^2**

(d) BH_4^- (H–B⁻–H with two further H on B) B: 4 σ, **sp^3**

3.3.6 Bond Lengths

Atoms in a molecule vibrate along the internuclear axis. **Bond length** between two atoms is therefore an *average distance* between them.

Bond lengths can vary depending on the **hybridization** of atoms or the **bond order** (i.e. the number of bonds between two atoms).

Effect of Hybridization

The hybridization of atoms influences bond lengths. **The greater the s character of the hybrid orbital of atom** (such as carbon), **the shorter is the bond length**. This is because s electrons are held closer to the nucleus than p electrons.

Bond lengths are generally expressed in the units of **Angstrom (Å)** or **nanometre (nm)** [$1\ Å = 10^{-10}\ m = 10^{-1} \times 10^{-9}\ m = 0.1\ nm$].

The **C–H bond lengths** (in Å) in ethane, ethylene and acetylene are as follows:

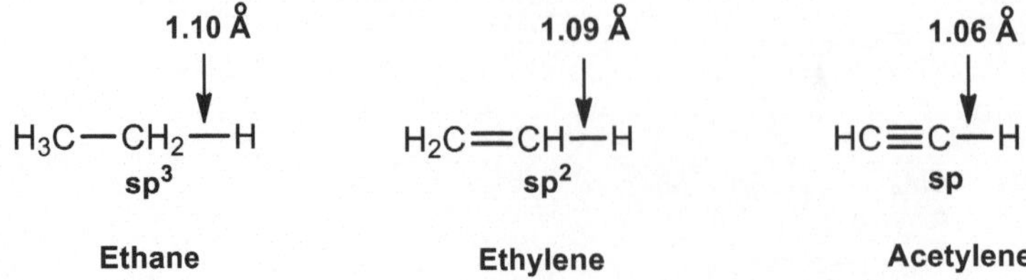

The **sp** hybrid orbital of carbon (50% s) gives the *shortest* C–H bond (1.06 Å) in acetylene (ethyne), while the **sp**^{3} hybrid orbital (25% s) forms the *longest* C–H bond (1.10 Å) in ethane.

Effect of Bond Order

Bond order is the number of bonds between two atoms. A single bond has a bond order of 1, a double bond has a bond order of 2 and a triple bond has a bond order of 3. The bond order also influences bond lengths.

Bond lengths become **shorter** with **increasing bond order** (i.e. increasing number of bonds) because the constituent atoms are held more tightly.

Thus the carbon-carbon triple bond (C≡C) in acetylene is *shorter* than the carbon-carbon double bond (C=C) in ethylene which, in turn, is *shorter* than the carbon-carbon single bond (C–C) in ethane.

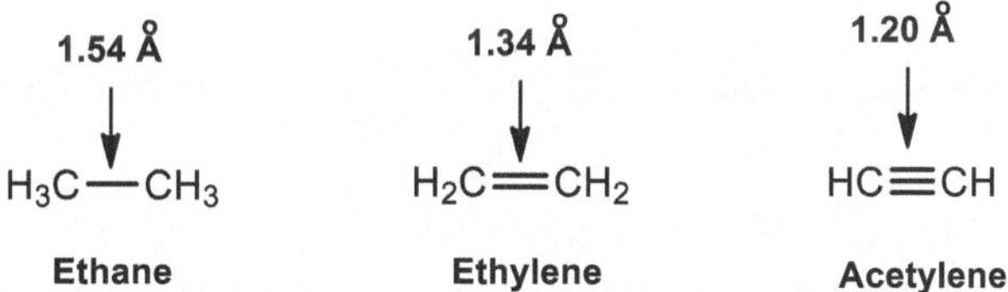

3.3.7 Bond Energies

The energy of a bond is a form of **potential energy** that arises from attractive forces between atoms.

Bond energies can be specified in two ways. The energy necessary to cleave a **specific bond** is called the **bond dissociation energy (BDE)**, whereas the **average of all similar bond energies** is termed the **bond energy (BE)**.

The conventional unit for BDE or BE is **kcal mol**$^{-1}$, but the SI unit is **kJ mol**$^{-1}$ (1 cal = 4.184 J).

Consider **methane** (CH_4). The **bond dissociation energy (BDE)** of an *individual* C–H bond in methane differs, as shown below.

Reaction	BDE (kcal mol^{-1})
$H_3C—H \longrightarrow CH_3 + H$	102
$H_2C—H \longrightarrow CH_2 + H$	110
$HC—H \longrightarrow CH + H$	104
$C—H \longrightarrow C + H$	81
	Total = 397

Bond energy (BE) of C-H bond
= (397/4) kcalmol^{-1} = 99.3 kcal mol^{-1}

The **total energy** for the cleavage of the four C–H bonds is 397 kcal mol^{-1}. Therefore, the **bond energy (BE)** of a C–H bond in methane is (397/4) = 99.3 kcal mol^{-1}.

Note that for **diatomic molecules** such as halogens (X_2, X = F, Cl, Br, I) and hydrogen halides (HX), the bond dissociation energy and the bond energy are the same.

Effect of Bond Length

There is usually a correlation between the bond energy and the bond length. The shorter the bond length, the stronger is the bond or higher is the bond energy.

The bond energies of some common bonds are as follows:

Bond	Bond energy (kcal mol^{-1})	Bond	Bond energy (kcal mol^{-1})	Bond	Bond energy (kcal mol^{-1})
C—H	99	C—C	83	C—O	86
		C=C	146	C=O	179
		C≡C	200		

For single bonds (C–H, C–C and C–O), the shortest C–H bond has the highest bond energy (99 kcal mol^{-1}).

A carbon-carbon double bond (C=C) is *shorter* than a carbon-carbon single bond (C–C), and the double bond is stronger. However, **the bond energy of C=C** (146 kcal mol^{-1}) is **less than twice the bond energy of C–C** (83 kcal mol^{-1}). This is because a double bond consists of a σ bond and a π bond, but a π bond is weaker than a σ bond. From the bond energy data for C=C and C–C, the energy of a π bond can be calculated. The **π bond energy** is (146 – 83) = 63 kcal mol^{-1}.

A carbon-carbon triple bond (C≡C) is *shorter* than a carbon-carbon double bond (C=C), and the triple bond is stronger (200 kcal mol^{-1}).

In ketones, the carbon-oxygen double bond (C=O) has a high bond energy of 179 kcal mol^{-1}, which indicates that **C=O is stronger than C=C**. The carbonyl group (C=O) is **polar** and here the situation is slightly different: the bond energy of C=O is more than twice the bond energy of C–O single bond.

Heat of Reaction (ΔH^0)

In a chemical reaction, **some bonds are broken and some new bonds are formed**. Bond breaking requires *addition* of energy, while bond formation *releases* energy.

The energy released or absorbed during the reaction is called the heat of reaction. It is a measure of the change in enthalpy (ΔH) as the reactants are converted into products. Note that H stands for enthalpy or heat content.

The standard enthalpy change (ΔH^0) can be estimated from bond energies of bonds being broken and formed during the reaction:

$$\Delta H^0 = \text{Bond energies}_{(\text{broken})} - \text{Bond energies}_{(\text{formed})}$$

(The superscript zero (0) implies that reactants and products are in their standard states at a pressure of 1 atm).

To illustrate, consider the following reaction:

$$CH_4 + Cl_2 \rightarrow CH_3Cl + HCl$$

In this reaction, C–H and Cl–Cl bonds are broken, and C–Cl and H–Cl bonds are formed. Their bond energies are as follows:

Bonds broken	Bond energy (kcal mol^{-1})	Bonds formed	Bond energy (kcal mol^{-1})
C—H	99	C—Cl	81
Cl—Cl	58	H—Cl	103

The heat of reaction (ΔH^0) is estimated as

$$\Delta H^0 = [(99 + 58) - (81 + 103)] \text{ kcal mol}^{-1} = -27 \text{ kcal mol}^{-1}$$

Now, $\Delta H^0 = H^0_{\text{products}} - H^0_{\text{reactants}} = -27 \text{ kcal mol}^{-1}$

or, $H^0_{\text{reactants}} = H^0_{\text{products}} + 27 \text{ kcal mol}^{-1}$

This implies that the reaction gives products with release of heat energy (27 kcal mol^{-1}), and the reaction is called **exothermic**. Note that **ΔH^0 is negative for an exothermic reaction**. In an exothermic reaction, the products are of lower enthalpy (or energy) than the reactants and hence are **more stable**, and the reaction is **favourable**.

If ΔH^0 for a reaction is **positive**, the reaction is called **endothermic**.

PROBLEM 3.2: Estimate the heat of reaction for the following reaction, and indicate whether the reaction is exothermic or endothermic.

$$H_2C{=}CH_2 + H_2 \longrightarrow H_3C{-}CH_3$$

Ethylene → **Ethane**

[Given, bond energies (kJ mol^{-1}): H–H 436, C–H 414, C–C 347, C=C 610]

Answer

Bonds broken: C=C, H–H; bonds formed: C–C, C–H, C–H.

$$\Delta H^0 = [(610 + 436) - (347 + 414 + 414)] \text{ kJ mol}^{-1} = -129 \text{ kJ mol}^{-1}$$

Since ΔH^0 is negative, the reaction is exothermic.

Note: C=C is not completely broken; it is the π bond which breaks. However, breaking of C=C and formation of C–C accounts for breaking of the π bond.

3.3.8 Bond Angles

The bond angles associated with sp^3, sp^2 and sp hybridization of the central atom are 109°28′, 120° and 180° respectively. Thus **bond angles increase with the increasing s character** in the hybrid orbital from sp^3 (25% s) to sp^2 (33.3% s) to sp (50% s). In other words, **bond angles increase with the decreasing p character** in the hybrid orbital from sp^3 (75% p) to sp^2 (66.7% p) to sp (50% p). These angles are expected when the groups around the central atom are identical. However, the bond angles associated with an *unsymmetrically substituted central atom* often deviate a little from those ideal values.

Consider, for example, **methyl chloride** (CH_3Cl) in which H–C–H angles are 110°, whereas the H–C–Cl angles are 108°, indicating a small deviation from the ideal tetrahedral (sp^3) angle for CH_4.

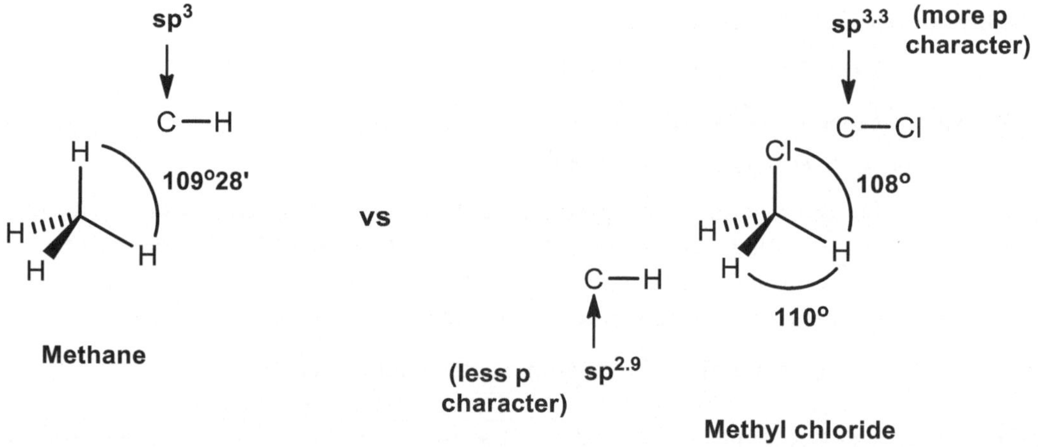

sp^3 → C—H; 109°28′

Methane

vs

$sp^{3.3}$ (more p character) → C—Cl; 108°; 110°

C—H ← $sp^{2.9}$ (less p character)

Methyl chloride

This can be explained using **modified hybrid orbitals** for the carbon atom. Besides *integer* combination of atomic orbitals (sp^3, sp^2 or sp), hybrid orbitals can also be derived from **noninteger combination of atomic orbitals**. In general, hybrid orbitals can be designated as **sp^n** where *n* is a positive integer or a fraction.

In CH_3Cl, the hybridization of carbon is not exactly sp^3. *Since Cl is more electronegative than C, the electron density in the C–Cl bond will be attracted more towards Cl.* Thus, for the **C–Cl bond, C uses a hybrid orbital with more p character** (p electrons are further from the nucleus than are s electrons), and consequently **C–H bonds are formed with hybrid orbitals of less p character**.

Increase in more p character (or less s character) in the hybrid orbital leads to a **decrease** of H–C–Cl bond angle to 108° from the ideal sp^3 angle of 109°28′, and a decrease in p character (or more s character) in the hybrid orbital for C–H bonds leads to an **increase** of H–C–H bond angle to 110°. The hybrid orbitals of carbon are approximately $sp^{3.3}$ for the C–Cl bond and $sp^{2.9}$ for the C–H bonds. The four hybrid orbitals of carbon are therefore not equivalent.

PROBLEM 3.3: The C–N–C bond angles in trimethylamine [$(CH_3)_3N$] are 108°. Explain using the concept of hybridization.

Answer

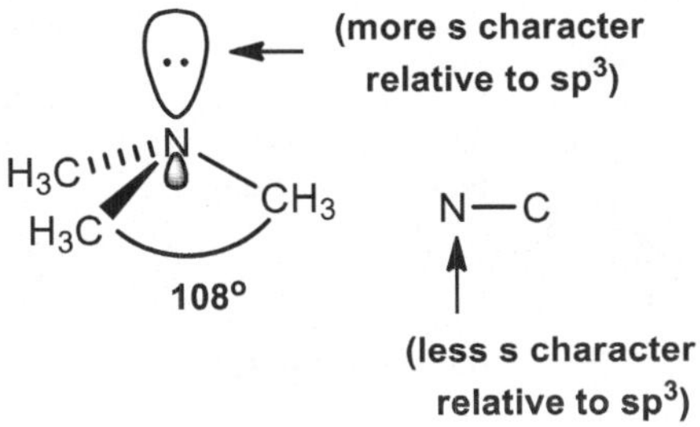

(more s character relative to sp^3)

H_3C N CH_3 H_3C 108° N—C

(less s character relative to sp^3)

Trimethylamine

The C–N–C angles in trimethylamine are slightly smaller than tetrahedral. This can be explained as follows:

The lone pair of electrons on N is close to the nucleus because they are not extended outward in bonding to other atom. Thus N uses a hybrid orbital with *more s character* (relative to sp^3) to accommodate the lone pair, since s electrons are closer to the nucleus than are p electrons. Consequently, *less s character* (or more p character) is associated with the *nitrogen hybrid orbitals* for forming the N–C bonds, resulting in a slight decrease in bond angle.

Valence Shell Electron-pair Repulsion (VSEPR) Theory

The bond angles around the central atom can also be explained using a simple theory called the **valence shell electron-pair repulsion (VSEPR) theory**. Valence electron pairs of the central atom can be electron pairs shared in covalent bonds, called **bond pairs** and the electron pairs that are unshared, called **lone pairs**.

An electron pair (bond pair or lone pair) will tend to be as far apart as possible from other electron pairs because of **repulsion** between electrons. A lone pair of electrons is closer to the nucleus than a bond pair of electrons, and hence lone pairs would exert more electron repulsion. The repulsive interactions are in the order:

Bond pair/bond pair < bond pair/lone pair < lone pair/lone pair

To illustrate, consider bond angles in **methane** (CH_4), **ammonia** (NH_3) and **water** (H_2O).

For **methane**, only a *tetrahedral arrangement* will allow four C–H bond pairs to have the *maximum possible separation* that will minimize their mutual electron repulsion. Any other arrangement, say, a square planar orientation, will place the electron pairs closer together. Therefore, H–C–H bond angles in methane are 109°28′, the tetrahedral angle.

In **ammonia**, the electron repulsion between the lone pair on nitrogen and the bond pair of N–H bond is *greater* than the bond pair/bond pair repulsion. As a result, H–N–H angles are smaller (107°) than the ideal tetrahedral angle (109°28′).

In **water**, the H–O–H bond angle is 105° which is even smaller than the bond angle in ammonia. This is attributed to the *much greater* lone pair/lone pair repulsion between two lone pairs on the oxygen atom as compared to the bond pair/bond pair repulsion between two O–H bonds.

109°28'

sp^3
(more lone pair/ bond pair repulsion)
(less bond pair/ bond pair repulsion)
107°

(large lone pair/ lone pair repulsion)
sp^3 sp^3
(small bond pair/ bond pair repulsion)
105°

Methane Ammonia Water

Next, consider bond angles in **methyl chloride**, **propane** and **ethylene**.

In the case of **methyl chloride** (CH_3Cl), the electronegative Cl attracts electron density away from the carbon atom. As a result, the electron repulsion due to the C–Cl bond pair will be less, resulting in a smaller H–C–Cl angle (108°) than H–C–H angle (110°).

For **propane**, the C–C–C bond angle is 112.4° which is larger than the H–C–H angle (106.1°) on the central carbon. This is because there is greater repulsion between two C–C bonds than between two C–H bonds.

In **ethylene**, the C–C–H angle (121.5°) is larger than the H–C–H angle (117°). Since the double bond has *two* electron pairs, repulsions by two electron pairs are evidently greater than those by one electron pair of the single bonds, resulting in a larger C–C–H angle.

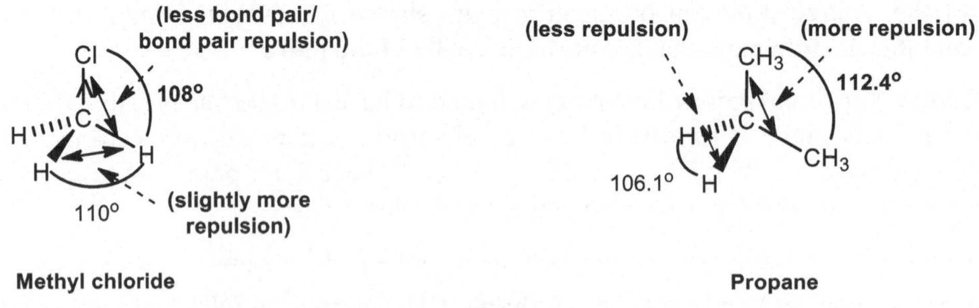

Methyl chloride Propane

121.5° (more repulsion)
117°
(less repulsion)

Ethylene

3.4 CONJUGATION AND DELOCALIZATION

Organic compounds can contain two or more double bonds. When two double bonds are separated by *only one single bond*, the two double bonds are said to be **conjugated** (see also Section 2.3.2). The simplest **conjugated diene** is **1,3-butadiene**.

If one of the two double bonds is a C=O double bond, the compound is called a **conjugated carbonyl compound**. An example is **acrolein** (or acraldehyde).

Sketetal representation	***Examples***
C=C—C=C	H_2C=CH—CH=CH_2
Conjugated diene	**1,3-Butadiene**
C=C—C=O	H_2C=CH—CH=O
Conjugated carbonyl compound	**Acrolein**

Two double bonds are **not conjugated** when they are separated by *more than one single bond* or when both double bonds are *directly attached to an atom*. In the former case, the two double bonds are said to be **isolated** and in the latter case, they are said to be **cumulated** (see also Section 2.3.2).

An example of an isolated diene is **1, 4-pentadiene**, and the simplest cumulated diene is **allene**.

H_2C=CH—CH_2—CH=CH_2	H_2C=C=CH_2
1,4-Pentadiene	**Allene**
(isolated diene)	**(cumulated diene)**
Not conjugated	***Not conjugated***

In **nonconjugated** dienes, each π bond is *restricted between two carbon atoms only*, similar to the π bond in ethylene, and is said to be **localized**. The situation is different in **conjugated** systems when the *π bonds can spread over more than two atoms*, and the π bonds are said to be **delocalized**.

Orbital Pictures

In order to understand **π-conjugation (or delocalization of π electrons)**, we will use simple pictorial descriptions called **orbital pictures**. In drawing orbital pictures, conjugated molecules are shown in their *proper geometry* and the use of **p orbitals with no phase sign** will suffice (see below).

These orbital pictures are *not* molecular orbitals. It is necessary to use *p orbitals with phase sign* while representing *π molecular orbitals* of conjugated molecules, which will show in-phase overlaps, out-of-phase overlaps and nodes. However, molecular orbitals of conjugated systems and their representations are beyond the scope of our discussion.

3.4.1 π bond–π bond conjugation: Delocalization of π electrons over four atoms

Consider **1,3-butadiene** containing two π bonds. An orbital picture of 1,3-butadiene is shown below.

Each carbon atom of the diene is bonded to three atoms, and is **sp^2-hybridized.** Therefore, the molecule has a **planar geometry** (all four carbon atoms and six hydrogen atoms lie in the same plane).

The **unhybridized p orbitals** on the carbon atoms are **perpendicular** to the planar framework, and they are **mutually parallel.**

The **two π bonds** are normally formed by the lateral overlap of p orbitals on C1 and C2, and on C3 and C4. It also appears that a **lateral overlap is possible between the p orbitals on C2 and C3** as they are parallel to each other.

As a result, the two π bonds and four π electrons of butadiene are **delocalized over the whole length of four carbon atoms.** This delocalization is a stabilizing phenomenon and confers **extra stability** to the molecule.

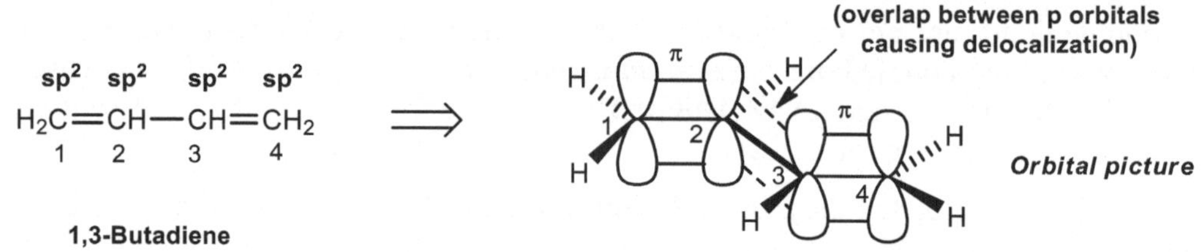

sp^2 sp^2 sp^2 sp^2
$H_2C{=}CH{-}CH{=}CH_2$
1 2 3 4

1,3-Butadiene ⟹ **Orbital picture**

(overlap between p orbitals causing delocalization)

The p orbitals are perpendicular to the molecular plane, and are mutually parallel

Bond Lengths

The **carbon-carbon bond lengths** of butadiene show that the two terminal carbon-carbon double bonds have the same length (1.34 Å) as the double bond of ethylene, but the **central C–C bond** between C2 and C3 is considerably **shorter** (1.47 Å) than the single bond in ethane (1.54 Å).

The shortening of the central C2–C3 bond has been attributed to a **partial π bond character** as a result of delocalization and also to the **σ overlap of two sp^2 orbitals** (in contrast to sp^3–sp^3 overlap in ethane) as bond lengths become shorter with increase of s character in the sp^2 hybrid orbitals.

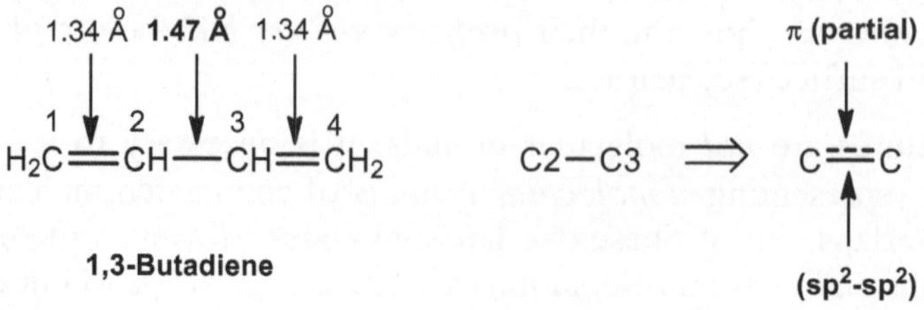

1.34 Å **1.47 Å** 1.34 Å

$\overset{1}{H_2C}{=}\overset{2}{CH}{-}\overset{3}{CH}{=}\overset{4}{CH_2}$

1,3-Butadiene

C2−C3 ⟹ C$\overset{\pi \text{ (partial)}}{\underset{(sp^2\text{-}sp^2)\ \sigma}{=}}$C

π (partial)

(sp^2-sp^2) σ

3.4.2 π bond–p orbital conjugation: Delocalization of π electrons over three atoms

Conjugation is possible between a **double bond** and a **p orbital** on an adjacent atom that is separated from the double bond by *only one* single bond.

Consider, for example, **methyl vinyl ether**. Here the oxygen has two σ bonds (bonded to two carbon atoms) and two lone pairs; the hybridization of oxygen is thus expected to be sp^3. But an sp^3 orbital is *not* parallel to the π bond and hence is not effective for conjugation.

To achieve a **favourable conjugation**, **oxygen** becomes **sp^2-hybridized** with an electron pair in a **p orbital** that is parallel to the **π bond**, as shown below in the orbital picture of the molecule. The **π bond–p orbital conjugation** leads to the delocalization of **four π electrons** (two in the π bond and two in the p orbital) over **three atoms** (two carbons and one oxygen). The other lone pair on oxygen in an sp^2 orbital is not shown.

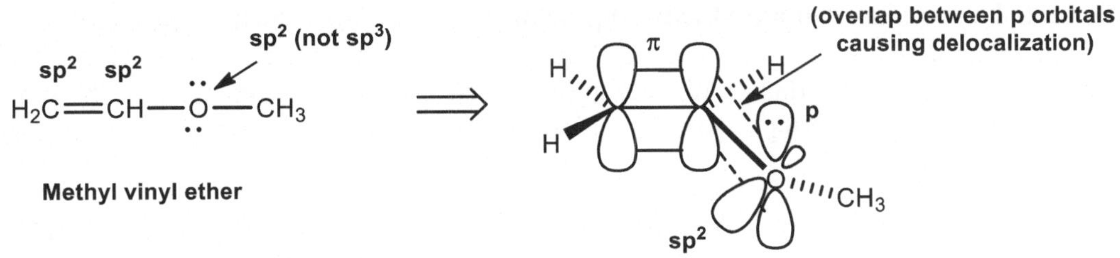

Orbital picture

PROBLEM 3.4: Draw orbital pictures for allyl cation and allyl anion showing π–p conjugation. How many π electrons are delocalized in each case?

Answer

In both allyl cation and allyl anion, each carbon is sp^2-hybridized, and the cationic or anionic carbon has a p orbital. The p orbital in allyl cation is vacant, whereas the p orbital in allyl anion is completely filled with two electrons.

The π–p conjugation leads to delocalization of two π electrons over three carbon atoms in the case of allyl cation, and delocalization of four π electrons over three carbon atoms in the case of allyl anion.

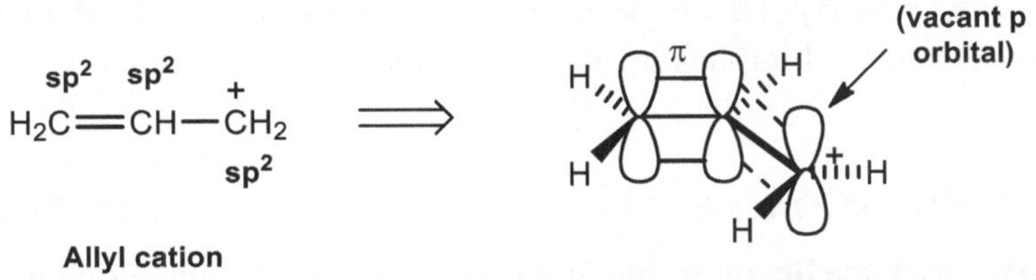

Delocalization of 2π electrons

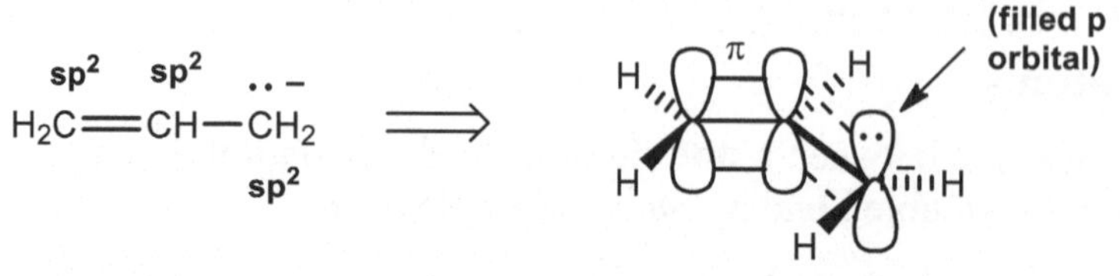

Allyl anion

Delocalization of 4π electrons

3.4.3 Orthogonal π Bonds: No Conjugation or Delocalization

Consider, for example, **allene** ($H_2C=C=CH_2$) in which two double bonds are attached to the same carbon atom.

The **central carbon (C2)** of allene is bonded to two atoms and is **sp-hybridized**, whereas **two terminal carbons (C1 and C3)**, each bonded to three atoms, are **sp^2-hybridized**. The C=C=C chain in allene is **linear** since the central carbon is sp.

The central carbon has **two unhybridized p orbitals (p_y and p_z)**, which are **perpendicular** to each other and also to the linear carbon chain. As shown below, one of the two p orbitals (p_z) on C2 overlaps with the unhybridized p orbital (p_z) on C1 and the other p orbital (p_y) on C2 overlaps with the unhybridized p orbital (p_y) on C3 to form **two π bonds**.

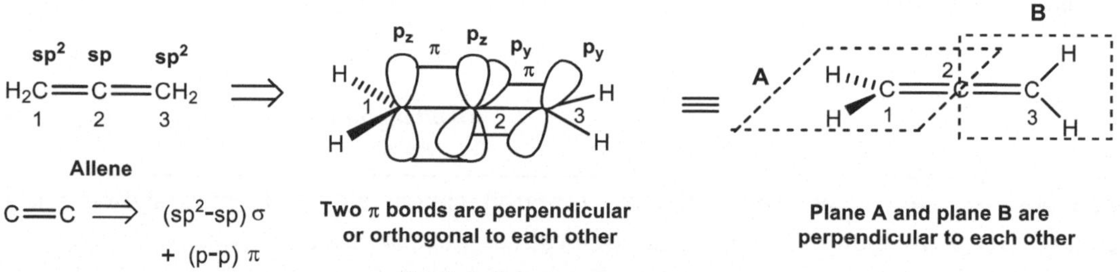

Two π bonds are perpendicular or orthogonal to each other

Plane A and plane B are perpendicular to each other

Orbital picture

The arrangement of the p orbitals in allene clearly indicates that the **two π bonds are perpendicular or orthogonal to each other, and hence no π–π delocalization is possible**. Remember that for delocalization to occur, the two π bonds need to be parallel.

The sp^2 plane containing C1, C2 and two H atoms at C1 (plane A) is **perpendicular** to the sp^2 plane containing C2, C3 and two H atoms at C3 (plane B).

The C=C **bond length** in allene is 1.31 Å which is slightly **shorter** than the C=C bond length in ethylene (1.34 Å). This is because the σ bond in the allene double bond arises from sp^2–sp overlap (sp orbital has greater s character) as compared to sp^2–sp^2 σ overlap in ethylene.

3.5 CYCLIC CONJUGATION: AROMATICITY AND ANTIAROMATICITY

Benzene (C_6H_6) is a **cyclic** molecule with alternate single and double bonds. Three double bonds in benzene are therefore **conjugated**. Each carbon atom (bonded to three other atoms) is **sp^2-hybridized**, and the benzene ring is **planar** with all carbon and hydrogen atoms lying in the same plane.

The C–C–C bond angles in benzene are the normal sp^2 angle of 120°, which is the same as the geometric angles of 120° of a regular hexagon. As such, the planar geometry of benzene becomes highly favourable.

An **orbital picture** of benzene is shown below. In benzene, the **six unhybridized p orbitals**, each containing one electron, are **perpendicular** to the plane of the ring, and make a **cyclic array** of p orbitals. The p orbital of any one carbon atom can overlap equally well with the two adjacent p orbitals leading to a continuous **cyclic p orbital overlap**. This leads to **cyclic delocalization of six π electrons** and imparts **exceptional stability** to the benzene molecule.

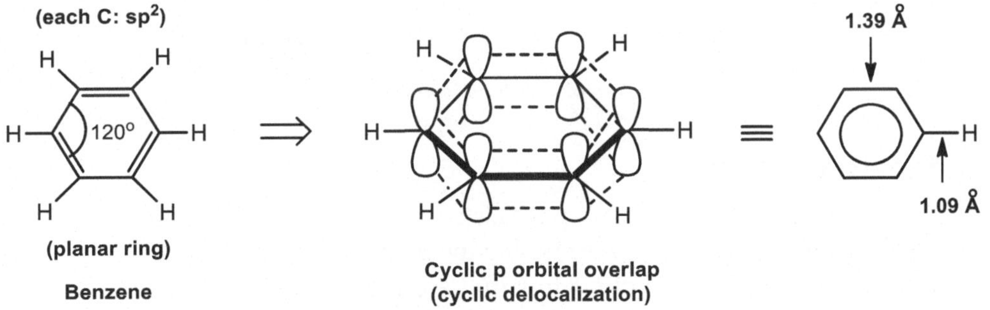

Orbital picture

Because of cyclic p orbital overlap, the carbon–carbon bonds in benzene are expected to be neither single bonds nor double bonds. This is exactly what is found experimentally by spectroscopic measurements. **All carbon-carbon bonds of benzene are of equal length (1.39 Å)**, being *intermediate* between the length of a *single bond* between sp^2-hybridized carbon atoms (1.47 Å) and that of a C=C *double bond* (1.34 Å).

Aromaticity and antiaromaticity

Due to **cyclic conjugation or delocalization, benzene** becomes especially stable. It is **more stable** than its open-chain analogue, **1,3,5-hexatriene**. This extra stabilization is called **aromaticity**. Thus benzene is **aromatic**.

Since the aromaticity or aromatic character of benzene arises from cyclic conjugation, we might **assume** that all rings that contain alternating double bonds would have the similar aromatic character as benzene. But this assumption is *not correct always*.

For example, **cyclobutadiene** that has a **cyclic conjugation** of two double bonds is an extremely unstable compound. Cyclobutadiene is **less stable** than its acyclic analogue, **1,3-butadiene**. This destabilization as a result of cyclic conjugation is opposite to aromaticity, called **antiaromaticity**.

Thus, **cyclic conjugation may lead to stabilization or destabilization**. Some rings (like benzene) are stabilized by cyclic conjugation and become **aromatic**, while others (like cyclobutadiene) are destabilized by cyclic conjugation and become **antiaromatic**.

Stability:

>

($H_2C{=}CH{-}CH{=}CH{-}CH{=}CH_2$)

Benzene **1,3,5-Hexatriene**

Benzene is aromatic

Stability:

<

($H_2C{=}CH{-}CH{=}CH_2$)

Cyclobutadiene **1,3-Butadiene**

Cyclobutadiene is antiaromatic

3.5.1 The Hückel Rules

In 1931 Erich Hückel, a German physicist, investigated the *aromaticity* of **cyclic conjugated polyenes** (such as benzene) using molecular orbital calculations.

Hückel predicted that monocyclic rings that are planar, and provide a p orbital on each atom would be aromatic if they contain $(4n + 2)$ π electrons, where n is an integer (0, 1, 2, 3 and so on). This is called the Hückel $(4n + 2)$ rule.

The **Hückel (4*n* + 2) rule** is thus stated as:

> Planar, fully conjugated, monocyclic systems that contain $(4n + 2)$ π electrons ($n = 0, 1, 2, 3, \ldots$) will be aromatic.

Benzene is *planar* and *monocyclic* (contains one ring), and has a *p orbital on each carbon atom*. The number of π electrons in benzene is **6** (2 from each of the three double bonds). Since the number 6 is a Hückel number ($4n + 2$, $n = 1$), **benzene is aromatic**. The six π electrons in benzene are called the **aromatic sextet**.

On the other hand, **cyclobutadiene** is also *planar*, *monocyclic*, and has a *p orbital on each carbon*. It contains **4** π electrons (2 from each of the two double bonds). But 4 is *not* a Hückel $(4n + 2)$ number. Therefore, cyclobutadiene is *not* aromatic; it is **antiaromatic** (see above).

However, **4** is a **4*n* number** ($n = 1$). So there is another rule called the **Hückel (4*n*) rule**, which predicts antiaromaticity.

The **Hückel (4*n*) rule** is stated as:

> Planar conjugated monocyclic systems that contain $(4n)$ π electrons ($n = 1, 2, \ldots$) will be antiaromatic.

Thus, in consistent with the Hückel $(4n)$ rule, cyclobutadiene is antiaromatic.

3.5.2 Molecular Orbital Energy Diagram: Frost Circle

The Hückel's molecular orbital calculations are beyond our scope, but we can draw an **energy diagram** for the relative energies of π **molecular orbitals** of conjugated monocyclic systems (such as benzene and cyclobutadiene) in a relatively simple way, known as a **Frost circle**.

Consider, for example, **benzene** as a conjugated monocyclic system. Benzene has six *p* orbitals (one from each carbon). Therefore, **six π molecular orbitals** (ψ_1–ψ_6) must be formed, since the number of atomic orbitals combined is equal to number of molecular orbitals formed. A **Frost circle** would enable us to draw quickly an energy diagram for the six π molecular orbitals of benzene.

To make a **Frost circle**, we simply draw a **circle** and **inscribe** in it a regular polygon such as a **hexagon** for benzene, so that *one* of its corners is at the **bottom** (pointing down). **Each point where the polygon (hexagon) touches the circle corresponds to an energy level of a π molecular orbital of benzene.**

Figure 3.5 shows the construction of a molecular orbital energy diagram for benzene with the help of a Frost circle.

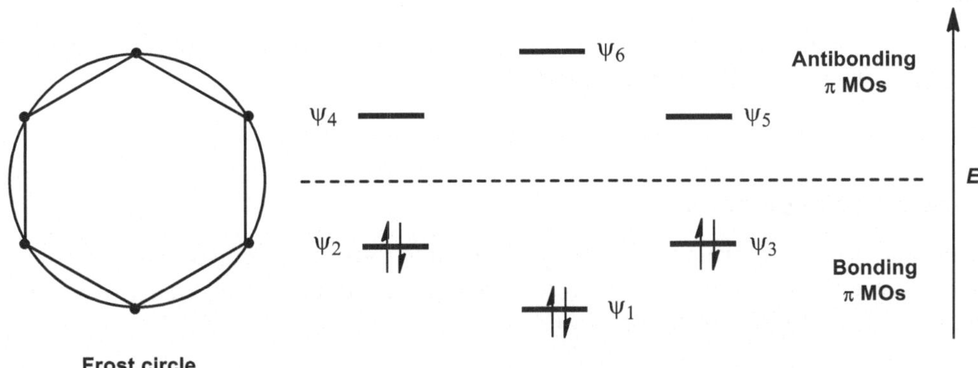

Figure 3.5: Frost circle and molecular orbital energy diagram for benzene.

The six points where the hexagonal ring touches the circle represent the locations of **six energy levels for six π MOs of benzene**. These energy levels are drawn to the right of the circle for clarity.

The energy levels *lower* than the centre of the circle correspond to the **bonding π molecular orbitals (ψ_1–ψ_3)**, while those *higher* than the centre of the circle correspond to the **antibonding π molecular orbitals (ψ_4–ψ_6)**. A **horizontal line** through the centre of the circle divides the bonding molecular orbitals from the antibonding molecular orbitals.

Since benzene has **six π electrons**, the lowest energy bonding orbital (ψ_1) and two degenerate bonding orbitals above it (ψ_2 and ψ_3) are completely filled. Therefore, **the π electron configuration of benzene is $\psi_1^2\psi_2^2\psi_3^2$**.

This electronic structure of benzene indicates that all three of the low-energy bonding orbitals are completely filled with electrons. This arrangement of six π electrons in benzene represents a **closed shell configuration** of π electrons, similar to the closed shell configuration of an inert gas atom. **Since benzene has 6 π electrons in a closed shell, it is exceptionally stable and aromatic.**

In general, the reason for the exceptional stability or aromaticity is that the **Hückel number $4n + 2$ gives a closed shell arrangement of π electrons.** Note that a closed shell configuration of cyclic π systems implies that all bonding molecular orbitals (and nonbonding orbitals, if present) are completely filled.

Note: Cyclobutadiene has a bonding π molecular orbital (ψ_1), two degenerate (equal energy) nonbonding π molecular orbitals (ψ_2 and ψ_3) and an antibonding π molecular orbital (ψ_4) (draw a Frost circle to verify). A π molecular orbital that falls on the horizontal line of the Frost energy diagram is a nonbonding molecular orbital; it has the same energy as the energy of a p atomic orbital. The π electron configuration of cyclobutadiene containing four π electrons is $\psi_1^2\psi_2^1\psi_3^1$, which indicates that the two nonbonding orbitals are half-filled, *not* completely filled. Thus cyclobutadiene *does not* represent a closed shell configuration, and is not aromatic.

3.5.3 Applications of the Hückel ($4n + 2$) Rule

The Hückel ($4n + 2$) rule can be applied to determine whether a molecule or an ionic species (cation or anion) is aromatic or not.

Monocyclic systems may also contain one or more heteroatoms (N, O or S), and if they are aromatic, they are called **heteroaromatic** compounds.

Aromatic systems with six π electrons (the aromatic sextet) are most common.

Heteroaromatic compounds

Pyrrole and **pyridine** are two heterocyclic compounds, each containing one nitrogen atom. As described below, both pyrrole and pyridine obey the Hückel ($4n + 2$) rule, where $n = 1$. Therefore, they are **aromatic**.

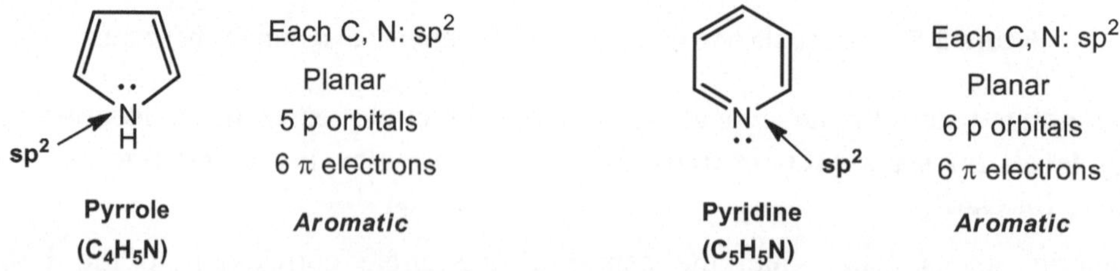

In **pyrrole**, the *N* atom is bonded to three atoms and it has one lone pair, but the hybridization of *N* **is sp^2** (instead of sp^3) so that cyclic delocalization can take place. With sp^2-hybridized carbons and nitrogen, pyrrole becomes **planar**, and there is a *cyclic array of 5 overlapping p orbitals*. The delocalized π system of pyrrole contains **6 electrons** (4 from the two π bonds and 2 from the nitrogen lone pair), and hence **pyrrole is aromatic** obeying the Hückel ($4n + 2$) rule, where $\boldsymbol{n = 1}$.

It should be noted that the nirogen lone pair in pyrrole contributes to the aromatic sextet.

On the other hand, the N atom in **pyridine** has two σ bonds and one lone pair, and the hybridization of N is **sp²** with its lone pair in an sp² orbital. All five carbon atoms and the nitrogen atom in pyridine are sp² with a p orbital on each atom. Thus pyridine is **planar**, and has a *cyclic array of 6 overlapping p orbitals*. Since the number of delocalized π electrons is **6** (from the three π bonds), **pyridine** obeys the Hückel ($4n + 2$) rule where $\boldsymbol{n} = \mathbf{1}$, and is **aromatic**.

In contrast to pyrrole, **the nitrogen lone pair in pyridine does not contribute to the aromatic sextet**.

Aromatic Charged Species

The Hückel ($4n + 2$) rule is also applicable to the **charged species** such as cyclopentadienyl anion, cycloheptatrienyl cation and cyclopropenyl cation.

Cyclopentadienyl anion is analogous to pyrrole, where an anionic carbon replaces the nitrogen atom. The anionic carbon is **sp²-hybridized** and has a **lone pair in a p orbital.** The *cyclic overlap of 5 p orbitals* in cyclopentadienyl anion gives a delocalized π system with **6 electrons**. Hence, cyclopentadienyl anion is **aromatic** obeying the Hückel ($4n + 2$) rule, where $\boldsymbol{n} = \mathbf{1}$.

Cycloheptatrienyl cation (also called **tropylium cation**) is a charged species in which the cationic carbon is bonded to three atoms (two carbons and one hydrogen). It is **sp²-hybridized** with an **empty p orbital** on the cationic carbon. Thus the tropylium cation is **planar**, and has a *cyclic array of 7 overlapping p orbitals* (one on each carbon). The delocalized π system contains **6 electrons** (6 from the three π bonds and none from the empty p orbital). Hence the **tropylium cation** obeys the Hückel ($4n + 2$) rule where $\boldsymbol{n} = \mathbf{1}$, and is **aromatic**.

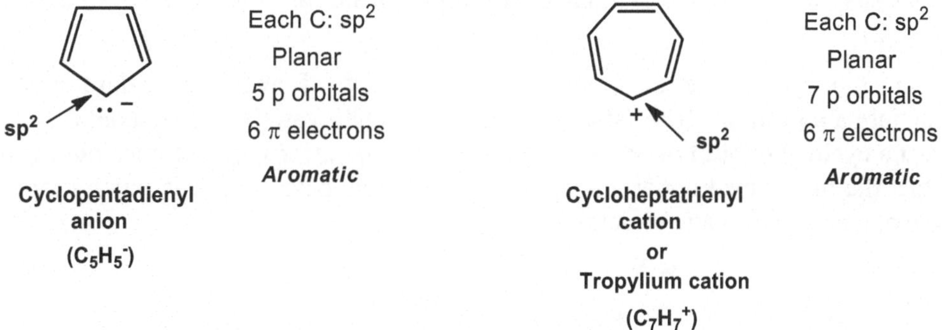

Cyclopropenyl cation (or **cyclopropenium cation)** has a three-membered ring which is necessarily **planar**. The cationic carbon is **sp²-hybridized** with an **empty p orbital** on it. Thus cyclopropenyl cation has a *cyclic array of 3 overlapping p orbitals* (one on each carbon). The delocalized π system contains **2 electrons** (2 from the π bond and none from the empty p orbital).

Besides the number 6, the **number 2** is also a Hückel ($4n + 2$) number, where $\boldsymbol{n} = \mathbf{0}$. Hence **cyclopropenyl cation is aromatic**.

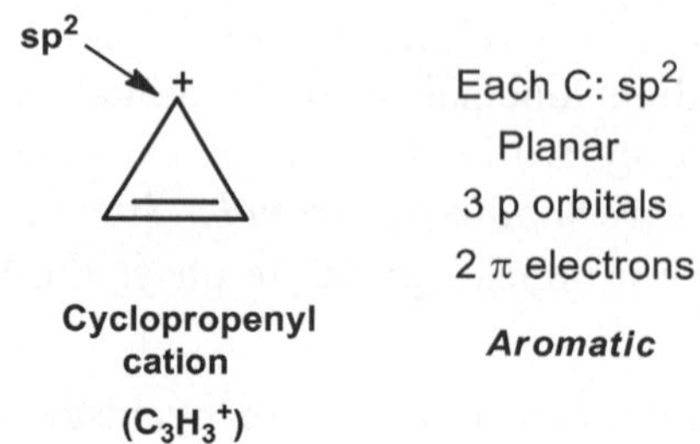

Annulenes

Annulenes (C_nH_n) are monocyclic systems having alternating double bonds. The **ring size** of annulenes is denoted by a *number in brackets*, such as [4]annulene, [6]annulene, [8]annulene and [10]annulene.

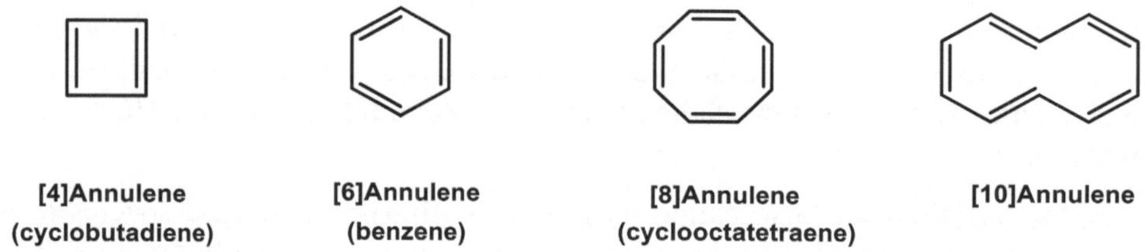

[4]Annulene is cyclobutadiene which is **antiaromatic**, and **[6] annulene** is benzene which is **aromatic**.

[8]Annulene is **cyclooctatetraene**. All carbon atoms in cyclooctatetraene are **sp^2-hybridized**. But a **planar structure** of cyclooctatetraene with **8 π electrons** (from the four π bonds) would be **antiaromatic** and *highly unstable* in terms of the **Hückel (4*n*) rule** ($n = 2$).

However, the cyclooctatetraene molecule is capable of avoiding antiaromaticity by bending into a **nonplanar 'tub' shape** when the p orbitals cannot overlap continuously, i.e. there is **no cyclic p orbital overlap**. As a result, **cyclooctatetraene** becomes **nonaromatic**. Note that the tub-shaped nonaromatic structure of cyclooctatetraene is much more stable than the planar antiaromatic structure.

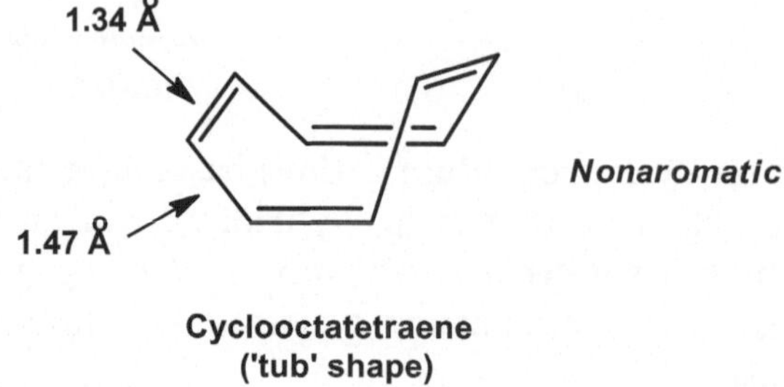

Because of the lack of conjugation, the carbon-carbon bond lengths in cyclooctatetraene are not equal but alternately long (1.47 Å) and short (1.34 Å) for the typical C–C and C=C bonds.

[10]Annulene is a monocyclic system with ten carbon atoms which are **sp^2-hybridized**. Thus the molecule is **expected** to be planar and aromatic with 10 π electrons, since the number 10 is a Hückel number ($4n + 2$, $n = 2$).

However, there is a severe **steric interaction** between the **two internal H atoms** in the centre of the [10]annulene ring. As a result, the ring is **prevented from being planar**. Since the ring is *not* planar, the **Hückel rule does not apply,** and hence **[10]annulene** is **nonaromatic**. Remember that planarity is required for an effective cyclic conjugation or p orbital overlap.

H H

Not planar due to steric interference of two internal H atoms

Nonaromatic

[10]Annulene

Naphthalene

If the two internal H atoms of [10]annulene are replaced by a carbon-carbon bond, it changes to **naphthalene**. Naphthalene is **planar** because the steric interference is eliminated. But the Hückel rule is applicable to a *monocyclic system*. So the question is: **Does the Hückel rule apply to naphthalene which is bicyclic?**

Although bicyclic, naphthalene can be regarded as a **monocyclic** system with an array of **10 p orbitals along the perimeter of the naphthalene ring** (ignoring the central C–C bond). A cyclic p orbital overlap is thus possible and the **Hückel rule is applicable.** Therefore, **naphthalene** with 10 π electrons is **aromatic** since the number 10 is a Hückel ($4n + 2$) number, where $n = 2$.

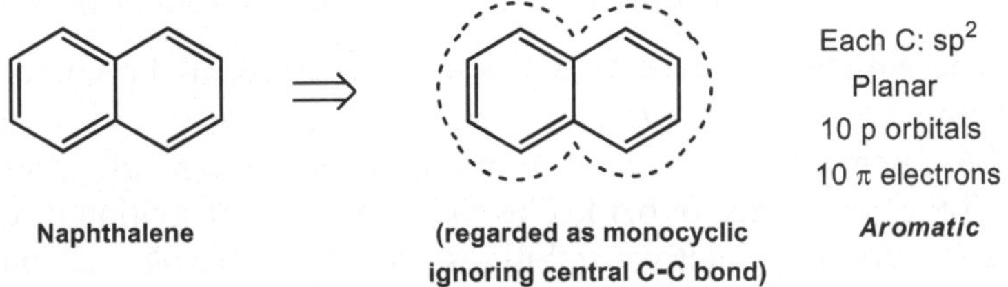

PROBLEM 3.5: Is the Hückel rule applicable to the following compounds? Hence predict whether each of them is aromatic or nonaromatic.

Answer

In each compound, *one* of the ring carbons is sp^3-hybridized and lacks a p orbital. Therefore, they do not have a cyclic array of p orbitals, and the π electrons cannot be delocalized over the entire ring because of the intervening sp^3-hybridized carbon. The Hückel rule is *not* applicable, and each of the compounds is **nonaromatic**.

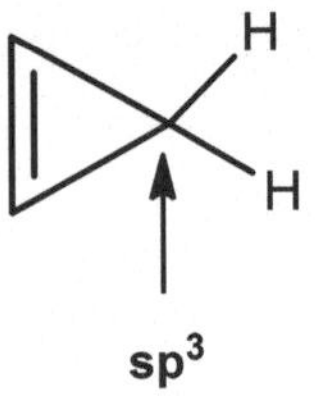

H H sp^3 **Nonaromatic**   H H sp^3 **Nonaromatic**   H H sp^3 **Nonaromatic**

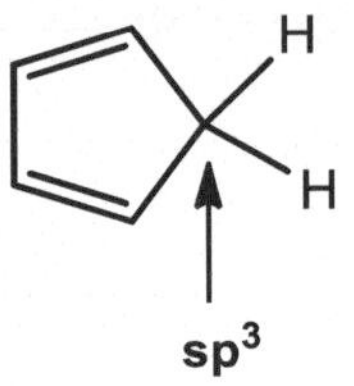

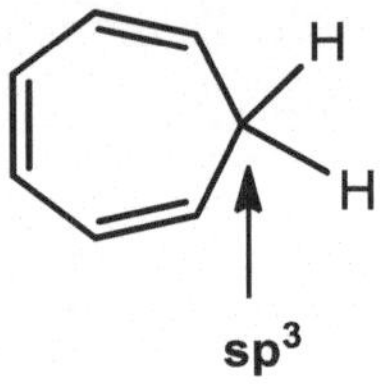

Note: Don't be misled just by counting the number of π electrons, i.e. 2 in cyclopropene, 4 in cyclopentadiene and 6 in cycloheptatriene to predict aromaticity. First, ensure whether the Hückel rule is applicable or not, based on cyclic conjugation or delocalization. If applicable, then count the number of π electrons to determine aromaticity or antiaromaticity.

3.6 RESONANCE

The two most important theories for describing bonding in organic molecules are the molecular orbital (MO) theory and the valence bond (VB) theory.

In this section, we will introduce the **valence bond method**, which approaches the problem by developing the concept of **resonance**. There are several ways in which π *electrons* and *lone pair electrons* can be distributed in a σ *bonding framework* in organic molecules, and each possible distribution can be represented by a proper **Lewis structure**. Such Lewis structures are called **resonance structures or canonical structures**.

No one resonance structure represents the *actual* structure of the compound but each resonance structure has some *contribution* to the actual structure. The **actual or real structure** of the compound is regarded as a *weighted average* of different resonance structures, called the **resonance hybrid**. The weighing or contribution of each resonance structure depends on their relative stabilities, *the most stable resonance structure contributing most.*

It is important to note that **resonance structures are hypothetical but the resonance hybrid is real**. The resonance hybrid represents the actual or real molecule.

The *energy of the actual molecule* (i.e. resonance hybrid) is *lower* than the *energy of any individual resonance structure.* Thus the actual molecule is **stabilized by resonance**. The **difference** between the energy of the actual molecule (or resonance hybrid) and the energy estimated for the *most stable* resonance structure is called the **resonance energy (RE)**.

Resonance energy is thus a measure of stabilization of an organic compound by resonance; **the greater the RE, the more stable the compound is.**

3.6.1 Rules for resonance

Resonance structures are Lewis structures that can be drawn using curved arrows (see Section 2.2.3). The double-headed arrows (↔) are used to *interrelate* the contributing resonance structures.

The following rules give a general guidance for drawing resonance structures and evaluating their importance.

1. ***All of the resonance structures must be proper Lewis structures.*** For example, a Lewis structure in which N has five bonds is not proper.

(N has 5 bonds)

Not a proper Lewis structure

2. ***The positions of the nuclei of atoms must be the same in all of the structures.*** This implies that only *electrons are allowed to move but not the atoms.*

For example, the structures **I** and **II** are two resonance structures of acetone but not the structure **III**, in which a hydrogen atom has moved from carbon to oxygen.

Resonance structures of acetone:

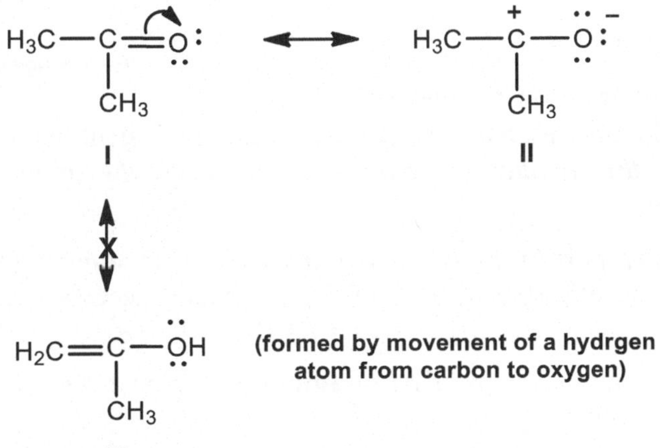

I

II

III

(formed by movement of a hydrgen atom from carbon to oxygen)

Not a resonance structure

3. ***Electron pairing must be conserved in all resonance structures.*** For example, the resonance structures of 1,3-butadiene are represented by the structures **IV–VI**, of which the structures **V** and **VI** are equivalent. The structure **VII** is, however, *not* a resonance structure of 1,3-butadiene because the single electrons in it have the same spin orientation, and hence are not paired.

Resonance structures of 1,3-butadiene:

$H_2C{=}CH{-}CH{=}CH_2 \longleftrightarrow H_2\overset{+}{C}{-}CH{=}CH{-}\overset{\cdot\cdot}{\underset{}{C}}H_2 \longleftrightarrow H_2\overset{\cdot\cdot}{C}{-}CH{=}CH{-}\overset{+}{C}H_2$

IV **V** **VI**

$\updownarrow$ ✗

$H_2\dot{C}{-}CH{=}CH{-}\dot{C}H_2$ **(two unpaired electrons with parallel spins)**

VII

Not a resonance structure

4.

(a) ***Equivalent resonance structures contribute equally and lead to a large resonance stabilization.*** For example, benzene is mostly represented by two equivalent resonance structures **VIII** and **IX** (known as Kekulé structures), and has a large resonance energy (see Section 3.6.2). The resonance hybrid is often drawn by inscribing a circle in the hexagon.

Benzene:

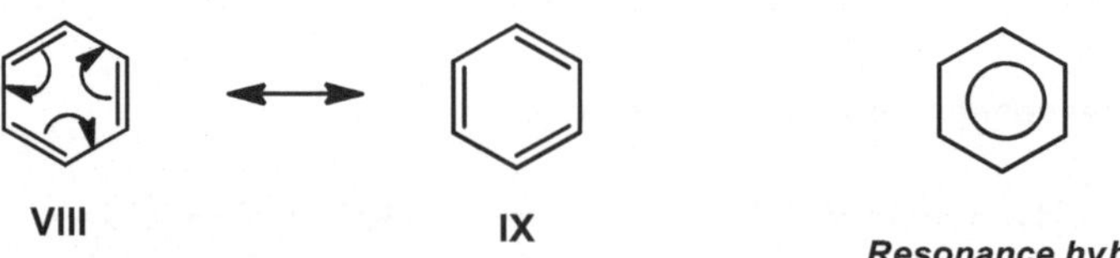

VIII **IX** ***Resonance hybrid***

Two equivalent resonance structures

(b) If the contributing resonance structures are not equivalent but are similar in energy, ***the greater the number of resonance structures, the greater is the resonance stabilization.***

5. ***All atoms taking part in the resonance must lie in a plane or nearly so.*** The reason for planarity is the effective overlap of the p orbitals necessary for delocalization or resonance.

For example, the amino group (NH_2) in **aniline** is coplanar with the benzene ring, and thus in resonance with it.

Aniline:

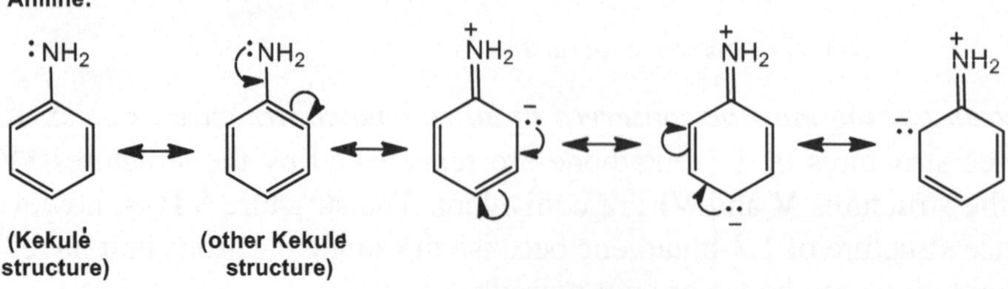

Five resonating structures

In orbital terms, the p orbital of nitrogen overlaps continuously with the π system of the benzene ring, thereby causing **delocalization** of the nitrogen lone pair into the benzene ring.

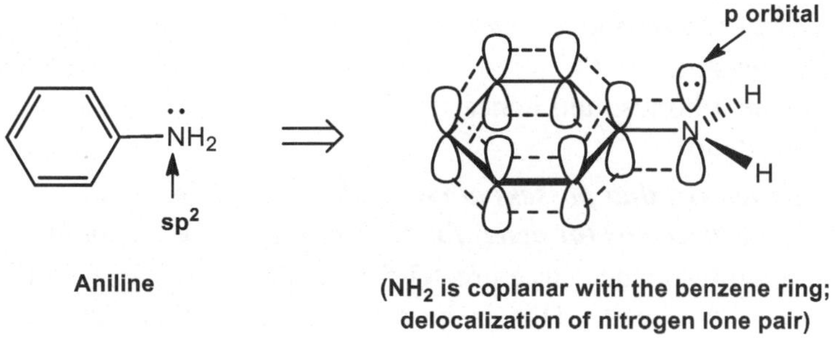

Aniline

(NH_2 is coplanar with the benzene ring; delocalization of nitrogen lone pair)

Steric inhibition of resonance

Consider, for example, ***N,N*,2,6-tetramethylaniline**. Due to **steric interactions** of the CH_3 groups at 2,6 positions of the aromatic ring with the CH_3 groups on nitrogen, the dimethylamino group [$N(CH_3)_2$] is rotated and forced **out of planarity** with the benzene ring. As a result, the resonance of the dimethylamino group with the ring is inhibited. This is known as the **steric inhibition of resonance**.

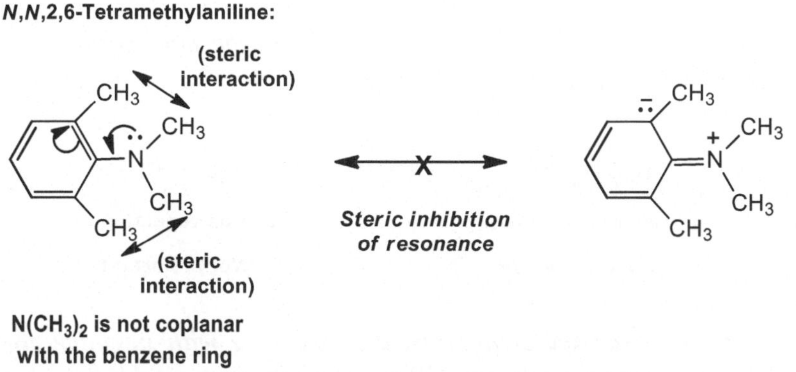

***N,N*,2,6-Tetramethylaniline:**

$N(CH_3)_2$ is not coplanar with the benzene ring

In orbital terms, the p orbital of nitrogen is *not* parallel with the π system of the benzene ring, thus there is **no delocalization** of the nitrogen lone pair into the benzene ring.

***N,N*,2,6-Tetramethylaniline**

$N(CH_3)_2$ is not coplanar with the benzene ring

No delocalization of nitrogen lone pair

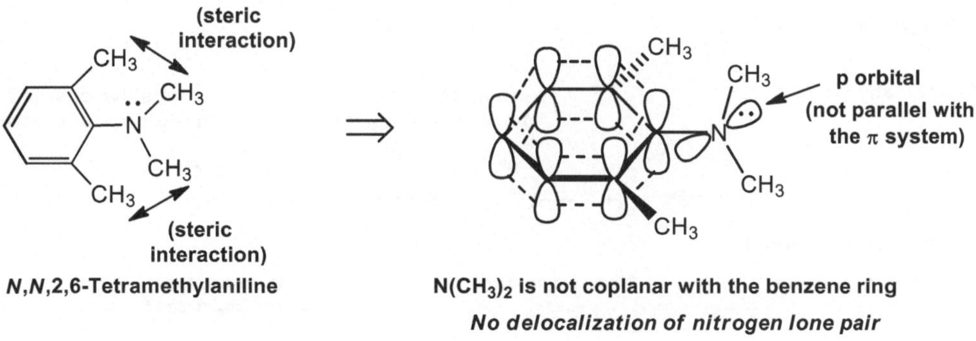

6. ***Each resonance structure contributes in proportion to its stability, the more stable a structure is, the greater is its contribution to the resonance hybrid.*** The relative stabilities of resonance structures can be decided as follows:

(a) ***The more covalent bonds a resonance structure has, the more stable it is.***
For example, the structure **IV** in the resonance structures of 1,3-butadiene (see above) has more covalent bonds, and hence it is the most stable resonance structure and is the major contributor.

(b) ***Resonance structures that possess atoms with a complete valence shell of electrons, i.e. eight valence electrons (or octet) for C, N and O, are more stable.***
For example, the resonance structure **XI** for the cationic species is more stable than the resonance structure **X**. This is because the carbon has six valence electrons (sextet) in **X**, while it has eight valence electrons (octet) in **XI**.

(sextet) (octet) (octet) (octet)

$CH_3\overset{+}{C}H-\ddot{O}CH_3 \longleftrightarrow CH_3CH=\overset{+}{\ddot{O}}CH_3$

X — ***Less stable*** **XI** — ***More stable***

(c) ***Stability of resonance structure decreases with charge separation.***
For example, the charge-separated structure of ethylene is not important.

$H_2C=CH_2 \longleftrightarrow H_2\overset{+}{C}-\overset{-}{C}H_2$

(no charge separation) ***Most important*** **(charge separation)** ***Not important***

(d) ***Between charge-separated structures, the structure with negative charge located on the more electronegative atom will be relatively more stable.***
For example, consider the following resonance structures for formaldehyde. The structure with no charge separation is most important. However, a charge-separated structure with negative charge on oxygen is more important than that with a positive charge on oxygen.

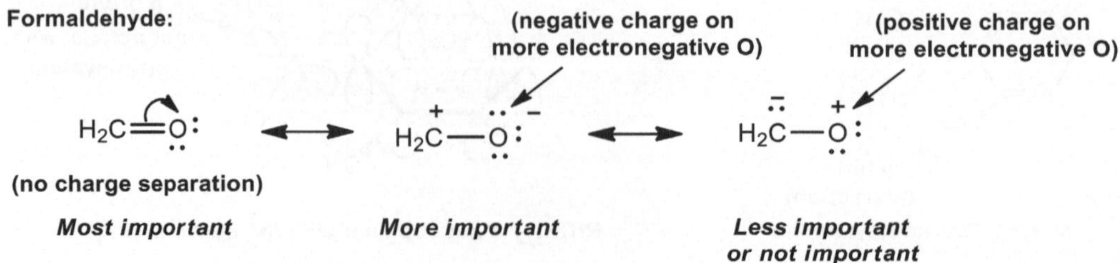

PROBLEM 3.6: Justify or criticise the following representations of resonance:

(a) 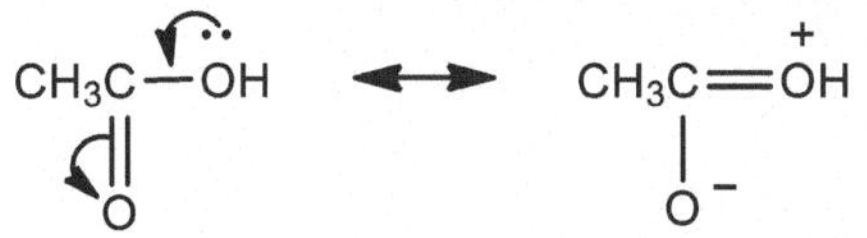 $CH_3C(=O)-\ddot{O}H \longleftrightarrow CH_3C(-O^-)=\overset{+}{O}H$

(b) 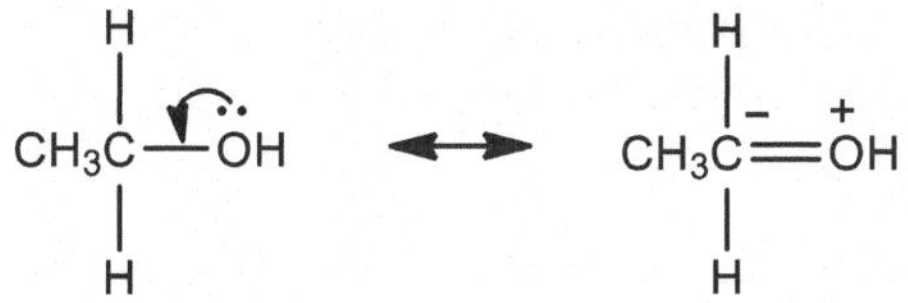 $CH_3CH_2-\ddot{O}H \longleftrightarrow CH_3\overset{-}{C}H_2=\overset{+}{O}H$

(c) $H_2C=CH-\overset{+}{C}H_2 \longleftrightarrow H_2\overset{+}{C}-CH=CH_2$

(d)  $H_2\overset{-}{C}-C(=O)-CH_3 \longleftrightarrow H_2C=C(-O^-)-CH_3$

(e) 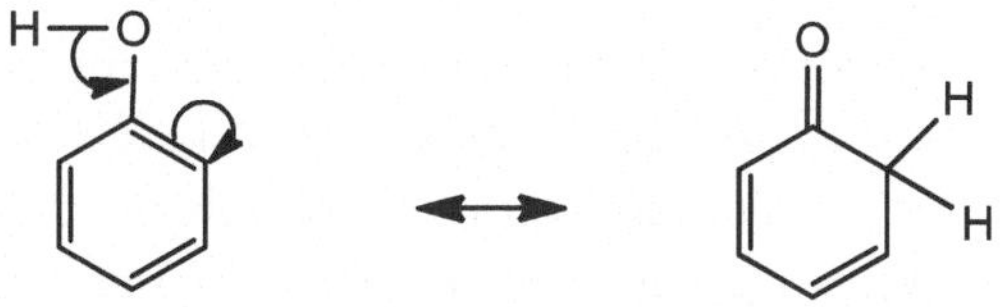 H–O (phenol) $\longleftrightarrow$ O, H, H (cyclohexa-2,4-dienone)

Answer

(a) Both resonance structures have the same number of covalent bonds and make significant contribution to the resonance hybrid. However, the structure on the left is the more important contributor because the other structure exhibits charge separation and is less stable.

(b) Wrong representation: the structure on the right is not a proper resonance structure as one carbon has five bonds.

(c) The two resonance structures are equivalent and contribute equally to the hybrid.

(d) Both resonance structures have the same number of covalent bonds and are significant contributors. However, the structure on the right is the more significant contributor as the negative charge is on the more electronegative O atom.

(e) Wrong representation: The two structures are not resonance structures. They are isomers (tautomers) formed by the movement of H atom from O to C. The structure on the left is the enol form and the structure on the right is the keto form.

3.6.2 Resonance Energy of Benzene: An Estimate from Heat of Hydrogenation Data

Benzene is a resonance hybrid of two equivalent Kekulé structures (see above). The resonance energy (RE) of benzene is therefore given by

$$\text{Resonance energy} = E_{\text{Kekulé}} - E_{\text{hybrid}}$$

where, $E_{\text{Kekulé}}$ and E_{hybrid} denote the energy of the Kekulé resonance structure and the energy of the resonance hybrid (or actual benzene) respectively.

Heat of hydrogenation

One of the methods for estimating the resonance energy of benzene uses the heat of hydrogenation data. When an alkene is hydrogenated in the presence of a metal catalyst (such as Pt), it is converted into an alkane with the release of heat energy. The energy so released is called the **heat of hydrogenation**.

The heat of hydrogenation of benzene (i.e. resonance hybrid) can be determined **experimentally**. But the heat of hydrogenation of the Kekulé resonance structure (which is not real) cannot be determined experimentally, and is therefore **calculated** with reference to a **model** compound. A simple model is **cyclohexene**.

Estimation of resonance energy

Cyclohexene has *one* double bond, which can be hydrogenated to **cyclohexane**. The heat of hydrogenation was measured experimentally and found to be **28.6 kcal mol^{-1}**.

The **Kekulé structure** represents a *hypothetical* molecule that contains *three* double bonds in a six-membered ring. The heat of hydrogenation for the Kekulé structure cannot be measured experimentally, but we can predict the hypothetical molecule to liberate (3 × 28.6) kcal mol^{-1} if it were hydrogenated to produce **cyclohexane**. Thus the **calculated** heat of hydrogenation for the Kekulé structure is **85.8 kcal mol^{-1}**.

The heat of hydrogenation of the **resonance hybrid** (i.e. actual benzene) was measured experimentally, and the **observed** value is **49.8 kcal mol^{-1}**.

These results are represented in Figure 3.6.

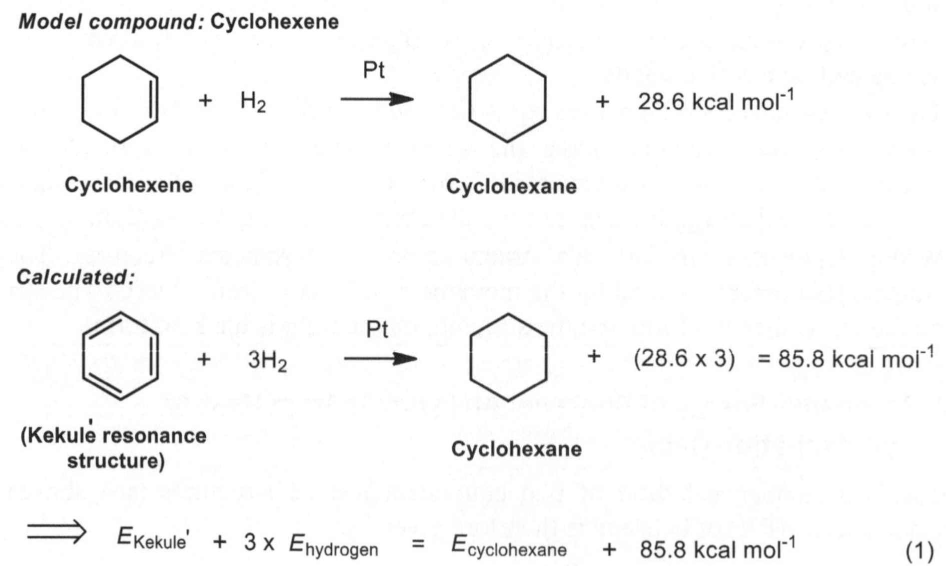

Observed:

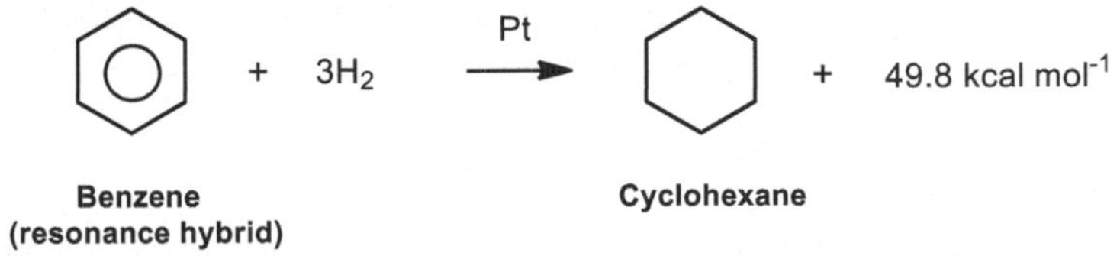

$$\Longrightarrow \quad E_{\text{hybrid}} + 3 \times E_{\text{hydrogen}} = E_{\text{cyclohexane}} + 49.8 \text{ kcal mol}^{-1} \qquad (2)$$

Subtracting eqn (2) from eqn (1):

$$E_{\text{Kekule'}} - E_{\text{hybrid}} = (85.8 - 49.8) \text{ kcal mol}^{-1}$$

or, **Resonance energy = 36 kcal mol^{-1}**

Figure 3.6: Resonance energy of benzene from heat of hydrogenation data using cyclohexene as a model compound.

The **difference** in the calculated and observed values gives the **resonance energy (RE)** of benzene as **36 kcal mol^{-1}**. This large value of RE implies that benzene is highly stabilized by resonance.

3.7 HYPERCONJUGATION

Besides π-conjugation which involves π bond–π bond or π bond–p orbital overlap, there is another type of conjugation where a **σ bond** participates in conjugation or delocalization with an *adjacent* **p orbital** or **π bond**. This type of conjugation (**σ–p** or **σ–π**) is called **hyperconjugation**.

The σ bond involved is usually a **C–H** bond. The overlap that occurs between the σ bond and the p orbital (or π bond) is generally *weak*; however, this is important in many cases, for example, in carbocations and alkenes.

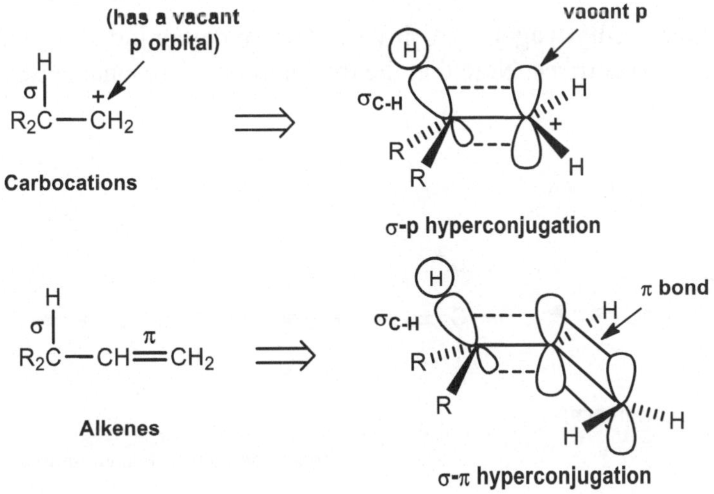

For **carbocations**, hyperconjugation involves **two electrons** (two from a C–H σ bond and no electron from the empty p orbital), whereas in the case of **alkenes**, hyperconjugation involves **four electrons** (two from a C–H σ bond and two from the π bond).

Resonance representation

The **resonance representation** of hyperconjugation for carbocations and alkenes are given below. The resonance structure in which there is no bond between the carbon and hydrogen is called a **hyperconjugative structure**. Hyperconjugation is sometimes called *no-bond resonance*.

Carbocations:

$$R_2C(H){-}\overset{+}{C}H_2 \longleftrightarrow R_2C{=}CH_2 \;\; H^+ \text{ (no bond)}$$

Hyperconjugative structure

Alkenes:

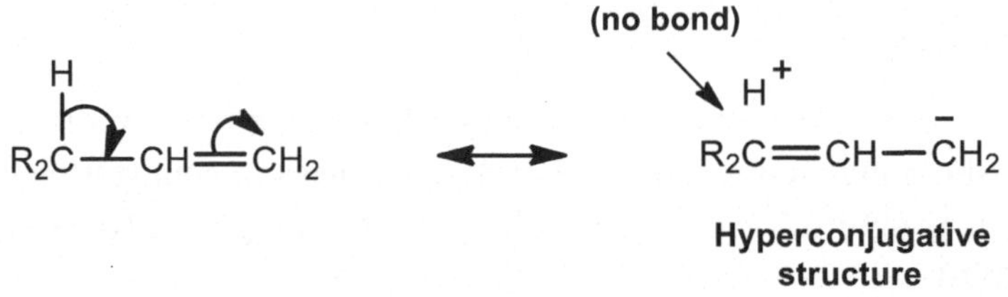

$$R_2C(H){-}CH{=}CH_2 \longleftrightarrow R_2C{=}CH{-}\overset{-}{C}H_2 \;\; H^+ \text{ (no bond)}$$

Hyperconjugative structure

Examples

Consider **ethyl cation** and **propene**. The carbon atom *adjacent* to C^+ or C=C is called the **α-carbon** and the hydrogens attached to α-carbon are called **α-hydrogens**.

In each case, **three α-hydrogens** are involved in hyperconjugation which give **three hyperconjugative structures**. Note that the total number of resonance structures is 4.

Ethyl cation:

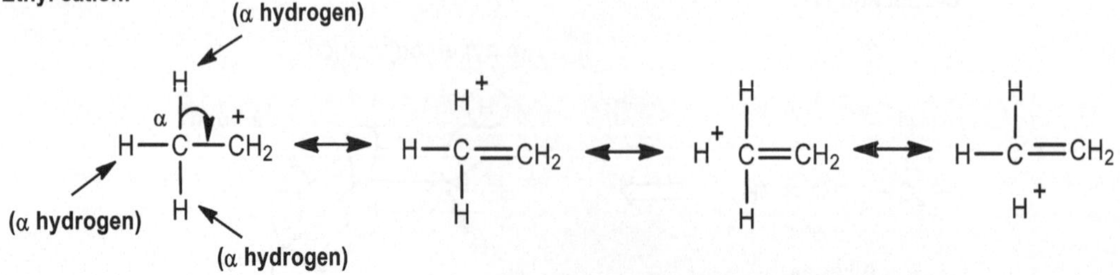

$$H_3C{-}\overset{+}{C}H_2 \longleftrightarrow H^+\;H_2C{=}CH_2 \longleftrightarrow H^+\;H_2C{=}CH_2 \longleftrightarrow H^+\;H_2C{=}CH_2$$

(α hydrogen) (α hydrogen) (α hydrogen)

Three hyperconjugative structures

Propene:

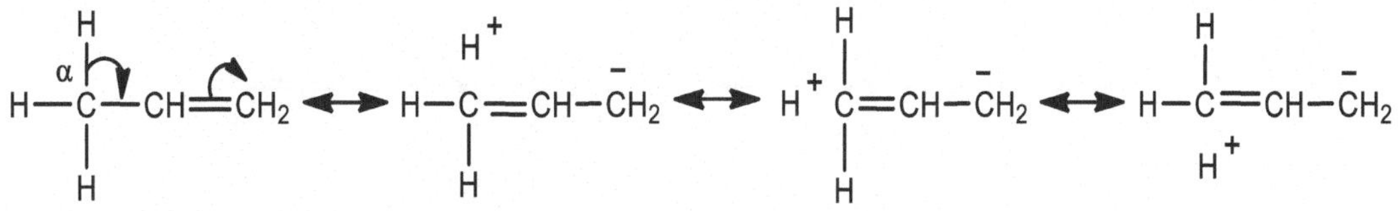

$$H_3C{-}CH{=}CH_2 \longleftrightarrow H^+\ H_2C{=}CH{-}\overset{-}{C}H_2 \longleftrightarrow H^+\ H_2C{=}CH{-}\overset{-}{C}H_2 \longleftrightarrow H^+\ H_2C{=}CH{-}\overset{-}{C}H_2$$

(three α hydrogens)

Three hyperconjugative structures

Relative stabilities of carbocations (primary, secondary and tertiary)

Although hyperconjugation is less effective than π-conjugation, it is believed to have played important roles in many situations, for example, in the relative stabilities of primary (1°), secondary (2°) and tertiary (3°) carbocations.

A **primary carbocation** has a positive charge on a primary carbon, a **secondary carbocation** has a positive charge on a secondary carbon, and a **tertiary carbocation** has a positive charge on a tertiary carbon. Alkyl groups are able to donate electron density to the carbocation centre through **hyperconjugation**, thereby stabilizing it by delocalizing the positive charge.

There are *three* hyperconjugable α-hydrogens for the **primary carbocation** ($CH_3CH_2^+$) which give **three hyperconjugative structures**. Similarly, *six* hyperconjugable α-hydrogens for the **secondary carbocation** [$(CH_3)_2CH^+$] gives **six hyperconjugative structures**, and *nine* hyperconjugable α-hydrogens for the *tertiary* carbocation [$(CH_3)_3C^+$] gives **nine hyperconjugative structures**.

According to the rules of resonance, the greater the number of hyperconjugative structures, the greater is the stability.

Therefore, the stability order of carbocations is:

$$(CH_3)_3C^+ > (CH_3)_2CH^+ > CH_3CH_2^+$$

or, tertiary (3°) > secondary (2°) > primary (1°)

A primary (1°) carbocation: Ethyl carbocation

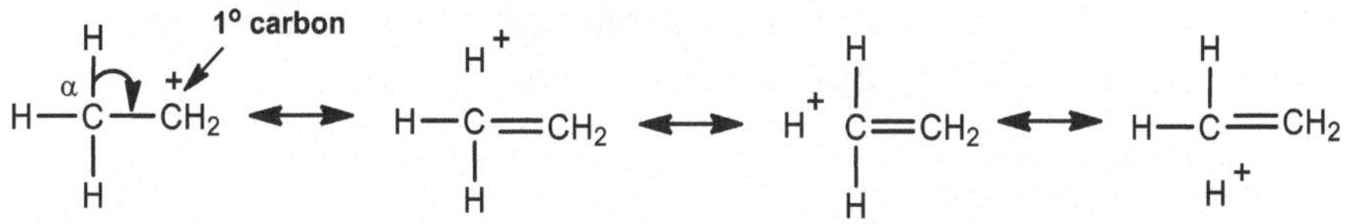

$$H_3C{-}\overset{+}{C}H_2 \longleftrightarrow H^+\ H_2C{=}CH_2 \longleftrightarrow H^+\ H_2C{=}CH_2 \longleftrightarrow H^+\ H_2C{=}CH_2$$

1° carbon

[$CH_3CH_2^+$]

(3 hyperconjugable H's)

Three hyperconjugative structures

A secondary (2°) carbocation: Isopropyl carbocation

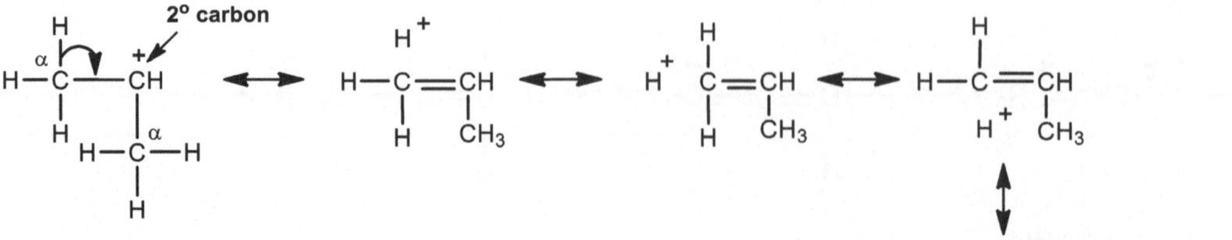

2° carbon

$[(CH_3)_2CH^+]$

(6 hyperconjugable H's)

Three more hyperconjugative structures

Six hyperconjugative structures

A tertiary (3°) carbocation: *tert*-Butyl carbocation

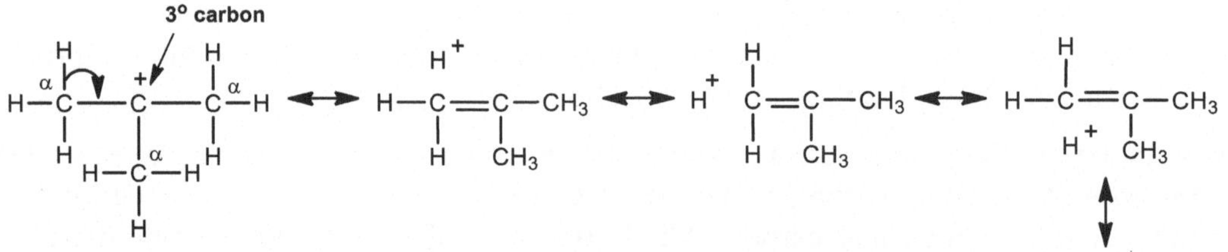

3° carbon

$[(CH_3)_3C^+]$

(9 hyperconjugable H's)

Six more hyperconjugative structures

Nine hyperconjugative structures

Stability order: $(CH_3)_3C^+ > (CH_3)_2CH^+ > CH_3CH_2^+$

3° 2° 1°

Note: Besides C–H bonds, C–C and C–Si σ bonds can also take part in hyperconjugation and play important roles in certain cases.

3.8 INDUCTIVE EFFECTS

In a **σ-bond** between a carbon atom and another different atom, the electrons are *not* shared equally between them but are attracted towards the more electronegative atom. This creates a **polarization** of the σ bond with **a partial negative charge on the more electronegative atom**.

Consider a C–X σ bond where X is more electronegative than C. Here the electrons in the σ bond are attracted toward X, which gives a partial negative charge (δ-) on X.

Similarly, in a C–Y σ bond where C is more electronegative than Y, the electrons in the σ bond are attracted toward C with a partial negative charge (δ-) on it.

(atom (X) is more electronegative than C)

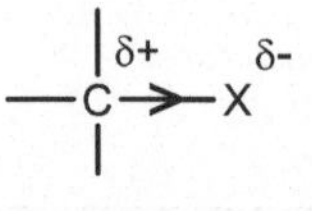

δ+ δ-

Electrons attracted to X

(atom (Y) is less electronegative than C)

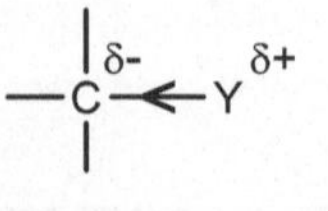

δ- δ+

Electrons attracted to C

For example, in **methyl chloride** (CH_3Cl), the C–Cl bond is polarized, whose negative end is Cl (more electronegative).

In **ethyl chloride** (CH_3CH_2Cl), the polarization of the C–Cl bond can **induce** polarization of the adjacent C–C σ bond. As a result, the distant methyl C will get a partial positive charge (δδ+) which is smaller than δ+ on the C of the C–Cl bond.

$$\overset{\delta+}{H_3C}-\overset{\delta-}{Cl}$$

Methyl chloride

$$\overset{\delta\delta+}{H_3C}-\overset{\delta+}{CH_2}-\overset{\delta-}{Cl}$$

Ethyl chloride

(δδ+ is smaller than δ+)

The polarization of a bond by the influence of a group or by an adjacent polar bond is known as the **inductive effect.**

The inductive effect of a group is most powerful on the carbon atom to which it is *directly* bonded, and carbon atoms *further away* are affected to some degree. The effect on successive carbon atoms however falls off with their increasing distance. This is illustrated with **butyl chloride** as an example.

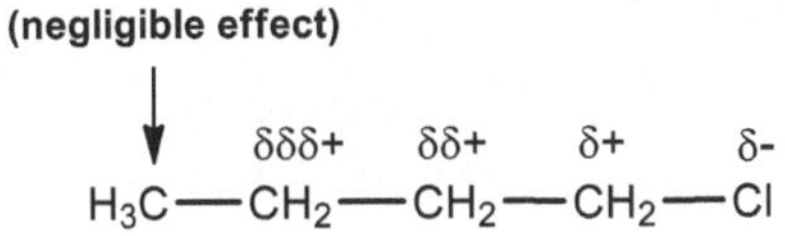

(negligible effect) ↓

$$H_3C-\overset{\delta\delta\delta+}{CH_2}-\overset{\delta\delta+}{CH_2}-\overset{\delta+}{CH_2}-\overset{\delta-}{Cl}$$

(δδδ+ is smaller than δδ+, which, in turn, is smaller than δ+)

Butyl chloride

The inductive effect operates through σ bonds. Based on inductive effects (*I*), groups can be classified as electron-donating (+*I*) or electron-withdrawing (–*I*), relative to hydrogen.

Electron-donating (+I) groups

The following are some common electron-donating (+*I*) groups:

Electron-donating (+*I*) groups: M (metals such as Li or Mg), O^-, CO_2^-, R (alkyl)

The (+*I*) groups include metal atoms of low electronegativity (e.g. Li, Mg), groups with a negative charge (O^- or CO_2^-) because the electrons of the σ bond is repelled towards the carbon by the negative charge, and alkyl groups (R).

Electron-withdrawing (–I) groups

Most groups are electron-withdrawing (–*I*) and the following are some examples:

Electron-withdrawing (–*I*) groups: NR_3^+, NR_2, NO_2, OH, OR, F, Cl, Br, I
CHO, COR, CO_2H, CO_2R, CN
$C(R){=}CR_2$, Ar, $C{\equiv}CR$

The (–*I*) groups include groups containing nitrogen, oxygen or halogen atoms, carbonyl, and nitrile groups.

Alkenyl and aryl groups (e.g. $CH{=}CH_2$, C_6H_5) with sp^2 carbons, and alkynyl groups (e.g. $C{\equiv}CH$) with sp carbons are also –*I*. This is because sp^2 and sp carbons are more electronegative than sp^3 carbons. *As s electrons are closer to the nucleus than p electrons, they are attracted more strongly by the nucleus*, and a hybrid orbital with more s character becomes more electronegative. Thus the **electronegativity** of carbon is the order: **sp> sp^2> sp^3.**

Note: Alkyl groups (e.g. CH_3) are commonly regarded as +*I* groups, although these may be electron-withdrawing (–*I*) in some cases.

3.9 MESOMERIC EFFECTS

Electrons can move via π-bonds by conjugation or resonance. The electron transfer by π-conjugation (or π-resonance) is called the **mesomeric effect**.

The mesomeric effect operates through π bonds. Based on mesomeric effects (*M*), functional groups can be classified as electron-donating (+*M*) or electron-withdrawing (–*M*).

Electron-donating (+M) groups

The following are some common electron-donating (+*M*) groups:

Electron-donating (+*M*) groups: OH, OR, NH_2, NR_2, F, Cl, Br, I

A +*M* group usually contains a *lone pair* of electrons on O, N or halogen atom, which is involved in π-resonance. The electron donation by the group to a π bond produces a **negative charge on a carbon atom** in the molecule. This is illustrated with the following examples:

+M
$H_2C{=}CH{-}\ddot{O}CH_3$ ⟷ $H_2\bar{C}{-}CH{=}\overset{+}{O}CH_3$

Methyl vinyl ether **(carbanion centre)**

+M
$H_2C{=}CH{-}\ddot{Cl}:$ ⟷ $H_2\bar{C}{-}CH{=}\overset{+}{Cl}:$

Vinyl chloride **(carbanion centre)**

+M
$H_2C{=}CH{-}\ddot{N}(CH_3)_2$ ⟷ $H_2\bar{C}{-}CH{=}\overset{+}{N}(CH_3)_2$

***N,N*-Dimethylvinylamine** **(carbanion centre)**

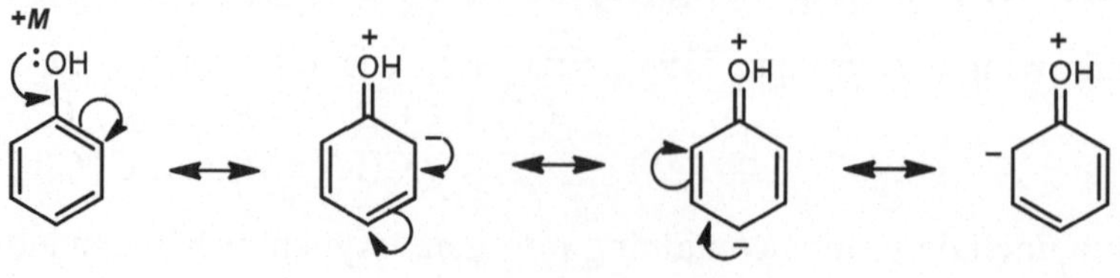

+M
:OH ⟷ $\overset{+}{O}H$ ⟷ $\overset{+}{O}H$ ⟷ $\overset{+}{O}H$

Phenol **(carbanion centres at two *ortho* and one *para* positions)**

Note that in drawing resonance structures, lone pairs on atoms are not always explicitly shown (see Section 2.2.3).

Electron-withdrawing (–M) groups

The following are some examples of electron-withdrawing (–*M*) groups:

Electron-withdrawing (–*M*) groups: CHO, COR, CO_2H, CO_2R, CN, NO_2

A **–*M* group** usually contains a π bond involving a heteroatom (O or N), such as carbonyl group (C=O) in aldehydes, ketones, acids or esters, nitrile group (CN), and nitro group (NO_2).

The electron withdrawal by the group via π-resonance creates a **positive charge on a carbon atom** in the molecule. The following examples illustrate this.

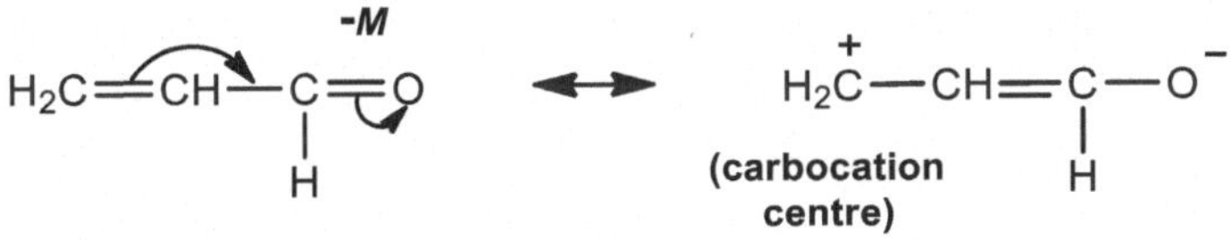

Acrolein (acraldehyde)

(carbocation centre)

Acrylonitrile

(carbocation centre)

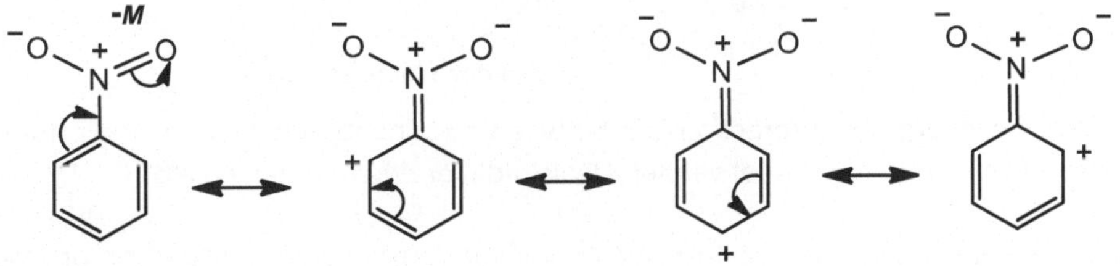

($C_6H_5NO_2$)

Nitrobenzene

(carbocation centres at two *ortho* and one *para* positions)

Notes:

1. The mesomeric effect is generally stronger than the inductive effect. Mesomeric effects can also operate over longer distances via conjugation.
2. A particular group may exert both inductive and mesomeric effects. If both effects are electron-donating (or electron-withdrawing), they are reinforced. For example, NO_2 is –*I* and –*M*; thus NO_2 acts as a powerful electron-withdrawing group.
3. Mesomeric and inductive effects of a group can also act in opposition, i.e. one effect is electron-donating while the other is electron-withdrawing. In this situation, it is the stronger effect that generally prevails. For example, OH is –*I* but +*M*, where + *M* > –*I*. Thus OH acts as a net electron-donating group.

3.10 STERIC EFFECTS

Besides inductive and mesomeric effects, the **size of atoms or groups** also influences the structure and reactivity of organic molecules. The spatial interactions between the substituent groups that are **not bonded** to each other are called the **steric effects**. These **nonbonded interactions are repulsive** in nature.

Every atom has an 'effective' size with respect to other atoms to which it is not bonded. This effective size of an atom is called its **van der Waals radius**. At infinite distance, the energy of interaction between two nonbonded atoms is zero.

As the two atoms approach each other, an **attractive force** operates which leads to a lowering of energy. **But if the atoms are forced closer together, a repulsive force arises** from the interactions between the filled orbitals of one atom and the filled orbitals of the other nonbonded atom.

A plot of energy (E) as a function of internuclear distance (r) is shown in Figure 3.7.

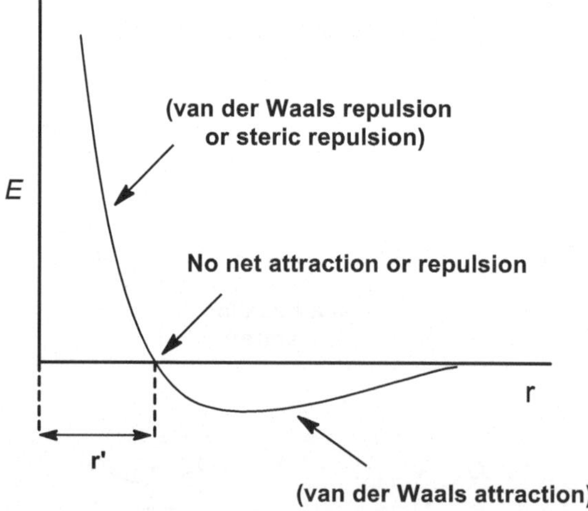

Figure 3.7: The energy of interaction (E) between two nonbonded atoms as a function of internuclear distance (r). r′ = sum of vander Waals radii for two nonbonded atoms.

The distance between two nonbonded atoms (when there is no net attraction or repulsion) is equal to the **sum of their van der Waals radii**.

If the **distance** between two specified atoms is **less than the sum** of their van der Waals radii, they will **repel** each other. Such repulsion due to the crowding of atoms is known as the **van der Waals repulsion or steric repulsion**.

Note: The van der Waals radius of an atom is different from its atomic radius. For example, the van der Waals radius of hydrogen atom is 1.25 Å, whereas its atomic radius is 0.53 Å.

SUPPLEMENTARY PROBLEMS

3.7 What is the hybridization of carbon, nitrogen and oxygen atoms in the following molecules?

(a) $H_2C{=}CH{-}C{\equiv}N$

Acrylonitrile

(b) $H_2C{=}C{=}O$

Ketene

[*Hints*: count the number of σ bonds and lone pairs (not shown). If the sum is 4, 3 and 2, the hybridization is sp^3, sp^2 and sp respectively]

3.8 The H–O–H bond angle in water is 104.5° but the H–S–H bond angle in H_2S is 92°. What does this indicate about the bonding in water and H_2S?

[*Hint*: O is hybridized but S is not hybridized]

3.9 Draw curved arrows to show the conjugation in the following compounds:

(a) CO_2CH_3 N CH_3 (b) CO_2CH_3 N CH_3 (c) CO_2CH_3 N CH_3

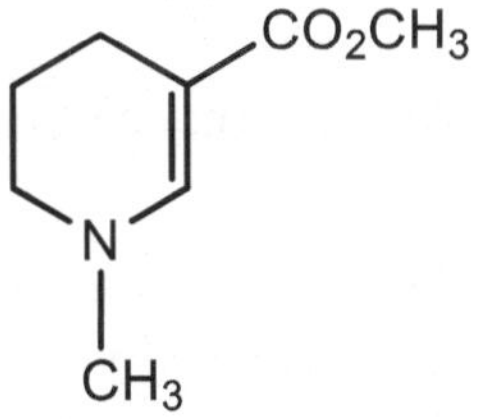

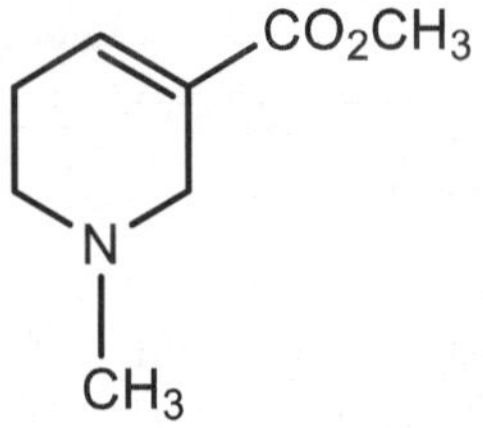

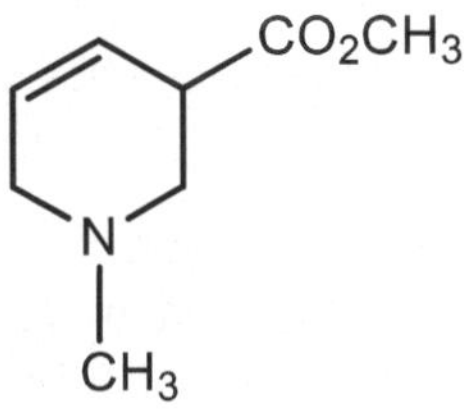

[*Hints*: in (a), nitrogen lone pair, C=C and C=O are in conjugation; in (b), C=C and C=O are conjugated; in (c), only ester conjugation]

3.10 Draw orbital pictures for (a) an amide ($RCONH_2$) and (b) a nitro compound (RNO_2) showing π–p conjugation in each case.

3.11 The bond lengths for *o*- and *p*-nitro groups in picryl iodide are 1.45 Å and 1.35 Å respectively. Explain why the bond lengths are different.

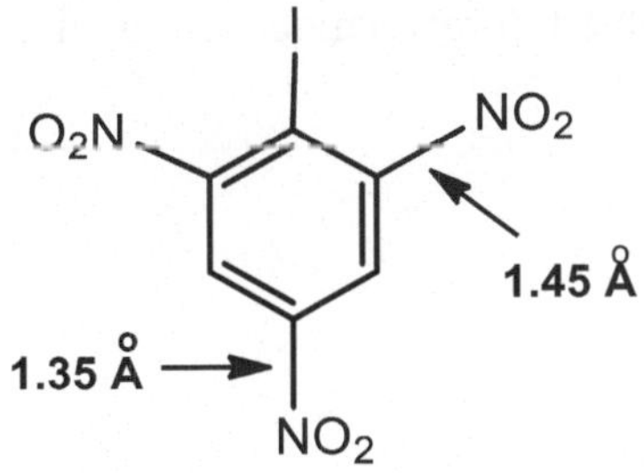

I O_2N NO_2 1.45 Å 1.35 Å NO_2

Picryl iodide

[*Hint*: resonance of the *p*-nitro group with the benzene ring gives a partial double bond character to the C–N bond and hence its bond length is shorter; for *o*-nitro groups, there is steric inhibition of resonance due to a large iodine atom]

3.12 Draw resonance structures of the carbonyl group in cycloheptatrienone (tropone) and cyclopentadienone, and hence predict which carbonyl group has longer bond length.

Cycloheptatrienone (tropone) **Cyclopentadienone**

[*Hints*: the charged resonance structure for tropone is aromatic, while that for cyclopentadienone is antiaromatic]

3.13 Can you apply the Hückel rule to determine the aromaticity of azulene? If so, predict whether it is aromatic or not. Also draw resonance structures in support of your prediction.

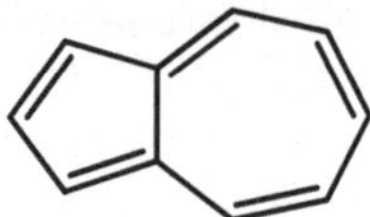

Azulene

[*Hints*: azulene has a cyclic array of p orbitals like a monocyclic system; charged resonance structures give aromatic character to each ring]

3.14 Both pyrrole and pyridine have nitrogen atom with a lone pair. Is the nitrogen lone pair involved in resonance for both or one of them or none? Show by drawing their resonance structures.

Pyrrole **Pyridine**

[*Hints*: the N lone pair is involved in resonance for pyrrole, but not for pyridine]

CHAPTER 4

Stereochemistry

IN THIS CHAPTER

- Constitutional isomers and stereoisomers
- Representation of three-dimensional molecules: Flying wedge and Fischer projection formulas; rules for drawing equivalent representations
- Chirality and symmetry
- Chiral and achiral molecules, optical activity and specific rotation
- Enantiomers, racemic forms, nonracemic forms and enantiomeric excess
- Stereocentres or chiral centres: Molecules with one and two (unlike/like) chiral centres; chiral and achiral diastereomers; *erythro*/*threo* and *meso* nomenclature
- Chiral centre configuration: The *R*/*S* system and the D/L system
- Chiral molecules without a chiral centre; biological importance of chiral molecules
- Stereoisomerism in alkenes; *cis*/*trans* and *E*/*Z* nomenclatures
- Stereoisomerism in cycloalkanes; *cis* and *trans* diastereomers
- Conformation of acyclic and cyclic molecules: Eclipsed, staggered, *gauche*, *anti* and chair conformations

In this chapter, we will look at the **structure** of organic molecules in **three-dimensions**. We begin by classifying **isomers** that are different molecules having the same molecular formula.

4.1 CLASSIFICATION OF ISOMERS

Isomers are broadly classified into constitutional isomers and stereoisomers.

4.1.1 Constitutional Isomers

Isomers that differ in the bonding connectivity of atoms are called **constitutional isomers**. They can be divided into three types depending on how the atoms are connected to each other.

(a) Constitutional isomers that differ in the type of the carbon chain are called **chain isomers**. The carbon chain may be a **straight chain**, a **branched chain** or a **closed chain (ring)**.

For example, **butane** and **isobutane** of the same molecular formula C_4H_{10} are chain isomers. Butane has a *straight carbon chain*, whereas isobutane has a *branched carbon chain*.

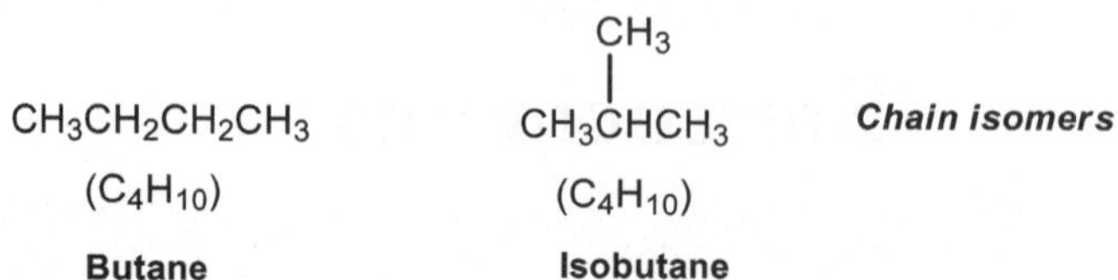

(b) Constitutional isomers that differ in the location (or position) of a substituent or a functional group along a carbon chain or ring are called **regioisomers** (regio from region). They are also known as *position isomers*.

The regioisomers are illustrated with the following examples.

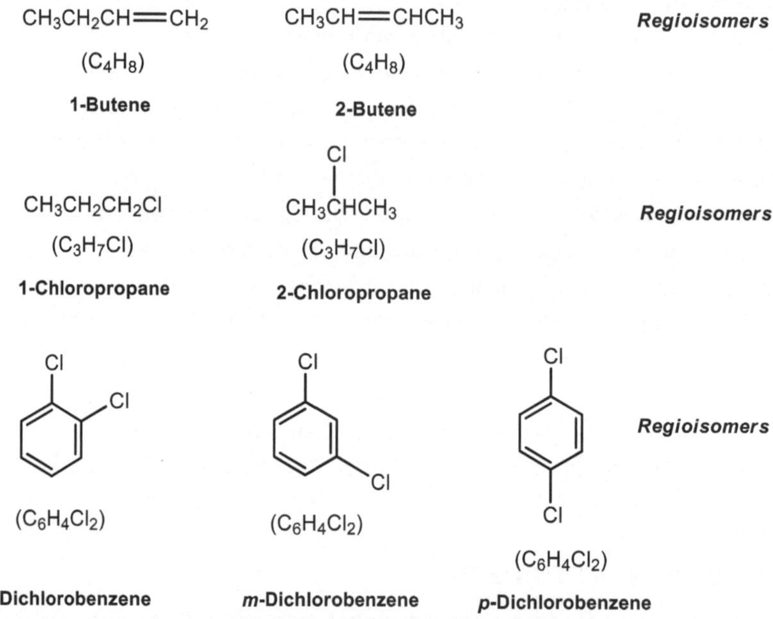

1-Butene and **2-butene** are regioisomers as they differ in the position of the double bond along the carbon chain. Similarly, **1-chloropropane** and **2-chloropropane** are regioisomers because the location of Cl atom is different in the two isomers. The three isomers ***o*-, *m*- and *p*-dichlorobenzene** differ in the relative positions of two Cl atoms on the benzene ring, and are therefore regioisomers.

(c) Constitutional isomers that have different functional groups are called **functional group isomers.**

For example, **ethanol** and **dimethyl ether** are functional group isomers because ethanol has an alcohol functional group (OH) while dimethyl ether has an ether linkage (–O–).

$CH_3CH_2—OH$	$CH_3—O—CH_3$	***Functional group isomers***
(C_2H_6O)	(C_2H_6O)	
Ethanol	**Dimethyl ether**	

4.1.2 Stereoisomers

Stereoisomers are different from constitutional isomers. **Isomers that have the same atom connectivity but different orientations of atoms in space are called stereoisomers.** They are further divided into two kinds: (a) enantiomers and (b) diastereomers.

(a) **Enantiomers are stereoisomers that are mirror images of each other**. These mirror image molecules are *not* identical, because they cannot be superimposed onto each other.

(b) **Diastereomers are stereoisomers that are *not* mirror images of each other**. Therefore, diastereomers are stereoisomers that are *not* enantiomers.

The classification of isomers is summarized in Figure 4.1.

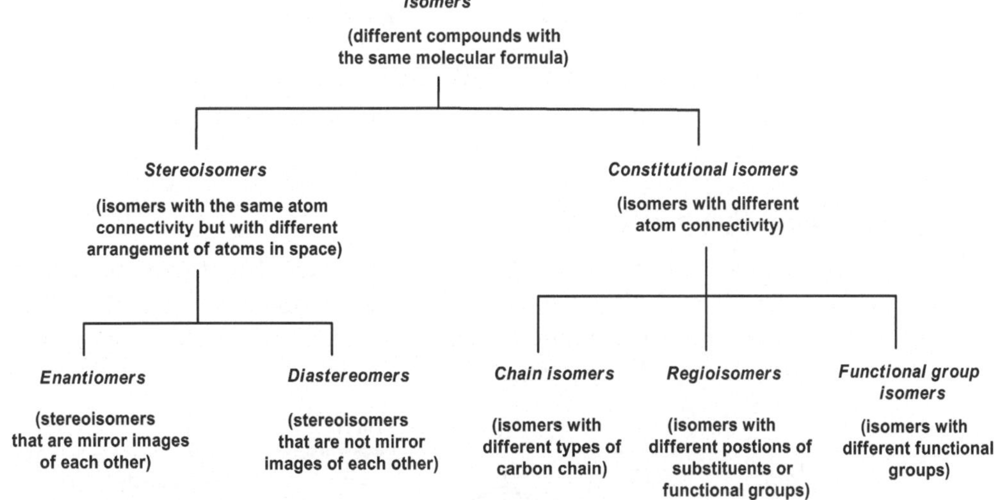

Figure 4.1: Classification of isomers.

4.2 REPRESENTATION OF THREE-DIMENSIONAL MOLECULES

To study stereochemistry, it is important to represent or draw **three-dimensional molecules** on a **two-dimensional paper**.

For this, **molecular models** are of great help. In fact, you need to play with these models, sometimes messing things up, but in the process you do get a feel for how stereochemistry works.

In this section, we will describe two types of representations of three-dimensional molecules, namely, **flying wedge formulas** and **Fischer projection formulas**.

4.2.1 Flying wedge formulas

A **tetrahedral carbon** has four bonds. Of these, a *maximum* of two bonds can lie on a plane (say, the plane of the paper), when one bond (third) is above the plane and the other (fourth) is below the plane (use a molecular model).

A **line bond** indicates a bond that is in the same plane of the paper. A **wedged bond** indicates a bond that is above the plane of the paper, while a **hashed bond** indicates a bond that is below the plane of the paper. This type of stereochemical representation is called the **flying wedge formula**.

For example, a flying wedge formula of **bromochlorofluoromethane** (CHFClBr) is given here.

(CHFClBr)

Bromochlorofluoromethane **A flying wedge formula**

In this flying wedge formula, the C–H and C–F bonds (line bonds) are in the plane of the paper, the C–Cl bond (a wedged bond) is above the plane, and the C–Br bond (a hashed bond) is below the plane.

Drawing equivalent flying wedge formulas

The flying wedge formula (**I**) for CHFClBr can be converted to two other equivalent flying wedge formulas (**II and III**) as shown below (use a molecular model). **II** has only one bond in the plane of the paper, while **III** has no bonds in the plane of the paper. The formula **I** is the most common flying wedge representation.

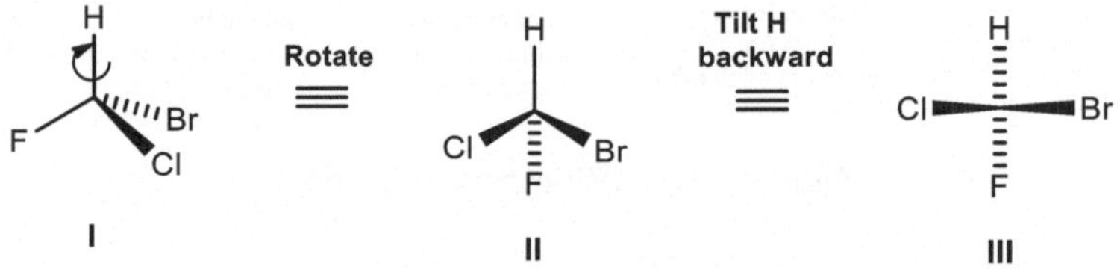

Rotate **Tilt H backward**

I II III

Three equivalent flying wedge formulas

Flying wedge formulas for two tetrahedral carbons

A flying wedge formula can be drawn for molecules with *two* tetrahedral centres linked by a C–C bond.

Consider, for example, **2,3-dichlorobutane** ($CH_3CHClCHClCH_3$). As shown below, two tetrahedral units at C2 and C3 can be artificially combined to obtain a flying wedge formula for the molecule with respect to the central C2–C3 bond.

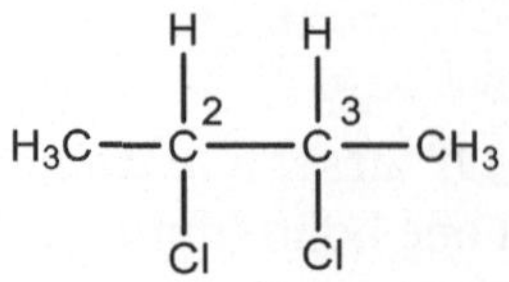

2,3-Dichlorobutane

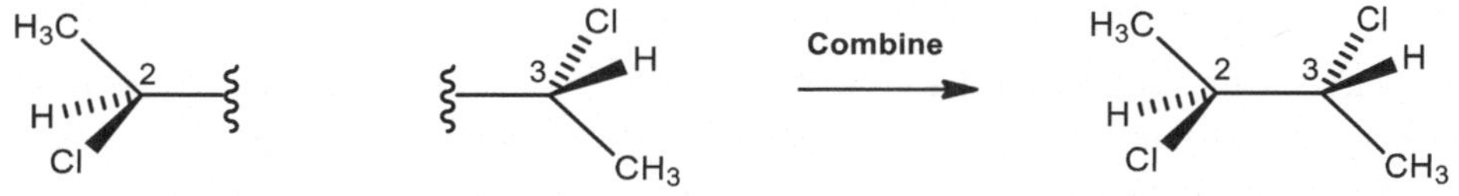

A tetrahedral arrangement at C2

A tetrahedral arrangement at C3

Combine

A flying wedge formula for 2,3-dichlorobutane

In this representation, **the carbon chain including the C2–C3 bond lies in the plane of the paper**. For *each* tetrahedral carbon C2 (or C3), two bonds are in the plane of the paper, one bond is above the plane and the other is below the plane (use a model).

4.2.2 Fischer Projection Formulas

A **Fischer projection formula** is a convenient two-dimensional formula that represents a three-dimensional molecule.

To draw a Fischer projection, a three-dimensional model is taken as a reference. In this reference model, as we view the tetrahedral carbon, two horizontal bonds must be in the front (above the plane) and two vertical bonds must be in the rear (below the plane).

The **projections** of these four bonds on the plane of the paper give a **Fischer projection formula**. These projections can be thought of as their shadows in the plane by shining a light through the molecular model.

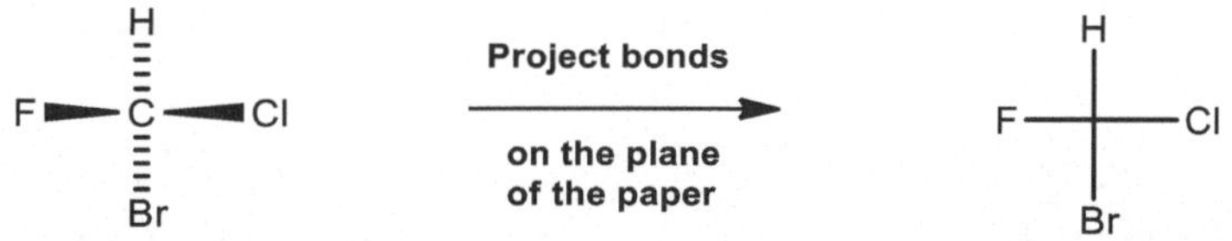

Three-dimensional reference model

Project bonds on the plane of the paper

Fischer projection formula

Notably, the **horizontal lines** and the **vertical lines** in a Fischer projection do not represent the actual orientation of the bonds in space, but correspond to *above the plane* and *below the plane* bonds respectively.

To **transform** *any* flying wedge formula into a Fischer projection, the molecule is to be viewed in a direction such that *the requirements of the reference model are fulfilled,* i.e. two horizontal groups are towards the observer (in the front), and two vertical groups are away from the observer (in the rear).

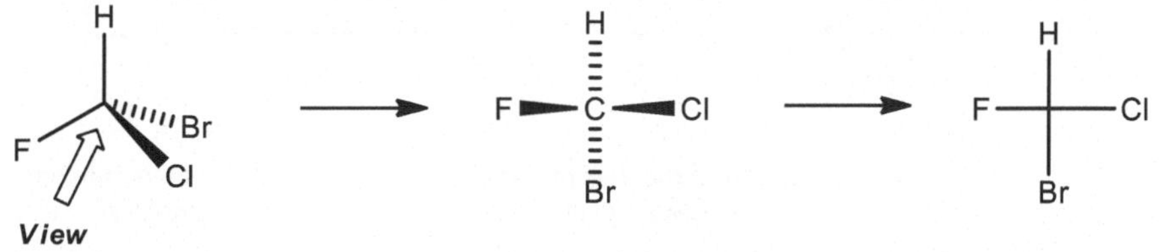

View

Flying wedge formula

Another Flying wedge formula

Fischer projection formula

A **Fischer projection formula** can also be drawn for molecules *with two tetrahedral arrangements of groups* around a C–C bond. In this case, the **reference** three-dimensional formula has horizontal bonds in the front and vertical bonds in the rear with respect to the C–C bond in the plane.

For example, a Fischer projection formula for **2,3-dichlorobutane** is shown below.

CH_3, H—C—Cl, H—C—Cl, CH_3 → CH_3, H—+—Cl, H—+—Cl, CH_3

Flying wedge formula | **Fischer projection formula**

Rules for drawing equivalent Fischer projection formulas

A given Fischer projection formula can be transformed into other equivalent Fischer projections representing the same molecule, using the following **rules**.

Consider a general case, where A, B, C and D are the four groups attached to the tetrahedral carbon.

1. Rotate *any three* groups on a Fischer projection while holding one group fixed. This gives an equivalent Fischer projection.

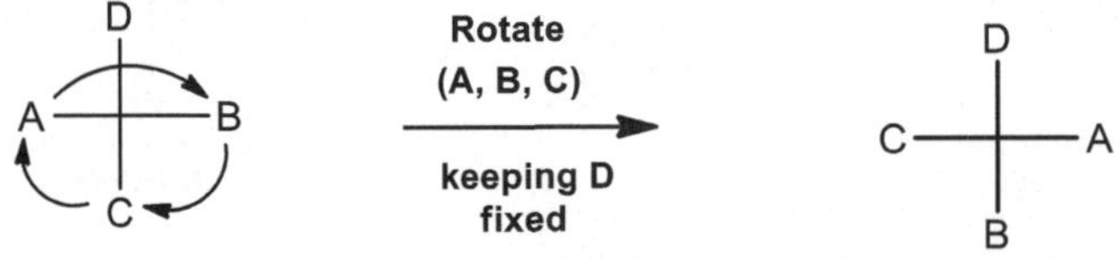

Equivalent Fischer projection

Same molecule

2. Exchange *any two* groups on a Fischer projection. An even number of exchanges (2 exchanges) give an equivalent Fischer projection, while an odd number of exchanges (1 or 3 exchanges) give a different Fischer projection.

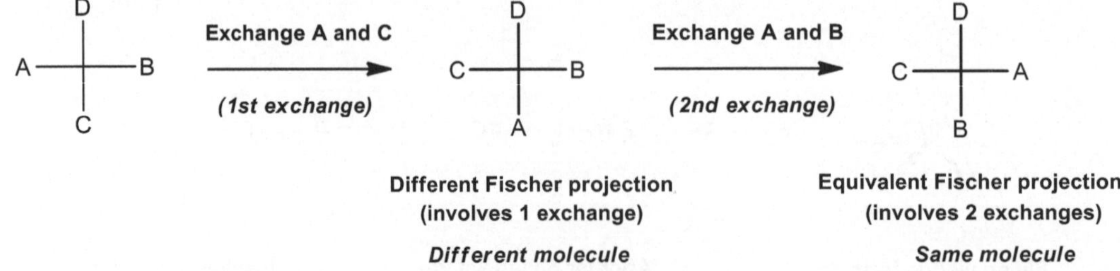

Different Fischer projection (involves 1 exchange)

Different molecule

Equivalent Fischer projection (involves 2 exchanges)

Same molecule

3. Rotate a Fischer projection by 180° in plane, which gives an equivalent Fischer projection. Rotating a Fischer projection by 90° in plane gives a different Fischer projection.

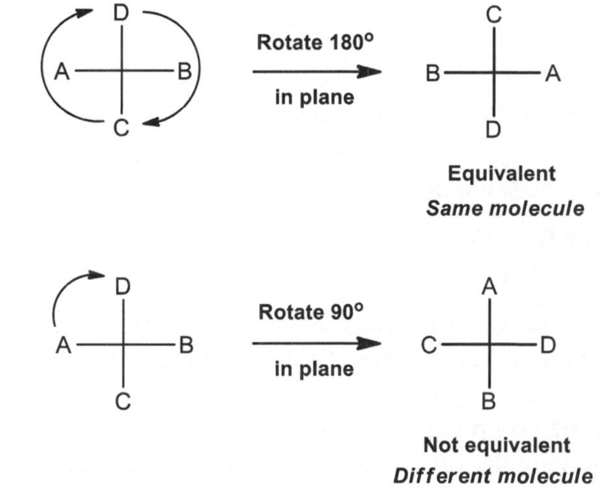

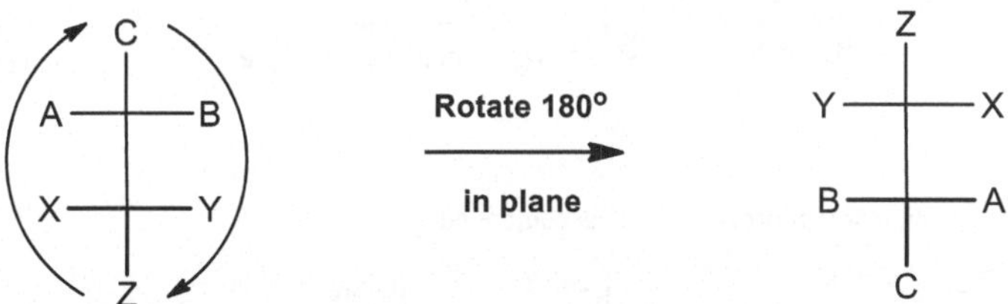

Two tetrahedral centres; one centre attached to groups (A, B, C) and the other attached to groups (X, Y, Z)

Equivalent Fischer projection
Same molecule

PROBLEM 4.1: Predict whether the two structures in each pair represent identical or different compounds.

(a) CH_3, H, Cl, Br and Cl, Br, H, CH_3

(b) CH_3, HO, H, C_2H_5 and H, HO, CH_3, C_2H_5

(c) CO_2H, H_2N, H, CH_3 and NH_2, H, CO_2H, CH_3

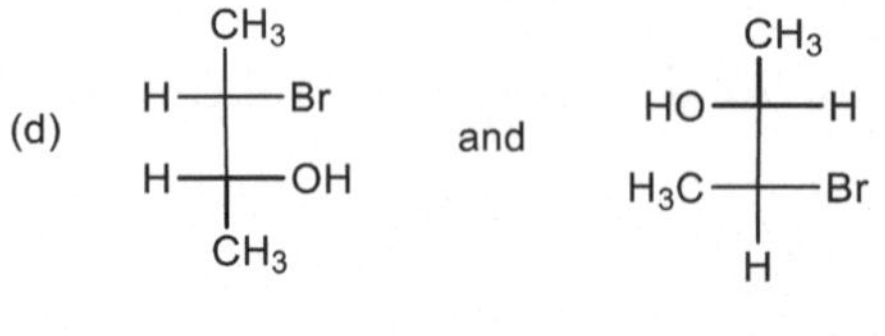

(d) Fischer projections: CH3 / H–Br / H–OH / CH3 and CH3 / HO–H / H3C–Br / H

Answer

The rules for Fischer projections (rotating three groups while holding one fixed or making two exchanges) also apply to flying wedge formulas, which give an equivalent flying wedge formula representing the same or identical molecule.

If two Fischer projections or flying wedge formulas are not so related or related by one or three exchanges, they are different.

In (a), the two structures are different because the first structure can be converted to the second structure by 3 (odd number) exchanges. An odd number of exchanges give different flying wedge structures representing different molecules. In (b), (c) and (d), the two structures represent the same compound because the first structure can be converted to the second structure by applying the rules that give equivalent structures.

(a) The two structures represent different compounds

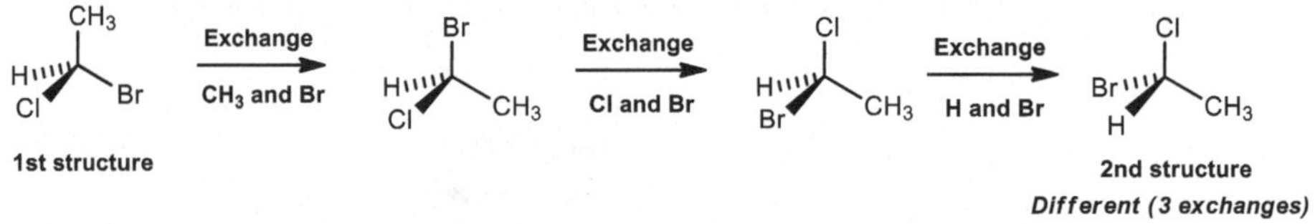

Exchange CH_3 and Br → Exchange Cl and Br → Exchange H and Br

1st structure — 2nd structure *Different (3 exchanges)*

(b) The two structures represent identical compound

view

1st structure → Rotate (H, CH_3, C_2H_5) keeping OH fixed → 2nd structure *Same molecule*

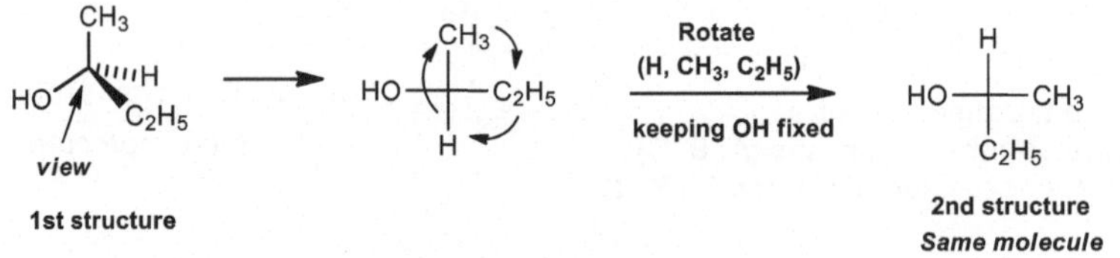

(c) The two structures represent identical compound

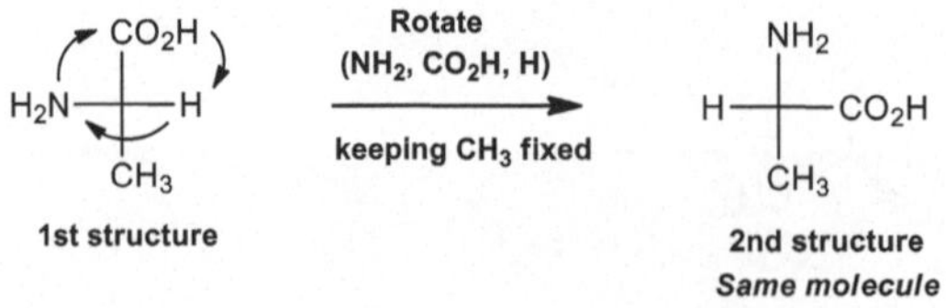

1st structure → Rotate (NH_2, CO_2H, H) keeping CH_3 fixed → 2nd structure *Same molecule*

(d) The two structures represent identical compound

1st structure → Rotate (H, CH_3, Br) keeping $CHOHCH_3$ fixed → *Equivalent structure* → Rotate 180° in plane → 2nd structure *Same molecule*

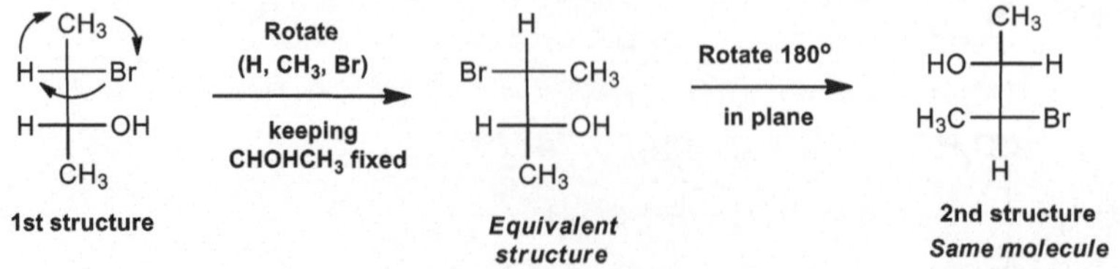

4.3 ENANTIOMERS AND CHIRALITY

Enantiomers are stereoisomers that are mirror images of each other. These mirror images are **nonsuperimposable**. Thus, a molecule of an enantiomer and its mirror image are **different** molecules; the mirror image molecule represents a second enantiomer.

In general, **any structure that is not superimposable on its mirror image is called chiral** (pronounced 'ky-ral'). Here the 'structure' is not limited to just molecules but applies to any everyday object.

The word **chiral** comes from the Greek word *cheir*, meaning 'hand'. Thus the term chiral is used to describe the way by which a left hand is related to a right hand.

If we view our left hand in a mirror, the mirror image will be a right hand. Now we try to **superimpose** our left hand with its mirror image (i.e. right hand), when their *similar sides* are facing in the *same direction*. We will find that they are **nonsuperimposable**. This is why a left glove does not fit a right hand and a right glove does not fit a left hand. This property of the hands is called **handedness** or **chirality**.

By definition, chirality is the nonsuperimposability of an object or a molecule with its mirror image.

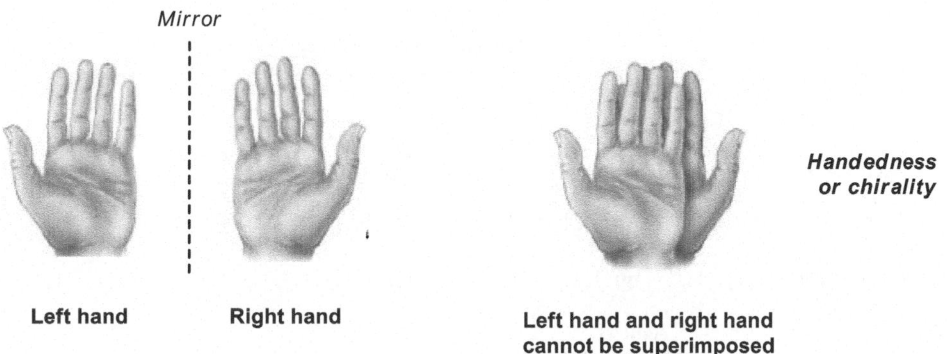

Enantiomers possess chirality, and the molecules of enantiomers are chiral. Notably, **chirality is the necessary and sufficient condition for the existence of enantiomers.**

Molecules that are **superimposable** on their mirror images are said to be **achiral**. An achiral molecule *cannot* have an enantiomer because its mirror image is identical to itself.

Chiral molecules

Consider, for example, **2-chlorobutane** ($CH_3CH_2CHClCH_3$). In this compound, the four atoms or groups attached to C2 (which is sp^3-hybridized) are arranged in a tetrahedral fashion.

If we construct a model for 2-chlorobutane (**I**) and the model of its mirror image (**II**), and then try to superimpose them, we will find that they are *not* superimposable. Therefore, **2-chlorobutane is chiral, and it has two enantiomers**.

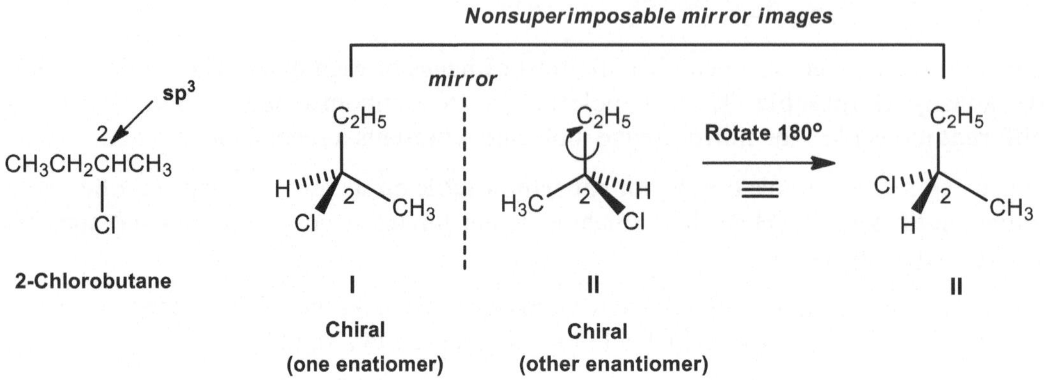

Nonsuperimposable mirror images

2-Chlorobutane

Chiral (one enatiomer) — I; Chiral (other enantiomer) — II

Achiral molecules

Consider **2-chloropropane** ($CH_3CHClCH_3$) in which the four atoms or groups attached to C2 (sp^3) are arranged in a tetrahedral fashion.

If we construct models for 2-chloropropane and its mirror image (**III** and **IV**), and try to superimpose them, we will find that they are superimposable, and hence identical. Therefore, **2-chloropropane is achiral, and it has no enantiomers**.

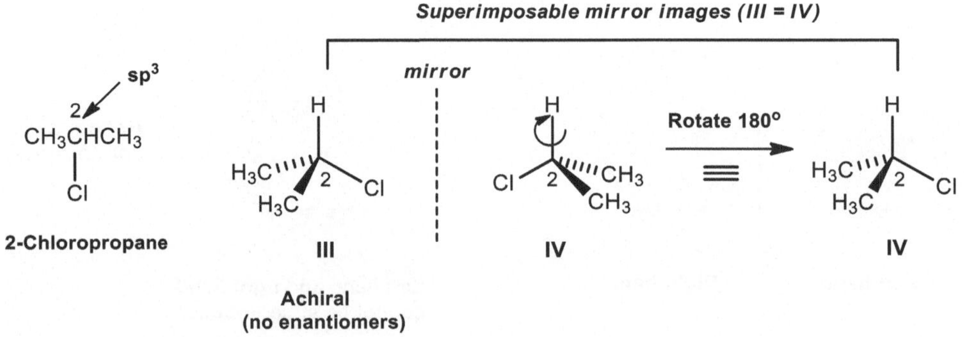

Superimposable mirror images (III = IV)

2-Chloropropane

Achiral (no enantiomers)

4.4 CHIRALITY AND SYMMETRY

The essential difference between a chiral and an achiral molecule is their **symmetry**. Two common symmetry elements are *plane of symmetry* and *centre of symmetry*.

A molecule that has a plane of symmetry or a centre of symmetry or both is achiral, whereas a chiral molecule has no plane of symmetry or no centre of symmetry.

Plane of symmetry

A plane of symmetry is defined as an imaginary plane that divides the molecule into two halves such that one half is the mirror image of the other half. A plane of symmetry is also called a mirror plane, which may pass through *some* atoms or *no* atoms.

We have already seen that 2-chloropropane is achiral but 2-chlorobutane is chiral. These results are consistent with their symmetry properties as described below.

2-Chloropropane has a **plane of symmetry**. Therefore, it is **achiral**. Note that the *mirror plane* passes through Cl, C2 and H, and bisects the molecule in such a way that the two halves of the molecule are mirror images of each other.

In contrast, **2-chlorobutane** does *not* possess a **plane of symmetry**, and hence it is **chiral**.

Plane of symmetry

(remains in left half) (remains in right half)

Cl H H_3C 2 CH_3

Two halves are mirror images of each other

2-Chloropropane (achiral)

Not plane of symmetry

(left half) (right half)

Cl H H_3C 2 CH_2CH_3

Two halves are not mirror images of each other

2-Chlorobutane (chiral)

Centre of symmetry

A centre of symmetry is a specific point within a molecule through which it is possible to project every atom in a structure to an equivalent position on the opposite side of the centre.

Consider, for example, a cyclic molecule, ***trans*-1,3-dimethylcyclobutane**. In this molecule, *all similar atoms or groups* (such as two CH_3 groups or two H atoms at C1 and C3) are *equidistant* from the centre and on the diametrically *opposite sides* of the centre.

Thus the molecule has a **centre of symmetry** at the centre of the ring, and hence it is **achiral**.

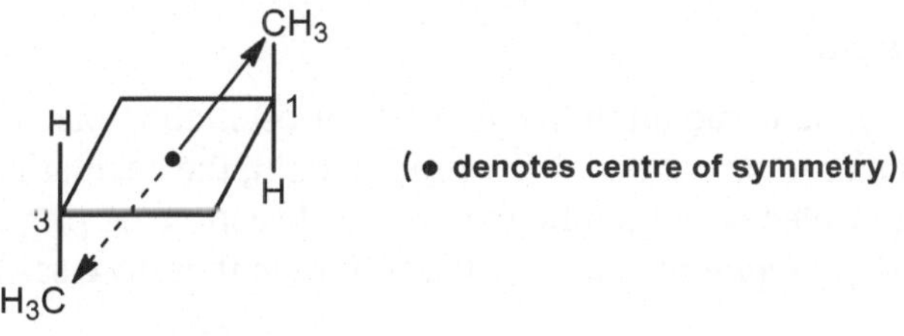

***trans*-1,3-Dimethylcyclobutane (achiral)**

PROBLEM 4.2: Using symmetry properties, predict whether the following molecules are chiral or achiral.

CH_3 / H—C—Cl / H—C—Cl / CH_3

H_3C Br H H Br CH_3

Answer

The first molecule possesses a plane of symmetry. Notably, the mirror plane does not pass through any atom. The molecule is achiral.

The second molecule has a centre of symmetry. As shown, two Br atoms are equidistant from the centre and on the diametrically opposite sides of the centre. Similar are the cases for two H atoms and for two CH_3 groups. It is also achiral.

Plane of symmetry

Centre of symmetry

4.5 ENANTIOMERS AND OPTICAL ACTIVITY

Enantiomers are two different compounds, but they have identical melting points, boiling points, solubilities and all other physical properties, **except one**.

If we pass a beam of **plane-polarized light** through a sample of an enantiomer, we will find that the *plane* of polarization is *rotated*. **The rotation of plane-polarized light is known as optical activity**, and the enantiomers are said to be **optically active**.

The difference is the two enantiomers will rotate the plane of plane-polarized light in **opposite directions** although in equal amounts. If the rotation of the plane-polarized light is in a **clockwise** direction (or to the right), it is said to be **positive (+)**, and if the rotation is anticlockwise (or to the left), it is said to be **negative (–)**. The enantiomer with (+) rotation is also called **dextrorotatory** (from the Latin *dexter*, meaning right), and the enantiomer with (–) rotation is called **levorotatory** (from the Latin *laevus*, meaning left).

Plane-polarized light

Light is an electromagnetic radiation that consists of oscillating electric and magnetic fields mutually perpendicular to one another. In ordinary light, the oscillations of the *electric field* occur in all possible planes perpendicular to the direction of propagation. The same is true for the oscillating *magnetic field*. Ordinary light is therefore said to be **unpolarized**.

When ordinary light passes through a plane-polarizing filter called **polarizer**, the electric field of the *emerging* light oscillates **only in one plane**, with the magnetic field oscillations in a perpendicular plane. Such light is called the **plane-polarized light**.

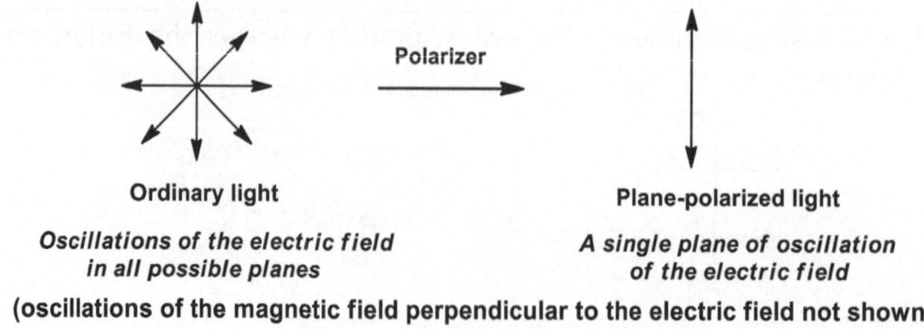

Oscillations of the electric field in all possible planes

A single plane of oscillation of the electric field

(oscillations of the magnetic field perpendicular to the electric field not shown)

Polarimeter and optical rotation

The instrument that is used to measure the *rotation* of plane-polarized light (called optical rotation) is known as **polarimeter**. The polarimeter has a **light source** (usually a sodium lamp), a **polarizer**, a **sample cell** containing a solution of the substance, and a **detector** that indicates by how much the light is rotated.

A schematic diagram for the measurement of **optical rotation (α)** using a polarimeter is shown in Figure 4.2.

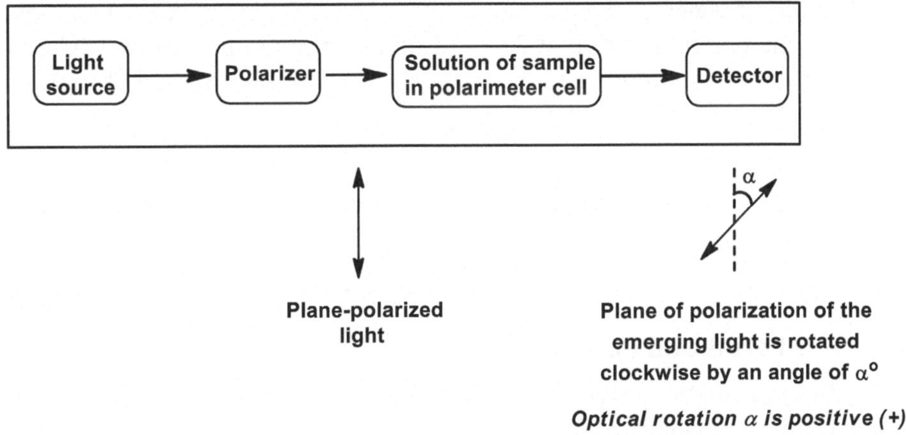

Figure 4.2: A schematic diagram for the measurement of optical rotation by a polarimeter.

The optical rotation (α) is measured in **degrees**, and the degree scale in a polarimeter ranges from 0° to 180° with positive and negative values in the scale.

4.5.1 Specific Rotation

The **magnitude of optical rotation** (α) depends on the number of enantiomer molecules being encountered by the plane-polarized light and hence on the **length of the cell** (l) and the **concentration of the solution** (c).

The measured optical rotation is usually standardized by a quantity called **specific rotation [α]**, which is defined as

$$[\alpha] = \frac{\alpha}{l \times c}$$

where α is the observed rotation in degrees (deg), l is the length of the cell in decimetres (dm), and c is the concentration of the solution in grams per millilitre (g/mL or g/cm^3). The length of a typical polarimeter cell is10 cm = 1 dm.

The **unit of [α]** is given by

$$[\alpha] = \frac{\alpha}{l \times c} = \frac{\text{deg cm}^3}{10 \text{ cm} \times \text{g}} = 10^{-1} \text{ deg cm}^2 \text{ g}^{-1}$$

But the *unit* of specific rotation [α] is generally *omitted*; for example, the specific rotation of an enantiomer of 2-butanol is denoted as +13.5, *not* +13.5°.

The specific rotation $[\alpha]$ also depends on the **wavelength of light** and the **temperature**, which are usually indicated as *subscript* and *superscript* of $[\alpha]$ respectively. Thus $[\boldsymbol{\alpha}]_{\boldsymbol{D}}^{\mathbf{25}}$ indicates a specific rotation using the wavelength of the emission line (*D*-line) of a sodium lamp at 25°*C*. The wavelength of *D*-line is 589 nm (1 nm = 10^{-9} m = 10^{-7} cm = 10 Å).

PROBLEM 4.3: A solution of sucrose (cane sugar) is prepared by dissolving 20 g of sucrose in 100 mL of water, and a portion of this solution is placed in a 10-cm polarimeter cell to measure the optical rotation at 25°C using a sodium lamp (*D*-line). The observed rotation of the sucrose solution is +13.3°. Calculate the specific rotation of sucrose.

Answer

Here, α = +13.3°; l = 10 cm = 1dm; c = (20/100) g/mL = 0.2 g/cm^3 (1 mL = 1 cm^3)

Therefore, $[\alpha]_D^{25} = \frac{\alpha}{l \times c} = \frac{+13.3}{1 \times 0.2} = +66.5$

4.5.2 Racemic Forms and Resolution

An enantiomer is **optically active**, and is described as (+) or (–) enantiomer based on its behaviour toward plane-polarized light.

The optical rotation of a *mixture* of (+) and (–) enantiomers in a 1 : 1 molar proportion will be *zero* because of exact cancellation of (+) and (–) rotations. An **equimolar mixture of two enantiomers** is therefore **optically inactive**, and is called a **racemic form, racemate or racemic mixture**.

All molecules in a racemic form are chiral but there is *equal* number of molecules with **opposite chirality**. A racemic form is thus said to be **optically inactive by external compensation** and is often indicated as (±). For example, a racemic form of 2-chlorobutane is indicated as **(±)-2-chlorobutane**, while its two enantiomers are indicated as (+)-2-chlorobutane and (–)-2-chlorobutane.

Resolution

Since the enantiomers have identical boiling points and identical solubilites in ordinary solvents, the constituent enantiomers of a racemic form cannot be separated by conventional methods as used for ordinary mixtures, such as distillation and crystallization. Thus the separation of eneantiomers requires special procedures. **The process of separation of a racemic form into the component enantiomers is called resolution.**

Historically, in 1848, Louis Pasteur first separated a racemic form of sodium ammonium salt of tartaric acid. He fortunately observed that two types of crystals were present in the salt, and he separated those using tweezers and a magnifying glass. The solution of one type of crystal was found to be *dextrorotatory*, and the solution of the other type of crystal was found to be *levorotatory*. Thus Pasteur effected the *separation of two optically active forms of tartaric acid*, and discovered the phenomenon of enantiomerism.

However, few organic compounds crystallize into separate crystals of enantiomers as do the salts of tartaric acid, and Pasteur's method is not applicable generally. The **general methods for resolution** of racemic forms are based on their *interactions with another chiral compound*, which lead to the formation of a **mixture of diastereomers**.

Diastereomers have different physical properties, i.e. different melting points and boiling points, different solubilities, and so on. Thus *a mixture of diastereomers* (usually salts and complexes) can be easily *separated by conventional methods* such as distillation and crystallization. Thereafter, the enantiomers are released from the separated diastereomers by simple chemical methods.

However, the methods of resolution of racemates are beyond our scope and will not be described here.

4.5.3 Nonracemic Forms and Enantiomeric Excess

A mixture of *unequal* amounts of two enantiomers is called a **nonracemic form**, which is **optically active**.

A nonracemic form that contains a greater proportion of a particular enantiomer is said to be **enantiomerically enriched (or enantioenriched)** with respect to that enantiomer. However, if a nonracemic form of a chiral substance consists of a *single* enantiomer, it is called **enantiomerically pure (or enantiopure)**.

Enantiomeric excess (ee)

The enantiomeric composition of a nonracemic form is denoted by a term called **enantiomeric excess (ee)**. The enantiomeric excess is defined as **the excess of one enantiomer over the other** and is expressed in percentage (%).

There are two *extreme* values of ee. An **enantiomerically pure** sample has a single enantiomer, and hence its **ee = 100%**. On the other hand, a **racemic form** has 50% (+)-enantiomer and 50% (–)-enantiomer, and therefore its **ee = 0%**.

Now, consider a nonracemic sample which is **dextrorotatory (+)** and has **ee = 50%**. This implies that (+) enantiomer has an excess of 50% over (–) enantiomer. Therefore, the enantiomeric composition of the nonracemic sample

= 75% (+)-enantiomer + 25% (–)-enantiomer

= 50% (+)-enantiomer + 25% (+)-enantiomer + 25% (–)-enantiomer

= 50% (+)-enantiomer + 50% (±)

Since a racemic form (±) has no net optical rotation, it is the 50% (+)-enantiomer that will contribute to the observed optical rotation of the nonracemic sample.

In general, **enantiomeric excess (ee)** is given by

$$\%\ \text{ee} = (x_1 - x_2) \times 100$$

where x_1 is the mole fraction of one enantiomer and x_2 is the mole fraction of the second enantiomer. Remember that mole fraction = number of moles/total number of moles.

Experimental determination of ee

An experimental method for determining an enantiomeric excess (ee) is based on optical rotations, and the ee can be calculated as

$$\% \textbf{ ee} = \frac{[\alpha]_{\textbf{obs}}}{[\alpha]_{\textbf{pure}}} \times \textbf{100}$$

where $[\alpha]_{obs}$ = observed specific rotation and $[\alpha]_{pure}$ = specific rotation of the pure enantiomer.

PROBLEM 4.4: A nonracemic sample of 2-butanol showed a specific rotation of + 8.11. The specific rotation of pure (–)-2-butanol is –13.52. Determine the enantiomeric excess of the sample.

Answer

Two enantiomers of 2-butanol have opposite but equal rotations. Therefore, specific rotation of pure (+)-2-butanol is +13.52.

Here $[\alpha]_{obs}$ = +8.11 and $[\alpha]_{pure}$ = +13.52. Hence, ee = (8.11/13.52) × 100% = 60%.

Thus the dextrorotatory sample has 80% (+) and 20% (–), i.e. 60% (+) and 40% (±).

4.6 CHIRAL CENTRES OR STEREOGENIC CENTRES

An atom (usually carbon) which is attached to four different atoms or groups is called a chiral centre or a stereogenic centre. The atoms or groups attached to the chiral centre are generally referred to as **ligands**. Notably, 'stereogenic centre' is a more general term; as such it is defined in Section 4.6.1.

A chiral centre is usually a carbon atom but it can be other atoms such as nitrogen, phosphorus or sulphur. An older term for the chiral centre is the *asymmetric carbon*, which has now become obsolete.

Consider **2-butanol** ($CH_3CH_2CHOHCH_3$). The molecule has four carbon atoms, but it is **C2** which is a **chiral centre** because C2 is attached to four different ligands (OH, CH_2CH_3, CH_3 and H). The chiral centre is conventionally labelled with an asterisk (*).

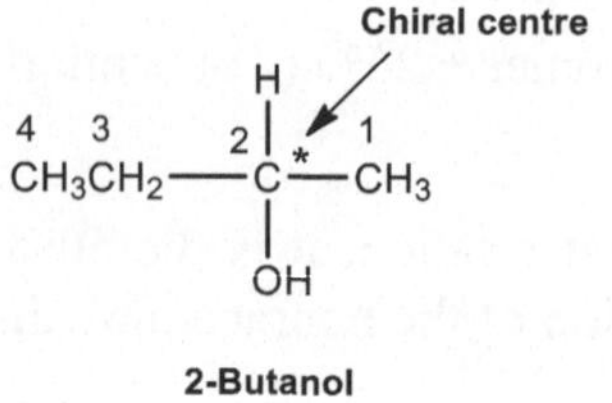

Chiral centre

4 3 2 1
CH_3CH_2—C*(H)(OH)—CH_3

2-Butanol

Ligands attached to C2:

OH, CH_2CH_3, CH_3, H

Four different ligands

4.6.1 Single Chiral Centres: Enantiomers but No Diastereomers

A molecule with a single chiral centre has *no* plane of symmetry or centre of symmetry, and it must be **chiral**. This means that the structure of the molecule and its mirror image structure are nonsuperimposable. So they constitute two stereoisomers that are enantiomers. Therefore, **an organic compound that contains a single chiral centre has only two enantiomers.**

Consider, for example, **2-butanol** which has a single chiral centre (C2). It has two enantiomers (**I** and **II**):

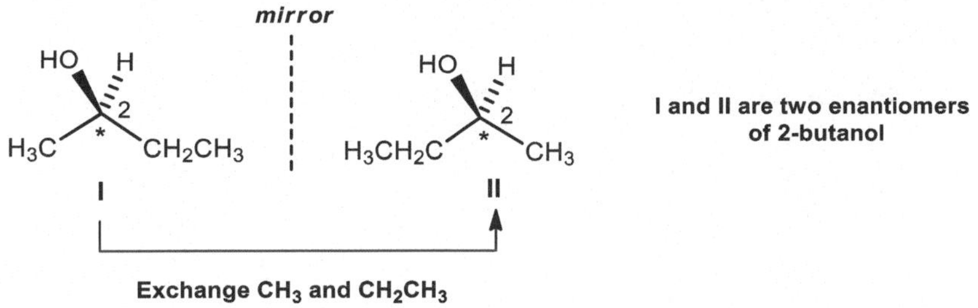

mirror

I and II are two enantiomers of 2-butanol

Exchange CH_3 and CH_2CH_3

It is of note that an exchange or interchange of two ligands (CH_3 and CH_2CH_3) at C2 converts the stereoisomer **I** into a different stereoisomer **II**. This is why the chiral centre C2 is called **stereogenic**.

In general, a **stereogenic centre (or stereocentre)** is defined as an atom (usually carbon) bearing four ligands, such that an exchange of *any two* ligands generates a stereoisomer.

Thus two enantiomers of 2-butanol can be drawn by an exchange of any two ligands at C2. For example, an exchange of H and OH gives the following representation of two enantiomers:

Exchange OH and H

2-Butanol (one enantiomer)

2-Butanol (other enantiomer)

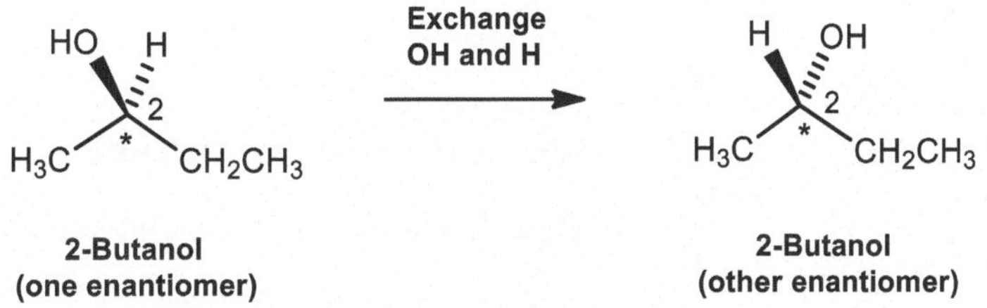

Another representation of two enantiomers of 2-butanol

As **enantiomers**, both stereoisomers of 2-butanol are **optically active**. An equimolar mixture of two enantiomers will give a **racemic form**.

Thus 2-butanol is available in three forms, namely, (+)-2-butanol, (−)-2-butanol, and racemic- or (±)-2-butanol.

PROBLEM 4.5: Draw Fischer projections for the stereoisomers of lactic acid and alanine:

Lactic acid

Alanine

Answer

Lactic acid is an α-hydroxy acid, whereas alanine is an α-amino acid. Each of these compounds contains a single chiral centre (C2).

In lactic acid, the different ligands attached to C2 are OH, CO_2H, CH_3 and H, while the ligands in alanine are NH_2, CO_2H, CH_3 and H. Each compound has two stereoisomers that are enantiomers.

It should be noted that an exchange of two ligands converts one enantiomer to the other enantiomer.

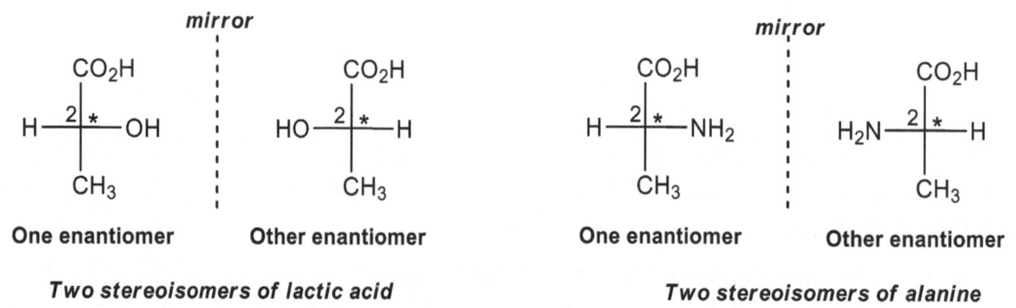

4.6.2 Two *unlike* Chiral Centres: Chiral Diastereomers and Enantiomers

Organic molecules may contain more than one chiral centre. Consider **3-bromo-2-butanol** which has **two chiral centres** (C2 and C3):

3-Bromo-2-butanol

Ligands attached to C2: OH, $CHBrCH_3$, CH_3, H

Ligands attached to C3: Br, $CHOHCH_3$, CH_3, H

Two chiral centres (C2 and C3) are unlike

The four different ligands attached to C2 are **OH, $CHBrCH_3$, CH_3 and H**, and those attached to C3 are **Br, $CHOHCH_3$, CH_3 and H**. Since the set of four ligands at C2 is *different* from the set of four ligands at C3, they are **unlike** chiral centres.

Drawing stereoisomers using Fischer projections

The most convenient way of drawing molecules with more than one chiral centre is with **Fischer projections**. The possible stereoisomers of **3-bromo-2-butanol** are determined as follows.

First, we draw arbitrarily a Fischer projection (labelled **A**) of the molecule. An exchange of two ligands at *one* of the two chiral centres (say, C2) then produces a Fischer projection **B**, which is **not the mirror image of A**. Therefore, the structures **A** and **B** represent two **diastereomers**.

The structure **A** or **B** does not possess a plane of symmetry, and both are **chiral**. So there are **two chiral diastereomers** of 3-bromo-2-butanol.

CH_3 / H—2—OH / H—3—Br / CH_3
A

Exchange H, OH at C2 →

CH_3 / HO—2—H / H—3—Br / CH_3
B

A and B represent diastereomers
Not mirror images

No plane of symmetry ***Chiral*** (A)

No plane of symmetry ***Chiral*** (B)

Since the diastereomer **A** is **chiral**, it has a nonsuperimposable mirror image **A′**. Thus **A** and **A′** are **two enantiomers** of the chiral diastereomer. Similarly, the diastereomer **B** has a nonsuperimposable mirror image **B′**, and they constitute **another pair of enantiomers**.

So there are **four possible stereoisomers** of 3-bromo-2-butanol (**A, A′, B, B′**). Each stereoisomer is **optically active**.

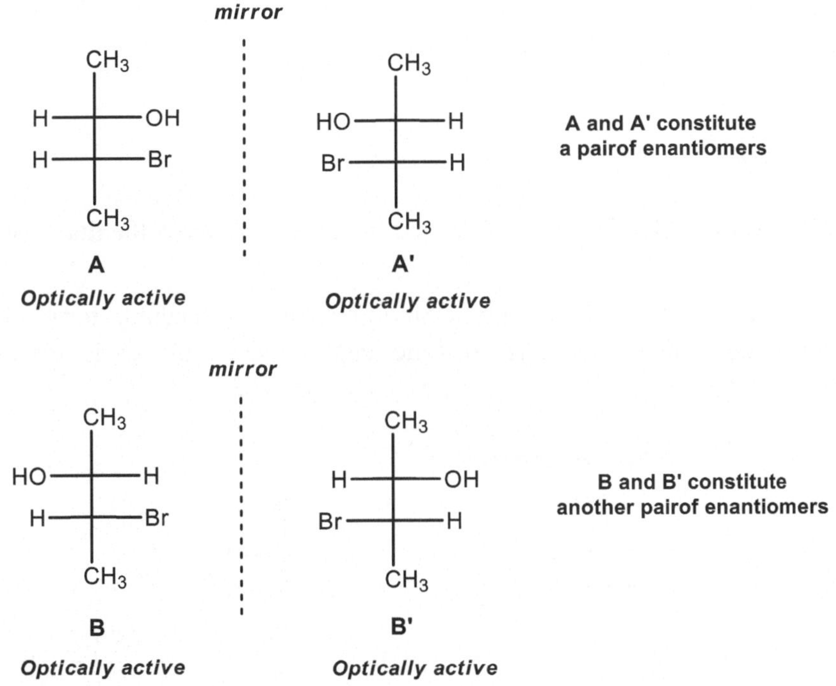

Note: An exchange of ligands at C3 in **A** gives the chiral diastereomer **B′** whose mirror image is **B**, thereby giving the same enantiomeric pair (**B** and **B′**). Thus, we get the same set of four stereoisomers.

In general, if a compound contains *n* unlike chiral centres, the total number of stereoisomers will be $\mathbf{2^n}$. In conformity with this, 3-bromo-2-butanol has $2^2 = 4$ stereoisomers.

Erythro/threo Nomenclature for diastereomers

A simple method for designating diastereomers uses the ***erythro/threo*** **nomenclature**. This nomenclature is based on the Fischer projection formulas of **erythrose** and **threose**, which are four-carbon sugars with two *unlike* chiral centres.

If a diastereomer has identical ligands on the **same side** of the Fischer projection as in erythrose, it is called ***erythro***, and if on the **opposite sides** as in threose, it is called ***threo***.

This nomenclature is illustrated with a general compound (RCHXCHXR′) having a heteroatom-containing ligand (X) as follows:

CHO; H—OH; H—OH; CH_2OH

Erythrose

R; H—X; H—X; R'

erythro

(X = heteroatom-containing ligand)

CHO; HO—H; H—OH; CH_2OH

Threose

R; X—H; H—X; R'

threo

Thus, the **diastereomer A** of 3-bromo-2-butanol is ***erythro***, whereas the **diastereomer B** is ***threo***.

These designations also apply to any enantiomer or the racemic form of a chiral diastereomer. Therefore, **A** is *erythro* (one enantiomer) and **A′** is *erythro* (other enantiomer).

Chiral diastereomers:

CH_3; H—OH; H—Br; CH_3

A

***erythro*-3-Bromo-2-butanol**

CH_3; HO—H; H—Br; CH_3

B

***threo*-3-Bromo-2-butanol**

Enantiomers of a chiral diastereomer:

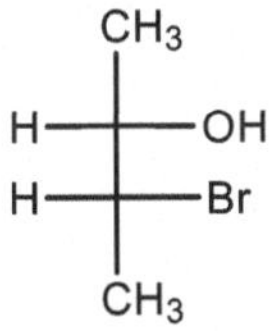

A

***erythro*-3-Bromo-2-butanol**

(one enantiomer)

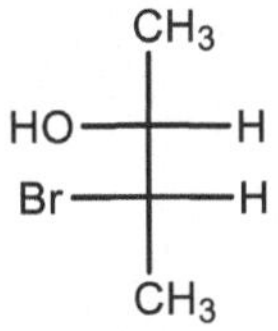

A'

***erythro*-3-Bromo-2-butanol**

(other enantiomer)

4.6.3 Two like chiral centres: Chiral diastereomer and enantiomers; achiral or meso diastereomer

Consider **2,3-butanediol.** Here the two chiral centres C2 and C3 are ***like*** because they carry the *same* set of four ligands **(OH, CO_2H, $CHOHCO_2H$, H)**.

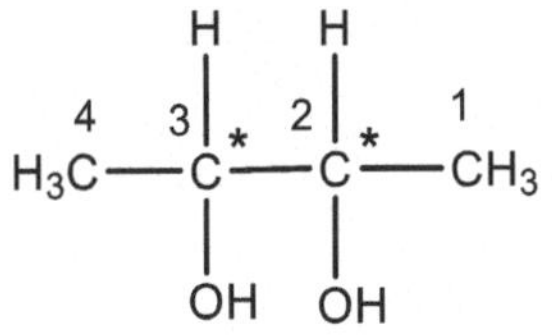

2,3-Butanediol

Ligands attached to C2: OH, $CHOHCH_3$, CH_3, H

Ligands attached to C3: OH, $CHOHCH_3$, CH_3, H

Two chiral centres (C2 and C3) are like

As in the case of two unlike chiral centres, starting from a Fischer projection **C** of 2,3-butanediol, an exchange of two ligands (H, OH) at *one* of the two chiral centres (say, C2) gives a Fischer projection **D**, when **C** and **D** represent **two diastereomers**. But there is a *difference*.

The diastereomer **C** has no plane of symmetry and is **chiral**, whereas the diastereomer **D** has a plane of symmetry and is therefore **achiral**. Note that the mirror plane does not pass through any atoms.

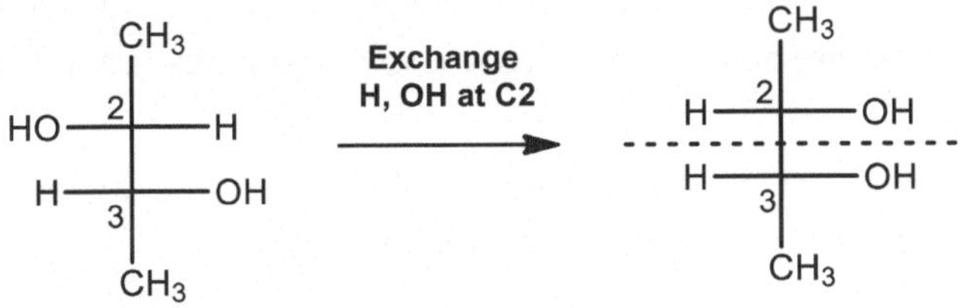

Plane of symmetry

C and D represent diastereomers

Not mirror images

C **D**

No plane of symmetry

Chiral ***Achiral***

The **chiral** diastereomer can exist as **two enantiomers** (**C** and **C′**) that are **optically active**. In contrast, the **achiral** diasteromer (**D**) is **optically inactive**. Note that the diastereomer **D** is superimposable on its mirror image because the mirror image structure rotated by 180° in plane gives the original structure **D**.

mirror

CH3, HO–H, H–OH, CH3 | CH3, H–OH, HO–H, CH3

C *Optically active* | **C'** *Optically active*

C and C' are two enantiomers

mirror

CH3, H–OH, H–OH, CH3 | CH3, HO–H, HO–H, CH3

D *Optically inactive* *meso*

(D and its mirror image are superimposable)

Rotate 180° in plane

CH3, H–OH, H–OH, CH3

D

It should be noted that the molecule represented by the structure **D** is **achiral** even though it contains two **chiral centres**. Thus *chiral centres* and *chiral molecules* are distinct terms.

Molecules that contain chiral centres but are achiral are called ***meso*** compounds. Therefore, the diastereomer D is designated *meso.*

To summarize, 2,3-butanediol has *three* stereoisomers comprising two optically active stereoisomers (**C** and **C′**) and an optically inactive *meso* stereoisomer (**D**).

PROBLEM 4.6 How many stereoisomers are possible for tartaric acid (HO_2CCHOH $CHOHCO_2H$)? Write their Fischer projections and indicate optically active and optically inactive isomers. Also draw a flying wedge formula for the optically inactive isomer.

Answer

The ligands attached to the two chiral centres of tartaric acid are the same (OH, CO_2H, $CHOHCO_2H$, H). Thus tartaric acid has two *like* chiral centres, and therefore has three stereoisomers of which two are optically active isomers (enantiomers) and one is an optically inactive isomer (*meso*).

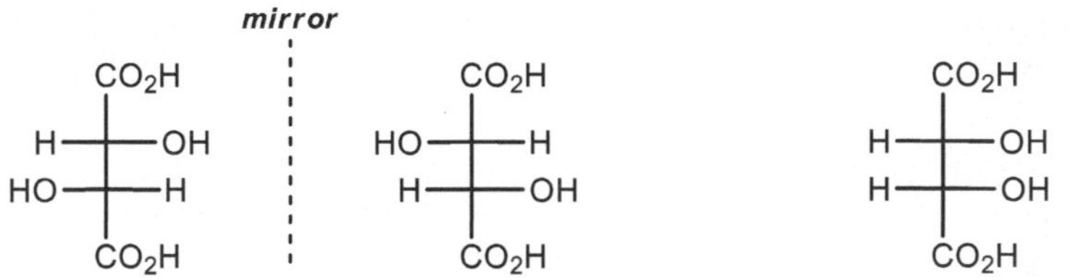

(one enantiomer) ***Optically active*** | **(other enantiomer)** ***Optically active*** | ***meso* isomer** ***Optically inactive***

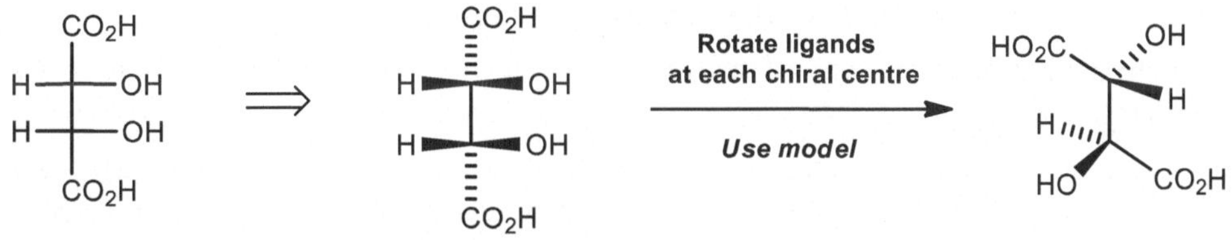

meso*-Tartaric acid** | ***Flying wedge formula | ***Another Flying wedge formula***

4.7 CONFIGURATION TO CHIRAL CENTRES: THE *R/S* SYSTEM

The **arrangement** of atoms or groups (called ligands) in space around a **chiral centre** is called its **configuration**.

Any chiral centre can have **two** possible configurations. These configurations can be specified by a system of nomenclature called the ***R/S* system**, which was devised by three chemists, R.S. Cahn, C.K. Ingold and V. Prelog. This system, also known as the **Cahn-Ingold-Prelog (CIP) system**, is now widely used to assign *R* or *S* configurations to chiral centres.

The CIP system involves two steps: (a) determination of order of priority (or preference) for ligands; (b) assignment of *R* or *S* configuration.

4.7.1 Determination of Priority Order for Ligands

Each of the four ligands attached to a chiral centre is assigned a **priority *a*, *b*, *c* or *d***, where the priority follows the order: $\boldsymbol{a > b > c > d}$. Thus the priority *a* denotes the highest priority while the priority *d* denotes the lowest priority.

The following **rules** can be used to determine the **priority order** of four ligands at a chiral centre.

1. Priority is based on the atomic number of the atoms *directly attached* to the chiral centre.

   Atoms of higher atomic number will have higher priority.

Example: Bromochlorofluoromethane (CHFClBr)

The highest priority *a* goes to Br which has the highest atomic number, and the lowest priority *d* goes to H which has the lowest atomic number. The overall priority order of the ligands is **Br > Cl > F > H**.

(*d*)
H
(*c*) F—C*—Br (*a*)
Cl (*b*)

Bromochlorofluoromethane

Ligands attached to the chiral centre: Br, Cl, F and H

Atomic number: Br > Cl > F > H

Overall priority order: Br > Cl > F > H
(*a*) (*b*) (*c*) (*d*)

2. When the *atom directly attached to the chiral centre* (the first atom) is the same, the next atoms (second atoms) of the ligands are examined.

   The ligand with *any* second atom of higher atomic number will have higher priority.

   If second atoms are the same, then the next outward atoms (third atoms) are examined similarly. If still there is no decision, the fourth atoms and so on are examined.

Example: 2-Chlorobutane ($CH_3CH_2CHClCH_3$)

The four ligands attached to the chiral centre are **Cl, CH_2CH_3, CH_3 and H**.

Cl has the highest atomic number of the four atoms (Cl, C, C and H) *directly attached* to the chiral centre, and therefore **Cl** is assigned the **highest priority *a***. On the other hand, **H** has the lowest atomic number and is given the **lowest priority *d***.

The first atom in both CH_2CH_3 and CH_3 is a C atom. Therefore, the **second atoms** of these ligands are to be examined to assign priority.

In CH_2CH_3, the second atoms are a set of one C atom and two H atoms **(C, H, H)**, whereas in CH_3, the second atoms are a set of three H atoms **(H, H, H)**. Since C has a higher atomic number than H, **CH_2CH_3 has the higher priority *b* over CH_3 (priority *c*)**.

Thus the overall priority order of the four ligands is **Cl > CH_2CH_3 > CH_3 > H**.

(*c*)
Second atoms in CH_3: (H, H, H)

H—C—C*(Cl (*a*), H (*d*))—C—CH_3

(*b*)
Second atoms in CH_2CH_3: (C, H, H)

(C, H, H) > (H, H, H)

(1 = first atom, 2 = second atom in a ligand)

Overall priority order: Cl > CH_2CH_3 > CH_3 > H
(*a*) (*b*) (*c*) (*d*)

3. When a ligand has a **double bond** between two atoms, it is treated as if there were **two single bonds** to ***both*** **atoms**.

Similarly, a **triple bond** between two atoms is treated as if there were **three single bonds** to ***both*** **atoms**.

These are schematically represented as,

A═B **is treated as** A—B (with A bonded to B below, B bonded to A below)

(A has two A-B bonds; B has two B-A bonds)

A≡B **is treated as** A—B (with A bonded to B above and below, B bonded to A above and below)

(A has three A-B bonds; B has three B-A bonds)

Example: Lactic acid ($CH_3CHOHCO_2H$)

The four ligands attached to the chiral centre are **OH, CO_2H, CH_3 and H**.

As the first atom, O has the highest atomic number, and therefore **OH** is assigned the **highest priority *a***. On the other hand, **H** has the **lowest priority *d***.

The first atom in both CO_2H and CH_3 is a C atom. Therefore, the **second atoms** of these ligands are to be examined to assign priority.

To do this, a C=O is treated as if the C atom has two C–O bonds and the O atom has two O–C bonds. Thus, in **CO_2H**, C has three C–O bonds including the C–OH bond. The second atoms in CO_2H are therefore **(O, O, O)**. In comparison, the second atoms in **CH_3** are **(H, H, H)**. Hence **CO_2H has the higher priority *b* over CH_3 (priority *c*)**.

The overall priority order of the four ligands is **OH > CO_2H > CH_3 > H**.

(*a*) OH; H_3C—C*—C═O (OH); H (*d*)

Lactic acid

—CH_3 ⟹ —C(2 H)(2 H)—H(2) **(*c*) (H, H, H)**

—C(═O)OH ⟹ —C(2 O)(C)—O 2, 2 O—H **(*b*) (O, O, O)**

Overall priority order: **OH > CO_2H > CH_3 > H**
(*a*) (*b*) (*c*) (*d*)

PROBLEM 4.7: Which ligand in the following pair has higher priority?

—CH═CH_2 and —CH$(CH_3)_2$

(vinyl) **(isopropyl)**

Answer

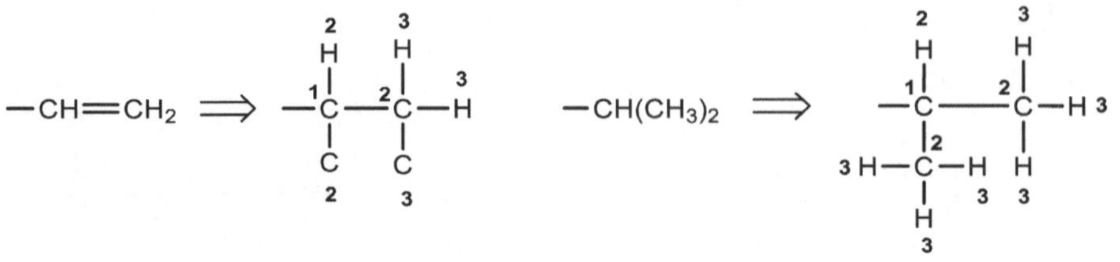

$-CH=CH_2 \Rightarrow$ [expanded structure] $\quad -CH(CH_3)_2 \Rightarrow$ [expanded structure]

2nd atoms: (C, C, H)	=	2nd atoms: (C, C, H)
3rd atoms: (C, H, H)	>	3rd atoms: (H, H, H), (H, H, H)

Priority order: $-CH=CH_2 > -CH(CH_3)_2$

4.7.2 Assignment of *R* or *S* Configuration

Once the priority order of four ligands attached to a chiral centre is established, the *R* or *S* configuration is assigned as follows.

View the molecule so that the **ligand with the lowest-priority *d*** is pointed toward the **back**. Then **trace a path** of the remaining three ligands in **decreasing priority** from the ligand of first-priority ***a*** to the ligand of second-priority ***b*** to the ligand of third-priority ***c***.

If ***a* → *b* → *c* traces clockwise**, the configuration is ***R*** (from the Latin *rectus*, meaning right), and if ***a* → *b* → *c* traces anticlockwise**, the configuration is ***S*** (from the Latin *sinister*, left).

Example 1: 2-Chlorobutane (two enantiomers)

The priority order of four ligands attached to the chiral centre is Cl > CH_2CH_3 > CH_3> H (see above).

Two enantiomers of 2-chlorobutane are shown in flying wedge formulas. The enantiomer on the left is assigned ***R*** configuration since Cl(*a*) → CH_2CH_3(*b*) → CH_3(*c*) traces clockwise, with H(d) pointed toward the back. Similarly, the enantiomer on the right is assigned ***S*** configuration because Cl(*a*) → CH_2CH_3(*b*) → CH_3(*c*) traces anticlockwise with H(d) in the back.

Two enantiomers of 2-chlorobutane:

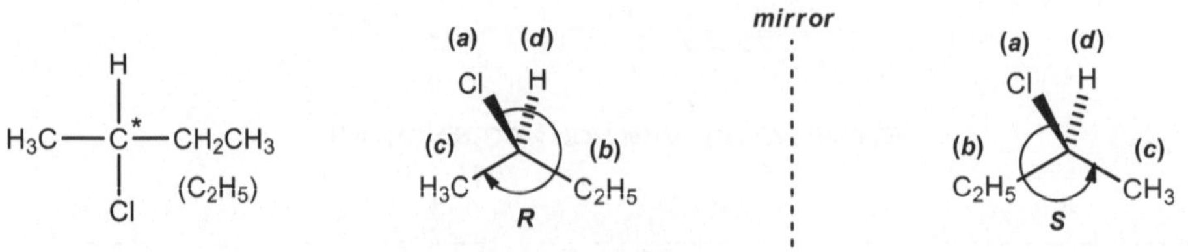

Note: An exchange of two ligands (e.g. CH_3 and C_2H_5) at the chiral centre changes the configuration from *R* to *S* or *S* to *R*. This is true for any chiral centre in molecules having two or more chiral centres.

Example 2: Bromochlorofluoromethane (one enantiomer)

Consider a flying wedge formula for an enantiomer of bromochlorofluoromethane (CHFClBr) in which the **lowest-priority ligand (H) is *not* in the back**.

To assign *R* or *S*, the molecule can be rotated so that the lowest-priority ligand is oriented to the **back**. Another alternative way to do this is to make *two* exchanges of ligands that give the identical structure. In *one* of the two exchanges, the lowest-priority ligand is placed in the **back**, as shown below.

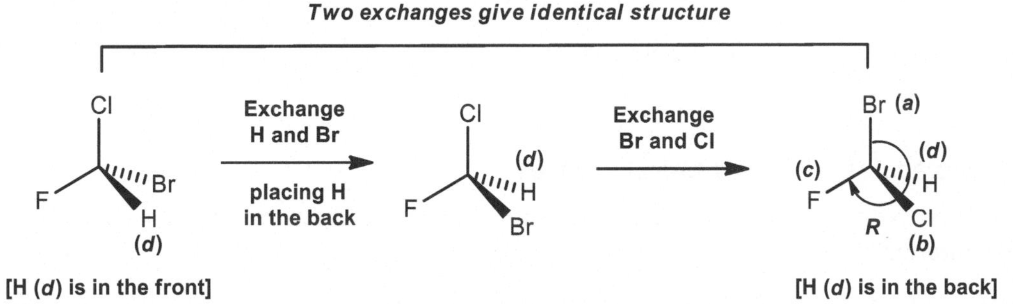

Two exchanges give identical structure

Cl, F, Br, H (*d*) — **Exchange H and Br** — **placing H in the back** → Cl, F, (*d*) H, Br — **Exchange Br and Cl** → Br (*a*), (*c*) F, (*d*) H, Cl (*b*), *R*

[H (*d*) is in the front] **[H (*d*) is in the back]**

(*R*)-Bromochlorofluoromethane

Assignment of R/S configuration in Fischer projection formulas

In the Fischer projection formula, a **vertical** ligand is actually oriented toward the **back** while a **horizontal** ligand is actually pointed toward the **front**. Two cases can arise:

(a) Ligand of the lowest priority *d* is on the vertical line (oriented toward the back)

A clockwise array of the remaining three ligands in descending priority $a \to b \to c$ gives the *R* configuration. But if $a \to b \to c$ is anticlockwise, the configuration is *S*.

(b) Ligand of the lowest priority *d* is on the horizontal line (oriented toward the front)

Here the ligand of priority *d* can be placed on the vertical line by *rotating three ligands* keeping one fixed or by making *two exchanges*, and then the above rule can be applied to assign *R* or *S* configuration.

However, we can use the following **short-cut**: with ***d*** on the **horizontal line**, the **opposite assignment** is taken to be correct, i.e. if $\boldsymbol{a \to b \to c}$ is **clockwise**, the configuration is ***S***, and if $\boldsymbol{a \to b \to c}$ is **anticlockwise**, the configuration is ***R***.

Example: Lactic acid (two enantiomers)

Consider the Fischer projections of two enantiomers of lactic acid, as shown below. Note that *one exchange* of two ligands (CH_3, H) gives the other enantiomer.

In the first enantiomer, *d* is on the vertical line, and $a \to b \to c$ is clockwise. Therefore, the configuration is *R*.

In the second enantiomer, *d* is on the horizontal line, and $a \to b \to c$ is clockwise. Therefore, the opposite assignment is correct, i.e the configuration is *S*.

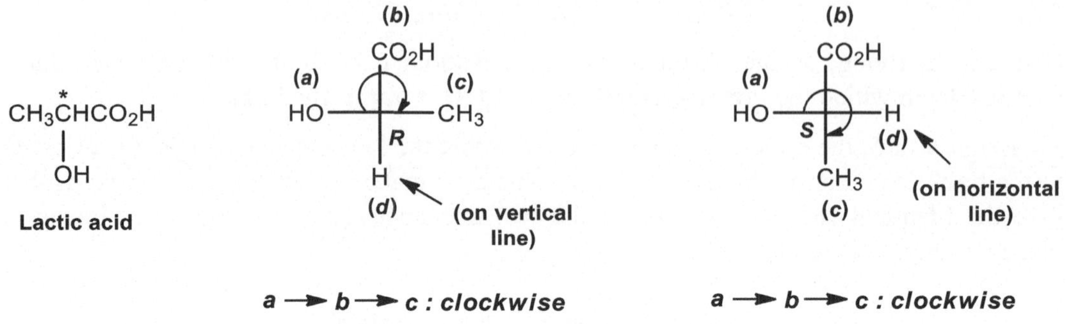

$CH_3\overset{*}{C}HCO_2H$ (with OH on the starred carbon)

Lactic acid

(b) CO_2H; (a) HO; (c) CH_3; *R*; H (d) (on vertical line)

a → b → c : clockwise

(*R*)-Lactic acid

(b) CO_2H; (a) HO; H (d) (on horizontal line); *S*; CH_3 (c)

a → b → c : clockwise

(*S*)-Lactic acid

4.7.3 Molecules with Two Chiral Centres

If a molecule contains two or more chiral centres, *each* chiral centre can be assigned *R* or *S* configuration.

Consider, for example, **tartaric acid** ($HO_2CCHOHCHOHCO_2H$) which has **two *like* chiral centres**. The four ligands attached to each chiral centre are **OH, CO_2H, $CHOHCO_2H$ and H.**

The priority order of the four ligands is decided as shown below.

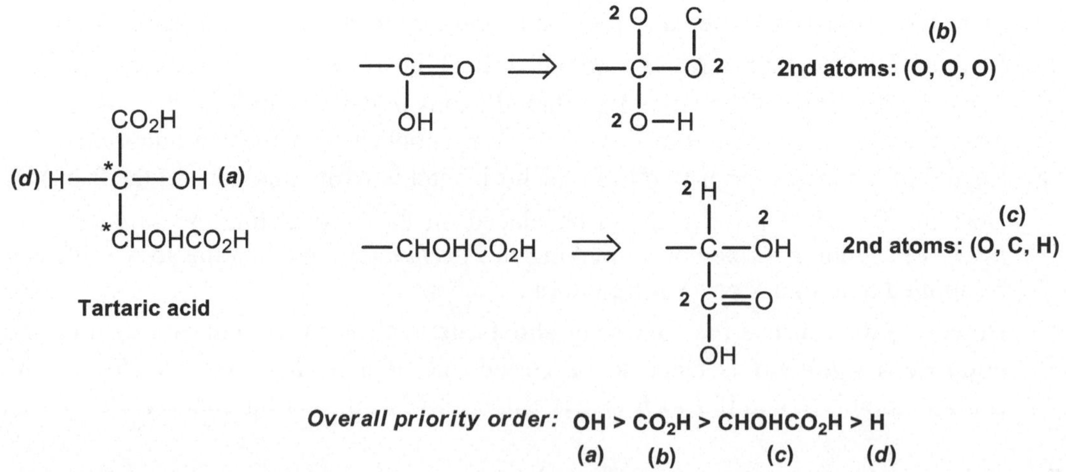

CO_2H; (d) H—$\overset{*}{C}$—OH (a); $^{*}CHOHCO_2H$

Tartaric acid

—C(=O)OH ⇒ —C(O)(O)(O—H) **2nd atoms: (O, O, O)** (b)

—$CHOHCO_2H$ ⇒ —C(H)(OH)(C=O, OH) **2nd atoms: (O, C, H)** (c)

Overall priority order: **OH > CO_2H > $CHOHCO_2H$ > H**
(a) (b) (c) (d)

Assignment of R/S Configurations

Tartaric acid has three stereoisomers comprising two enantiomers and a *meso* isomer (see problem 4.6). First, we consider an **enantiomer A** of tartaric acid and assign *R* or *S* configuration to the chiral centres C2 and C3 by treating them **separately**.

As shown below, the enantiomer **A** is designated **(2*R*, 3*R*)** or simply **(*R*,*R*)** for the two like chiral centres.

(*d* on the horizontal line) (*b*) CO_2H (*d*) H—C2(*R*)—OH (*a*) $CHOHCO_2H$ (*c*)

Consider C2

a → *b* → *c* : *anticlockwise*

CO_2H
H—2—OH
HO—3—H
CO_2H

A

(one enantiomer)

Consider C3

(*c*) $CHOHCO_2H$ (*a*) HO—C3(*R*)—H (*d*) CO_2H (*b*) (*d* on the horizontal line)

a → *b* → *c* : *anticlockwise*

⇒

CO_2H
H—2 *R*—OH
HO—3 *R*—H
CO_2H

(2*R*,3*R*) or (*R*,*R*)

CO_2H
H—2 *R*—OH
HO—3 *R*—H
CO_2H

A

(one enantiomer)

Exchange H, OH at C3 →

CO_2H
H—2 *R*—OH
H—3 *S*—OH
CO_2H

B

***meso* isomer**

(2*R*,3S) or (*R*,*S*)

Exchange H, OH at C2 and at C3 →

CO_2H
HO—2 *S*—H
H—3 *S*—OH
CO_2H

A'

Other enantiomer

(2*S*,3S) or (*S*,*S*)

We do not need to repeat the entire procedure for assigning the *R*/*S* configurations to the other stereoisomers. The *R*/*S* configuration can be assigned directly by **changing *R* to *S* (or *S* to *R*)** for an act of **one exchange of two ligands** at a chiral centre (see the note in Section 4.7.2).

An exchange of two ligands at *one* of the two chiral centres gives the *meso* diastereomer. Thus an exchange of two ligands (H, OH) at C3 gives the *meso* isomer **B** which is designated **(2*R*,3*S*)** or (***R***,***S***).

However, an exchange of two ligands at *each* chiral centre gives the other enantiomer. Thus the other enantiomer **A′** is designated **(2*S*,3*S*)** or (***S***,***S***) by an exchange of two ligands (H, OH) at both C2 and C3. Note that two enantiomers have opposite configurations at all chiral centres.

4.7.4 Configuration and Optical Rotation

Two enantiomers of the *same* chiral compound have **opposite configurations** (*R*) and (*S*) at a chiral centre, and they have **opposite optical rotations** (+) and (–). For example, (*R*)-2-butanol shows (+) rotation, while (*S*)-2-butanol shows (–) rotation.

However, two *different* chiral compounds with the same configuration (say, *R*) may have different sign of optical rotation. For example, **(*R*)-2-methyl-1-butanol** is **dextrorotatory (+)**, whereas **(*R*)-1-chloro-2-methylbutane** is **levorotatory (–)**.

CH_2OH ; H_3C ; H ; CH_2CH_3

(*R*)-(+)-2-Methyl-1-butanol

CH_2Cl ; H_3C ; H ; CH_2CH_3

(*R*)-(-)-1-Chloro-2-methylbutane

Thus the sign of optical rotation (say, +) as such cannot predict the configuration (*R* or *S*) for a chiral compound. It was not until 1951 when Bijvoet determined the **configuration of (+)-tartaric acid** by x-ray crystallographic technique to be **(*R,R*)**. For the same compound (tartaric acid), the configuration of the other enantiomer, i.e. **(–)-tartaric acid** is therefore **(*S,S*)**.

The configuration of ***meso*-tartaric acid** is **(*R,S*)**. Since tartaric acid has two *like* chiral centres, the contributions of optical rotation due to (*R*) and (*S*) chiral centres are **equal** in magnitude but **opposite** in sign. Therefore, in *meso*-tartaric acid, the contributions of (+) and (–) rotations cancel each other. The ***meso*-tartaric acid** is then said to be **optically inactive by internal compensation**.

This is in contrast to the **racemic** form which is optically inactive by external compensation (see Section 4.5.2).

PROBLEM 4.8: Draw structures (as indicated) for the following molecules:

(a) (*R*)-2-Butanol (flying wedge)
(b) (2*S*,3*R*)-3-Bromo-2-butanol (Fischer projection)

Answer

(a) The ligand (H) with the lowest-priority *d* is placed in the back, and the other three ligands with decreasing priorities *a*, *b* and *c* are arranged in a clockwise manner.

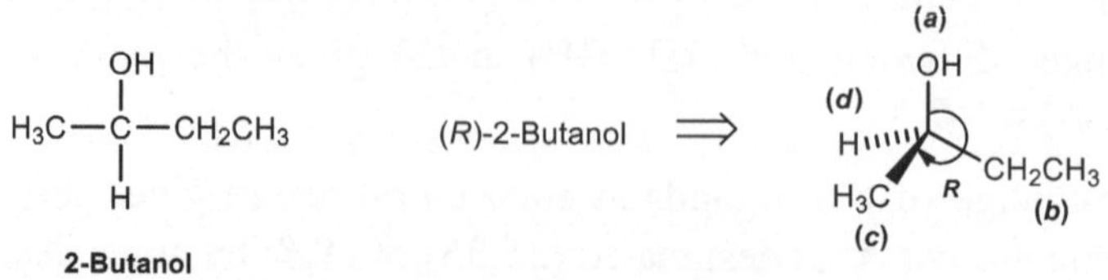

OH ; H_3C—C—CH_2CH_3 ; H

2-Butanol

Priority order of ligands:
OH > CH_2CH_3 > CH_3 > H

(*R*)-2-Butanol ⟹

(*a*) OH ; (*d*) H ; CH_2CH_3 (*b*) ; H_3C (*c*) ; *R*

(b) A part Fischer projection is drawn for the chiral centre C2 (*S* configuration) as well as for the chiral centre C3 (*R* configuration), with the carbon chain along the vertical line. The combination of the two parts gives the complete Fischer projection.

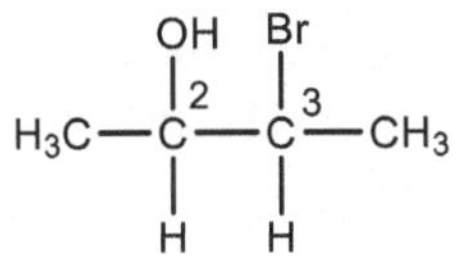

3-Bromo-2-butanol

Priority order of ligands at C2:
OH > $CHBrCH_3$ > CH_3 > H

Priority order of ligands at C3:
Br > $CHOHCH_3$ > CH_3 > H

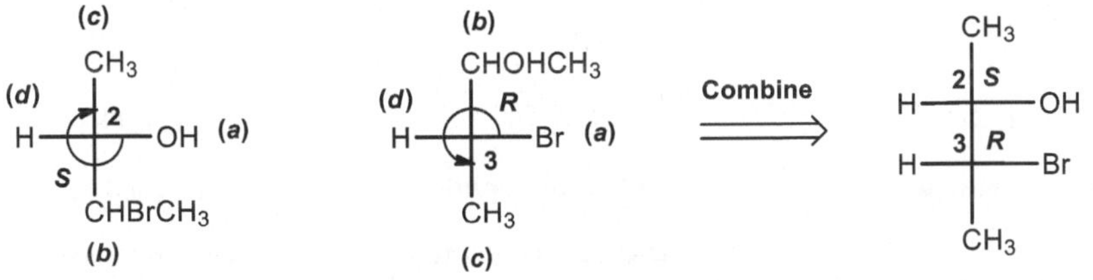

Combine

(2*S*,3*R*)-3-Bromo-2-butanol

4.8 CONFIGURATION TO CHIRAL CENTRES: THE D/L SYSTEM

The **D/L system** of nomenclature for configuration at a chiral centre is limited in scope but is still used for **α-amino acids** and **carbohydrates**.

The nomenclature is **defined in the Fischer projection formula** in which the **vertical line** denotes a main carbon chain with **C1** (as per the IUPAC numbering) at the **top**.

α-Amino acids

For α-amino acids, the **configuration** is **D** when **NH_2** at the chiral centre is placed on the **right side** of the horizontal line, and the **configuration** is **L** when **NH_2** is on on the **left side**.

Consider, for example, two enantiomers of **alanine** (2-aminopropanoic acid). In the Fischer projection, CO_2H is at the top of the vertical line. One enantiomer is D-alanine in which NH_2 is on the right side, and the other enantiomer is L-alanine in which NH_2 is on the left side.

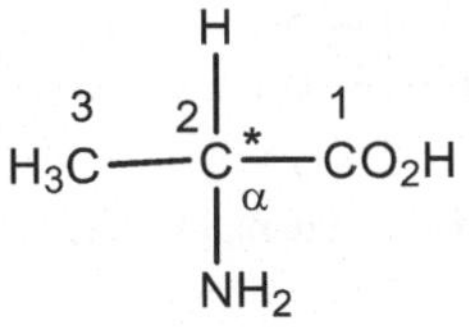

Alanine
(an α-amino acid)

IUPAC: 2-Aminopropanoic acid

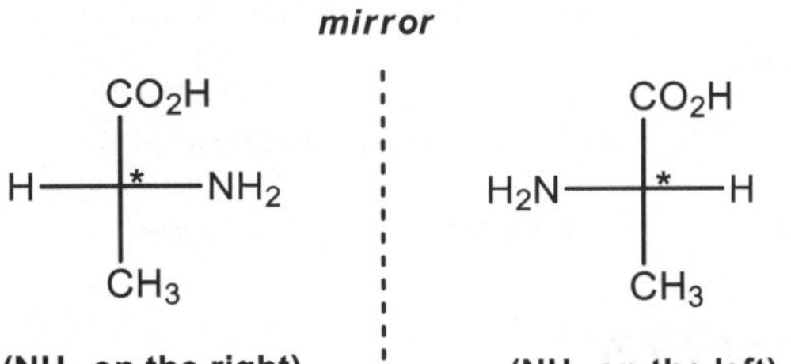

mirror

(NH_2 on the right)

D-Alanine

(NH_2 on the left)

L-Alanine

Carbohydrates

Carbohydrates or sugars are polyhydroxy aldehydes or ketones that may contain one or more chiral centres.

The **simplest carbohydrate** with a single chiral centre is **glyceraldehyde**. Similar to α-amino acids, the configuration of glyceraldehyde is **D** when **OH** at the chiral centre is on the **right**, and the configuration is **L** when **OH** is on the **left**.

1 CHO | H—2C*—OH | 3 CH_2OH
Glyceraldehyde

1 CHO | H—*—OH | CH_2OH
(OH on the right)
D-Glyceraldehyde

mirror

1 CHO | HO—*—H | CH_2OH
(OH on the left)
L-Glyceraldehyde

For carbohydrates containing **more than one chiral centre**, the D or L configuration is assigned on the basis of the **OH** group at the **highest numbered chiral centre**.

To illustrate, consider the two enantiomers each of **erythrose** (two chiral centres C2 and C3), **ribose** (three chiral centres C2, C3 and C4) and **glucose** (four chiral centres C2, C3, C4 and C5). The D or L configuration is assigned based on the OH at the highest numbered chiral centre, which is C3, C4 and C5 for erythrose, ribose and glucose respectively.

1 CHO | 2 CHOH* | 3 CHOH* | 4 CH_2OH
Erythrose

CHO | H—OH | H—3—**OH** | CH_2OH
(3-OH on the right)
D-Erythrose

CHO | HO—H | **HO**—3—H | CH_2OH
(3-OH on the left)
L-Erythrose

1 CHO | 2 CHOH* | 3 CHOH* | 4 CHOH* | 5 CH_2OH
Ribose

CHO | H—OH | H—OH | H—4—**OH** | CH_2OH
(4-OH on the right)
D-Ribose

CHO | HO—H | HO—H | **HO**—4—H | CH_2OH
(4-OH on the left)
L-Ribose

1 CHO | 2 CHOH* | 3 CHOH* | 4 CHOH* | 5 CHOH* | 6 CH_2OH
Glucose

CHO | H—OH | HO—H | H—OH | H—5—**OH** | CH_2OH
(5-OH on the right)
D-Glucose

CHO | HO—H | H—OH | HO—H | **HO**—5—H | CH_2OH
(5-OH on the left)
L-Glucose

Tartaric acid

The D or L configuration of tartaric acid is assigned based on the position of **OH** at the chiral centre **C3** (the top carbon being C1).

As shown below, **D-tartaric acid** has OH at C3 on the right corresponding to **(*S*,*S*)- or (–)-tartaric acid**, whereas **L-tartaric acid** has OH at C3 on the left corresponding to **(*R*,*R*)- or (+)-tartaric acid**. (see Sections 4.7.3 and 4.7.4)

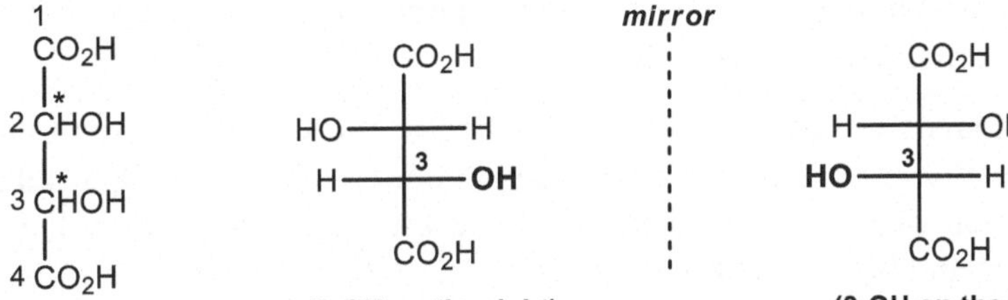
1
CO2H
2 CHOH
3 CHOH
4 CO2H
CO2H
HO
H
H
3
OH
CO2H
mirror
CO2H
H
OH
HO
3
H
CO2H

Tartaric acid | (3-OH on the right) D-Tartaric acid = (*S*,*S*)-(-)-Tartaric acid | (3-OH on the left) L-Tartaric acid = (*R*,*R*)-(+)-Tartaric acid

PROBLEM 4.9: Assign *R* or *S* configuration to (a) L-alanine and (b) D-glyceraldehyde.

Answer

(a) NH_2 has the highest priority *a* because nitrogen (directly attached to the chiral centre) has the highest atomic number. The overall priority order is $NH_2 > CO_2H > CH_3 > H$. As shown below, L-alanine has *S* configuration.

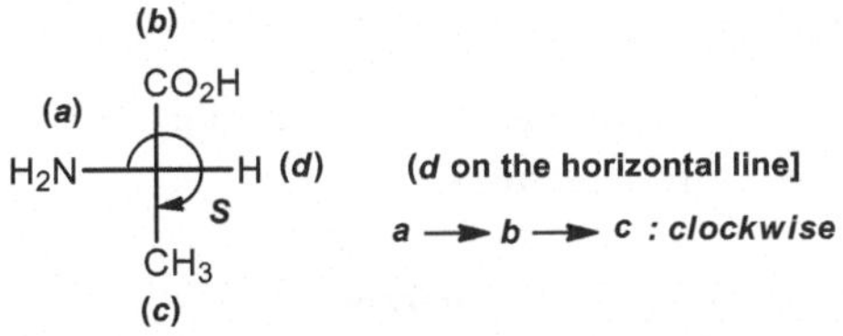
(b)
CO2H
(a)
H2N
H (d)
S
CH3
(c)
(d on the horizontal line]
a → b → c : clockwise

L-Alanine = (*S*)-Alanine

(b) OH has the highest priority *a*, and CHO has higher priority *b* over CH_2OH (priority *c*). The overall priority order is $OH > CHO > CH_2OH > H$.
As shown below, D-glyceraldehyde has *R* configuration.

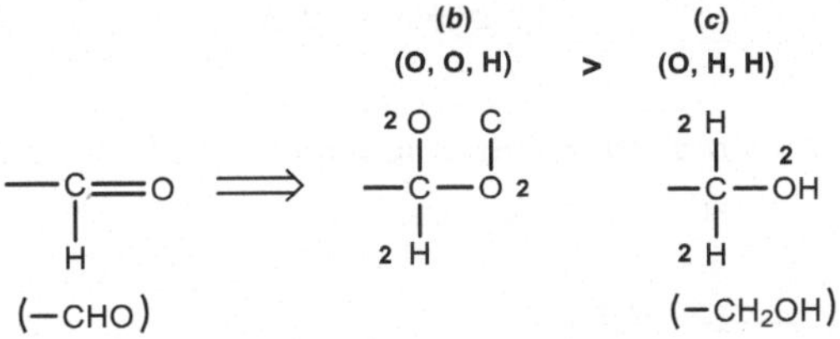
(b)
(O, O, H)
>
(c)
(O, H, H)
(–CHO)
(–CH2OH)

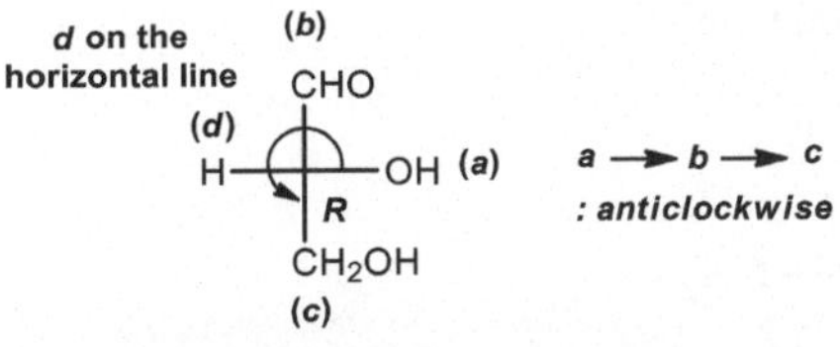
d on the
horizontal line
(b)
CHO
(d)
H
OH (a)
R
CH2OH
(c)
a → b → c
: anticlockwise

D-Glyceraldehyde = (*R*)-Glyceraldehyde

4.9 CHIRAL MOLECULES WITHOUT A CHIRAL CENTRE

Chiral centre and **chiral molecule** are **distinct** terms. The presence of a chiral centre is a common but *not* an essential attribute that can make a molecule chiral.

Chiral compounds are known that do not contain chiral centres. For example, **allenes** that do not possess a chiral centre can be **chiral.**

In an allene, the ligands (atoms or groups) attached to the terminal carbon atoms of the C=C=C system lie in *perpendicular planes* (see Section 3.4.3). If **two ligands** attached to *each* terminal carbon of C=C=C unit in an allene are **different**, the molecule **lacks a plane of symmetry** and becomes **chiral**.

There are two general cases when allenes become chiral:

Case I:

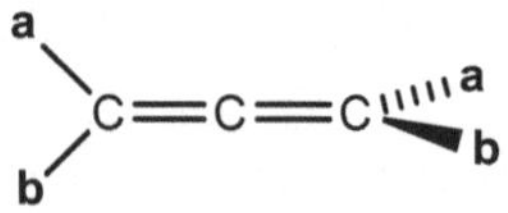

a, b C=C=C a, b

(a and b are different)

(no plane of symmetry)

Chiral

Case II:

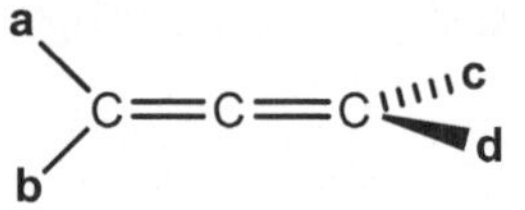

a, b C=C=C c, d

(a and b are different; c and d are different)

(no plane of symmetry)

Chiral

Example: 1,3-Dimethylallene

1,3-Dimethylallene ($CH_3CH{=}C{=}CHCH_3$) is an example for Case I (a = CH_3, b = H). The allene is **chiral** and it has two **enantiomers**.

mirror

H_3C, H C=C=C H, CH_3

(one enantioner)

H, H_3C C=C=C CH_3, H

(other enantioner)

1,3-Dimethylallene

PROBLEM 4.10: Which of the following allenes is chiral or achiral?

(a) Cl—CH=C=CH—CH_3

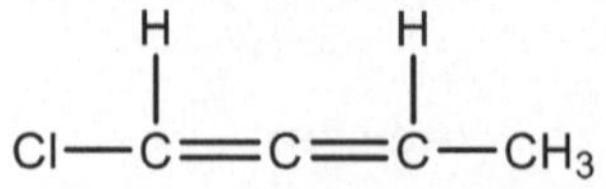

1-Chloro-3-methylallene

(b) H—CH=C=CH—CH_3

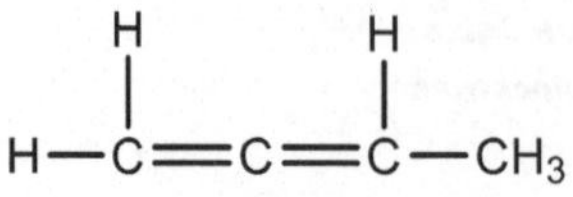

1-Methylallene

Answer

(a) The allene is **chiral** because the two ligands attached to *each* terminal carbon atom are different ($CH_3 \neq$ H, H $\neq$Cl). It has two stereoisomers (enantiomers).

(b) The allene is **achiral** because there are two identical ligands (H, H) at one of the two terminal carbon atoms. It has no stereoisomers.

4.10 BIOLOGICAL IMPORTANCE OF CHIRAL MOLECULES

Chiral molecules are ubiquitous in nature. Molecules that make up plants and animals are mostly **chiral**. Interestingly, nature is highly **selective** and cares for only **one enantiomer** of chiral compounds.

Naturally occurring proteins are derived from **L-amino acids**, *not* from enantiomeric D-amino acids. **Natural sugars** are **D-sugars**, *not* L-sugars. They include D-ribose and D-2-deoxyribose that occur in RNA (ribonucleic acid) and DNA (deoxyribonucleic acid) respectively.

Differences in chirality may also be manifest in many ways. The subtle differences in the way the molecules of enantiomers are arranged in three-dimensional space can be recognized by our *nose receptors*.

For example, the ***R*** or **(+) enantiomer of limonene** is responsible for the **smells of oranges**, while the ***S*** or **(–) enantiomer** is responsible for **smells of lemons**.

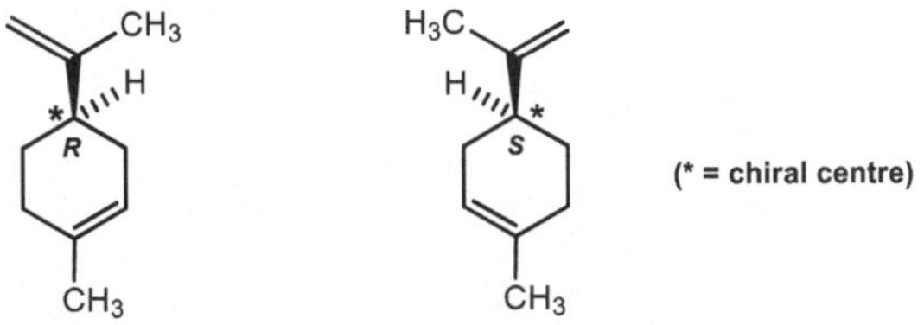

CH_3 H * *R* CH_3 H_3C H * *S* CH_3 (* = chiral centre)

(*R*)-(+)-Limonene
Smells of oranges

(*S*)-(-)-Limonene
Smells of lemons

In pharmaceutical industry, a **chiral drug** that contains a **single enantiomer** rather than a racemate is of much importance. In some cases, a drug has been marketed as a racemate for years, even though only one enantiomer of the drug is the active constituent.

For example, the **(*S*) enantiomer** of the anti-inflammatory drug **ibuprofen** functions as an **active drug**, whereas the (*R*) enantiomer is essentially inactive.

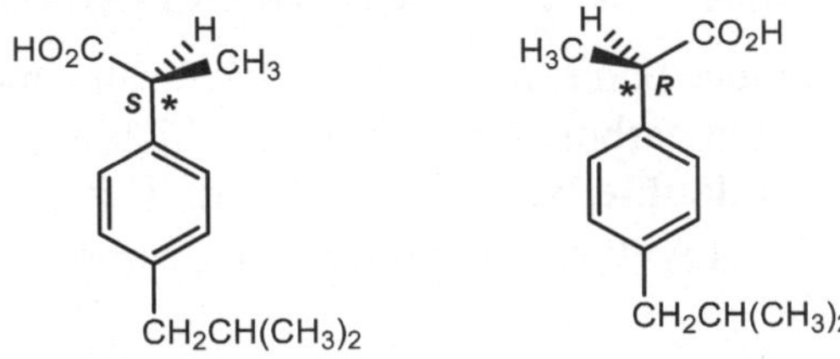

HO_2C H CH_3 *S* * $CH_2CH(CH_3)_2$ H H_3C CO_2H * *R* $CH_2CH(CH_3)_2$

(*S*)-Ibuprofen
Active drug

(*R*)-Ibuprofen
Inactive

Sometimes, the use of the right enantiomer of a drug is crucial and even can be a matter of life and death.

For example, *only* **one enantiomer** of the drug **dopa** (3,4-**d**ihydr**o**xy**p**henyl**a**lanine), called **L-dopa** or **(*S*)-dopa** is effective for the treatment of Parkinson's disease. The other enantiomer, D-dopa or (*R*)-dopa is not only ineffective, but also quite toxic.

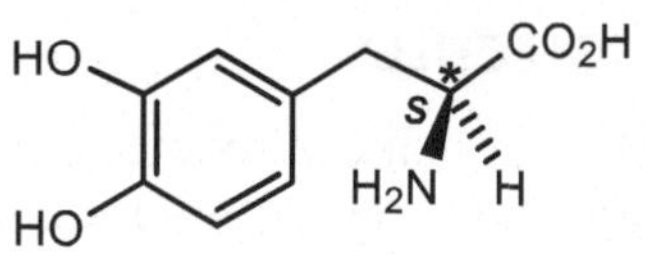

L-Dopa or (*S*)-Dopa

Effective for Parkinson's disease

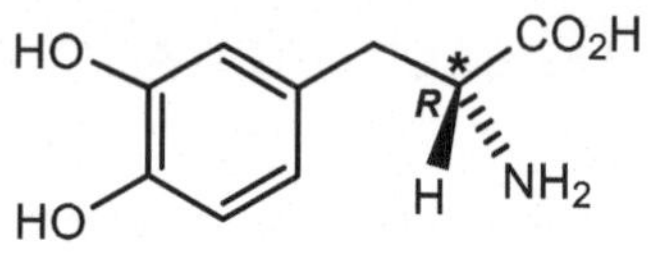

D-Dopa or (*R*)-Dopa

Toxic

An extremely harmful effect of a drug is known as the *thalidomide tragedy*. **Thalidomide** as a **racemic mixture** was prescribed in the late 1950s and early 1960s for the treatment of morning sickness in pregnant women. It was discovered in 1963 that the drug was responsible for the cause of horrible birth defects in thousands of children.

It is the ***R*** **enantiomer** of the drug that can control the morning sickness. But unfortunately the *S* enantiomer present in the racemic mixture caused terrible side effects.

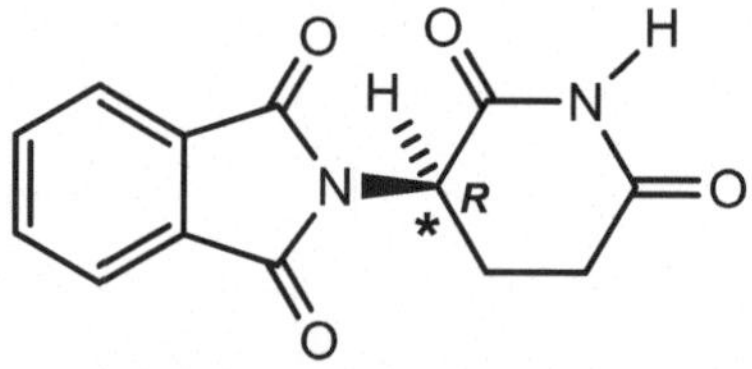

(*R*)-Thalidomide

In general, a single enantiomer of a chiral compound is involved in biological recognition and response. Thus making a chiral compound as a single enantiomer is of extreme importance. The synthesis of one enantiomer **selectively** over another, although difficult, is possible, which is known as the **asymmetric synthesis**.

4.11 STEREOISOMERS OF ALKENES: *CIS* AND *TRANS* DIASTEREOMERS

The carbon atoms of a double bond are sp^2-hybridized, and molecules of alkenes are trigonal planar. Unlike carbon-carbon single bonds, which are free to rotate, **rotation around the carbon-carbon double bond is restricted**. This is because the **rotation** of one carbon of the double bond with respect to the other carbon atom **causes the π bond to break**.

A rotation of 90° will break the π bond completely as there is no net overlap between the two p orbitals.

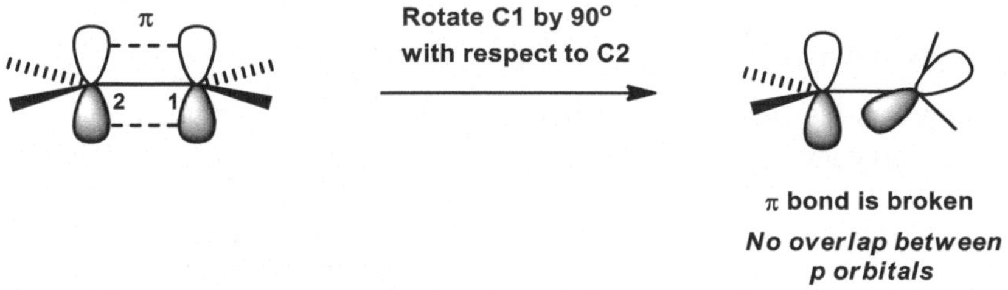

Rotate C1 by 90° with respect to C2

π bond is broken
No overlap between p orbitals

As a result of restricted rotation, double bonds become rigid, and the ligands (atoms or groups) attached to the double bond are fixed in space. Thus alkenes have the possibility of having stereoisomers.

cis-trans Stereoisomerism

Consider, for example, **2-butene**. There are *two* possible **stereoisomers**.

The **stereoisomer** in which two CH_3 ligands (or two H atoms) are on the **same side** of the double bond is called the ***cis*** (from the Latin, meaning on this side), and the **stereoisomer** in which two identical ligands are on **opposites sides** of the double bond is called the ***trans*** (from the Latin, meaning across).

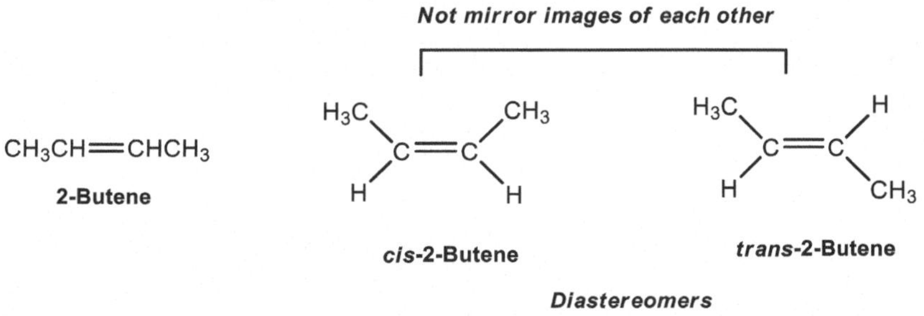

Not mirror images of each other

$CH_3CH{=}CHCH_3$

2-Butene

cis-2-Butene **_trans_-2-Butene**

Diastereomers

The *cis* and *trans* stereoisomers of 2-butene are not mirror images of each other, and are therefore diastereomers.

Both ***cis*-2-butene** and ***trans*-2-butene** have a plane of symmetry, which is the plane of the molecule. Therefore, they are **achiral** and **optically inactive**.

(The *cis–trans* isomerism is also known as the *geometrical isomerism* due to the difference in geometric locations of certain atoms or groups in these isomers. But the term geometrical isomerism is now out of date).

In general, **for an alkene to exhibit *cis–trans* isomerism, *each* of the two carbon atoms of the double bond must bear *two different ligands*.**

However, two pairs of ligands on the two alkene carbon atoms may be the same or different. For 2-butene, both alkene carbon atoms bear the same ligand pair (CH_3 and H).

No cis–trans stereoisomerism

If **one** or **both carbon atoms** of the double bond bear **two identical groups**, the ***cis–trans* isomerism is not possible**. Thus all double bonds do not lead to *cis–trans* stereochemistry.

Consider, for example, **1-butene** in which the alkene carbon (C1) bears two identical H atoms. Therefore, 1-butene cannot exist as a *cis* or *trans* stereoisomer.

Another example is **isobutene** which has no *cis*/*trans* stereoisomers. In this case, both alkene carbons carry identical ligands (C1 has two H atoms and C2 has two CH_3 groups).

Same structures

$\overset{2}{C}H_3CH_2CH{=}\overset{1}{C}H_2$

1-Butene

(two identical H atoms at C1)

$(CH_3CH_2)(H)C{=}C(H)(H) \equiv (H)(CH_3CH_2)C{=}C(H)(H)$

No *cis-trans* isomerism

$CH_3\overset{2}{C}(CH_3){=}\overset{1}{C}H_2$

Isobutene

(two identical H atoms at C1; two identical CH_3 ligands at C2)

$(H_3C)(H_3C)C{=}C(H)(H)$

No *cis-trans* isomerism

PROBLEM 4.11: Which of the following alkenes can exist as *cis* and *trans* diastereomers? Draw their structures.

(a) 1,1,2-Trichloroethene
(b) 1-Chloro-1-butene

Answer

(a) One of the two alkene carbon atoms bears two identical ligands (Cl, Cl). Therefore, 1,1,2-trichloroethane cannot exist as a *cis* or *trans* diastereomer.

$(Cl)(Cl)\overset{1}{C}{=}\overset{2}{C}(H)(Cl)$

C1 is attached to two identical ligands (Cl, Cl)

No cis-trans isomerism

(b) Each of the two carbon atoms of the double bond has two different ligands: one alkene carbon (C1) is attached to Cl and H, while the other (C2) is attached to CH_2CH_3 and H. Thus 1-chloro-1-butene can exist as *cis* and *trans* diastereomers.

$CH_3CH_2\overset{2}{C}H{=}\overset{1}{C}HCl$

At C1: H, Cl (different)

At C2: H, CH_2CH_3 (different)

$(CH_3CH_2)(H)C{=}C(Cl)(H)$

(two H atoms on the same side)

***cis*-1-Chloro-1-butene**

$(CH_3CH_2)(H)C{=}C(H)(Cl)$

(two H atoms on opposite sides)

***trans*-1-Chloro-1-butene**

4.11.1 The *E/Z* Nomenclature

The *cis-trans* nomenclature is applicable when two *identical ligands* are placed on the same side or opposite sides of the double bond. But if **four ligands** attached to the double bond (C=C) are all **different**, the *cis-trans* nomenclature cannot be applied.

Consider, for example, **1-bromo-1-chloropropene** in which four different ligands (CH_3, H, Cl, Br) are attached to the double bond. Clearly, this alkene has **two stereoisomers** because each carbon atom of the double bond is attached to two different ligands, but these stereoisomers cannot be designated as *cis* or *trans*.

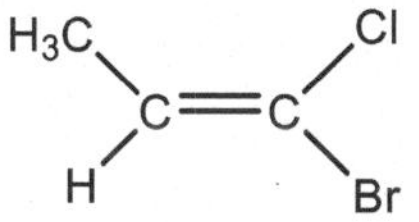

H_3C, H; Cl, Br — C=C

The four ligands are nonidentical;
the *cis-trans* nomenclature does not apply

1-Bromo-1-chloropropene

To assign the stereochemistry of alkenes, a **general system of nomenclature** has been proposed. This is called the ***E/Z* system**, which is based on the Cahn-Ingold-Prelog (CIP) system for the **priority of ligands** (see Section 4.7.1).

The priority order of two ligands attached to *each* alkene carbon is determined; thus the ligands are assigned the **higher priority (*h*)** and the **lower priority (*l*)**.

If **two ligands of higher priority** are on the **same side** of the double bond, the alkene is designated ***Z*** (from the German *zusammen*, meaning together). But if **two ligands of higher priority** are on **opposite sides** of the double bond, the alkene is designated ***E*** (from the German *entgegen*, meaning opposite).

Examples

The following examples illustrate how the stereochemistry of alkenes is assigned as *E* or *Z*. Note that (*Z*)-2-butene and (*E*)-2-butene correspond to *cis*-2-butene and *trans*-2-butene respectively.

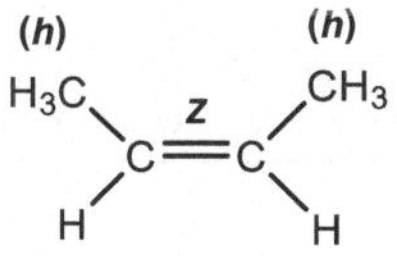

(*h*) H_3C, H; (*h*) CH_3, H — C=C, *Z*

(*Z*)-2-Butene
(*cis*-2-butene)

Priority: **CH_3 > H**
(*h*) (*l*)

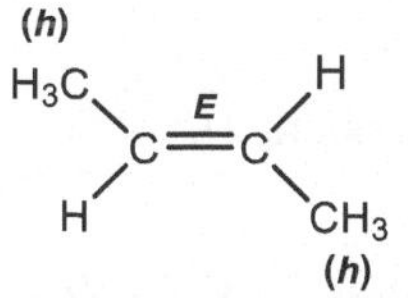

(*h*) H_3C, H; H, CH_3 (*h*) — C=C, *E*

(*E*)-2-Butene
(*trans*-2-butene)

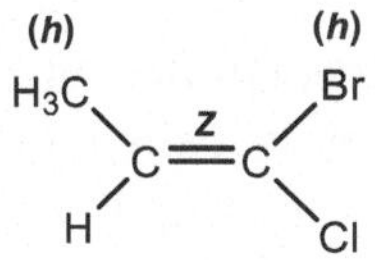

(*h*) H_3C, H; (*h*) Br, Cl — C=C, *Z*

(*Z*)-1-Bromo-1-chloropropene

Priority:
CH_3 > H
Br > Cl

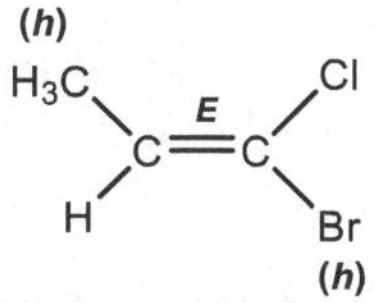

(*h*) H_3C, H; Cl, Br (*h*) — C=C, *E*

(*E*)-1-Bromo-1-chloropropene

PROBLEM 4.12: Assign *E*/*Z* designations to the following molecules:

(a) $H_3C(H)C{=}N{-}OH$

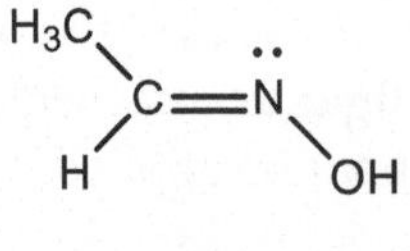

(b) $H_3C(H)C{=}C(H){-}C(H){=}C(CH_3)H$

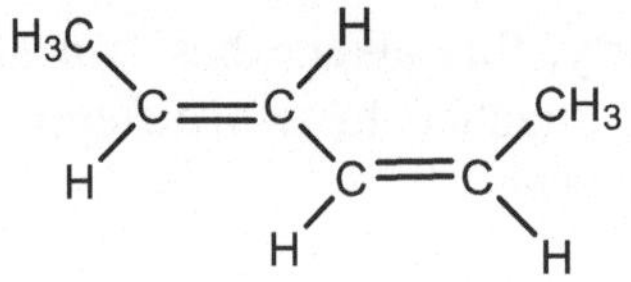

Answer

(a) The *E*/*Z* nomenclature also applies to the stereochemistry of C=N double bond. Here the lone pair on nitrogen has the *least priority* as it may be considered as an imaginary atom with atomic number zero (0). The oxime is designated (*E*).
(An older nomenclature for oximes is *syn*/*anti* based on the relationship between the OH and the alkyl group; accordingly, the oxime is *anti*).

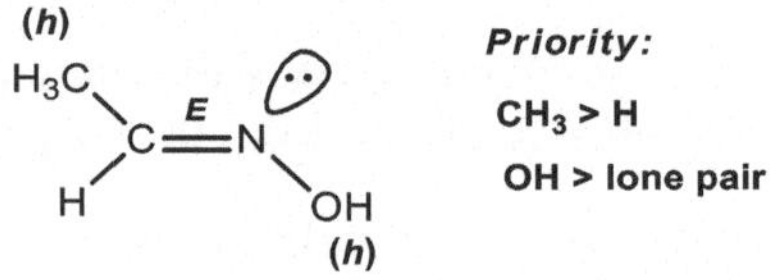

Priority:

CH_3 > H

OH > lone pair

(*E*)-Acetaldoxime

(b) The *E*/*Z* nomenclature for a diene is assigned for *each double bond* treating it separately, as shown below. The stereoisomer of 2,4-hexadiene is designated (2*E*,4*Z*).

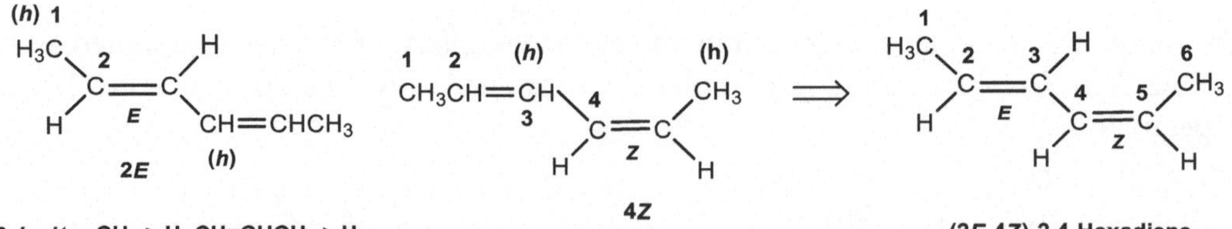

2*E*   4*Z*   (2*E*,4*Z*)-2,4-Hexadiene

Priority: CH_3 > H; $CH{=}CHCH_3$ > H

4.12 STEREOISOMERS OF CYCLOALKANES: *CIS* AND *TRANS* DIASTEREOMERS

A **cycloalkane ring** has *two* distinct **faces or sides**. As a result, **two substituents** on a cycloalkane can be either on the **same side** of the ring or on **opposite sides** of the ring. Thus **cycloalkanes** have the possibility of having **stereoisomers**.

Consider, for example, **1,2-dimethylcyclohexane**. The **stereoisomer** in which the two CH_3 groups are on the *same side* of the cyclohexane ring is called the ***cis***, and the **stereoisomer** in which the CH_3 groups are on *opposite sides* of the ring is called the ***trans***.

The *cis* and *trans* stereoisomers of cycloalkanes are generally drawn in the **Haworth projection formulas** or in the **perspective or three-dimensional formulas**.

Figure 4.3 shows *cis* and *trans* isomers of 1,2-dimethylcyclohexane in both formulas.

Haworth projection formulas:

Perspective or three-dimensional formulas:

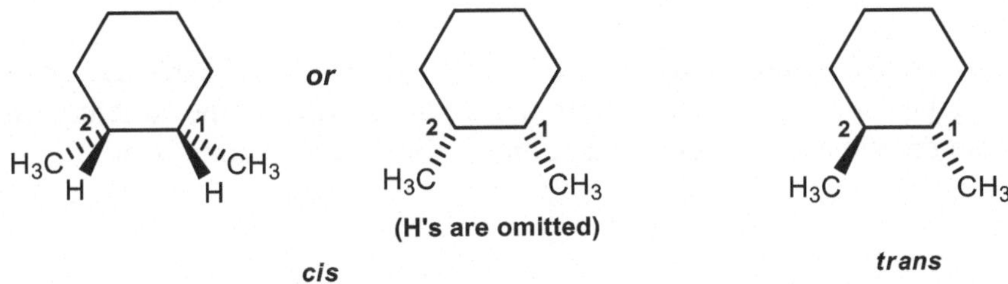

Figure 4.3: Haworth projection formulas and perspective formulas for *cis* and *trans* stereoisomers of 1,2-dimethylcyclohexane. In the perspective formulas, H atoms are usually not shown.

The ***cis* and *trans* stereoisomers** are **not mirror images** of each other, and hence they are **diastereomers**. Thus *cis*-1,2-dimethylcyclohexane and *trans*-1,2-dimethylcyclohexane are two diastereomers of 1,2-dimethylcyclohexane.

Not mirror images of each other

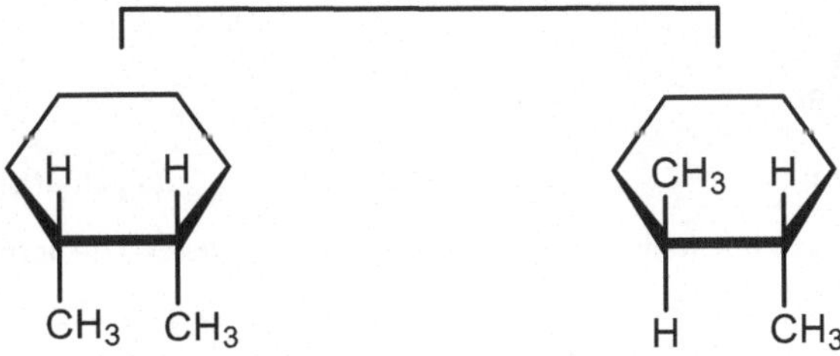

***cis*-1,2-Dimethylcyclohexane** ***trans*-1,2-Dimethylcyclohexane**

Diastereomers

The *cis* diastereomer has a plane of symmetry. Therefore, it is achiral and optically inactive.

In contrast, the *trans* diastereomer has no plane of symmetry and is chiral. It has two enantiomers, each of which is optically active.

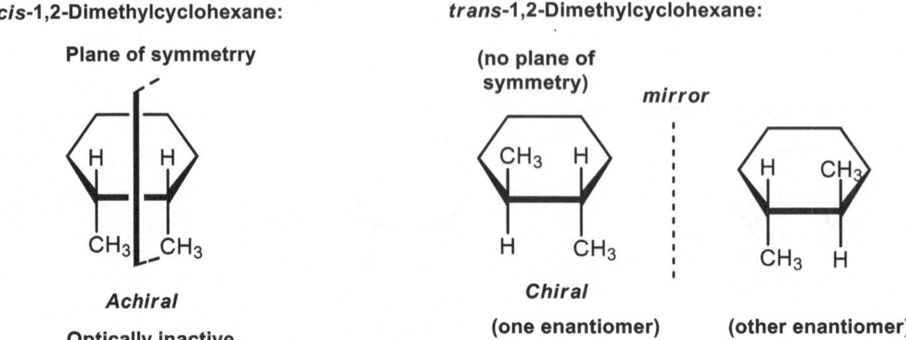

Note: Unlike *cis-* or *trans*-2-butene, *cis-* or *trans*-1,2-dimethylcyclohexane contains **chiral centres** (C1 and C2). The two chiral centres are *like* because they bear the same set of four different ligands [$-CH(CH_3)CH_2CH_2CH_2CH_2$, $-CH_2CH_2CH_2CH_2CH(CH_3)$, $-CH_3$, $-H$]. Since *cis*-1,2-dimethylcyclohexane is achiral and optically inactive, it is also called *meso*.

PROBLEM 4.13: Draw structures of *cis*-1,3-dibromocyclopentane and *trans*-1,3-dibromocyclopentane in perspective formulas. Indicate whether the structures are chiral or achiral. Also draw the structures of enantiomers for the chiral compound.

Answer

cis-1,3-Dibromocyclopentane has a plane of symmetry. Therefore, it is achiral and optically inactive. The *cis* isomer is also called *meso* isomer.

On the other hand, *trans*-1,3- dibromocyclopentane has no plane of symmetry and is chiral. It exists as a pair of enantiomers.

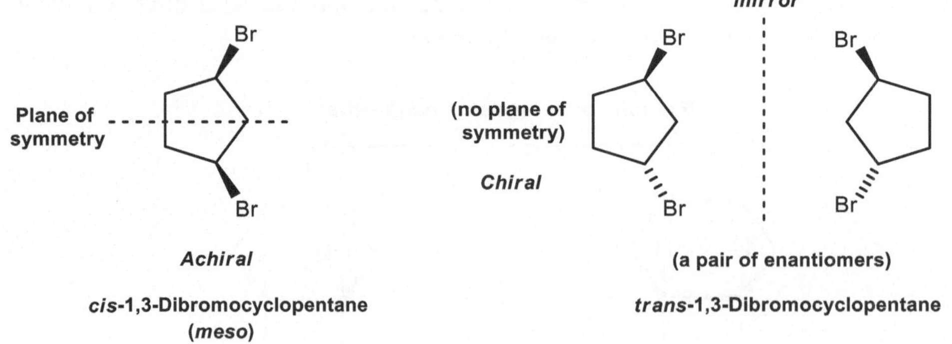

4.13 CONFORMATION OF ORGANIC MOLECULES

A **C–C single bond** is a σ bond that is **cylindrically symmetrical** about the line joining the two nuclei. Rotation about the C–C single bond does not affect the σ overlap, and C–C single bonds are therefore capable of rotating.

Rotation about the C–C single bond implies that *atoms or groups at one end of the single bond* are **rotated** *with respect to atoms or groups at the other end.* This leads to different arrangements of atoms or groups in space. **The arrangements of atoms or groups in space that result from rotation about single bonds are called conformations**.

A convenient way to draw different conformations is to use **Newman projection formulas**. Figure 4.4 shows how a Newman projection formula of a general molecule can be drawn from its three dimensional flying wedge formula.

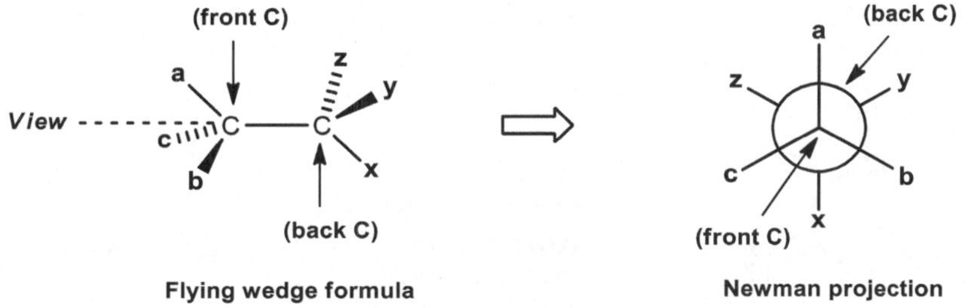

Figure 4.4 Drawing a Newman projection for a general molecule. Three groups (**a**, **b** and **c**) are bonded to front carbon, and three groups (**x**, **y** and **z**) are bonded to back carbon.

To draw the **Newman projection**, the three-dimensional molecule is **viewed** from one end directly **along the C–C bond axis.**

The **front carbon atom** of the C–C bond is represented by a **central point** in the Newman formula from which **three bonds** of the front carbon are drawn. The **back carbon atom** of the C–C bond is depicted as a **circle** from whose perimeter **three bonds** of the back carbon emerge. Notably, the C–C bond that we are looking along is not visible in the Newman projection.

We shall see in the following sections that Newman projections of different conformations can be drawn in a straightforward manner just by *rotating one carbon at a time*, either the front carbon or the back carbon.

4.13.1 Conformations of Ethane: Eclipsed and Staggered

The simplest molecule with a C–C single bond is **ethane**. The **rotation** of one methyl group with respect to the other by **any angle** (even a fraction of a degree) gives a **different conformation**. Thus an **infinite number of conformations** are possible.

All these conformations are *not* of equal energy or stability. The two most important conformations of ethane are called the **eclipsed conformation** and the **staggered conformation**.

In an **eclipsed conformation**, the C–H bonds on the front carbon and the C–H bonds on the back carbon are **directly opposite** to each other, i.e. they **eclipse** each other. Thus the **angle** between the eclipsed C–H bonds is **0 degrees**.

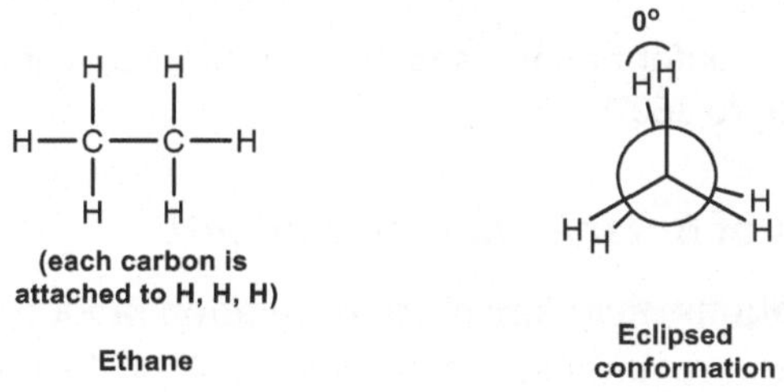

Note that in the eclipsed conformation, the H atoms on the back carbon cannot be seen. In order to see them, they are *artificially rotated* a little in the drawing.

Starting from the eclipsed conformation, a **rotation** of the back carbon by an angle of **60°** generates a **staggered conformation**, which allows a maximum possible separation of the C–H bonds on the front carbon and the back carbon.

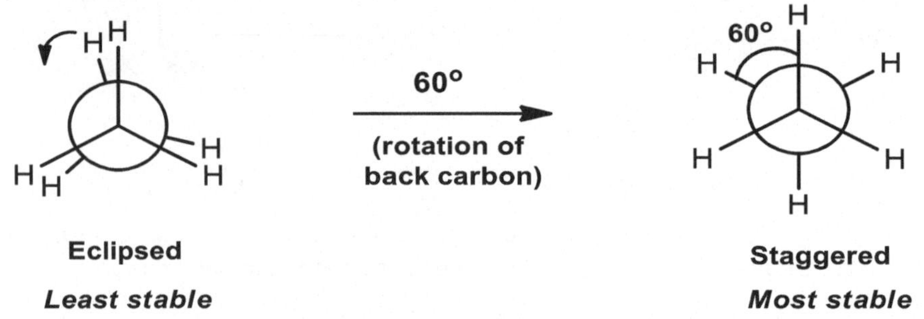

60°

(rotation of back carbon)

Eclipsed

Least stable

Staggered

Most stable

The staggered conformation is the **most stable** conformation of ethane, while the eclipsed conformation is the **least stable** conformation of ethane. At any instant, most of the molecules will be in the staggered conformation.

Figure 4.5 shows that in a complete rotation circuit (0° to 360°), there are **three eclipsed conformations** of equal energy and **three staggered conformations** having the same energy.

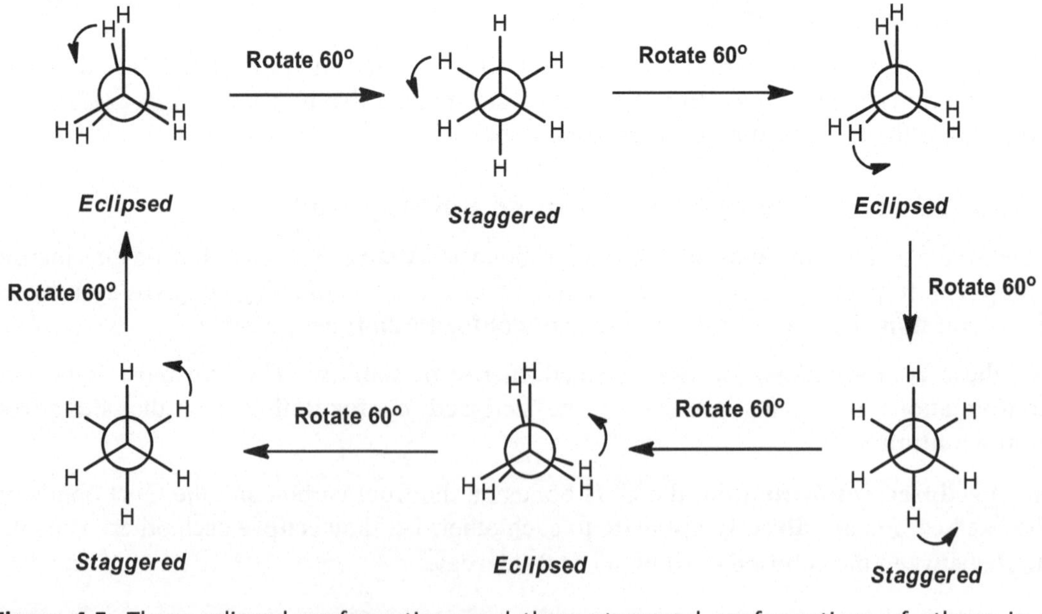

Eclipsed — Rotate 60° → Staggered — Rotate 60° → Eclipsed — Rotate 60° → Staggered — Rotate 60° → Eclipsed — Rotate 60° → Staggered — Rotate 60° → Eclipsed

Figure 4.5: Three eclipsed conformations and three staggered conformations of ethane in a complete rotation circuit (0° to 360°).

4.13.2 Conformations of Butane: *gauche* and *anti*

With **butane**, conformations may arise from rotation about a **terminal C–C bond** (C1–C2 or C3–C4) or from rotation about the **central C–C bond** (C2–C3).

Conformations due to rotation about the C1–C2 bond

Conformations from **rotation about C1–C2 bond** in butane are very similar to those of ethane. Here the eclipsed or staggered conformation has an ethyl group (CH_2CH_3) in place of a H atom for ethane.

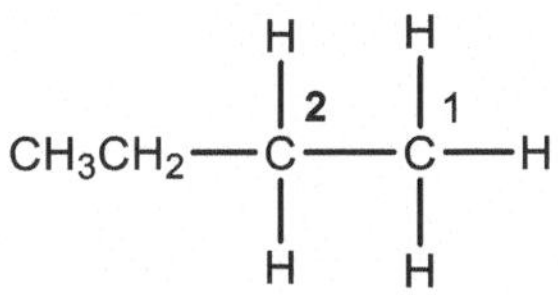

Butane

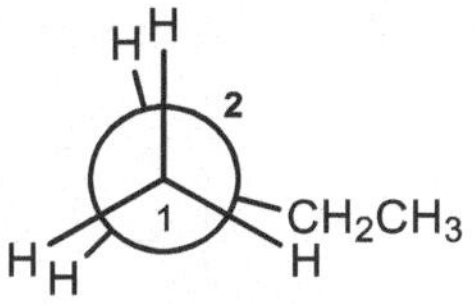

Eclipsed

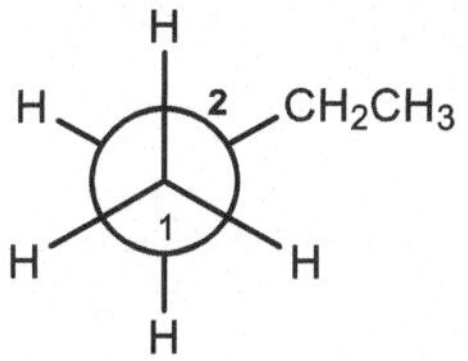

Staggered

Conformations due to rotation about the C2–C3 bond

For butane, conformations arising from rotation about the central C2–C3 bond are more important. Here the situation is more complex than that with ethane.

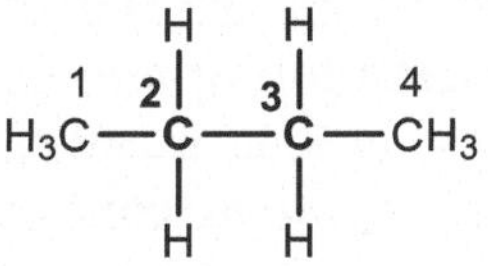

C2 (or C3) is attached to CH_3, H and H

Butane

Unlike ethane, the three eclipsed conformations of butane are *not* of equal energy, and the three staggered conformations are also *not* of the same energy.

The **three eclipsed conformations** of butane consist of a conformation **I** of *highest energy* and two conformations **II** and **III** of *equal energy*, as shown below.

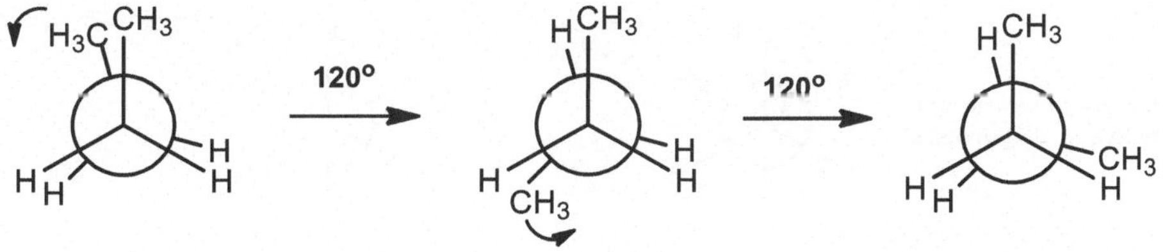

120°

120°

I
Eclipsed

II
Eclipsed

III
Eclipsed

Highest energy

II and III are of equal energy

A **120° rotation** converts one eclipsed conformation into another eclipsed conformation (cf. Figure 4.5). The eclipsed conformation **I** has the highest energy because it involves a severe eclipsing of two large methyl groups (CH_3/CH_3 eclipsing). The eclipsed conformations **II** and **III** have the same energy because *each* involves CH_3/H eclipsing interactions that are less severe than CH_3/CH_3 eclipsing.

The **staggered conformations** of butane are of lower energy, and are most important of all the conformations. At any instant, butane exists mostly in staggered conformations, and they are called **conformers**.

The **three** staggered conformations of butane consist of a conformer **IV** of *lowest energy* and two conformers **V** and **VI** of equal energy, as shown below. A **120° rotation** converts one staggered conformation into another staggered conformation.

CH_3 H H H H CH_3 — 120° → CH_3 H CH_3 H H H — 120° → CH_3 H_3C H H H H

IV **Staggered** ***anti*** **V** **Staggered** ***gauche*** **VI** **Staggered** ***gauche***

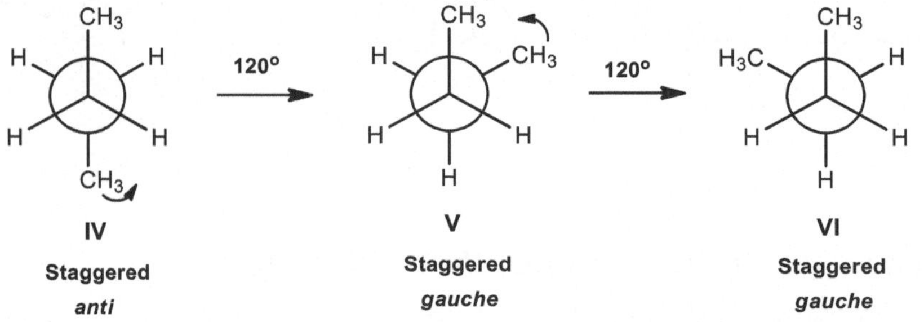

In the staggered conformation **I**, the two methyl groups are 180 degrees apart from each other, and this conformation is called ***anti*** **conformation** or ***anti*** **conformer**. In the other staggered conformations **V** and **VI**, the two methyl groups are 60 degrees apart from each other, and they are called ***gauche*** **conformations** or ***gauche*** **conformers**.

The ***anti*** **conformer (I)** is **most stable** because there is **no repulsive steric interaction** between the two methyl groups as they are far apart.

But in a ***gauche*** **conformer (V or VI)**, the methyl groups are close enough that they interact **sterically** with each other. This kind of steric or van der Waals interaction in a staggered conformation is known as the ***gauche*** **interaction**. The repulsive van der Waals interaction acts to increase the energy of a conformer. Therefore, **a *gauche* conformer is less stable than *anti* conformer**.

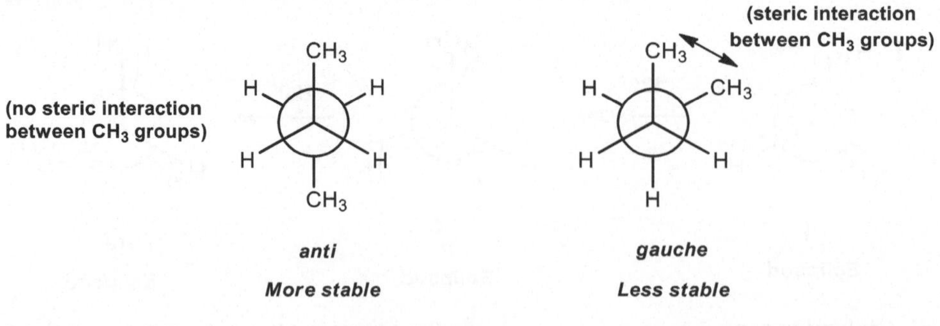

(no steric interaction between CH_3 groups) CH_3 H H H H CH_3 — ***anti*** ***More stable***

(steric interaction between CH_3 groups) CH_3 H CH_3 H H H — ***gauche*** ***Less stable***

The **energy difference** between a *gauche* conformer and *anti* conformer is 0.9 kcal mol^{-1}, and **butane** exists as a **mixture of *anti* (70%) and *gauche* (30%)**. However, *anti* and *gauche* conformers are **not separable** at normal temperatures because they are rapidly interconverted by rotation about the single bond.

Note: The two *gauche* conformers (V and VI) are **mirror images** of each other, and they are **not superimposable**. Thus the *gauche* conformer is **chiral**. The two *gauche* forms are different molecules (called conformational enantiomers). They have the same energy. On the other hand, the *anti* conformer (IV) has a centre of symmetry; it is **achiral**. The *anti* and *gauche* forms are not mirror images of each other, and they represent conformational diastereomers.

4.13.3 Conformations of Cyclohexane: Chair Conformation and Ring Inversion

Cyclohexane is typically drawn *as if it were planar*. But all carbon atoms of cyclohexane are **sp³**-hybridized, and the cyclohexane ring constituted of six C–C bonds is actually **not planar**. A nonplanar structure of cyclohexane ring represents a **conformation**.

The **most stable** conformation of cyclohexane is the **chair conformation** in which the bond angles are close to a tetrahedral angle of 109.5°. This conformation is so called because it looks like a reclining chair.

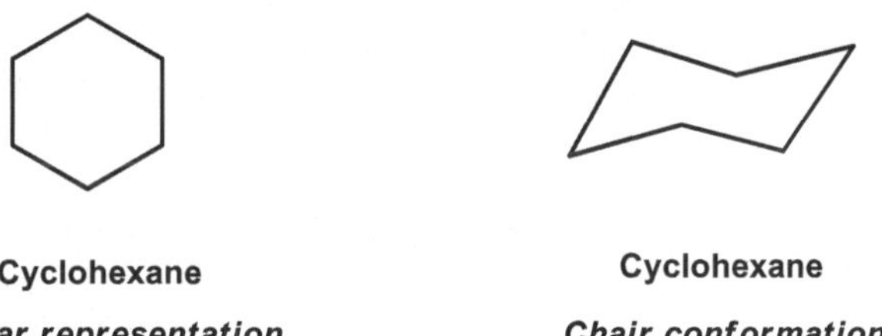

Cyclohexane — ***Planar representation***

Cyclohexane — ***Chair conformation***

Other conformations for cyclohexane include boat, twist-boat and half-chair conformations, which are beyond our scope at this level.

Drawing chair conformation

Drawing a chair conformation can be somewhat difficult at first. It will be easier if the chair is drawn with *parts* rather than with one continuous line, as shown below.

First, **draw two lines** (1, 2) from one end (like a *V*). Next, **draw two parallel lines** (3, 4) of equal length. Finally, add the **last two lines** (5, 6) (like an inverted *V*) so that these lines are parallel to the first pair of lines (5 is parallel to 2; 6 is parallel to 1).

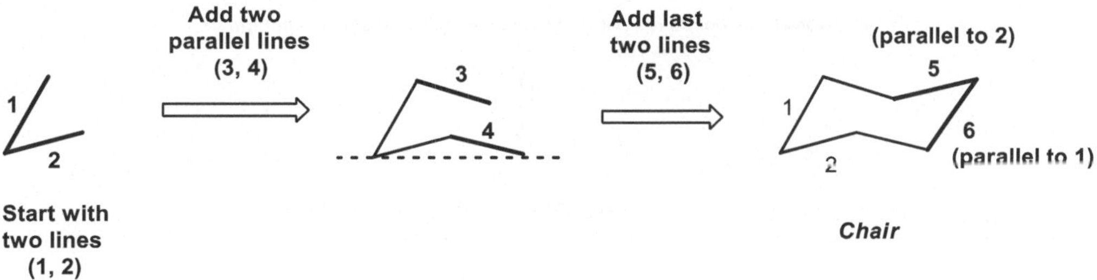

Start with two lines (1, 2)

Chair

Axial and equatorial hydrogen atoms

The **chair conformation** of cyclohexane contains two different kinds of hydrogen atoms: **axial hydrogen atoms** and **equatorial hydrogen atoms**.

Axial hydrogen atoms are hydrogen atoms that are oriented in a direction parallel to an *imaginary axis* which is perpendicular to the average plane of the ring. There are **six** axial hydrogen atoms that stick **up** and **down** alternately.

At a **carbon** atom on the chair that points **up**, the **axial hydrogen** is **up**; similarly, at a **carbon** that points **down**, the **axial hydrogen** is **down**.

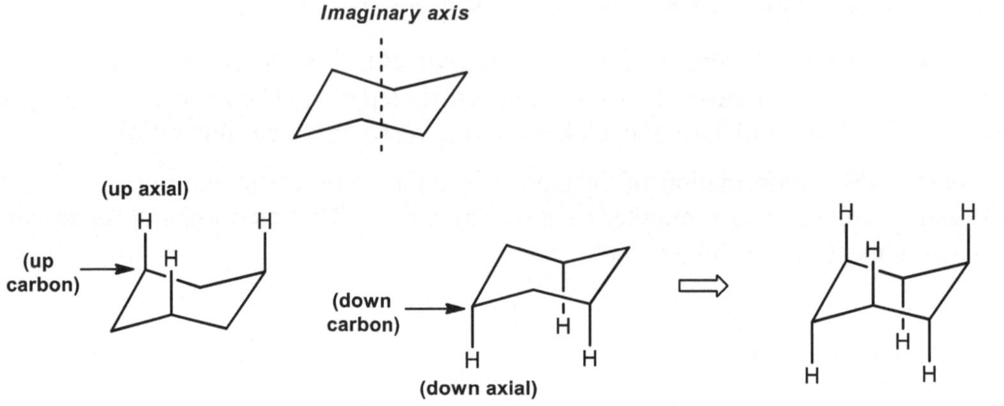

Imaginary axis

(up axial)

(up carbon)

(down carbon)

(down axial)

Three axial hydrogens (up)

Three axial hydrogens (down)

All six axial hydrogens (alternately up and down)

Six other hydrogen atoms, one on each carbon, that lie roughly in the plane of the ring are called **equatorial**, by analogy with the equator of the earth. The **equatorial hydrogen atoms** slope **up** and **down** alternately with respect to the mean plane of the ring.

Notably, the **equatorial C–H bonds** on the *diagonally opposite carbon atoms* are not only mutually parallel but they are also parallel to the intervening C–C bonds.

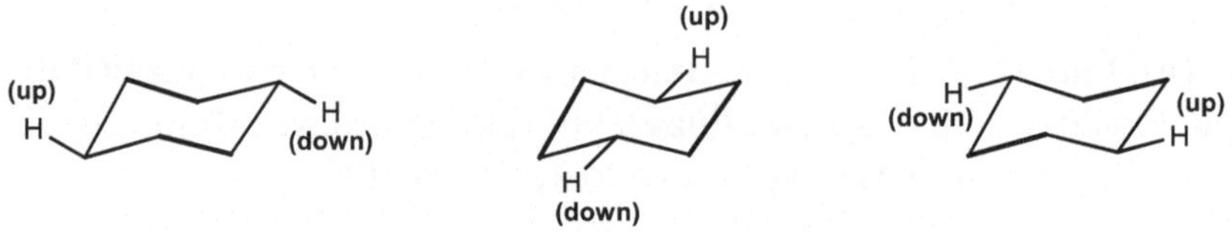

The equatorial C-H bonds are parallel to the C-C bonds that are highlighted

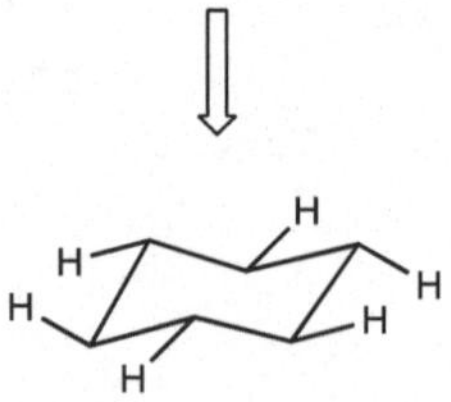

All six equatorial hydrogens (alternately up and down)

The following is the **complete chair conformation** with all axial and equatorial hydrogen atoms:

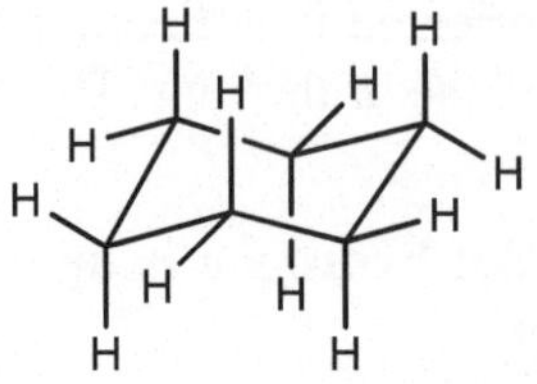

Complete chair conformation with all axial and equatorial hydrogens

Ring flipping or ring inversion

A chair conformation of cyclohexane can readily **flip** to an alternative chair conformation. This process is called **ring flipping**. The ring flipping occurs by rotating bonds when **up** carbons go **down**, and **down** carbons go **up**.

In effect, the **ring flipping inverts the chair** and it is also called **ring inversion**. (Try yourself the process of ring inversion with the help of molecular model).

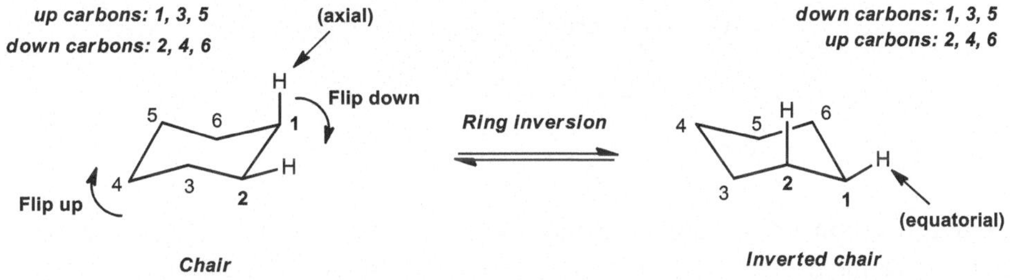

up carbons: 1, 3, 5
down carbons: 2, 4, 6
(axial)
Flip down
Flip up
Chair
Ring inversion
down carbons: 1, 3, 5
up carbons: 2, 4, 6
(equatorial)
Inverted chair

As shown, an axial hydrogen at C1 becomes equatorial and an equatorial hydrogen at C2 become axial as a result of ring inversion. This is true for all axial and equatorial bonds. Thus, **the ring inversion causes all axial hydrogens to become equatorial and all equatorial hydrogens to become axial.**

SUPPLEMENTARY PROBLEMS

4.14 Predict whether the two structures in each of the following pairs represent identical compounds, enantiomers or diastereomers.

(a) and

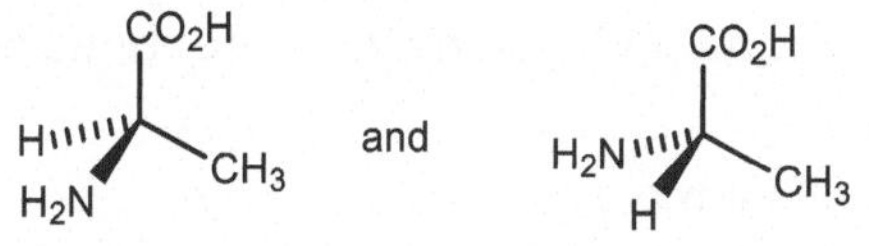

(b) and

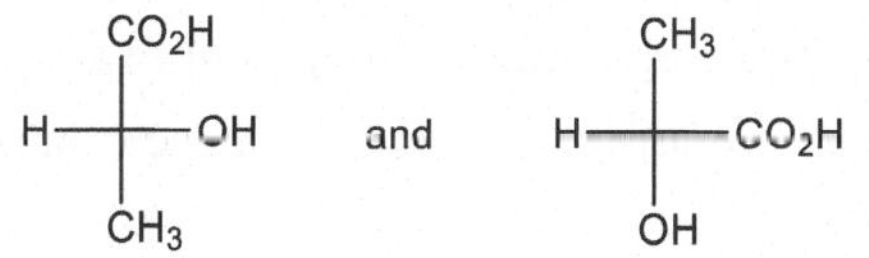

(c) and

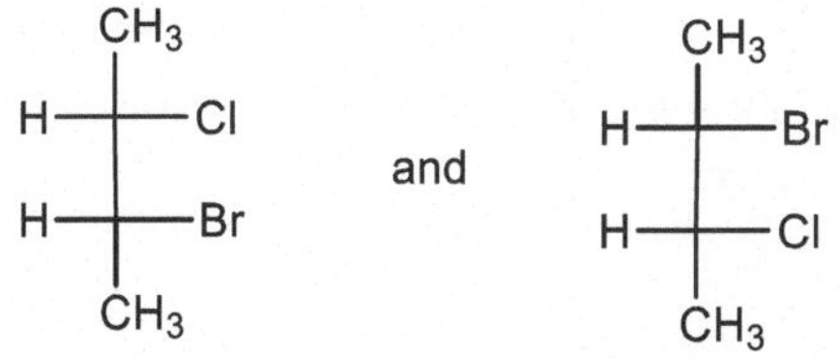

(d) and

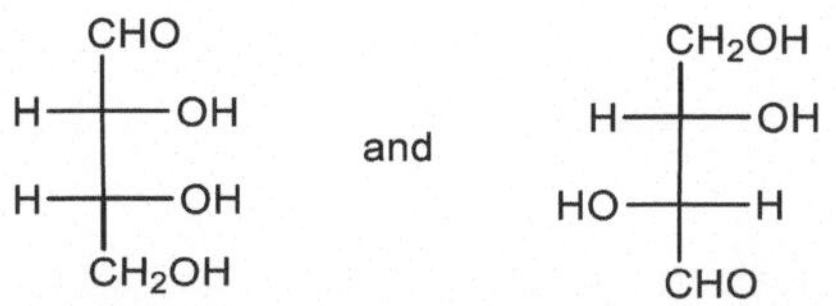

[*Ans:* (a) enantiomers (b) identical (c) enantiomers (d) diastereomers]

4.15 What is the minimum number of carbon atoms that an alkane needs to become chiral? How many chiral alkanes are possible with this minimum number of carbon atoms? Draw flying wedge formulas for the enantiomers of each chiral alkane.

[*Hints:* consider four ligands: hydrogen, methyl, ethyl and propyl (or isopropyl); two possible chiral alkanes]

4.16 Which ligand in each of the following pairs has higher priority?

(a) $-CH{=}CH_2$ and $-C{\equiv}CH$

(b) $-C_6H_5$ (phenyl) and $-C(CH_3)_3$

4.17 A hydrocarbon of molecular formula C_6H_8 exists as a pair of enantiomers. Draw flying wedge formulas for the enantiomers, and assign the *R* or *S* configuration to the chiral centre.

4.18 Draw Fischer projections for the two diastereomers of 2,3-dibromopentane. Designate them by the *erythro*/*threo* nomenclature. Indicate whether each diastereomer is chiral or achiral. Draw the enantiomers (if any) and indicate the total number of stereoisomers possible for 2,3-dibromopentane.

4.19 Oxidation of *R*-(+)-glyceraldehyde with HgO gives (−)-glyceric acid. What is the configuration of the product (*R* or *S*)? Comment on the stereochemical result.

[*Ans: R*; configuration is retained if no bond to the chiral centre is broken in the reaction]

4.20 Both *cis* and *trans* isomers of 1,4-dimethylcyclohexane are optically inactive. Explain.

[*Hints:* look for a plane of symmetry in each isomer]

4.21 Draw Newman projections for a *gauche* conformer and the *anti* conformer of ethylene glycol ($HOCH_2CH_2OH$). Which of the conformers is more stable and why?

[*Hints:* consider intramolecular hydrogen bonding for the relative stability of the conformers]

4.22 Draw two chair conformations for the ring inversion of methylcyclohexane.

[*Hints:* one chair has the axial methyl group and the other chair has the equatorial methyl group]

CHAPTER 5

Physical Properties, Acidity and Basicity

IN THIS CHAPTER

- Intermolecular forces: Ion-ion, ion-dipole, dipole-dipole, hydrogen bond, van der Waals forces and polarizability
- Melting point and boiling point; solubility, solvent polarity, polar protic and polar aprotic solvents
- Dipole moment and its prediction: σ, π and lone pair moments, group moments, resultant or net dipole moment
- Acids and bases: Brønsted-Lowry theory, acidity, basicity and pK_a; Lewis theory
- Comparing acidities of organic compounds: Carboxylic acids, alcohols, phenols, hydrocarbons, carbonyl compounds (aldehydes, ketones, esters) and β-dicarbonyl compounds
- Comparing basicities of organic compounds: Bases with different basic atoms, anionic oxygen bases, aliphatic and aromatic amines

A **physical property** of an organic compound does not involve a change in its composition, and is a *characteristic* of the compound. All organic compounds have distinct physical properties such as melting point, boiling point, solubility or dipole moment.

Suppose you intend to perform a synthesis of an organic compound. The first experimental observation you would like to note is whether the product is a solid or liquid at room temperature and 1-atm pressure.

How does a solid differ from a liquid? The packing of molecules in the solid crystalline state is *well-ordered*, while it is *more random* in the liquid state. The forces between molecules, called **intermolecular forces**, bind the molecules together. A large amount of thermal energy is needed to break up the intermolecular forces, which leads to the changes of phases from solid to liquid and liquid to gas.

The temperature at which solid → liquid transition occurs is called the **melting point**, and the temperature at which liquid → gas transition occurs is called the **boiling point**. These are characteristic physical properties of organic compounds, which are useful to identify them.

Another useful physical property is the **solubility**, which refers to the dissolution of an organic compound in a liquid solvent. The process of dissolution involves the destruction of ordered crystalline structure of the solid and the formation of more disordered arrangements of molecules in solution. If the product of a synthesis is a solid, it can be isolated by *crystallization* based on different solubilities of the substance in different solvents.

A covalent bond between two atoms of different electronegativity is *polar* because the atom with greater electronegativity draws the electron pair closer to it. This gives a partial positive charge (δ+) to the less electronegative atom and a partial negative charge (δ–) to the more electronegative atom, thereby forming a *dipole*. Any diatomic molecule with a polar bond has a **dipole moment**. However, the dipole moment is a *vector quantity*; thus polyatomic molecules with polar bonds may or may not have dipole moments depending on the *vector sum* of the individual bond moments. The dipole moment of an organic compound can be measured experimentally.

A **chemical property** involves a chemical change or reaction. **Acidity** and **basicity** are two chemical properties that are extremely important. Almost every reaction in organic chemistry involves acid–base chemistry. Common acids such as hydrochloric acid (HCl) and sulphuric acid (H_2SO_4) are inorganic acids, and common bases such as sodium hydroxide (NaOH) and potassium hydroxide (KOH) are inorganic bases. In comparison, organic acids or bases are generally much weaker. In this chapter, we will describe organic acids and bases, and their relative acidity or basicity based on the structure of compounds.

5.1 INTERMOLECULAR FORCES

The intermolecular forces that bind molecules or ions together are attractive forces. These include ion–ion forces, ion–dipole forces, dipole–dipole forces, hydrogen bonds, and van der Waals forces.

Ion–ion forces

Ion–ion forces are electrostatic forces that occur between positive and negative ions.

For example, in **crystalline sodium acetate** (CH_3CO_2Na), each sodium ion (Na^+) is surrounded by negatively charged acetate ions ($CH_3CO_2^-$), and each acetate ion is surrounded by positively charged sodium ions. Such forces that hold the positive and negative ions together in the crystal lattice are very strong.

Ion–dipole forces

Ion–dipole forces occur between an ion and a dipolar molecule.

For example, when sodium acetate is dissolved in water, **positive sodium ions** are surrounded by **polar water molecules** in three dimensions with the negative end of water dipole towards the positive sodium ion. Sodium ions are thus **solvated** by water molecules by ion–dipole forces.

Note that the water dipole contains a partial positive charge (δ+) at one end and a partial negative charge at the other end (δ–) in contrast to a full positive charge (+) on sodium ion.

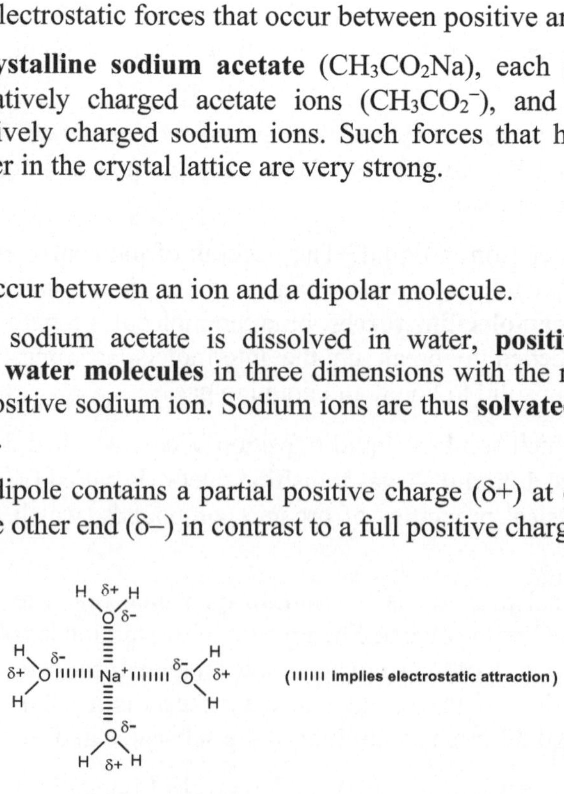

(IIIIII implies electrostatic attraction)

Ion-dipole interactions

The negative acetate ions are also solvated by water molecules; here the solvation occurs primarily by forming strong hydrogen bonds (see later).

Dipole–dipole forces

Dipole–dipole forces occur between molecules that have a *permanent* dipole.

For example, **acetone** (CH_3COCH_3) has a permanent dipole because the carbonyl group (C=O) is highly polarized. In the **liquid state**, dipole-dipole attractive forces occur between acetone molecules when the negative end of one molecule is oriented towards the positive end of another.

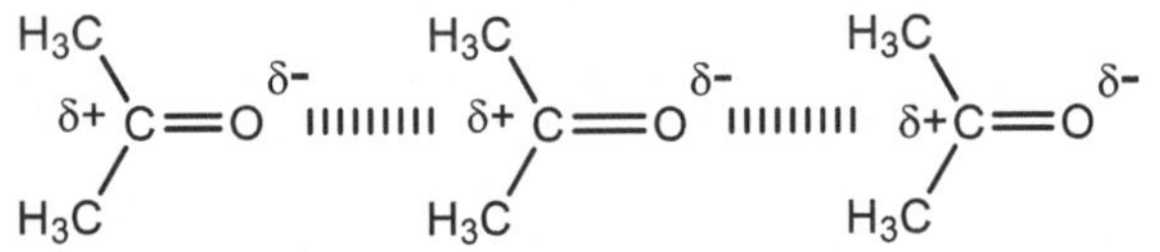

Dipole-dipole interactions

Hydrogen bond

A **hydrogen bond** is a kind of strong dipole-dipole attraction that occurs between a **hydrogen atom** (usually bonded to highly electronegative atoms such as O, N or F) and the **unshared electron pair or lone pair** on other such electronegative atoms. The hydrogen atom is positively polarized and the electronegative atom is negatively polarized.

As shown below, molecules of **methyl alcohol** (CH_3OH) can form hydrogen bonds (indicated by dotted lines) to each other. A hydrogen bond is much stronger than ordinary dipole-dipole attractions but weaker than a covalent bond.

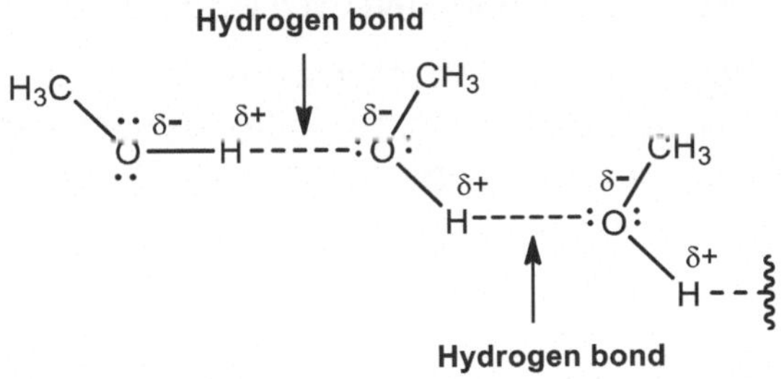

Van der Waals forces

Nonpolar molecules can associate with each other by attractive intermolecular forces called **van der Waals forces**. The vander Waals forces are also known as **London forces** or **dispersion forces**. These forces are responsible for nonpolar compounds (such as methane) to become a liquid or solid at low temperatures.

In a nonpolar molecule, the average distribution of electronic charge is uniform over a period of time. But as electrons move, the electronic charge may not be uniformly distributed *at any given instant*. As a result, a small **temporary dipole** will be created. This temporary dipole in one molecule can then **induce** an opposite dipole in a surrounding molecule and thereby create an attractive force between them.

Notably, the temporary dipoles would change constantly as electrons move, but the net result is to produce an attraction between nonpolar molecules.

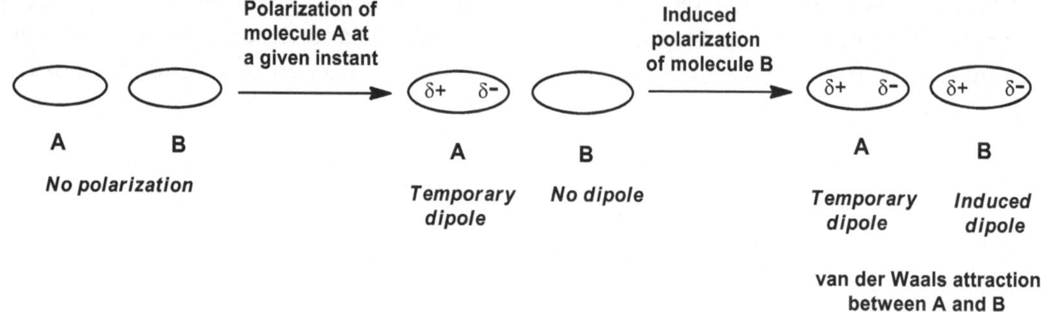

An important factor that influences the magnitude of van der Waals forces is the relative polarizability of the atoms involved. Polarizability refers to the ability of the electrons of an atom to distort in a changing electric field.

When the *valence electrons* of an atom are relatively *far* from the nucleus, they are loosely held, and such atoms have *greater* polarizability.

For example, **polarizability** of halogen atoms increases in the order **F < Cl < Br < I**. Since iodine atom is largest in the series, its valence electrons are most loosely held, and it has the highest polarizability. On the other hand, the smallest fluorine atom has the lowest polarizability because its valence electrons are most tightly held by the nucleus.

5.2 MELTING POINT, BOILING POINT AND SOLUBILITY

Melting point and boiling point are two characteristic physical properties of organic compounds, which are used to identify them. Solubility is another useful physical property. Based on different solubility behaviour, organic compounds can be isolated or purified by crystallization.

5.2.1 Melting Point

Ionic compounds have high melting points. For example, sodium acetate that has a highly ordered lattice structure of sodium ions and acetate ions held together by strong ion–ion forces has a high melting point of 324°C. This is because a large amount of thermal energy is needed to break up the ordered structure of the crystalline solid into the disordered open structure of the liquid.

Covalent organic compounds generally have much lower melting points. An important factor that influences the melting points of many organic compounds is the compactness and rigidity of their individual molecules.

More symmetrical and less symmetrical molecules

Organic molecules that are more symmetrical have higher melting points. This is because more symmetrical molecules would pack more closely in the crystalline state. Consequently, attractive intermolecular forces between molecules are greater, and melting points are higher.

Branching of chains can give highly **symmetrical** structures that lead to abnormally high melting points.

Consider, for example, two isomeric alkanes, **2,2,3,3-tetramethylbutane** and **octane**. Since 2,2,3,3-tetramethylbutane has a highly branched and symmetrical structure, its melting point (mp) is abnormally high (94-97°C) compared to the straight chain and less symmetrical octane (mp, –57°C).

Similarly, between two isomeric alcohols (*tert*-butanol and butanol), ***tert*-butanol** which has a branched and more symmetrical structure has much higher melting point (25°C) than the straight chain and less symmetrical **butanol** (mp, –90°C).

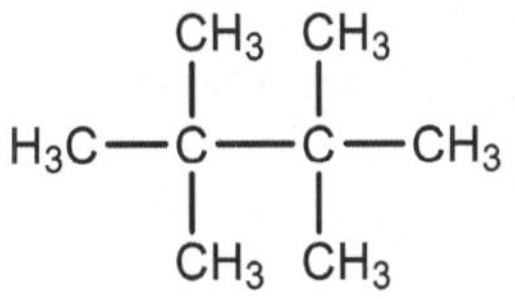

$(CH_3)_3C{-}C(CH_3)_3$

2,2,3,3-Tetramethylbutane
(mp, 94-97°C)

$CH_3CH_2CH_2CH_2CH_2CH_2CH_2CH_3$

Octane
(mp, -57°C)

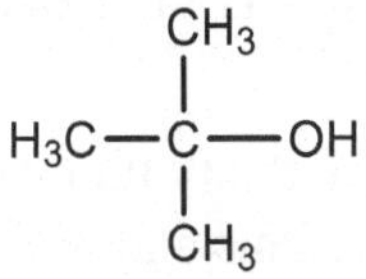

$H_3C{-}C(CH_3)_2{-}OH$

***tert*-Butanol**
(mp, 25°C)

$CH_3CH_2CH_2CH_2OH$

Butanol
(mp, -90°C)

Cycloalkanes also have much higher melting points than the corresponding alkanes because cycloalkanes have **more compact and symmetrical structures**.

For example, **cyclohexane** melts at 6.5°C whereas **hexane** melts at –95°C.

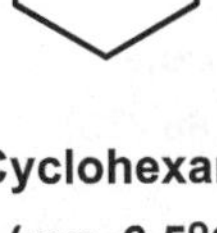

Cyclohexane
(mp, 6.5°C)

$CH_3CH_2CH_2CH_2CH_2CH_3$

Hexane
(mp, -95°C)

Straight chain alkanes

For **straight chain alkanes**, there is no smooth increase in melting points with increase in molecular weight.

To illustrate, **propane** has a lower melting point (–188°C) than **ethane** (mp, –183°C) but **butane** (mp, –138°C) melts higher than ethane.

CH_3CH_3	$CH_3CH_2CH_3$	$CH_3CH_2CH_2CH_3$
Ethane	**Propane**	**Butane**
(mp, -183°C)	**(mp, -188°C)**	**(mp, -138°C)**

Ethane melts higher than propane ***Butane melts higher than ethane***

Alkane chains with even number of carbon atoms are packed more closely in the solid state than alkane chains with odd number of carbon atoms. So, ethane with *two* carbon atoms melts higher than propane with *three* carbon atoms.

However, since both ethane and butane have even number of carbon atoms, there is an increase in melting point with increasing molecular weight from ethane to butane.

5.2.2 Boiling Point

The boiling point of a liquid refers to the liquid–vapour transition. The **boiling point** is the **temperature** at which the *vapour pressure* is equal to the *pressure of the atmosphere* above the liquid.

Therefore, **boiling point depends on the pressure at which it is measured**. The normal boiling point of a liquid is its boiling point at 1-atm pressure.

There are several factors or forces that can operate to influence the boiling point of a liquid. These include van der Waals attractions, dipole–dipole attractions and hydrogen bond.

Role of van der Waals attractions

Nonpolar compounds such as **alkanes** usually boil at low temperatures because the **intermolecular van der Waals attractions** between molecules are weak. Methane, ethane, propane and butane are gases at normal temperature (25°C) and pressure (1 atm).

Boiling points of straight chain alkanes increase with increasing molecular weight.

For example, the boiling point (bp) of pentane is 36.1°C whereas the boiling point of hexane is 68.7°C.

$CH_3CH_2CH_2CH_2CH_3$	$CH_3CH_2CH_2CH_2CH_2CH_3$
Pentane	**Hexane**
(bp, 36.1°C)	**(bp, 68.7°C)**

As hexane molecules are *larger* in size, they have *greater surface areas* than pentane molecules. Thus, intermolecular van der Waals attractions are larger for hexane, and hence greater thermal energy (higher temperature) is required for hexane molecules to escape the liquid surface.

For isomeric alkanes, branching of the carbon chain lowers the boiling point.

For example, hexane boils at 68.7°C, 2-methylpentane with *one branch* boils lower at 60.3°C, and 2,2-dimethylbutane with *two branches* boils still lower at 49.7°C.

$CH_3CH_2CH_2CH_2CH_2CH_3$ — **Hexane** **(bp, 68.7°C)**

$CH_3CH(CH_3)CH_2CH_2CH_3$ — **2-Methylpentane** **(bp, 60.3°C)**

$CH_3C(CH_3)_2CH_2CH_3$ — **2,2-Dimethylbutane** **(bp, 49.7°C)**

Branching of the chain makes a molecule more compact with a *reduced surface area.* Therefore, intermolecular van der Waals attractions between the branched molecules are reduced.

The strength of the van der Waals forces decreases more with increasing branching. This leads to the lowering of boiling point progressively from hexane to 2-methylpentane to 2,2-dimethylbutane.

Role of dipole–dipole attractions

Aldehydes and ketones have higher boiling points than alkanes of the *same* molecular weight.

For example, acetone and butane have the same molecular weight ($M = 58$) but acetone has higher boiling point (56.1°C) than butane (–0.5°C).

$CH_3C(=O)CH_3$ — **Acetone** **(bp, 56.1°C)**

$CH_3CH_2CH_2CH_3$ — **Butane** **(bp, -0.5°C)**

The carbonyl group (C=O) in acetone is a strongly polar group. Thus **intermolecular dipole–dipole attractive forces** operate between acetone molecules in the liquid state (see Section 5.1). These dipole–dipole forces are *stronger* than van der Waals forces between butane molecules, and therefore lead to a higher boiling point for acetone.

Role of hydrogen bonding

A **hydrogen bond** is usually **much stronger** than other types of intermolecular forces.

Consider ethyl alcohol and dimethyl ether that are isomers. Ethyl alcohol has a much higher boiling point (78.5°C) than dimethyl ether (–24.9°C).

Stronger interaction

Intermolecular hydrogen bond

CH_3CH_2–O–H- - -O(CH_2CH_3)–H

Ethyl alcohol
(bp, 78.5°C)

Weaker interaction

Dipole-dipole attraction

$(H_3C)_2O^{\delta-}$ ||||||| $^{\delta+}CH_3$–$O^{\delta-}$–CH_3

Dimethyl ether
(bp, -24.9°C)

Molecules of ethyl alcohol can associate by forming **strong intermolecular hydrogen bonds** since a hydrogen atom is bonded to a highly electronegative oxygen atom (O–H). But dimethyl ether has no O–H group and hydrogen bonding does not occur between molecules of dimethyl ether. The intermolecular forces that occur between molecules of dimethyl ether are *dipole-dipole attractions*, which are much weaker than hydrogen bonds. As a result, ethyl alcohol boils at a much higher temperature than dimethyl ether.

PROBLEM 5.1: The fluorocarbon C_5F_{12} (M = 288) has much higher molecular weight than pentane C_5H_{12} (M = 72), yet the fluorocarbon has a lower boiling point (29°C) than pentane (36.1°C). Explain.

Answer

The strengths of intermolecular van der Waals forces depend on the relative polarizability of the atoms involved. The fluorine atom has a very low polarizability which results in very small van der Waals attractions between the fluorocarbon molecules. Thus the fluorocarbon has much lower boiling point than that expected for its high molecular mass.

PROBLEM 5.2: Which compound in each of the following pairs would you expect to have the higher boiling point? Explain.

(a) Dipropylamine and triethylamine
(b) Propyl alcohol and ethylene glycol

Answer

(a) Dipropyl amine [$(CH_3CH_2CH_2)_2NH$] and triethylamine [$(CH_3CH_2)_3N$] are isomeric. Since triethylamine has a branched structure, it has a smaller surface area, resulting in lower van der Waals attractions. Therefore, triethylamine has a lower boiling point (90°C) than dipropylamine (110°C).

$CH_3CH_2CH_2NHCH_2CH_2CH_3$

Dipropylamine
(bp, 110°C)

$CH_3CH_2N(CH_2CH_3)CH_2CH_3$

Triethylamine
(bp, 90°C)

(b) Propyl alcohol ($M = 60$) and ethylene glycol ($M = 62$) have similar molecular weights but propyl alcohol has one OH group whereas ethylene glycol has two OH groups. Intermolecular hydrogen bonds can occur in both compounds.

However, one molecule of propyl alcohol can form *one* intermolecular hydrogen bond, whereas one molecule of ethylene glycol can form *two* intermolecular hydrogen bonds. Therefore, a greater thermal energy is required to break more hydrogen bonds in order to separate ethylene glycol molecules. Thus ethylene glycol has a higher boiling point (197°C) than propyl alcohol (97.2°C).

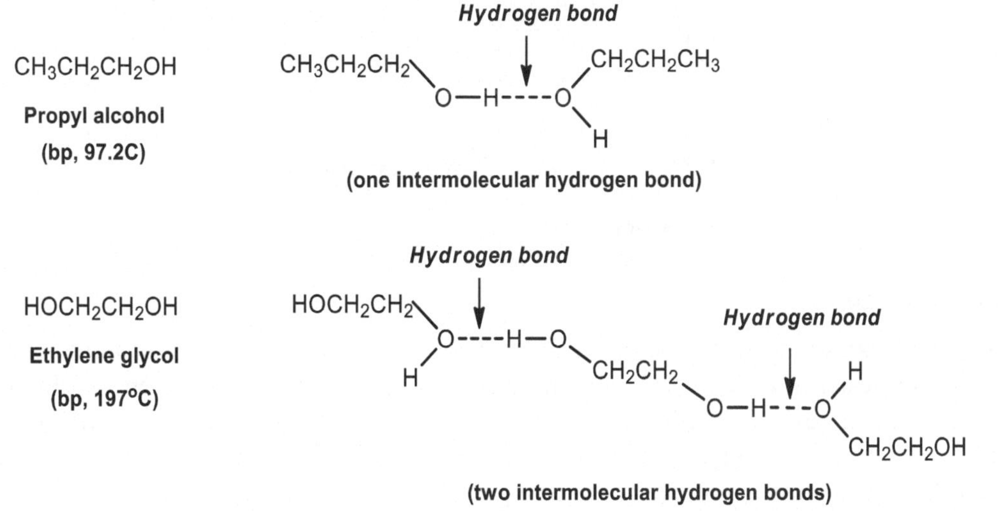

5.2.3 Solubility

The **solubility** of an organic compound in a given solvent is also governed by **intermolecular forces**. Usually, the **solute** is a solid or a liquid, and the **solvent** is a liquid. The solubility of a liquid compound in a liquid solvent is often referred to as **miscibility**.

Polar and nonpolar solvents

The polarity of a solvent is usually measured by a parameter called the **dielectric constant (ε)**, which indicates how well the solvent can accommodate charged species. As the dielectric constant of a solvent increases, the electrostatic attraction between the ions of the solute decreases and the solute has a greater tendency to ionize. **The higher the value of the dielectric constant, the more polar the solvent is**.

Generally, solvents with $\varepsilon > 5$ are termed **polar**, and solvents with $\varepsilon < 5$ are termed **nonpolar**.

A *rule of thumb* for ascertaining solubilities is: **like dissolves like**. This implies that ionic or polar compounds tend to dissolve in polar solvents, whereas nonpolar compounds tend to dissolve in nonpolar solvents.

Polar protic and polar aprotic solvents

When a polar solvent has a hydrogen atom that is capable of forming hydrogen bond, it is called a **polar protic solvent**. For example, **water** is a polar protic solvent; it is highly polar ($\varepsilon = 78.5$) and is capable of forming strong hydrogen bonds.

On the other hand, if the polar solvent does not possess a hydrogen atom that can participate in hydrogen bonding, it is called a **polar aprotic solvent**. An example is ***N,N*-dimethylformamide (DMF)**; it is polar ($\varepsilon = 37$) but cannot form hydrogen bond as no hydrogen atom in the molecule is attached to an electronegative oxygen or nitrogen atom.

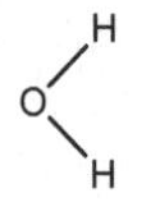

Water
(ε = 78.5)

Polar protic solvent

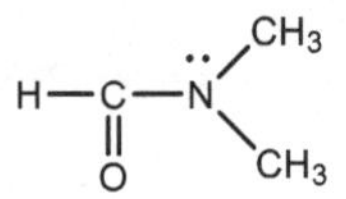

N,N-Dimethylformamide (DMF)
(ε = 37)

Polar aprotic solvent

Ionic compounds

Ionic compounds such as sodium acetate dissolve in water. Positive sodium ions are solvated by water molecules through ion–dipole forces with the negative end of the water dipole oriented towards the positive ion. Negative acetate ions are solvated by water molecules through hydrogen bonds. The energies supplied by these intermolecular forces in the dissolution process are great enough to overcome the strong intermolecular ion–ion forces in the crystal lattice.

Alcohols

Alcohols (ROH) have a polar OH group which can form hydrogen bond with water. With a relatively **small alkyl chain**, as in methyl alcohol, ethyl alcohol, propyl alcohol or butyl alcohol, the alcohol becomes soluble (or miscible) with water.

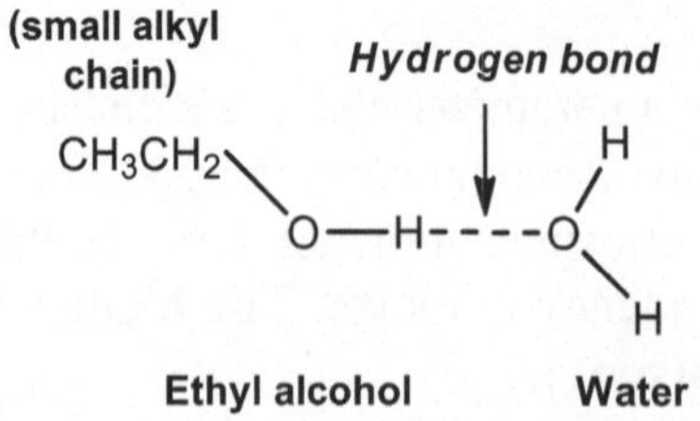

(small alkyl chain)

Hydrogen bond

Ethyl alcohol

Water

Ethyl alcohol is soluble in water through hydrogen bonding

If the alkyl chain of the alcohol is long, it becomes much less soluble in water. The long alkyl chains are said to be hydrophobic (water avoiding).

Consider, for example, **decyl alcohol** [$CH_3(CH_2)_8CH_2OH$] which has a *long 10-carbon chain*. Decyl alcohol is **insoluble in water**. Here, although the OH group is **hydrophilic** (water loving), the very large decyl group ($C_{10}H_{21}$) is **hydrophobic** which imparts insolubility of the compound in water.

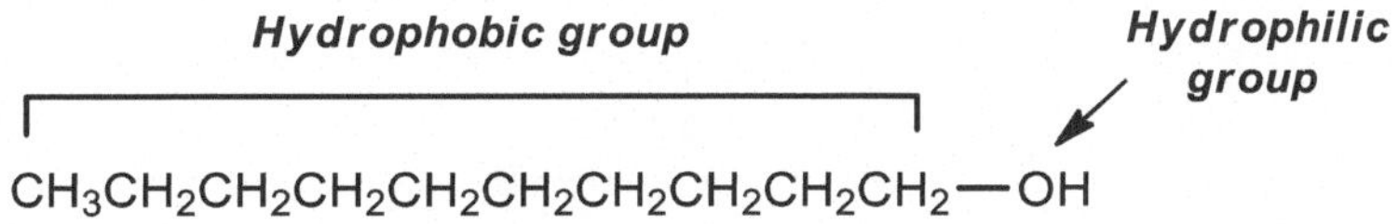

Hydrophobic group **Hydrophilic group**

$CH_3CH_2CH_2CH_2CH_2CH_2CH_2CH_2CH_2CH_2—OH$

Decyl alcohol
(insoluble in water)

The reason why nonpolar alkyl chains avoid an aqueous environment is generally attributed to what is called the **hydrophobic effect**. Accommodation of a nonpolar chain in water requires a **more ordered** arrangement of water molecules around the chain in comparison to the **less ordered** arrangement of water molecules in the bulk water.

Entropy is associated with the relative order of a system: **the more random or disordered a system is, the greater is its entropy**. Thus changes from disorder to order (or less order to more order) are accompanied with a *decrease of entropy*, which is thermodynamically *unfavourable.*

For nonpolar groups in water, the change from less order to more order for water molecules results in an **unfavourable entropy change** which is primarily responsible for the hydrophobic effect.

Ethers, aldehydes and ketones

The oxygen atom of **ethers** is also able to form hydrogen bonds with water and the *low molecular weight ethers* (such as diethyl ether) are **soluble** in water.

The carbonyl group in **aldehydes and ketones** is polar and the carbonyl oxygen can form hydrogen bonds to molecules of water. Thus *low molecular weight aldehydes and ketones* (such as acetaldehyde and acetone) are **soluble** in water.

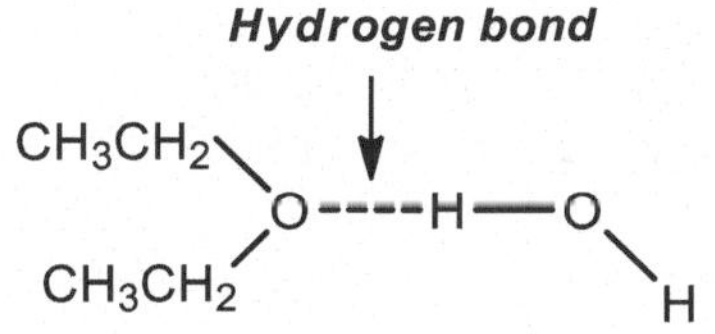

Hydrogen bond

$(CH_3CH_2)_2O$----H—O—H

Diethyl ether
(soluble in water)

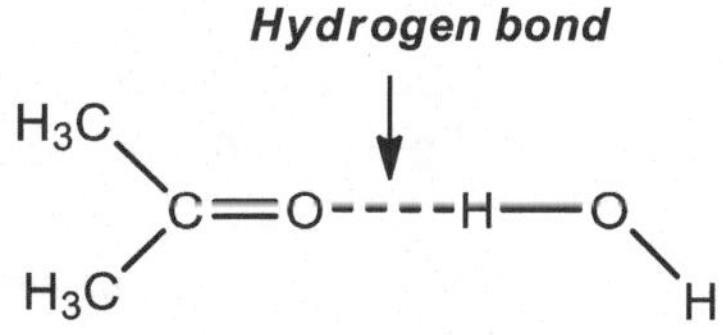

Hydrogen bond

$(H_3C)_2C=O$----H—O—H

Acetone
(soluble in water)

Amines

Amines are also moderately polar and the nitrogen atom of amines can form hydrogen bonds to molecules of water. As a result, primary, secondary and tertiary amines with *small alkyl chains* are **soluble** in water.

Examples of such amines are methylamine (CH_3NH_2), ethylamine ($CH_3CH_2NH_2$), dimethylamine [$(CH_3)_2NH$] and trimethylamine [$(CH_3)_3N$].

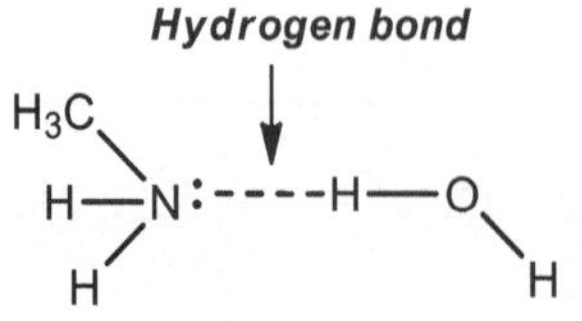

Hydrogen bond

H_3C–NH(H):----H—O–H

Methylamine (soluble in water)

Alkanes, alkenes and alkynes

Alkanes, alkenes and alkynes are almost **completely insoluble** in water because they are nonpolar and cannot form hydrogen bonds. They generally dissolve in **nonpolar solvents or solvents of low polarity** such as benzene (ε = 2.3), carbon tetrachloride (ε = 2.2), chloroform (ε = 4.8), and other hydrocarbons.

5.3 DIPOLE MOMENT

A polar covalent bond between two atoms represents a **dipole** with a partial positive charge (δ+) on the less electronegative atom and a partial negative charge (δ–) on the more electronegative atom. The polar bond has a dipole moment.

The **dipole moment (μ)** is defined as the product of the magnitude of the charge (q) and the distance (d) separating the charges:

$$\mu = q \times d$$

Typically, the charges are on the order of 10^{-10} in electrostatic units (esu) and the distances are on the order of 10^{-8} cm.

Therefore, the dipole moments are on the order of 10^{-18} esu cm. **The unit, 10^{-18} esu cm, is called the debye (D),** named after the scientist Peter J. W. Debye.

The dipole moment is a **vector quantity**; it has both magnitude and direction. The **direction** of the dipole vector is indicated by the following arrow symbol:

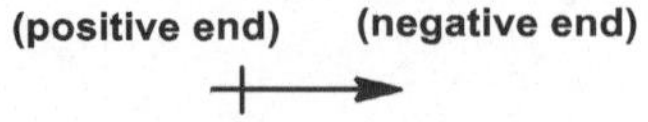

(positive end) **(negative end)**

$+\!\!\longrightarrow$

Direction of the dipole vector

The crossed end is the positive end and the arrow head is the negative end. *The vector will point from the positive to the negative end.*

To illustrate, consider HCl which is a diatomic molecule. The dipole moment (μ) of HCl is 1.08 D. The direction of the dipole moment vector is shown as

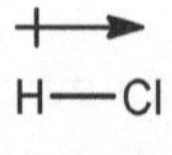

$+\!\!\longrightarrow$

H—Cl

(μ = 1.08 D)

5.3.1 Predicting Dipole Moments of Organic Molecules

Organic molecules may contain two or more polar bonds. Dipole moments of such molecules can be predicted. These are illustrated with the following cases.

Bond moments and group moments

The dipole moment of a polar bond is called the **bond moment**. When a molecule contains two or more polar bonds; the dipole moment of the molecule is determined by the **resultant or vector sum** of all individual bond moments.

Consider, for example, **carbon tetrachloride (CCl_4)**. The molecule has four C–Cl bonds, and is **tetrahedral** in shape. Since chlorine is more electronegative than carbon, each C–Cl bond is polar and has a bond moment. The four bond moments are vectors of **equal magnitude** but are oriented in **different directions** in a tetrahedral fashion.

As a result, the net dipole moment of a group of *three* C–Cl bonds, called the **group moment** of CCl_3 group, is equal in magnitude but opposite in direction to the **bond moment** of the remaining C–Cl bond. Since the group moment of CCl_3 and bond moment of C–Cl are opposed and cancel each other, the **CCl_4 molecule has no resultant dipole moment ($\mu = 0$).**

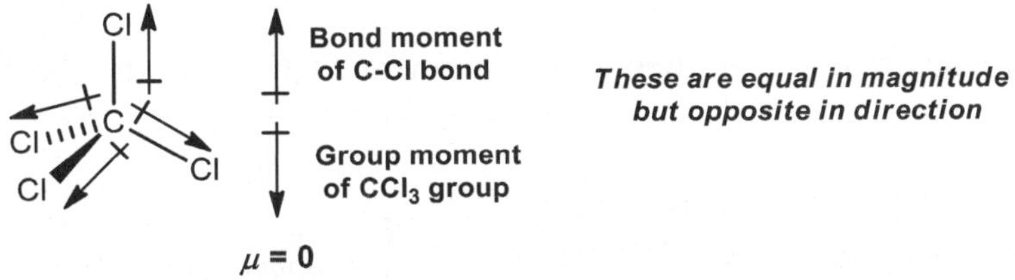

In contrast, **chloroform ($CHCl_3$)** has a dipole moment of 1.0 D. Like CCl_4, $CHCl_3$ is also **tetrahedral**. A C–H bond is not appreciably polar since the electronegativity of carbon is slightly more than that of hydrogen. Thus the contribution of the C–H bond to the dipole moment of $CHCl_3$ is small. On the other hand, the group moment of CCl_3 (which is equal to the bond moment of C Cl) contributes to a large extent.

The bond moment of C–H acts in the **same direction** as the group moment of CCl_3; they reinforce each other and are additive. Therefore, the **$CHCl_3$ molecule has a finite dipole moment ($\mu = 1.0$ D).**

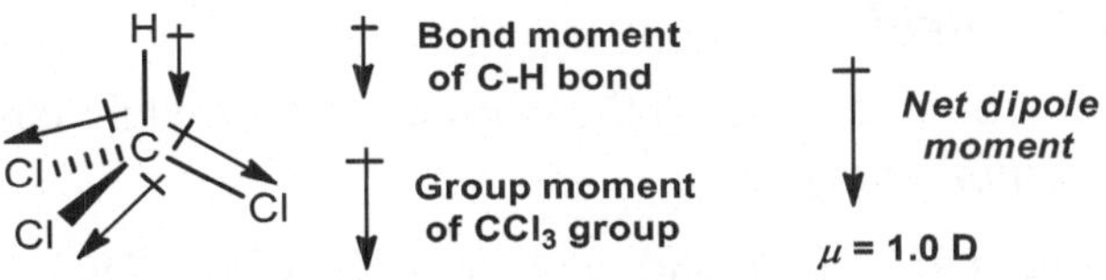

PROBLEM 5.3: Although F is more electronegative than Cl, the dipole moment of CH_3F (1.82 D) is less than that of CH_3Cl (1.94 D). Explain.

Answer

The magnitude of dipole moment depends on both charge and distance ($\mu = q \times d$)). The C–F bond in CH_3F is shorter than the C–Cl bond in CH_3Cl. The shorter bond distance tends to decrease the value of the dipole moment for CH_3F, even though the amount of negative charge is more on F than on Cl in CH_3Cl.

Moments due to lone pairs

Lone pairs of electrons can make significant contributions to the dipole moment of molecules. A lone pair has no atom attached to it that can partially neutralize its negative charge by electron pair sharing. Thus **the lone pair contributes a large moment which is directed away from the central atom.**

In **ammonia (NH_3)**, the **lone pair** on N is in an **sp^3** orbital. The N–H bond is polar as nitrogen is more electronegative than hydrogen. The resultant moment of *three* N–H bonds that are arranged in a tetrahedral-like (pyramidal) fashion is the group moment of NH_3 group. **The group moment of NH_3 group and the contribution of moment from the lone pair are in the same direction, and reinforce each other.**

In contrast, in **NF_3**, the resultant moment of *three* N–F bonds (N is less electronegative than F) **opposes** the moment of the lone pair.

Therefore, **NF_3 has a much smaller dipole moment (0.24 D) than NH_3 (1.46 D)**. The direction of the dipole moment vector of NF_3 is *indeterminate*.

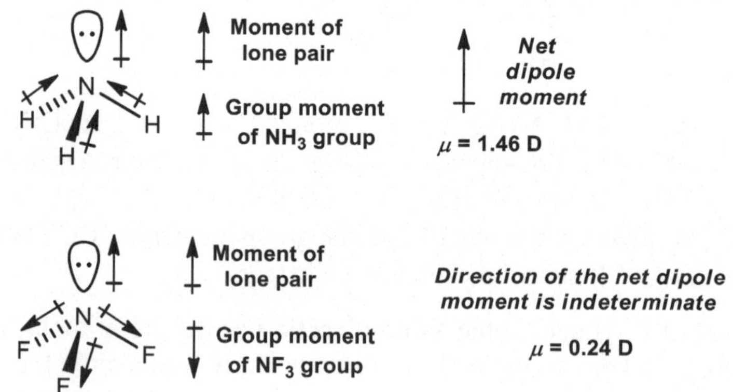

σ Moments and π moments

The bond moment of a σ bond is called the **σ moment (μ_σ)** while the bond moment of a π bond is called the **π moment (μ_π)**.

Consider ethyl chloride (CH_3CH_2Cl) and vinyl chloride ($CH_2{=}CHCl$).

In **ethyl chloride**, the electron withdrawing inductive effect (*–I*) of Cl creates the C–Cl bond dipole. The major contributing factor toward the dipole moment of ethyl chloride is the **bond moment of C(sp^3)–Cl σ bond.**

In **vinyl chloride**, the carbon atom attached to chlorine is **sp²**-hybridized, which is more electronegative than sp³ carbon in ethyl chloride. As a result, the Cl atom in C(sp²)–Cl σ bond is less negatively polarized. Therefore, the **σ moment (μ_σ)** for the C–Cl bond in vinyl chloride is smaller than that in ethyl chloride. Further, in vinyl chloride, Cl can exert an electron donating $+M$ effect which contributes a moment due to the π bond between C and Cl. The **π moment (μ_π) acts in the opposite direction to μ_σ**. The C–Cl bond in vinyl chloride is also *shorter* (due to some π character and more s character of sp² carbon) that tends to decrease its dipole moment. Thus **the dipole moment of vinyl chloride (1.4 D) is much lower than that of ethyl chloride (2.10 D).**

Since $-I > +M$ for Cl, Cl is net electron-withdrawing. Hence the direction of μ_σ determines the **direction of the net dipole moment** of vinyl chloride.

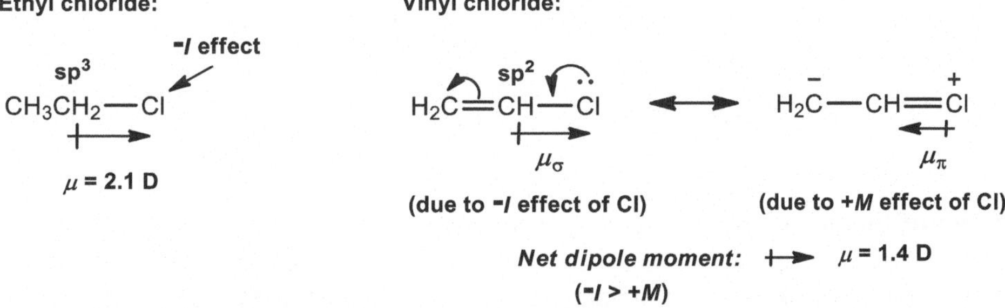

For similar reasons, the dipole moment of **chlorobenzene** (1.7 D) is smaller than that of **cyclohexyl chloride** (2.2 D).

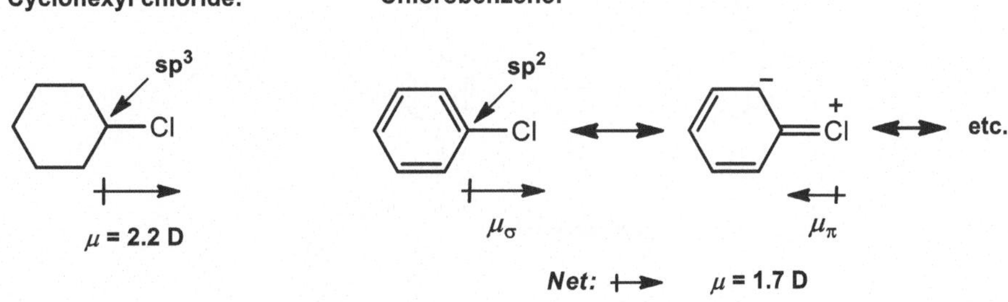

Next, compare dipole moments of **aniline** and **nitrobenzene**. The dipole moment of aniline is 1.5 D whereas the dipole moment of nitrobenzene is 4.0 D. Here the two dipole moment vectors act in opposite directions.

$:NH_2$ $\mu = 1.5$ D

NO_2 $\mu = 4.0$ D

Aniline

Nitrobenzene

These results can be explained as follows. For **aniline**, the NH_2 group has $-I$ and $+M$ effects, but $+M > -I$. As a result, $\mu_\pi > \mu_\sigma$, and the dipole moment vector of aniline is pointed *toward the ring*. In contrast, for **nitrobenzene**, the NO_2 group has $-I$ and $-M$ effects; thus **μ_σ and μ_π reinforce** each other, and the dipole moment vector for nitrobenzene is pointed *away from the ring*.

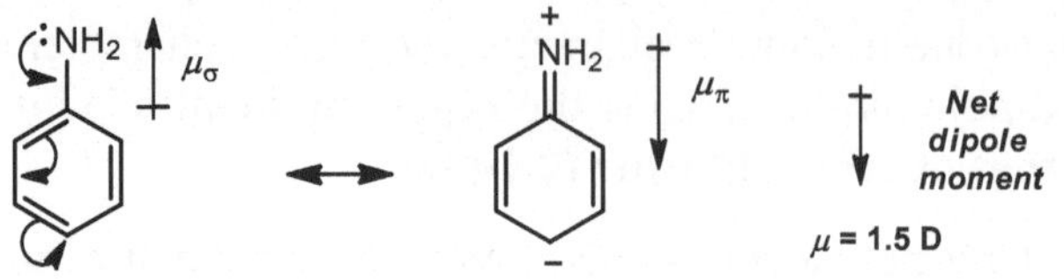

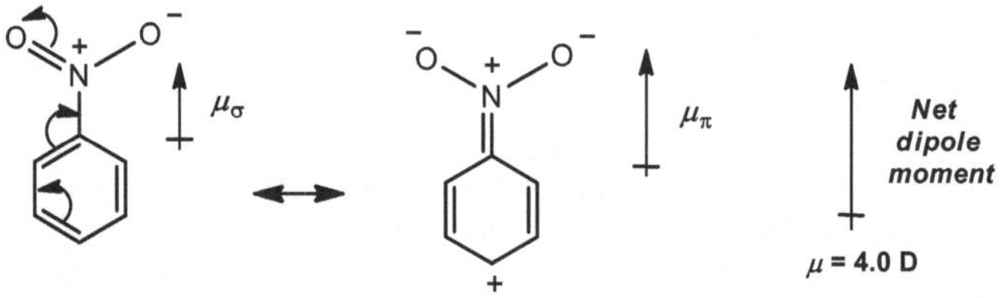

Net dipole moment

μ = 1.5 D

Net dipole moment

μ = 4.0 D

In ***p*-nitroaniline**, the dipole moment vectors due to NH_2 and NO_2 groups act in the same direction and reinforce each other. But, interestingly, **the dipole moment of *p*-nitroaniline (6.2 D) is greater than the sum of the values for aniline and nitrobenzene (5.5 D)**. This is because the NH_2 and NO_2 groups can enter into resonance, thereby increasing the extent of polarization as well as the distance separating the charges.

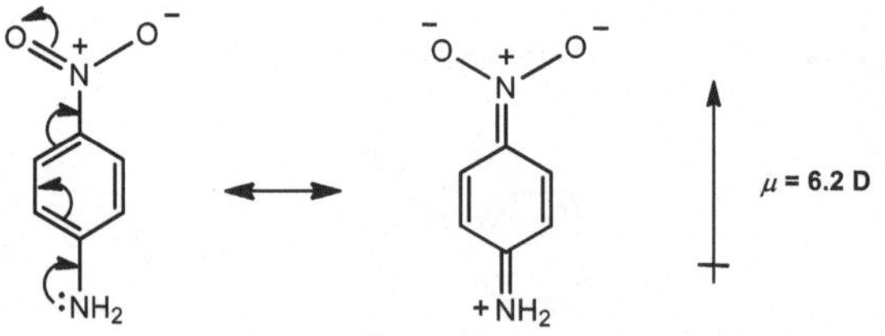

μ = 6.2 D

***p*-Nitroaniline**

Angle between bond moments

If two bond moments μ_1 and μ_2 are inclined at an angle θ, the resultant dipole moment (μ) is given by

$$\mu = \sqrt{\mu_1^2 + \mu_2^2 + 2\mu_1\mu_2 cos\theta}$$

Consider dipole moments of three disubstituted benzenes: ***o*-dichlorobenzene, *m*-dichlorobenzene and *p*-dichlorobenzene**. The angle between two C–Cl bond dipoles can be determined from the *geometry* of the molecules. For the *o*- *m*- and *p*-isomer, the angle is 60°, 120° and 180° respectively.

o-Dichlorobenzene (θ = 60°)

m-Dichlorobenzene (θ = 120°)

p-Dichlorobenzene (θ = 180°)

If the bond moment of a C–Cl bond is *x* D, the dipole moment (in D) of the three isomers can be calculated using the above expression as shown below.

Here, $\mu_1 = \mu_2 = x$. Therefore,

$$\mu_{ortho} = \sqrt{x^2 + x^2 + 2x^2 cos 60^o} = \sqrt{3x^2} = \sqrt{3}\,x$$

$$\mu_{meta} = \sqrt{x^2 + x^2 + 2x^2 cos 120^o} = \sqrt{x^2} = x$$

$$\mu_{para} = \sqrt{x^2 + x^2 + 2x^2 cos 180^o} = 0$$

Thus the dipole moment decreases in the order: ***o*-dichlorobenzene > *m*-dichlorobenzene > *p*-dichlorobenzene**. Experimentally, the dipole moments are found to be 2.54 D for *ortho*, 1.48 D for *meta* and 0 for *para*.

It may also be noted that the *p*-isomer has no dipole moment because the two vectors are opposed and cancel each other.

5.4 ACIDS AND BASES

Acids and bases are quite familiar terms. We know that lemons and oranges are acidic, while soaps and bleach are basic.

But how could we define acids and bases? There are two important theories, namely, the **Brønsted-Lowry theory** and the **Lewis theory**, which are currently used for defining whether a molecule is an acid or a base.

5.4.1 The Brønsted-Lowry theory

According to the Brønsted-Lowry theory, an acid is a proton donor and a base is a proton acceptor.

When an **acid (HA)** loses a proton (H^+), it forms a species (ion or molecule) called the **conjugated base (cB)** of the acid. Similarly, the species that is formed when a **base (B)** accepts a proton (H^+) is called the **conjugated acid (cA)** of the base.

Overall, the reaction between an acid (HA) and a base (B) gives a mixture of a conjugate base (cB) and a conjugate acid (cA).

$$\underset{\textbf{Acid}}{HA} \rightleftharpoons H^+ + \underset{\textbf{Conjugate base (cB)}}{A^-}$$

$$\underset{\textbf{Base}}{B:} + H^+ \rightleftharpoons \underset{\textbf{Conjugate acid (cA)}}{BH^+}$$

Overall:

$$\underset{\textbf{(acid)}}{HA} + \underset{\textbf{(base)}}{B:} \rightleftharpoons \underset{\textbf{(cB)}}{A^-} + \underset{\textbf{(cA)}}{BH^+}$$

In comparison to *strong* **inorganic acids** (e.g. HCl) which are completely ionized in water, **organic acids** are relatively *much weaker*. For example, **acetic acid** (CH_3CO_2H) dissolved in water gives an equilibrium that lies more to the side of the undissociated acid. At 25°C, only about 1% of acetic acid molecules are dissociated in a 0.1 M aqueous solution of acetic acid.

$$\underset{\textbf{Strong acid}}{HCl} + H_2O \longrightarrow Cl^- + H_3O^+ \quad \textbf{(complete ionization of HCl)}$$

$$\underset{\textbf{Weak acid}}{CH_3CO_2H} + H_2O \rightleftharpoons CH_3CO_2^- + H_3O^+ \quad \textbf{(equilibrium lies to the left)}$$

Here water acts as a base since it accepts a proton from the acid.

Notes: (1) A molecule can be an acid *only with reference to* another molecule. Similarly, a molecule can be a base relative to another molecule. The reference molecule is usually water. Generally, any molecule that is more acidic than water is considered an acid and any molecule that is more basic than water is regarded as a base.

(2) In practice, whether a molecule acts as an acid or a base depends on what is present in the reaction mixture along with it. For example, nitric acid (which we know as a very strong acid) actually acts as a base in the presence of a more acidic sulphuric acid (see nitration of benzene in Chapter 13). Thus acids and bases are more general terms. *Water can act both as an acid and a base.*

5.4.2 Acidity, Dissociation Constant (K_a) and pK_a

The dissociation of an acid (HA) in water is represented as

$$HA + H_2O \rightleftharpoons A^- + H_3O^+$$

Therefore, the **equilibrium constant (K_{eq})** is given by

$$K_{eq} = \frac{[A^-][H_3O^+]}{[HA][H_2O]}$$

Since the concentration of water is essentially constant for dilute solutions, K_{eq} [H_2O] is *constant*, which is called the **acid dissociation constant (K_a)** for the acid.

$$K_a = K_{eq}[H_2O] = \frac{[A^-][H_3O^+]}{[HA]}$$

(acid dissociation constant)

A larger value of K_a implies more dissociation and the acid is stronger, while a smaller value of K_a means less dissociation and the acid is weaker.

Acidity or acid strength of an organic compound is generally expressed in terms of **pK_a** which is defined as

$$pK_a = -\log K_a$$

This implies that when K_a is large, the value of pK_a is small. So there is an *inverse* relationship between the value of the pK_a and the strength of the acid: **the smaller the value of the pK_a, the stronger is the acid.**

5.4.3 Ionization of water: pH and pK_a

Ionization of water occurs as,

$$H_2O + H_2O \rightleftharpoons H_3O^+ + HO^-$$

Notably, the proton (H^+) resulting from ionization is not free in water; it exists mainly as H_3O^+.

pH

The pH is the negative logarithm of the concentration of H_3O^+, that is, pH = $-\log[H_3O^+]$. At 25°C, pure water is only slightly dissociated and the concentration of H_3O^+ is extremely low: $[H_3O^+] = 10^{-7}$ M.

Therefore, the **pH of pure water** is

$$pH = -\log[H_3O^+] = -\log 10^{-7} = 7$$

An **aqueous solution** of a compound is said to be **neutral** at pH 7. Below pH 7, there is more H_3O^+ ($[H_3O^+] > 10^{-7}$ M) and the solution is **acidic**. Above pH 7, there is less H_3O^+ ($[H_3O^+] < 10^{-7}$ M) and the solution is **basic**.

pK$_a$

Since the dissociation of pure water is extremely small, the concentration of undissociated water in the ionization equilibrium is essentially constant. The dissociation also gives equal amounts of H_3O^+ and HO^-, i.e. $[H_3O^+] = [HO^-] = 10^{-7}$ M.

The mass of 1 L of water = 1000 g; therefore, **$[H_2O]$ = (1000/18) M = 55.5 M**. The **dissociation constant (K_a) of water** is then given by

$$K_a = \frac{[H_3O^+][HO^-]}{[H_2O]} = \frac{10^{-7} \times 10^{-7}}{55.5}$$

Therefore, the **pK_a of pure water** is

$$pK_a = -\log K_a = -\log(10^{-14}/55.5) = 15.7$$

5.4.4 Basicity and pK_a

The **basicity or the strength of a base (B)** in water is determined by the following equilibrium:

$$B\!: + H_2O \rightleftharpoons BH^+ + HO^-$$

The greater the concentration of the conjugate acid (BH^+), the stronger is the base (B).

Usually, acidity and basicity are measured in terms of a **unified pK_a scale**. Thus we consider the following equilibrium in terms of the conjugate acid (BH^+) in water:

$$BH^+ + H_2O \rightleftharpoons B\!: + H_3O^+$$

K_a and pK_a are then given by

$$K_a = \frac{[B\!:][H_3O^+]}{[BH^+]} \quad \text{and} \quad pK_a = -\log K_a$$

If the concentration of BH^+ is greater, K_a is smaller or pK_a is larger. So there is a *direct* relationship between the value of the pK_a and the strength of a base: the larger the value of the pK_a, the stronger is the base.

It should be noted that pK_a is a measure of the acid strength of the conjugate acid (cA) of the base. **The weaker the conjugate acid (smaller K_a or larger pK_a), the stronger is the base.**

5.4.5 The Lewis theory

A more general definition of **acids** has been put forward by G. N. Lewis, which includes compounds (such as BF_3) that have *no* protons to donate. Instead, they can accept electrons.

According to the Lewis theory, an acid is an electron-pair acceptor and a base is an electron-pair donor.

Bases are similar in the Brønsted-Lowry theory and the Lewis theory since a base must donate a pair of electrons to accept a proton. For example, **ammonia** (NH_3) donates an electron pair to make a bond with a proton. NH_3 is therefore a base in terms of both Lewis and Brønsted-Lowry definitions.

$$H_3N\!: \; + \; H^+ \longrightarrow H_3\overset{+}{N}{-}H$$

Lewis base or Brpnsted base → **Ammonium ion**

Lewis acids have an electron-deficient atom that can accept an electron pair. **Boron trifluoride** (BF_3) and **aluminium chloride** ($AlCl_3$) are common Lewis acids.

BF_3 is usually used as its complex with ether, called 'etherate'. Diethyl ether, as a Lewis base, donates a pair of electrons to the electron-deficient B atom having a sextet of electrons.

Similarly, $AlCl_3$ in which Al has a sextet of electrons can accept an electron pair from a Lewis base (NH_3) to form a Lewis acid-Lewis base complex.

$$BF_3 + :O(C_2H_5)_2 \longrightarrow F_3\overset{-}{B}-\overset{+}{O}(C_2H_5)_2$$

Lewis acid **Lewis base** **Lewis acid-Lewis base complex (called BF_3-etherate)**

$$AlCl_3 + :NH_3 \longrightarrow Cl_3\overset{-}{Al}-\overset{+}{N}H_3$$

Lewis acid **Lewis base** **Lewis acid-Lewis base complex**

Halides of transition metals in which metal atoms have vacant orbitals can accept electron pairs, and therefore act as **Lewis acids**. Two examples are **ferric chloride** ($FeCl_3$) and **zinc chloride** ($ZnCl_2$).

$FeCl_3$ can accept an electron pair from a Cl atom of Cl_2 to form a Lewis acid-Lewis base complex. Similarly, $ZnCl_2$ can accept an electron pair from the O atom of an alcohol (ROH) to form a complex.

$$:\ddot{Cl}-\ddot{Cl}: + FeCl_3 \longrightarrow :\ddot{Cl}-\overset{+}{Cl}-\overset{-}{Fe}Cl_3$$

Lewis base **Lewis acid** **Lewis acid-Lewis base complex**

$$R-\ddot{O}(H): + ZnCl_2 \longrightarrow R-\overset{+}{O}(H)-\overset{-}{Zn}Cl_2$$

Lewis base **Lewis acid** **Lewis acid-Lewis base complex**

5.5 RELATIVE ACIDITIES OF ORGANIC COMPOUNDS

In general, the strength of an acid depends on the stability of its conjugate base. **Strong acids have stable conjugate bases.**

An acid that has a more stable conjugate base will be more acidic than an acid that has a less stable conjugate base. The relative strengths of two or more acids can be determined thus: **the more stable the conjugate base, the stronger is the acid.**

Factors affecting the stability of conjugate base

The conjugate base of an acid is usually negatively charged. So the question is: *what structural features contribute to stabilizing negative charges?*

Generally, a negative charge on an **electronegative atom** makes an anion stable.

However, the most important feature that stabilizes a negative charge is the **delocalization or spreading of the negative charge over several atoms.**

Delocalization is possible when the **groups** attached to the negatively charged atom are **electron-withdrawing** through **inductive effect (–*I*)** or by **mesomeric effect (–*M*)**.

The following are two general cases for the dispersal or delocalization of a negative charge in the conjugate base:

(a) the negatively charged atom (A) in the conjugate base is attached to a –*I* group (X)
(b) the negatively charged atom (A) in the conjugate base is attached to a –*M* group (Y=Z).

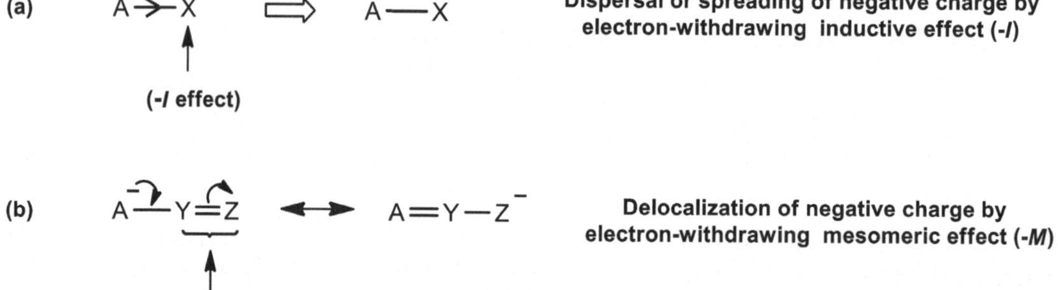

5.5.1 Carboxylic Acids

Carboxylic acids (RCOOH) and alcohols (ROH) both possess O–H group that is capable of donating a proton. But **a carboxylic acid is a much stronger acid than an alcohol.**

For example, **acetic acid** (pK_a = 4.75) is much more acidic than **ethanol** (pK_a = 15.9). (Remember that a smaller pK_a value indicates a stronger acid.) Notably, alcohols have a pK_a value similar to that of water (pK_a = 15.7).

The **conjugate base** of acetic acid is **acetate ion**, which is stabilized by both electron-withdrawing –*I* and –*M* effects of the carbonyl group (C=O). The two resonance structures of acetate ion are *equivalent*, and therefore contribute **large resonance stabilization**.

On the other hand, the **conjugate base** of ethanol is **ethoxide ion**, in which no significant inductive effect operates to disperse the negative charge, and the ethoxide ion has *no* resonance structures.

Consequently, the conjugate base of acetic acid is much **more stable** than that of ethanol. Hence, acetic acid is a much **stronger acid** than ethanol.

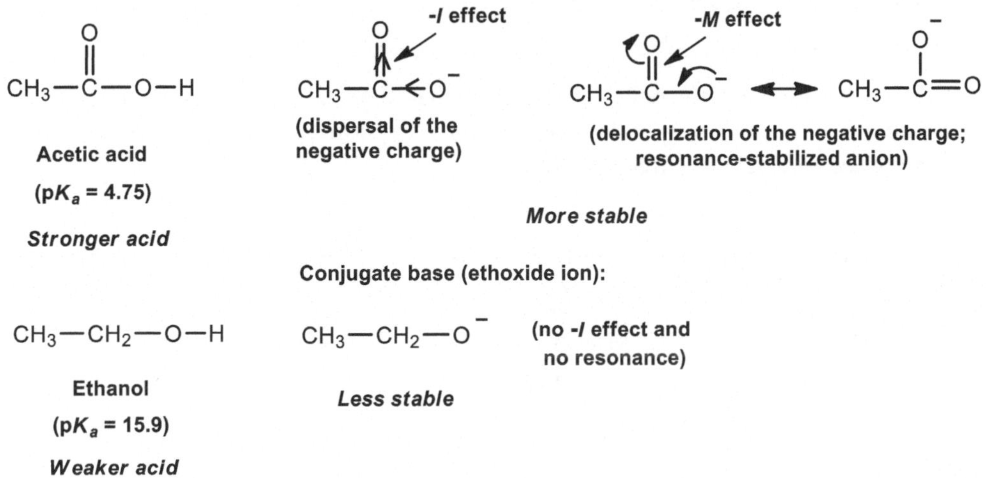

Relative acidities of aliphatic carboxylic acids

Acetic acid (pK_a = 4.75) **is a weaker acid than formic acid** (pK_a = 3.77). In the conjugate base (acetate ion), the methyl group (CH_3) exerts an electron donating $+I$ effect which tends to concentrate the negative charge, whereas in the conjugate base (formate ion), no such effect operates. As a result, the conjugate base of acetic acid is **less stable** than that of formic acid, and hence acetic acid is a weaker acid.

	Conjugate base:		Conjugate base:
$CH_3—CO_2H$	$CH_3 \rightarrow CO_2^-$	$H—CO_2H$	$H—CO_2^-$
Acetic acid	**+*I* effect**	**Formic acid**	***More stable***
(pK_a = 4.75)	***Less stable***	**(pK_a = 3.77)**	
Weaker acid		***Stronger acid***	

For **halogen-substituted carboxylic acids**, the acidity increases with increasing $-I$ effect of the halogen. The $-I$ effect increases with increasing electronegativity.

For example, **fluoroacetic acid** (FCH_2CO_2H) is a **stronger acid** than **chloroacetic acid** ($ClCH_2CO_2H$) since F is more electronegative (more electron-withdrawing) than Cl.

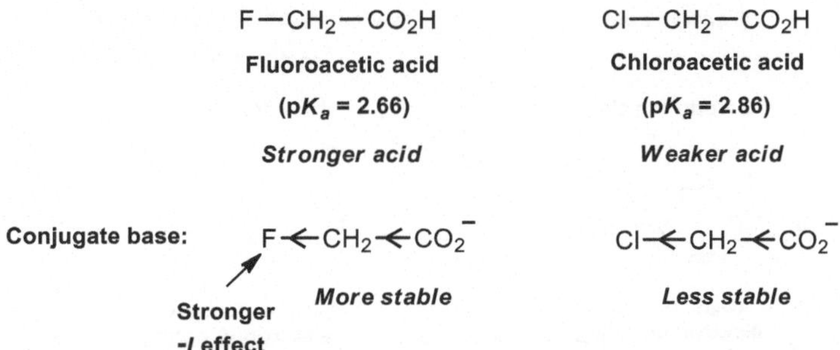

An increase in the number of halogen substituents further promotes the acidity. Thus the acidity increases in the order:

$ClCH_2CO_2H < Cl_2CHCO_2H < Cl_3CCO_2H$.

In the series, **trichloroacetic acid** (Cl_3CCO_2H) has a small pK_a value (0.65) and is the strongest acid because the conjugate base anion of trichloroacetic acid is highly stabilized by the combined –*I* effect of three Cl atoms.

Cl—CH_2—CO_2H | Cl—CH(Cl)—CO_2H | Cl—C(Cl)(Cl)—CO_2H

	Chloroacetic acid	Dichloroacetic acid	Trichloroacetic acid
pK_a	2.86	1.30	0.65
			Strongest acid

Cl←C(←Cl)(←Cl)—CO_2^-

Conjugate base of trichloroacetic acid is highly stabilized by -*I* effects of three Cl atoms

Relative acidities of aromatic carboxylic acids

Aromatic carboxylic acids are generally **stronger acids** than their saturated analogues. For example, **benzoic acid** (pK_a = 4.20) **is a stronger acid than cyclohexanecarboxylic acid** (pK_a = 4.87).

An unsaturated sp^2 carbon with more s character is *less electron donating* (or *more electron withdrawing*) than a saturated sp^3 carbon with less s character. Thus **phenyl group is less electron donating than cyclohexyl group**.

Since an electron-donating group tends to concentrate and destabilize the negative charge, the **conjugate base of benzoic acid** is relatively **less destabilized (or more stabilized)** relative to the conjugate base of cyclohexanecarboxylic acid. Therefore, benzoic acid is more acidic than cyclohexanecarboxylic acid.

	C_6H_5—CO_2H	C_6H_{11}—CO_2H
	Benzoic acid	**Cyclohexanecarboxylic acid**
	(pK_a = 4.20)	**(pK_a = 4.87)**
	Stronger acid	***Weaker acid***
Conjugate base:	C_6H_5—CO_2^- (sp^2)	C_6H_{11}—CO_2^- (sp^3)
	(C_6H_5 is less electron donating)	**(C_6H_{11} is more electron donating)**
	More stable	***Less stable***

The presence of an electron-withdrawing group on the aromatic ring increases the acidity. For example, *p*-nitrobenzoic acid (pK_a = 3.43) is a stronger acid than benzoic acid (pK_a = 4.20).

The NO_2 group is powerfully electron-withdrawing. It exerts a strong –*I* effect to disperse the negative charge in the conjugate base of *p*-nitrobenzoic acid. Further, its electron withdrawing –*M* effect creates a positive charge at the *para* position, which causes a dispersal of the negative charge of the CO_2^-, thereby stabilizing the conjugate base. Thus the **conjugate base of *p*-nitrobenzoic acid is much more stable than that of benzoic acid**. Hence *p*-nitrobenzoic acid is a stronger acid.

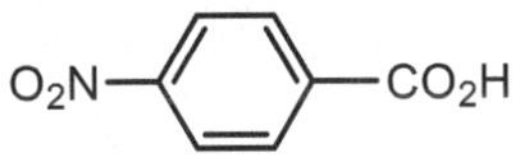

O_2N–C_6H_4–CO_2H ***Stronger acid than benzoic acid***

***p*-Nitrobenzoic acid**
(pK_a = 3.43)

Conjugate base:

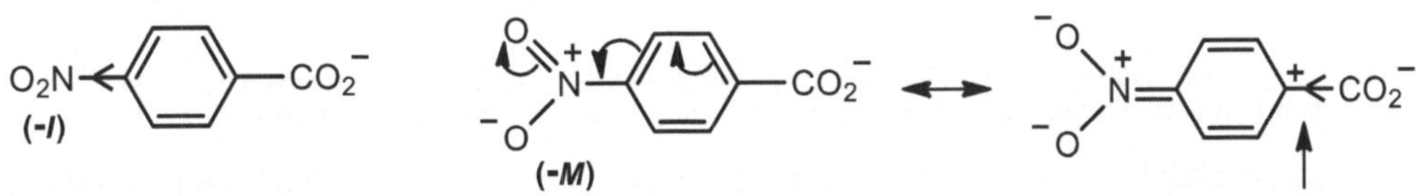

O_2N **(-*I*)** CO_2^- ; CO_2^- **(-*M*)** ⟷ CO_2^-

(stabilized by dispersal of negative charge) **(stabilized by dispersal of negative charge)**

Both -*I* and -*M* effects of NO_2 group stabilize the conjugate base pf *p*-nitrobenzoic acid; it is more stable than the conjugate base of benzoic acid

In contrast, ***p*-hydroxybenzoic acid** (pK_a = 4.58) is a **weaker acid** than benzoic acid. The OH group is –*I* but +*M*. Since the electron donating +*M* effect is dominant (+*M* > –*I*), the OH group is ***net* electron-donating**, which tends to inhibit the delocalization of the negative charge and thereby destabilize the conjugate base. As a result, the **conjugate base of *p*-hydroxybenzoic acid is less stable than that of benzoic acid**. Therefore, *p*-hydroxybenzoic acid is a weaker acid.

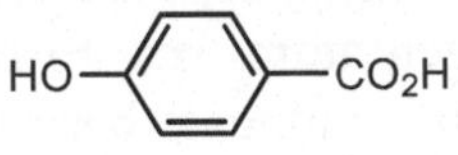

HO–C_6H_4–CO_2H ***Weaker acid than benzoic acid***

***p*-Hydroxybenzoic acid**
(pK_a = 4.58)

Conjugate base:

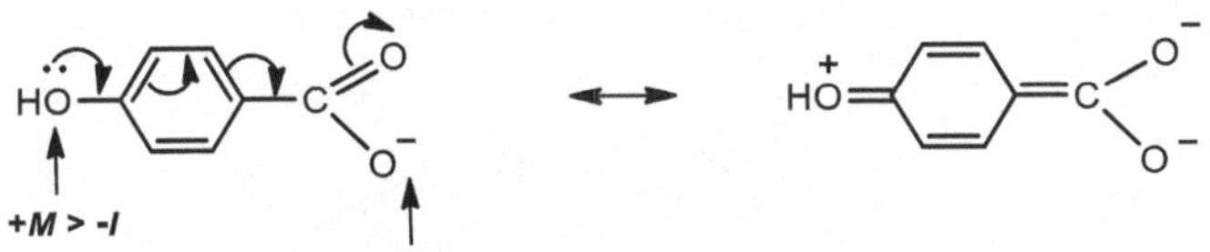

HO **+*M* > -*I*** ⟷ $H\overset{+}{O}$=

(delocalization of the negative charge is inhibited)

The stronger +*M* effect of OH destabilizes the conjugate base of *p*-hydroxybenzoic acid; it is less stable than the conjugate base of benzoic acid

Interestingly, ***o*-hydroxybenzoic acid (salicylic acid)** with a pK_a of 2.98 is a much **stronger acid** than ***p*-hydroxybenzoic acid**. It is even *stronger* than benzoic acid (pK_a = 4.20).

This is because the **conjugate base of the salicylic acid** is greatly stabilized by **intramolecular hydrogen bonding** between OH and CO_2^-.

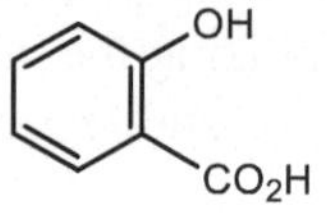

OH

CO_2H

Salicylic acid

(pK_a = 2.98)

Stronger acid
than p-hydroxybenzoic acid or benzoic acid

Conjugate base:

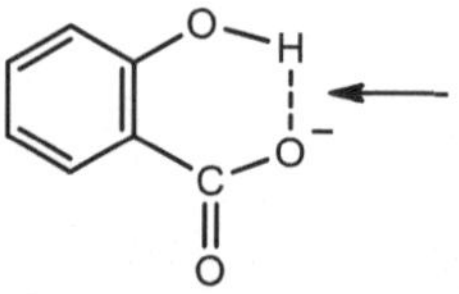

Intramolecular hydrogen bond

The conjugate base of salicylic acid is highly stabilized by intramolecular hydrogen bonding; it is much more stable than the conjugate base of *p*-hydroxybenzoic acid or benzoic acid

This is an example that illustrates the role of intramolecular hydrogen bonding in determining the relative strengths of acids.

PROBLEM 5.4: Sulphonic acids (RSO_3H) are much stronger acids than carboxylic acids (RCO_2H). Explain.

Answer

The electron withdrawing –*I* effect of the sulphone group (–SO_2–) in sulphonic acids is much stronger than the carbonyl group (–CO–) in carboxylic acids, which increases the acidity of sulphonic acids.

More importantly, the negative charge in the conjugate base of a sulphonic acid (sulphonate anion) is delocalized over *three* oxygen atoms due to the –*M* effect of the sulphone group, whereas the negative charge in the conjugate base of a carboxylic acid (carboxylate anion) is delocalized over *two* oxygen atoms. Thus, the conjugate base of a sulphonic acid is much strongly stabilized relative to the conjugate base of a carboxylic acid, and hence sulphonic acids have much higher acidity.

Conjugate base of a carboxylic acid:

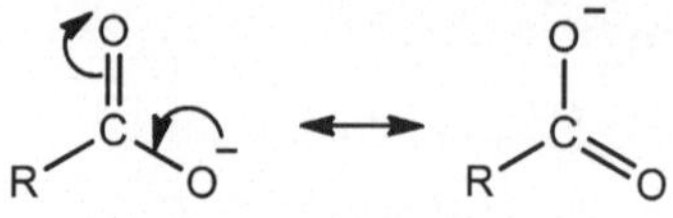

(negative charge is delocalized over two oxygen atoms)

Less stable

Conjugate base of a sulphonic acid:

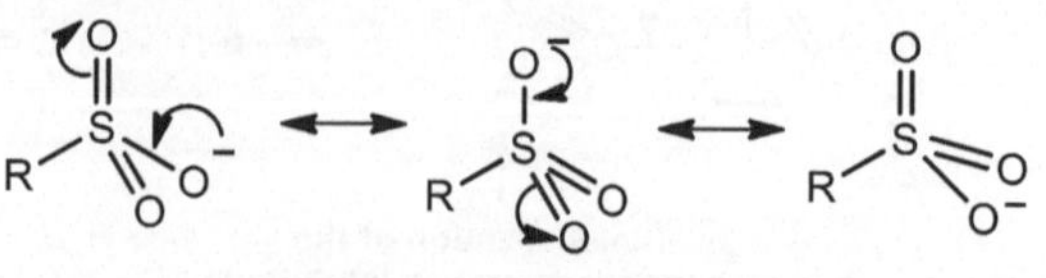

(negative charge is delocalized over three oxygen atoms)

More stable

An example of a sulphonic acid is methanesulphonic acid (CH_3SO_3H) which has a negative pK_a value (–1.9). This indicates that the equilibrium for dissociation of the sulphonic acid is highly favourable to the side of the dissociated ion. This is in contrast to acetic acid (CH_3CO_2H), a carboxylic acid where the equilibrium lies to the side of the undissociated acid (see Section 5.4.1).

Sulphonic acids are used as strong acid catalysts in organic synthesis.

5.5.2 Phenols

Phenols and alcohols both possess O–H groups, but **phenols are much stronger acids than alcohols**.

Phenol itself (pK_a = 10) is **more acidic** than **cyclohexanol** (pK_a = 18). Since phenyl group is less electron donating than cyclohexyl group (see above), the **conjugate base of phenol** (phenoxide ion) is relatively **more stabilized** than the conjugate base anion of cyclohexanol. Further, there is delocalization of the negative charge of phenoxide ion into the aromatic ring by resonance, while no such delocalization is possible in the anion of cyclohexanol. As a result, phenol is a stronger acid than cyclohexanol.

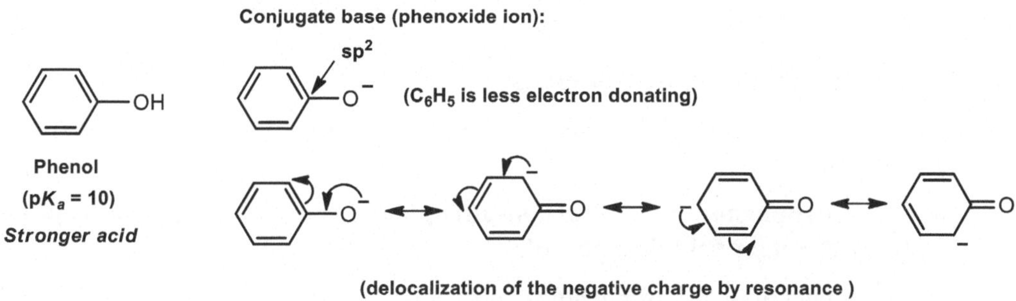

Conjugate base (anion of cyclohexanol):

sp^3

OH

O$^-$

(C_6H_{11} is more electron donating)

Cyclohexanol
(pK_a = 18)

Weaker acid

(no delocalization of the negative charge by resonance)

Less stable

However, phenols are less acidic than carboxylic acids. For example, Phenol itself (pK_a = 10) is a weaker acid than benzoic acid (pK_a = 4.20).

In the phenoxide ion (conjugate base of phenol), the negative charge on oxygen is delocalized onto the carbon atoms of the aromatic ring, whereas in the benzoate ion (conjugate base of benzoic acid), the negative charge on oxygen is delocalized onto the other electronegative oxygen atom. As a result, the **conjugate base of phenol** is relatively

less stabilized than the conjugate base of benzoic acid. Hence phenol is a weaker acid than benzoic acid.

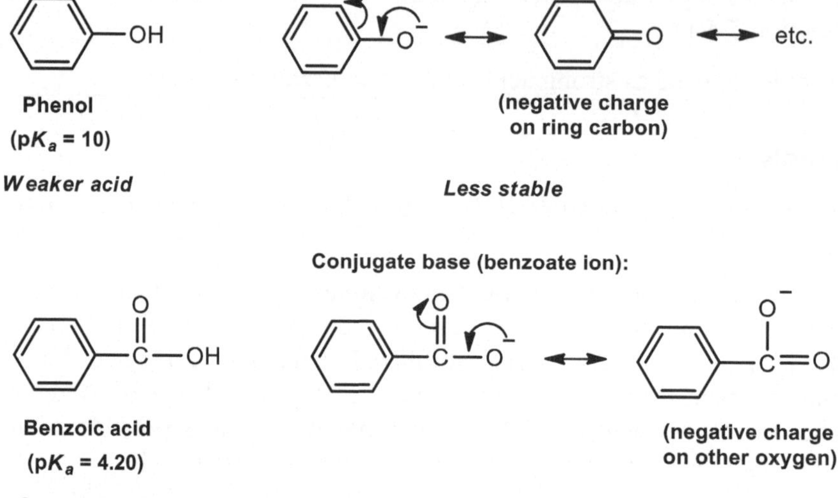

Relative acidities of substituted phenols

Introduction of an electron-withdrawing substituent into the aromatic ring increases the acidity of phenols.

For example, ***o*-nitrophenol** (pK_a = 7.23), ***m*-nitrophenol** (pK_a = 8.35) or ***p*-nitrophenol** (pK_a = 7.14) is a **stronger acid** than phenol.

However, *o*-nitrophenol or *p*-nitrophenol is more acidic than *m*-nitrophenol. This can be explained as follows.

The NO_2 group can stabilize the conjugate base through $-I$ and $-M$ effects. The $-I$ effect can operate from all positions, while the $-M$ effect is operative when the NO_2 group is in the *o*- or *p*-position. Therefore, the **conjugate base of the *o*- or *p*-isomer** is stabilized by both $-I$ and $-M$ effects, and is **more stable** than the conjugate base of the *m*-isomer which is stabilized *only* by the $-I$ effect.

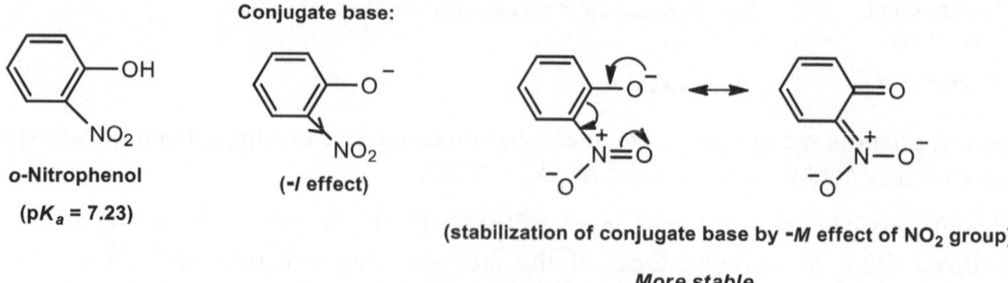

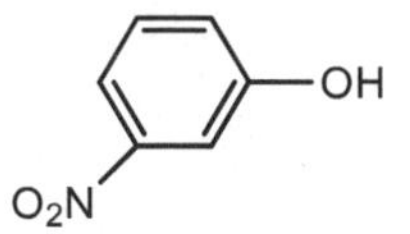

m-Nitrophenol
($pK_a = 8.35$)

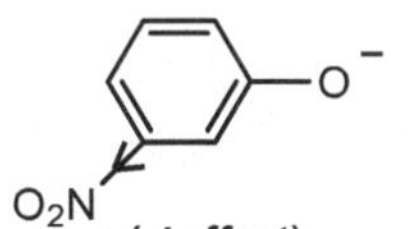

(-*I* effect)

Less stable

(-*M* effect of NO_2 group is not operative)

p-Nitrophenol
($pK_a = 7.14$)

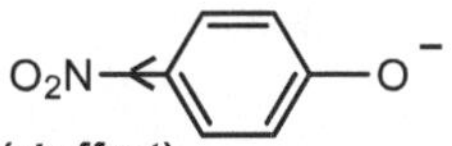

(-*I* effect)

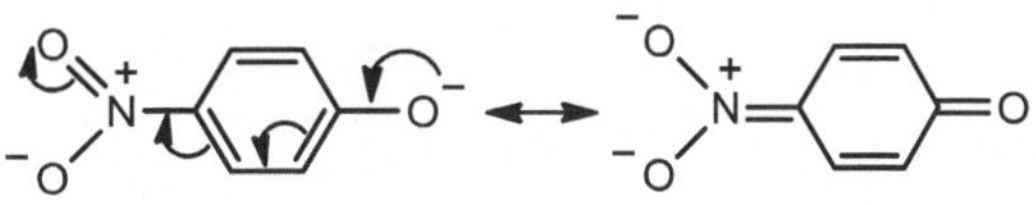

(stabilization of conjugate base by -*M* effect of NO_2 group)

More stable

The introduction of two or more NO_2 groups into the *ortho* and *para* positions promotes the acidity of phenols considerably. Thus **2,4,6-trinitrophenol** (pK_a = 1.0) is **a very strong acid**, called **picric acid.**

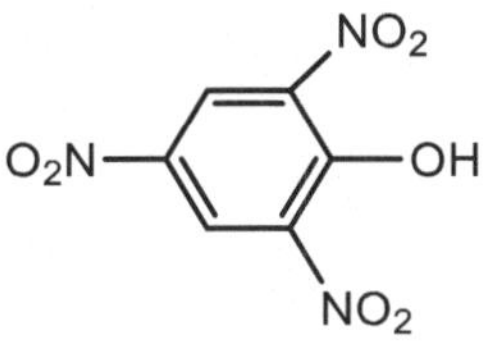

2,4,6-Trinitrophenol
(picric acid)

($pK_a = 1.0$)

5.5.3 Hydrocarbons: The Effect of Hybridization on Acidity

Acetylene (pK_a = 25) is **more acidic** than **ethylene** (pK_a = 44), which in turn, is **more acidic** than **ethane** (pK_a = 50).

The stability of conjugate base anion increases when the negative charge is on the more electronegative atom. A carbon with more s character draws the bonding electrons close to its nucleus, i.e. it is more electronegative. Thus the **electronegativity increases in the order: sp^3 C (25% s) < sp^2 C (33% s) < sp C (50% s).**

A hybrid orbital with more s character will be lower in energy; thus **the energy of hybrid orbitals is in the order: sp < sp^2 < sp^3.**

In the conjugate base of acetylene, the negative charge is on the most electronegative sp carbon, and the lone pair of the anion is in the sp hybrid orbital of lowest energy. Thus **the conjugate base anion of acetylene is most stable.**

In the conjugate base of ethylene, the negative charge is on the sp^2 carbon, which is less electronegative than sp carbon. Here the lone pair is in a higher energy sp^2 orbital. Therefore, the **conjugate base of ethylene is less stable than that of acetylene.**

In the conjugate base of ethane, the negative charge is on the least electronegative sp^3 carbon and the lone pair resides in the highest energy sp^3 hybrid orbital. Therefore, the **conjugate base of ethane is least stable.**

The more stable the conjugate base is, the greater is the acidity. Hence the acidity order of the hydrocarbons is: acetylene > ethylene > ethane.

$HC{\equiv}CH$ (pK_a = 25) $H_2C{=}CH_2$ (pK_a = 44) $H_3C{-}CH_3$ (pK_a = 50)

Relative stability of conjugate base:

$HC{\equiv}\ddot{C}^-$ (sp) > $H_2C{=}\ddot{C}H^-$ (sp^2) > $H_3C{-}\ddot{C}H_2^-$ (sp^3)

Relative acidities of cyclopentadiene and acetylene

Cyclopentadiene is much **more acidic** (pK_a = 16) than acetylene (pK_a = 25) although cyclopentadiene has the C(sp^3)–H bond, whereas acetylene has the C(sp)–H bond.

This is because the **conjugate base anion of cyclopentadiene** with six π electrons is **aromatic** according to the Hückel rule, and is therefore **much more stabilized** relative to the conjugate base of acetylene.

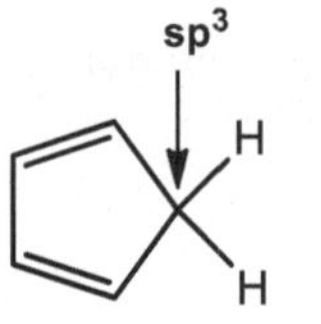

sp^3 H H

$HC{\equiv}CH$

Cyclopentadiene **(pK_a = 16)** ***Stronger acid***

Acetylene **(pK_a = 25)** ***Weaker acid***

Conjugate base: sp^2 $\ddot{C}H^-$ **6 π electrons; aromatic**

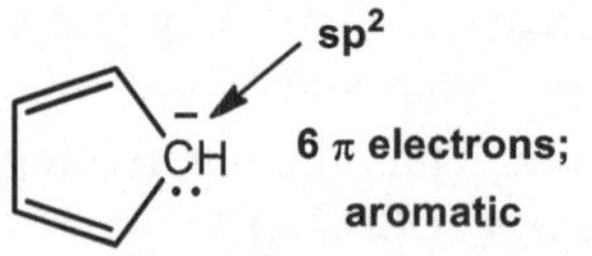

sp $HC{\equiv}\ddot{C}^-$

Cyclopentadienyl anion ***More stable***

Acetylene ion ***Less stable***

PROBLEM 5.5: Explain the following order of acidities:

(a) $H_3C{-}OH > H_3C{-}NH_2 > H_3C{-}CH_3$

(b) $H{-}F < H{-}Cl < H{-}Br < H{-}I$

Answer

(a) These refer to the acidities of O–H, N–H and C–H bonds. The negative charge in the conjugate base anion prefers to be accommodated on a more electronegative atom.

Electronegativity increases across a *horizontal row* of the periodic table from left to right: C < N < O. Therefore, a negative charge is more stable on oxygen than on nitrogen; similarly, a negative charge is more stable on nitrogen than on carbon. As a result, an alcohol (CH_3OH) is more acidic than an amine (CH_3NH_2), which in turn is more acidic than an alkane (CH_3CH_3).

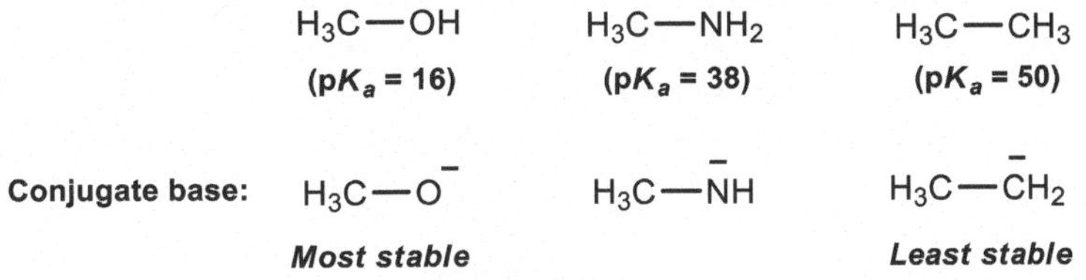

	$H_3C—OH$	$H_3C—NH_2$	$H_3C—CH_3$
	(pK_a = 16)	**(pK_a = 38)**	**(pK_a = 50)**
Conjugate base:	$H_3C—O^-$	$H_3C—\bar{N}H$	$H_3C—\bar{C}H_2$
	Most stable		***Least stable***

(b) HX (X = F, Cl, Br, I) are hydrides of halogen atoms that lie on a *vertical column* of the periodic table. In this case, **the strength of the bond to the hydrogen is the dominating factor to determine the relative acidity**.
The strength of the H–X bond *decreases* with increase in size: H–F > H–Cl > H–Br > H–I. Therefore, even though fluorine is a more electronegative atom than chlorine, HCl is more acidic than HF because H–Cl bond is much weaker (103.1 kcal mol^{-1}) than H–F bond (135.9 kcal mol^{-1}). The weakest bond is the H–I bond, and hence HI is the strongest acid.

	H—F	H—Cl	H—Br	H—I
	(pK_a = 3.2)	**(pK_a = -7)**	**(pK_a = -9)**	**(pK_a = -10)**
Bond energy (kcal mol^{-1}):	**135.9**	**103.1**	**87.4**	**71.4**

5.5.4 Carbonyl Compounds: The Acidity of the α Hydrogens

Carbonyl group (C=O) is a strong electron withdrawing group, which promotes acidity of hydrogen atoms on carbon atoms adjacent to it. The carbon atom adjacent to the carbonyl group is called the **α carbon**, and hydrogen atoms attached to the α carbon are called the **α hydrogens**.

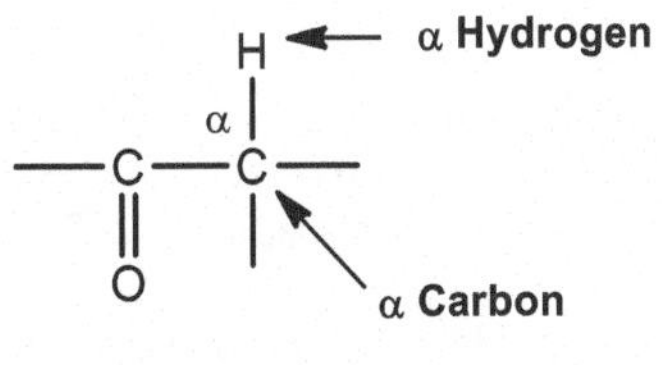

α Hydrogen

α Carbon

The α hydrogens of carbonyl compounds are unusually acidic; they are much more acidic than the hydrogen atom attached to the triple bond in alkynes.

For example, **acetaldehyde** (pK_a = 16.7) and **acetone** (pK_a = 19.3) are **more acidic** than **acetylene** (pK_a = 25). Note that acetaldehyde (CH_3CHO) has *three* α hydrogens, while acetone (CH_3COCH_3) has *six* α hydrogens.

Relative acidities of aldehydes and ketones

The acidity of the α hydrogens of carbonyl compounds is attributed to the **relative stability of their conjugated bases by resonance** when the negative charge in the conjugate base anion is delocalized.

Between acetaldehyde and acetone, **acetone has a lower acidity**. The conjugated bases of both acetaldehyde and acetone are stabilized by resonance. However, acetone has an *additional* electron donating (+*I*) methyl group which tends to decrease the stability of its conjugate base relative to the conjugate base of acetaldehyde. Hence acetone is less acidic than acetaldehyde.

Conjugate base:

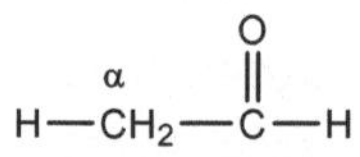

$H—CH_2—C(=O)—H$ (α at CH_2)

Acetaldehyde

(pK_a = 16.7)

More acidic

$\ddot{C}H_2—C(=O)—H \longleftrightarrow H_2C{=}C(O^-)—H$

More stable

$H—CH_2—C(=O)—CH_3$ (α at CH_2 and CH_3)

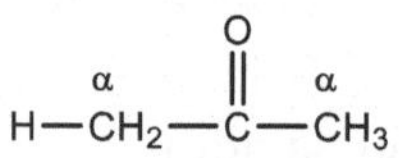

Acetone

(pK_a = 19.3)

Less acidic

$\ddot{C}H_2—C(=O)\leftarrow CH_3 \longleftrightarrow H_2C{=}C(O^-)—CH_3$

(+*I* effect)

Less stable

Relative acidities of ketones and esters

The α hydrogens of esters are less acidic than those of ketones. For example, ethyl acetate (pK_a = 25.1) is less acidic than acetone (pK_a = 19.3).

This is because the **ester carbonyl** attached to a stronger electron donating (+*M*) alkoxy group (OC_2H_5) is **less electron withdrawing** than a keto carbonyl. Due to the +*M* effect of the ethoxy group, delocalization of the negative charge of the conjugate base of ethyl acetate becomes less effective. As a result, the **conjugate base of ethyl acetate** is **less stable** than the conjugate base of acetone. Hence ethyl acetate is a weaker acid.

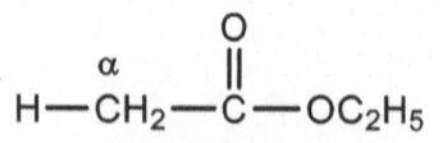

$H—CH_2—C(=O)—OC_2H_5$ (α at CH_2)

Ethyl acetate

(pK_a = 25.1)

Less acidic

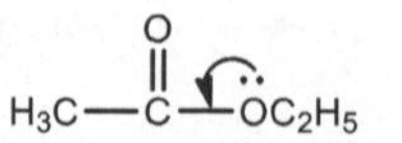

$H_3C—C(=O)—\ddot{O}C_2H_5$

+*M* effect

Stronger

Ester carbonyl

(less electron withdrawing)

$H—CH_2—C(=O)—CH_3$ (α at CH_2)

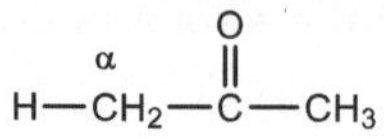

Acetone

(pK_a = 19.3)

More acidic

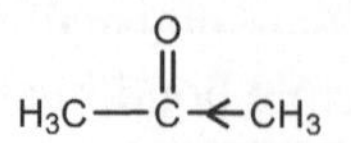

$H_3C—C(=O)\leftarrow CH_3$

+*I* effect

Weaker

Keto carbonyl

(more electron withdrawing)

Conjugate base of ester:

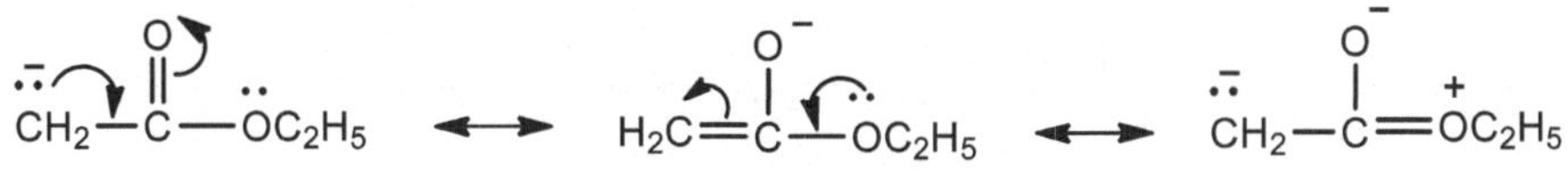

$\ddot{C}H_2-C(=O)-\ddot{O}C_2H_5 \longleftrightarrow H_2C=C(O^-)-\ddot{O}C_2H_5 \longleftrightarrow \ddot{C}H_2-C(O^-)=\overset{+}{O}C_2H_5$

(+*M* effect of OC_2H_5 tends to create a carbanion centre, thereby diminishes delocalization of the negative charge on carbon)

Less stable than the conjugate base of acetone

Acidity of β-dicarbonyl compounds

If an α hydrogen is located on a carbon atom between two carbonyl groups, its acidity is enhanced further. Compounds having a carbon atom flanked by two carbonyl groups are called β-dicarbonyl compounds.

Two examples are **ethyl acetoacetate** (a β-keto ester) and **diethyl malonate** (a malonic ester) that are widely used in organic synthesis.

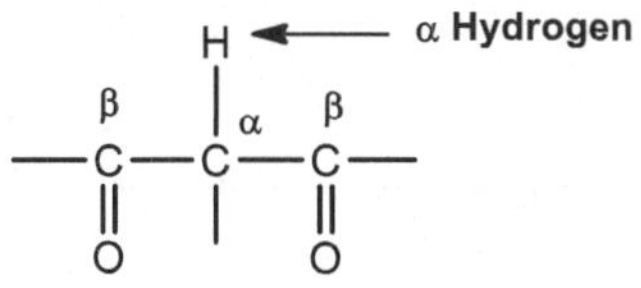

$-\overset{\beta}{C}(=O)-\overset{\alpha}{C}H(-)-\overset{\beta}{C}(=O)-$ (H ← α Hydrogen)

β-Dicarbonyl compounds

Examples

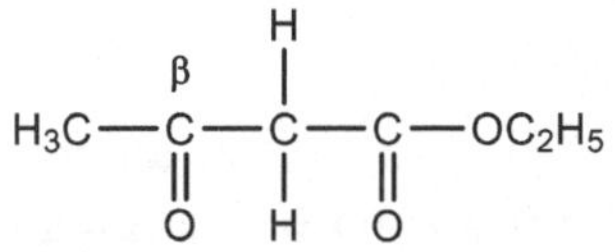

$H_3C-\overset{\beta}{C}(=O)-CH_2-C(=O)-OC_2H_5$

Ethyl acetoacetate (a β-keto ester)

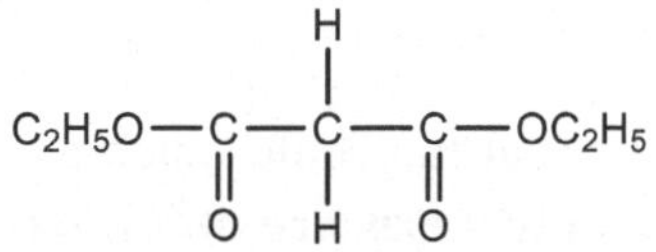

$C_2H_5O-C(=O)-CH_2-C(=O)-OC_2H_5$

Diethyl malonate (a malonic ester)

Both **ethyl acetoacetate** ($pK_a = 10.7$) and **diethyl malonate** ($pK_a = 12.9$) are **more acidic** than **acetone**. This is because the conjugate base of ethyl acetoacetate or diethyl malonate is stabilized by delocalization of the negative charge with *two* carbonyl groups, whereas the conjugate base of acetone is stabilized through a *single* carbonyl group.

Between ethyl acetoacetate and diethyl malonate, **ethyl acetoacetate is more acidic**. The conjugate base of ethyl acetoacetate involves delocalization through a more electron withdrawing keto carbonyl and a less electron withdrawing ester carbonyl, while for the conjugate base of diethyl malonate, the delocalization involves two less electron withdrawing ester carbonyls.

Due to the more effective delocalization by a keto carbonyl compared to an ester carbonyl, the **conjugate base of ethyl acetoacetate** becomes **more stable** than the conjugate base of diethyl malonate. Hence ethyl acetoacetate has a greater acidity.

Conjugate base:

$H_3C-C(=O)-CH(H)-C(=O)-OC_2H_5$

Ethyl acetoacetate
(pK_a = 10.7)

$H_3C-C(=O)-\ddot{C}H^{-}-C(=O)-OC_2H_5 \longleftrightarrow CH_3-C(O^-)=CH-C(=O)-OC_2H_5 \longleftrightarrow H_3C-C(=O)-CH=C(O^-)-OC_2H_5$

More stable

(involves resonance with a more electron withdrawing keto carbonyl and a less electron withdrawing ester carbonyl)

$C_2H_5O-C(=O)-CH(H)-C(=O)-OC_2H_5$

Diethyl malonate
(pK_a = 12.9)

$C_2H_5O-C(=O)-\ddot{C}H^{-}-C(=O)-OC_2H_5 \longleftrightarrow C_2H_5O-C(O^-)=CH-C(=O)-OC_2H_5 \longleftrightarrow C_2H_5O-C(=O)-CH=C(O^-)-OC_2H_5$

Less stable

(involves resonance with two less electron withdrawing ester carbonyls)

5.6 RELATIVE BASICITIES OF ORGANIC COMPOUNDS

A base is a proton acceptor that contains an atom with at least one unshared electron pair (lone pair) for bonding with a proton. The **basic atom or site** is commonly a **nitrogen** or **oxygen atom**.

Bases include neutral molecules such as amines (RNH_2), as well as anions such as amide ion (NH_2^-). **Anionic bases are much stronger than their neutral analogues:**

$NH_2^- >> NH_3$; $HO^- >> H_2O$

Basicities of anionic bases with different basic sites

Amongst anionic bases that contain different basic atoms, the basicity increases with decreasing electronegativity of the basic atom.

For example, the order of basicities of the following anionic bases is:

$\mathbf{R_3C^- > R_2N^- > RO^-}$ (electronegativity: $C < N < O$)

This is because a negative charge on the *less electronegative atom* is relatively *less stable*, and therefore has a more tendency to neutralize its negative charge by accepting a proton.

The **conjugate acids** of these bases are R_3CH (alkanes), R_2NH (amines) and ROH (alcohols), whose acidities decrease in the order: ROH > R_2NH > R_3CH (see Problem 5.5a).

Thus the stronger the conjugate acid, the weaker is the base.

Basicities of anionic bases with the same basic site

In general, the basicity order arises from the relative destabilization of the negative charge of the anionic bases. **The greater the destabilization of the negative charge, the stronger is the basicity**. This is because a less stable anion has a greater tendency to neutralize its negative charge by combining with a proton.

In other words, since an anion is stabilized by the dispersal of the negative charge, a less stable anion has a greater concentration of negative charge, i.e. it is more electron-rich. A lone pair on the more electron-rich basic atom is then more available for protonation. Thus, **the greater the availability of a lone pair of electrons for protonation, the higher is the basicity**.

To illustrate, consider the following anionic bases containing the same basic oxygen atom, and their basicity order:

$$(CH_3)_3CO^- > CH_3O^- > C_6H_5O^- > CH_3CO_2^-$$

Between *t*-butoxide ion [$(CH_3)_3CO^-$] and methoxide ion (CH_3O^-), the electron donating $+I$ effect of *three* CH_3 groups in the *t*-butoxide ion tend to concentrate the negative charge on oxygen and destabilize it more. In other words, a lone pair on the more electron-rich oxygen in the *t*-butoxide ion is more available for protonation. Therefore, ***t*-butoxide ion** is a **stronger base** than **methoxide ion**.

The **phenoxide ion** ($C_6H_5O^-$) is a **weaker base** than CH_3O^- because the negative charge in the phenoxide ion is stabilized by delocalization into the aromatic ring (see also Section 5.5.2), and is less available for protonation.

The **acetate ion** ($CH_3CO_2^-$) is the **weakest base** because delocalization of the negative charge is more effective with two equivalent resonance structures (see also Section 5.5.1), and hence is least available for protonation.

***t*-Butoxide ion**

(negative charge is more concentrated on oxygen; most available for protonation)

Strongest base

Methoxide ion

Phenoxide ion

(delocalization of negative charge; less available for protonation)

Acetate ion

(more effective delocalization of negative charge; least available for protonation)

Weakest base

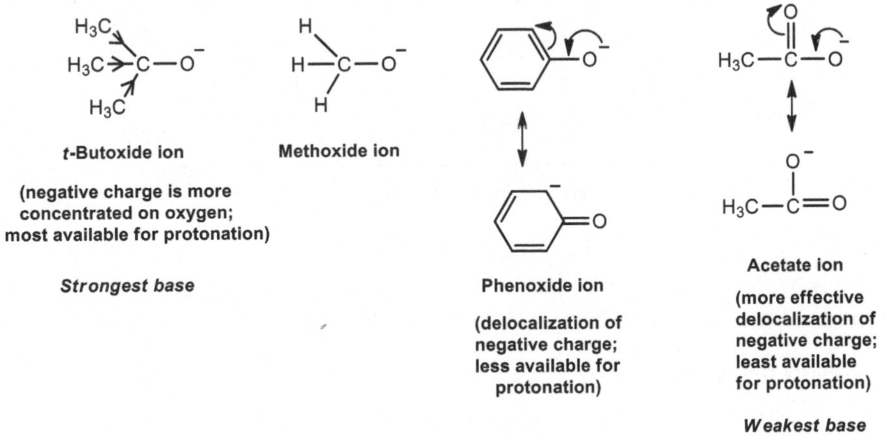

5.6.1 Aliphatic Amines

Aliphatic amines are the most common organic nitrogen bases. They are somewhat stronger bases than ammonia ($pK_a = 9.26$).

Basicity of aliphatic amines in aqueous medium

The basicities of primary amines (RNH_2), secondary amines (R_2NH) and tertiary amines (R_3N) in aqueous medium reveal that secondary amines are the strongest bases.

For example, the basicities of methylamine (1°), dimethylamine (2°) and trimethylamine (3°) in water are given by

	$CH_3\ddot{N}H_2$	$(CH_3)_2\ddot{N}H$	$(CH_3)_3\ddot{N}$
	Methylamine (1°)	**Dimethylamine (2°)**	**Trimethylamine (3°)**
Basicity in water:	**p*K*$_a$ = 10.64**	**p*K*$_a$ = 10.77**	**p*K*$_a$ = 9.80**
		Strongest base	***Weakest base***

(Remember that a higher pK_a value indicates a stronger base.)

The basicity order of amines can be explained on the basis of the relative stability of the conjugate acid of the base: **the greater the stability of the conjugate acid, the stronger is the base.**

In aqueous solution, *two* factors can influence the stability of the **positively charged** conjugate acid. These are **electron donating +*I* effect of the alkyl group**, and **solvation through hydrogen bonding**.

The electron donating +*I* effect of CH_3 group tends to stabilize the conjugate acid by dispersing its positive charge. The greater the number of CH_3 groups attached to nitrogen, the greater is the dispersal of the positive charge, and the higher is the stability. Therefore, the **+*I* effect** tends to stabilize conjugate acids of methylamine, dimethylamine and trimethylamine in the following order:

Stability of conjugate acid:

$$H_3C \rightarrow \overset{+}{N}H_3 \quad < \quad (H_3C \rightarrow)_2\overset{+}{N}H_2 \quad < \quad (H_3C \rightarrow)_3\overset{+}{N}H$$

Increasing stability by the +I effect of methyl groups

On the other hand, the more H atoms attached to nitrogen in the conjugate acid, the greater is the solvation by water through hydrogen bonding. Therefore, **solvation** tends to stabilize conjugate acids of methylamine, dimethylamine and trimethylamine in the following order:

Stability of conjugate acid:

$$CH_3\overset{+}{N}(H\text{---}OH_2)_3 \quad > \quad (CH_3)_2\overset{+}{N}(H\text{---}OH_2)_2 \quad > \quad (CH_3)_3\overset{+}{N}\text{–}H\text{---}OH_2$$

Decreasing stability by solvation through hydrogen bonding

Two factors thus act in opposition, and the stability of the conjugate acid is determined by a balance of these factors.

In the *net*, the **conjugate acid of dimethylamine (2°)** is the **most stable**, since both $+I$ effect and solvation effect contribute significantly. But poorer solvation of the conjugate acid of trimethylamine more than counteracts the combined $+I$ effects of three methyl groups, and as a result, the **conjugate acid of trimethylamine (3°)** is the **least stable**.

Therefore, the basicity order of the amines in aqueous medium is:

$(CH_3)_2NH > CH_3NH_2 > (CH_3)_3N$

Basicity of aliphatic amines in gas phase

In the gas phase, the solvation effect cannot operate. Therefore, the basicity order of primary, secondary and tertiary amines in the gas phase will be determined only by the $+I$ effect, and is given by

Basicity in gas phase: $(CH_3)_3N > (CH_3)_2NH > CH_3NH_2$
(3°) (2°) (1°)

Note: If basicity of amines is measured in a solvent such as **chlorobenzene** in which solvation by hydrogen bonding cannot take place, the basicity order of the amines will be the *same* as in the gas phase: $(CH_3)_3N > (CH_3)_2NH > CH_3NH_2$.

Basicity of amides

Amides ($RCONH_2$) are much less basic than amines (RNH_2), and are practically *neutral*. This is because the lone pair on nitrogen in an amide is effectively delocalized into the carbonyl group, and hence is much less available for protonation.

$$R-C(=O)-\ddot{N}H_2 \longleftrightarrow R-C(-O^-)=\overset{+}{N}H_2$$

(delocalisation of nitrogen lone pair)

An example is **acetamide** (CH_3CONH_2) which is a **very weak base** with a small pK_a value of 0.5.

PROBLEM 5.6: Explain the following:

(a) Guanidine [$HN{=}C(NH_2)_2$] is a strong base (pK_a = 13.6).

(b) A nitrile is a weaker base than an amine.

Answer

(a) Protonation of guanidine is highly favourable because the resulting cation (conjugate acid) is greatly stabilized involving three equivalent resonance structures. Resonance is also possible in guanidine but it is less effective since two of the resonance structures involve a separation of charge.
Thus guanidine behaves as a strong base.

Resonance structures of guanidine:

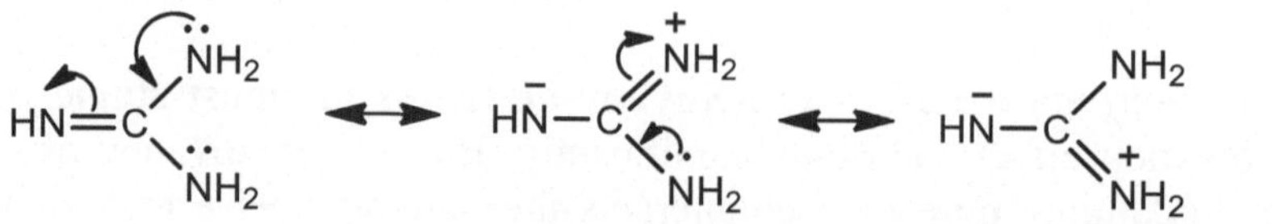

(charge separation) (charge separation)

Less stabilization

Resonance structures of conjugate acid:

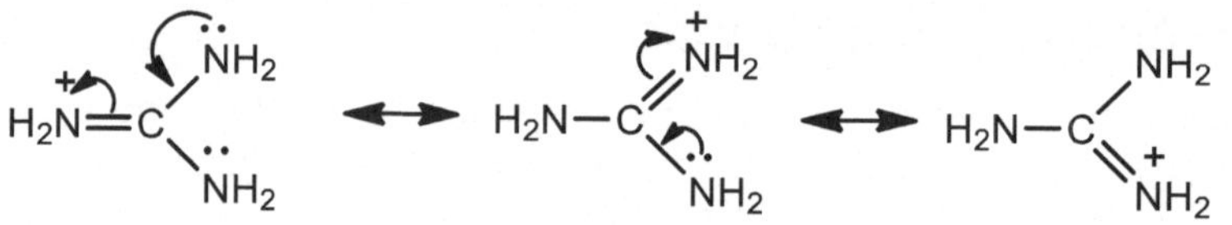

Three equivalent resonance structures

Highly stabilized

(b) The nitrogen atom in a nitrile is sp-hybridized, which is more electronegative than the sp^3 hybridized nitrogen atom in an amine. Therefore, the nitrogen lone pair in a nitrile is more tightly held and not quite available for protonation. Thus a nitrile is less basic than an amine.

An example of a nitrile is acetonitrile (CH_3CN) which has a negative pK_a value of –4.3.

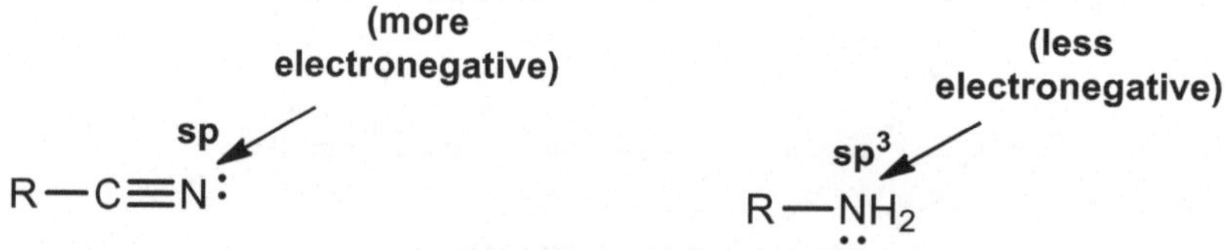

(more electronegative) **(less electronegative)**

A nitrile **An amine**

Less basic ***More basic***

5.6.2 Aromatic Amines

Aromatic amines are much weaker bases than the corresponding aliphatic amines.

For example, **aniline** (pK_a = 4.62) is a very **weak base** compared with **cyclohexylamine** (pK_a = 10.68). This is because the nitrogen lone pair in aniline is delocalized into the ring and is much less available for protonation. Also, the resonance stabilization of aniline as a result of delocalisation of the nitrogen lone pair is lost on protonation (by forming its conjugate base).

In contrast, no delocalisation of nitrogen lone pair occurs in the case of cyclohexylamine, and the lone pair is more available for protonation. Further, since there is no resonance, the formation of the conjugate acid involves no loss of resonance energy, and is more favoured.

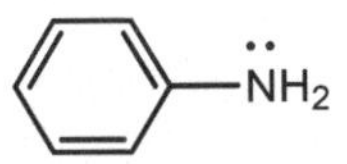

Aniline
(pK_a = 4.62)
Weaker base

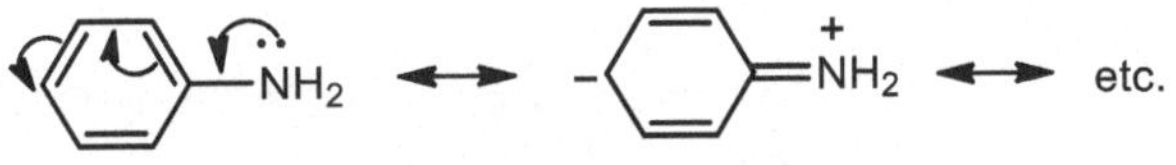

Delocalization of nitrogen lone pair in aniline; lone pair is less available for protonation

Resonance stabilization of aniline

Considering conjugate acid: $\overset{+}{N}H_3$ **(loss of resonance energy)**

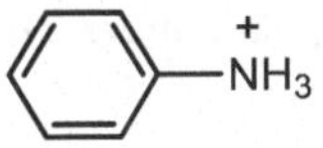

Less favoured

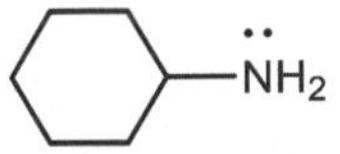

Cyclohexylamine
(pK_a = 10.68)
Stronger base

No delocalization of nitrogen lone pair in cyclohexylamine; lone pair is more available for protonation

No resonance stabilization

Considering conjugate acid: $\overset{+}{N}H_3$ **(no loss of resonance energy)**

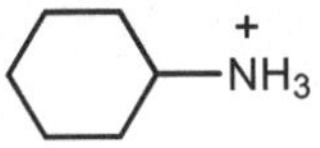

More favoured

Effect of substituents on basicity

An electron withdrawing group on the aromatic ring decreases the basicity of aromatic amines, while an electron donating group increases the basicity.

NO_2 is a powerful **electron withdrawing group** through inductive effect (–*I*) and mesomeric effect (–*M*).

In *p*-niroaniline, the nitrogen lone pair is more effectively delocalized into the aromatic ring, making it almost unavailable for protonation. Thus *p*-niroaniline (pK_a = 0.98) is a **much weaker base** than **aniline**.

O_2N **(-*I*)** **(-*M*)**

***p*-Nitroaniline**
(pK_a = 0.98)
Very weak base

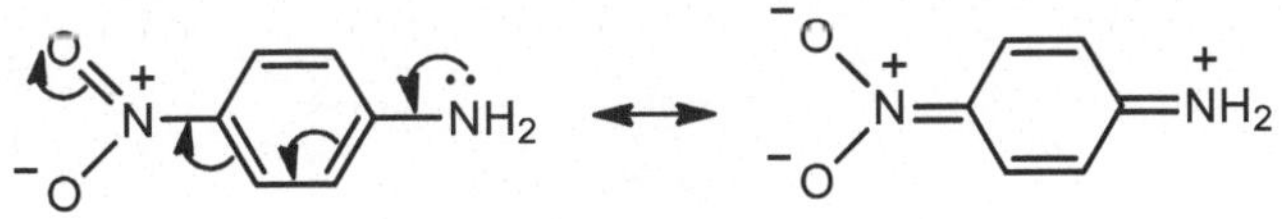

More effective delocalization of nitrogen lone pair; lone pair is much less available for protonation

The substituent OH is electron withdrawing by inductive effect (–*I*), but electron donating by a stronger mesomeric effect (+*M*). Therefore, OH is ***net* electron donating**.

In *p*-hydroxyaniline, OH can exert its +*M* effect by conjugation with the aromatic ring and create a negative charge on the carbon atom adjacent to NH_2 group, which tend to inhibit the delocalisation of the nitrogen lone pair. Thus ***p*-hydroxyaniline (or *p*-aminophenol)** is a **stronger base** (pK_a = 5.30) than **aniline**.

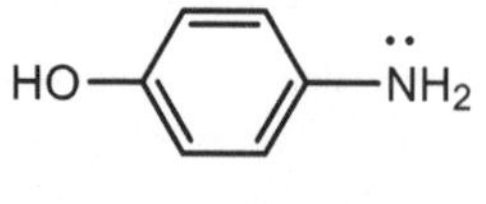

HO—C6H4—$\ddot{N}H_2$; HO—C6H4—$\ddot{N}H_2$ (+M) ⟷ $H\overset{+}{O}$=C6H4=$\ddot{N}H_2$

p-Hydroxyaniline (or p-aminophenol)

(pK_a = 5.30)

Stronger base than aniline

(negative charge adjacent to NH_2 diminishes delcalization of nitrogen lone pair)

Lone pair is more available for protonation than the lone pair in aniline

---

PROBLEM 5.7: Pyridine (pK_a = 5.21) is a much stronger base than pyrrole (pK_a = −0.27). Explain.

Answer

In pyridine, the nitrogen lone pair is not involved in resonance or aromatic delocalization, whereas in pyrrole, the nitrogen lone pair is involved in the aromatic sextet. As a result, the nitrogen lone pair in pyridine is much more available for protonation than that in pyrrole. Thus pyridine is more basic than pyrrole.

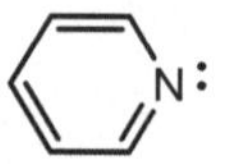

N: ; NH ⟷ $\overset{+}{N}H$

(aromatic; no delocalizaion of nitrogen lone pair)

Pyridine

More basic

(aromatic; delocalisaion of nitrogen lone pair)

Pyrrole

Less basic

SUPPLEMENTARY PROBLEMS

5.8 Isopropyl bromide has a lower boiling point than propyl bromide. Explain.

$CH_3CH_2CH_2Br$; $CH_3CH(Br)CH_3$

Propyl bromide (bp, 71°C)

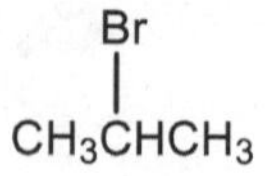

Isopropyl bromide (bp, 59.4°C)

5.9 Both phenol and succinimide are insoluble in water but soluble in aqueous sodium hydroxide. Explain.

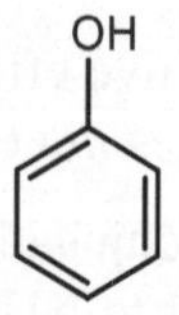

OH ; O, NH, O

Phenol

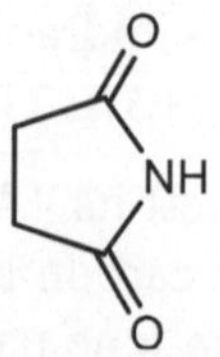

Succinimide

5.10 Account for the following observations:

(a) CO_2 has no dipole moment but SO_2 has a dipole moment of 1.6 D

(b) Dipole moment of CH_2Cl_2 (1.6 D) is greater than that of $CHCl_3$ (1.0 D)

(c) Dipole moment of $CH_2{=}CHCHO$ (3.04 D) is greater than that of CH_3CH_2CHO (2.73 D)

5.11 Arrange the following compounds in decreasing order of their dipole moments:

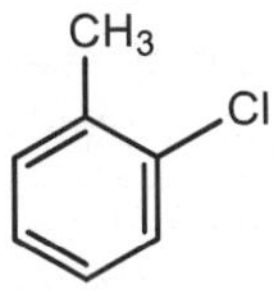

o-Chlorotoluene 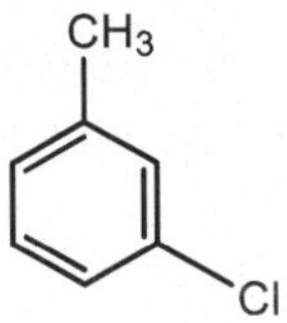 *m*-Chlorotoluene  *p*-Chlorotoluene

[*Ans:* *p*-chlorotoluene (1.9 D) > *m*-chlorotoluene (1.78 D) > *o*-chlorotoluene (1.3 D)]

5.12 Methanethiol (CH_3SH) is a stronger acid than methanol (CH_3OH). Explain.

5.13 Which compound in each of the following pairs is more acidic? Give your reasoning.

(a) CHF_3 and $CHCl_3$

(b) *m*-Chlorophenol and *p*-Chlorophenol

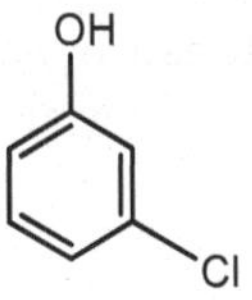

[*Ans:* (a) $CHCl_3 > CHF_3$; (b) *m*-chlorophenol > *p*-chlorophenol]

5.14 Which compound in each of the following pairs is more basic? Explain.

(a) 2,4,6-Trinitroaniline and 2,4,6-Trinitro-*N,N*-dimethylaniline

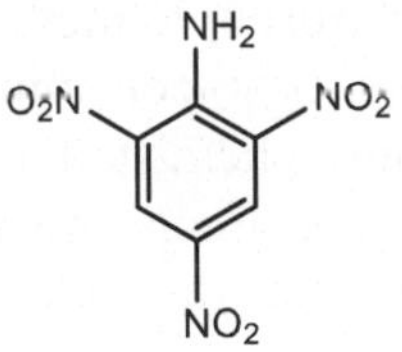

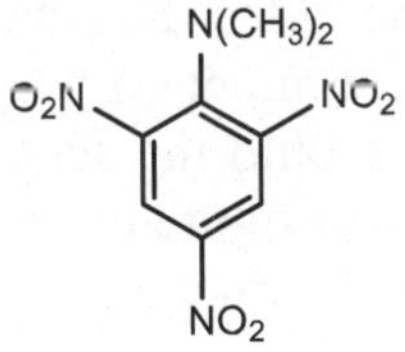

(b) Pyridine and Piperidine

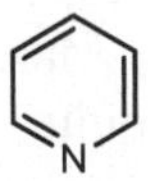

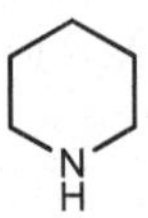

[*Hints:* (a) consider resonance and steric inhibition of resonance; (b) consider hybridization of nitrogen]

CHAPTER 6

Spectroscopy

IN THIS CHAPTER

- Introduction to spectroscopy
- Mass spectrometry (MS): Mass spectrum (molecular ion and base peak), isotopic peaks, nitrogen rule, determination of molecular formula, double bond equivalent (DBE)
- Ultraviolet and visible (UV–Vis) spectroscopy: UV spectrum (λ_{max} and ε_{max}), chromophores, electronic transitions ($\pi \rightarrow \pi^*$ and $n \rightarrow \pi^*$), effect of conjugation, red shift and colour
- Infrared (IR) spectroscopy: IR spectrum (functional group and fingerprint regions), Hooke's law, stretching frequencies, characteristic IR absorptions for C–H, O–H, N–H, C=C, C≡C, C≡N and C=O
- Nuclear magnetic resonance (NMR) spectroscopy: Introduction and theory
- Proton NMR (^{1}H NMR) spectroscopy: ^{1}H NMR spectrum (chemical shift, peak integration and proton counting), shielding and deshielding, chemical shifts of different types of protons, splitting of signals, illustrative examples
- Carbon NMR (^{13}C NMR) spectroscopy: ^{13}C NMR spectrum (broad band proton decoupling), chemical shifts of different types of carbons, illustrative examples
- Determination of structure using spectral data

Physical properties and chemical reactivity of an organic compound are governed by its structure. But how do we know *what structures organic molecules actually have?* It is **spectroscopy** that can provide valuable information about the structure of a compound.

As a structure-analysing tool, several spectroscopic techniques are now widely used by modern organic chemists. Back in the olden days when spectroscopic methods were not available, the method of structure determination was tedious, time-consuming and even unreliable. With spectroscopy coming onto the scene, organic chemists have got a tool that is amazingly powerful, and that can figure out the structure of any organic molecule from the simple to the large and complex molecules like enzymes that catalyse chemical reactions in the biological cell.

By definition, spectroscopy involves light and its interactions with molecules. Spectroscopic methods include ultraviolet and visible (UV-Vis) spectroscopy, infrared (IR) spectroscopy and nuclear magnetic resonance (NMR) spectroscopy. Mass spectrometry (MS) is usually *included* in spectroscopic techniques, but it is quite different from spectroscopy because it does not involve light at all. This technique is aptly called mass spectrometry, *not* mass spectroscopy.

To determine the structure of an organic compound, first we would like to know the **molecular mass** and **atomic composition (or molecular formula)** of the compound. In the early work, a *quantitative analysis* of carbon, hydrogen, nitrogen and halogens led to the *percentage composition* and *empirical formula* of a compound, and *molecular formula* was then derived by determining the molecular weight.

Nowadays, **the molecular mass is determined by mass spectrometry**, which gives a value with incredible accuracy. The measured molecular mass refers to the *relative molecular mass* or *molecular weight* (*M*). The correct **molecular formula** of a compound can also be deduced from mass spectrometry.

While mass spectrometry can determine the molecular mass (and also some of the structures), **UV-Vis spectroscopy** can provide information on π bond conjugation, and **IR spectroscopy** can give us an idea of what kinds of functional groups are present in the molecule. **NMR spectroscopy** is the most versatile and powerful method that can show us reliably how the hydrogen and carbon atoms are distributed and arranged within the molecule, i.e. how the molecule is pieced together.

The following is a **summary** of what each spectroscopic method tells us about the molecular structure:

Spectroscopic method	Structural information
Mass spectrometry (MS)	Molecular mass and composition
Ultraviolet-visible (UV-Vis) spectrocopy	π-Conjugation
Infrared (IR) spectroscopy	Functional groups
Nuclear magnetic resonance (NMR) spectroscopy	Distribution of H and C atoms

In this chapter, we will introduce briefly each of the above methods, describe how each method works, and discuss how we can analyse each kind of spectrum to obtain the structural information. Finally, we will see how all pieces of information can be put together to determine the structure of an organic molecule.

6.1 MASS SPECTROMETRY (MS)

A mass spectrometer converts an organic molecule to positively charged ions, separates them on the basis of their mass-to-charge (m/z) ratios, and records the relative amount (or abundance) of each ion. A plot of ion abundance versus m/z is called the mass spectrum.

Mass spectrometry is performed in the *gas phase*. A small organic sample is introduced into a high-vacuum chamber where it is vaporized by heating under low pressure. Molecules in the gaseous state are then bombarded with a beam of high-energy electrons. As a result of **electron bombardment**, *one* of the electrons of molecules is dislodged, resulting in the formation of a positively charged ion, called the **molecular ion**.

Mass spectrometry involving this type of electron bombardment is called the **electron impact mass spectrometry (EIMS)**.

$$\underset{\textbf{Molecule}}{M} + \underset{\textbf{Electron}}{e^-} \longrightarrow \underset{\textbf{Molecular ion}}{M^{\bullet +}} + 2e^-$$

Radical cation

The positively charged molecular ion contains an *unpaired electron*; it is therefore a radical. Thus the **molecular ion** belongs to a general class of ions called **radical cations**.

The electron bombardment can eject an *electron* from the **σ bond**, **π bond** or **lone pair (lp)** of the molecule to form the **molecular ion**. This is illustrated with alkane, alkene, ketone and amine.

$$\underset{\textbf{Alkane}}{R\overset{\sigma}{-}R} \equiv R\,{:}\,R \xrightarrow{e^-} \underset{\textbf{Molecular ion}}{R \,{}^{\bullet+}\, R} + 2e^-$$

$$\underset{\textbf{Alkene}}{R_2C\overset{\pi}{=}CR_2} \equiv R_2C\overset{\bullet\bullet}{-}CR_2 \xrightarrow{e^-} \underset{\textbf{Molecular ion}}{R_2C\overset{\bullet+}{-}CR_2} + 2e^-$$

$$\underset{\textbf{Ketone}}{R_2C{=}\ddot{O}{:}\ (lp)} \xrightarrow{e^-} \underset{\textbf{Molecular ion}}{R_2C{=}\overset{\bullet+}{O}{:}} + 2e^- \qquad \underset{\textbf{Amine}}{R{-}\ddot{N}H_2\ (lp)} \xrightarrow{e^-} \underset{\textbf{Molecular ion}}{R{-}\overset{\bullet+}{N}H_2} + 2e^-$$

Since the molecular ion has a mass that is essentially the same as the mass of the molecule, mass spectrometry can be used to determine the **molecular mass**.

In most cases, the molecular ion is a *highly energetic* species. Some of the molecular ions stay intact, but others *fragment* spontaneously into smaller neutral radicals and cations (called daughter ions). Only charged particles (molecular ions or daughter ions) can be detected and recorded in a mass spectrometer.

Mass spectrum

A **mass spectrum**, which is a plot of relative ion abundance versus *m*/*z*, is usually presented as a bar graph. The most intense peak in the mass spectrum is called the **base peak**, and this is arbitrarily assigned an intensity of 100%. The intensities of all other peaks are reported as **percentages of the base peak** (i.e. as **relative abundance**).

The base peak can represent the molecular ion or a fragment ion. The molecular ion and most fragments often have a charge of +1 (z =1). The m/z axis then represents the molecular mass of the charged species.

Figure 6.1 shows a bar graph representation of the mass spectrum of **pentane (C_5H_{12})**. Here the base peak represents a fragment at m/z 43, and the small peak at m/z 72 is obviously the molecular ion (the molecular weight of pentane = 72). Notably, the fragment ion at m/z 43 corresponds to the daughter ion $C_3H_7^+$.

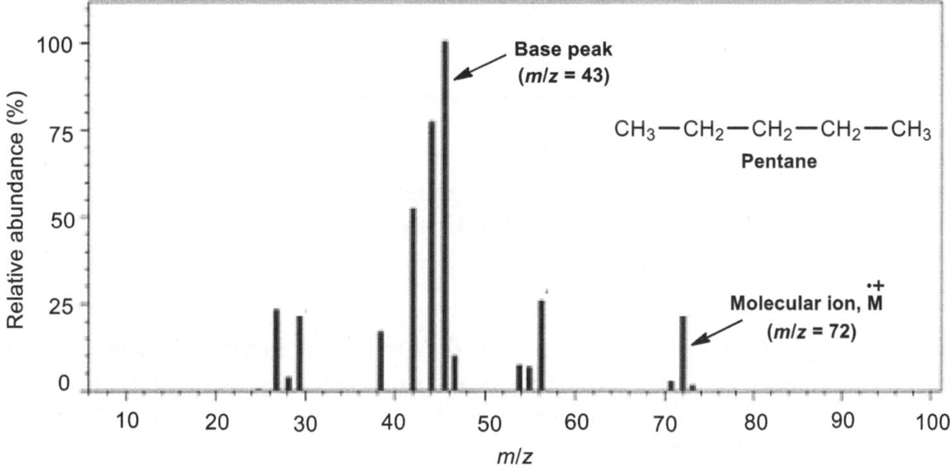

Figure 6.1: The mass spectrum of pentane.

Note: A mass spectrum usually contains a number of peaks because of fragment ions. The fragmentation of molecular ion can take place in a variety of ways and the resulting fragments may undergo further fragmentation. An analysis of the fragmentation pattern can provide important information on the structure of the parent molecule. However, this discussion is beyond our scope.

6.1.1 Molecular Ion and Isotopic Peaks

Many elements have more than one isotope, of which one of them is usually the most abundant. For example, the natural abundance of ^{12}C is 98.9%, while the natural abundance of ^{13}C is only 1.1%.

In a mass spectrum, besides the molecular ion peak, **isotopic peaks** are also observed, but these peaks are generally **much less intense** than the molecular ion peak because of the low natural abundance of those isotopes.

M + 1 isotopic peak

The **principal heavier isotope** of carbon, hydrogen and nitrogen is **^{13}C, ^{2}H** (deuterium) and **^{15}N** respectively, which is **one mass unit greater** than their most common isotope ^{12}C, ^{1}H and ^{14}N.

Thus, compounds containing C, H and N can give rise to an **isotopic peak** at **M + 1**, where **M** indicates the **molecular ion peak**. Note that the *symbol* (·+) for the molecular ion is *omitted* for simplicity.

The natural abundances of ^{13}C, ^{2}H and ^{15}N isotopes that give rise to M + 1 isotopic peaks are shown below.

M + 1 Isotopic peak

Element	Isotope	Natural abundance (%)
Carbon	^{13}C	1.1
Hydrogen	^{2}H	0.016
Nitrogen	^{15}N	0.36

It is seen that the natural abundance of ^{2}H (deuterium) is extremely small (0.016%) and the natural abundance of ^{15}N (0.36%) is much smaller than that of ^{13}C (1.1%). Thus **^{13}C is the most important contributor to the M + 1 peak.**

M + 2 isotopic peak

For oxygen, sulphur, chlorine and bromine, the **principal heavier isotope** is **^{18}O, ^{34}S, ^{37}Cl** and **^{81}Br** respectively, which is **two mass units greater** than their most common isotope ^{16}O, ^{32}S, ^{35}Cl and ^{79}Br. The presence of these elements in a compound can give rise to an **isotopic peak** at **M + 2.**

The ratio of natural abundances of isotopes for oxygen, sulphur, chlorine and bromine is the following:

M + 2 Isotopic peak

Element	Isotopes	Ratio of natural abundance
Oxygen	^{16}O, ^{18}O	99.8 : 0.2
Sulphur	^{32}S, ^{34}S	95.8 : 4.2
Chlorine	^{35}Cl, ^{37}Cl	75.8 : 24.2
Bromine	^{79}Br, ^{81}Br	50.5 : 49.5

Since ^{18}O has a very low natural abundance (0.2%), its contribution to the M + 2 peak is almost negligible. Therefore, **the presence of M + 2 peaks indicates the presence of S, Cl or Br in the molecule.**

A characteristic **3:1 ratio of M and M + 2 peaks** suggests the presence of **one chlorine atom** in the molecule (since the ratio of abundance of ^{35}Cl and ^{37}Cl is nearly 3:1).

Similarly, a characteristic **1:1 ratio of M and M + 2 peaks** indicates the presence of **one bromine atom** in the molecule (the ratio of abundance of ^{79}Br and ^{81}Br being 1:1).

6.1.2 The Nitrogen Rule

A molecule containing carbon, hydrogen and oxygen atoms has a relative molecular mass of *even number*. If the molecule contains nitrogen atoms, the relative molecular mass can be *odd* or *even*.

The nitrogen rule states that a molecule containing odd number of nitrogen atoms will have an odd molecular mass and a molecule containing even number of nitrogen atoms will have an even molecular mass.

The rule can be rationalized as follows. A neutral nitrogen atom can bear *three* hydrogen atoms, whereas a carbon atom can bear *four* hydrogen atoms. Therefore, for **every substitution of a carbon atom by a nitrogen atom**, there is a **loss of one hydrogen atom or one mass unit**. Since a molecule containing carbon and hydrogen atoms has even molecular mass, the presence of odd number of nitrogen atoms would lead to an odd molecular mass (because even – odd = odd). Similarly, the presence of even number of nitrogen atoms would give an even molecular mass (even – even = even).

Thus, if the molecular mass is **odd**, the compound contains **odd number of nitrogen atoms**. But if the molecular mass is **even**, the compound contains either **no nitrogen atoms or even number of nitrogen atoms**.

The following examples illustrate the nitrogen rule. *M* denotes relative molecular mass or molecular weight.

$CH_3CH_2NH_2$	$H_2NC(=NH)NH_2$	$H_2NCH_2CH_2NH_2$	CH_3CH_2OH
(*M* = 45; odd)	**(*M* = 59; odd)**	**(*M* = 60; even)**	**(*M* = 44; even)**
1 N atom	**3 N atoms**	**2 N atoms**	**No N atom**
Odd	***Odd***	***Even***	***Zero***

6.1.3 Determination of Molecular Formula

The molecular formula of an organic compound can be determined from its mass spectrum using the following procedure:

1. Estimate the number of carbon atoms from the relative abundance of M + 1 peak (expressed as the percentage of the intensity of M peak) and the natural abundance of ^{13}C:

$$\text{Number of C atoms} = \frac{\text{Relative abundance of M + 1 (as the percentage of M)}}{\text{Natural abundance of } ^{13}\text{C}} = \frac{\text{Relative abundance of M + 1 (as the percentage of M)}}{1.1}$$

2. Determine the presence of nitrogen atoms using the nitrogen rule.
3. Determine the presence of a sulphur, chlorine or bromine atom from the ratio of M and M + 2 peaks, which is approximately 23:1 for S, 3:1 for Cl, and 1:1 for Br.

4. Calculate the molecular formula by determining the number of H atoms, and adding the appropriate number of O atoms, if necessary.

Example 1

Consider the following mass spectral data for an unknown compound:

m/z	27	28	29	39	41	42	43	44	72	73	74
Relative abundance (%)	59.0	15.0	54.0	23.0	60.0	12.0	79.0	100	73.0	3.3	0.2

The mass spectrum shows the molecular ion (M) peak, fragment peaks, as well as M +1 and M + 2 peaks. The *m/z* value for the molecular ion must be *higher* than any of its fragments. Therefore, **the *m/z* values of 72, 73 and 74 indicate M, M + 1 and M + 2 peaks respectively**.

	m/z	Relative abundance (%)
M →	72	73.0
M + 1 →	73	3.3
M + 2 →	74	0.2
	44	100.0 (base peak)

Note that the molecular ion peak at *m/z* 72 is *not* the base peak; rather a fragment peak at *m/z* 44 is the base peak.

1. To determine the number of C atoms, M + 1 peak needs to be expressed as the percentage of M peak:

$$\text{Relative abundance of M + 1 (as \% of M)} = \frac{\text{Relative abundance of M + 1}}{\text{Relative abundance of M}} \times 100 = \frac{3.3}{73} \times 100 = 4.5$$

$$\text{Therefore, number of C atoms} = \frac{4,5}{1,1} = 4$$

2. Since M at *m/z* 72 is **even**, the compound contains either no N atom or even number of N atoms. But with 2 N atoms (smallest even number), the formula C_4N_2 gives a molecular weight = 76, which is greater than the molecular weight of the unknown compound. Therefore, the compound contains **no N atom**.
3. With M + 2 = 0.2%, S, Cl and Br are absent but O may be present.
4. If the molecule contains 4 C atoms and H atoms only, the number of H atoms = 72 – 48 = 24, but C_4H_{24} is impossible. So the molecule contains **one O atom** (consistent with M + 2 = 0.2%). Thus the number of **H atoms** = 72 – 48 – 16 = **8**.

Therefore, the compound has the molecular formula C_4H_8O.

Example 2

Consider the following mass spectral data for an unknown compound:

m/z	14	15	18	28	29	42	43	44	73	74	75
Relative abundance (%)	8.0	38.6	16.3	39.7	23.4	46.6	10.7	100	86.1	3.2	0.2

The m/z value for the molecular ion must be higher than any of its fragments. Thus the molecular ion (M) peak is at m/z 73, and M + 1 and M + 2 peaks are at m/z values of 74 and 75 respectively.

1. The M + 1 peak as the percentage of M peak is determined and the number of C atoms is then calculated.

$$\text{Relative abundance of M + 1 (as \% of M)} = \frac{3.2}{86.1} \times 100 = 3.7$$

$$\text{Number of C atoms} = \frac{3.7}{1.1} = 3$$

   (Note that the number of C atoms is taken to be the nearest whole number.)
2. Since M at m/z 73 is **odd**, the molecule has **one N atom** (odd number). It cannot contain 3 N atoms (the next higher odd number), because the formula C_3N_3 gives a molecular weight of 78 which is greater than the molecular weight of the unknown compound.
3. With M + 2 = 0.2%, S, Cl and Br are absent but O may be present.
4. If the molecule contains 3 C atoms, 1 N atom and H atoms only, the number of H atoms = 73 – 36 – 14 = 23, but $C_3H_{23}N$ is impossible. So the molecule contains one O atom, when the number of H atoms = 23 – 16 = 7.

   Therefore, the molecular formula of the compound is C_3H_7NO.

   Note: High resolution mass spectrometry can determine the molecular mass with an accuracy of several digits after the decimal point. Since actual masses of different atoms are *not* integers, an accurate measurement of the molecular mass can distinguish two or more different molecules. The determination of molecular formula is thus possible (see Problem 6.2).

PROBLEM 6.1: An organic compound gives the molecular ion (M) peak at m/z 122 and a peak of nearly equal relative abundance at m/z 124. Suggest the molecular formula of the compound.

Answer

The peak at m/z 124 is the M + 2 peak. A characteristic 1:1 ratio of M and M + 2 peaks indicates the presence of one Br atom (atomic mass = 79 amu) in the molecule. The residual mass (122 – 79 = 23) corresponds to the alkyl group C_2H_5. Therefore, the molecular formula is C_2H_5Br.

PROBLEM 6.2: An unknown organic compound could be either $C_6H_{12}O$ or $C_4H_4O_3$. High resolution mass spectrometry of the compound gives an actual mass of 100.08871.

Identify the correct molecular formula of the compound. (Given: atomic mass (in amu), $^{1}H = 1.00782$, $^{12}C = 12.00000$, $^{16}O = 15.99491$)

Answer

Relative molecular mass corresponding to each molecular formula is calculated as:

$C_6H_{12}O = 6(12.00000) + 12(1.00782) + 15.99491 = 100.08875$
$C_4H_4O_3 = 4(12.00000) + 4(1.00782) + 3(15.99491) = 100.01601$

The relative molecular masses corresponding to the two molecular formulas can be distinguished by an extremely accurate mass determination by high resolution mass spectrometry. Since the measured mass (100.08871) conforms to the calculated molecular mass of 100.08875, the correct molecular formula of the unknown compound is $C_6H_{12}O$.

6.1.4 Double Bond Equivalent (DBE)

Molecular formula itself can provide some information on the structure of the molecule.

The general formula of *alkanes* is C_nH_{2n+2}, whereas the general formula of *alkenes* (one double bond) or *cycloalkanes* (one ring) is C_nH_{2n}. Thus there is a **loss of two H atoms** for the presence of **a double bond** or **a ring** in the molecule.

A double bonds of any kind (such as C=C, C=O, C=N, etc.) or a ring of any size is called a double bond equivalent (DBE).

The general formula of *alkynes* is C_nH_{2n-2}, which indicates the *loss of four H atoms* relative to alkanes. Thus **a triple bond has DBE = 2**.

The possible structural information from DBE = 0, 1 or 2 is summarized here.

Double bond equivalent (DBE)	Possible structural information
0	**Saturated**
1	**1 double bond or 1 ring**
2	**2 double bonds or 1 triple bond or 1 double bond and 1 ring or 2 rings**

Benzene has *one ring* and *three double bonds*, therefore, it has **DBE = 4**. So the presence of a benzene ring in an organic compound contributes a DBE of 4.

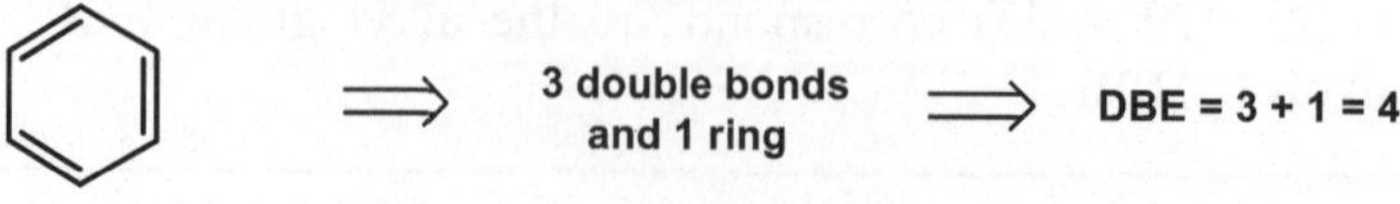

Benzene

General formula

It is possible to calculate the double bond equivalent (DBE) from the molecular formula of an organic compound. **The presence of oxygen atoms does not make any difference** in the calculation, but **the presence of nitrogen or halogen atoms does.**

For a general formula $C_aH_bN_cO_dX_e$ (X = halogen), the DBE can be calculated from the following expression:

Molecular formula: $C_aH_bN_cO_dX_e$ (X = halogen)

$$\mathrm{DBE} = (a + 1) - \frac{(b - c + e)}{2}$$

To illustrate, consider the following examples:

Example 1

$C_4H_9NO_2$ DBE = (4 + 1) − (9 - 1)/2 = 1

(a = 4, b = 9, c = 1)

Example 2

$C_6H_4NO_2Cl$ DBE = (6 + 1) − (4 - 1 + 1)/2 = 5

(a = 6, b = 4, c = 1, e = 1)

6.1.5 Other Mass Spectrometric Techniques

We have introduced a mass spectrometric technique that ionizes organic molecules by bombardment with a beam of high-energy electrons. This is called the **electron impact mass spectrometry (EIMS)**. A problem with the EIMS is that the high energy of the bombarding electrons might cause the molecule to fragment completely in some cases, so as to lose all traces of the molecular ion.

This problem can be avoided by using a number of other 'soft' ionization techniques, such as **chemical ionization (CI)** and **electrospray ionization (ESI)**.

In chemical ionization, an organic molecule is protonated in the gas phase to form a **cation** (not a radical cation as in EI), and the mass spectrum gives the $[M + H]^+$ or M + 1 peak (one mass unit higher than the molecular mass).

In electrospray ionization, an aerosol of organic molecules is ionized, and ionization in the presence of sodium ions gives the $[M + Na]^+$ or M + 23 peak.

It is also possible to perform mass spectrometry of **biomolecules** such as proteins and nucleic acids using 'soft' ionization techniques. These include **electrospray ionization (ESI), matrix-assisted laser desorption ionization (MALDI)**, and **fast atom bombardment (FAB)** mass spectrometry. These techniques have been useful for the analysis of amino acid sequence of proteins, nucleotide sequence of nucleic acids, enzyme-substrate complexes, and drug-receptor interactions.

6.2 THE ELECTROMAGNETIC SPECTRUM

Ultraviolet and visible (UV–Vis), infrared (IR), and nuclear magnetic resonance (NMR) spectroscopy involves interaction of molecules with light.

Light is **electromagnetic radiation**, which is described in terms of its **wavelength (λ)** or its **frequency (ν)**. The product of wavelength and frequency gives the **velocity (*c*)** of light:

$$c = \lambda\nu$$

It is of note that all electromagnetic radiations travel through vacuum at the same velocity ($c = 3 \times 10^8$ m s^{-1}).

The **energy (*E*)** of a quantum of electromagnetic radiation is *directly proportional* to its frequency (ν) or *inversely proportional* to its wavelength (λ):

$$E = h\nu = \frac{hc}{\lambda}$$

where h = Planck's constant. The value of h is 6.63×10^{-34} J s.

The equation implies that the higher the frequency, the higher is the energy or the longer the wavelength, the lower is the energy.

Regions of the electromagnetic spectrum

Organic molecules can absorb electromagnetic radiation in different regions of wavelength (or frequency) of the electromagnetic spectrum, which leads to **absorption spectroscopy**.

The different regions of the electromagnetic spectrum with respect to ultraviolet, visible, infrared and radio waves, *in terms of wavelength*, are shown in Figure 6.2.

The **wavelengths** are generally expressed in **nanometre (nm)** (1 nm = 10^{-9} m = 10×10^{-10} m = 10 Å).

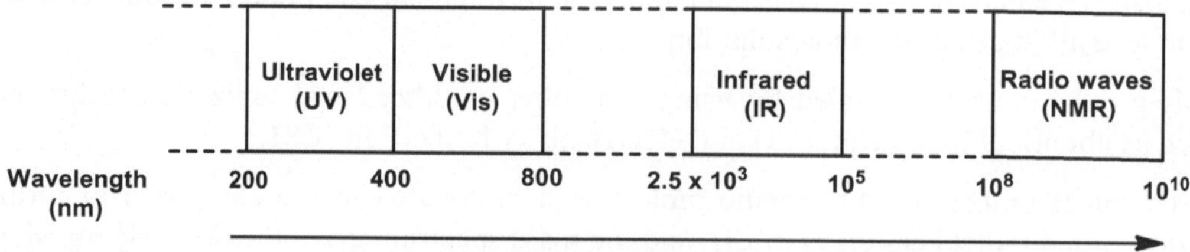

Figure 6.2: Ultraviolet and visible (UV–Vis), infrared (IR) and radio waves (NMR) regions of the electromagnetic spectrum in terms of wavelength (nm). 1 nm = 10^{-9} m.

The **energy** of electromagnetic radiation decreases in the order: **UV > Vis > IR > Radio waves.**

The UV–Vis spectroscopy and the IR spectroscopy involve absorptions of electromagnetic radiation in their respective regions, while the NMR spectroscopy involves absorption of electromagnetic radiation in radio waves region.

6.3 ULTRAVIOLET AND VISIBLE (UV–VIS) SPECTROSCOPY

In UV–Vis spectroscopy, the wavelengths of electromagnetic radiation are usually expressed in nm (1 nm = 10 Å). The **UV region is 200–400 nm**, and the **visible region is 400–800 nm**.

When electromagnetic radiation in the UV or visible region passes through a sample of organic molecules, a portion of the radiation is absorbed. The absorption of radiation causes the excitation or promotion of an *electron* from a molecular orbital of *lower energy* to another molecular orbital of *higher energy*. This process is called an **electronic transition**, and the UV/Vis spectra are often called the **electronic spectra**.

The instrument, called the **UV–Vis spectrophotometer**, measures the amount of light absorbed at each wavelength of the radiation. Due to absorption of light, **the intensity of transmitted light (I)** emerging from the sample will be **less** than **the intensity of incident light (I_0)** striking the sample.

The absorption of light at a particular wavelength is estimated as the **absorbance (A)**, which is defined as,

$$A = \log_{10}(\frac{I_0}{I})$$

An older term for absorbance is known as *optical density* (OD).

UV spectrum

A plot of absorbance (A) at each wavelength versus wavelengths of the region gives a graph, called the **UV/Vis spectrum**.

Figure 6.3 shows a typical **UV absorption spectrum** of **isoprene** (2-methyl-1,3-butadiene) in methanol. The spectrum exhibits a broad absorption peak. **The wavelength corresponding to the absorption peak or absorption maximum is reported as λ_{max}.** Isoprene gives a λ_{max} of 222 nm.

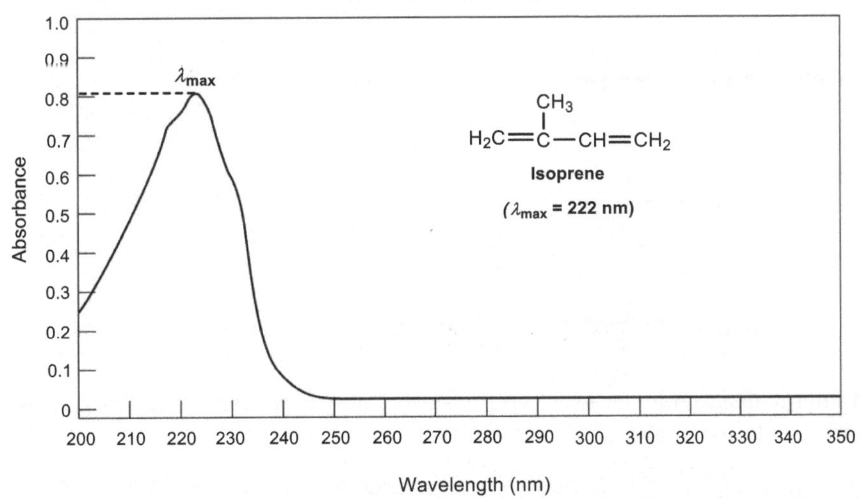

Figure 6.3: The UV absorption spectrum of isoprene.

Lambert-Beer expression

The **absorbance (A)** is related to the **molar concentration (c)** of the sample and the **length (l)** of the path of the light through the sample according to the **Lambert-Beer expression**:

$$A = \varepsilon c l$$

$$\text{or, } \varepsilon = \frac{A}{cl}$$

where ε = **molar absorptivity** or **molar extinction coefficient**, c = concentration in mol L^{-1}, and l = path length (i.e. length of the sample cell) in cm. The **unit of ε is L mol^{-1} cm^{-1}**.

The **intensity (or strength)** of the absorption at λ_{max} (the wavelength of maximum absorption) is reported as $\boldsymbol{\varepsilon_{max}}$.

Electronic transitions

On absorption of UV light, an electron from a bonding or nonbonding molecular orbital is excited to a higher-energy antibonding molecular orbital.

A **lone pair** of electrons on an atom in a molecule resides in a **nonbonding molecular orbital**, which has *higher energy* than the energy of a bonding molecular orbital but has *lower energy* than the energy of an antibonding molecular orbital.

Two types of electron excitations are most common. These are:

(a) Excitation of an electron from a bonding π molecular orbital to an antibonding π* molecular orbital, called the **π → π* transition**

(b) Excitation of an electron from a nonbonding molecular orbital (n) to an antibonding π* molecular orbital, called the ***n* → π* transition**.

These transitions are diagrammatically shown in Figure 6.4.

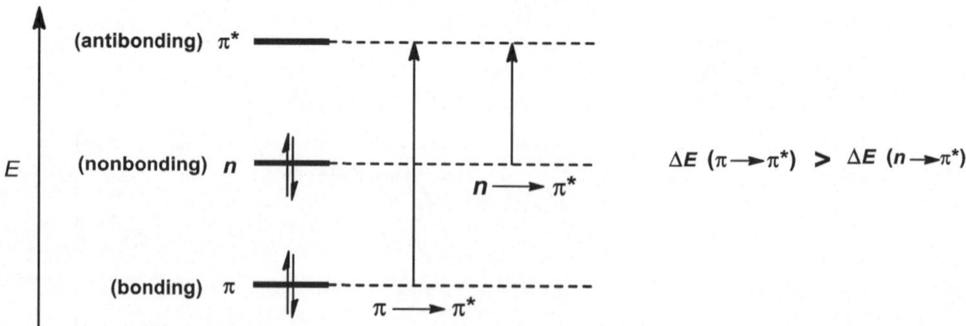

Figure 6.4 Energy diagram for π → π* and n → π* transitions in UV spectroscopy. Nonbonding electrons are usually lone pairs.

The energy diagram shows that the energy difference (ΔE) for π → π* transition is greater than the energy difference for n → π* transition, i.e. ΔE (π → π*) > ΔE (n → π*).

In general, if the energy difference between two energy levels is greater, the absorption of radiation occurs at a shorter wavelength (since ΔE is *inversely* proportion to wavelength, λ). Therefore, **π → π* transition (larger energy difference) leads to a shorter λ_{max} compared to a λ_{max} for *n* → π* transition (smaller energy difference)**.

The molar absorptivity (ε_{max}) denotes the intensity of the absorption peak, which depends on the probability of the electronic transition.

The π → π* transition has a high probability and is said to be **'allowed'**; it gives a **large value of ε_{max}**. But the ***n* → π*** transition has a low probability and is said to be **'forbidden'**; it gives a **small value of ε_{max}**.

Examples

Consider an alkene such as **ethylene** ($H_2C{=}CH_2$). It has π and π* molecular orbitals. The energy difference between these orbitals is so large that the π → π* transition for ethylene occurs at a **very short wavelength** below the UV region (**λ_{max} = 171 nm**). The **intensity** of the absorption is, however, **very high** (**ε_{max} = 15,530**) because the transition is 'allowed'.

Next, consider a ketone such as **acetone** (CH_3COCH_3). It has a carbon-oxygen double bond (C=O) and two lone pairs on the oxygen atom. Therefore, acetone possesses π, π* and *n* (nonbonding) orbitals. Both *n* → π* and π → π* transitions are now possible.

Acetone shows a broad absorption peak (**λ_{max} = 280 nm**) in the UV region with a very low intensity (**ε_{max} = 15**). This absorption is due to the ***n* → π*** transition, which leads to a greater λ_{max} (due to a smaller energy difference) and a very small ε_{max} (as the transition is 'forbidden').

The high-energy π → π* transition for acetone gives a **λ_{max} at 160 nm** below the UV region with a high **ε_{max} of 18,000.**

The above results for ethylene and acetone are summarized here.

Compound	Transition	λ_{max} (nm)	ε_{max}
$H_2C{=}CH_2$	π → π*	171	15,530
$(H_3C)_2C{=}\ddot{O}:$	π → π*	160	18,000
	n → π*	280	15

Chromophore

In UV–Vis absorption spectroscopy, **any functional group that absorbs electromagnetic radiation is called a chromophore**. For example, the carbonyl group (C=O) is a chromophore.

The term chromophore comes from the Greek *chromophoros*, meaning colour carrier, and was originally applied to a group that imparts colour to a compound.

6.3.1 Effect of conjugation: Red shift

Like simple alkenes, nonconjugated dienes such as 1,4-pentadiene (CH_2=$CHCH_2CH$=CH_2) usually gives λ_{max} *below* 200 nm.

However, compounds containing conjugated double bonds have λ_{max} *above* 200 nm. For example, **1,3-butadiene** (CH_2=CH–CH=CH_2) in which two double bonds are conjugated, has a $\boldsymbol{\lambda_{max}}$ **at 217 nm**.

Another example is **isoprene** (2-methyl-1,3-butadiene) which shows a slightly greater $\boldsymbol{\lambda_{max}}$ **at 222 nm** (see Figure 6.3).

It is a general principle that conjugation shifts the absorption to longer wavelength. The greater the number of conjugated double bonds, the larger is the value of λ_{max}.

The shift to longer wavelength is called a **bathochromic shift or red shift**. The term red shift is used in analogy with a shift toward the red end (longest wavelength) of the visible spectrum (VIBGYOR). It may be mentioned that a shift to shorter wavelength is called a hypsochromic shift or blue shift.

Theory

The shift of absorption of electromagnetic radiation to a longer wavelength as a result of conjugation can be explained using an energy diagram as shown in Figure 6.5.

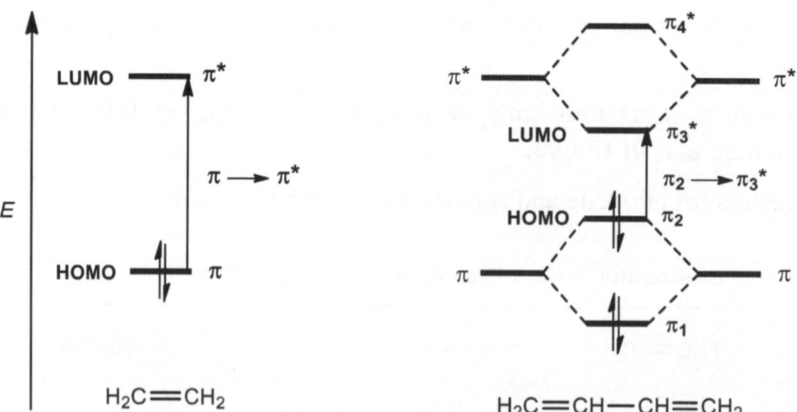

Figure 6.5: Relative energies of π molecular orbitals of a simple alkene (ethylene) and a conjugated diene (1,3-butadiene), and the HOMO/LUMO transitions involved in UV spectroscopy. HOMO = highest occupied molecular orbital; LUMO = lowest unoccupied molecular orbital.

A simple alkene such as ethylene (CH_2=CH_2) has a bonding π orbital and an antibonding π* orbital. Two electrons of the π bond occupy the lower energy π orbital. So the π electron configuration of ethylene is $\pi^2\pi^{*0}$, i.e. π is the occupied molecular orbital and π* is the unoccupied molecular orbital. In general terminology, **π** is called the highest occupied molecular orbital (**HOMO**) and **π*** is called the lowest unoccupied molecular orbital (**LUMO**). (Remember that π orbital has higher energy than σ orbital but π* orbital has lower energy than σ*orbital; cf. Figure 3.4).

In a conjugated diene such as 1,3-butadiene ($CH_2{=}CH{-}CH{=}CH_2$), the two bonding π orbitals of separate alkene groups can combine to form **two new bonding molecular orbitals π_1 and π_2**. Similarly, the two antibonding π* orbitals of separate alkene groups can combine to form **two new antibonding molecular orbitals π_3^* and π_4^***. Four π electrons of the diene occupy the bonding orbitals π_1 and π_2. So the π electron configuration of 1,3-butadiene is $\pi_1^2\pi_2^2$. Thus π_2 is the occupied orbital of higher energy; it is called the highest occupied molecular orbital (**HOMO**). The antibonding orbitals π_3^* and π_4^* are unoccupied or vacant. Of these, π_3^* has the lower energy; it is called the lowest unoccupied molecular orbital (**LUMO**).

From the energy diagram, it can be seen that a π → π* type of transition, **π_2 (HOMO) → π_3^* (LUMO)**, is now possible for 1,3-butadiene as a result of conjugation. The *energy difference* between the HOMO (π_2) and the LUMO (π_3^*) for 1,3-butadiene is *much less* than the energy difference between the HOMO (π) and the LUMO (π*) for ethylene:

$$\Delta E\ (\pi_2 \rightarrow \pi_3^*) < \Delta E\ (\pi \rightarrow \pi^*)$$

Therefore, 1,3-butadiene absorbs at a longer wavelength in the UV region (λ_{max} = 217 nm) than ethylene does (λ_{max} = 171 nm).

In general, conjugation lowers the HOMO/LUMO energy difference and thereby shifts the absorption to longer wavelength.

PROBLEM 6.3: Methyl vinyl ketone ($CH_2{=}CHCOCH_3$) shows the following absorption maxima in its UV spectrum. Identify the transition responsible for each absorption peak.

λ_{max} 219 nm, ε_{max} 3600

λ_{max} 324 nm, ε_{max} 24

Answer

$H_2C{=}CH{-}C(CH_3){=}O$

Methyl vinyl ketone

Methyl vinyl ketone is a conjugated ketone in which the carbonyl group (C=O) is conjugated with a carbon-carbon double bond (C=C).

Two kinds of transitions are possible: π → π* and *n* → π*. The π → π* transition for the conjugated system occurs in the UV region but at a shorter wavelength than the *n* → π* transition. The former is more intense ('allowed' transition) than the latter ('forbidden' transition). Thus, the first peak (λ_{max} 219 nm, ε_{max} 3600) is due to the π → π* transition, and the second peak (λ_{max} 324 nm, ε_{max} 24) is due to the *n* → π* transition.

PROBLEM 6.4: Two compounds **A** and **B** have the same molecular formula C_6H_8. Both compounds react with 2 molar equivalents of H_2 in the presence of Pt catalyst to yield cyclohexane. Compound **A** shows an absorption peak at 256 nm, whereas compound **B** shows no absorption maximum beyond 200 nm. Propose structures for **A** and **B**.

Answer

For the molecular formula C_6H_8, DBE = (6 + 1) – (8/2) = 3.

Since both **A** and **B** on hydrogenation give cyclohexane, each compound contains a six-membered ring. Again, as both compounds require 2 moles of H_2 for complete hydrogenation, they may contain 2 double bonds or 1 triple bond. But a triple bond requires a linear arrangement of four carbons in a six-membered ring, which would lead to a highly strained ring. Therefore, the presence of a triple bond is unlikely.

Hence **A** and **B** both contain a six-membered ring with two double bonds, consistent with DBE = 3. Since **A** gives a λ_{max} at 256 nm in the UV region, it is a conjugated diene (1,3-cyclohexadiene). In contrast, **B** shows no λ_{max} in the UV region (above 200 nm), and hence it is a nonconjugated diene (1,4-cyclohexadiene).

+ 2 H_2 —Pt→

A

(conjugated,
λ_{max} 256 nm)

1,3-Cyclohexadiene

+ 2 H_2 —Pt→

B

(nonconjugated,
λ_{max} < 200 nm)

1,4-Cyclohexadiene

6.3.2 Aromatic Compounds

The **benzene ring** of aromatic compounds contains a conjugated π system that can act as a **chromophore**.

For benzene itself, several $\pi \rightarrow \pi^*$ transitions are possible, which give rise to absorption bands at 184, 204 and 254 nm.

Substituents such as NH_2 and OH cause a shift of λ_{max} to the longer wavelength (red shift).

For example, the **λ_{max} at 204 nm** for benzene is shifted to a **λ_{max} at 230 nm** for **aniline**. This is because the nitrogen lone pair in aniline can enter into conjugation with the benzene ring, resulting in extended conjugation which shifts the absorption maximum to the longer wavelength. (The λ_{max} at 254 nm for benzene also shifts to a λ_{max} at 280 nm for aniline.)

Interestingly, **in aqueous HCl, aniline shows a λ_{max} at 203 nm**, similar to benzene. This happens because aniline is protonated in aqueous HCl and the nitrogen lone pair is no longer available for conjugation with the benzene ring. As a result, the anilinium cation behaves like benzene.

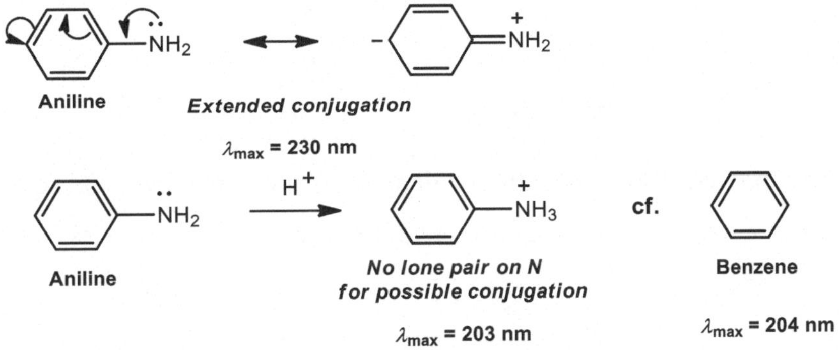

Aniline ⟷ Extended conjugation

λ_{max} = 230 nm

Aniline $\xrightarrow{H^+}$ $-\overset{+}{N}H_3$ cf. Benzene

No lone pair on N for possible conjugation

λ_{max} = 203 nm

Benzene λ_{max} = 204 nm

Phenol shows an absorption peak at a longer wavelength at **210 nm** than benzene (204 nm) because the oxygen lone pair can enter into conjugation with the benzene ring. Notably, the $+M$ effect of OH is weaker than the $+M$ effect of NH_2, causing a smaller red shift for phenol compared to that for aniline.

In **aqueous NaOH**, phenol loses a proton and is converted into **phenoxide ion** with a full negative charge on oxygen. Thus a much more effective conjugation is possible in phenoxide ion (there is no separation of charge in its resonance structures), and the λ_{max} shifts further to a longer wavelength of **235 nm**.

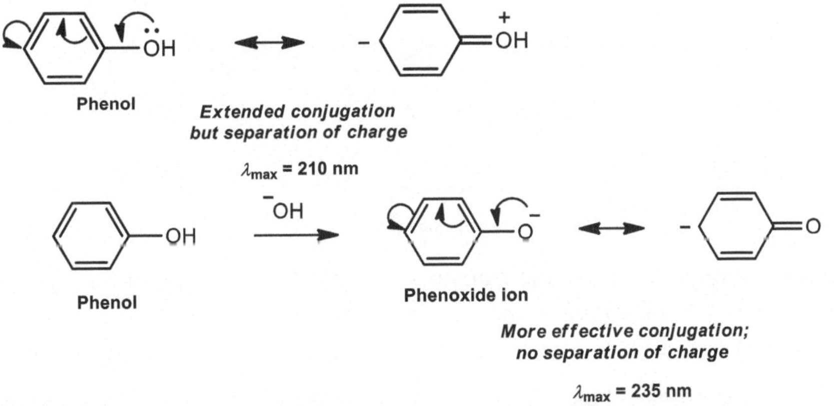

Phenol ⟷ Extended conjugation but separation of charge

λ_{max} = 210 nm

Phenol $\xrightarrow{^-OH}$ Phenoxide ion

More effective conjugation; no separation of charge

λ_{max} = 235 nm

6.3.3 Visible Spectrum and Colour

If a compound has a colour, it must absorb light in the visible region (400–800 nm) of the electromagnetic spectrum.

When the compound absorbs all wavelengths of visible light, it will be **black**. On the other hand, if the compound reflects all wavelengths, it will be **white**.

A few basic colours of light absorbed and their complementary colours (i.e. observed colours) are shown here.

Light absorbed		Colour observed
λ (nm)	Colour	(complementary colour)
450	Blue	Yellow
550	Green	Red
600	Yellow	Blue
700	Red	Green

Thus a compound that absorbs blue light will appear yellow in colour, and so on.

It is of note that a *range of wavelengths* around the indicated wavelength is associated with a colour of the light.

Examples

If a compound has enough conjugated double bonds, the absorption will shift to much longer wavelength in the visible region, i.e. the compound will absorb visible light and it will be coloured.

***β*-Carotene** is a natural pigment found in carrots and tomatoes. It has 11 conjugated double bonds, and it gives an absorption maximum (λ_{max}) at 497 nm in the visible region. This light absorbed by *β*-carotene has a blue-green colour. Thus, *β*-carotene has the complementary colour of blue-green, which is **orange-red**.

(Me = CH_3)

***β*-Carotene** (λ_{max}, **497 nm)**

Orange-red

Many aromatic compounds have their absorption maximum (λ_{max}) in the UV region, but the *range of their light absorption can extend into the visible region*. As a result, these compounds appear in colour.

For example, ***p*-nitrophenol** has a λ_{max} at 320 nm in the UV region but its end absorption, though small, occurs above 400 nm in the visible region. Thus the compound has a **pale yellow** colour (complementary to a bluish colour).

O_2N–C₆H₄–OH

λ_{max} **= 320 nm, with a slight end absorption above 400 nm**

***p*-Nitrophenol**

Pale yellow

6.4 INFRARED (IR) SPECTROSCOPY

Bonds in molecules are not fixed, but are dynamic. Bonds are constantly vibrating in a variety of ways, and there are specific or quantized vibrational energy levels in a molecule.

The energy differences between vibrational energy levels are relatively *small*, and so **low energy infrared radiation** (compared to UV or visible light) can cause excitation of the molecule from a *vibrational level of lower energy* to a *vibrational level of higher energy*. Since vibrational transitions are involved, the **IR spectroscopy** is called the **vibrational spectroscopy**.

However, not all molecular vibrations result in the absorption of infrared radiation. Vibrations that occur with the absorption of infrared light are called IR active. **In order for a vibration to be IR active, the dipole moment of the molecule must change as the vibration occurs**.

Infrared spectra are absorption spectra. But unlike UV absorption spectra, **IR spectra** do not generally indicate the wavelength (λ), but use the **inverse of wavelength (λ^{-1}** or **$1/\lambda$)**, called the **wavenumber**. The **unit** of wavenumber is normally taken to be **cm^{-1}**.

Since wavelength and frequency are inversely related, **wavenumber ($\bar{\nu}$) refers to frequency in cm^{-1}**, *not* the usual frequency (ν) in cycles per second (s^{-1}) or hertz (Hz).

The general relationship between wavenumber ($\bar{\nu}$) and frequency (ν) in Hz is given by

$$\text{Wavenumber } (\bar{\nu}) = \frac{1}{\lambda} = \frac{1}{c/\nu} = \frac{\nu}{c}$$

Stretching and bending vibrations

Bonds in a molecule vibrate as if these bonds were tiny *springs* connecting the atoms. Two atoms joined by a covalent bond can undergo vibration where the atoms move back and forth along the bond. Such vibrations are called **stretching vibrations**.

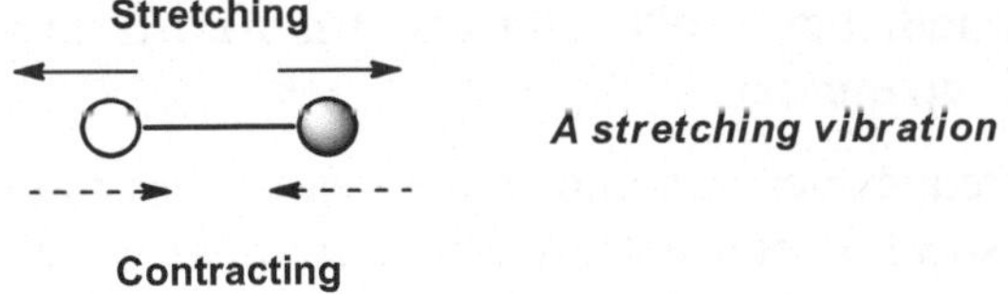

A stretching vibration

Stretching vibrations change (increase or decrease) *bond lengths* and the assigned bond length between two atoms is the average length resulting from these vibrations.

Any set of three atoms in molecules can undergo **stretching vibrations** as well as **bending vibrations (or deformations)**.

In bending vibrations, bond lengths remain constant but *bond angles* change (increase or decrease). The assigned bond angle is an average. For a set of three atoms, several kinds of stretching and bending vibrations are possible.

Of the several cases, only a **stretching vibration**, an **in-plane bending vibration** (called scissoring) and an **out-of-plane bending vibration** (called twisting) are shown here.

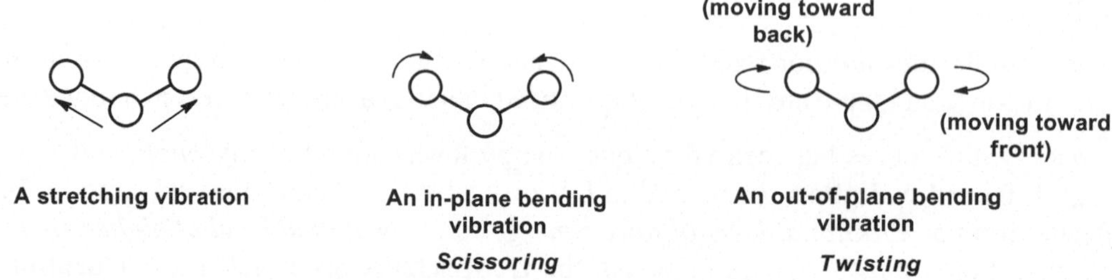

To stretch a spring (or a bond) requires more energy than to bend it. Thus, **the stretching frequency of a bond is greater than its bending frequency**.

Functional group and finger print regions

Typically, the whole IR region is 4000–500 cm^{-1}.

The **stretching vibrations** for organic molecules generally correspond to the IR region **4000–1500 cm^{-1}**. This region of the IR spectrum is highly useful for detecting the presence of functional groups (including C–H bonds), and is often called the **functional group region**. The presence of an IR peak in this region is almost always a definite indication that some particular functional group is present in the molecule. Similarly, the absence of a particular IR peak in this region implies that the concerned functional group is absent.

Infrared spectra also contain many absorption peaks which are not open to interpretation. These *unassigned peaks* are associated with complex bending vibrations and some single-bond (C–C, C–N, C–O) stretching vibrations. This complex pattern of absorption bands is uniquely characteristic of each molecule, and serves as a *fingerprint* of the molecule. This region extends from **1500 to 500 cm^{-1}**, and is often called the **fingerprint region** that can provide a proof of identity for the molecule.

IR spectrum

In IR spectroscopy, a beam of infrared radiation (**4000–500 cm^{-1}**) is passed through a sample in an instrument called the **IR spectrometer**.

The sample absorbs some specific IR frequencies and allows the rest to pass through it (transmitted light). The detector of the spectrometer detects the *transmitted frequencies*, which can also reveal the values of the absorbed frequencies.

The intensity of transmission or transmittance (T) of a sample is defined as

$$T = \frac{I}{I_0}.$$

where I is the intensity of transmitted light and I_0 is the intensity of incident light.

The **IR spectrum** is a plot of **transmittance** (in percentage) versus **wavenumbers** (in cm^{-1}).

Figure 6.6 shows a typical IR spectrum of **hexane**. Since hexane has only C–H and C–C bonds, the spectrum is relatively simple. The **functional group region** shows a characteristic C–H stretching frequency below 3000 cm^{-1} (2940–2880 cm^{-1}). The **fingerprint region** is also indicated in the spectrum.

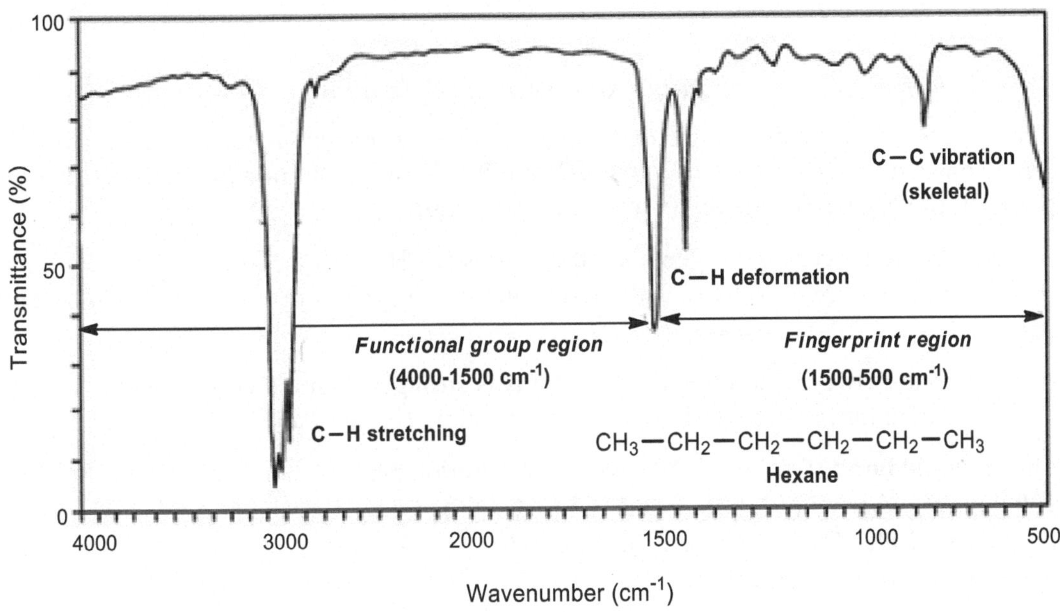

Figure 6.6: The IR spectrum of hexane.

The relationship between absorbance (*A*) and transmittance (*T*) is given by

$$A = \log_{10}\left(\frac{I_0}{I}\right) = \log_{10}\left(\frac{1}{T}\right)$$

Thus transmittance bears an inverse and logarithmic relationship to absorbance. This is why the **IR spectrum** in the transmission mode appears **inverted or upside down** compared to the UV spectrum in the absorption mode.

6.4.1 Hooke's Law and Stretching Frequencies

In physics, Hooke's law is used to describe the frequency of vibrations for a ball and spring system. Since bonds behave like springs, **Hooke's law** can be applied to calculate the vibrational stretching frequency (ν) of a bond, which is given by

$$\nu = \frac{1}{2\pi}\sqrt{\frac{k}{\mu}}$$

where k = force constant or strength of the bond (or spring), and μ = reduced mass.

Therefore, the vibrational stretching frequency in wavenumbers ($\bar{\nu}$) is:

$$\bar{\nu} = \frac{\nu}{c} = \frac{1}{2\pi c}\sqrt{\frac{k}{\mu}}$$

The **reduced mass (μ)** is a function of the masses (m_1 and m_2) of the two atoms or balls, and is given by

$$\mu = \frac{m_1 m_2}{m_1 + m_2}$$

Stretching frequencies for single, double and triple bonds

The vibrational stretching frequency ($\bar{\nu}$) is *directly proportional* to the square root of force constant (k) or strength of a bond. Therefore, **stronger bonds give IR absorption peaks in the higher frequency (wavenumber) region**.

A **single bond** between hydrogen and other atoms (C–H, N–H or O–H) is very short and strong, and therefore absorbs IR radiation in the very high frequency region (**4000–2500 cm^{-1}**).

A triple bond between carbon and other atoms is stronger than the corresponding double bond, which in turn is stronger than the corresponding single bond.

Thus a **triple bond** (C≡C or C≡N) gives an absorption peak at a higher frequency (**2500-2000 cm^{-1}**), whereas a **double bond** (C=C, C=N or C=O) absorbs at a relatively lower frequency (**2000–1500 cm^{-1}**). A **single bond** (C–C, C–N or C–O) absorbs at a much lower frequency (**< 1500 cm^{-1}**).

The four important regions of the IR spectrum are summarized here.

Bonds/functional groups	Wavenumber (cm^{-1})
Bonds to hydrogen: C—H N—H O—H	4000-2500
Triple bonds: C≡C C≡N	2500-2000
Double bonds: C=C C=N C=O	2000-1500
Single bonds: C—C C—N C—O	< 1500

Effect of reduced mass on stretching frequencies

The vibrational stretching frequency ($\bar{\nu}$) is *inversely proportional* to the square root of reduced mass (μ). Therefore, **bonds with smaller reduced mass give IR absorption peaks at higher frequency**.

A bond with a lighter atom has a smaller reduced mass that a bond with a heavier atom. To illustrate, compare **O–H** bond and **O–D** bond.

For the O–H bond, the reduced mass (using relative atomic mass or atomic weight) = $[(16 \times 1)/(16 + 1)] = 0.94$.

For the O–D bond, the reduced mass (in relative mass units) = [(16 × 2)/(16 + 2)] = 1.77.

Therefore, **O–H bond** with a smaller reduced mass has a **higher stretching frequency** than O–D bond.

Bond	Reduced mass (in relative mass units)	Stretching frequency (cm^{-1})
O—H	0.94	3600
O—D	1.77	2600

It should be noted that force constants for O–H and O–D bonds are nearly the same as they are chemically similar.

PROBLEM 6.5: Calculate the approximate vibrational stretching frequency (in cm^{-1}) of the C–H bond using Hooke's law. Given: $k = 5 \times 10^5$ g s^{-2}, $m_H = 1.6 \times 10^{-24}$ g, $m_C = 20 \times 10^{-24}$ g.

Answer

The stretching frequency (in cm^{-1}) is calculated as follows. The reduced mass is determined from the expression $\mu = (m_1 m_2)/(m_1 + m_2)$, taking the *actual* masses of C and H atoms.

$$\bar{\nu} = \frac{1}{2\pi c}\sqrt{\frac{k}{\mu}} = \frac{1}{2 \times 3.14 \times 3 \times 10^{10}\ \text{cm s}^{-1}}\sqrt{\frac{5 \times 10^5\ \text{g s}^{-2}}{(20 \times 10^{-24}\ \text{g})(1.6 \times 10^{-24}\ \text{g})/(21.6 \times 10^{-24}\ \text{g})}}$$

$= 3100\ cm^{-1}$

6.4.2 Aliphatic and Aromatic Hydrocarbons: C–H Stretching

All hydrocarbons give IR absorption peaks associated with **C–H stretching** vibrations in the **2800–3300 cm^{-1}** region.

The position of the IR peak depends on the strength of the C–H bond, which in turn depends on the hybridization of the carbon (sp^3, sp^2 or sp) bearing the hydrogen atom. *The greater the s character in the hybrid orbital, the stronger is the bond.*

The **bond strength** increases in the order:

$C(sp^3)–H < C(sp^2)–H < C(sp)–H$

The C–H stretching peaks for $C(sp^3)$–H bonds in **alkanes or alkyl groups** occur at the lower frequency **below 3000 cm^{-1}** (2800–3000 cm^{-1}) (see Figure 6.6), whereas the C–H stretching peaks for $C(sp^2)$–H bonds in **alkenes or aromatic rings** occur at the relatively higher frequency **above 3000 cm^{-1}** (3000–3100 cm^{-1}).

The C–H stretching vibrations for C(sp)–H bonds in terminal **alkynes** occur at the highest frequency **around 3300 cm^{-1}**.

C–H stretching frequencies (cm^{-1})

Alkanes		Alkenes and aromatics		Alkynes	
$C(sp^3)$—H	2800-3000	$C(sp^2)$—H	3000-3100	C(sp)—H	~3330

Examples

Consider **toluene** and **1-hexyne**.

The IR spectrum of **toluene** gives the characteristic C–H stretching peak for CH_3 (alkyl group) at 2925 cm^{-1} (< 3000 cm^{-1}), as well as the C–H absorption peak of the **aromatic hydrogens** (Ar–H) at 3090 cm^{-1} (> 3000 cm^{-1}).

1-Hexyne gives the characteristic **≡C–H** peak at 3320 cm^{-1} (~ 3300 cm^{-1}) along with C–H stretching frequencies for **CH_3 and CH_2** groups at 2970 and 2880 cm^{-1} (< 3000 cm^{-1}).

Toluene

Group	C-H stretching (cm^{-1})
CH_3	2925
Ar—H	3090

1-Hexyne

Group	C-H stretching (cm^{-1})
≡C–H	3320
CH_3, CH_2	2970, 2880

6.4.3 Alcohols, Carboxylic Acids and Amines: O–H and N–H Stretchings

The O–H bond is *stronger* than N–H bond, which is *stronger* than C–H bond. Therefore, the **N–H stretching** occurs at a high frequency at about 3400 cm^{-1} compared to a C–H stretching. The **O–H stretching frequency** is higher still at around 3600 cm^{-1}.

Bond	IR frequency (cm^{-1})	Intensity
O—H	3500-3600	Strong
N—H	3300-3400	Medium

The **intensity of O–H stretching peak is stronger** than the intensity of N–H stretching peak. This is because stretching of O–H bond produces a **large change in dipole moment**, since O–H bond is more polarized than N–H bond (oxygen being more electronegative

than nitrogen). The N–H and O–H absorptions are also *stronger* than C–H absorptions because C–H bonds are only weakly polarized.

Alcohols

Alcohols (ROH) give the characteristic O–H stretching frequency in their IR spectra. When an **alcohol** is present as a very **dilute solution** in CCl_4, it gives a sharp O–H absorption peak at about 3600 cm^{-1}. But in a **concentrated solution**, the sharp peak is replaced by a broad band in the 3200–3550 cm^{-1} region.

Group	Stretching frequency (cm^{-1})
Alcohols (ROH)	
O—H (free) **(dilute solution)**	3600 **(sharp peak)**
O—H (hydrogen bonded) **(concentrated solution)**	3200-3550 **(broad band)**

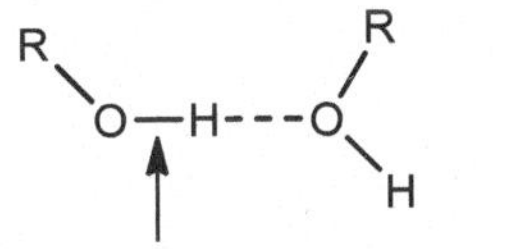

R—O—H---O(R)—H

Bond length increases due to hydrogen bonding

The sharp peak at 3600 cm^{-1} is due to O–H stretching for 'free' alcohol molecules, since these molecules are widely separated in a very dilute solution, and therefore *not* associated by intermolecular hydrogen bonding.

But in a concentrated solution, alcohol molecules are associated through **intermolecular hydrogen bonding**. Hydrogen bonding involves a lengthening of the original O–H bond. As a result, the O–H bond is weakened, and its force constant is reduced causing a lowering of the O–H stretching frequency.

At some intermediate concentrations, when both free and hydrogen bonded alcohol molecules are present, the IR spectra give a sharp band at 3600 cm^{-1} in addition to the broad band at 3350 cm^{-1}.

Carboxylic acids

Carboxylic acids (RCOOH) show the characteristic O–H stretching frequency as a very **broad band** in the 2500–3000 cm^{-1} region. This is attributed to OH groups in hydrogen-bonded **dimers** with two strong hydrogen bonds between the carbonyl oxygen atom of one molecule and the acidic hydrogen atom of the other molecule.

Group	Stretching frequency (cm^{-1})
Carboxylic acids (RCOOH)	
O—H (hydrogen bonded)	2500-3000 **(broad band)**

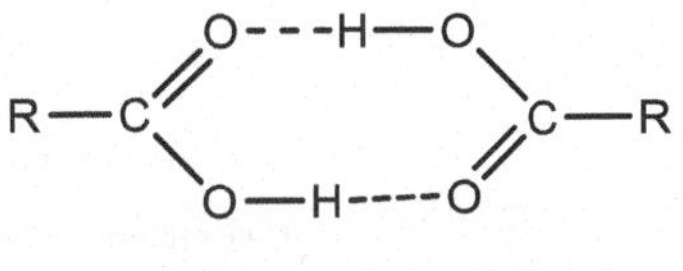

R—C(=O---H—O)(—O—H---O=)C—R

Hydrogen-bonded dimer

Amines

Primary amines (RNH_2) and **secondary amines** (R_2NH) give N–H absorption peaks in the 3300–3400 cm^{-1}.

IR spectroscopy can distinguish between primary, secondary and tertiary amines.

Primary (1°) amines give **two sharp peaks** due to *coupling* of two N–H stretching vibrations, whereas **secondary** (2°) amines give only **one peak** as they have one N–H bond. But **tertiary** amines (R_3N) have no N–H bond, and hence give **no peak** in this region.

Group	Stretching frequency (cm^{-1})
Amines (RNH_2, R_2NH, R_3N)	
N—H	3300-3400 **(1°: 2 peaks; 2°: 1 peak 3°: no peak)**

6.4.4 Alkenes, Aromatics, Alkynes and Nitriles: C=C, C≡C and C≡N Stretchings

The following are the stretching frequencies for double or triple bonds in alkenes, aromatics, alkynes and nitriles.

Group	Stretching frequency (cm^{-1})
Alkenes	
C=C	1620-1680
Aromatics	
C⩦C	1450-1600
Alkynes	
C≡C	2100-2250
Nitriles	
C≡N	~2250

Alkenes

Besides C–H stretching vibrations (see Section 6.4.2), **alkenes** can also give absorption peaks in the **1620–1680 cm^{-1}** region that result from **C=C stretching vibrations.**

These absorptions are usually weak, and are *not* detected at all if the double bond is symmetrically substituted. **For symmetrically substituted alkenes, C=C stretching is IR inactive** because no dipole moment change is associated with this vibration.

$H_2C=CH_2$ $(H_3C)_2C=C(CH_3)_2$

(symmetrical alkenes)

C=C stretching is IR inactive

$H_2C=C(CH_2CH_3)(CH_3)$

(unsymmetrical alkene)

C=C stretching is IR active

Aromatic compounds

The **benzene ring** of aromatic compounds (aromatics) has carbon-carbon bonds with a partial double bond character (C⋯C). Thus aromatic compounds can show **stretching vibrations of carbon-carbon bonds** with a set of *weak* absorption peaks in the **1450–1600 cm^{-1}** region.

These absorptions occur in addition to the characteristic C–H stretching frequency of aromatic hydrogens (see Section 6.4.2).

Alkynes

Alkynes can give **C≡C stretching** peaks in the **2100–2200 cm^{-1}** region, in addition to C–H stretching frequencies at about 3300 cm^{-1} (see Section 6.4.2).

However, **C≡C stretching is IR inactive for symmetrically substituted alkynes** when no change in dipole moment is associated with this vibration.

H—C≡C—H H_3C—C≡C—CH_3 H—C≡C—CH_2CH_3

(symmetrical alkynes)   **(unsymmetrical alkyne)**

C≡C stretching is IR inactive   **C≡C stretching is IR active**

Nitriles

Nitriles have a carbon-nitrogen triple bond, and they show characteristic **C≡N stretching frequency** at about **2250 cm^{-1}**.

Nitrile group (CN) itself is **unsymmetrical** and **polar**. Thus, a change in dipole moment is associated with the C≡N stretching vibration, and it is always **IR active**.

6.4.5 Carbonyl Compounds: C=O Stretching

Carbonyl compounds include aldehydes, ketones, carboxylic acids, esters and amides. Carbonyl group (C=O) is a strongly polar group which gives a strong and prominent absorption peak associated with **C=O stretching vibration** in the **1630-1780 cm^{-1}** region.

IR spectroscopy is particularly useful to *detect the presence of carbonyl group* in organic compounds.

Group	Stretching frequency (cm^{-1})
Carbonyl compounds	
C=O **(aldehydes)**	1690-1740
C=O **(ketones)**	1680-1750
C=O **(carboxylic acids)**	1710-1780
C=O **(esters)**	1735-1750
C=O (amides)	1630-1690

Simple and conjugated ketones

Simple **ketones** (e.g. acetone) absorb near **1715 cm^{-1}**.

Conjugation of the carbonyl group (C=O) with a C=C or a benzene ring **lowers the frequency of C=O stretching** by about 30 cm^{-1}.

Due to conjugation or resonance, carbonyl group assumes a more *single bond character*. As a result, the carbonyl bond is weakened, and its force constant is reduced. This leads to a shift of carbonyl stretching vibration to lower frequency.

R—C(=O)—R $\bar{\nu}_{C=O}$ 1715 cm^{-1}

R—CH=CH—C(=O)—R ⟷ R—$\overset{+}{C}$H—CH=C(O^-)—R (single bond) $\bar{\nu}_{C=O}$ 1685 cm^{-1}

Conjugation lowers carbonyl stretching frequency

Carboxylic acid derivatives

Carboxylic acid derivatives such as acid chlorides, esters and amides are best distinguished by the difference in IR stretching frequency of the C=O group.

In **acid chlorides**, the electron withdrawing inductive effect (–*I*) of Cl dominates over its +*M* effect and tends to increase the double bond character of the carbonyl group, which strengthens the carbonyl bond. As a result, **C=O stretching frequency shifts to a higher frequency (~1815 cm^{-1})**.

In **esters**, the inductive electron withdrawal (–*I*) and mesomeric electron donation (+*M*) of OR group are more or less balanced. In some cases, the –*I* effect slightly dominates, whereas in others, +*M* effect slightly dominates. **Esters give carbonyl stretching peaks at a lower frequency (~1745 cm^{-1}) than acid chlorides.**

In **amides**, the +*M* effect of NH_2 group strongly dominates and lowers the double bond character of the carbonyl group, which greatly weakens the carbonyl bond. Thus, **amides give carbonyl stretching peaks at a much lower frequency (~1650 cm^{-1}).**

(double bond) R—C(=O)→Cl (–*I* > +*M*) ⟷ R—$\overset{+}{C}$(O^-)→Cl

More important contributor | (destabilized by –*I* effect of Cl)

Acid chlorides $\bar{\nu}_{C=O}$ 1815 cm^{-1}

R—C(=O)→OR

(–*I* and +*M* effects of OR are more or less balanced)

Esters $\bar{\nu}_{C=O}$ 1745 cm^{-1}

R—C(=O)—NH_2 ⟷ R—C(O^-)=$\overset{+}{N}H_2$ (single bond)

(+*M* effect of NH_2 strongly dominates) | *More important contributor*

Amides $\bar{\nu}_{C=O}$ 1650 cm^{-1}

PROBLEM 6.6: Explain the following trends of C=O stretching frequencies for cyclic ketones:

	Cyclohexanone	Cyclopentanone	Cyclobutanone	Cyclopropanone
$\bar{\nu}_{C=O}$	1715 cm^{-1}	1745 cm^{-1}	1780 cm^{-1}	1810 cm^{-1}

Answer

Cyclohexanone absorbs at 1715 cm^{-1}, and the C=O stretching vibration shifts progressively to higher frequency from cyclohexanone → cyclopentanone → cyclobutanone → cyclopropanone. The increment of frequency is about 30 cm^{-1} from one compound to the next in the series. This can be explained in terms of **bond angle strain.**

The bond angle at the carbonyl group of cyclohexanone is the normal sp^2 angle of 120° as for simple acyclic ketones, and their carbonyl absorptions are similar. Small ring ketones introduce bond angle strain, as the bond angles in small rings are less than the optimum bond angle of 120°.

The C–C–C bond angles (θ) inside five-, four- and three-membered rings are 108°, 90° and 60° respectively. Therefore, the C–C–O bond angle (α) *outside* the ring will increase progressively as the ring size decreases. From the geometry, $\theta + 2\alpha = 360°$. Thus, the C–C–O bond angle (α) for cyclopentanone, cyclobutanone and cyclopropanone is 126°, 135° and 150° respectively.

As the C–C–O bond angle increases, s character in the C–O bond increases. Consequently, the C=O bond is shortened and strengthened, and the C=O stretching frequency increases.

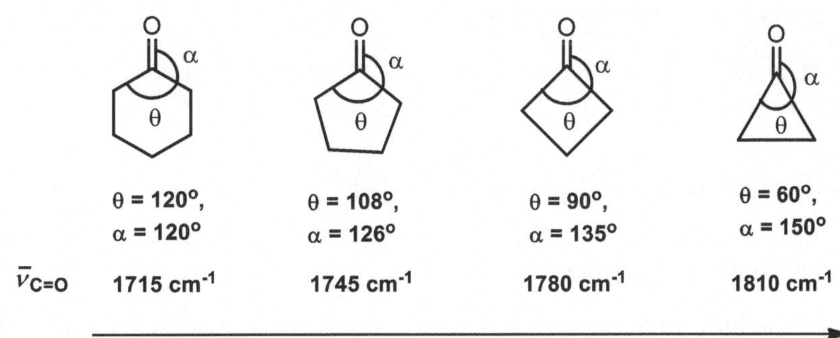

6.4.6 Summary of Infrared Absorptions

The characteristic IR **stretching frequencies** for some common groups are summarized in Table 6.1.

Table 6.1: Characteristic IR absorptions of some common groups

	Group	Stretching frequency (cm^{-1})	Absorption intensity
Alkanes	C—H	2800-3000	Medium to strong
Alkenes	C=C	1620-1680	Weak to medium
	=C—H	3000-3100	Medium
Aromatics	C⚌C	1450-1600	Weak
	Ar—H	3000-3100	Medium
Alkynes	C≡C	2100-2250	Medium
	≡C—H	~3330	Strong
Alcohols	O—H	3500-3600	Strong
Amines	N—H	3300-3400	Medium
Nitriles	C≡N	2220-2260	Medium
Aldehydes, ketones, acids, esters, and amides			
	C=O	1630-1780	Strong
	Aldehydes	1690-1740	
	Ketones	1680-1750	
	Carboxylic acids	1710-1780	
	Esters	1735-1750	
	Amides	1630-1690	

Absorption **intensities** are indicated approximately as strong, medium or weak. A stronger absorption implies a more intense absorption.

The IR spectra of unknown compounds can be interpreted by recognizing where the IR absorptions of these groups occur.

6.5 NUCLEAR MAGNETIC RESONANCE (NMR) SPECTROSCOPY

Nuclear magnetic resonance (NMR) spectroscopy allows us to 'see' the nuclei and their chemical or electronic environment in a molecule.

However, some nuclei cannot be seen in NMR experiments; they are called **NMR inactive**. For example, the carbon nucleus (**^{12}C**) is NMR inactive.

But the nucleus of the hydrogen atom (**^{1}H or proton**) and the nucleus of the carbon-13 isotope (**^{13}C**) are **NMR active**.

The **^{1}H NMR (proton NMR or PMR)** is often the most useful NMR spectroscopy to determine the structure of organic molecules. Besides the ^{1}H NMR, we will also introduce briefly the **^{13}C NMR**.

Theory of NMR spectroscopy

The theory of NMR spectroscopy is quite complicated, and the details are beyond our scope. We give here only a simplified version of how NMR works.

The nuclei that have **magnetic properties** are NMR active. The property of a nucleus responsible for its magnetic properties is the **nuclear spin** or the **nuclear spin quantum number (*I*)**.

The nuclei for which $\boldsymbol{I > 0}$ are **magnetic**, and they are NMR active. On the other hand, the nuclei for which $\boldsymbol{I = 0}$ are **nonmagnetic**, and they are NMR inactive.

The nuclear spin (*I*) is associated with atomic number and mass number of the nuclei:

Atomic number	Mass number	Spin quantum number (*I*)
Even	Even	0
Odd or Even	Odd	$\frac{1}{2}, \frac{3}{2}, ...$
Odd	Even	1, 2, ...

Thus, **^{12}C** nucleus with atomic number 6 (even) and mass number 12 (even) has $\boldsymbol{I = 0}$, and hence is **nonmagnetic**. But **^{1}H** nucleus with atomic number 1 (odd) and mass number 1 (odd) or **^{13}C** nucleus with atomic number 6 (even) and mass number 13 (odd) has nuclear spin $\boldsymbol{I = ½}$, and is **magnetic**.

The deuterium nucleus **^{2}D** with atomic number 1 (odd) and mass number 2 (even) is also **magnetic ($\boldsymbol{I = 1}$)**.

A **^{1}H or ^{13}C nucleus** has both electric charge and spin, and behaves as a tiny spinning bar magnet (cf. any spinning charged body that generates a magnetic field). Under the influence of a **strong external magnetic field**, the magnetic nucleus can take up different orientations with respect to the applied field. In general, **the number of possible orientations is given by ($\boldsymbol{2I + 1}$)**.

Thus, a ^{1}H nucleus ($I = ½$) or ^{13}C nucleus ($I = ½$) has two possible orientations ($2I + 1 = 2 \times ½ + 1 = 2$): the axis of the spin is either aligned with (parallel to) the applied field or against (antiparallel to) the applied field.

The parallel spin is called **α spin** and the antiparallel spin is called **β spin**.

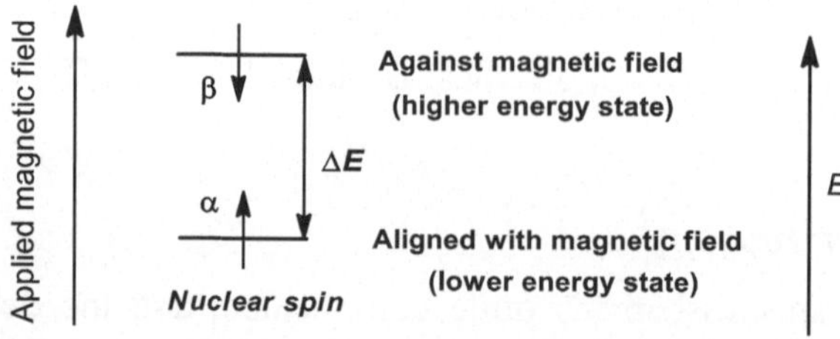

The nuclear spin (α) parallel to the applied field has a lower energy than the nuclear spin (β) antiparallel to the applied field. Therefore, a ^{1}H or ^{13}C nucleus in an external magnetic field has two energy levels.

However, the **energy difference (Δ*E*)** between these levels is very small, and low-energy electromagnetic radiation of **radio-wave frequency** is needed to flip the nucleus from the lower energy state to the higher energy state.

The flipping of α spin to β spin occurs when the nucleus absorbs energy of the correct radiofrequency (ν), which is given by

$$\Delta E = h\nu$$

Under this condition, the nucleus and the radiofrequency radiation are said to be in **resonance** (hence the term **nuclear magnetic resonance**; this resonance must not be confused with resonance structures).

The **NMR spectrometer** uses a powerful electromagnet to produce a strong magnetic field, and irradiates the sample of an organic compound with electromagnetic radiation in the radiofrequency region.

The sample is dissolved in a suitable solvent such as $CDCl_3$, placed in a narrow tube, and put inside the electromagnet. Then the sample is irradiated with a short **pulse** of radiofrequency energy. Some nuclei absorb the energy and are excited or promoted to the higher energy level. When the pulse finishes, the nuclei return or **relax** back to the lower energy state, and the radiation that comes out is detected by the radiofrequency receiver.

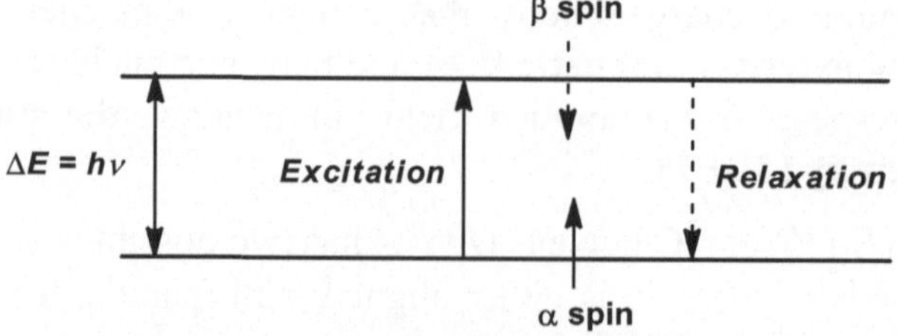

After extensive computation, the results are displayed in the form of an **NMR spectrum**, which is typically a plot of intensity (amount of radiation detected) versus frequency of radiation.

In NMR, the frequency scale (in Hz) is replaced with a scale in **'parts per million' (ppm)** (see Section 6.6.1).

The **energy difference** (ΔE) between the two states depends on the **strength of the applied magnetic field** (B_0) and the **type of the magnetic nucleus** (^{1}H or ^{13}C), and is given by,

$$\Delta E = \frac{h}{2\pi}\gamma B_0$$

where γ= magnetogyric ratio, which is a characteristic of the particular nucleus.

Thus, the stronger the magnetic field, the greater the energy difference between the two states, and higher frequency (higher energy) radiation is needed.

The exact **frequency (ν)** at which the nucleus resonates, is related to the **applied field (B_0)** as

$$\Delta E = h\nu = \frac{h}{2\pi}\gamma B_0$$

$$\text{or, } \nu = \frac{\gamma B_0}{2\pi}$$

This implies that at a magnetic field strength (B_0) of 4.7 tesla (T), the **resonance frequency (ν) of protons** will be **200 megahertz (MHz)**, where 1 MHz = 1 million Hz = 10^6 Hz.

At the same field strength (B_0 = 4.7 T), the **^{13}C** nuclei would resonate at a different frequency of **50.3 MHz** because its γ value is different.

6.6 ^{1}H NMR (PROTON NMR) SPECTROSCOPY

All protons in a molecule do not resonate at the same frequency in a given magnetic field. This is because electrons surround the nuclei (protons), and the electronic environments around the nuclei are not the same. **The exact resonance frequency of a proton depends on what sort of electronic environment it is in.**

Protons that differ in their electronic or chemical environment are said to be chemically **non-equivalent**, and these non-equivalent protons give **different signals** (i.e. peaks at different frequencies) in the ^{1}H NMR spectrum.

In contrast, the **protons** that are chemically **equivalent** will give a **single NMR signal**.

6.6.1 The Chemical Shift

In NMR spectroscopy, the resonance frequency of protons in organic molecules is measured with respect to the frequency at which the protons of a standard reference sample resonate.

The **reference sample** most often used is **tetramethylsilane (TMS)**. TMS has 12 *equivalent* protons that give a strong *single* signal. TMS is used as a reference because its protons resonate at a frequency *at one extreme* which is lower than the frequencies of the vast majority of protons in organic molecules.

CH_3

H_3C—Si—CH_3

CH_3

Tetramethylsilane (TMS)

The resonance frequency of protons of an organic sample is measured relative to the frequency of protons in TMS. The difference in frequency is called the chemical shift:

Chemical shift (in Hz) = $\nu_{sample} - \nu_{TMS}$

where ν_{sample} and ν_{TMS} are the frequencies of protons (in Hz) for the sample and TMS respectively.

The delta (δ) scale

Since the resonance frequency of a proton depends on the external applied magnetic field ($\nu = \frac{\gamma B_0}{2\pi}$), **the chemical shift of a proton (in Hz) is also proportional to the applied magnetic field**.

NMR spectrometers with different magnetic field strengths are commercially available. Therefore, if an organic sample is run on different NMR machines, the chemical shift values in Hz will vary. It is thus desirable to express chemical shifts in a form that is *independent* of NMR machines with different external field strengths.

An NMR machine of magnetic field strength of 9.4 T is referred to as a **400 MHz NMR spectrometer**, because that is the frequency at which the protons in the reference sample resonate at this magnetic field strength. This frequency is called the **operating frequency** of the NMR spectrometer.

Since chemical shift values are very small, protons of an organic sample resonate at frequency values close to the operating frequency of the NMR machine.

In order to express chemical shifts independent of field strengths of different NMR spectrometers, **the chemical shift (in Hz) is divided by the operating frequency (ν_0) of the NMR spectrometer (in Hz)**. As chemical shifts are typically < 5000 Hz and the operating frequency is in the order of MHz (million Hz or 10^6 Hz), the fractions from the division are very small. Therefore, the **fractions are multiplied with a million (10^6)** to express the chemical shift values conveniently in **parts per million (ppm)**.

The chemical shift scale in ppm is called the delta (δ) scale, where

$$\delta\ (\text{ppm}) = \frac{\nu_{sample}\ (\text{Hz}) - \nu_{TMS}\ (\text{Hz})}{\nu_0\ (\text{Hz})} \times 10^6 = \frac{\nu_{sample}\ (\text{Hz}) - \nu_{TMS}\ (\text{Hz})}{\nu_0\ (\text{MHz})}$$

(ν_0 = operating frequency of the NMR spectrometer)

For the reference sample TMS, **δ = 0 ppm**. Almost all other proton signals occur at *higher* δ values with respect to this signal, typically in the range of **δ = 0–10 ppm**.

¹H NMR Spectrum

A typical ¹H NMR spectrum of **methyl acetate** is shown in Figure 6.7. The spectrum exhibits *two* NMR signals for two kinds of methyl protons.

Note that the δ scale runs to the left from the TMS signal at δ = 0, i.e. backwards from the usual style.

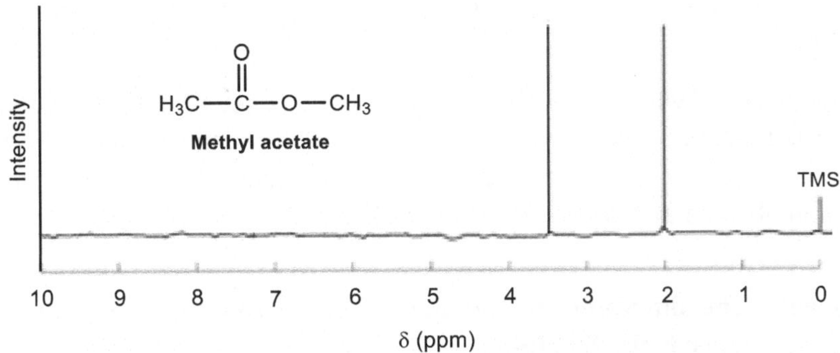

Figure 6.7: The ^{1}H NMR spectrum of methyl acetate. Chemical shift values are expressed in δ (ppm).

6.6.2 Shielding and Deshielding of Protons

Unlike a *bare* proton, a proton in a molecule is surrounded by **electrons**. In an external magnetic field, the electrons are *induced* to circulate and thereby set up a tiny electric current. This current produces its own magnetic field, called the **induced magnetic field** (**B_i**), which *opposes* the **applied magnetic field** (**B_0**), as shown in Figure 6.8.

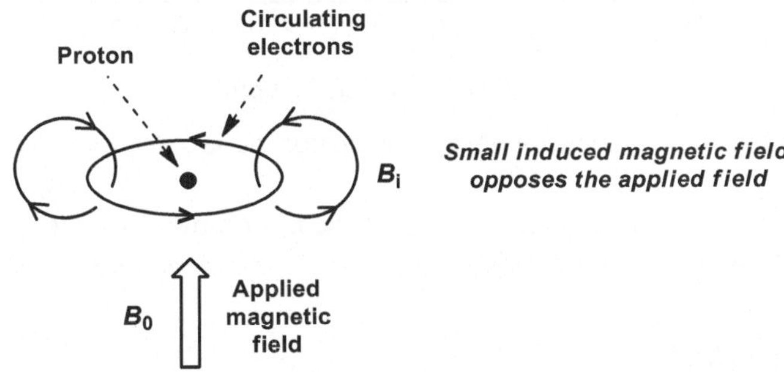

Figure 6.8: Shielding of a proton by surrounding electrons in an external magnetic field.

As a result, the **actual or effective magnetic field** (**$B_{effective}$**) sensed by the **proton** is slightly **less** than the applied magnetic field:

$$B_{effective} = B_0 - B_i$$

The electrons are then said to **shield** the proton from the external magnetic field.

Upfield and downfield proton signals: Relative chemical shifts (δ, ppm)

A **shielded proton** will require the *external magnetic field* to be made slightly *larger* in order to compensate for the small induced field. This implies that **shielding causes the proton signal to appear upfield**.

All protons in a molecule are not shielded to the same extent. Some protons are more shielded and some are less shielded. In relative terms, the less shielded protons are said to be **deshielded** with respect to the more shielded protons. The **deshielded protons** then appear **downfield** with respect to shielded protons.

Thus, **shielding shifts the proton signal upfield, whereas deshielding shifts the proton signal downfield.**

In the modern pulsed NMR spectrometer, the external magnetic field is kept constant. Thus a shielded proton senses a slightly smaller field than the applied field. As a result, the energy difference (ΔE) between α and β spin states becomes slightly smaller and the shielded proton absorbs at a *lower frequency* (ν), thereby giving a *smaller chemical shift* (δ).

Thus, **the greater the shielding, the smaller is the δ value**. Similarly, **the greater the deshielding, the larger is the δ value**.

Figure 6.9 shows the different ways of describing proton signals and their chemical shifts.

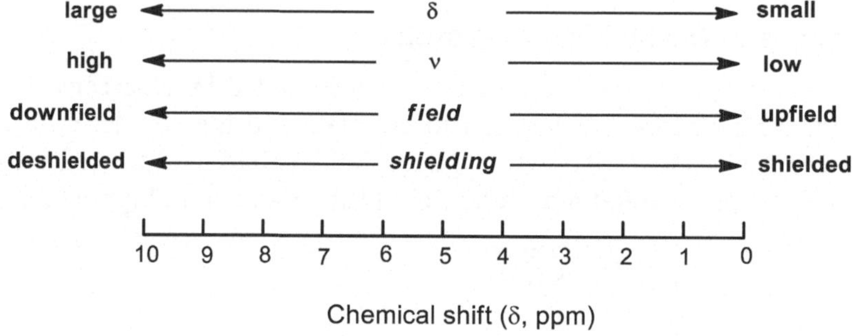

Figure 6.9: Different ways of describing proton signals and chemical shifts.

Notably, **upfield** implies *low* frequency (ν) and *small* chemical shift (δ), while **downfield** implies *high* ν and *large* δ.

Electronegativity effects

The extent to which a proton is shielded depends on the *relative electron density* around the proton. If there is more electron density near the nucleus (proton), there will be greater shielding.

Electronegative groups withdraw electron density from the C–H σ bond and *decrease* the shielding of the nucleus; the δ value will be then *larger*.

For example, the chemical shift (δ, ppm) of methyl protons in **methyl halides** increases with increasing electronegativity of the attached halogen atom (I < Br < Cl < F).

The greater the electronegativity, the smaller is the shielding, and the larger is the δ value.

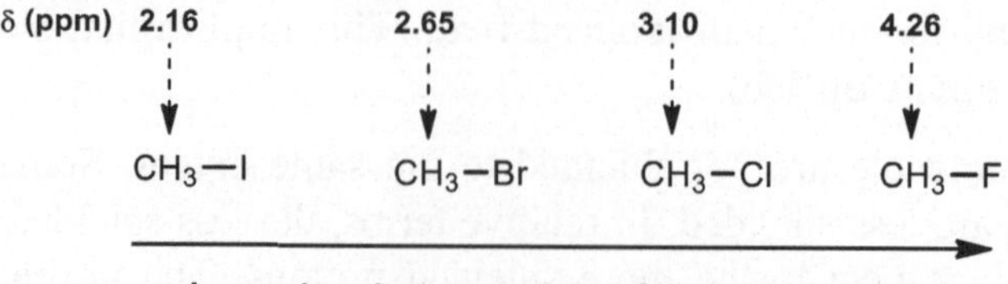

In contrast, **electropositive groups** near the nucleus will increase the surrounding electron density, and hence *increase* the shielding of the proton.

For example, consider **tetramethylsilane** (TMS), $(CH_3)_4Si$. In TMS, **Si is electropositive** and it pushes electron density *into* the CH_3 groups and exerts a powerful shielding effect.

As a result, the CH_3 protons of TMS resonate at a *very low frequency* or a **very low δ (defined as zero).** Since four CH_3 groups of TMS are chemically equivalent, the 12 protons of TMS give a *single NMR signal.*

Anisotropic shielding and deshielding

Besides σ electrons, π electrons also cause shielding or deshielding of protons.

When placed in a magnetic field, the π electrons in the *aromatic ring*, *alkenes*, *alkynes*, *carbonyl compounds* or *other π systems* are induced to circulate. This circulation of π electrons generates an **induced** magnetic field.

The induced field can *oppose* the applied field in *certain zones around the π clouds*, called the **diamagnetic shielding**, while in *some other zones*, the induced field can *reinforce* the applied field, called the **paramagnetic deshielding**. The magnetic effects caused by the circulating π electrons are thus said to be **anisotropic**, i.e. not operating equally through space. (The term 'anisotropic' is opposite to the term 'isotropic' which means operating equally through space.)

The shielding or deshielding of a proton depends on its **location** in the diamagnetic or paramagnetic zone.

Figure 6.10 illustrates diamagnetic shielding and paramagnetic deshielding associated with circulation of π electrons in the **benzene ring**.

Benzene
(six H atoms are equivalent)

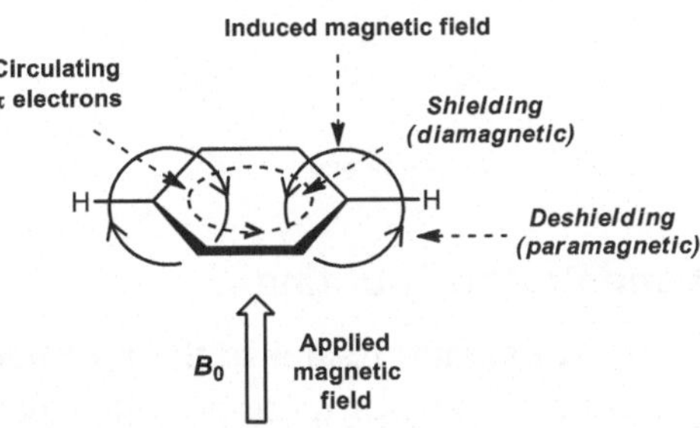

Figure 6.10: Anisotropic shielding and deshielding associated with aromatic ring current.

In benzene, π electrons are delocalized over the aromatic ring. In the presence of the applied field (B_0), the π electrons are induced to circulate around the ring, which set up a substantial electric current, called the **ring current**. This produces an *induced* magnetic field that **reinforces** the applied field **outside the ring**, and **opposes** the applied field **inside the ring**.

Thus the induced field is **diamagnetic** (opposing B_0) inside the ring, and is **paramagnetic** (reinforcing B_0) outside the ring.

Aromatic protons around the periphery of the ring fall in the paramagnetic zone and experience *enhanced* magnetic field, and hence are **strongly deshielded**. Because of this deshielding effect, the chemical shifts of aromatic protons occur at **high** δ values. For example, the chemical shift of protons of **benzene** occurs at **δ 7.27**.

It may be noted that there are lots of magnetic lines of force associated with the induced field but only representative lines of force are drawn in Figure 6.10.

The case of [18]annulene

In some aromatic systems, a few protons can be directed *inside* the ring, while others are directed outside the ring. The **internal protons** are located in the diamagnetic zone, and hence are **very strongly shielded**. They resonate at a **very low** δ value.

For example, consider **[18]annulene** which is aromatic (check using the Hückel rule). In [18]annulene, 12 equivalent protons are *outside* the ring, while 6 equivalent protons are *inside* the ring.

The 12 protons resonate at **δ 8.9**, similar to the δ value for six aromatic protons of benzene. But the 6 internal protons that fall in the diamagnetic zone and are very strongly shielded, resonate at δ **−1.8**, which is even lower than δ = 0 for TMS.

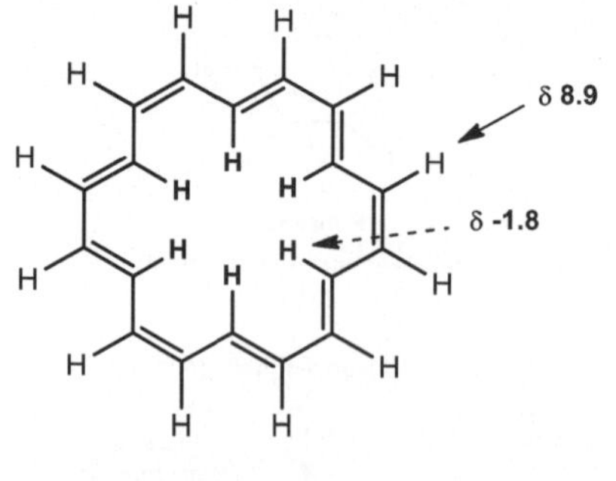

δ 8.9

δ -1.8

[18]Annulene

6.6.3 Peak Integration and Proton Counting

The **intensity** of a proton peak (or signal) relates to the **number of protons** that the peak represents. The *relative intensities* of two or more proton peaks thus give the *ratio* of the number of protons causing those peaks.

The intensities are proportional to the **peak areas** that are measured automatically by an electronic integrator as a part of the NMR spectrometer. The results are displayed as **integration curves or traces** over each NMR signal.

The heights of the integration curves are proportional to the number of protons, and therefore, the ratio of heights gives the ratio of protons in the peaks.

Figure 6.11 shows the integration curves over two signals for ***tert*-butyl methyl ether**.

One signal is for **three** methyl protons and the other for **nine** *tert*-butyl protons. Measurements of the heights of the integration lines show that **the ratio of heights = 1:3 = 3:9**, which is the ratio of the number of methyl protons and *t*-butyl protons.

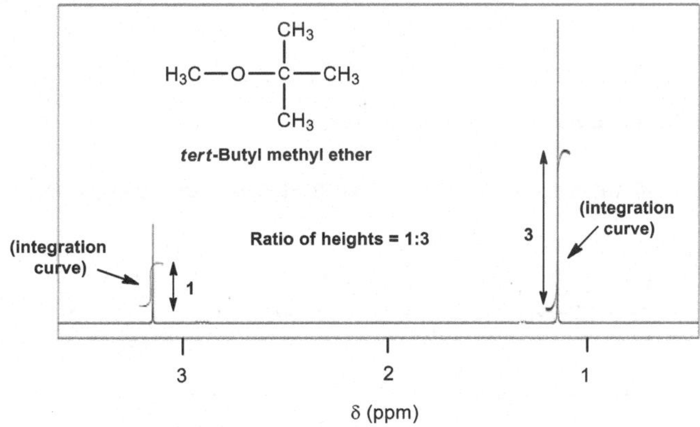

Figure 6.11: Integration of ^{1}H NMR signals for *tert*-butyl methyl ether and determination of the ratio of protons.

6.6.4 Chemical Shifts and Molecular Structure

Chemically equivalent protons that are in the same electronic environment give **only one ^{1}H NMR signal** with a characteristic chemical shift (δ).

The **number of signals** in the ^{1}H NMR spectrum of an organic compound depends on the **number of different types of protons** in the molecule. Each type of protons consists of one or more hydrogen atoms that are chemically equivalent.

Chemical shifts indicate the **positions of different peaks or signals** in the ^{1}H NMR spectrum, and chemical shift values of most protons lie downfield (to the left) of TMS in the **δ (ppm) scale.**

Chemical shifts of different types of protons

Certain generalizations can be made to account for the chemical shift values of protons in different molecular environments:

1. **The more hydrogen atoms are on a carbon, the greater is the shielding** because hydrogen is less electronegative than carbon. Consequently, the chemical shift is further upfield and **δ is lower.**

—C—H (methine) —CH_2 (methylene) —CH_3 (methyl)

Shielding increases; δ decreases

2. Electron withdrawal by adjacent electronegative groups has a deshielding effect, and the chemical shift is further downfield and δ is higher.

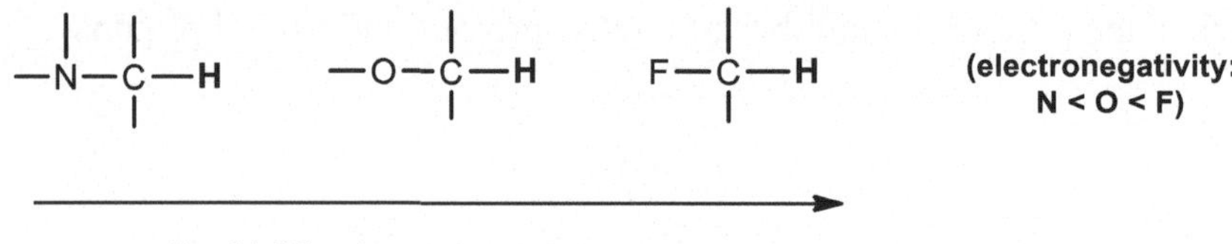

Deshielding increases; δ increases

3. Double bonds or aromatic π systems have a strong deshielding effect, and δ increases in the order:

—C—C—H —C=C—H C_6H_5—H

Deshielding increases; δ increases

4. **Acetylenic protons have lower δ relative to alkene protons**, although sp carbon is more electronegative than sp^2 carbon. This occurs because circulation of π electrons of the triple bond causes **shielding** of acetylenic protons that fall in the diamagnetic zone.

—C=C—H —C≡C—H

Shielding increases; δ decreases

Table 6.2 shows approximate chemical shifts of different types of protons in organic molecules.

Table 6.2: Approximate chemical shifts of different types of protons

Type of proton	Chemical shift (δ, ppm)	Type of proton	Chemical shift (δ, ppm)
Alkyl		**Vinylic**	
Methyl, RCH_3	0.8-1.0	$R_2C{=}C(R){-}H$	4.6-5.0
Methylene, R_2CH_2	1.2-1.4		
Methine, R_3CH	1.4-1.7	**Aromatic**	
Allylic		ArH	7.0-9.0
$R_2C{=}C(R){-}CH_3$	1.6-1.9	**Aldehyde**	
		$RC(=O)H$	9.5-9.6
Benzylic			
$ArCH_3$	2.2-2.5	**Alcohol hydroxyl**	
Ketone		ROH	1-5[a]
$RC(=O)CH_3$	2.1-2.6	**Amino**	
		RNH_2	1-3[a]
Alcohol/Ether		**Carboxylic**	
$HOCH_3$ / $ROCH_3$	3.3-4.0	$RC(=O)OH$	10-13[a]

[a] Chemical shifts vary in different solvents and with concentration

Example

Consider 1,4-dimethylbenzene (*p*-xylene).

The molecule has two equivalent methyl groups; thus the **six methyl protons** are equivalent (labelled *a*). The **four aromatic protons** are also equivalent (labelled *b*). Therefore, 1,4-dimethylbenzene gives **two ^{1}H NMR signals** for two types of protons.

The **protons (*a*)** are benzylic in nature and resonate at **δ 2.30** (6H), while **protons (*b*)** are aromatic and resonate at **δ 7.05** (4H).

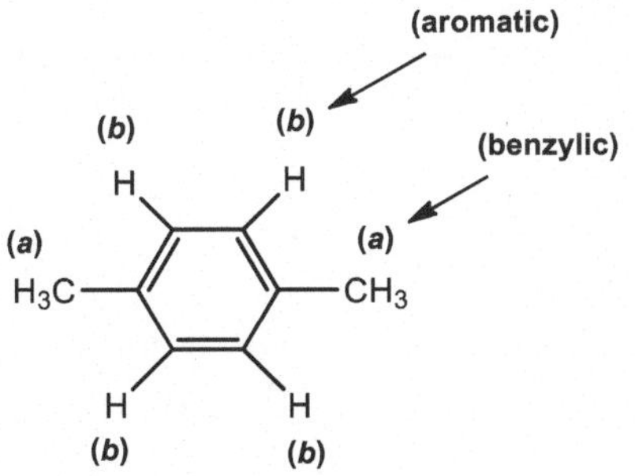

(aromatic)

(benzylic)

(*b*) H, (*b*) H, (*a*) H_3C, (*a*) CH_3, H (*b*), H (*b*)

1,4-Dimethylbenzene (*p*-xylene)

Two signals

protons (*a*) : δ 2.30 (6H)

protons (*b*) : δ 7.05 (4H)

PROBLEM 6.7: How many ^{1}H NMR signals would you expect from each of the following compounds? Predict their relative chemical shifts.

(i) $(CH_3)_3COCH_3$
(ii) $CH_3CO_2CH_3$
(iii) $CH_3COCH_2CH_3$

Answer

(i)

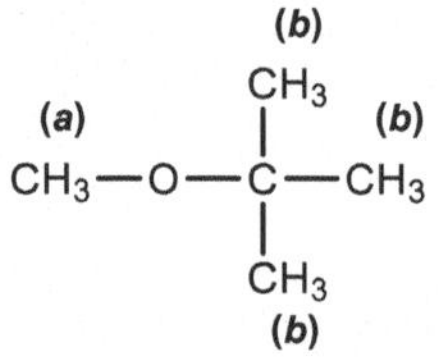

$$\underset{(a)}{CH_3}-O-\overset{\overset{(b)}{CH_3}}{\underset{\underset{(b)}{CH_3}}{C}}-\overset{(b)}{CH_3}$$

***t*-Butyl methyl ether**

Chemical shift: protons (*a*) > protons (*b*)
(δ 3.2) (δ 1.2)

The three CH_3 groups of $C(CH_3)_3$ in *t*-butyl methyl ether are equivalent. So there are two types of protons (labelled *a* and *b*), which give **two ^{1}H NMR signals** (see Figure 6.11). Methyl protons (*a*) are next to electronegative oxygen and are therefore more downfield than methyl protons (*b*).

Therefore, chemical shift of (*a*) protons is higher than that of (*b*) protons. The (*a*) protons (3H) resonate at δ 3.2 while (*b*) protons (9H) resonate at δ 1.2.

(ii)

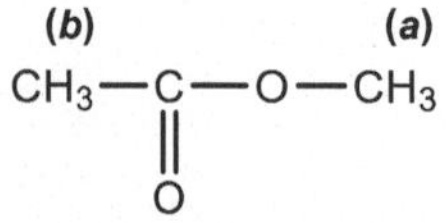

$$\overset{(b)}{CH_3}-\underset{\underset{O}{\|}}{C}-O-\overset{(a)}{CH_3}$$

Methyl acetate

Chemical shift: protons (*a*) > protons (*b*)
(δ 3.6) (δ 2.0)

Methyl acetate gives **two ^{1}H NMR signals** (see Figure 6.7) for two types of methyl protons. Protons (*a*) next to electronegative oxygen are more downfield than methyl protons (*b*) next to carbonyl group. Therefore, (*a*) protons (3H) have a higher chemical shift at δ 3.6 than (*b*) protons (3H) at δ 2.0.

Note: Protons (*a*) in methyl acetate are slightly more downfield at δ 3.6 than protons (*a*) in *t*-butyl methyl ether at δ 3.2. This is because protons (*a*) in methyl acetate are more deshielded by an additional electron-withdrawing carbonyl group (C=O).

(iii)

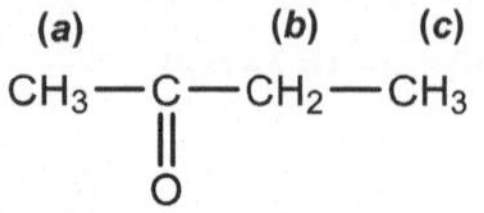

$$\overset{(a)}{CH_3}-\underset{\underset{O}{\|}}{C}-\overset{(b)}{CH_2}-\overset{(c)}{CH_3}$$

Butanone

Chemical shift: protons (*b*) > protons (*a*) > protons (*c*)
(δ 2.4) (δ 2.1) (δ 1.2)

Butanone gives **three ^{1}H NMR signals** for three types of protons. Methyl protons (*a*) and methylene protons (*b*) are both *next* to electron-withdrawing carbonyl group, but since *methylene* protons are generally more downfield than *methyl* protons, the chemical shift of CH_2 protons (*b*) is slightly higher than that of CH_3 protons (*a*). The CH_3 protons (*c*) are farthest from C=O; therefore, they shift upfield with a lowest chemical shift value.

The δ (ppm) values of (*b*), (*a*) and (*c*) protons are 2.4, 2.1 and 1.2 respectively.

6.6.5 Splitting of Signals in ^{1}H NMR

The signals in the ^{1}H NMR spectrum are *not always* single peaks. Most spectra show **single peaks** (called **singlets**) and **groups of peaks** (called **multiplets**).

Consider, for example, **1,1,2-trichloroethane**. The molecule has two types of protons (*a* and *b*) which produce two chemical shift signals (δ 5.8 and δ 4.0).

However, the more downfield **signal at δ 5.8 from the proton (*a*)** is split into *three* peaks (**triplet**), while the **signal at δ 4.0 from the protons (*b*)** is split into *two* peaks (**doublet**), as shown in Figure 6.12.

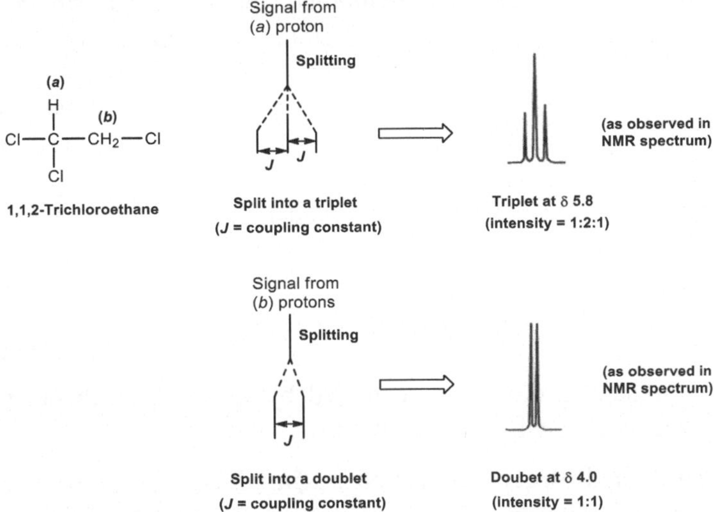

Figure 6.12: Splitting of signals for 1,1,2-trichloroethane.

The **signal splitting** arises from the phenomenon called the **spin-spin coupling**. The theory of spin-spin coupling is beyond our scope.

However, it is worthwhile to mention that the **coupling** refers to the **mutual magnetic influences** between non-equivalent protons (*a* and *b*), which are transmitted **via the electrons of the intervening bonds** between the non-equivalent hydrogen atoms.

The separation between the peaks in a *doublet*, *triplet* or a *multiplet* is called the **coupling constant (*J*)**, which is expressed in **frequency units (Hz)**.

Since coupling is caused entirely by internal forces in the molecule, the **coupling constant (*J*) is independent of applied magnetic fields**, and its measured value will be the *same* in NMR spectrometers operating at different radiofrequencies (say, at 60 MHz or 400 MHz).

Equivalent protons: No splitting of signals

Chemically equivalent protons *do not* couple. Thus, **signal splitting is not observed** for three equivalent protons of a **methyl group** or six equivalent protons of **benzene**.

(three equivalent protons)

No signal splitting

(δ 7.3, singlet)

(six equivalent protons)

No signal splitting

Effect of the number of intervening σ bonds between non-equivalent protons

Signal splitting (as observed for 1,1,2-trichloroethane) is *most common* when the non-equivalent protons (H_a and H_b) are on adjacent atoms and separated by three σ bonds.

(H_a and H_b are nonequivalent and separated by 3 σ bonds)

Signal splitting occurs

1,1,2-Trichloroethane

When the non-equivalent protons are separated by **more than three σ bonds**, signal splitting is very weak and usually *not* observed.

Thus **methyl acetate** and ***t*-butyl methyl ether** (in which *a* and *b* protons are separated by 5 single or σ bonds) show **no signal splitting** in their ^{1}H NMR spectra (see Figures 6.7 and 6.11). All the peaks are observed as **singlets**.

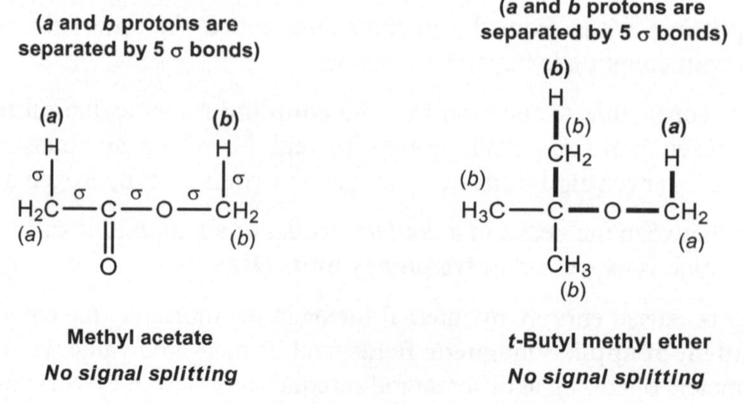

Methyl acetate
No signal splitting

***t*-Butyl methyl ether**
No signal splitting

The n +1 rule

In signal splitting, the **number of peaks** (or **multiplicity**) is related to the **number of neighbouring protons**. The number of peaks is given by $\boldsymbol{n+1}$, where n is the number of equivalent neighbouring protons. This is known as the $\boldsymbol{n+1}$ **rule**.

Consider 1,1,2-trichloroethane (see Figure 6.12). **Protons (*b*)** in CH_2Cl group have **one neighbouring proton** in Cl_2CH group, and therefore appear as a **doublet** ($n + 1 = 1 + 1 = 2$). But **proton (*a*)** in Cl_2CH group has **two neighbouring protons** in CH_2Cl group and appears as a **triplet** ($n + 1 = 2 + 1 = 3$).

The **intensity** of two peaks in a doublet is roughly equal (1:1), called a **1:1 doublet**. But the **intensity** of three peaks in a triplet is in the ratio of 1:2:1, called a **1:2:1 triplet.**

The splitting pattern according to the $n + 1$ rule, where $n = 1, 2, 3$ and 4, and the relative intensities are given below.

Number of neighbouring protons (*n*)	Number of peaks (*n* + 1)	Splitting patteren	Relative intensity
0	1	singlet (s)	1
1	2	doublet (d)	1:1
2	3	triplet (t)	1:2:1
3	4	quartet (q)	1:3:3:1
4	5	quintet (quin)	1:4:6:4:1

Illustrative Examples

1. *Ethyl Bromide*

Figure 6.13 shows the proton NMR spectrum of ethyl bromide (CH_3CH_2Br). The spectrum exhibits two chemical shift positions for **methyl protons (*a*) at δ 1.7** and for **methylene protons (*b*) at δ 3.4**.

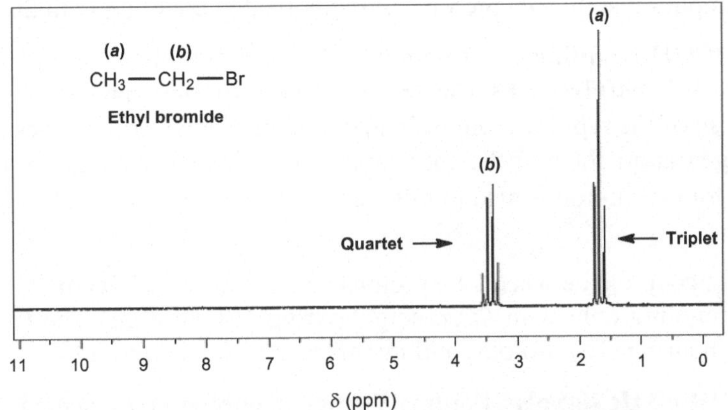

Figure 6.13: The ^{1}H NMR spectrum of ethyl bromide.

The **protons (*a*)** appear as a **triplet** (three peaks with relative intensities 1:2:1) since it has *two* neighbouring CH_2 protons, while the **protons (*b*)** appear as a **quartet** (four peaks with relative intensities 1:3:3:1) since it has *three* neighbouring CH_3 protons.

The **^{1}H NMR signals** of ethyl bromide are thus represented as: **δ 1.7 (3H, t)** and **δ 3.4 (2H, q)**.

It should be noted that the chemical shift (δ) reflects the central or middle position of the relevant multiplet (triplet or quartet).

2. *Ethyl Alcohol*

The proton NMR spectrum of *ordinary* ethyl alcohol (CH_3CH_2OH) is shown in Figure 6.14. The spectrum exhibits three chemical shifts for **CH_3 (*a*), CH_2 (*b*)** and **OH (*c*) protons** at **δ 1.26, δ 3.7 and δ 2.6** respectively.

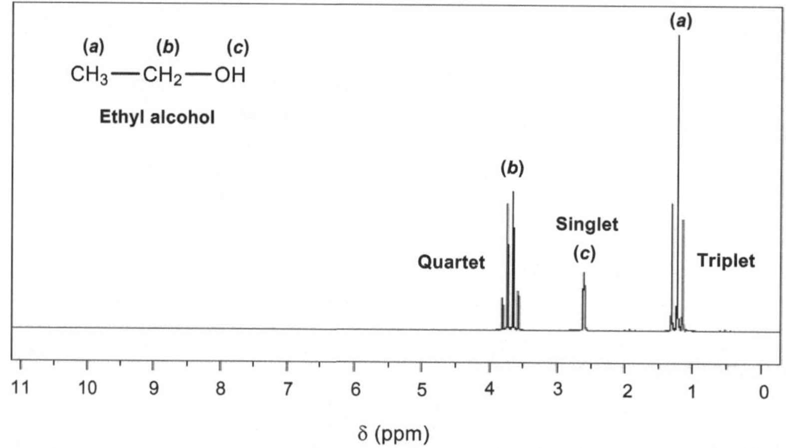

Figure 6.14: The ^{1}H NMR spectrum of ordinary ethyl alcohol.

In *ordinary alcohol*, in which traces of water or acidic impurities are present, the **OH proton** is observed as a **singlet** (single peak) even though it has *two neighbouring CH_2 protons*. This implies that the OH proton is *not* involved in coupling with the CH_2 protons.

The **absence of OH coupling** is attributed to the **fast proton exchange** among ethanol molecules, which is **catalysed by the traces of acid or base** present in *ordinary* ethyl alcohol. Because of the rapid proton exchange, one particular proton does not reside on a particular oxygen atom for a sufficiently long time for the coupling to take place. The chemical shift for OH proton is also highly variable, sometimes quoted as δ ~5 (see Table 6.2).

Since there is no coupling between CH_2 protons and OH proton, **CH_2 protons (*b*)** appear as a **quartet** by coupling only with *three* neighbouring CH_3 protons. The **CH_3 protons (*a*)** have *two* neighbouring CH_2 protons, and therefore appear as a **triplet**.

Therefore, the **^{1}H NMR signals** of ordinary ethyl alcohol are represented as: **δ 1.26 (3H, t), δ 2.6 (1H, s)** and **δ 3.7 (2H, q)**.

Note: Protons that couple with each other have peaks that are separated by the **same coupling constant (*J*)**. Thus the value of *J* for the CH_3 triplet and for the CH_2 quartet is the same ($J = 8$ Hz).

The OH coupling

When the NMR spectrum is recorded with *very pure* **ethyl alcohol**, the expected **triplet for OH proton** is observed due to **OH–CH_2 coupling** (coupling of OH proton with *two* neighbouring CH_2 protons).

On the other hand, the **CH_2 signal is split into a multiplet of eight peaks.** This can be explained by applying the ***n* + 1 rule** to two sets of protons (CH_3 and OH) **separately.**

The CH_2 signal is first split into a *quartet* by three neighbouring CH_3 protons, and then *each peak* of the quartet is further split into a *doublet* by the OH proton. The resulting pattern is a **multiplet** of eight peaks, called a **quartet of doublet (qd).**

Thus the **^{1}H NMR signals** of very pure ethyl alcohol are represented as: **δ 1.26 (3H, t)**, **δ 2.6 (1H, t)** and **δ 3.7 (2H, qd).**

Hydrogen–deuterium exchange

If the NMR spectrum is run with a *solution* of **ethyl alcohol in D_2O**, the **OH signal disappears.**

This is because D_2O exchanges rapidly with OH of the alcohol as follows:

$$\underset{}{CH_3CH_2O—H} + \underset{(D_2O)}{D—O—D} \longrightarrow CH_3CH_2O—D + \underset{(HOD)}{H—O—D}$$

The rapid **hydrogen-deuterium exchange** gives CH_3CH_2OD and HOD; thereby removing the OH proton peak of the alcohol. But at the same time a **new signal for HOD** appears at δ 4.7.

This experiment in D_2O can be used to identify the presence of OH group in organic molecules.

PROBLEM 6.8: In which of the following compounds does signal splitting occur in the ^{1}H NMR spectrum? When splitting is observed, give the splitting pattern of each signal.

(i) $ClCH_2CH_2Cl$

(ii) $Cl_2CHCHClCHCl_2$

Answer

(i) **No splitting** is observed for 1,2-dichloroethane because the two CH_2 groups are equivalent. There is only one NMR signal with a single peak (singlet) for the four protons.

(ii)

(*a*)	(*b*)	(*a*)
Cl_2CH—	—CH(Cl)—	—$CHCl_2$

1,1,2,3,3-Pentachloropropane

Chemical shift:	protons (*a*) >	proton (*b*)
	(δ 6.1)	(δ 4.5)
	Doublet	**Triplet**

1,1,2,3,3-Pentachloropropane has two different kinds of protons (*a* and *b*), which give rise to two NMR signals; the signal of protons (*a*) being more downfield (higher δ) due to the presence of two electronegative Cl atoms.

The two **protons (*a*)** appear as a **doublet** since each has one neighbouring proton, while the **proton (*b*)** appears as a **triplet** because it has two equivalent coupling neighbours. The coupling constant (*J*) is the same for the doublet and the triplet.

6.6.6 Application of ^{1}H NMR in Medicine

^{1}H NMR spectroscopy has been widely employed not only in chemistry, but also in medicine. An important application of ^{1}H NMR spectroscopy in medical diagnosis is known as the **magnetic resonance imaging (MRI)**.

In this technique, the patient is the sample. A portion of the patient's body is placed inside a powerful donut-shaped magnet and irradiated with radiofrequencies using the pulse method to excite the protons in the tissues under investigation.

The results are translated into an image showing contrasts between different regions in soft tissues. These contrasts in magnetic resonance imaging are inherently much better than that produced by X-ray techniques, and are used to detect any abnormalities, such as tumour growth and bone injury.

6.7 ^{13}C NMR SPECTROCOPY

^{13}C NMR (carbon-13 NMR) spectroscopy is similar to ^{1}H NMR.

^{13}C NMR can provide information on the different types of carbon atoms in organic molecules. Equivalent carbon atoms show a single peak, while non-equivalent carbons give separate peaks.

^{13}C isotope has a natural abundance of only 1.1% (^{12}C has a natural abundance of about 99%, but is nonmagnetic). Thus ^{13}C NMR is not as sensitive as ^{1}H NMR, and more samples and more time are needed to record ^{13}C NMR spectra.

Broad band proton decoupling

The **probability** of two ^{13}C atoms to be *adjacent* to each other in a molecule is **very small** (1.1% × 1.1%, i.e. about 1 in 10000). Therefore, **^{13}C–^{13}C coupling is minimal**, and the absence of such type of splitting simplifies the ^{13}C spectrum.

^{13}C NMR signals can however be split into multiplets by protons attached to carbon atoms. But this ^{1}H–^{13}C coupling can be eliminated by choosing instrumental parameters, known as **broad band proton decoupling**.

In this way, all types of proton–carbon interactions can be decoupled from the carbon signals. As a result, **each type of carbon produces only one peak (singlet)**. Most ^{13}C spectra are recorded in this simplified decoupled mode.

Note: It is possible to obtain ^{13}C NMR spectra that can identify whether a carbon is primary (CH_3), secondary (CH_2), tertiary (CH) or quaternary (C). One such spectrum is called the DEPT (distortionless enhanced polarization transfer) ^{13}C spectrum.

Chemical shifts

The **chemical shift range** for ^{13}C NMR is quite large, typically **δ = 0–200 ppm**. The reference standard is tetramethylsilane (TMS), to which δ is assigned as 0 ppm.

TMS has four equivalent carbon atoms. Since silicon is more electropositive, carbon atoms in TMS are quite electron-rich, and therefore highly shielded, so that almost all other ^{13}C signals occur downfield (at higher δ values) with respect to the TMS signal.

Like ^{1}H NMR, the chemical shift of a ^{13}C nucleus depends on the relative electron density around it. **Electron-withdrawing groups** cause **deshielding** of the ^{13}C nucleus and its signal appears **downfield (higher δ)**.

But **electron-donating groups** cause **shielding** of the ^{13}C nucleus and cause the ^{13}C signal to appear **upfield (lower δ)**.

The ^{13}C NMR spectra can be divided into four major regions as follows:

Type of carbon atom	Chemical shift (δ, ppm)
sp^3 ↓ R—CH_2—R	0-50
sp^3 next to O ↓ R—CH_2—OR'	50-100
sp^2 carbons C=C or (benzene ring)	100-150
sp^2 next to O ↓ C=O	150-200

Illustrative examples

1. *1-Chloro-2-propanol*

Figure 6.15 shows the ^{13}C NMR spectrum of 1-chloro-2-propanol ($CH_3CHOHCH_2Cl$). The spectrum exhibits **three peaks (singlets)** at δ 20, δ 51 and δ 67.

Since the compound contains *three* nonequivalent carbons C1 (*a*), C2 (*b*) and C3 (*c*), it gives three ^{13}C NMR peaks or signals.

The lower the electron density around a carbon, the less that carbon will be shielded, and its signal will appear more downfield (higher δ). The oxygen atom of OH group is most electronegative and hence is most electron-withdrawing. Thus **C2 (*b*)** bearing the OH group is **most deshielded** and its signal appears **most downfield at δ 67**.

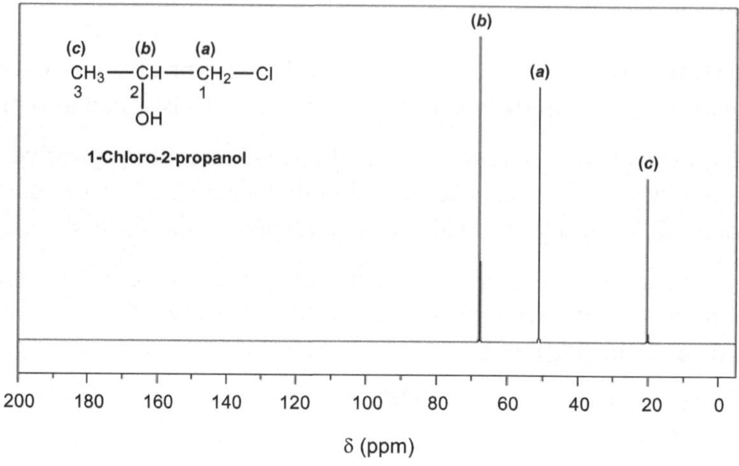

Figure 6.15: The ^{13}C NMR spectrum of 1-chloro-2-propanol.

Chlorine is less electronegative than oxygen and **C1 (*a*)** bearing Cl gives the signal which shifts **upfield at δ 51.**

The methyl carbon **C3 (*c*)** with no electronegative group attached to it gives the signal that is **further upfield at δ 20.**

2. *Adipic acid*

The ^{13}C NMR spectrum of adipic acid (hexanedioic acid) is shown in Figure 6.16. Adipic acid ($HO_2CCH_2CH_2CH_2CH_2CO_2H$) contains *six* carbons, but it gives only **three ^{13}C NMR peaks** at δ 24, δ 33 and δ 174.

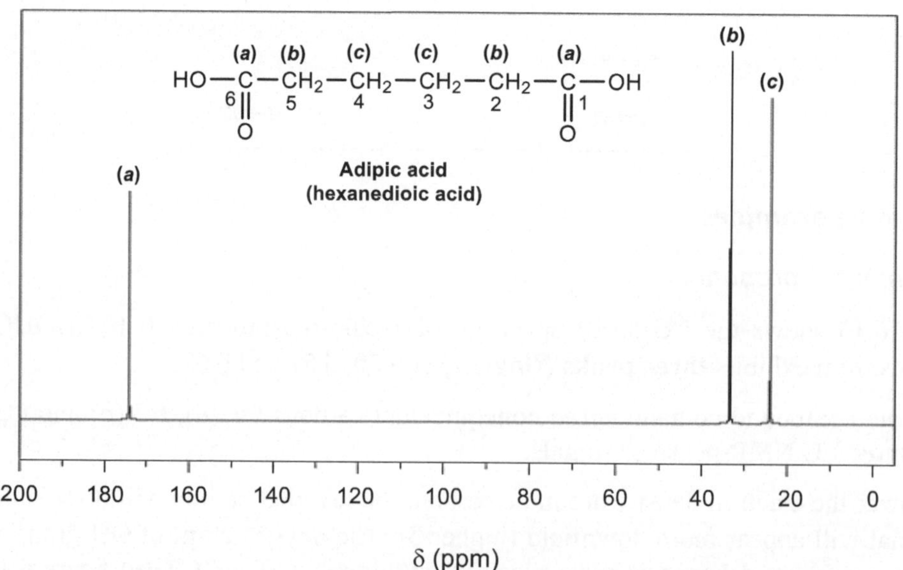

Figure 6.16: The ^{13}C NMR spectrum of adipic acid (hexanedioic acid).

Because of the symmetry of the structure of the molecule, two carboxyl (CO_2H) groups are equivalent, and therefore **carbonyl carbons C1/C6 (*a*)** give a single peak, which is **most downfield at δ 174.**

Similarly, C2 and C5 carbons are equivalent, and so are C3 and C4 carbons.

Since C2/C5 carbons (*b*) are next to the electron-withdrawing CO_2H group, they are **more deshielded** than C3/C4 carbons (*c*). Therefore, **C2/C5 (*b*)** give signal that is more downfield **at δ 33** than the signal for **C3/C4 (c) at δ 24.**

PROBLEM 6.9: Paracetamol (*p*-hydroxyacetanilide) is a familiar analgesic and antipyretic drug. How many signals would you expect from paracetamol in its broad band proton-decoupled ^{13}C NMR spectrum?

Answer

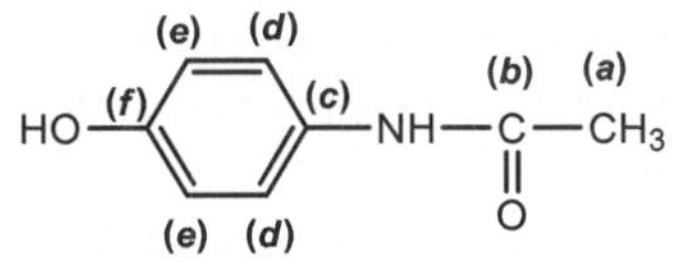

Paracetamol

In paracetamol, the two *ortho* carbon atoms with respect to $NHCOCH_3$ are equivalent, so are the two *ortho* carbons with respect to OH.

Thus, paracetamol has six different types of carbons: methyl carbon (*a*), carbonyl carbon (*b*), ring carbon (*c*), *ortho* carbons (*d*), *ortho* carbons (*e*), and ring carbon (*f*). Hence paracetamol gives six ^{13}C NMR signals.

6.8 DETERMINATION OF STRUCTURE USING SPECTRAL DATA

Spectroscopic methods can provide sufficient information for the determination of **structure** of organic molecules.

Determining structure from spectral data might appear to be intimidating at first, but what it really needs is a lot of practice, and ultimately it can be fun. However, the big picture of spectroscopic solutions to structural problems is beyond our scope.

Here we give an introduction as to how we can extract important pieces of structural information from **molecular formula**, **IR** and **NMR** data, and then putting the pieces together to arrive at the molecular structure.

The steps are as follows:

(a) Calculate the double bond equivalent (DBE) from molecular formula
(b) Determine the major functional groups from IR data
(c) Find the fragments that fit with the NMR data with respect to the number of protons, chemical shifts and splitting patterns
(d) Establish the structure by putting all fragments together

Problem 1

Propose a structure for an organic compound of molecular formula $\mathbf{C_6H_{12}O}$ that gives the following IR and ^{1}H NMR spectral data (δ in ppm):

IR: 1710 cm^{-1} (strong peak)

^{1}H NMR: δ 1.15 (9H, singlet)

δ 2.1 (3H, singlet)

Solution:

(a) The molecular formula, $C_6H_{12}O$, gives

DBE = (6 + 1) – (12/2) = 1

This indicates that the molecule has either **a double bond or a ring.**

(b) The *strong* IR peak at **1710 cm^{-1}** is characteristic of a carbonyl group. Therefore, the molecule contains a **C=O** group, which accounts for **DBE = 1** as well as the **oxygen atom** in the molecular formula.

(c) The ^{1}H NMR spectrum gives two peaks at δ 1.15 and δ 2.1, each of which is a **singlet (s)**. Because there is *no* signal splitting, protons for the two peaks are *not* on adjacent carbon atoms.

The peak at **δ 1.15** is typical of an alkyl group (see Table 6.2), and its nine equivalent protons **(9H)** can be accounted for by a *tertiary*-butyl group, **–C(CH₃)₃**.

The peak at **δ 2.1** is also indicative of an alkyl group in which its protons are deshielded somewhat, and the three equivalent protons **(3H)** suggest a methyl group, **–CH₃**, *next* to the electron-withdrawing carbonyl (C=O).

Since the NMR data provide no peaks attributable to an aldehyde proton at δ ~ 10, the compound is a **ketone.**

(d) The **three fragments [C=O, C(CH₃)₃, CH₃]** account for the molecular formula ($C_6H_{12}O$), and there is *only one* way to put the fragments together. This gives the following structure for the organic compound:

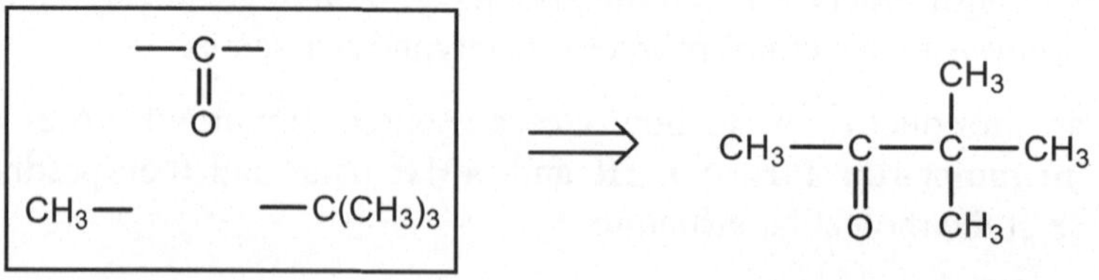

–C(=O)– CH₃– –C(CH₃)₃ ⟹ CH₃–C(=O)–C(CH₃)₂–CH₃

Fragments

***t*-Butyl methyl ketone (3,3-dimethyl-2-butanone)**

The compound is ***t*-butyl methyl ketone** (3,3-dimethyl-2-butanone).

Finally, the proposed **structure** of the compound is **confirmed** by assigning the spectral data to different groups:

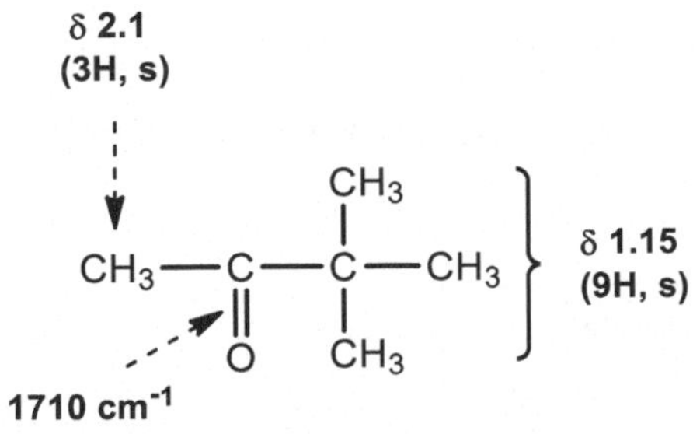

δ 2.1
(3H, s)

$$CH_3-\underset{\underset{O}{\|}}{C}-\overset{\overset{CH_3}{|}}{\underset{\underset{CH_3}{|}}{C}}-CH_3$$

δ 1.15 (9H, s)

1710 cm^{-1}

Problem 2

Suggest a structure for an organic compound of molecular formula $\mathbf{C_4H_8O_3}$ that gives the following IR and ^{1}H NMR spectral data (δ in ppm). Note that s = singlet, t = triplet, and q = quartet.

IR: 2500–3000 cm^{-1} (broad)
1715 cm^{-1}

^{1}H NMR: δ 1.27 (3H, t)
δ 3.66 (2H, q)
δ 4.13 (2H, s)
δ 10.95 (1H, s)

Solution:

(a) The molecular formula ($C_4H_8O_3$) gives **DBE = (4 + 1) – (8/2) = 1**, which indicates that **a double bond or a ring** is a part of the structure.

(b) The IR absorption at **1715 cm^{-1}** is indicative of a **C=O** group and accounts for the DBE = 1. The *broad* IR peak in **2500-3000 cm^{-1}** region probably indicates an **OH** group.

(c) In ^{1}H NMR, a peak at **δ 10.95** indicates a highly deshielded proton, and the one-proton singlet **(1H, s)** suggests a **carboxylic proton** (CO_2H). Thus, the compound contains a carboxylic acid group, **–COOH**, which is also consistent with the IR data for C=O and OH groups.

The characteristic NMR pattern of a three-proton triplet **(3H, t)** at δ 1.27 and a two-proton quartet **(2H, q)** at δ 3.66 is indicative of an ethyl group (**CH_3CH_2–**) (cf. Figure 6.13).

The deshielded quartet (q) at **δ 3.66** suggests that CH_2 of the ethyl group is *attached to an electron-withdrawing oxygen atom* (O). Thus, the compound contains an ethoxy group, **CH_3CH_2O–**. This oxygen atom, together with two oxygen atoms of carboxyl, accounts for all three oxygen atoms of the molecule.

The NMR data also show the presence of a peak at **δ 4.13** as a two-proton singlet **(2H, s)**. This indicates a methylene group, **CH_2**, which has *no* adjacent proton neighbours.

Further, these methylene protons are *more deshielded* (δ 4.13) than the methylene protons (δ 3.66) of the ethoxy group. The greater deshielding at δ 4.13 suggests that the **CH_2** is attached to *both electron-withdrawing O atom and electron-withdrawing COOH group*.

(d) Putting the three fragments (CH_3CH_2O, CH_2, COOH) together gives the following structure for the organic compound ($C_4H_8O_3$):

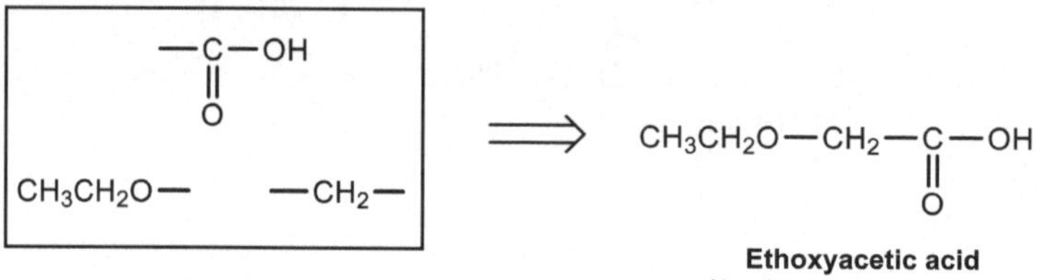

$$\boxed{\begin{array}{c} -\underset{\underset{O}{\parallel}}{C}-OH \\ CH_3CH_2O- \quad\quad -CH_2- \end{array}} \Longrightarrow CH_3CH_2O-CH_2-\underset{\underset{O}{\parallel}}{C}-OH$$

Fragments

Ethoxyacetic acid (2-ethoxyethanoic acid)

The compound is **ethoxyacetic acid** (2-ethoxyethanoic acid).

Finally, the proposed **structure** of the compound is **confirmed** by assigning the spectral data to different groups:

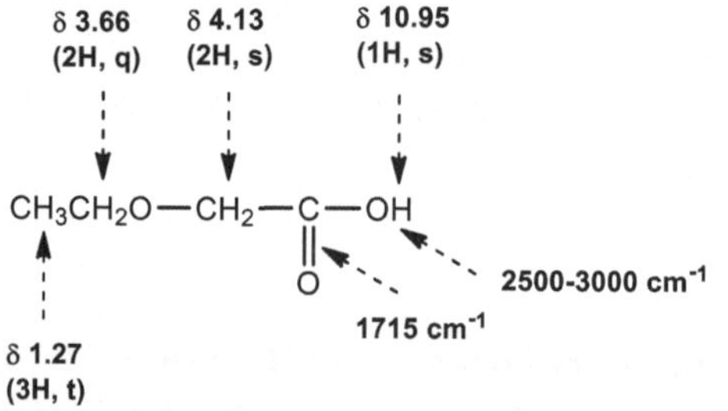

$$\overset{\delta\,1.27\,(3H,\,t)}{CH_3}\overset{\delta\,3.66\,(2H,\,q)}{CH_2}O-\overset{\delta\,4.13\,(2H,\,s)}{CH_2}-\underset{\underset{O}{\parallel}}{C}-\overset{\delta\,10.95\,(1H,\,s)}{OH}$$

δ 3.66 (2H, q); δ 4.13 (2H, s); δ 10.95 (1H, s); δ 1.27 (3H, t); C=O: 1715 cm^{-1}; O–H: 2500-3000 cm^{-1}

SUPPLEMENTARY PROBLEMS

6.10 Two diastereomers of stilbene ($C_6H_5CH{=}CHC_6H_5$) have their λ_{max} values at 278 and 294 nm. Identify the diastereomers.

6.11 Three compounds **A**, **B** and **C** have the same molecular formula C_5H_6. Each reacts with 3 moles of H_2 to give pentane. Compounds **B** and **C** give a precipitate with ammoniacal $AgNO_3$ but compound **A** does not. Compounds **A** and **B** show λ_{max} near 230 nm but compound **C** shows no absorption maximum beyond 200 nm. What are the structures of **A**, **B** and **C**?

[*Ans:* **A**: $CH_2{=}CH{-}C{\equiv}CCH_3$; **B**: $CH_3CH{=}CH{-}C{\equiv}CH$; **C**: $CH_2{=}CHCH_2C{\equiv}CH$]

6.12 How would you distinguish the following isomeric alkenes using IR spectroscopy?

$$(CH_3)_2C{=}C(CH_3)_2 \qquad\qquad (CH_3)_2CH(CH_3)C{=}CH_2$$

2,3-Dimethyl-2-butene **2,3-Dimethyl-1-butene**

6.13 Concentrated solutions of ethyl alcohol (CH_3CH_2OH) and ethylene glycol ($HOCH_2CH_2OH$) give broad bands near 3350 cm^{-1}. On dilution with CCl_4, the IR spectrum of ethylene glycol does not change, but the spectrum of ethyl alcohol shows a sharp peak at 3600 cm^{-1} replacing the broad band. Explain.

[*Hints:* intermolecular *versus* intramolecular hydrogen bonding]

6.14 How many signals do you expect in the ^1H NMR spectrum of 1,3-dichloropropane ($ClCH_2CH_2CH_2Cl$)? Predict the multiplicity of each signal.

[*Ans:* two signals: (4H, triplet) and (2H, quintet)]

6.15 How many signals would you expect in the ^{13}C NMR spectrum of butanone ($CH_3COCH_2CH_3$)? Predict their relative chemical shifts.

[*Ans:* four signals. chemical shifts: C2 (carbonyl carbon) > C3 > C1 > C4]

6.16 The IR and NMR absorption peaks for two compounds along with their molecular formulas are given below. The chemical shift (δ) values are expressed in ppm. Suggest a structure for each compound.

(a) Molecular formula: C_4H_8O
IR: 1720 cm^{-1} (strong peak)
^1H NMR: δ 1.05 (3H, t), 2.13 (3H, s), and 2.47 (2H, q)

(b) Molecular formula: $C_6H_{12}O_2$
IR: 1745 cm^{-1}
^1H NMR: δ_H 1.20 (9H, s) and 3.67 (3H, s)
^{13}C NMR: δ_C 179, 52, 39, 27

[*Ans:* (a) $CH_3CH_2COCH_3$ (b) $(CH_3)_3CCO_2CH_3$]

6.17 Compound **X** has a strong IR absorption at 1730 cm^{-1}, and it gives two singlet peaks at δ 1.08 and δ 9.48 in its ^1H NMR spectrum. On reaction with $NaBH_4$, **X** produces compound **Y** which has a broad IR absorption at 3200-3500 cm^{-1}. Reaction of **Y** with NaH (a base) followed by treatment with CH_3I gives compound **Z**, which shows a molecular ion peak at *m*/*z* 102 in its mass spectrum. Suggest the structures for compounds **X**, **Y** and **Z**.

[*Ans:* **X**: $(CH_3)_3CCHO$; **Y**: $(CH_3)_3CCH_2OH$; **Z**: $(CH_3)_3CCH_2OCH_3$]

PART-II

Reactions of Organic Compounds

CHAPTER 7

Introduction to Organic Reactions: Classification and Mechanisms

IN THIS CHAPTER

- Conventions of writing organic reactions
- Classification of reactions: Addition, elimination, substitution and rearrangement
- Classification of mechanisms: Ionic, radical and pericyclic
- Nucleophiles: Lone pair, π and σ nucleophiles; ambident nucleophiles
- Electrophiles: Lewis acid, π bond and σ bond electrophiles
- Drawing ionic mechanisms: Illustrative examples
- Reactive intermediates: Carbocations and carbanions
- Concerted and multistep reactions: Ionic mechanisms
- Reaction thermodynamics: Equilibrium constant (K) and standard free energy change (ΔG^0)
- Reaction kinetics: Rate constant (k) and order, transition structure (TS) and molecularity, free energy of activation ($\Delta G^{\ddagger}$), free energy diagrams and Hammond postulate

7.1 INTRODUCTION

A chemical reaction depicts the conversion of **reactants** into **products**, and is written as,

$$A + B \longrightarrow C + D$$

where A and B are reactants, and C and D are products.

For organic reactions, one of the reactants must be an organic compound, and the other reacting component may be an inorganic or organic species (molecule or ion). The *organic reactant* that undergoes structural or functional group change is called the **substrate**, and the *other reactant* which reacts with the substrate is called the **reagent**. The reaction is then represented as,

$$\text{Substrate} + \text{Reagent} \longrightarrow \text{Products}$$

Besides substrates and reagents, organic reactions may also involve **solvents** or **catalysts**.

There are different ways by which organic reactions can be written. Usually, solvents, catalysts, reaction conditions (e.g. temperature), and sometimes reagents are written over or under the arrow. **Loss of small molecules** like water, salts or gaseous products are sometimes written with a **negative sign over or under the arrow**. The equations representing organic reactions are often not balanced.

Figure 7.1 illustrates some conventions of writing organic reactions.

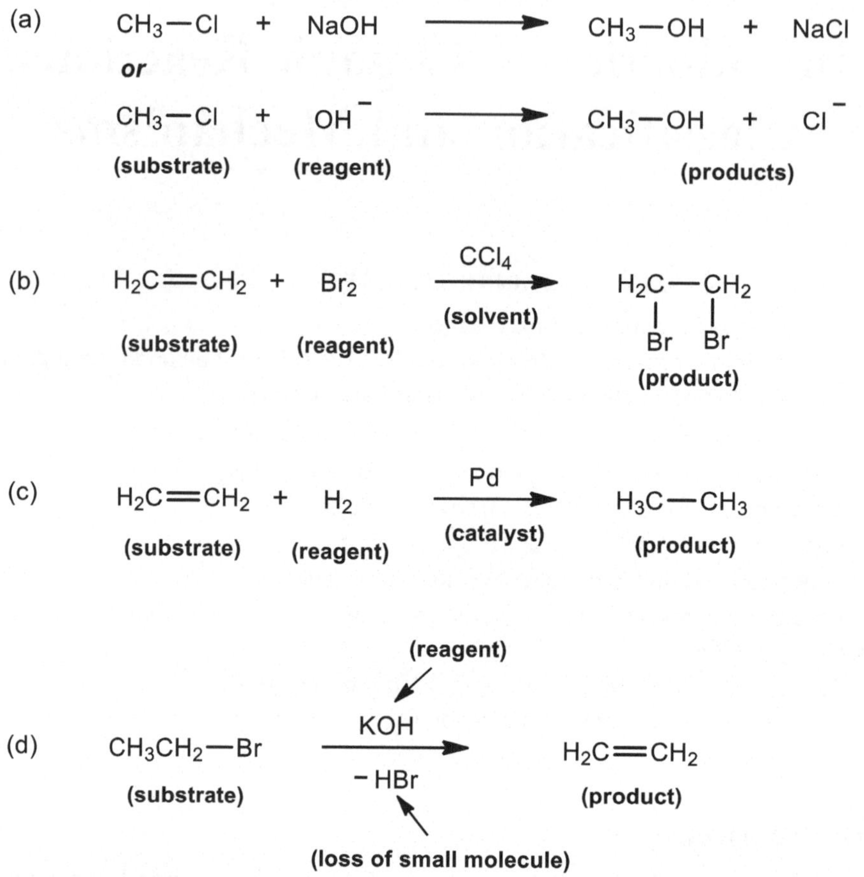

(a) CH_3-Cl + NaOH ⟶ CH_3-OH + NaCl

or

CH_3-Cl + OH^- ⟶ CH_3-OH + Cl^-

(substrate) **(reagent)** **(products)**

(b) $H_2C{=}CH_2$ + Br_2 $\xrightarrow{CCl_4}$ $H_2C(Br)-CH_2(Br)$

(substrate) **(reagent)** **(solvent)** **(product)**

(c) $H_2C{=}CH_2$ + H_2 $\xrightarrow{Pd}$ H_3C-CH_3

(substrate) **(reagent)** **(catalyst)** **(product)**

(d) CH_3CH_2-Br $\xrightarrow[-\,HBr]{KOH}$ $H_2C{=}CH_2$

(reagent)

(substrate) **(product)**

(loss of small molecule)

Figure 7.1: Some conventions (a–d) of writing organic reactions.

Note: Organic substrates, reagents, solvents, catalysts or products can be intrinsically toxic and hazardous to use. Thus attention has also been directed to design chemical reactions that can lessen or eliminate such hazards on human health and the environment. This is what is called the **green chemistry**.

Green chemistry is the design of chemical processes that reduce or eliminate the use and production of hazardous substances. Besides using safer chemicals with little or no toxicity, the concept of green chemistry envisages maximizing **atom economy** when the final product contains the maximum number of atoms of starting materials. This ensures a waste of few or no atoms so that there is no need to treat or clean up the byproducts. The green chemistry also seeks to increase the **energy efficiency** when the reactions can be run at room temperature and pressure wherever possible.

Importantly, green chemistry reduces pollution at its *source* by minimizing the hazards associated with starting materials, solvents and products.

7.2 CLASSIFICATION OF REACTIONS

There are countless reactions in organic chemistry. However, almost all organic reactions fall into four classes. They are *addition*, *elimination*, *substitution*, and *rearrangement*.

7.2.1 Addition

Addition reactions are characteristic reactions of **unsaturated substrates** that contain double or triple bonds. Double bonded or triple bonded atoms can accept additional atoms or groups from reagents, without exceeding their maximum valence.

Addition reaction of an *unsaturated substrate* with a reagent gives a single *saturated product*. In the process, **a π bond in the substrate is replaced by two new σ bonds**.

Addition reactions commonly involve addition to carbon-carbon double bonds (C=C) and addition to carbon-oxygen double bonds (C=O).

Addition to C=C

Alkenes possess C=C double bonds. Reagents such as hydrogen halides (HCl, HBr or HI) can add to C=C of alkenes to form **alkyl halides**.

$$-\overset{|}{\underset{|}{C}}\overset{\pi}{=}\overset{|}{\underset{|}{C}}- \quad + \quad H-X \quad \longrightarrow \quad -\overset{H}{\underset{|}{\overset{\sigma|}{C}}}-\overset{X}{\underset{|}{\overset{|\sigma}{C}}}-$$

Alkene **(X = Cl, Br, I)** **Alkyl halide**

Example

$$H-\underset{H}{\underset{|}{C}}=\underset{H}{\underset{|}{C}}-H \quad + \quad H-Cl \quad \longrightarrow \quad H-\overset{H}{\overset{|}{\underset{H}{\underset{|}{C}}}}-\overset{Cl}{\overset{|}{\underset{H}{\underset{|}{C}}}}-H$$

Ethylene **Ethyl chloride**

Addition to C=O

Aldehydes and ketones contain a carbonyl group (C=O). Addition of a reagent such as hydrogen cyanide to C=O produces a **cyanohydrin**.

$$-\underset{|}{C}\overset{\pi}{=}O \quad + \quad H-CN \quad \longrightarrow \quad -\overset{CN}{\overset{\sigma|}{\underset{|}{C}}}-O\overset{\sigma}{-}H$$

Aldehyde/ketone **Cyanohydrin**

Example

$$H_3C-\underset{H}{\underset{|}{C}}=O \quad + \quad H-CN \quad \longrightarrow \quad H_3C-\overset{CN}{\overset{|}{\underset{H}{\underset{|}{C}}}}-OH$$

Acetaldehyde **Acetaldehyde cyanohydrin**

7.2.2 Elimination

Elimination reactions are the *opposite* of addition reactions. In an elimination reaction, a small molecule is eliminated from a *saturated substrate* under the action of the reagent to give an *unsaturated product* with a double bond. Here, **two σ bonds in the substrate are replaced by a new π bond**.

$$\text{—}\overset{\overset{\text{H}}{\sigma|}}{\underset{|}{\text{C}}}\text{—}\overset{\overset{\text{X}}{|\sigma}}{\underset{|}{\text{C}}}\text{—} \xrightarrow[\text{(loss of HX)}]{-\text{HX}} \text{—}\underset{|}{\text{C}}\overset{\pi}{=}\underset{|}{\text{C}}\text{—}$$

Dehydrohalogenation

The **elimination of hydrogen halide** from an alkyl halide with a strong base produces an alkene. This elimination reaction is called **dehydrohalogenation**.

For example, the elimination of HBr from isopropyl bromide in the presence of KOH produces propylene.

$$\underset{\text{Isopropyl bromide}}{H_2\overset{\overset{H}{|}}{C}-\overset{\overset{Br}{|}}{\underset{\underset{H}{|}}{C}}-CH_3} \xrightarrow[\text{Heat }(\Delta)]{KOH} \underset{\text{Propylene}}{H_2C=\underset{\underset{H}{|}}{C}-CH_3} + HBr$$

Dehydration

The **loss of water** from an alcohol with a strong acid gives an alkene. This elimination is called **dehydration**.

For example, elimination of water from isopropyl alcohol in the presence of H_2SO_4 produces propylene.

$$\underset{\text{Isopropyl alcohol}}{H_2\overset{\overset{H}{|}}{C}-\overset{\overset{OH}{|}}{\underset{\underset{H}{|}}{C}}-CH_3} \xrightarrow[\Delta]{H_2SO_4} \underset{\text{Propylene}}{H_2C=\underset{\underset{H}{|}}{C}-CH_3} + H_2O$$

7.2.3 Substitution

A **substitution reaction** involves the replacement of a group of the substrate by another group supplied by the reagent. In this case, **a σ bond in the substrate is replaced by a new σ bond**.

Substitution reactions can take place at both *saturated* and *unsaturated* carbon atoms.

Substitution at a saturated carbon

A substitution reaction at a **saturated (sp^3) carbon** is the reaction of an **alkyl halide** with *hydroxide ion* to form an **alcohol**. In this reaction, the halogen (X) of the substrate is replaced by OH. A reagent such as NaOH supplies the HO^- ion.

$$\underset{\text{Alkyl halide}}{-\overset{|}{\underset{|}{C}}\overset{sp^3}{\underset{\sigma}{-}}X} + \underset{(NaOH)}{OH^-} \longrightarrow \underset{\text{Alcohol}}{-\overset{|}{\underset{|}{C}}\underset{\sigma}{-}OH} + X^-$$

Example

$$\underset{\text{Methyl chloride}}{H_3C-Cl} + OH^- \longrightarrow \underset{\text{Methyl alcohol}}{H_3C-OH} + Cl^-$$

Substitution at an unsaturated carbon (C=O)

There are two important types of substitution reactions that take place at an **unsaturated (sp^2) carbon**. One type involves substitution at a carbon atom of the **carbonyl group (C=O)**.

An example is the hydrolysis of **esters**, when an alkoxy group (OR) is replaced by OH to produce **carboxylic acids**.

$$\underset{\text{Ester}}{-\overset{O}{\overset{\|}{C}}\overset{sp^2}{\underset{\sigma}{-}}OR} + OH^- \longrightarrow \underset{\text{Carboxylic acid}}{-\overset{O}{\overset{\|}{C}}\underset{\sigma}{-}OH} + {}^-OR$$

Example

$$\underset{\text{Methyl acetate}}{H_3C-\overset{O}{\overset{\|}{C}}-OCH_3} + OH^- \longrightarrow \underset{\text{Acetic acid}}{H_3C-\overset{O}{\overset{\|}{C}}-OH} + {}^-OCH_3$$

Substitution at an unsaturated carbon (aromatic substitution)

The other type of substitution at an unsaturated carbon involves substitution at a carbon atom of the benzene ring, called the **aromatic substitution**.

An example is the *chlorination* of **benzene**, when a hydrogen atom of benzene is replaced by a chlorine atom to form **chlorobenzene**.

$$\text{Benzene (C–H }\sigma\text{, sp}^2\text{)} + Cl_2 \xrightarrow{FeCl_3} \text{Chlorobenzene (C–Cl }\sigma\text{)}$$

Benzene

Chlorobenzene

7.2.4 Rearrangement

A **rearrangement reaction** involves a reorganization of the carbon skeleton of the substrate to form a product with a *modified* carbon skeleton.

For example, when an alkene (3,3-dimethyl-1-butene) is heated with a strong acid, it **rearranges** to a new alkene (2,3-dimethyl-2-butene) involving the **movement of a methyl group from one carbon to another**. The strong acid supplies H^+ ion, and acts as a catalyst.

$$H_3C-C(CH_3)_2-CH=CH_2 \xrightarrow{H^+} H_3C-C(CH_3)=C(CH_3)-CH_3$$

3,3-Dimethyl-1-butene

2,3-Dimethyl-2-butene

Rearranged alkene

Note: Organic reactions include **oxidation and reduction** but they do not represent separate classes of reactions. They reflect the changes which accompany addition, elimination or substitution.

7.3 MECHANISMS OF REACTIONS

A mechanism for an organic reaction is a detailed, step-by-step description of the events that take place on the molecular level as the reactants are converted into products. It is not actually possible to observe the molecular events directly, since molecules are too small. A mechanism is thus a theory that can be proposed based on experimental facts and chemical intuition.

Often, *more than one* mechanism for a reaction can be thought of. But **a plausible or reasonable mechanism** is the one that is consistent with all experimental results. If future experiments contradict the proposed mechanism, then the mechanism needs to be changed or modified so as to be consistent with the new experimental facts.

Mechanisms help us organize the vast body of organic reactions and make them understandable. There are *three* types of basic mechanisms, namely, **ionic, radical and pericyclic** mechanisms.

7.3.1 Ionic Mechanism

On the molecular level, a chemical reaction always involves the **breaking and making of bonds**.

In general, a covalent bond may break in two possible ways. One possibility is the unsymmetrical breaking or heterolytic cleavage of the bond.

Heterolytic Bond Cleavage

The type of bond breaking when one fragment takes away both electrons of the bond, leaving the other fragment with no electrons of the bond is called the **heterolytic cleavage (or heterolysis)**. The heterolytic bond cleavage leads to the formation of charged fragments or **ions**.

$$\text{A—B} \equiv \text{A : B} \xrightarrow{\text{Heterolytic bond cleavage}} \underbrace{\text{A}^{+} + \text{:B}^{-}}_{\text{Ions}}$$

It should be noted that the movement of an *electron pair* is indicated by a **curved arrow** (see Section 2.2.3).

Heterolytic Bond Formation

Bond making or bond formation is the *reverse* of bond breaking. In **heterolytic bond formation**, ions (anion and cation) combine to form a covalent bond.

The anion possesses a lone pair of electrons and is electron-rich, whereas the cation is electron-deficient. The heterolytic bond formation is represented using **a curved arrow**. Importantly, the curved arrow has a **direction**; it must point from the electron-rich centre to the electron-deficient centre, as shown below.

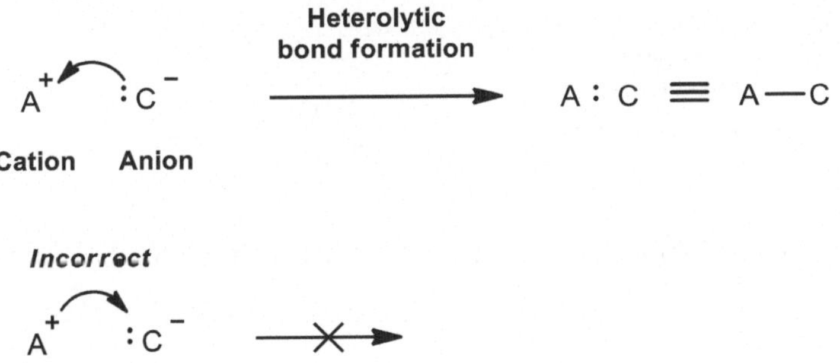

Note: The overall reaction involving the heterolytic cleavage of A–B bond and heterolytic formation of A–C bond as shown above is represented by

$$\text{A—B} + \text{:C}^{-} \longrightarrow \text{A—C} + \text{:B}^{-}$$

Ionic Mechanisms

Mechanisms that involve bond cleavage and bond formation in a heterolytic fashion are called ionic mechanisms. They proceed by the *movement of electrons as electron pairs*.

To illustrate, consider the following substitution reaction that involves an ionic mechanism:

Substitution reaction:

$(CH_3)_3C—Cl$ + ^-OH ⟶ $(CH_3)_3C—OH$ + Cl^-

***t*-Butyl chloride** **_t_-Butyl alcohol**

Ionic mechanism:

$(CH_3)_3C—Cl$ —*Heterolytic bond cleavage*⟶ $(CH_3)_3C^+$ + Cl^- **(step 1)**

$(CH_3)_3C^+$ + ^-OH —*Heterolytic bond formation*⟶ $(CH_3)_3C—OH$ **(step 2)**

Note that lone pairs on oxygen are not shown. Here a curved arrow is drawn from the – charge on the oxygen atom in step 2 (see Section 2.2.3).

7.3.2 Radical Mechanism

Another possibility of bond breaking is the symmetrical breaking or homolytic cleavage of the bond.

Homolytic Bond Cleavage

In homolytic bond cleavage, each fragment takes away one of the electrons of the bond. Thus the homolytic cleavage leads to the formation of fragments with *unpaired electrons*, called **radicals.**

A—B ≡ A : B —**Homolytic bond cleavage**⟶ A· + ·B

Radicals

Note that the movement of a *single electron* is denoted by **a half-headed arrow (or fishhook arrow).**

Homolytic Bond Formation

Homolytic bond formation can occur between two radicals and is represented by **fishhook arrows**. The fishhook arrows point towards each other to make a new two-electron bond.

A· ·C —**Homolytic bond formation**⟶ A : C ≡ A—C

Radical **Radical**

Note: The overall reaction involving the homolytic bond cleavage and homolytic bond formation as shown above is represented by

$$A{-}B + {}^{\bullet}C \longrightarrow A{-}C + {}^{\bullet}B$$

Radical Mechanisms

Mechanisms that involve bond cleavage and bond formation in a homolytic fashion are called radical mechanisms. They proceed by the *movement of single electrons*.

To illustrate, consider the following substitution reaction that proceeds by a radical mechanism:

Substitution reaction:

$$\underset{\textbf{Methane}}{H_3C{-}H} + Cl_2 \xrightarrow{\text{Light}} \underset{\textbf{Methyl chloride}}{H_3C{-}Cl} + HCl$$

Radical mechanism:

$$Cl{-}Cl \xrightarrow[\textit{bond cleavage}]{\textit{Homolytic}} Cl^{\bullet} + Cl^{\bullet} \quad \textbf{(step 1)}$$

$$Cl^{\bullet} \quad H{-}CH_3 \xrightarrow{\textit{Homolytic}} H{-}Cl + H_3C^{\bullet} \quad \textbf{(step 2)}$$

bond formation ***bond cleavage***

$$H_3C^{\bullet} \quad Cl{-}Cl \xrightarrow{\textit{Homolytic}} H_3C{-}Cl + Cl^{\bullet} \quad \textbf{(step 3)}$$

bond formation ***bond cleavage***

7.3.3 Pericyclic Mechanism

Bond breaking and bond making in a reaction can occur simultaneously in a cyclic fashion. Mechanisms that involve simultaneous (or concerted) bond breaking and bond making in a single step with the cyclic movement of electrons are called pericyclic mechanisms.

In this mechanism, no ionic or radical intermediates are involved. **Curved arrows** are also used to depict a pericyclic mechanism.

To illustrate, consider the following addition reaction between 1,3-butadiene (a conjugated diene) and ethylene (an alkene) to form cyclohexene (a cyclic product). The reaction, known as the **Diels-Alder reaction**, involves a pericyclic mechanism with the *cyclic movement of electrons*.

The movement of electrons can be *clockwise* or *anticlockwise*.

Diels-Alder reaction:

$H_2C{=}CH{-}CH{=}CH_2$ + $H_2C{=}CH_2$ $\equiv$ + $\xrightarrow{\Delta}$

1,3-Butadiene Ethylene Cyclohexene

Pericyclic mechanism:

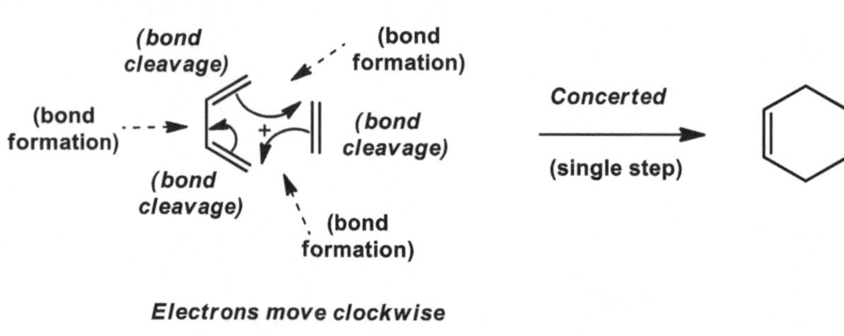

OR

Concerted

(single step)

Electrons move anticlockwise

7.4 IONIC REACTIONS: NUCLEOPHILES AND ELECTROPHILES

Most organic reactions proceed by ionic mechanisms, and they are called **ionic reactions**. In ionic reactions, *substrates* as well as *reagents* are divided into two groups based on their electronic character. They are called **nucleophiles and electrophiles**.

7.4.1 Nucleophiles

Nucleophiles are electron-rich species that can donate a pair of electrons to an electron-deficient atom to form a covalent bond. The word nucleophile comes from nucleus (the positive part of an atom) + phile (from the Greek word *philein* meaning loving). So nucleophile means 'nucleus loving', i.e. it seeks a positive centre.

The electron pair that can be donated by a nucleophile may be a *lone pair*, a *σ bond* or a *π bond.*

Lone pair nucleophiles

Lone pair nucleophiles are the most common type of nucleophiles, which can donate a lone pair of electrons from an atom, called the **nucleophilic atom (or nucleophilic site)**. These nucleophiles are either *negatively charged ions (anions)* or *neutral molecules*.

If the nucleophilic atom is oxygen, it is sometimes called an *oxygen nucleophile*, if nitrogen, a *nitrogen nucleophile*, and so on.

The following are some examples of **lone pair nucleophiles** that contain O, N and X (halogen) atom as the nucleophilic site.

Lone pair nucleophiles

Anions: $Nu:^-$ *or* Nu^- **Neutral molecules:** Nu:

Anion	Nucleophilic atom or site
HO^- (hydroxide)	O
RO^- (alkoxide)	O
X^- (halide)	X

X = F, Cl, Br, I

O has 3 lone pairs, and X has 4 lone pairs, none of which is shown

Neutral molecule	Nucleophilic atom or site
$H_2O:$ (water)	O
$R\ddot{O}H$ (alcohol)	O
$\ddot{N}H_3$ (ammonia)	N
$R_3\ddot{N}$ (amine)	N

O has 2 lone pairs, one of which is shown

Note that lone pairs are not shown for the anionic nucleophiles; a negative charge is indicative of a lone pair.

Some nucleophiles possess lone pairs or negative charge on two *different* atoms, each of which is capable of donating a pair of electrons to an electron-deficient site. Such nucleophiles with *two nucleophilic sites* are called **ambident nucleophiles**.

Two examples are **cyanide ion** (CN^-) and **nitrite ion** (NO_2^-).

Ambident nucleophiles

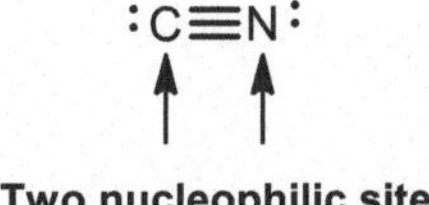

Two nucleophilic sites (C and N) in cyanide ion

$O{=}\ddot{N}{-}O^-$

Two nucleophilic sites (N and O) in nitrite ion

Note that two O atoms in the nitrite ion are equivalent because of resonance (draw resonance structures).

π Bond nucleophiles

Molecules without lone pairs can also act as nucleophiles. A pair of electrons in a **π bond** can be donated to an electron-deficient site to form a covalent bond. These nucleophiles that can donate π electrons with concomitant cleavage of π bonds are called **π bond nucleophiles** or simply, **π nucleophiles**.

Alkenes and aromatic rings are the common π nucleophiles.

(donates π electrons)

$H_2C{=}CH_2$

Ethylene

An alkene

(donates π electrons)

Benzene

An aromatic compound

σ Bond nucleophiles

When a σ bond occurs between carbon and an electropositive atom M (M = Li, B, Al or Mg), a pair of electrons in the **σ bond** (C–M) can be donated to an electron-deficient site to form a covalent bond. Similarly, a pair of electrons in the H–M σ bond can also be donated. Such nucleophiles that can donate σ electrons with concomitant cleavage of σ bonds are called **σ bond nucleophiles** or simply, **σ nucleophiles**.

Grignard reagents (R–MgX), alkyllithiums (R–Li), sodium borohydride (Na^+ BH_4^-) and lithium aluminium hydride (Li^+ AlH_4^-) are common examples of σ nucleophiles.

***σ* bond nucleophiles**

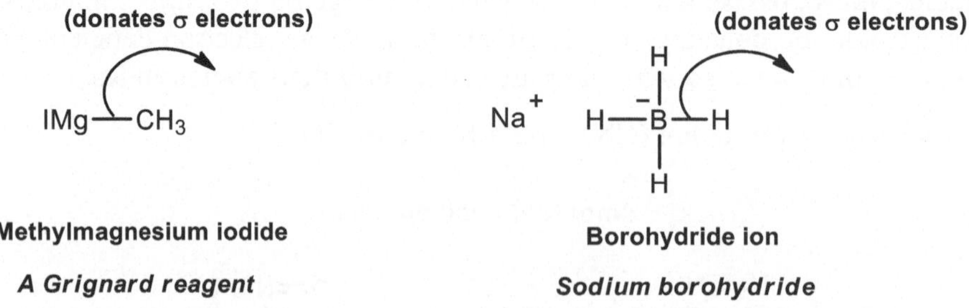

Methylmagnesium iodide

A Grignard reagent

Borohydride ion

Sodium borohydride

7.4.2 Electrophiles

Electrophiles are electron-deficient species that can accept a pair of electrons from an electron-rich site. Electrophile means 'electron loving', i.e. it seeks a negative charge or a pair of electrons.

In electrophiles, an atom can be electron-deficient by having a *sextet* of electrons, or by a *cleavage of a π bond or a σ bond.* Such an atom that accepts an electron pair is called the **electrophilic atom or electrophilic site.**

Lewis acid electrophiles

Electrophiles that possess atoms usually with *sextet* of electrons and an *empty p orbital* resemble Lewis acids. Such electrophiles are called **Lewis acid electrophiles**. A Lewis acid electrophile may be a *positively charged ion* or a *neutral molecule.*

Carbocations (R_3C^+) and aluminium chloride ($AlCl_3$) are common Lewis acid electrophiles. However, the simplest Lewis acid electrophile is the hydrogen cation (H^+), which has no electron in the 1s orbital.

Lewis acid electrophiles

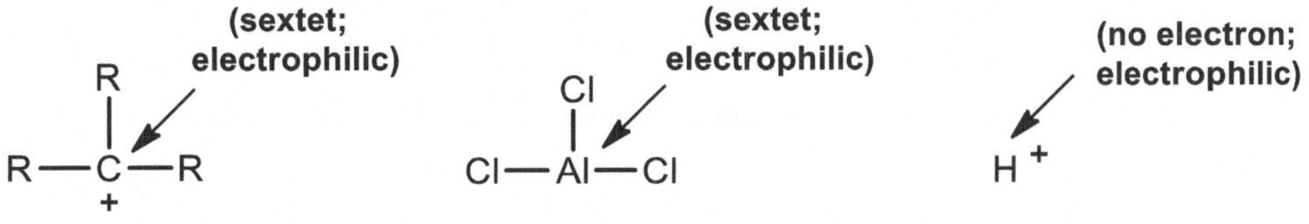

(sextet; electrophilic)
R—C(R)—R, +

Carbocations

(sextet; electrophilic)
Cl—Al(Cl)—Cl

Aluminium chloride

(no electron; electrophilic)
H^+

Hydrogen ion

π Bond electrophiles

Molecules with a **double or triple bond to an electronegative atom** such as **C=O, C=N** or **C≡N** can act as electrophiles. In these electrophiles, the *less electronegative atom* of the double or triple bond is the **electrophilic site.**

In aldehydes or ketones, the **carbon atom of the carbonyl group (C=O) is electrophilic.** Although the carbonyl carbon has an octet of electrons, it can become electron-deficient with a sextet of electrons by *breaking* the carbon-oxygen π bond, and hence can accept a pair of electrons. Such electrophiles such as acetone in which the electron-deficiency of the electrophilic atom is caused by the *breaking of a π bond* are called **π bond electrophiles.**

π bond electrophiles

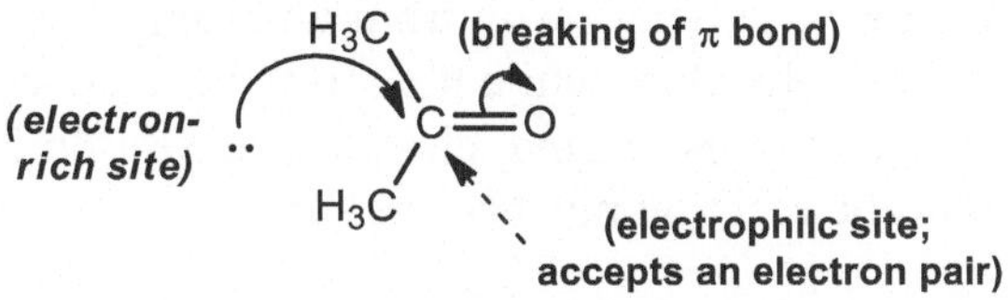

H_3C (breaking of π bond)
(electron-rich site)
C=O
H_3C
(electrophilc site; accepts an electron pair)

Acetone

A ketone

σ Bond electrophiles

It is also possible to create an electron-deficient atom in a molecule by the breaking of a **σ bond to an electronegative atom**. As the electrophilic atom accepts a pair of electrons, *the σ bond is broken*. Such electrophiles are called **σ bond electrophiles**.

For example, an alkyl halide such as methyl chloride (CH_3–Cl) is a σ bond electrophile; the breaking of the C–Cl σ bond makes the carbon atom electrophilic. Also note that, since Cl is more electronegative than C, the carbon atom of the C–Cl bond acquires a partial positive charge showing its electrophilic character.

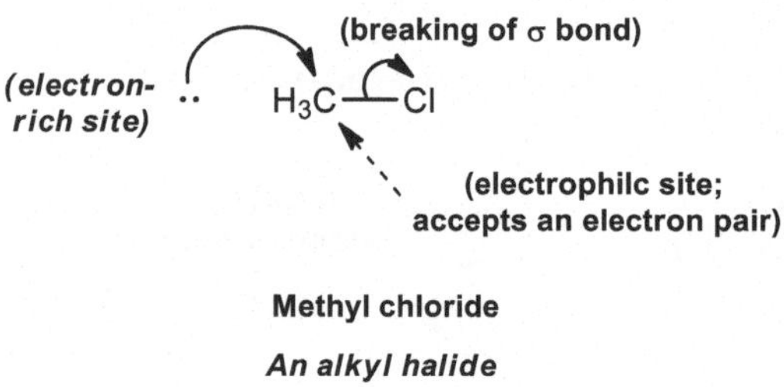

7.4.3 Drawing Ionic Mechanisms

Ionic mechanisms involve combination of nucleophiles with electrophiles. A new covalent bond is formed when *electrons flow from a nucleophile to an electrophile.* A curved arrow is used to represent the electron flow (movement of a pair of electrons). The mechanism is drawn with the help of curved arrows.

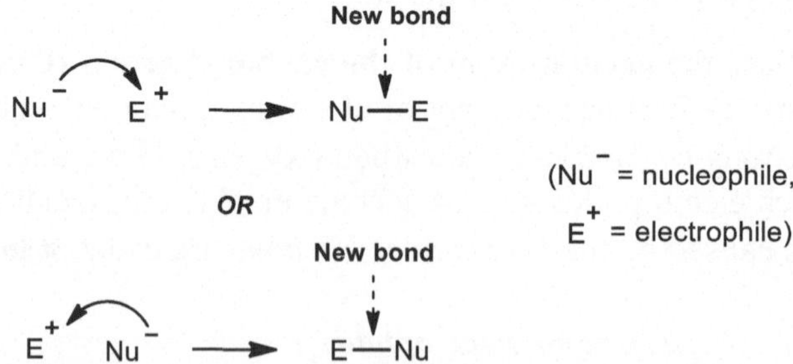

Remember that a curved arrow moving *away* from an atom gives it a *positive* charge in addition to any charge held by the atom while the arrow moving *towards* an atom gives it a *negative* charge over and above its original charge (see Section 2.2.3). Importantly, the *net* charge before and after the reaction must be the same.

Example 1

Chloride ion (Cl^-) combines with ***tert*-butyl cation** [$(CH_3)_3C^+$] to form ***tert*-butyl chloride**. A new C–Cl bond is formed.

New bond
Cl^- (lone pair nucleophile) + $^+C(CH_3)_3$ (Lewis acid electrophile) ⟶ $Cl{-}C(CH_3)_3$
t-Butyl chloride

Here Cl^- is a lone pair nucleophile and $(CH_3)_3C^+$ is a Lewis acid electrophile. As the curved arrow moves away from the Cl atom, the Cl atom gets a positive charge which cancels its negative charge, as shown in the product. On the other hand, the curved arrow moves toward the C atom and gives it a negative charge which cancels its positive charge. The net charge before and after the combination is zero.

Example 2

Isobutene $[(CH_3)_2C{=}CH_2]$ reacts with H^+ to give a ***t*-butyl carbocation** $[(CH_3)_3C^+]$. A new C–H bond is formed.

$H_3C{-}C(CH_3){=}CH_2$ (π bond nucleophile) + H^+ (Lewis acid electrophile) ⟶ $H_3C{-}\overset{+}{C}(CH_3){-}CH_2{-}H$
(single bond) New bond
t-Butyl carbocation

Isobutene is a π bond nucleophile which uses two electrons of the π bond to combine with H^+ (a Lewis acid electrophile) to form a C–H bond. As a result, C=C is converted to a C–C single bond in the carbocation. Since the curved arrow moves away from the carbon bearing two methyl groups, it gets a positive charge. On the other hand, the curved arrow moves toward the H atom and cancels its positive charge.

Example 3

Acetone $[(CH_3)_2C{=}O]$ combines with **cyanide ion** (CN^-) to form an **oxyanion species**. A new C–C bond is formed.

$H_3C{-}C(CH_3){=}O$ (π bond electrophile) + ^-CN (lone pair nucleophile) ⟶ $H_3C{-}C(O^-)(CH_3){-}CN$
New bond
Oxyanion species

The lone pair nucleophile CN^- uses its lone pair or negative charge on carbon to make the new C–C bond. As acetone is a π bond electrophile, the π bond of the C=O group is broken and the O atom gets a negative charge. **Two curved arrows** are needed here; one arrow begins with the negative charge on carbon and the other arrow terminates on the oxygen atom.

Note that in a sequence of curved arrows, it is only necessary to look at the *initial* and *terminal* atoms for the assignment of *new charges*. Thus the initial carbon has no charge since the arrow moves away from the carbon carrying a negative charge, and the terminal oxygen has a negative charge since the other arrow moves toward it.

Example 4

Ammonia (NH_3) reacts with **methyl chloride** (CH_3Cl) to form **methyl ammonium ion** ($CH_3NH_3^+$) and chloride ion (Cl^-). A new C–N bond is formed.

$$H_3\ddot{N} \quad H_3C{-}Cl \longrightarrow H_3\overset{+}{N}{-}CH_3 + Cl^-$$

(lone pair nucleophile) (σ bond electrophile) New bond; Methyl ammonium ion

The nucleophile NH_3 uses the lone pair on N to form the new C–N bond. Since CH_3–Cl is a σ bond electrophile, the C–Cl bond is broken resulting in the formation of Cl^- ion. Note that in the sequence of *two* curved arrows, the N atom gets a positive charge as the arrow moves away from it, and the Cl atom gets a negative charge as the other arrow moves toward it.

Example 5

Acetaldehyde ($CH_3CH{=}O$) reacts with **methylmagnesium iodide** (CH_3MgI) to give the compound shown below. A new C–C bond is formed.

$$H_3C{-}\overset{O}{\overset{\|}{C}}{-}CH_3 \quad H_3C{-}MgI \longrightarrow H_3C{-}\overset{OMgI}{\overset{|}{C}}{-}CH_3 \text{ (with } CH_3 \text{ below)}$$

(π bond electrophile) (σ bond nucleophile) New bond; Product

CH_3MgI has a C–Mg (metal) bond and acts as a σ bond nucleophile, while CH_3CHO acts as a π bond electrophile. CH_3MgI uses two electrons of the C–Mg σ bond to form the new C–C bond. The sequence of *two* arrows gives Mg a positive charge since an arrow moves away from it, and O gets a negative charge as the second arrow moves toward it. Note that O^- is associated with IMg^+ as OMgI.

7.5 REACTIVE INTERMEDIATES: CARBOCATIONS AND CARBANIONS

In ionic reactions, *heterolysis of a bond to a carbon atom* can produce a cation with a positive charge on the carbon atom, called a **carbocation** or an anion with a negative charge on the carbon atom, called a **carbanion**.

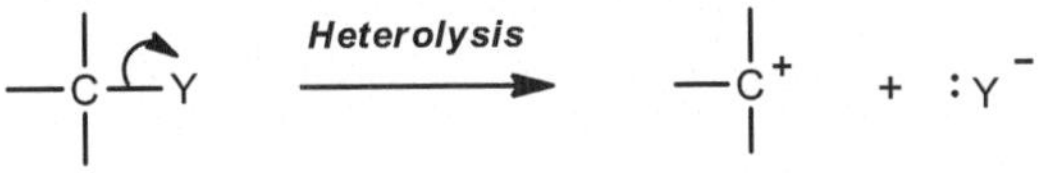

Heterolysis

Carbocation

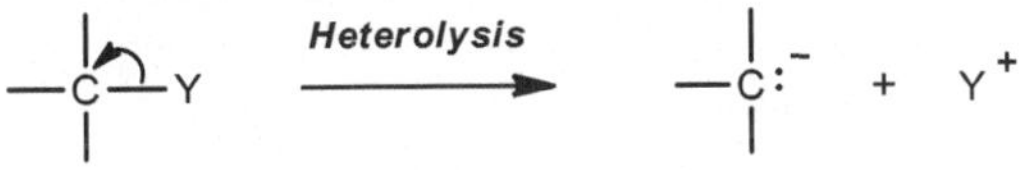

Heterolysis

Carbanion

Most carbocations and carbanions are short-lived and highly reactive. They form as **reactive intermediates** that cannot be isolated.

Carbocations

Carbocations are generally tricoordinate, and the positively charged carbon atom is **sp^2** hybridized. Thus carbocations possess a **planar** geometry and contain a **vacant p orbital**. They are electron-deficient with the cationic carbon having a *sextet* of electrons, and therefore act as **electrophiles**.

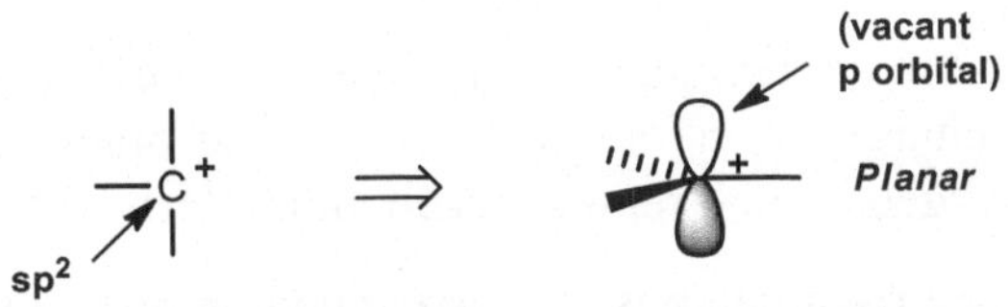

(vacant p orbital)

Planar

sp^2

Carbocation

Electrophile

Carbanions

Carbanions are also tricoordinate, with the anionic carbon possessing a **lone pair** of electrons.

In simple carbanions, the negative charge is *not* conjugated, and the negatively charged carbon is **sp^3** hybridized. These carbanions are therefore **pyramidal** with a lone pair of electrons in an sp^3 orbital. They can donate a pair of electrons, and act as **nucleophiles**.

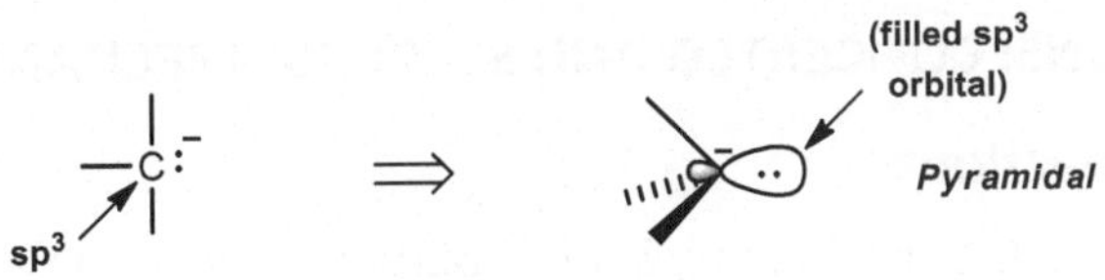

(filled sp^3 orbital)

Pyramidal

sp^3

Carbanion

Nucleophile

In **conjugated carbanions**, the negative charge is usually conjugated with a carbonyl (C=O) group, when the hybridization of the anionic carbon changes from **sp^3 to sp^2**. This is because the planarity permits delocalization of the negative charge by extended p orbital overlap with the C=O group, and thereby stabilizing the carbanion. The conjugated carbanions then become **planar**.

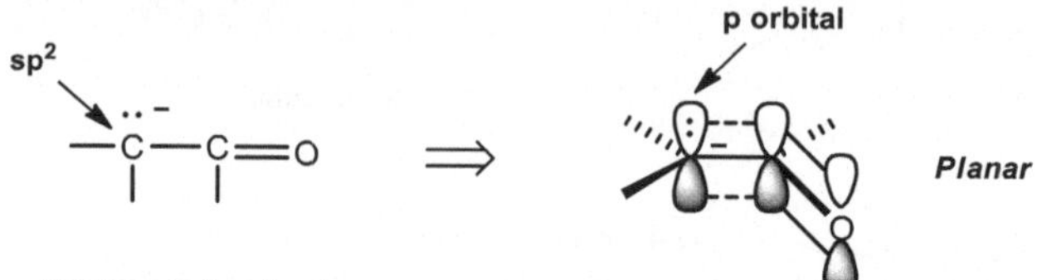

sp²

Conjugated carbanion

p orbital

Planar

Delocalization of negative charge stabilizes the carbanion

These stabilized conjugated carbanions can form when a carbonyl compound having **α hydrogens** is treated with a base. For example, acetone forms an **α-carbanion** (a conjugated carbanion) when treated with hydroxide ion.

$$H_2C(H)\text{—}C(=O)\text{—}CH_3 + {}^-OH \rightleftharpoons H_2\ddot{C}^-\text{—}C(=O)\text{—}CH_3 + H_2O$$

Acetone

α-Carbanion

Conjugated carbanion

Due to conjugation, the α-carbanion is **stabilized by resonance**. In one resonance structure, the negative charge is on carbon, while in the other resonance structure, the negative charge is on oxygen, which represents an anion derived from enol.

These resonance-stabilized α-carbanions are also called **enolate ions**. They can behave as **ambident nucleophiles** with two nucleophilic sites (carbon and oxygen).

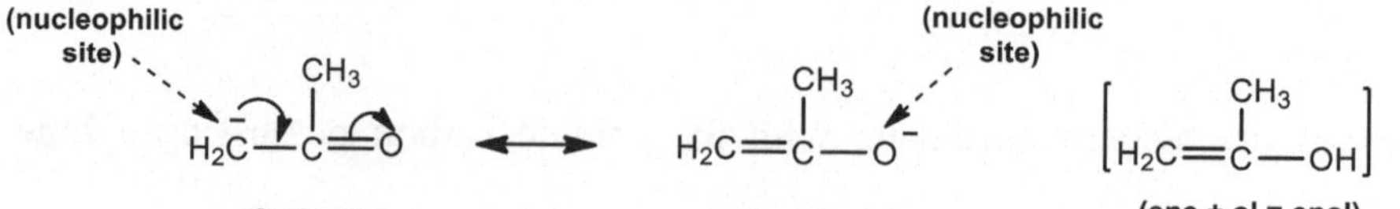

(nucleophilic site)

$$H_2\bar{C}\text{—}C(CH_3)=O \longleftrightarrow H_2C=C(CH_3)\text{—}O^- \qquad [H_2C=C(CH_3)\text{—}OH]$$

(nucleophilic site)

α-Carbanion **Anion of enol** **(ene + ol = enol)**

α-Carbanion or enolate ion

(ambident nucleophile)

7.6 IONIC REACTIONS: CONCERTED AND STEPWISE MECHANISMS

7.6.1 Concerted Reactions

Ionic reactions in which bond-breaking and bond-making occur *simultaneously* (at the same time) and involve a *single* step without any intermediate are called **concerted reactions**.

Consider a substitution reaction between methyl chloride and hydroxide ion to produce methanol and chloride ion. The hydroxide ion is supplied by aqueous sodium hydroxide.

Reaction: $CH_3Cl + OH^- \text{ (NaOH)} \longrightarrow CH_3OH + Cl^-$

Methyl chloride **Methanol**

Experimental evidences suggest that the reaction proceeds through a **concerted mechanism**. The C–Cl bond breaking and C–O bond making are simultaneous and occur in a single step. *Two curved arrows* represent the mechanism:

Mechanism:

$HO^- \; H_3C{-}Cl \xrightarrow{\text{Concerted}} HO{-}CH_3 + Cl^-$

(nucleophile) **(electrophile)**

7.6.2 Multistep Reactions

Ionic reactions that proceed through the formation of one or more **intermediates** and involve *more than one step* are called **multistep reactions**. These are illustrated with two examples.

Example 1

Isobutene reacts with HCl to produce ***t*-butyl chloride**. This is an addition reaction of an alkene.

Reaction: $H_3C{-}C(CH_3){=}CH_2 + HCl \longrightarrow H_3C{-}C(CH_3)_2{-}Cl$

Isobutene ***t*-Butyl chloride**

HCl is an acid which dissociates into H^+ and Cl^-. The mechanism involves *two* steps. In **step 1**, isobutene (a nucleophile) combines with H^+ (an electrophile) to form a **carbocation** (*t*-butyl cation) as an *intermediate*. The intermediate *t*-butyl carbocation is a *new* electrophile, which combines with Cl^- (a nucleophile) in **step 2** to give *t*-butyl chloride as the product. The steps are indicated under the respective reaction arrows (→).

Mechanism:

$H_3C{-}C(CH_3){=}CH_2 \;\; H^+ \xrightarrow[\text{Step 1}]{} H_3C{-}C^+(CH_3)_2 \xrightarrow[\text{Step 2}]{Cl^-} H_3C{-}C(CH_3)_2{-}Cl$

Isobutene **Carbocation (*t*-butyl cation)** ***Intermediate*** ***t*-Butyl chloride**

Example 2

***t*-Butyl alcohol** reacts with concentrated HCl to form ***t*-butyl chloride**. This is a substitution reaction of an alcohol.

The mechanism involves *three* steps. In **step 1**, *t*-butyl alcohol (a nucleophile) combines with H^+ (an electrophile) to form a **protonated alcohol** (*t*-butyloxonium ion) as the *first intermediate*. **Step 2** involves the *heterolysis* of the C–O bond of the protonated alcohol to form a **carbocation** (*t*-butyl cation) as the *second intermediate* and water. The carbocation (a new electrophile) then combines with Cl^- (a nucleophile) in **step 3** to give *t*-butyl chloride as the product.

7.7 REACTION THERMODYNAMICS: THE EQUILIBRIUM CONSTANT (*K*) AND THE STANDARD FREE ENERGY CHANGE (ΔG^0)

Reaction thermodynamics tells us whether a reaction is **favourable** or **unfavourable**, based on the *extent* of the reaction, i.e. how completely the reactants are converted into products.

In principle, all chemical reactions are reversible, and the reactions can be written as equilibrium processes:

$$\mathrm{A} \rightleftharpoons \mathrm{B}$$

where A is the reactant and B is the product. The *position of equilibrium* is expressed by the **equilibrium constant (*K*)** which is given by

$$K = \frac{[\mathbf{B}]_{\mathbf{eq}}}{[\mathbf{A}]_{\mathbf{eq}}}$$

where $[B]_{eq}$ is the concentration of product B at equilibrium and $[A]_{eq}$ is the concentration of reactant A at equilibrium.

If $K > 1$, the concentration of B will be larger than the concentration of A at equilibrium, and **the equilibrium favours the product side**. For an efficient conversion of reactants to products, a *high value* of *K* is necessary.

Relationship between equilibrium constant and standard free energy change

There is an important relationship between the **equilibrium constant (*K*)** and the **standard free energy change** denoted by ΔG^0 (read delta G zero) for the reaction:

$$\Delta G^0 = -RT \ln K = -2.303\, RT \log K$$

where

ΔG^0 = standard free energy (G^0) of products – standard free energy (G^0)of reactants;

R = gas constant (1.987 cal K^{-1} mol^{-1}); T = absolute temperature (in K)

Note that *G* denotes the free energy (or the Gibbs free energy), while the standard free energy is symbolized with superscript zero as G^0. By standard free energies, we mean that the products and reactants are taken in their standard states, i.e. 1 atmosphere (atm) for a gas and 1 molar (*M*) for a solution.

The **standard free energy change (ΔG^0)** for a reaction indicates whether the reaction is **favourable** or **not**.

- If ΔG^0 is **negative** (the standard free energy of products is lower than the standard free energy of reactants), $K > 1$ (from the equation), and the reaction is **favourable**.
- If ΔG^0 is **positive** (the standard free energy of products is higher than the standard free energy of reactants), $K < 1$ (from the equation), and the reaction is **unfavourable**.

Exergonic and Endergonic Reactions

A reaction that occurs with a *negative* free energy change ($\Delta G^0 < 0$) is called **exergonic**, while a reaction that proceeds with a *positive* free energy change ($\Delta G^0 > 0$) is called **endergonic**. This is illustrated here with the help of free energy diagrams:

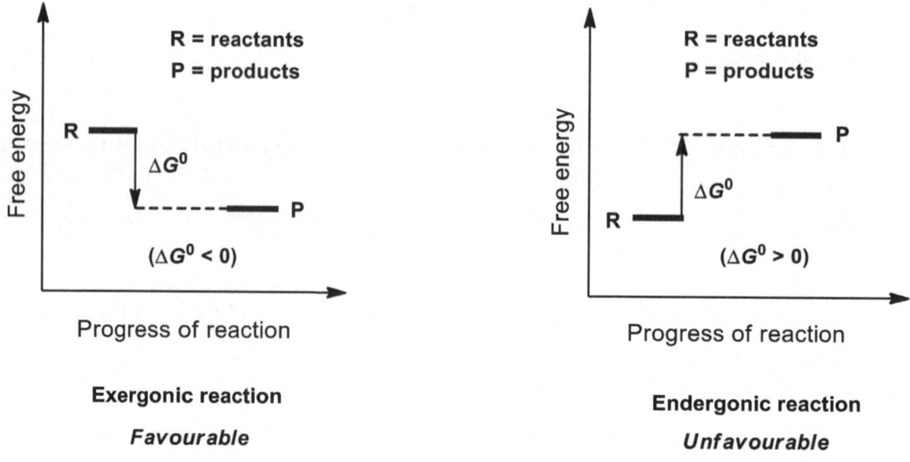

Exergonic reaction

Favourable

Endergonic reaction

Unfavourable

For an exergonic reaction, if $\Delta G^0 \ll 0$, the reaction is highly favourable. For example, the reaction between methyl chloride and hydroxide ion in aqueous solution at 60°C is **highly favourable** ($\Delta G^0 = -24$ kcal mol^{-1}).

$$CH_3Cl + OH^- \xrightarrow[60^\circ C]{H_2O} CH_3OH + Cl^- \qquad \Delta G^0 = -24 \text{ kcal mol}^{-1}$$

This becomes more clear if we calculate K using the equation $\Delta G^0 = -2.303\ RT \log K$. We get

$$\log K = \frac{-\Delta G^0}{2.303\ RT} = \frac{24000 \text{ cal mol}^{-1}}{2.303 \times 1.987 \text{ cal K}^{-1} \text{ mol}^{-1} \times 333 \text{ K}} = 15.75$$

or, $K = 5.6 \times 10^{15}$

As the equilibrium constant (K) is extremely large, the reaction practically proceeds to completion, and hence is highly favourable.

The standard free energy change (ΔG^0) for a reaction is related to the standard enthalpy change (ΔH^0) and the standard entropy change (ΔS^0), and is given by,

$$\boldsymbol{\Delta G^0 = \Delta H^0 - T\Delta S^0}$$

Enthalpy Change (ΔH^0)

We have already learned (Section 3.3.7) that ΔH^0 is associated with energy changes involved in bonds broken and bonds formed in a reaction. If stronger bonds are formed in the products than those broken in the reactants, **ΔH^0 is negative** ($\Delta H^0 < 0$) and the reaction is **exothermic**. But if weaker bonds are formed in the products than those broken in the reactants, ***ΔH^0 is* positive** ($\Delta H^0 > 0$) and the reaction is **endothermic**.

In the equation $\Delta G^0 = \Delta H^0 - T\Delta S^0$, the enthalpy term ($\Delta H^0$) has a positive sign. This implies that a negative value of ΔH^0 will contribute to making ΔG^0 negative, and will tend to favour the reaction.

Entropy Change (ΔS^0)

Entropy change (ΔS^0) is associated with the changes in the relative order of a system (see Section 5.2.3). **The more the disorder of a system, the greater is its entropy**.

Consider the following general reaction:

$$A \rightarrow B + C$$

Here *one* molecule reacts to produce *two* molecules. As the increase in the number of molecules is associated with more disorder, **the entropy change (ΔS^0)** for the reaction is **positive**.

In general, if the number of molecules of products is greater than the number of molecules of reactants, the entropy change is positive ($\Delta S^0 > 0$).

In the equation $\Delta G^0 = \Delta H^0 - T\Delta S^0$, the term ($T\Delta S^0$) has a negative sign; thus a positive value of ΔS^0 (from order to disorder) will contribute to making ΔG^0 negative, and will tend to favour the reaction.

Note: For most reactions at ordinary temperatures, $T\Delta S^0$ is small compared with ΔH^0; thus ΔH^0 often determines whether a reaction is favourable or unfavourable.

7.8 REACTION KINETICS: THE RATE CONSTANT (*K*) AND THE FREE ENERGY OF ACTIVATION ($\Delta G^{\ddagger}$)

Reaction kinetics refers to the **rate** of the reaction and tells us whether a reaction is **fast** or **slow**.

Reaction kinetics and reaction thermodynamics are independent of one another. A favourable equilibrium constant (*K*) does not necessarily guarantee a fast reaction. Although many reactions are found to be favourable both thermodynamically and kinetically, it is possible to have a fast but unfavourable reaction or a slow but favourable reaction.

7.8.1 Rate Equation and Order

The **rate of a reaction** depends on temperature and can be determined experimentally at a particular temperature by measuring the *decrease in concentration of the reactants (or the increase in concentration of the products) with time*. Experimental results indicate that the **rate depends on the concentration of the reactants**.

Consider the following reaction:

$$CH_3Cl + OH^- \longrightarrow CH_3OH + Cl^-$$

Measurement of the reaction rate reveals that the rate depends on *both* the concentration of CH_3Cl and the concentration of OH^-.

If the concentration of *either* CH_3Cl *or* OH^- is made **two times**, the rate is **doubled**, and if **both** concentrations are **doubled**, the rate increases by **four-fold**. This means that the rate of the reaction is directly proportional to the concentrations of both CH_3Cl and OH^-.

Thus, the **rate equation** is given by

Rate $\propto$ $[CH_3Cl]\ [OH^-]$

or, Rate = $k\ [CH_3Cl]\ [OH^-]$

where k is the proportionality constant, called the **rate constant**.

Order of a Reaction

An *experimental* quantity, called the **order**, is determined from the rate equation. The order of a reaction is defined as the **sum of the powers of the concentration of the reactants** in the rate equation.

Since for the above reaction, the power of $[CH_3Cl]$ is 1 and that of $[OH^-]$ is 1, the order of the reaction is 1 + 1 = 2. The reaction is said to be **second order**.

Sometimes, the order is specified with respect to each reactant. Thus, the reaction is first order with respect to CH_3Cl and first order with respect to OH^-. The reaction is second order overall.

It should be noted that order may be a positive integer, a fraction or even zero.

7.8.2 Transition Structure (TS), Molecularity and the Free Energy of Activation

In a chemical reaction, the reactants have to cross a barrier, called a **free energy barrier**, to become products. This means that the *reactants first go up an energy hill, reach the top of the hill, and then go downhill to products*. This will be true, even if the reactant has a higher free energy than the product ($\Delta G < 0$).

If there *were* no energy barrier, the reactant would immediately be converted into the product and it could not exist!

Transition Structure

The arrangement of the reacting species at the top of the free energy barrier is called the **transition structure (TS)**. The transition structure has an extremely short lifetime. It does not last longer than the time required for one molecular vibration (~10^{-12} s), and cannot be isolated.
Consider the above reaction between methyl chloride (CH_3Cl) and hydroxide ion (OH^-). This is a one-step, concerted reaction (see Section 7.6.1).
The reaction involves the breaking of the C–Cl bond and making of the C–OH bond. The **transition structure (TS)** is thought to be an **arrangement of atoms in which the nucleophile (HO^-) and the leaving group (Cl) are both partially bonded to the carbon atom**. In this arrangement, the C–Cl bond is about to break and the C–O bond is about to form in a concerted fashion. The TS is often drawn in square brackets with a superscript double-dagger (‡), as shown below.

$$HO^- \ \ H_3C\text{—}Cl \longrightarrow \left[\overset{\delta-}{HO}\text{- - -}CH_3\text{- - -}\overset{\delta-}{Cl}\right]^{\ddagger} \longrightarrow HO\text{—}CH_3 + Cl^-$$

Reactants — **Transition structure (TS)** — **Products**

Molecularity

The concept of TS relates to the mechanism of the reaction. A *theoretical* quantity, called the **molecularity**, is derived from the transition structure. The molecularity of a reaction is defined as the **number of species (molecule or ion) involved in the TS of the reaction**.

Since in the above reaction, *both* CH_3Cl and OH^- participate in the TS, the molecularity of the reaction is 2, and the reaction is said to be **bimolecular**.

It is of note that the bimolecular mechanism is consistent with the observed second order kinetics of the reaction (see Section 7.8.1).

Free energy Diagram

The difference in free energy between the reactants and the transition structure is called the free energy of activation, denoted by $\Delta G^{\ddagger}$ (read delta G double-dagger).

The **rate of a reaction** depends on the **free energy of activation ($\Delta G^{\ddagger}$)**, *not* on ΔG^0. A reaction is **fast** if the free energy barrier ($\Delta G^{\ddagger}$) is **low**, and it is **slow** if $\Delta G^{\ddagger}$ is **high**.

The free energy changes along the reaction pathway from **reactants → TS → products** can be represented graphically. **The plot of free energy versus the reaction coordinate is called the free energy diagram.** The reaction coordinate is a parameter that denotes the progress of the reaction.

Figure 7.2 shows a free energy diagram for the above reaction involving reactants, TS, and products. The free energy of activation ($\Delta G^{\ddagger}$) is clearly indicated as the free energy difference between the TS and the reactants.

Note that the free energy of the products is lower than the free energy of the reactants (see Section 7.7), and this difference in free energy (ΔG^0) is also indicated in the diagram.

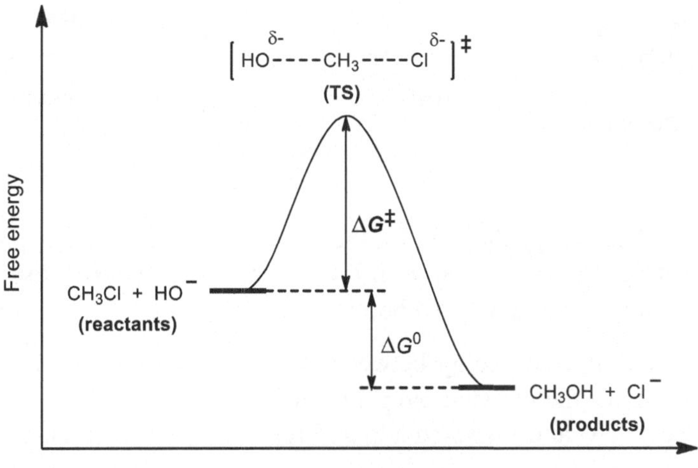

Figure 7.2: A free energy diagram for the one-step reaction between methyl chloride and hydroxide ion.

7.8.3 Multistep Reactions and the Rate-Determining Step

Multistep reactions involve two or more steps with unstable intermediates. If *one* of the steps is *intrinsically slower* than all the others, then the rate of the overall reaction is controlled or limited by the *rate of the slowest step.* The **slowest step** of the reaction is called **the rate-determining (r/d) step.**

Consider a general **two-step reaction** which involves *one* intermediate. Assume that step 1 is intrinsically slow and step 2 is fast, which means that the rate constant (k_1) for step 1 is much smaller than the rate constant (k_2) for step 2.

$$\text{Reactants} \xrightarrow[\text{slow}]{k_1} \text{Intermediate} \xrightarrow[\text{fast}]{k_2} \text{Products} \qquad k_1 < k_2$$

Step 1 ***Step 2***

Here, **step 1 is the rate-determining (r/d) step.** *Each step involves a transition structure.* Therefore, the two-step reaction involves **two transition structures (TS_1 and TS_2).**

On a free energy diagram, the step involving the formation of **highest-energy TS** indicates the **r/d step.**

Free Energy Diagram for a Two-Step Reaction

Consider the two-step reaction between isobutene with HCl (see example 2 in Section 7.6.2).

Step 1 involves the formation of a carbocation intermediate which is **slow**, and **step 2** involves the **fast** reaction of the reactive carbocation with the nucleophile (Cl^-). **The formation of carbocation (step 1) is the slow, r/d step.**

$$(CH_3)_2C{=}CH_2 \;\; H{-}Cl \xrightarrow[\text{Step 1}]{\text{slow}} (CH_3)_3C^+ + Cl^- \xrightarrow[\text{Step 2}]{\text{fast}} (CH_3)_3C{-}Cl$$

Reactants **Carbocation (intermediate)** **Product**

Figure 7.3 shows a free energy diagram for the reaction. The free energy profile shows two maxima, representing TS_1 associated with formation of the **carbocation in the r/d step** and TS_2 for formation of the new C–Cl bond.

$\Delta G_1^\ddagger$ is the difference in free energy between TS_1 and reactants, and represents the **free energy of activation** for the **r/d step (step 1)**. On the other hand, $\Delta G_2^\ddagger$ is the difference in free energy between TS_2 and intermediate, and represents the **free energy of activation** for **step 2.**

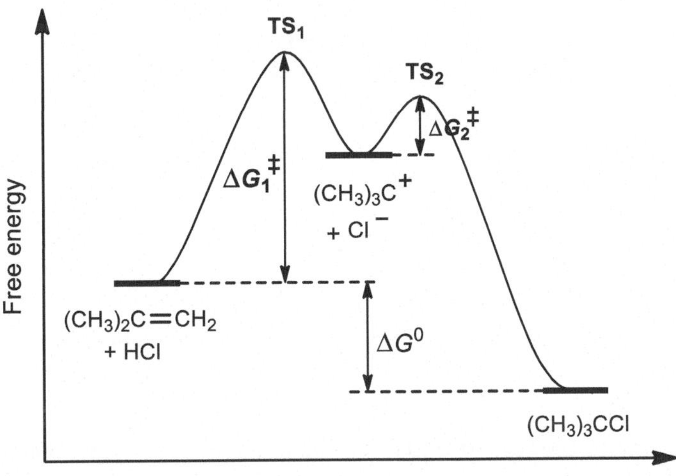

Figure 7.3: A free energy diagram for a two-step reaction between isobutene and HCl.

Note that the carbocation is a highly unstable species, and therefore has *higher* free energy than either reactants or product. The carbocation and Cl^- constitute the intermediate, i.e. reactants for step 2. The free energy difference between the product and the reactants (ΔG^0) is also indicated in the diagram.

Molecularity for a Multistep Reaction

The **molecularity** for a two-step or multistep reaction refers to the **TS of the r/d step**. In the above two-step reaction, two species (isobutene and HCl) participate in the r/d step (step 1), and the mechanism of the reaction is said to be **bimolecular**.

The Hammond postulate

The information about a transition structure cannot be obtained directly. An intuitive way of visualizing the TS is given by a postulate proposed by Hammond. **The Hammond postulate states that the transition structure resembles the species which is closer to it in free energy**.

In the above two-step reaction, TS_1 in the r/d step is nearer to the carbocation intermediate (not to the reactants) in free energy. Thus TS_1 resembles the carbocation, and **the stability of TS_1 is directly related to the stability of the carbocation**.

If the carbocation can be stabilized, the TS_1 will also get stabilized with lower free energy. This will lead to a **lowering** of the free energy barrier ($\Delta G_1^‡$), and hence an **increase** in reaction rate.

As shown below, the reaction between isobutene and H^+ (from HCl) can lead to **two different carbocations** in step 1:

$(CH_3)_2C{=}CH_2$ H^+ —**Step 1**→ $(CH_3)_2\overset{+}{C}{-}CH_2H$

3° Carbocation
More stable

H^+ $(CH_3)_2C{=}CH_2$ —**Step 1**↛ $(CH_3)_2C(H){-}\overset{+}{C}H_2$

1° Carbocation
Less stable

These two carbocations are not of equal stability: the **tertiary (3°) carbocation is more stable than the primary (1°) carbocation**. Therefore, *TS_1 resembles the more stable 3° carbocation*, which leads to a lower free energy of activation. Thus, the reaction proceeds preferentially through the **more stable 3° carbocation** in step 1 of the reaction.

CHAPTER 8

Alkenes, Alkynes and Conjugated Dienes: Addition Reactions

IN THIS CHAPTER

- Electrophilic and radical addition to alkenes
- Addition of hydrogen halides to alkenes: Markovnikov addition
- Addition of hydrogen bromide to alkenes in the presence of peroxide: Anti-Markovnikov addition
- Addition of water to alkenes: Acid-catalysed hydration, oxymercuration-demercuration, and hydroboration-oxidation
- Addition of halogens to alkenes and stereochemistry of bromine addition
- Epoxidation and dihydroxylation of alkenes
- Oxidative cleavage and ozonolysis of alkenes
- Reduction of alkenes: Heats of hydrogenation and relative stabilities
- Reactions of alkynes: Addition of hydrogen halides, halogens and water; oxidative cleavage and reduction
- Reactions of conjugated dienes with hydrogen halides and halogens: 1,2- and 1,4-additions
- Pericyclic addition of conjugated dienes: The Diels-Alder reaction

The most important reactions of alkenes, alkynes or conjugated dienes are **addition reactions** (see Section 7.2.1). These addition reactions proceed *mostly* by **ionic mechanisms**, while some others involve a **radical** or **pericyclic mechanism**.

8.1 ELECTROPHILIC ADDITION TO ALKENES

Alkenes contain a carbon-carbon double bond (C=C) that can act as a *π nucleophile* (see Section 7.4.1) by donating two electrons of the π bond to an *electrophile* supplied by the adding *reagent*. The reaction follows a **stepwise ionic mechanism**.

The addition reaction is initiated by an attack of the electrophile to the double bond of an alkene, which is normally the slow and rate-determining (r/d) step of the reaction. This mechanism of addition to alkenes is called an **electrophilic addition**.

To illustrate, consider a general reaction between an **alkene** and the **reagent (E–Nu)**. The reagent may dissociate into an electrophilic species (E^+) and a nucleophilic species (Nu^-) *or* it may act as a σ bond electrophile by breaking the E–Nu bond.

Figure 8.1 shows the mechanism of electrophilic addition. The mechanism involves *two* steps. In **step 1**, the electrophile (E^+ or E–Nu) reacts with the π bond of alkene to form a **carbocation intermediate**. This step is *slow* and *rate-determining (r/d)*. In **step 2** which is *fast*, the carbocation as a new electrophile reacts with the nucleophile (Nu^-) to give the addition product. Note that each step involves combination of a nucleophile with an electrophile.

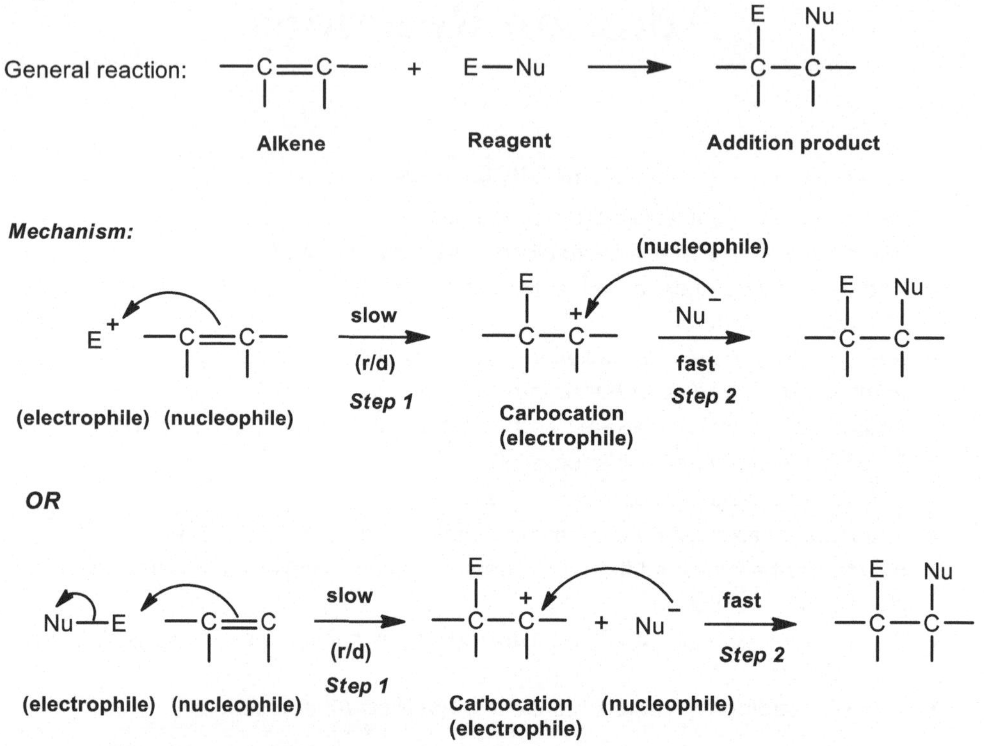

General reaction: —C=C— + E—Nu ⟶ E, Nu on —C—C—

Alkene Reagent Addition product

Mechanism:

E^+ —C=C— (electrophile) (nucleophile) —slow (r/d) *Step 1*→ Carbocation (electrophile) + Nu^- (nucleophile) —fast *Step 2*→ addition product

OR

Nu—E —C=C (electrophile) (nucleophile) —slow (r/d) *Step 1*→ Carbocation (electrophile) + Nu^- (nucleophile) —fast *Step 2*→ addition product

Figure 8.1: A general mechanism for electrophilic addition to alkenes.

8.2 ADDITION OF HYDROGEN HALIDES TO ALKENES: MARKOVNIKOV ADDITION

The addition of a **hydrogen halide** (HF, HCl, HBr or HI) to an alkene gives an **alkyl halide**. This *addition reaction* is called the **hydrohalogenation** of alkenes.

The order of reactivity of hydrogen halides follows their order of acidity (i.e. their ability to donate H^+): HI > HBr > HCl > HF.

Although HF can add to alkenes to produce alkyl fluorides, its use is extremely hazardous and is avoided whenever possible.

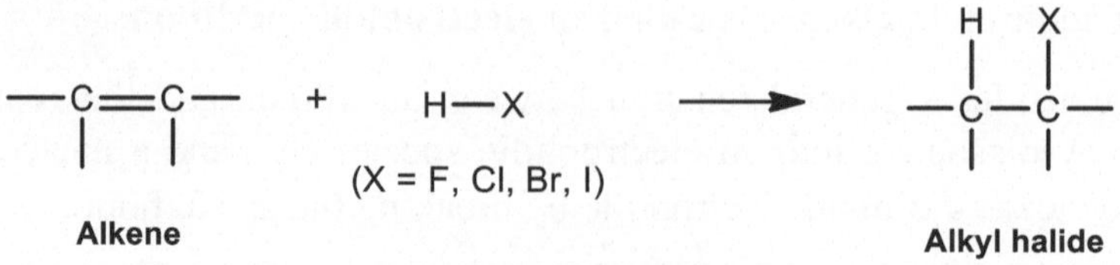

—C=C— + H—X (X = F, Cl, Br, I) ⟶ H, X on —C—C—

Alkene Alkyl halide

Most alkenes are *unsymmetrical* when different groups are attached to one or both ends of the double bond. The addition of HX (X = Cl, Br or I) to an unsymmetrical alkene can take place in **two possible directions or orientations** to give a mixture of two products.

However, one product is usually formed exclusively or predominantly. **Addition reactions that occur in one direction selectively are said to be regioselective** (regio from the Latin word *regionem* meaning direction).

For example, the addition of HBr to **propene** is regioselective and leads to **2-bromopropane** as the predominant or major product.

Regioselective addition:

$$CH_2{=}CHCH_3 + H{-}Br \longrightarrow \underset{(CH_3CHBrCH_3)}{\overset{H\quad Br}{CH_2{-}CHCH_3}} \quad \textbf{2-Bromopropane (Major product)}$$

$$CH_2{=}CHCH_3 + H{-}Br \not\longrightarrow \underset{(BrCH_2CH_2CH_3)}{\overset{Br\quad H}{CH_2{-}CHCH_3}} \quad \textbf{1-Bromopropane (Negligible product)}$$

Propene

The Russian Chemist Vladimir Markovnikov in 1870 proposed a rule for addition reactions of alkenes based on the experimental results then available. This rule is known as the **Markovnikov rule.** The Markovnikov rule states that **the addition of a hydrogen halide (HX) to an alkene proceeds in such a way that the H atom adds to the carbon atom of the double bond already bearing the greater number of hydrogen atoms.**

Addition reactions that obey the Markovnikov rule are called **Markovnikov additions** and the product is called the **Markovnikov product**. Thus, the above addition reaction of propene with HBr is a Markovnikov addition, which gives 2-bromopropane as the Markovnikov product.

(C with more H atoms)

$$\underset{\uparrow\,H\quad\uparrow\,Br}{H_2C{=}CHCH_3} \xrightarrow{HBr} \overset{Br}{CH_3{-}CHCH_3}$$

Markovnikov product

8.2.1 Ionic Mechanism for Markovnikov Addition

The mechanism for **electrophilic addition** can explain the basis for the Markovnikov rule. Initial attack of the electrophile (H^+ or H–X) to the double bond (cf. step 1 in Figure 8.1) *controls* the overall direction of addition or regioselectivity. The reaction proceeds favourably in that direction which produces a **more stable carbocation** as the intermediate. This direction leads to the **Markovnikov product.**

Example 1

Consider again the addition of HBr to propene. The reaction proceeds via the formation of a secondary (2°) carbocation that is *more stable* than a primary (1°) carbocation formed in an alternative pathway. The observed product is therefore 2-bromopropane, which is the Markovnikov product.

$CH_3CH{=}CH_2$ H—Br → $CH_3\overset{+}{C}H{-}CH_3$ Br^- → $CH_3CH(Br){-}CH_3$

Propene — 2° Carbocation, *More stable* — Observed product

or

Br—H $CH_3CH{=}CH_2$ ⇸ Br^- $CH_3CH_2{-}\overset{+}{C}H_2$ → $CH_3CH_2{-}CH_2Br$

Propene — 1° Carbocation, *Less stable* — Not formed

Example 2

The reaction of HCl with **isobutene** produces only ***t*-butyl chloride**. Here the reaction proceeds through the formation of tertiary (3°) carbocation which is *more stable* than a primary (1°) carbocation formed in an alternative pathway. The only product is *t*-butyl chloride as the Markovnikov product.

Reaction: $H_3C{-}C(CH_3){=}CH_2$ $\xrightarrow{HCl}$ $H_3C{-}C(CH_3)(Cl){-}CH_3$

Isobutene — *t*-Butyl chloride

Mechanism:

$H_3C{-}C(CH_3){=}CH_2$ (Isobutene) + H^+

⇸ $H_3C{-}CH(CH_3){-}\overset{+}{C}H_2$ (1° Carbocation, *Less stable*) $\xrightarrow{Cl^-}$ $H_3C{-}CH(CH_3){-}CH_2Cl$ (Not formed)

→ $H_3C{-}\overset{+}{C}(CH_3){-}CH_3$ (3° Carbocation, *More stable*) $\xrightarrow{Cl^-}$ $H_3C{-}C(CH_3)(Cl){-}CH_3$ (Only product, *Markovnikov product*)

Thus, in more general terms, Markovnikov rule can be stated as follows: In the electrophilic addition, the electrophile adds to the double bond so as to form the more stable carbocation as an intermediate.

PROBLEM 8.1: Give the structure of the product that would be obtained from the addition of ICl to 2-methyl-1-butene. Write a plausible mechanism to account for the formation of the product.

Answer

Cl is more electronegative than I; therefore the positive or electrophilic end of the reagent ICl is I atom which, according to Markovnikov rule, adds to the C atom with the greater number of H atoms. The product is 2-chloro-1-iodo-2-methylbutane.

Reaction: $CH_3CH_2-C(CH_3)=CH_2 \xrightarrow{ICl} CH_3CH_2-C(CH_3)(Cl)-CH_2I$

2-Methyl-1-butene → **2-Chloro-1-iodo-2-methylbutane**, ***Markovnikov product***

The reaction proceeds through the more stable 3° carbocation to give the product.

Mechanism:

$CH_3CH_2-C(CH_3)=CH_2 + \overset{\delta+}{I}-\overset{\delta-}{Cl} \longrightarrow CH_3CH_2-\overset{+}{C}(CH_3)-CH_2I \xrightarrow{Cl^-} CH_3CH_2-C(CH_3)(Cl)-CH_2I$

3° Carbocation (more stable)

8.3 ADDITION OF HYDROGEN BROMIDE TO ALKENES IN THE PRESENCE OF PEROXIDES: ANTI-MARKOVNIKOV ADDITION

When hydrogen bromide (HBr) is added to an alkene in the presence of a peroxide (RO–OR), the addition takes place in anti-Markovnikov fashion. This means that the H atom of HBr becomes attached to the carbon atom of the double bond already bearing the fewer number of hydrogen atoms.

An organic peroxide (RO–OR) is a derivative of hydrogen peroxide (H_2O_2 or H–O–O–H). The anti-Markovnikov addition is attributed to a **peroxide effect** on the addition of HBr to alkenes.

For example, consider the following reaction of **propene** with HBr in the presence of a peroxide. The product is **1-bromopropane**, which is an **anti-Markovnikov** product.

(C atom with fewer number of H atoms) ↓

$CH_3CH=CH_2 + HBr \xrightarrow[\text{(ROOR)}]{\text{Peroxide}} CH_3CH(H)-CH_2(Br)$ $(CH_3CH_2CH_2Br)$

Propene → **1-Bromopropane** ***anti-Markovnikov addition***

It has been observed that when HBr addition is carried out with air *excluded,* the Markovnikov addition occurs. But if air is *present*, the anti-Markovnikov addition takes place. This is because of the presence of some organic peroxides that are formed by the action of atmospheric oxygen on the alkenes.

8.3.1 Radical Mechanism for Anti-Markovnikov Addition

To account for the anti-Markovnikov addition of HBr in the presence of peroxides, a mechanism different from the ionic mechanism of Markovnikov addition has been proposed. This mechanism is called a **radical mechanism** (see Section 7.3.2) that involves a **radical chain reaction**, as shown in Figure 8.2.

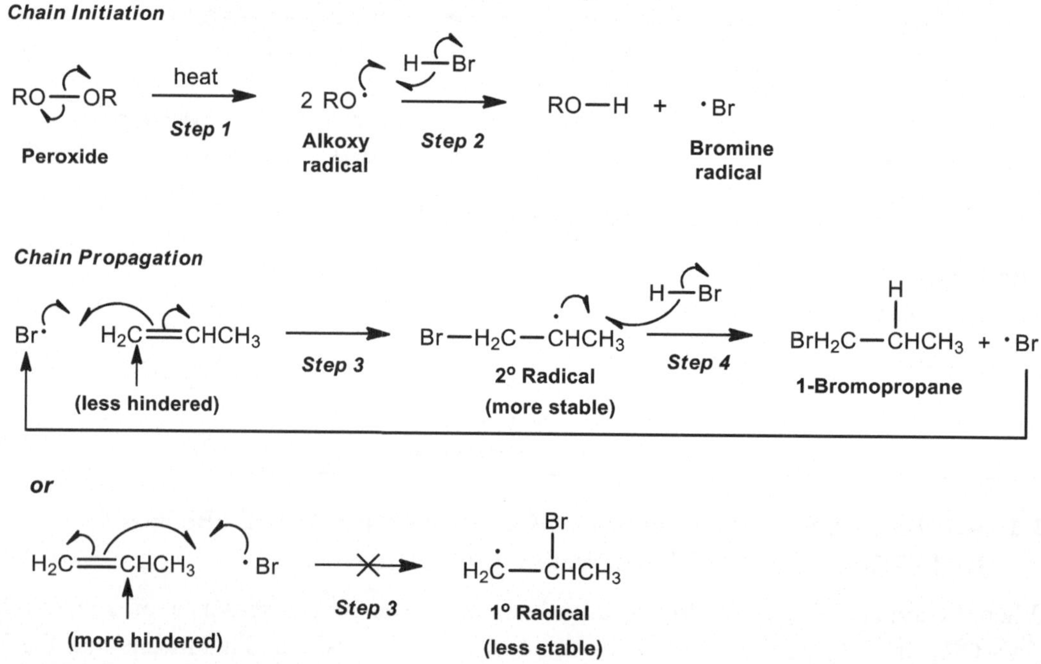

Figure 8.2: A radical chain mechanism for the anti-Markovnikov addition of HBr to propene in the presence of a peroxide.

The radical chain reaction involves a **chain initiation sequence** and a **chain propagation sequence**. The chain initiation sequence consists of two steps. The O–O bond of peroxides is weak and the chain initiation begins with the homolytic cleavage of O–O bond in **step 1** to form alkoxy radicals (RO·), which then abstract H atom from HBr in **step 2** to generate the bromine atom or radical (Br·).

The chain propagation sequence also consists of two steps. In **step 3**, Br· radical adds to the double bond of propene to form a *more stable* secondary (2°) radical instead of a *less stable* primary (1°) radical. Note that the order of stability of radicals is similar to that of carbocations. The *steric factor* also favours the formation of the 2° radical. Since Br atom has a relatively large size, the attack of the Br atom on the *less substituted (or less hindered)* methylene (CH_2) carbon is more facile. In **step 4**, the 2° radical abstracts H atom from HBr to give the product (1-bromopropane). Importantly, this step *regenerates* Br· radical that participates again in step 3 and so on, leading to a **chain reaction**.

Note: The difference that results in Markovnikov or anti-Markovnikov addition is that a proton (electrophile) adds to the double bond to form a *more stable carbocation* in the ionic reaction, whereas a bromine atom (radical) adds to the double bond to give a *more stable radical* in the radical chain reaction.

No anti-Markovnikov Addition with HF, HCl or HI

The anti-Markovnikov addition does not occur with HF, HCl or HI even when peroxides are present.

With HF, the fluorine radical is *not* formed by the homolysis of H–F bond in the **initiation sequence** (cf. step 2 in Figure 8.2), because H–F bond is very strong. With HCl, the **second propagation step** (cf. step 4 in Figure 8.2) is endothermic and not favourable. With HI, the **first propagation step** (cf. step 3 in Figure 8.2) is endothermic and not favourable.

However, with HBr, both propagation steps (see Figure 8.2) are exothermic and favourable, and the reaction proceeds through a radical chain mechanism to give the anti-Markovnikov product.

8.4 ADDITION OF SULPHURIC ACID TO ALKENES

Alkenes dissolve in cold concentrated sulphuric acid by forming **alkyl hydrogen sulphates**. The addition of sulphuric acid to alkenes follows the **Markovnikov rule**.

For example, propene reacts with sulphuric acid to form isopropyl hydrogen sulphate.

$$\underset{\textbf{Propene}}{H_2C{=}CHCH_3} + \underset{(H_2SO_4)}{H{-}O{-}\overset{\overset{O}{\|}}{\underset{\underset{O}{\|}}{S}}{-}OH} \longrightarrow \underset{\substack{[(CH_3)_2CH\text{-}OSO_3H] \\ \textbf{Isopropyl hydrogen sulphate}}}{H_2\overset{\overset{H}{|}}{C}{-}\overset{\overset{OSO_3H}{|}}{C}HCH_3} \quad \textit{\textbf{Markovnikov addition}}$$

Alkyl hydrogen sulphates can be easily converted to **alcohols** by hydrolysis on heating with water. The overall reaction of adding sulphuric acid to an alkene followed by hydrolysis is equivalent to the addition of water to an alkene. Thus, propene on treatment with cold H_2SO_4 followed by hydrolysis produces isopropanol.

$$\underset{\textbf{Propene}}{H_2C{=}CHCH_3} \xrightarrow{\text{cold } H_2SO_4} \underset{\underset{OSO_3H}{|}}{CH_3CHCH_3} \xrightarrow[\text{heat}]{H_2O} \underset{\substack{\underset{OH}{|} \\ \textbf{Isopropanol}}}{CH_3CHCH_3} + H_2SO_4$$

Net addition of water (H-OH)

PROBLEM 8.2: Outline an industrial synthesis of ethyl alcohol from ethylene.

Answer

Ethylene is dissolved in concentrated H_2SO_4, then water is added and the mixture is heated. The product is ethyl alcohol.

$$\underset{\textbf{Ethylene}}{H_2C{=}CH_2} \xrightarrow{\text{conc. } H_2SO_4} \underset{\textbf{Ethyl hydrogen sulphate}}{CH_3CH_2OSO_3H} \xrightarrow[\text{heat}]{H_2O} \underset{\textbf{Ethyl alcohol}}{CH_3CH_2OH}$$

8.5 ADDITION OF WATER TO ALKENES

Water can be added to alkenes directly or indirectly to form alcohols. The addition of water to alkenes is called the **hydration** of alkenes. It is possible to direct the hydration to proceed as Markovnikov addition or anti-Markovnikov addition.

8.5.1 Acid-catalyzed Hydration: Markovnikov Addition

Water adds to alkenes in the presence of an acid catalyst to produce **alcohols**. Acids commonly used are dilute sulphuric acid or phosphoric acid; they provide H^+ ion as catalyst.

$$\underset{\textbf{Alkene}}{-\overset{|}{C}{=}\overset{|}{C}-} + \underset{(H_2O)}{H{-}OH} \xrightarrow{H^+} \underset{\textbf{Alcohol}}{-\overset{H}{\overset{|}{C}}-\overset{OH}{\overset{|}{C}}-}$$

The **mechanism** for the hydration of an alkene involves *three* steps. **Step 1** is the slow, rate-determining step for the electrophilic addition of proton (H^+) to the double bond to form a **carbocation**. In **step 2**, the carbocation combines with the nucleophile (H_2O) to form a **protonated alcohol**, which then loses a proton in **step 3** to produce the **alcohol** itself.

Mechanism:

$$\underset{\textbf{Alkene}}{-\overset{|}{C}{=}\overset{|}{C}-} \xrightarrow[\textbf{slow}\ \textit{\textbf{Step 1}}]{H^+} \underset{\textbf{Carbocation}}{-\overset{+}{C}-\overset{H}{\overset{|}{C}}-} \xrightarrow[\textbf{fast}\ \textit{\textbf{Step 2}}]{H_2\ddot{O}} \underset{\textbf{Protonated alcohol}}{-\overset{H-\overset{H}{\overset{|}{O^+}}}{\overset{|}{C}}-\overset{H}{\overset{|}{C}}-} \xrightarrow[\textbf{fast}\ \textit{\textbf{Step 3}}]{-H^+} \underset{\textbf{Alcohol}}{-\overset{OH}{\overset{|}{C}}-\overset{H}{\overset{|}{C}}-}$$

The acid-catalyzed hydration of alkenes follows the **Markovnikov rule**. The formation of a *more stable* carbocation accounts for the Markovnikov addition.

The following are some examples of acid-catalyzed hydration of alkenes:

$$H_2C{=}CH_2 + H_2O \xrightarrow{H_3PO_4} CH_3CH_2OH$$

Ethylene **Ethanol**

$$CH_3CH{=}CH_2 \xrightarrow[H_2SO_4]{H_2O} CH_3CH(OH)CH_3$$

Propene **Isopropanol** ***Markovnikov addition***

$$CH_3CH_2CH{=}CH_2 \xrightarrow[H_2SO_4]{H_2O} CH_3CH_2CH(OH)CH_3$$

1-Butene ***sec*-Butanol** ***Markovnikov addition***

PROBLEM 8.3: Give a plausible mechanism for the hydration of isobutene in the presence of dilute H_2SO_4.

Answer

The reaction proceeds through a more stable 3° carbocation to produce *t*-butyl alcohol.

$$(CH_3)_2C{=}CH_2 \ (\textbf{Isobutene}) \xrightarrow{H^+} \not\rightarrow (CH_3)_2CH{-}\overset{+}{C}H_2 \ (\textbf{1° Carbocation (less stable)})$$

$$(CH_3)_2C{=}CH_2 \xrightarrow{H^+} (CH_3)_3C^+ \ (\textbf{3° Carbocation (more stable)}) \xrightarrow{H_2\ddot{O}} (CH_3)_3C{-}\overset{+}{O}H_2 \xrightarrow{-H^+} (CH_3)_3C{-}OH \ (\textbf{\textit{t}-Butyl alcohol})$$

8.5.2 Oxymercuration-Demercuration: Markovnikov Hydration

Oxymercuration-demercuration is a two-step method for the *net* addition of water to alkenes.

Oxymercuration involves the addition of water and mercuric acetate [$Hg(OAc)_2$] to an alkene to form an **organomercury compound**. In this reaction, the solvent is usually tetrahydrofuran (THF).

Demercuration is accomplished by *reducing* organomercury compound with sodium borohydride ($NaBH_4$) in aqueous sodium hydroxide. Demercuration causes replacement of the acetoxymercuri group (HgOAc) with H, and produces an **alcohol**.

$$-\overset{|}{C}=\overset{|}{C}- + H_2O + Hg(OAc)_2 \xrightarrow[\textit{Oxymercuration}]{THF} -\overset{|}{\underset{OH}{C}}-\overset{|}{\underset{HgOAc}{C}}- + AcOH$$

Alkene; Mercuric acetate ($OAc = OCOCH_3$); Organomercury compound; Acetic acid (CH_3COOH)

Demercuration (reduction): $NaBH_4$, $NaOH$

$$\longrightarrow -\overset{|}{\underset{OH}{C}}-\overset{|}{\underset{H}{C}}- + Hg + AcONa$$

Alcohol; Sodium acetate

Net addition of water

Both oxymercuration and demercuration steps are simple to carry out at or below room temperature. The demercuration (reduction) is performed without isolating the organomercury compound. Oxymercuration-demercuration method is a useful laboratory procedure for the synthesis of alcohols, and the overall reaction gives alcohols in high yields (> 90%).

With *unsymmetrical* alkenes, oxymercuration is highly regioselective and proceeds by **Markovnikov addition**. Thus, oxymercuration-demercuration leads to a net Markovnikov addition of water to alkenes. The following are two examples:

$$CH_3CH_2CH_2CH{=}CH_2 \xrightarrow[H_2O/THF]{Hg(OAc)_2} CH_3CH_2CH_2\underset{OH}{CH}-\underset{HgOAc}{CH_2} \xrightarrow[OH^-]{NaBH_4} CH_3CH_2CH_2\underset{OH}{CH}CH_3$$

1-Pentene → 2-Pentanol

Markovnikov addition

1-Methylcyclohexene $\xrightarrow[H_2O/THF]{Hg(OAc)_2}$ (1-methyl-1-hydroxy-2-(HgOAc)cyclohexane: CH_3, OH, HgOAc) $\xrightarrow[OH^-]{NaBH_4}$ 1-Methylcyclohexanol (CH_3, OH)

Markovnikov addition

Oxymercuration-demercuration method has *advantages* over direct hydration of alkenes. For example, 3,3-dimethyl-1-butene tends to **rearrange** in the direct *acid-catalyzed hydration*, but the alkene is structurally stable and **does not rearrange** during *oxymercuration-demercuration*.

$$H_3C-C(CH_3)_2-CH{=}CH_2 \xrightarrow[H^+]{H_2O} H_3C-C(CH_3)(OH)-CHCH_3(CH_3)$$

3,3-Dimethyl-1-butene → **2,3-Dimethyl-2-butanol (major product)** ***Rearrangement***

$$H_3C-C(CH_3)_2-CH{=}CH_2 \xrightarrow[\text{2. NaBH}_4,\ OH^-]{\text{1. Hg(OAc)}_2,\ H_2O/THF} H_3C-C(CH_3)_2-CH(OH)CH_3$$

3,3-Dimethyl-1-butene → **3,3-Dimethyl-2-butanol** ***No rearrangement***

8.5.3 Hydroboration-Oxidation: Anti-Markovnikov Hydration

Hydroboration is the addition of borane (BH_3) to an alkene to form an **organoborane**. The addition of borane to the double bond occurs in a single step.

Hydroboration is usually carried out using diborane (B_2H_6) in tetrahydrofuran (THF). In THF, diborane dissociates into two molecules of borane which forms a Lewis acid-Lewis base complex with THF.

$$B_2H_6 \overset{THF}{\rightleftharpoons} 2\ BH_3$$

$$\text{THF (O:)} + BH_3 \longrightarrow \text{THF}\overset{+}{O}-\overset{-}{B}H_3 \equiv \text{THF:BH}_3$$

THF (Lewis base) **Borane (Lewis acid)** **Complex of BH_3 with THF**

Hydroboration:

$$\text{Alkene } -C{=}C- \;+\; \text{Borane } H-BH_2 \xrightarrow[(B_2H_6/THF)]{THF:BH_3} -\underset{H}{C}-\underset{BH_2}{C}-$$

Organoborane

With an *unsymmetrical* alkene, **BH_3 adds in such a way that the B atom is attached to the less substituted carbon atom of the double bond**. This orientation of borane addition results, in part, from the **steric factor** which favours the approach of the bulky boron reagent (THF:BH_3) to the *less substituted (i.e. less hindered)* carbon atom. The addition is **anti-Markovnikov** because the H atom of borane becomes attached to the carbon atom with the fewer number of hydrogen atoms.

The addition of BH_3 to an alkene first forms an **alkylborane**. But since the alkylborane has still B–H bonds, it adds to a second molecule of alkene to form a **dialkylborane**, which further reacts with a third molecule of alkene to finally produce a **trialkylborane**.

For example, the reaction of propene with borane proceeds as follows:

$H—BH_2$ + $CH_3CH{=}CH_2$ (Propene) (less substituted) $\xrightarrow[\textit{anti-Markovnokov addition}]{THF:BH_3}$ $CH_3CH_2—CH_2BH_2$ $[CH_3CH_2CH_2BH_2]$ **Propylborane** $\xrightarrow[\text{(second molecule)}]{H_2C{=}CHCH_3}$ $[(CH_3CH_2CH_2)_2BH]$ **Dipropylborane** $\xrightarrow[\text{(third molecule)}]{H_2C{=}CHCH_3}$ $[(CH_3CH_2CH_2)_3B]$ **Tripropylborane**

Oxidation of a trialkylborane with hydrogen peroxide in aqueous NaOH replaces the boron group with OH and produces an **alcohol**. Since the hydroboration is an anti-Markovnikov addition, the result of **hydroboration-oxidation** is an **anti-Markovnikov addition of water**.

The trialkylborane formed in the hydroboration reaction is not usually isolated, but is oxidized in the same reaction flask to an alcohol with alkaline hydrogen peroxide. The following scheme illustrates the **hydroboration-oxidation of propene**:

$$3\ CH_3CH{=}CH_2 \xrightarrow[\textit{Hydroboration}]{THF:BH_3} (CH_3CH_2CH_2)_3B \xrightarrow[NaOH]{H_2O_2} 3\ CH_3CH_2CH_2OH + Na_3BO_3$$

Propene → **Tripropylborane** → **Propyl alcohol** (*Oxidation*)

Net addition of water (anti-Markovnikov)

The following are two more examples of hydroboration-oxidation of alkenes:

$$CH_3CH_2CH_2CH{=}CH_2 \xrightarrow[2.\ H_2O_2/OH^-]{1.\ THF:BH_3} CH_3CH_2CH_2CH_2CH_2OH$$

1-Pentene → **1-Pentanol**

anti-Markovnikov addition of water

1-Methylcyclohexene $\xrightarrow[2.\ H_2O_2/OH^-]{1.\ THF:BH_3}$ **2-Methylcyclohexanol**

anti-Markovnikov addition of water

8.6 ADDITION OF HALOGENS TO ALKENES

Halogens add to alkenes to form **1,2-dihalides**, also called **vicinal-dihalides (*vic*-dihalides)**. This addition reaction is called the **halogenation** of alkenes. Bromine and chlorine are generally used in halogenation reactions.

$$\underset{\textbf{Alkene}}{>C{=}C<} + \underset{(X = Br,\ Cl)}{X{-}X} \longrightarrow \underset{\substack{\textbf{1,2-Dihalide}\\ (\textbf{\textit{vic}-Dihalide})}}{-\underset{X}{\overset{|}{C}}-\underset{X}{\overset{|}{C}}-}$$

Bromination of alkenes is generally carried out with **Br_2 in carbon tetrachloride (CCl_4)** at room temperature in the dark. The following are some examples of bromination and chlorination of alkenes:

Bromination:

$$\underset{\textbf{Ethylene}}{H_2C{=}CH_2} \xrightarrow[CCl_4]{Br_2} \underset{\substack{\textbf{Ethylene dibromide}\\ \textbf{(1,2-dibromoethane)}}}{H_2\underset{Br}{C}{-}\underset{Br}{C}H_2}$$

Bromination:

$$\underset{\textbf{Cyclohexene}}{C_6H_{10}} + Br_2 \xrightarrow{CCl_4} \underset{\textbf{1,2-Dibromocyclohexane}}{C_6H_{10}Br_2}$$

Chlorination:

$$\underset{\textbf{2-Butene}}{CH_3CH{=}CHCH_3} \xrightarrow{Cl_2} \underset{\textbf{2,3-Dichlorobutane}}{CH_3\underset{Cl}{C}H{-}\underset{Cl}{C}HCH_3}$$

Qualitative test for detection of carbon-carbon double bond (C=C)

The reaction of alkenes with Br_2 in CCl_4 is used as a **test** for the detection of carbon-carbon double bond (C=C) in an organic compound (see also Section 8.8.2). The **red-brown** solution of Br_2 in CCl_4 is slowly added to the suspected alkene. Since the alkene and the addition product are usually *colourless*, the **disappearance of the red-brown colour** of bromine indicates that addition has occurred, and C=C is present in the organic sample.

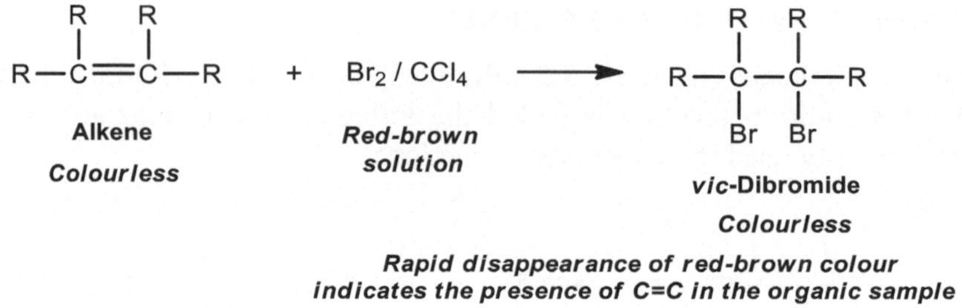

R R
R—C═C—R + Br_2 / CCl_4 ⟶ R—C(R)(Br)—C(R)(Br)—R

Alkene
Colourless

Red-brown solution

vic-Dibromide
Colourless

Rapid disappearance of red-brown colour indicates the presence of C=C in the organic sample

8.6.1 Mechanism of Bromine Addition: Anti Addition

The **mechanism** for addition of bromine to an alkene involves *two* steps, as shown in Figure 8.3.

As the bromine molecule approaches the alkene, the electrons in the Br–Br bond is repelled away from the π electrons of the alkene. As a result, the bromine molecule becomes *polarized* with the bromine atom nearer the double bond developing a partial positive charge. This polarized bromine molecule can then act as an electrophile by breaking the weak Br–Br bond.

In **step 1**, the attack by the alkene (nucleophile) to the polarized bromine (electrophile) gives an **intermediate**. Interestingly, this intermediate is **not** a normal carbocation but a three-membered ring cation called a **bromonium ion**, which is *more stable* than the alternative carbocation. Notably, the cyclic bromonium ion is a real intermediate in which a positively charged bromine atom is bonded to two carbon atoms. The two C–Br bonds are formed using two electron pairs: π bond and a lone pair on the Br atom.

In **step 2**, the bromide ion (Br^-) as a nucleophile attacks one of the two carbon atoms of the bromonium ion to give a ***vic*-dibromide**.

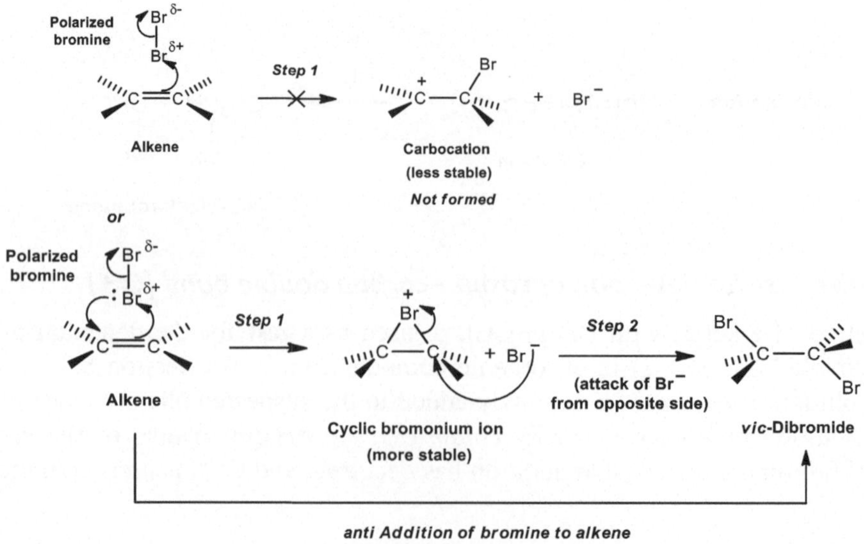

Polarized bromine
$Br^{\delta-}$
$Br^{\delta+}$
Alkene
Step 1
Carbocation
(less stable)
Not formed
+ Br^-

or

Polarized bromine
$Br^{\delta-}$
$Br^{\delta+}$
Alkene
Step 1
Cyclic bromonium ion
(more stable)
+ Br^-
Step 2
(attack of Br^- from opposite side)
vic-Dibromide

anti Addition of bromine to alkene

Figure 8.3: A mechanism for anti addition of bromine to alkenes via a cyclic bromonium ion intermediate.

Importantly, since one side of the original alkene molecule is occupied by the formation of bromonium ion, the nucleophile (Br^-) approaches from the *opposite* side. The result is the **anti addition** of bromine to the alkene.

8.6.2 Stereochemistry of Bromine Addition

Addition of bromine to 2-butene produces 2,3-dibromobutane. However, 2-butene can exist as two stereoisomers: ***cis*-2-butene** and ***trans*-2-butene**. So the question is: *what will be the stereochemistry of the product if cis and trans isomers of 2-butene are treated with bromine separately?*

The stereochemistry of products results from the **anti addition** of bromine to *cis*-2-butene and *trans*-2-butene, as shown in Figure 8.4. The anti addition can occur in *two* possible ways in each case.

***cis*-2-Butene** adds to Br_2 to give a product which is **chiral** because it has no plane of symmetry or centre of symmetry. The two possible anti additions give the chiral product as a mixture of enantiomers, i.e. **racemic-2,3-dibromobutane**.

On the other hand, the anti addition of bromine to ***trans*-2-butene** gives a product that is **achiral** because it has a centre of symmetry. Here, the two possible anti additions give rise to the same achiral product, i.e. ***meso*-2,3-dibromobutane**.

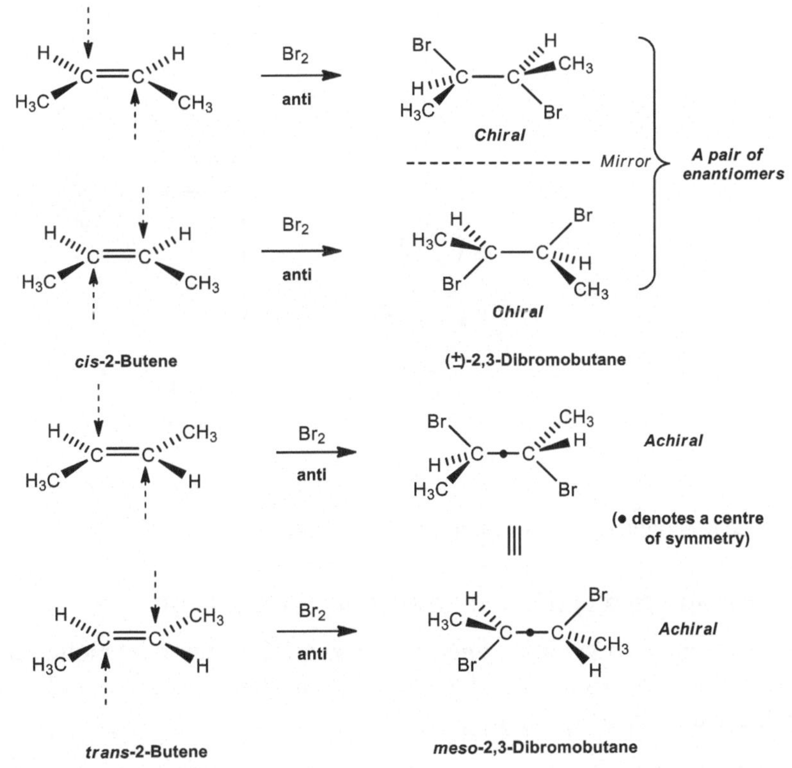

Figure 8.4: Stereochemistry of products resulting from two possible anti additions of bromine to *cis*-2-butene and *trans*-2-butene.

Thus, *cis* and *trans* isomers of 2-butene lead to products with different stereochemistry. These reactions in which stereoisomerically different reactants (such as *cis* and *trans*) give stereoisomerically different products (such as racemic and *meso*) are called **stereospecific reactions.**

Mechanisms

The mechanisms of bromine addition to *cis*-2-butene and *trans*-2-butene are delineated in Figure 8.5. The first step is the formation of a cyclic bromonium ion (cf. Figure 8.3). In the second step, attack of the nucleophile (Br^-) to either of the two carbon atoms of the bromonium ion (path *a* and path *b*) leads to the product.

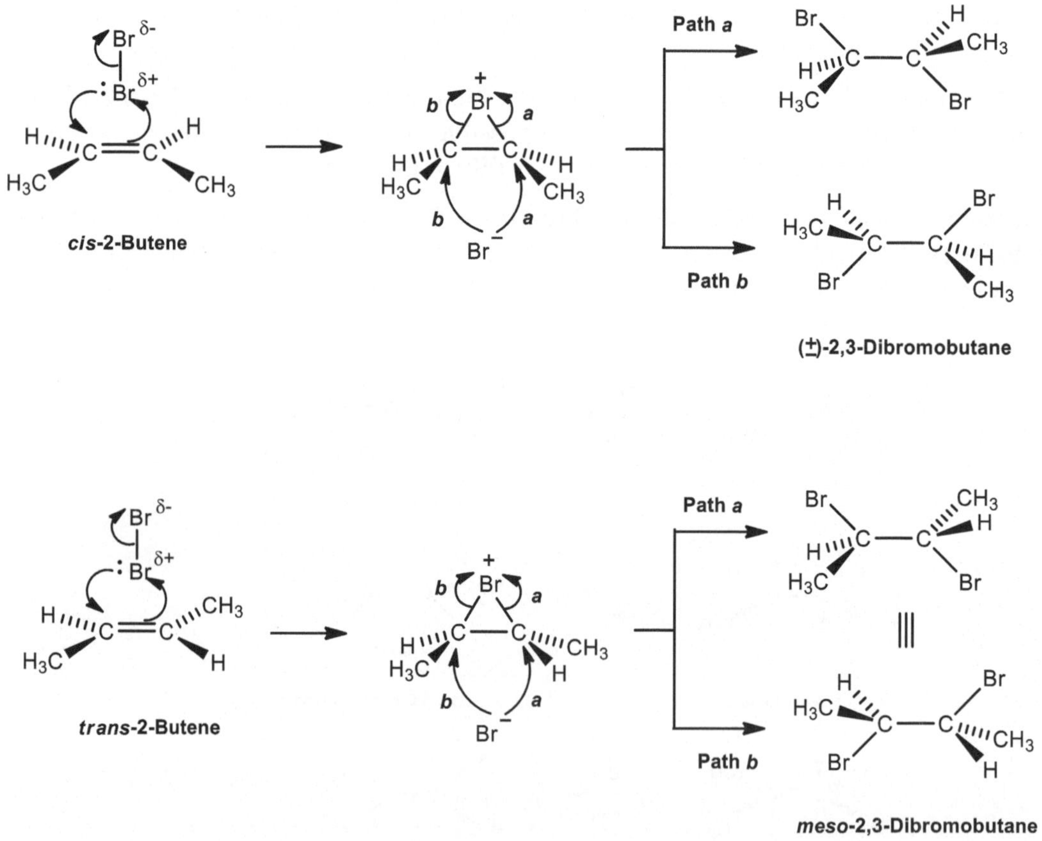

cis-2-Butene

(±)-2,3-Dibromobutane

trans-2-Butene

meso-2,3-Dibromobutane

Figure 8.5: Mechanisms for bromine addition to *cis*-2-butene and *trans*-2-butene.

Addition of bromine to a cyclic alkene such as cyclohexene

The addition of bromine to **cyclohexene** gives ***trans*-1,2-dibromocyclohexane**. Notably, 1,2-dibromocyclohexane can exist as two possible diastereomers (*cis* and *trans*), but the anti addition leads to only *trans* isomer.

The *trans* isomer is chiral having no plane or centre of symmetry, and is obtained as a mixture of two enantiomers, i.e. racemate.

Mirror

or

Br_2

anti

Br H

H Br

H Br

Br H

A pair of enantiomers

Cyclohexene

Chiral

Chiral

trans-1,2-Dibromocyclohexane (racemic)

8.6.3 Halohydrin Formation

If a **halogen** (Cl_2 or Br_2) reacts with an alkene in the presence of **water**, the *major* product is a halo alcohol called a **halohydrin** (a chlorohydrin from chlorine/water or a bromohydrin from bromine/water).

$$H_2C{=}CH_2 + Cl_2 \xrightarrow{H_2O} H_2C(Cl){-}CH_2(OH) + H_2C(Cl){-}CH_2(Cl)$$

Ethylene — Chlorohydrin (*Major product*) — *vic*-Dichloride (*Minor product*)

$$CH_3CH{=}CHCH_3 + Br_2 \xrightarrow{H_2O} CH_3CH(Br){-}CH(OH)CH_3 + CH_3CH(Br){-}CH(Br)CH_3$$

2-Butene — Bromohydrin (*Major product*) — *vic*-Dibromide (*Minor product*)

In the formation of a halohydrin (such as bromohydrin), **water as a nucleophile** attacks the cyclic bromonium ion formed by the reaction of an alkene with bromine (cf. Figure 8.3). The resulting species is a **protonated bromohydrin**, which then loses a proton to give the **bromohydrin** itself.

Br_2 Br^- H_2O $-H^+$

Alkene — Bromonium ion — Protonated bromohydrin — Bromohydrin

Since water is the *solvent*, water molecules are far greater in number than bromide ions, and therefore compete more favourably to attack the bromonium ion to give the bromohydrin as the **major** product.

It should be noted that bromohydrin formation can be regarded as the addition of the *components* of **hypobromous acid (HO–Br)** to alkenes. For an *unsymmetrical* alkene such as propene, the bromohydrin formation follows the **Markovnikov addition**. This means that the electrophilic Br atom (Br is less electronegative than O) adds to the carbon atom bearing the greater number of hydrogen atoms.

$$\underset{\textbf{Propene}}{CH_3CH{=}CH_2} \xrightarrow[\left(\overset{\delta-}{HO}-\overset{\delta+}{Br}\right)]{Br_2/H_2O} \underset{\textbf{Bromohydrin}}{CH_3\overset{OH}{\overset{|}{C}}H-\overset{Br}{\overset{|}{C}}H_2} \quad \textbf{\textit{Markovnikov addition}}$$

PROBLEM 8.4: Predict the possible products of the following reaction:

$$H_2C{=}CH_2 + Br_2 \xrightarrow[NaCl]{H_2O}$$

Answer

The electrophilic attack of Br_2 on ethylene gives a bromonium ion intermediate. The products then can arise from the attack on the bromonium ion by each of the three nucleophiles (Br^-, H_2O and Cl^-) present in the reaction mixture. The three possible products are 1,2-dibromoethane, 2-bromoethanol and 1-bromo-2-chloroethane.

$$\underset{\textbf{Ethylene}}{H_2C{=}CH_2} \xrightarrow{Br_2} H_2C\overset{\overset{+}{Br}}{\wedge}CH_2 \begin{cases} \xrightarrow{Br^-} \underset{\textbf{1,2-Dibromoethane}}{BrCH_2CH_2Br} \\ \xrightarrow{H_2O} \underset{\textbf{2-Bromoethanol}}{BrCH_2CH_2OH} \\ \xrightarrow{Cl^-} \underset{\textbf{1-Bromo-2-chloroethane}}{BrCH_2CH_2Cl} \end{cases}$$

8.7 OXIDATION OF ALKENES: EPOXIDATION

Oxidation of an alkene with a **peroxy acid** (or **peracid**) gives an **epoxide**. The oxidation reaction of alkenes to form epoxides is called the **epoxidation**.

The most commonly used peracid is *m*-chloroperoxybenzoic acid (*m*-CPBA), which is a stable crystalline solid. In the epoxidation reaction, **the peracid transfers an oxygen atom to the alkene**.

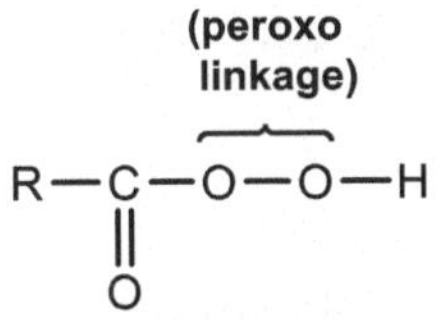

(peroxo linkage)

R—C(=O)—O—O—H

Peroxy acids (peracids)

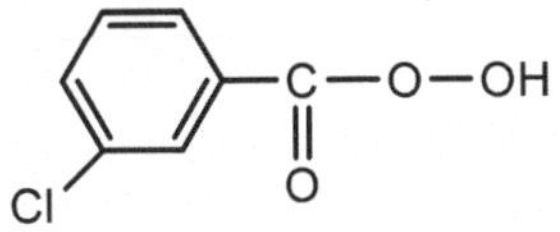

m-ClC₆H₄—C(=O)—O—OH   *A commonly used peracid*

***m*-CPBA (*m*-chloroperoxybenzoic acid)**

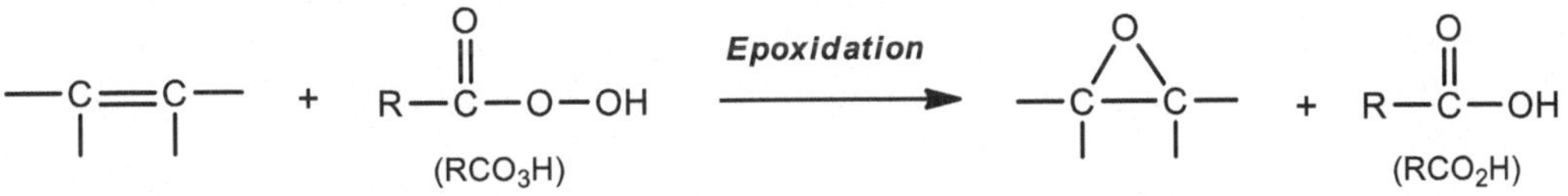

$$\text{—C=C—} + \underset{(RCO_3H)}{\text{R—C(=O)—O—OH}} \xrightarrow{\textit{Epoxidation}} \text{—C—C— (with O bridge)} + \underset{(RCO_2H)}{\text{R—C(=O)—OH}}$$

Alkene **Peracid** **Epoxide**

Epoxides are *cyclic ethers* containing a three-membered ring. They are also called **oxiranes**. The simplest epoxide (or oxirane) is **ethylene oxide** which is widely used in synthesis.

Ethylene oxide is industrially prepared by direct oxidation of ethylene with air or oxygen over a silver catalyst under pressure at 200–300°C.

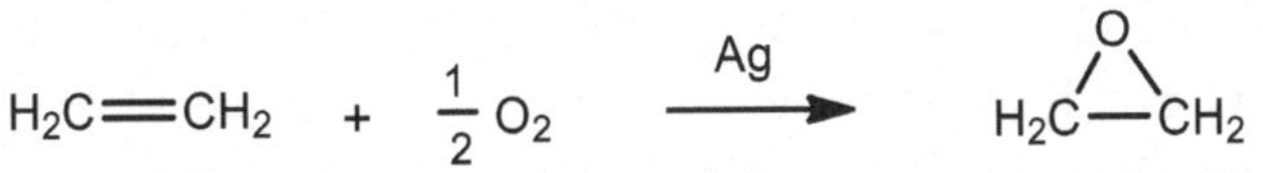

$$H_2C{=}CH_2 + \frac{1}{2}O_2 \xrightarrow{Ag} H_2C\overset{O}{—}CH_2$$

Ethylene **Ethylene oxide**

An epoxide or oxirane

Epoxides can be regarded as the *oxygen analogues* of bromonium ions, but unlike bromonium ions, they are quite stable. The following are some examples of epoxidation of alkenes with a peracid (*m*-CPBA):

$$CH_3CH{=}CH_2 \xrightarrow{m\text{-CPBA}} CH_3CH\overset{O}{—}CH_2$$

Propene **2-Methyloxirane (an epoxide)**

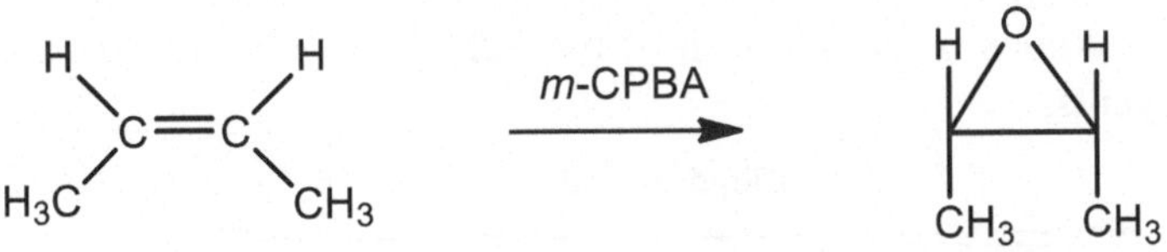

$$\text{(H)(H}_3\text{C)C=C(H)(CH}_3\text{)} \xrightarrow{m\text{-CPBA}} \text{epoxide (H, H on top; CH}_3\text{, CH}_3\text{ below)}$$

***cis*-2-Butene** ***cis*-2,3-Dimethyloxirane (a *cis*-epoxide)**

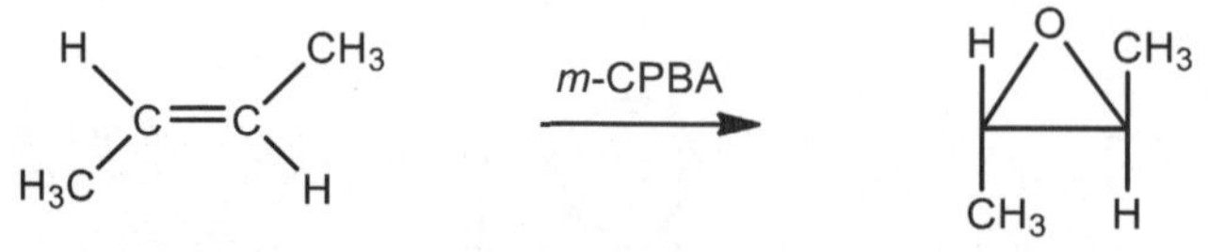

m-CPBA

trans-2-Butene

trans-2,3-Dimethyloxirane
(a *trans*-epoxide)

Notably, ***cis*-2-butene gives a *cis* epoxide** (*cis*-2,3-dimethyloxirane), whereas ***trans*-2-butene gives a *trans* epoxide** (*trans*-2,3-dimethyloxirane).

In the epoxidation of an alkene, the peracid acts as an **electrophile**; the O–O bond is weak and is easily broken by the nucleophilic attack of the alkene. The *electrophilic oxygen atom* is the oxygen atom which, upon nucleophilic attack, displaces a carboxylate (a good leaving group).

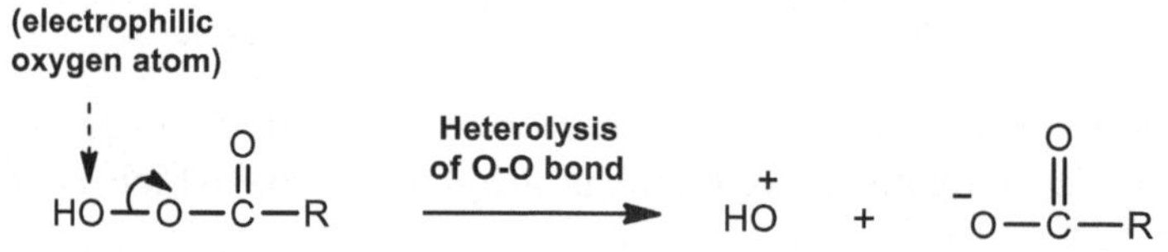

(electrophilic oxygen atom)

Heterolysis of O-O bond

Peracid

Carboxylate anion
(good leaving group)

The **mechanism** for epoxidation of an alkene is shown in Figure 8.6. The epoxidation involves a transfer of the electrophilic oxygen atom from the peracid to the alkene in a *single step*. Two C–O bonds are formed from two electron pairs: one pair from the π bond and the other pair from the O–H bond. The result is the formation of an epoxide. A carboxylic acid is also formed.

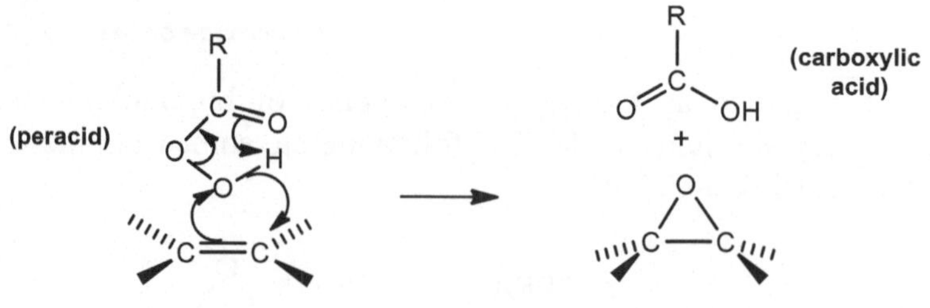

(peracid)

(carboxylic acid)

Alkene

Epoxide

Figure 8.6: Mechanism for epoxidation of an alkene with a peracid.

8.8 OXIDATION OF ALKENES: DIHYDROXYLATION

Dihydroxylation of alkenes is the addition of two OH groups across the double bond to form **1,2-diols** or **glycols**.

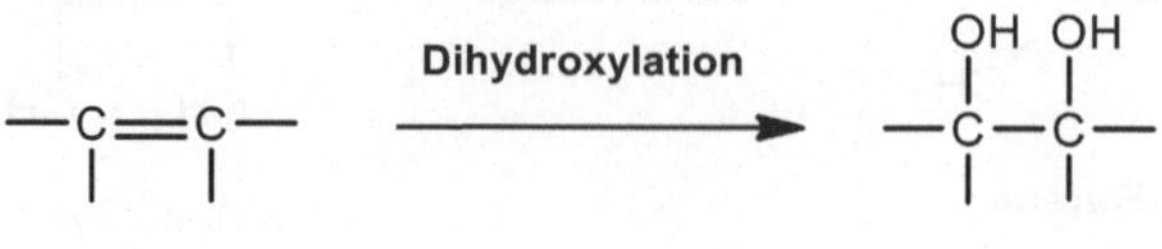

Dihydroxylation

Alkene

1,2-Diol
(or glycol)

Dihydroxylation can be carried out in several ways. In some cases, the two OH groups add to the *opposite* faces of an alkene called an **anti-dihydroxylation**, while in others, the two OH groups add to the *same* face of the alkene called a **syn-dihydroxylation**.

8.8.1 anti-Dihydroxylation

Oxidation of alkenes with a peracid produces **epoxides**, which can readily undergo acid-catalyzed hydrolysis to produce **1,2-diols**. The hydrolysis involves an opening of the strained three-membered ring of epoxides. Thus, **epoxidation of alkenes followed by hydrolysis** provides a suitable method for the **dihydroxylation of alkenes**.

RCH=CHR (**Alkene**) —*m*-CPBA, *Epoxidation*→ RHC—CHR with bridging O (**Epoxide**) —H_2O, H^+, *Hydrolysis*→ RHC(OH)—CHR(OH) (**1,2-Diol**)

The simplest 1,2-diol is **ethylene glycol**, the engine coolant widely used in automobiles. Ethylene glycol can be prepared by the hydrolysis of ethylene oxide in aqueous acid.

H_2C—CH_2 with bridging O (**Ethylene oxide**) + H_2O —H^+→ $H_2C(OH)$—$CH_2(OH)$ (**Ethylene glycol (1,2-ethanediol)**)

Mechanism for acid-catalysed hydrolysis of epoxides

The mechanism for acid-catalysed hydrolysis of an epoxide involves *three* steps. In **step 1**, the epoxide is protonated on oxygen. In **step 2**, water as a nucleophile attacks the protonated epoxide from the side *opposite* to the epoxide group to form a protonated diol, which then loses a proton in **step 3** to give the diol itself. The two OH groups become **anti** in the 1,2-diol product.

Epoxide —H^+, *Step 1*→ **Protonated epoxide** —*Step 2*, H_2O:→ **Protonated 1,2-diol** —*Step 3*, $-H^+$→ **anti-1,2-Diol**

It is therefore possible to perform anti-dihydroxylation of alkenes via epoxides.

Cyclic alkenes

Consider, for example, cyclohexene. The anti-dihydroxylation of cyclohexene can be achieved by epoxidation of cyclohexene with *m*-CPBA followed by hydrolysis in aqueous acid. The product is *trans*-1,2-cyclohexanediol.

m-CPBA / Epoxidation; H_2O, H^+ / Hydrolysis

Cyclohexene

trans-1,2-Cyclohexanediol

8.8.2 syn-Dihydroxylation

The **syn-dihydroxylation** of alkenes can be performed using *basic* potassium permanganate ($KMnO_4$) or osmium tetroxide (OsO_4).

Dihydroxylation with basic $KMnO_4$

Treatment of an alkene such as ethylene with cold, dilute and basic solution of potassium permanganate ($KMnO_4$) produces ethylene glycol (1,2-ethanediol).

$$H_2C{=}CH_2 + KMnO_4 \xrightarrow[OH^-]{H_2O} HOCH_2{-}CH_2OH + MnO_2$$

Ethylene — Ethylene glycol (1,2-ethanediol)

The dihydroxylation reaction with permanganate (MnO_4^-) proceeds through a **cyclic intermediate** which undergoes cleavage at the O–Mn bonds with aqueous base to give 1,2-diols. Since the two OH groups are formed on the *same* face of the original alkene, the reaction is a **syn-dihydroxylation**.

H_2O, OH^-; + MnO_2

Alkene — Cyclic intermediate — syn-1,2-Diol

Note: $KMnO_4$ is a strong oxidising agent, so it can cause further oxidation of the 1,2-diol product. The yields of the dihydroxylation reaction are often low, and the method has a limited synthetic utility.

Qualitative Test for Alkenes

We have already described the Br_2/CCl_4 test for alkenes (see Section 8.6). Dilute alkaline solution of $KMnO_4$ has also been used as a **qualitative test** for the detection of carbon-

carbon double bond. The *purple* $KMnO_4$ solution is slowly added to the suspected alkene. Rapid **disappearance of the purple colour** and the formation of a *dark brown* precipitate of manganese dioxide indicate a positive test for alkenes.

$$R_2C{=}CR_2 + KMnO_4/OH^- \longrightarrow R_2C(OH){-}C(OH)R_2 + MnO_2\downarrow$$

Alkene *Colourless* — *Purple solution* — *Colourless* — *Dark brown precipitate*

Rapid disappearance of purple colour and formation of a dark brown precipitate indicates a positive test for alkenes

Dihydroxylation with OsO_4

Alkenes react with osmium tetroxide (OsO_4) in pyridine to form a cyclic osmate ester, which is hydrolysed with aqueous $NaHSO_3$ to produce 1,2-diol.

The **cyclic osmate ester** is formed in a single step by a pericyclic mechanism. In this step, Os(VIII) in OsO_4 is converted to Os(VI) in the osmate ester. During hydrolysis, both OH groups are formed on the *same* face of the original alkene. The result is a **syn-dihydroxylation**.

Alkene + Osmium tetroxide *Os(VIII)* —Pyridine→ Cyclic osmate ester *Os(VI)* —$NaHSO_3$, H_2O→ syn-1,2-Diol + $(HO)_2OsO_2$ *Os(VI)*

For example, the syn-dihydroxylation of **cyclohexene** on treatment with OsO_4 followed by hydrolysis produces ***cis*-1,2-cyclohexanediol**.

Cyclohexene —OsO_4, Pyridine→ Osmate ester —$NaHSO_3$, H_2O→ *cis*-1,2-Cyclohexanediol

Note: The dihydroxylation reaction with OsO_4 is sufficiently mild and does not cause further oxidation of the diol, and gives high yields. However, the reagent is highly toxic and expensive. For this reason, the method has been modified by using a catalytic or small amount of OsO_4 in the presence of a co-oxidant such as *N*-methylmorpholine-*N*-oxide.

PROBLEM 8.5: Predict the stereochemistry of products formed in the following reactions:

cis-2-Butene $\xrightarrow[\text{2. NaHSO}_3/\text{H}_2\text{O}]{\text{1. OsO}_4\text{, pyridine}}$

trans-2-Butene $\xrightarrow[\text{2. NaHSO}_3/\text{H}_2\text{O}]{\text{1. OsO}_4\text{, pyridine}}$

Answer

The reaction of an alkene with OsO_4 followed by hydrolysis is a syn-dihydroxylation. The two possible pathways for syn-dihydroxylation of *cis*-2-butene give the same achiral product having a plane of symmetry, which is *meso*-2,3-butanediol. On the other hand, syn-dihydroxylation of *trans*-2-butene gives a chiral product with no plane of symmetry, and the two possible paths (*a* and *b*) lead to a mixture of enantiomers, i.e. racemic-2,3-butanediol.

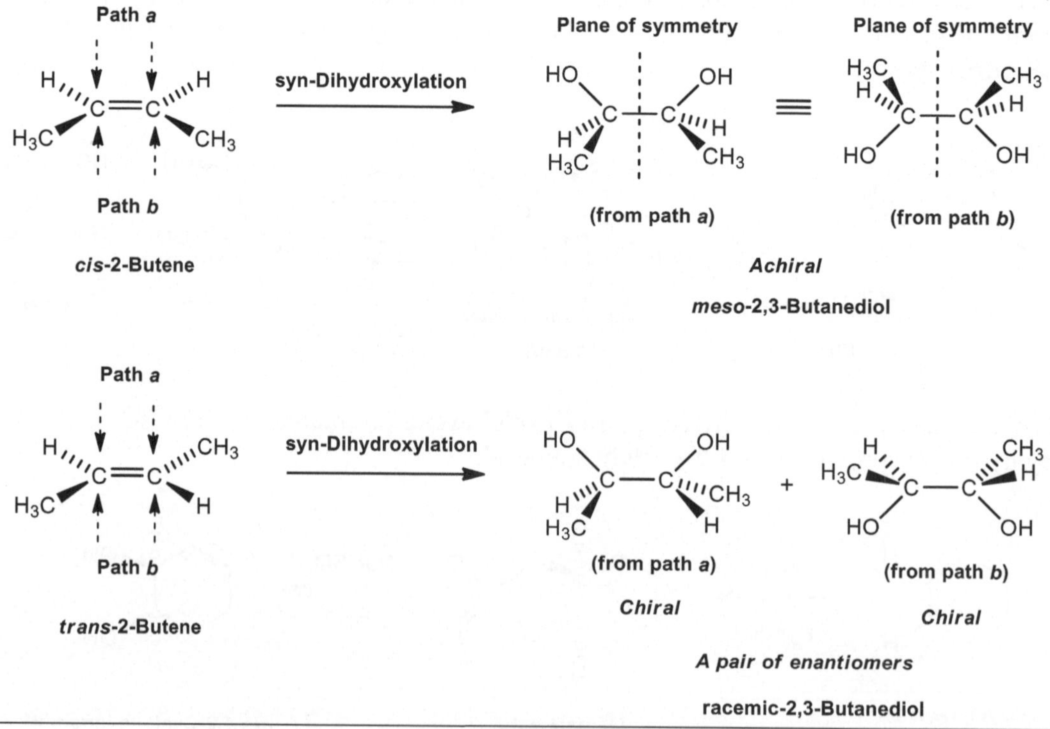

8.9 OXIDATIVE CLEAVAGE AND OZONOLYSIS OF ALKENES

It is possible to **cleave** a *double bond* of an alkene completely by **oxidation**. The process is called the **oxidative cleavage** of alkenes. The oxidative cleavage can be accomplished *via* the formation of a 1,2-diol intermediate.

Cold basic $KMnO_4$ solutions oxidize alkenes to 1,2-diols, but **hot basic $KMnO_4$ solutions** can further oxidize 1,2-diols with cleavage of the C–C σ bond. The result is the oxidative cleavage of alkenes.

The following are two examples of oxidative cleavage of alkenes by hot basic permanganate solutions:

$$CH_3CH{=}CHCH_3 \xrightarrow[OH^-,\ Heat]{KMnO_4} \left[CH_3CH(OH){-}CH(OH)CH_3\right] \longrightarrow 2\ H_3C{-}C(=O){-}O^- \xrightarrow{H^+} 2\ H_3C{-}C(=O){-}OH$$

2-Butene → Acetate ion → Acetic acid

$$(CH_3)_2C{=}CH_2 \xrightarrow[2.\ H^+]{1.\ KMnO_4,\ OH^-,\ heat} H_3C{-}C(CH_3){=}O + CO_2 + H_2O$$

Isobutene → Acetone

In the first example, *one mole* of **2-butene** is completely oxidized to *two moles* of **acetate ion**. Acidification of the mixture produces two moles of acetic acid. In the second example, **isobutene** has a terminal CH_2 group which is completely oxidized to **CO_2** and **H_2O**. The other product is **acetone** (a ketone). In general, a disubstituted C atom of a double bond (as in isobutene) is oxidized to C=O.

The **oxidative cleavage** of alkenes can also be performed in *two* steps. The alkene is first treated with **osmium tetroxide** (OsO_4) to form a 1,2-diol or glycol, which is then oxidized with **periodic acid** (HIO_4). The periodic acid oxidation leads to cleavage of the C–C bond between the two hydroxyl groups, which forms two C=O groups.

For example, **2-butene** gives two moles of **acetaldehyde**.

$$CH_3CH{=}CHCH_3 \xrightarrow[2.\ NaHSO_3/H_2O]{1.\ OsO_4,\ pyridine} CH_3CH(OH){-}CH(OH)CH_3 \xrightarrow{HIO_4} 2\ CH_3CH{=}O$$

2-Butene → 1,2-Diol or glycol → Acetaldehyde

This method is milder than the previous $KMnO_4$ method. In the OsO_4 method, 2-butene produces acetaldehyde but the $KMnO_4$ method gives acetic acid by further oxidation of acetaldehyde (see above).

8.9.1 Ozonolysis of Alkenes

The oxidative cleavage of alkenes is also possible by ozonolysis. This method is important for structure elucidation of alkenes.

Ozone (O_3) reacts with alkenes at low temperatures (–78°C) to form compounds called **ozonides**. Ozone is generated from oxygen by passage of an electric discharge through air and then bubbled through a solution of the alkene in dichloromethane (CH_2Cl_2).

Ozonides are quite unstable, and some are explosive. They are not isolated, but are converted to **carbonyl compounds** (aldehydes or ketones) by **reduction** with Zn and water. Other reducing agents such as dimethyl sulphide [$(CH_3)_2S$] and triphenylphosphine [$(C_6H_5)_3P$] can also be used in the reduction of ozonides.

$$R_2C{=}CHR + O_3 \xrightarrow[-78^{\circ}C]{CH_2Cl_2} \text{Ozonide}$$

Alkene **Ozonide**

Reduction: Zn/H_2O or $(CH_3)_2S$ or $(C_6H_5)_3P$

$$R_2C{=}O + O{=}CHR$$

Ketone/aldehyde

The overall process of forming the ozonides and their subsequent reduction is called the ozonolysis. Ozonolysis leads to the oxidative cleavage of alkenes, as shown below.

$$R_2C{=}CHR \xrightarrow[2.\ Zn/H_2O]{1.\ O_3,\ CH_2Cl_2,\ -78^{\circ}C} R_2C{=}O + O{=}CHR$$

Alkene *Ozonolysis* **Ketone** **Aldehyde**

The following examples illustrate the ozonolysis of some alkenes:

$$CH_3C(CH_3){=}CHCH_3 \xrightarrow[2.\ Zn/H_2O]{1.\ O_3,\ CH_2Cl_2,\ -78^{\circ}C} CH_3C(CH_3){=}O + CH_3CH{=}O$$

2-Methyl-2-butene **Acetone** **Acetaldehyde**

$$CH_3CH_2C(CH_3){=}CH_2 \xrightarrow[2.\ (CH_3)_2S]{1.\ O_3,\ CH_2Cl_2,\ -78^{\circ}C} CH_3CH_2C(CH_3){=}O + H_2C{=}O$$

2-Methyl-1-butene **Butanone** **Formaldehyde**

$$\text{Cyclohexene} \xrightarrow[2.\ (C_6H_5)_3P]{1.\ O_3,\ CH_2Cl_2,\ -78^{\circ}C} \text{(CH}_2)_4(CH{=}O)_2 \equiv HCCH_2CH_2CH_2CH_2CH \text{ (with C=O at each end)}$$

Cyclohexene **1,6-Hexanedial**

Treatment of ozonides with oxidising agents such as H_2O_2 gives **carboxylic acids,** whereas reducing the ozonide with more powerful reducing agents such as $NaBH_4$ produces **alcohols**.

$$CH_3CH{=}CHCH_3 \xrightarrow[\text{2. } H_2O_2]{\text{1. } O_3} 2\ CH_3C(OH){=}O$$

2-Butene → **Acetic acid**

$$CH_3CH{=}CHCH_3 \xrightarrow[\text{2. } NaBH_4]{\text{1. } O_3} 2\ CH_3CH_2OH$$

2-Butene → **Ethyl alcohol**

Structure Elucidation

Ozonolysis can be used as a **method for locating the double bond** in an alkene. After identifying the products (aldehydes or ketones), we **work backward** from the products to the reactant alkene that might have led to those products. This 'working backward' approach is indicated by an arrow symbol (⇒).

Example 1

An unknown alkene of molecular formula C_6H_{12} gives on ozonolysis acetone and propanal. Suggest a structure for the alkene.

Solution:

Acetone (CH_3COCH_3) is a three-carbon ketone and propanal (CH_3CH_2CHO) is also a three-carbon aldehyde. The two products result from the oxidative cleavage of C=C of the six-carbon alkene. Thus, working backward, the structure of the alkene is:

$$CH_3C(CH_3){=}O + O{=}CHCH_2CH_3 \Longrightarrow CH_3C(CH_3){=}CHCH_2CH_3$$

Acetone + **Propanal** ⇒ **2-Methyl-2-pentene** (***Alkene***)

Example 2

Suggest a structure of an unknown alkene (C_7H_{12}) which undergoes ozonolysis to give 6-oxoheptanal ($CH_3COCH_2CH_2CH_2CH_2CHO$).

Solution:

Since only one product is obtained and the product has the same number of carbon atoms as the reactant, a reasonable explanation is that the alkene has a double bond in a ring. Thus, working backward, the structure of the cycloalkene is

6-Oxoheptanal ⟹ **1-Methylcyclohexene**

Cyclic alkene

PROBLEM 8.6: Predict the structures of alkenes that would produce the following products on ozonolysis:

(a) Acetone only

(b) Cyclopentanone and formaldehyde

Answer

(a) Acetone has one C=O group. Therefore, 2 moles of acetone are produced from 1 mole of alkene. Working backward, the structure of alkene is

$$CH_3C(CH_3){=}O \; + \; O{=}C(CH_3)CH_3 \Longrightarrow CH_3C(CH_3){=}C(CH_3)CH_3$$

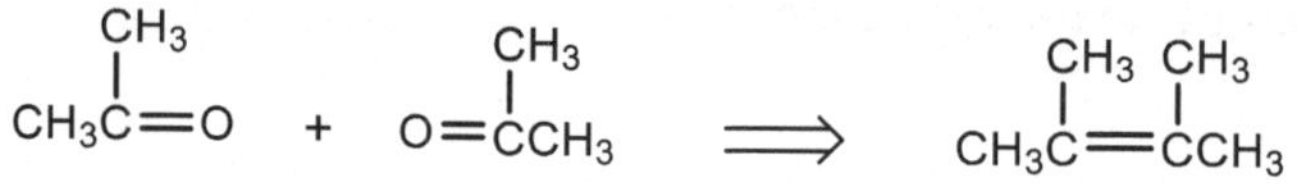

Acetone (2 moles) **2,3-Dimethyl-2-butene**

Alkene

(b) Working backward, the structure of alkene is

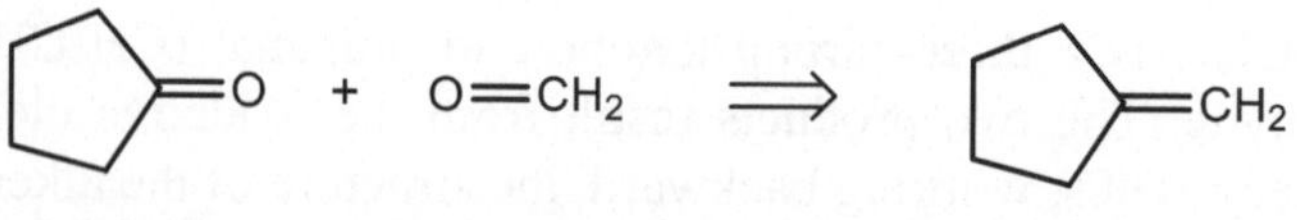

Cyclopentanone + **Formaldehyde** ($O{=}CH_2$) ⟹ **Methylenecyclopentane** (${=}CH_2$)

Alkene

8.10 REDUCTION OF ALKENES: CATALYTIC HYDROGENATION

Alkenes react with hydrogen in the presence of a variety of **metal catalysts** such as nickel, palladium or platinum to produce **alkanes**.

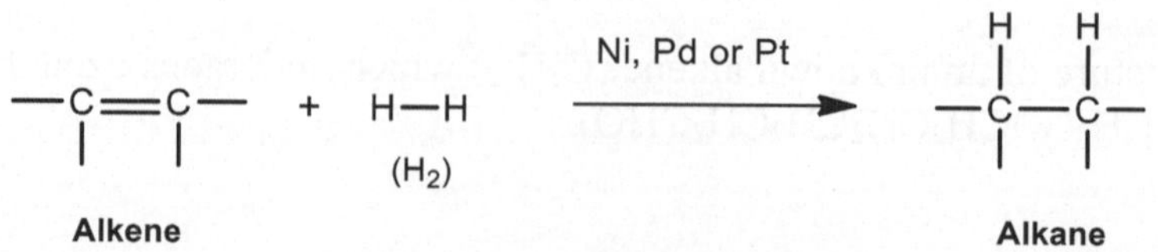

$$\text{—C=C—} \;+\; \text{H—H} \; (H_2) \xrightarrow{\text{Ni, Pd or Pt}} \text{—CH—CH—}$$

Alkene **Alkane**

This addition of hydrogen to an alkene in the presence of a metal catalyst represents a case of **reduction**. More frequently, this type of reduction of alkenes is called the

catalytic hydrogenation. The reaction is usually carried out by adding finely divided metal catalyst to the solution of an alkene and then passing hydrogen gas through the mixture under pressure in a special apparatus.

The following are some examples of catalytic hydrogenation of alkenes:

$$CH_3CH{=}CH_2 + H_2 \xrightarrow[C_2H_5OH,\ 25^\circ C,\ 50\ atm]{Ni} CH_3CH_2CH_3$$

Propene → Propane

$$(CH_3)_2C{=}CH_2 + H_2 \xrightarrow[C_2H_5OH,\ 25^\circ C,\ 50\ atm]{Ni} CH_3CH(CH_3)CH_3$$

Isobutene → Isobutane

Cyclohexene + H_2 $\xrightarrow[C_2H_5OH,\ 25^\circ C,\ 1\ atm]{Pt}$ Cyclohexane

In **catalytic hydrogenation**, the hydrogen molecule is *adsorbed* on the metal surface, and the two H atoms are delivered on the *same* side (or face) of the alkene. The catalytic hydrogenation is therefore a **syn addition**.

Alkene $\xrightarrow[Pt]{H_2}$ [Catalyst surface] $\xrightarrow{\text{syn addition}}$ Alkane

8.10.1 Heats of Hydrogenation: Relative Stabilities of Alkenes

The reaction of an alkene with hydrogen is *exothermic* and the heat evolved is called the **heat of hydrogenation**. The heat of hydrogenation is denoted by the enthalpy difference (ΔH^0):

$$\Delta H^0 = H^0_{\text{product}} - H^0_{\text{reactants}}$$

The product (alkane) has lower enthalpy than reactants (alkene and hydrogen), i.e. $H^0_{\text{product}} < H^0_{\text{reactants}}$. Therefore, ΔH^0 is negative for an exothermic reaction.

Enthalpy Diagram

Consider two alkenes (**alkene 1** and **alkene 2**) which give the *same* product on hydrogenation. Assume that alkene 1 evolves *smaller amount* of heat than alkene 2 on hydrogenation. Figure 8.7 shows an enthalpy diagram for hydrogenation of the alkenes.

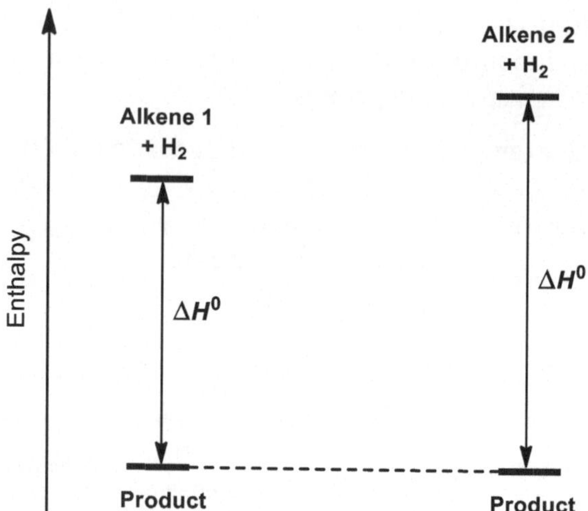

Figure 8.7: An enthalpy diagram for hydrogenation of two alkenes (alkene 1 and alkene 2). The alkenes give the same product on hydrogenation.

Since the product and one reactant (hydrogen) are the same in both cases, **alkene 1** with smaller value of ΔH^0 has **lower enthalpy** than alkene 2. *The lower the enthalpy, the greater is the stability*. Thus, when hydrogenation products are the same, **the alkene that evolves smaller amount of heat on hydrogenation has lower enthalpy, and hence is more stable**.

To illustrate, consider the ΔH^0 values for hydrogenation of 1-butene, *cis*-2-butene and *trans*-2-butene, which give the same product (butane).

$$\underset{\text{1-Butene}}{CH_3CH_2CH{=}CH_2} \xrightarrow[\text{Pt}]{H_2} CH_3CH_2CH_2CH_3 \qquad \Delta H^0 = -30.3 \text{ kcal mol}^{-1}$$

$$\underset{\textit{cis}\text{-2-Butene}}{(H)(H_3C)C{=}C(H)(CH_3)} \xrightarrow[\text{Pt}]{H_2} CH_3CH_2CH_2CH_3 \qquad \Delta H^0 = -28.6 \text{ kcal mol}^{-1}$$

$$\underset{\textit{trans}\text{-2-Butene}}{(H)(H_3C)C{=}C(CH_3)(H)} \xrightarrow[\text{Pt}]{H_2} \underset{\text{Butane}}{CH_3CH_2CH_2CH_3} \qquad \Delta H^0 = -27.6 \text{ kcal mol}^{-1}$$

Since **1-butene** evolves the greatest amount of heat ($\Delta H^0 = -30.3$ kcal mol^{-1}) on hydrogenation, it has the **highest enthalpy** and therefore it is the **least stable**. On the other hand, ***trans*-2-butene** evolves the least amount of heat ($\Delta H^0 = -27.6$ kcal mol^{-1}) and hence it has the **lowest enthalpy** and is the **most stable**.

So the order of stability obtained from the *experimental* heat of hydrogenation data is:

trans-2-butene > *cis*-2-butene > 1-butene.

In general, **a more substituted alkene has higher stability than a less substituted alkene**. Thus 1-butene with *one* alkyl group (CH_2CH_3) on the double bond is *less stable* than 2-butene with *two* alkyl groups (CH_3 and CH_3) on the double bond. This can be explained using **hyperconjugation** (see Section 3.7). 1-Butene with *two* α-hydrogens gives *two* hyperconjugative structures, whereas 2-butene with *six* α-hydrogens gives *six* hyperconjugative structures. Therefore, 2-butene is more stable than 1-butene by hyperconjugation.

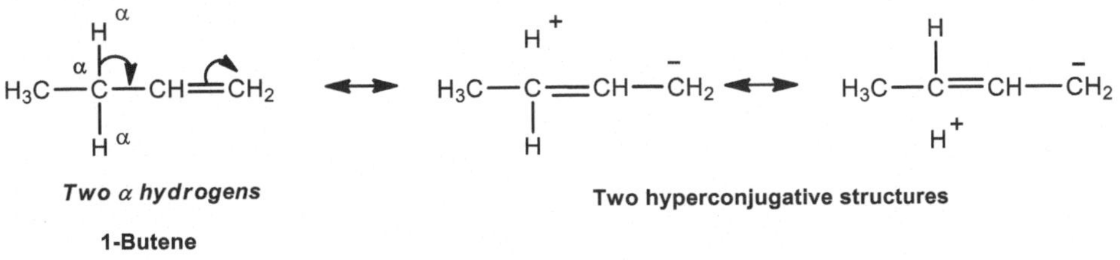

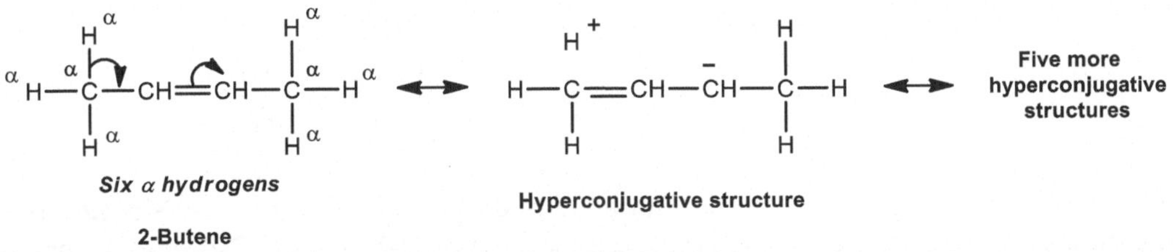

$H_3C-CH_2-CH=CH_2 \longleftrightarrow H_3C-CH=CH-\overset{-}{C}H_2\ (H^+) \longleftrightarrow H_3C-CH=CH-\overset{-}{C}H_2\ (H^+)$

Two α hydrogens **1-Butene** **Two hyperconjugative structures**

$H_3C-CH=CH-CH_3 \longleftrightarrow H-CH=CH-\overset{-}{C}H-CH_3\ (H^+) \longleftrightarrow$ **Five more hyperconjugative structures**

Six α hydrogens **2-Butene** **Hyperconjugative structure**

Between *cis* and *trans* stereoisomers of 2-butene, the relative stability can be explained using the **steric effects** (see Section 3.10). In the *cis* isomer, there is a repulsive steric interaction between two methyl groups on the *same* side of the double bond, whereas there is no such steric interaction between two methyl groups on the *opposite* sides of the double bond in the *trans* isomer. Therefore, *trans*-2-butene is more stable than *cis*-2-butene.

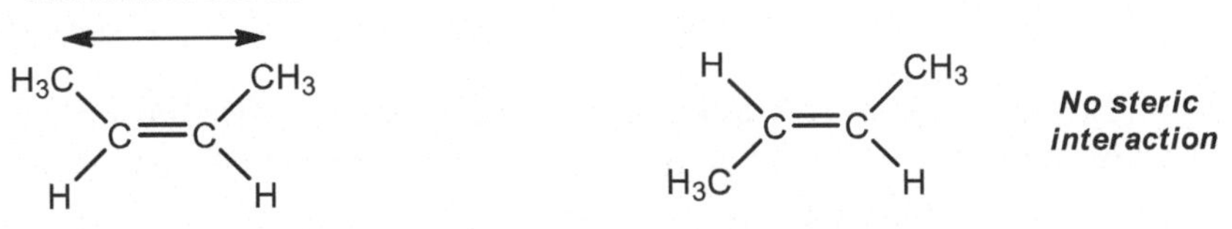

Steric interaction

No steric interaction

***cis*-2-Butene** ***trans*-2-Butene**

8.11 ADDITION OF HYDROGEN HALIDES TO ALKYNES

Alkynes can react with *one* or *two* molar equivalents of hydrogen halides (HX; X= Cl, Br). Addition of 1 mole of HX to an alkyne gives a **haloalkene**, while in the presence of 2 moles of HX, a **geminal (*gem*) dihalide** is formed. A *gem*-dihalide has two halogen atoms on the *same* carbon atom.

$$-C\equiv C- \xrightarrow[(X = Cl, Br)]{HX} -C(H)=C(X)- \xrightarrow{HX} -C(H)(H)-C(X)(X)-$$

Alkyne — Haloalkene — *gem*-Dihalide

The simplest alkyne, **acetylene** reacts with 1 mole of HCl to give **vinyl chloride**. Vinyl chloride is the monomer used in the preparation of polyvinylchloride (PVC) polymer.

The reaction of **2-butyne** with 1 mole of HCl gives **2-chloro-2-butene**, which is obtained as a *more stable* ***trans*** **stereoisomer**.

$$HC\equiv CH \xrightarrow{HCl} HC(H)=CH-Cl \text{ or } H_2C=CHCl$$

Acetylene — Vinyl chloride

$$H_3C-C\equiv C-CH_3 \xrightarrow{HCl} (H_3C)(H)C=C(Cl)(CH_3)$$

2-Butyne — *trans*-2-chloro-2-butene

With *unsymmetrical* alkynes, the addition of hydrogen halides follows **Markovnikov addition**. For example, **1-pentyne** reacts with 1 mole of HBr to form 2-bromo-1-pentene, and with 2 moles of HBr, it yields 2,2-dibromopentane. Both steps follow Markovnikov additions.

$$CH_3CH_2CH_2C\equiv CH \xrightarrow[\text{Markovnikov addition}]{HBr} CH_3CH_2CH_2C(Br)=CH_2 \xrightarrow[\text{Markovnikov addition}]{HBr} CH_3CH_2CH_2C(Br)_2-CH_3$$

1-Pentyne — 2-Bromo-1-pentene — 2,2-Dibromopentane

Alkynes can also undergo **anti-Markovnikov addition** of HBr in the presence of *peroxides*. For example, 1-pentyne adds to HBr to give 1-bromo-1-pentene when peroxides are present in the reaction mixture.

$$CH_3CH_2CH_2C\equiv CH \xrightarrow[\text{Peroxide}]{HBr} CH_3CH_2CH_2CH=CHBr$$

1-Pentyne — 1-Bromo-1-pentene

anti-Markovnikov addition

8.12 ADDITION OF HALOGENS TO ALKYNES

Alkynes can react with bromine or chlorine with *one* or *two* molar equivalents of halogen. For example, **2-butyne** reacts with 1 mole of bromine to form a **dibromoalkene** (2,3-dibromo-2-butene). The bromine addition is **anti**, similar to the addition of bromine to an alkene. Thus, the product is ***trans*-2,3-dibromo-2-butene**.

$$H_3C{-}C{\equiv}C{-}CH_3 + Br_2 \text{ (1 mol)} \xrightarrow[\text{anti addition}]{CCl_4} \text{(H}_3\text{C)(Br)C=C(Br)(CH}_3\text{)}$$

2-Butyne → ***trans*-2,3-Dibromo-2-butene**

The addition of 2 moles of bromine or chlorine to an alkyne produces a **tetrabromoalkane** or a **tetrachloroalkane**. This is illustrated with acetylene (or ethyne) and propyne.

$$HC{\equiv}CH \xrightarrow[CCl_4]{Br_2} HC(Br){=}CH(Br) \xrightarrow[CCl_4]{Br_2} HCBr_2{-}CHBr_2$$

Ethyne → **1,2-Dibromoethene** → **1,1,2,2-Tetrabromoethane**

$$CH_3C{\equiv}CH \xrightarrow{Cl_2} CH_3C(Cl){=}CH(Cl) \xrightarrow{Cl_2} CH_3CCl_2{-}CHCl_2$$

Propyne → **1,2-Dichloropropene** → **1,1,2,2-Tetrachloropropane**

8.13 ADDITION OF WATER TO ALKYNES

Alkynes add water readily when treated with an aqueous solution of **sulphuric acid** and **mercuric sulphate**. The addition reaction called hydration of alkynes is thus catalysed by mercuric ion (Hg^{2+}) and H^+. The initial product is a **vinylic alcohol** which is *unstable*, and it isomerizes (or rearranges) to a **ketone** or an **aldehyde**.

$$-C{\equiv}C- + H{-}OH \; (H_2O) \xrightarrow[H_2SO_4]{HgSO_4} \left[-C(H){=}C(OH)- \right] \xrightarrow{\text{Tautomerization}} -C(H)(H){-}C(=O)-$$

Alkyne + (H_2O) → **Vinylic alcohol (unstable)** *Enol form* → **Ketone or aldehyde** *Keto form*

The isomerization involves the movement of a H atom from oxygen to carbon with relocation of the double bond. This kind of isomerization or rearrangement is called the

keto-enol tautomerization (see later Section 15.1). The vinylic alcohol is the **enol** (en for alkene + ol for alcohol) form and the carbonyl compound (usually a ketone) is the **keto** form.

Acetylene itself undergoes addition of water to form **acetaldehyde** (an aldehyde), whereas *other alkynes* produce a *ketone*. For example, **2-butyne** adds water to produce **butanone**.

$$HC{\equiv}CH + H_2O \xrightarrow[H_2SO_4]{HgSO_4} [H_2C{=}CH(OH)] \longrightarrow H_3C{-}CH({=}O)$$

Acetylene → **Acetaldehyde**

$$H_3C{-}C{\equiv}C{-}CH_3 + H_2O \xrightarrow[H_2SO_4]{HgSO_4} [H_3C{-}C(H){=}C(OH){-}CH_3] \longrightarrow H_3C{-}CH_2{-}C({=}O){-}CH_3$$

2-Butyne → **Butanone**

The addition of water to alkynes follows the **Markovnikov rule**. Therefore, hydration of **terminal alkynes** (–C≡CH) proceeds by the attachment of H atom of water molecule to the terminal carbon atom bearing a hydrogen atom. The products are **ketones**, *not* aldehydes.

This is illustrated with hydration of **propyne** and **1-butyne**. Note that hydration of 1-butyne gives the same product (butanone) as produced by 2-butyne.

$$CH_3C{\equiv}CH + H_2O \xrightarrow[H_2SO_4]{HgSO_4} [H_3C{-}C(OH){=}CH(H)] \longrightarrow CH_3{-}C({=}O){-}CH_3$$

(OH and H add as shown: OH to the internal carbon, H to the terminal carbon.)

Propyne → ***Markovnikov addition*** → **Acetone**

$$CH_3CH_2C{\equiv}CH + H_2O \xrightarrow[H_2SO_4]{HgSO_4} [CH_3CH_2C(OH){=}CH_2] \longrightarrow CH_3CH_2C({=}O)CH_3$$

1-Butyne → ***Markovnikov addition*** → **Butanone**

8.14 OXIDATIVE CLEAVAGE OF ALKYNES

The C≡C triple bond of an alkyne can be completely cleaved with **basic potassium permanganate** or by **ozonolysis**.

The oxidative cleavage of a *terminal alkyne* (e.g. 1-butyne) produces a carboxylic acid and carbon dioxide, whereas *nonterminal alkynes* (such as 2-butyne and 2-pentyne) with internal triple bond produce only carboxylic acids.

$$CH_3CH_2C{\equiv}CH \xrightarrow[2.\ H^+]{1.\ KMnO_4,\ OH^-,\ heat} CH_3CH_2CO_2H + CO_2$$

1-Butyne **Propionic acid**

$$CH_3C{\equiv}CCH_3 \xrightarrow[2.\ H_2O]{1.\ O_3} 2\ CH_3CO_2H$$

2-Butyne **Acetic acid**

$$CH_3C{\equiv}CCH_2CH_3 \xrightarrow[2.\ H_2O]{1.\ O_3} CH_3CO_2H + CH_3CH_2CO_2H$$

2-Pentyne **Acetic acid** **Propionic acid**

8.15 REDUCTION OF ALKYNES: CATALYTIC HYDROGENATION

One or two molar equivalents of hydrogen can add to alkynes depending on the metal catalyst used.

In the presence of palladium or platinum catalyst, an alkyne usually adds *two* moles of hydrogen to form an **alkane**.

$$CH_3C{\equiv}CCH_3 \xrightarrow[Pd]{H_2} [CH_3CH{=}CHCH_3] \xrightarrow[Pd]{H_2} CH_3CH_2CH_2CH_3$$

2-Butyne **Butane**

It is possible to add only one mole of hydrogen when an alkyne is converted to an **alkene**, without further reduction to an alkane. This can be done using a modified metal catalyst such as **Lindlar's catalyst**. The Lindlar's catalyst consists of metallic palladium deposited on calcium carbonate, which is partially poisoned or deactivated with lead acetate and quinoline.

For example, **2-butyne** can be hydrogenated using Lindlar's catalyst to 2-butene. However, catalytic hydrogenation is a **syn addition**. Therefore, the product is ***cis*-2-butene**.

$$CH_3C{\equiv}CCH_3 + H_2 \xrightarrow[\text{Lead acetate, Quinoline}]{Pd/CaCO_3} (H_3C)(H)C{=}C(CH_3)(H)$$

2-Butyne ***Lindlar's catalyst*** ***cis*-2-Butene**

In general, a disubstituted alkyne gives a *cis*-alkene on catalytic hydrogenation in the presence of Lindlar's catalyst.

8.16 ADDITION OF HYDROGEN HALIDES AND HALOGENS TO CONJUGATED DIENES: 1,2- AND 1,4-ADDITIONS

The addition reactions of conjugated dienes exhibit special features. To illustrate, consider **1,3-butadiene**.

The addition of hydrogen bromide to 1,3-butadiene gives *two* products. One product is **3-bromo-1-butene** that results from the addition of HBr to a double bond between C1 and C2. This addition is called a **1,2-addition or direct addition**, which is expected. Note that the double bond between C1 and C2 and the double bond between C3 and C4 are equivalent.

The second product is **1-bromo-2-butene** that involves an addition across the ends (C1 and C4) of the conjugated diene, with *relocation* of the remaining double bond between the central carbon atoms (C2 and C3). This addition is called a **1,4-addition or conjugate addition**, which is unexpected.

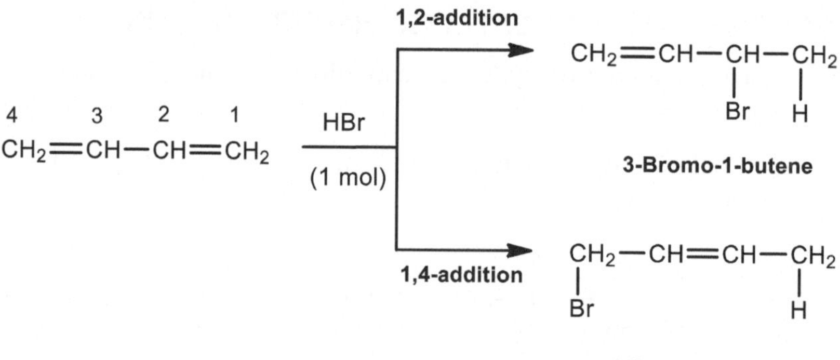

Mechanism

The mechanism for the addition of HBr to 1,3-butadiene involves *two* steps, as shown in Figure 8.8.

In **step 1**, a proton adds to a **terminal carbon atom** (C1) of the diene to form a **secondary and allylic carbocation**, which is stabilized by resonance. The proton *does not* add to an inner carbon atom (C2) that produces a *primary carbocation* which cannot be stabilized by resonance, and is much less stable.

In **step 2**, the nucleophile (Br^-) can combine with the allylic carbocation by attacking either at C2 or C4 to give the **1,2-addition product** and the **1,4-addition product**.

Attack at C2

$CH_2{=}CH{-}CH_2{-}\overset{+}{C}H_2$ **Not formed**

1° carbocation (less stable)

$\overset{4}{C}H_2{=}\overset{3}{C}H{-}\overset{2}{C}H{=}\overset{1}{C}H_2 \xrightarrow[\textit{Step 1}]{H^+}$

$\left[CH_2{=}CH{-}\overset{+}{C}H{-}CH_3 \longleftrightarrow \overset{+}{C}H_2{-}CH{=}CH{-}CH_3\right]$

Attack at C1

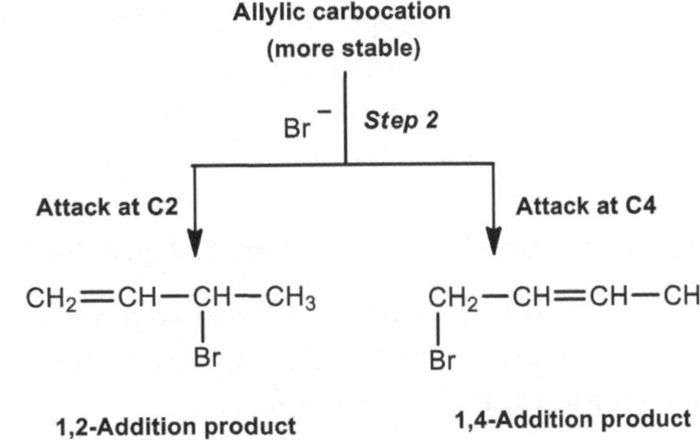

Figure 8.8: Mechanism for 1,2- and 1,4-additions of HBr to 1,3-butadiene.

Effect of Temperature

The relative amounts of 1,2- and 1,4-addition products depend on the temperatures at which the experiments are performed. At a **low temperature** (–80°C), the **1,2-addition product** predominates (80%), whereas at a **higher temperature** (40°C), the **1,4-addition product** is the major product (80%).

1,2-Addition product *1,4-Addition product*

$CH_2{=}CH{-}CH{=}CH_2$ —HBr→

-80°C: $CH_2{=}CH{-}CH(Br){-}CH_3$ (80%) + $CH_2(Br){-}CH{=}CH{-}CH_3$ (20%)

40°C: $CH_2{=}CH{-}CH(Br){-}CH_3$ (20%) + $CH_2(Br){-}CH{=}CH{-}CH_3$ (80%)

Relative rates of 1,2- and 1,4-additions

The **1,2-product** is formed **faster** than the 1,4-product. This is attributed to the **proximity effect**, which refers to the proximity or closeness of the Br^- ion to the C2 of the carbocation right after the protonation of the double bond takes place. Since the Br^- ion is *nearer* to C2, it can readily attack C2 than it can attack C4 which is farther apart, and the formation of the 1.2-product becomes faster.

$CH_2{=}CH{-}CH{=}CH_2$ (1,3-Butadiene) + H–Br → $CH_2{=}CH{-}\overset{+}{C}H{-}CH_3$ (2) ⟷ $\overset{+}{C}H_2{-}CH{=}CH{-}CH_3$ (4)

Nearer: Br^- attacks C2 — *Faster* → $CH_2{=}CH{-}CH(Br){-}CH_3$ **1,2-Product**

Farther apart: Br^- attacks C4 — *Slower* → $CH_2(Br){-}CH{=}CH{-}CH_3$ **1,4-Product**

When the product composition is determined by the *relative rates* of the reaction, the reaction is said to be under **kinetic control**. At low temperatures, the **1,2 product** which forms at a faster rate is the major product under kinetic control of the reaction, and is called the **kinetically controlled product**.

Relative stabilities of 1,2- and 1,4-products

The **1,4-product** is a **more substituted alkene** with two substituents, whereas the 1,2-product is a less substituted alkene with one substituent. Therefore, the **1,4-product is more stable than the 1,2 product**.

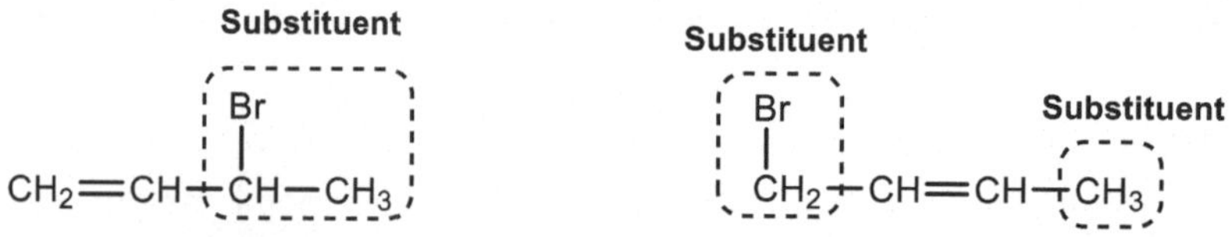

Substituent

$CH_2{=}CH{-}CH(Br){-}CH_3$

1,2-Product

(less substituted alkene)

Less stable

Substituent **Substituent**

$BrCH_2{-}CH{=}CH{-}CH_3$

1,4-Product

(more substituted alkene)

More stable

When the product composition is determined by the *relative stabilities* of the products, the reaction is said to be under **thermodynamic control**. At a higher temperature, the more stable **1,4 product** is the predominant product under thermodynamic or equilibrium control of the reaction, and is called the **thermodynamically controlled product**.

Note: The reaction is essentially irreversible at low temperatures and the major product is the product formed from the faster reaction. But at higher temperatures, the reaction is reversible and the major product is the more stable product at equilibrium.

Further Examples

1,3-Butadiene shows similar behaviour of 1,2- and 1,4-additions with other hydrogen halides (e.g. HCl) or halogens (such as Cl_2).

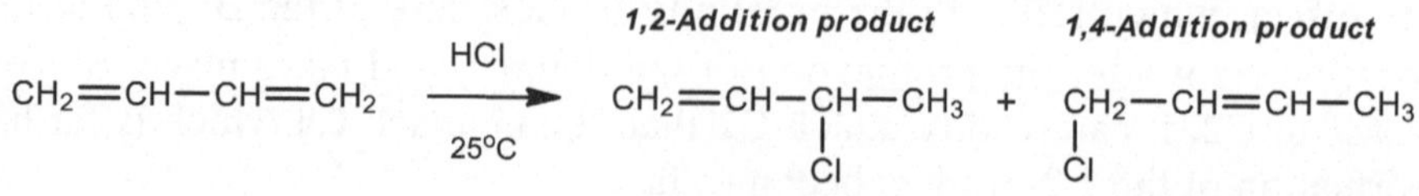

$$CH_2{=}CH{-}CH{=}CH_2 \xrightarrow[25^{\circ}C]{HCl} CH_2{=}CH{-}CH(Cl){-}CH_3 + ClCH_2{-}CH{=}CH{-}CH_3$$

1,2-Addition product: **3-Chloro-1-butene (78%)**

1,4-Addition product: **1-Chloro-2-butene (22%)**

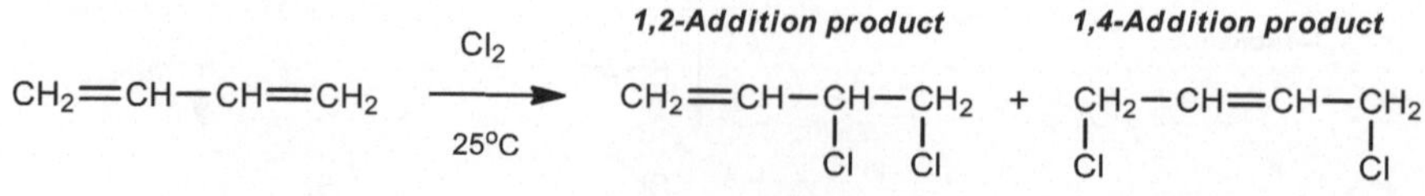

$$CH_2{=}CH{-}CH{=}CH_2 \xrightarrow[25^{\circ}C]{Cl_2} CH_2{=}CH{-}CH(Cl){-}CH_2Cl + ClCH_2{-}CH{=}CH{-}CH_2Cl$$

1,2-Addition product: **3,4-Dichloro-1-butene (60%)**

1,4-Addition product: **1,4-Dichloro-2-butene (40%)**

8.17 CYCLOADDITION REACTIONS OF CONJUGATED DIENES: THE DIELS-ALDER REACTION

Conjugated dienes can undergo addition reactions with **alkenes** to form addition products (or adducts) that are cyclic, called **cycloadducts**. Such reactions in which two reactant molecules add to each other to give a cycloadduct are called **cycloaddition reactions.**

An important and highly popular cycloaddition reaction is the **Diels-Alder reaction,** developed by two German chemists Otto Diels and Kurt Alder. The reaction has been highly versatile and greatly useful in organic synthesis, and they were awarded the Nobel Prize for Chemistry in 1950.

The two reactants in the Diels-Alder reaction are usually a conjugated **diene** and an alkene called a **dienophile** (which means diene loving). The product of the Diels-Alder reaction is a **cycloadduct**, which is a cyclic alkene (cyclohexene).

The simplest Diels-Alder reaction is the cycloaddition of **1,3-butadiene** with **ethylene** to form **cyclohexene**.

Reactants:

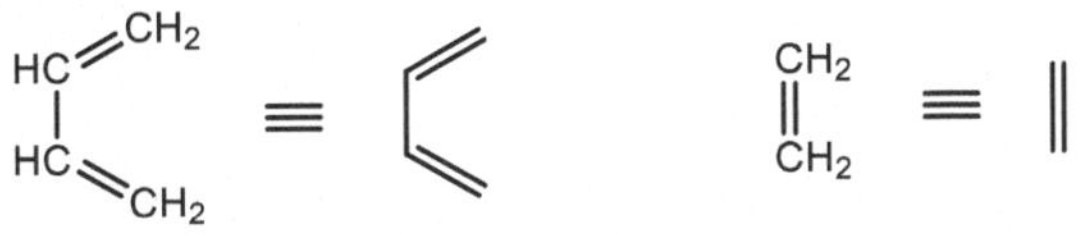

$HC(=CH_2)-HC=CH_2$ ≡ [skeletal structure] $CH_2=CH_2$ ≡ ‖

1,3-Butadiene **Ethylene**

Diels-Alder reaction:

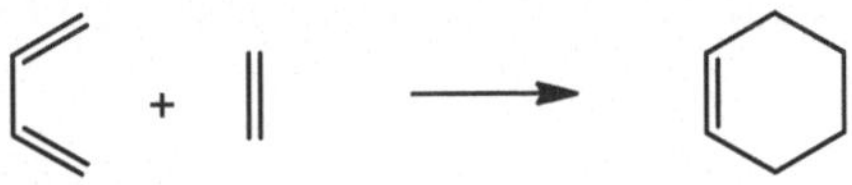

1,3-Butadiene + **Ethylene** ⟶ **Cyclohexene**

Diene ***Dienophile*** ***Cycloadduct***

Pericyclic Mechanism

The Diels-Alder reaction occurs by a pericyclic mechanism (see Section 7.3.3). It is a one-step, concerted process involving cyclic movement of electrons. In the process, three π bonds of the reactants are broken, and two new σ bonds and one new π bond are formed in the product.

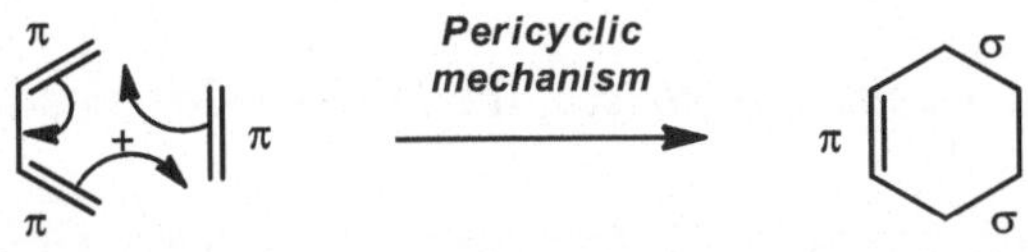

π π π ***Pericyclic mechanism*** ⟶ π σ σ

(three π bonds are broken) **(two new σ bonds and one π bond are formed)**

Reactivity

The Diels-Alder reaction between 1,3-butadiene and ethylene is very slow, and needs a temperature of 165°C and high pressure. However, the Diels-alder reaction becomes faster when electron-withdrawing groups are present in the dienophile.

For example, the Diels-Alder reaction between **1,3-butadiene** and **acrolein** (a dienophile containing an electron-withdrawing aldehyde carbonyl group) occurs under milder conditions at 150°C and normal pressure of 1 atm.

165°C
900 atm

1,3-Butadiene **Ethylene** **Cyclohexene**

Diene ***Dienophile***

O
H
150°C
1 atm
CHO

(CH_2=CH−CHO)

1,3-Butadiene **Acrolein** ***Cycloadduct***

Diene ***Dienophile***

The presence of electron-donating groups in the diene also makes the reaction faster. Thus, the reaction of **2,3-dimethyl-1,3-butadiene** (containing two electron-donating methyl groups) with **acrolein** is much faster and takes place at a much lower temperature of 30°C.

H_3C
H_3C
O
H
30°C
H_3C
H_3C
CHO

2,3-Dimethyl-1,3-butadiene **Acrolein** ***Cycloadduct***

Diene ***Dienophile***

Regioselectivity

With substituted dienes that are *unsymmetrical*, the Diels-Alder reaction is **regioselective** and obeys the **'*ortho-para*'** rule.

1-substituted butadienes give 1,2-disubstituted or **'*ortho*'** cycloadducts, whereas **2-substituted butadienes** give 1,4-disubstituted or **'*para*'** cycloadducts. The following are two examples.

CH_3 1 + O H Δ → CH_3 CHO

1-Methyl-1,3-butadiene **Acrolein** **'*ortho*' Cycloadduct**

1-substituted diene

H_3C 2 + O H Δ → CH_3 CHO

2-Methyl-1,3-butadiene **Acrolein** **'*para*' Cycloadduct**

2-substituted diene

Stereochemistry

In the Diels-Alder reaction, the addition occurs on the same face of the dienophile and also on the same face of the diene. So the Diels-Alder reaction is a **syn addition**. One consequence of this addition is that **the *cis* or *trans* stereochemistry of the dienophile is retained in the product.**

For example, **dimethyl maleate** (a dienophile with ***cis*** ester groups) reacts with 1,3-butadiene to give a ***cis*** **cycloadduct**. Similarly, the reaction of **dimethyl fumarate** (a dienophile with ***trans*** ester groups) with 1,3-butadiene produces a ***trans*** **cycloadduct.** These Diels-Alder reactions are therefore **stereospecific reactions** (cf. Section 8.6.2).

+ H CO_2Me *cis* H CO_2Me → H CO_2Me CO_2Me H

1,3-Butadiene **Dimethyl maleate (*cis* dienophile)** **(two CO_2Me groups are *cis*)** ***cis*** **Cycloadduct**

+ MeO_2C H *trans* H CO_2Me → H CO_2Me CO_2Me H

1,3-Butadiene **Dimethyl fumarate (*trans* dienophile)** **(two CO_2Me groups are *trans*)** ***trans*** **Cycloadduct**

PROBLEM 8.7: The Diels-Alder reaction of 1,3-butadiene with maleic anhydride in benzene occurs at 100°C with a quantitative yield of the product. Predict the structure of the product with stereochemistry.

Answer

Maleic anhydride has a ring structure and is a *cis* dienophile. Therefore, the product (cycloadduct) has two rings and a *cis* stereochemistry at the ring junction.

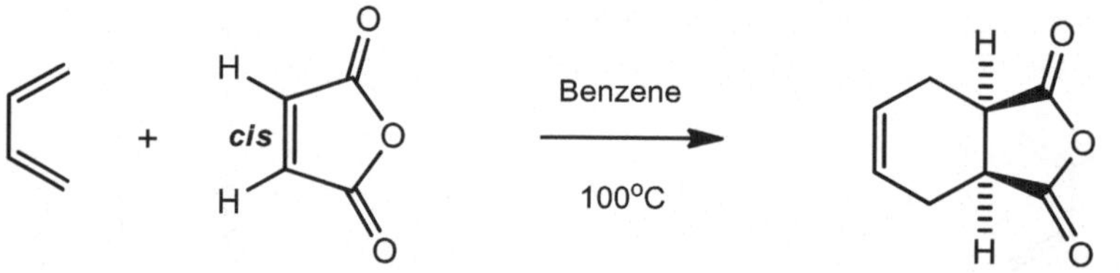

Benzene, 100°C

1,3-Butadiene ***Diene*** + **Maleic anhydride** ***Dienophile*** → **(two H atoms or two C=O groups in the junction of the rings are *cis*)** ***cis* Cycloadduct**

SUPPLEMENTARY PROBLEMS

8.8 Suggest the major product in each of the following reactions:

(a) 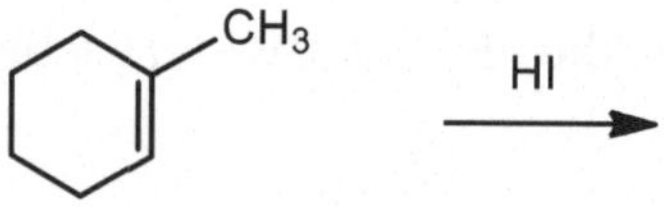 1-methylcyclohexene (CH_3) $\xrightarrow{HI}$

(b) 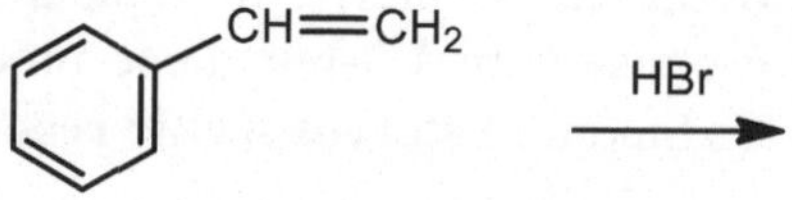 C_6H_5–$CH{=}CH_2$ $\xrightarrow{HBr}$

(c) $H_2C{=}CH{-}\overset{+}{N}(CH_3)_3\ I^-$ $\xrightarrow{HI}$

[*Hints:* (a) Markovnikov addition; (b) Markovnikov addition; (c) anti-Markovnikov addition (compare the relative stabilities of carbocation intermediates)]

8.9 How would you prepare the following alcohols from appropriate alkenes using the methods indicated?

(a) 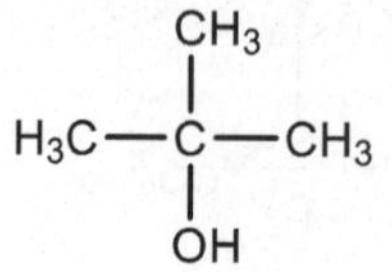 $H_3C{-}C(CH_3)(OH){-}CH_3$ (using oxymercuration-demercuration)

***t*-Butyl alcohol**

(b) 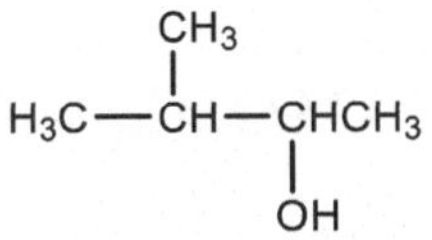 $H_3C—CH(CH_3)—CH(OH)CH_3$ (using hydroboration-oxidation)

3-Methyl-2-butanol

8.10 Give the structures of alkynes that would yield the following ketones on treatment with $HgSO_4$ and dilute H_2SO_4. When more than one alkyne is possible, suggest which of them is a better choice for the synthesis of the given ketone?

(a) 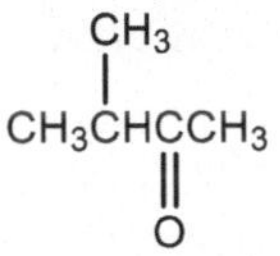 $CH_3CH(CH_3)C(=O)CH_3$

3-Methyl-2-butanone

(b) 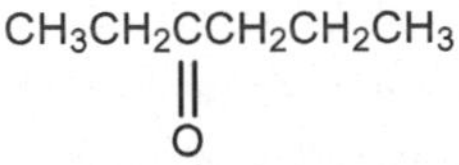 $CH_3CH_2C(=O)CH_2CH_2CH_3$

3-Hexanone

[*Ans:* (a) 3-methyl-1-butyne; (b) two alkynes: 3-hexyne and 2-hexyne; 3-hexyne is a better choice since it gives the given ketone as the only product while 2-hexyne gives the given ketone as one of the two possible products]

8.11 Which diene and dienophile would you use in a Diels-Alder reaction for the synthesis of each of the following compounds? Draw their structures in bond-line formulas.

(a) 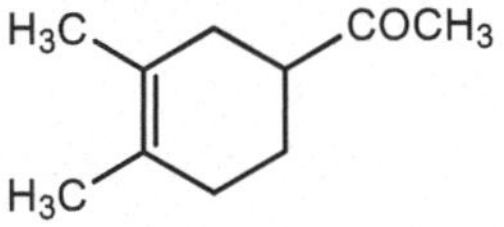 H3C, H3C, COCH3 (substituted cyclohexene)

(b) 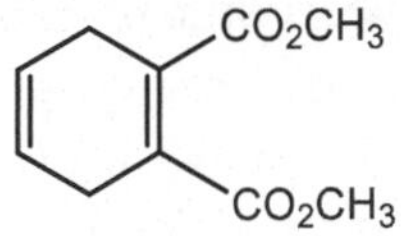 CO2CH3, CO2CH3 (substituted cyclohexadiene)

[*Ans:* The diene and the dienophile in each case can be derived using 'working backward' approach (see also Section 18.10).

(a) 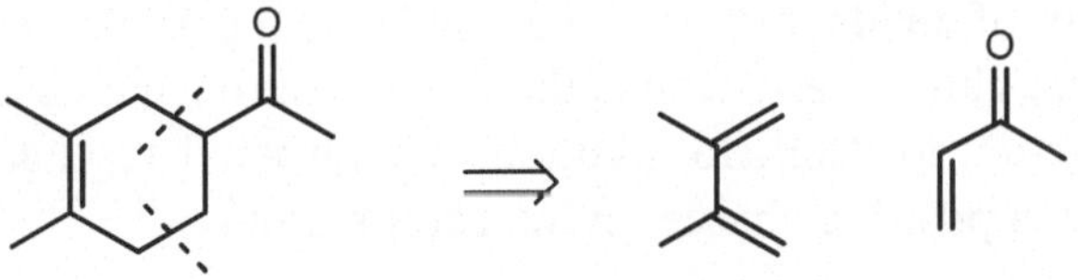 ⟹ **(diene)** **(dienophile)**

(b) 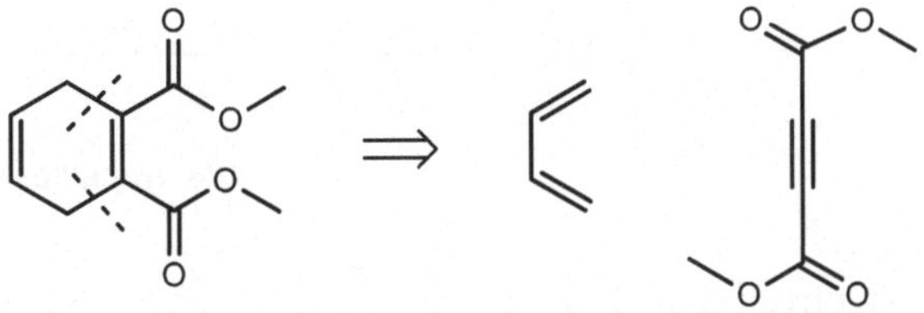 ⟹ **(diene)** **(dienophile)**

(a) diene: 2,3-dimethyl-1,3-butadiene; dienophile: methyl vinyl ketone (a substituted alkene)

(b) diene: 1,3-butadiene; dienophile: dimethyl acetylenedicarboxylate (a substituted alkyne)]

CHAPTER 9

Aldehydes, Ketones and Conjugated Carbonyl Compounds: Addition Reactions

IN THIS CHAPTER

- Nucleophilic addition to carbonyl group (C=O) and relative reactivity of aldehydes and ketones
- Addition of hydrogen cyanide to aldehydes and ketones: Cyanohydrins
- Addition of alcohols and water to aldehydes and ketones: Hemiacetals, acetals and hydrates
- Addition of derivatives of ammonia to aldehydes and ketones: Imines, oximes, hydrazones, semicarbazones and enamines
- Reduction of aldehydes and ketones to alcohols
- Wolff-Kishner reduction and Clemmensen reduction
- Addition of Grignard reagents and organolithium compounds to aldehydes and ketones
- Addition of ylides: The Wittig reaction
- Oxidation of aldehydes and ketones: The Baeyer-Villiger oxidation
- Addition of hydrogen cyanide and Grignard reagents to α,β-unsaturated aldehydes and ketones: Conjugate addition

9.1 NUCLEOPHILIC ADDITION TO THE CARBONYL GROUP (C=O)

The **carbonyl carbon atom** of an aldehyde or a ketone bears a partial positive charge, and hence is **electrophilic**. This arises because the more electronegative oxygen atom attracts the bonding electron pairs to itself. In addition, the carbonyl resonance causes a polarization of the π bond with a positive charge on the carbon atom.

Resonance structures:

$$\overset{\delta+}{C}=\overset{\delta-}{O} \qquad C=O \longleftrightarrow \overset{+}{C}-O^{-}$$

Electrophilic   Electrophilic

Since the carbonyl group is electrophilic, it is susceptible to an attack by a **nucleophile**. Thus a nucleophilic reagent can add to a C=O group of aldehydes and ketones. In these additions, the bond formation to the carbonyl carbon by the nucleophile is the most important change, and they are called **nucleophilic additions**.

The following are nucleophilic additions to aldehydes and ketones in general cases. In the addition of the nucleophilic reagent H–Nu to the C=O group, the nucleophilic part (Nu) is added to the carbonyl carbon and the H atom is added to the oxygen.

$$R{-}CH{=}O \;+\; H{-}Nu \;\longrightarrow\; R{-}CH(Nu){-}OH$$

Aldehyde (substrate) **Nucleophilic reagent** **Addition product**

$$R{-}C(R){=}O \;+\; H{-}Nu \;\longrightarrow\; R{-}C(R)(Nu){-}OH$$

Ketone (substrate) **Nucleophilic reagent** **Addition product**

Mechanism

The **nucleophilic addition** to the carbonyl group involves a **stepwise ionic mechanism** which can proceed in two general ways, as shown in Figure 9.1 and Figure 9.2.

Figure 9.1 shows the mechanism of nucleophilic addition with a *strong* nucleophile (Nu^-), which involves two steps.

In **step 1**, the nucleophile attacks the carbonyl carbon with breaking of the carbonyl π bond. The ability of the electronegative oxygen atom to accommodate a negative charge facilitates the nucleophilic attack at the carbonyl carbon. A new C–Nu bond is formed and the hybridization of carbon changes from sp^2 to sp^3, converting the *trigonal planar* aldehyde or ketone into a *tetrahedral* intermediate.

In **step 2**, the anionic oxygen accepts a proton from the reagent to form the **addition product**.

$$RR'C{=}O \;\xrightarrow[\text{Step 1}]{Nu^-}\; R{-}C(R')(Nu){-}O^- \;\xrightarrow[\text{Step 2}]{H{-}Nu}\; R{-}C(R')(Nu){-}OH \;+\; Nu^-$$

sp^2 — ***Trigonal planar*** — **Aldehyde (R' = H) or ketone**

sp^3 — ***Tetrahedral*** — **Intermediate**

Product

Figure 9.1: Mechanism of nucleophilic addition to aldehydes or ketones with a strong nucleophile.

The mechanism of nucleophilic addition to aldehyde or ketone with a *weak* nucleophile is shown in Figure 9.2 when the nucleophilic attack is facilitated by **acid catalysis**. The acid-catalysed mechanism involves three steps.

In **step 1**, the acid donates a proton to the carbonyl oxygen to form a **protonated carbonyl compound**. As a result, the carbonyl carbon becomes *more electrophilic* (one resonance structure has a full positive charge on carbon).

Step 2 involves the nucleophilic attack by HNu (the neutral nucleophile in acidic medium) to form a C–Nu bond in the *tetrahedral* **intermediate**.

In **step 3**, the loss of a proton gives the **addition product**.

R, R', C=O, H^+, Step 1, Aldehyde (R' = H) or ketone

Protonated carbonyl compound, Nu–H, Step 2

Tetrahedral intermediate, $-H^+$, Step 3, Product

Figure 9.2: Acid-catalyzed mechanism of nucleophilic addition to aldehyders or ketones with a weak nucleophile.

Reversibility

Unlike electrophilic additions to C=C which are *essentially irreversible*, **many nucleophilic additions to C=O are reversible**. Thus, a reversible arrow (⇌) is sometimes used to depict the addition reactions of aldehydes and ketones and their mechanisms.

In reversible additions, the products can be isolated if the equilibrium is favourable, i.e. if it lies to the side of the products. If the equilibrium lies far to the side of the reactants, the nucleophilic addition is unfavourable and for practical purposes, the reaction does not take place.

9.2 RELATIVE REACTIVITY OF ALDEHYDES AND KETONES

Aldehydes are usually **more reactive** than **ketones** in nucleophilic additions. This is attributed to both electronic and steric factors.

Electronic Factor

Alkyl groups are *electron donating* and therefore tend to partially neutralize the positive charge on the carbonyl carbon atom. As a result, the *carbonyl carbon* becomes *less electrophilic* and thus less susceptible to an attack by a nucleophile. **The more electrophilic the carbonyl carbon, the more reactive is the carbonyl compound.**

Aldehydes (except formaldehyde) have *one* electron-donating alkyl group, whereas ketones have *two* electron-donating alkyl groups. Thus the **carbonyl carbon in aldehydes** is more positive or **more electrophilic**, and the **carbonyl carbon in ketones** is less positive or **less electrophilic**.

The relative reactivity is: RCHO (aldehydes) > RR′CO (ketones)

(R, R' = electron-donating alkyl groups)

Aldehyde (carbonyl carbon is more electrophilic)

Ketone (carbonyl carbon is less electrophilic)

Relative reactivity: aldehyde > ketone

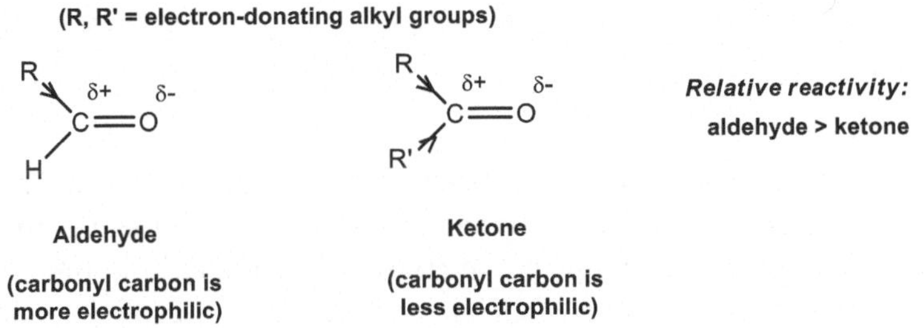

Formaldehyde has *no* electron-donating alkyl groups and hence is the **most reactive**. Thus, the order of reactivity is: **HCHO (formaldehyde) > RCHO (aldehydes) > RR′CO (ketones)**.

In general, an electron-donating substituent causes the carbonyl carbon to be less positive or less electrophilic but an electron-withdrawing substituent causes the carbonyl carbon to be more positive or more electrophilic.

Steric Factor

In the nucleophilic addition, the carbonyl carbon changes its hybridization from sp^2 to sp^3, and its bond angles change from about 120° to about 109°. Thus the substituents at the carbonyl carbon move closer together with **increase in steric crowding** in the tetrahedral intermediate.

For **aldehydes**, one of the substituents at the carbonyl carbon is a small hydrogen atom (for formaldehyde both substituents are hydrogen atoms), but for **ketones**, both substituents at the carbonyl carbon are larger alkyl groups. Therefore, the tetrahedral intermediate is *less crowded* for aldehydes than for ketones, and **nucleophilic addition to aldehydes is more favourable**.

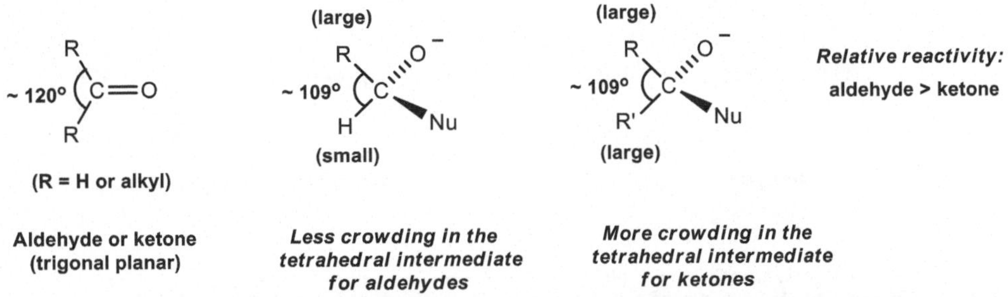

(R = H or alkyl)

Aldehyde or ketone (trigonal planar)

Less crowding in the tetrahedral intermediate for aldehydes

More crowding in the tetrahedral intermediate for ketones

Relative reactivity: aldehyde > ketone

The effect of steric crowding is *extremely small* in the case of **formaldehyde**. Thus, based on steric effects, the order of reactivity is **HCHO > RCHO > RR′CO**, as derived from the electronic factor.

PROBLEM 9.1: Arrange the following aldehydes in order of increasing reactivity toward nucleophilic addition:

C_6H_5-CHO | $H_3CO-C_6H_4-CHO$ | $O_2N-C_6H_4-CHO$

Benzaldehyde | ***p*-Methoxybenzaldehyde** | ***p*-Nitrobenzaldehyde**

Answer

Each aldehyde has an *aryl* substituent. In benzaldehyde, conjugation of the phenyl group with the carbonyl group diminishes the positive charge on the carbonyl carbon. In comparison, *p*-methoxyphenyl group in *p*-methoxybenzaldehyde is much more electron donating because the methoxy group (OCH_3) exerts a strong electron-donating $+M$ effect. In contrast, *p*-nitrophenyl group in *p*-nitrobenzaldehyde is much less electron donating than phenyl group because the nitro group (NO_2) exerts a strong electron-withdrawing $-M$ effect. Consequently, the carbonyl carbon is *least* electrophilic in *p*-methoxybenzaldehyde, and is *most* electrophilic in *p*-nitrobenzaldehyde.

Thus, the reactivity order is: *p*-methoxybenzaldehyde < benzaldehyde < *p*-nitrobenzaldehyde.

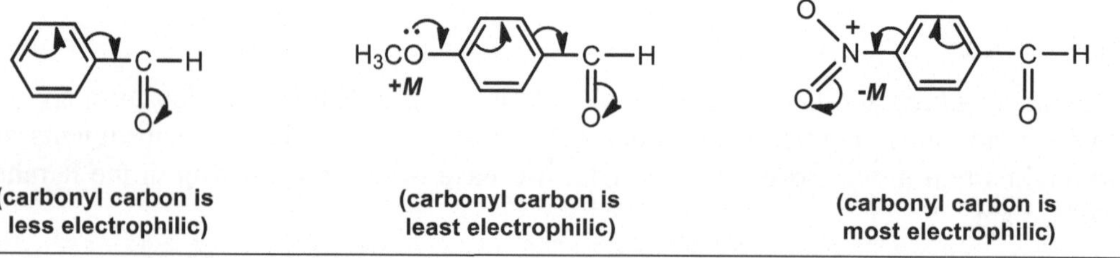

(carbonyl carbon is less electrophilic) | **(carbonyl carbon is least electrophilic)** | **(carbonyl carbon is most electrophilic)**

9.3 ADDITION OF HYDROGEN CYANIDE: CYANOHYDRINS

Addition of hydrogen cyanide (HCN) to **aldehydes and ketones** produces 2-hydroxynitriles, called **cyanohydrins**.

$$R-C(H)=O + HCN \longrightarrow R-C(H)(CN)-OH$$

Aldehyde → **Aldehyde cyanohydrin**

$$R-C(R')=O + HCN \longrightarrow R-C(R')(CN)-OH$$

Ketone → **Ketone cyanohydrin**

HCN is a very toxic gas. The addition of HCN itself also occurs very slowly because it is a poor nucleophile. Thus, an **aqueous sodium or potassium cyanide in an acidic medium** is usually used as a source of HCN. The cyanohydrin reaction is **reversible**.

For example, **acetaldehyde** reacts with sodium cyanide in aqueous sulphuric acid to form **acetaldehyde cyanohydrin** (2-hydroxypropanenitrile).

$$H_3C-CH{=}O + NaCN \xrightleftharpoons[]{H_2O,\ H_2SO_4} H_3C-CH(CN)-OH$$

Acetaldehyde

($CH_3CHOHCN$)

Acetaldehyde cyanohydrin (2-hydroxypropanenitrile)

Note: If the substituents (R, R′) in ketones are quite large, its carbonyl group becomes **sterically hindered**, and such ketones do not form cyanohydrins.

Mechanism

The mechanism of the cyanohydrin reaction involves two steps (cf. Figure 9.1). The cyanide ion (CN^-) is a *strong* **nucleophile** which can attack the carbonyl carbon of an aldehyde or ketone to form an intermediate **oxyanion or alkoxide**. The alkoxide is a strong base that abstracts a proton from solvent (water) or weakly acidic HCN to produce the **cyanohydrin**.

The **mechanism** is depicted below taking acetaldehyde as an example. (The steps are assumed to be *reversible*; however, reversible arrows (⇌) are not shown.)

$$H_3C-CH{=}O \xrightarrow{:\bar{C}\equiv N} H_3C-CH(CN)-O^- \xrightarrow{H-CN} H_3C-CH(CN)-OH + {}^-CN$$

Acetaldehyde — **An alkoxide** — **Acetaldehyde cyanohydrin**

Synthetic Applications

Cyanohydrins are useful intermediates in organic synthesis. A cyanohydrin has two functional groups – hydroxy and nitrile. The **nitrile group** of cyanohydrins can be easily hydrolysed to produce **α-hydroxy acids**.

For example, **lactic acid** (α-hydroxypropionic acid) can be prepared from **acetaldehyde** through a cyanohydrin intermediate.

$$CH_3CHO \xrightarrow{HCN} CH_3CH(OH)CN \xrightarrow[H_2O]{HCl} CH_3CH(OH)CO_2H + NH_4Cl$$

Acetaldehyde — **Acetaldehyde cyanohydrin** — **Lactic acid (α-hydroxypropionic acid)**

PROBLEM 9.2: How would you carry out the following conversions?

(a) Benzaldehyde (C_6H_5CHO) → Mandelic acid ($C_6H_5CH(OH)CO_2H$)

(b) Cyclohexanone → A β-aminoalcohol (1-(aminomethyl)cyclohexanol, OH and CH_2NH_2 on the same carbon)

Answer

(a) Mandelic acid is an α-hydroxy acid. Treatment of benzaldehyde with HCN and subsequent hydrolysis in aqueous acid gives mandelic acid.

$$C_6H_5CHO \xrightarrow{HCN} C_6H_5CH(OH)CN \xrightarrow[H^+]{H_2O} C_6H_5CH(OH)CO_2H$$

(b) The nitrile group of a cyanohydrin can be reduced to an amino group forming a β-aminoalcohol. Therefore, treatment of cyclohexanone with HCN followed by reduction with lithium aluminium hydride ($LiAlH_4$) accomplishes the desired conversion. An aqueous work-up is necessary to isolate the product.

Alternatively, the reduction can also be carried out by catalytic reduction (H_2/Ni).

$$\text{Cyclohexanone} \xrightarrow{HCN} \text{1-hydroxycyclohexane-CN} \xrightarrow[\text{2. } H_2O \text{ (work-up)}]{\text{1. } LiAlH_4} \text{1-hydroxycyclohexane-}CH_2NH_2$$

9.3.1 Addition of Hydrogen Chloride

Hydrogen chloride (HCl) also adds to the carbonyl group, but the addition product cannot be isolated. This is because the reaction is reversible and **the equilibrium lies far to the side of the reactants**. The equilibrium constant (K) is much less than 1.

$$R{-}CH{=}O + HCl \rightleftharpoons R{-}CH(Cl){-}OH \qquad K << 1$$

Aldehyde

Addition product cannot be isolated

This unfavourable equilibrium is attributed to the energies of the bonds broken and formed during the reaction. Calculation from the bond energy data shows that the addition of HCl is **less favourable** by 11 kcal mol^{-1} compared to the addition of HCN.

9.4 ADDITION OF ALCOHOLS: HEMIACETALS AND ACETALS

The addition of **one mole of alcohol** to an aldehyde or a ketone produces a **hemiacetal**. A hemiacetal has a hydroxy group (OH) and an alkoxy group (OR) attached to the same carbon atom. The hemiacetal formation is a **reversible process**.

$$\underset{\textbf{Acetaldehyde}}{H_3C{-}C({=}O){-}H} + \underset{\textbf{Methanol}}{CH_3O{-}H} \rightleftharpoons \underset{\textbf{A hemiacetal}}{H_3C{-}C(OH)(H){-}OCH_3}$$

$$\underset{\textbf{Acetone}}{H_3C{-}C({=}O){-}CH_3} + \underset{\textbf{Methanol}}{CH_3O{-}H} \rightleftharpoons \underset{\textbf{A hemiacetal}}{H_3C{-}C(OH)(CH_3){-}OCH_3}$$

Most *acyclic* hemiacetals are usually too unstable to isolate ($K < 1$). However, *cyclic* hemiacetals are often much more stable. For example, glucose exists primarily as a cyclic hemiacetal (see later Chapter 20).

Formation of Acetals

The addition of **two moles of alcohol** to an aldehyde or a ketone produces an **acetal**. An acetal has two alkoxy (OR) groups attached to the same carbon atom, i.e. it is a *gem*-diether. The acetal formation is also **reversible**.

The initially formed hemiacetal reacts with a second mole of the alcohol to produce an acetal and water.

$$\underset{\textbf{Aldehyde}}{R'{-}C({=}O){-}H} \overset{ROH}{\rightleftharpoons} \underset{\textbf{A hemiacetal}}{\left[R'{-}C(OH)(H){-}OR\right]} \overset{ROH}{\rightleftharpoons} \underset{\substack{\textbf{An acetal}\\ \textbf{(a \textit{gem}-diether)}}}{R'{-}C(OR)(H){-}OR} + H_2O$$

$$R'{-}\overset{\overset{O}{\|}}{C}{-}R'' \underset{}{\overset{ROH}{\rightleftharpoons}} \left[R'{-}\overset{\overset{OH}{|}}{\underset{\underset{R''}{|}}{C}}{-}OR \right] \overset{ROH}{\rightleftharpoons} R'{-}\overset{\overset{\mathbf{OR}}{|}}{\underset{\underset{R''}{|}}{C}}{-}\mathbf{OR} + H_2O$$

Ketone **A hemiacetal** **An acetal (a *gem*-diether)**

The acetal formation is an **acid-catalysed process** and occurs under **anhydrous conditions**, e.g. by passing **dry HCl gas** through a solution of aldehyde or ketone in an *anhydrous* alcohol. The reaction can be driven to completion by the *removal of water* (one of the products) from the reaction mixture as it is formed (La Chatelier principle).

The following are two examples:

$$CH_3CH_2\overset{\overset{O}{\|}}{C}H + 2\ C_2H_5OH \xrightarrow{HCl(g)} CH_3CH_2\overset{\overset{OC_2H_5}{|}}{C}H{-}OC_2H_5 + \underset{\text{(removed)}}{H_2O}$$

Propanal **Ethanol** **1,1-Diethoxypropane**
Acetal

$$CH_3CH_2\overset{\overset{O}{\|}}{C}CH_3 + 2\ C_2H_5OH \xrightarrow{HCl(g)} CH_3CH_2\overset{\overset{OC_2H_5}{|}}{\underset{\underset{CH_3}{|}}{C}}{-}OC_2H_5 + \underset{\text{(removed)}}{H_2O}$$

Butanone **Ethanol** **2,2-Diethoxybutane**
Acetal

A **1,2-diol** (such as ethylene glycol) can be used to prepare acetals that are cyclic. For example, treatment of a **ketone** with **one mole of ethylene glycol** in the presence of an acid catalyst produces a **cyclic acetal.**

Cyclic acetal formation is *highly favoured.*

$$\begin{matrix} H_2C{-}OH \\ | \\ H_2C{-}OH \end{matrix} + O{=}C\begin{matrix} R \\ \\ R' \end{matrix} \xrightarrow{H^+} \begin{matrix} H_2C{-}O \\ | \\ H_2C{-}O \end{matrix} C \begin{matrix} R \\ \\ R' \end{matrix} + H_2O$$

Ethylene glycol **Ketone** **Cyclic acetal**

Hydrolysis of acetals

The reverse process for acetal formation is the **hydrolysis of acetal**, which is also **acid-catalysed**. The hydrolysis of acetals occurs under **aqueous conditions**, e.g. with aqueous HCl.

$$R'-C(OR)(OR)-R'' \xrightarrow[H^+]{H_2O} R'-C(=O)-R'' + 2\,ROH$$

Acetal — Ketone — Alcohol

Hydrolysis of the cyclic acetal with **aqueous acid** regenerates the ketone and ethylene glycol.

Cyclic acetal (2,2-dimethyl-1,3-dioxolane) $\xrightarrow[H^+]{H_2O}$ CH_2OH-CH_2OH (Ethylene glycol) + $O=C(CH_3)_2$ (Acetone)

Cyclic acetal $\xrightarrow[OH^-]{H_2O}$ ✕ *No reaction*

Both acetal formation and hydrolysis of acetal are useful in synthesis.

Note: Acetals or cyclic acetals are *not hydrolysed in basic solutions*. Like ethers, they are unreactive toward bases. Thus acetals can be used as **protecting groups** for aldehydes or ketones from their undesired reactions in basic solutions (see later Section 12.8.1).

PROBLEM 9.3: Draw a mechanism for the formation of an acetal from an aldehyde (RCHO) and an alcohol (R′OH) in the presence of acid catalyst.

Answer

The mechanism involves an acid-catalysed **formation of a hemiacetal intermediate** and then an acid-catalysed **elimination of water** from the reaction of the hemiacetal with a second molecule of alcohol. (All steps are assumed to be reversible; however, reversible arrows (⇌) are not drawn for simplicity)

Aldehyde $R-CH=O \xrightarrow{H^+} R-CH=\overset{+}{O}H \xrightarrow[\text{Alcohol, First molecule}]{HO-R'} R-CH(OH)-\overset{+}{O}(H)-R' \xrightarrow{-H^+} R-CH(OR')-OH$ Hemiacetal

Hemiacetal $\xrightarrow{H^+} R-CH(OR')-\overset{+}{O}H_2 \xrightarrow[\text{Elimination of water}]{-H_2O} R-CH=\overset{+}{O}R' \xrightarrow[\text{Alcohol, Second molecule}]{R'-OH} R-CH(OR')-\overset{+}{O}(H)-R' \xrightarrow{-H^+} R-CH(OR')-OR'$ Acetal

---

9.4.1 Addition of Water: Hydrates

The addition of water to the carbonyl group of an aldehyde or a ketone produces a *gem*-diol, called a **hydrate**. The reaction can be catalysed by either acids or bases.

$$R-C(=O)-R' + H-OH\ (H_2O) \rightleftharpoons R-C(OH)_2-R'$$

Aldehyde (R' = H) or ketone **Hydrate (*gem*-diol)**

The hydration is reversible and the equilibrium is unfavourable for *most* aldehydes and ketones.

However, **formaldehyde** as a very reactive aldehyde exists primarily as its **hydrate** (methanediol) in aqueous solution.

$$H-C(=O)-H + H_2O \rightleftharpoons H-C(OH)_2-H \quad (K = 2 \times 10^3)$$

Formaldehyde $[CH_2(OH)_2]$ **Methanediol (hydrate)**

At room temperature, formaldehyde is a gas. It is generally used as a 37% aqueous solution known as **formalin**, in which it exists almost entirely as the hydrate ($K = 2 \times 10^3$).

9.5 ADDITION OF DERIVATIVES OF AMMONIA

The derivatives of ammonia (NH_3) include **primary amines** (RNH_2), **secondary amines** (R_2NH), **hydroxylamine** (NH_2OH) and **hydrazine** (NH_2NH_2), which can add to aldehydes and ketones.

9.5.1 Imines

Primary amines (RNH_2) add to the carbonyl group (C=O) of aldehydes or ketones to form compounds with a carbon-nitrogen double bond (C=N), called **imines** or **Schiff bases.** Imines are the nitrogen analogues of carbonyl compounds.

Imine formation is acid-catalysed. The initial product of addition is a **hemiaminal**, which is an aminoalcohol containing a hydroxy group (OH) and an amino group (NHR) attached to the same carbon atom.

The hemiaminal is **unstable** and undergoes **dehydration** (loss of water) to give the **imine**.

$$\underset{\text{Aldehyde/ketone}}{-\overset{O}{\overset{\|}{C}}-} + \underset{\text{1° Amine}}{\underset{(RNH_2)}{H-NHR}} \xrightarrow{H^+} \underset{\text{Hemiaminal}}{\left[-\overset{OH}{\overset{|}{C}}-\overset{H}{\overset{|}{N}}R\right]} \longrightarrow \underset{\text{Imine (Schiff base)}}{-C{=}NR} + H_2O$$

A reaction between two molecules that gives a larger molecule, usually with the elimination of a smaller molecule such as water, is called a **condensation reaction**. Thus **imine formation is a condensation reaction**. The reaction is reversible.

The following is an example of condensation of acetaldehyde with methylamine to form an imine:

$$\underset{\text{Acetaldehyde}}{CH_3CH{=}O} + \underset{\text{Methylamine}}{H_2N-CH_3} \overset{H^+}{\rightleftharpoons} \underset{\text{An imine}}{CH_3CH{=}NCH_3} + H_2O \qquad \textit{Condensation reaction}$$

Mechanism

The imine formation involves a **nucleophilic addition–elimination mechanism**. The amine (RNH_2) acts as a nucleophile which attacks the carbonyl group of an aldehyde or ketone to form the hemiaminal intermediate. Then an acid-catalysed elimination of water (dehydration) from the hemiaminal produces the imine, after the loss of a proton from the iminium ion.

$$\underset{\text{Aldehyde or ketone}}{-C{=}O} + \underset{\text{1° Amine}}{H_2\ddot{N}-R} \underset{}{\overset{\textit{Nucleophilic addition}}{\rightleftharpoons}} -\overset{H_2\overset{+}{N}-R}{\overset{|}{C}}-O^- \overset{\text{Transfer of } H^+ \text{ from N to O}}{\rightleftharpoons} \underset{\text{Hemiaminal}}{-\overset{HN-R}{\overset{|}{C}}-OH}$$

$$\underset{\text{Hemiaminal}}{-\overset{HN-R}{\overset{|}{C}}-OH} \overset{H^+}{\rightleftharpoons} -\overset{\ddot{N}HR}{\overset{|}{C}}-\overset{+}{O}H_2 \overset{\textit{Elimination of water}}{\rightleftharpoons} \underset{\text{Iminium ion}}{-\overset{H\overset{+}{N}-R}{\overset{\|}{C}}} + H_2O \overset{-H^+}{\rightleftharpoons} \underset{\text{Imine}}{-C{=}NR}$$

The reaction is most rapid at a pH of 4–5. At very low or very high pH, the imine formation is quite slow. At a low pH (high acidity), the protonation of the amine (a base) occurs to a considerable extent, thereby decreasing the concentration of the free amine which reacts as a nucleophile. On the other hand, at a high pH (low acidity), the dehydration of the hemiaminal is very slow because it requires acid catalysis. A pH of 4–5 is an effective compromise for the reaction to take place.

Stabilization of Imines

Imines are often not very stable and are easily hydrolysed back to aldehydes or ketones and primary amines. However, **imines in which the nitrogen atom bears an electronegative group are more stable than simple imines**. This is because the electronegative substituent (X) has a lone pair of electrons which can stabilize the imine double bond by conjugation or resonance.

$$-\overset{|}{C}=N-\ddot{X} \longleftrightarrow -\overset{-}{\underset{|}{C}}-N=\overset{+}{X}$$

(X= electronegative substituent)

Stabilized by resonance

Examples of such stable imines include oximes (X = OH), hydrazones (X = NH_2), and semicarbazones (X = $NHCONH_2$).

9.5.2 Oximes, Hydrazones and Semicarbazones

Oximes are imines in which the nitrogen atom carries an OH group. The reaction of an aldehyde or a ketone with hydroxylamine (NH_2OH) produces an oxime. Aldehydes give **aldoximes**, whereas ketones give **ketoximes**.

$$-\overset{|}{C}=O + H_2N-OH \longrightarrow -\overset{|}{C}=N-OH + H_2O$$

Aldehyde or ketone **Hydroxylamine** **Oxime (aldoxime or ketoxime)**

Example:

$$C_6H_5-\underset{H}{\underset{|}{C}}=O + H_2NOH \xrightarrow{-H_2O} C_6H_5-\underset{H}{\underset{|}{C}}=NOH$$

Benzaldehyde **Benzaldoxime**

Hydrazine (NH_2NH_2) and its aryl derivatives such as phenylhydrazine and 2,4-dinitrophenylhydrazine react with an aldehyde or a ketone to yield a **hydrazone**, **phenylhydrazone** and **2,4-dinitrophenylhydrazone** respectively.

$$-\overset{|}{C}=O + H_2N-NH_2 \longrightarrow -\overset{|}{C}=N-NH_2 + H_2O$$

Aldehyde or ketone **Hydrazine** **Hydrazone**

$$-\overset{|}{C}=O + H_2N-NHC_6H_5 \longrightarrow -\overset{|}{C}=N-NHC_6H_5 + H_2O$$

Aldehyde or ketone **Phenylhydrazine** **Phenylhydrazone**

$$\text{—C=O} + \text{H}_2\text{NNH—C}_6\text{H}_3(\text{NO}_2)_2 \xrightarrow{-\text{H}_2\text{O}} \text{—C=N—NH—C}_6\text{H}_3(\text{NO}_2)_2$$

Aldehyde or ketone **2,4-Dinitrophenylhydrazine** **2,4-Dinitrophenylhydrazone**

Example:

$$\text{C}_6\text{H}_{10}\text{=O} + \text{H}_2\text{NNHC}_6\text{H}_5 \xrightarrow{-\text{H}_2\text{O}} \text{C}_6\text{H}_{10}\text{=NNHC}_6\text{H}_5$$

Cyclohexanone **Phenylhydrazine** **Cyclohexanone phenylhydrazone**

Semicarbazide ($NH_2NHCONH_2$) is used to prepare a **semicarbazone**. It is the NH_2 group of the hydrazine moiety (NH_2NH) that is involved in the reaction. This is because it is much more nucleophilic than the NH_2 group of amide ($CONH_2$) in which the nitrogen lone pair is involved in resonance with the carbonyl group.

(more nucleophilic) **(less nucleophilic)**

$$\text{H}_2\text{N—NHC(=O)—NH}_2$$

Semicarbazide

$$\text{—C=O} + \text{H}_2\text{N—NHC(=O)NH}_2 \longrightarrow \text{—C=N—NHC(=O)NH}_2 + \text{H}_2\text{O}$$

Aldehyde or ketone **Semicarbazone**

Example:

$$\text{C}_6\text{H}_5\text{C(CH}_3\text{)=O} + \text{H}_2\text{NNHCONH}_2 \xrightarrow{-\text{H}_2\text{O}} \text{C}_6\text{H}_5\text{C(CH}_3\text{)=NNHCONH}_2$$

Acetophenone **Acetophenone semicarbazone**

Reagents as a Hydrochloride Salt of Ammonia Derivatives

The ammonia derivatives are usually employed as a hydrochloride salt. Thus the reagents are **hydroxylamine hydrochloride** ($HONH_3^+ Cl^-$), **hydrazine hydrochloride** ($H_2NNH_3^+ Cl^-$), **phenylhydrazine hydrochloride** ($C_6H_5NHNH_3^+ Cl^-$) and **semicarbazide hydrochloride** ($H_2NCONHNH_3^+ Cl^-$).

The salt is less easily oxidized by air than the free ammonia derivative, and so is best preserved. However, the *free derivative is the actual nucleophile* which is liberated from the salt in the presence of an aldehyde or ketone with a base, usually sodium acetate.

To illustrate, consider the formation of an **oxime** when acetone reacts with hydroxylamine hydrochloride and sodium acetate.

$$H_3C-C(CH_3)=O + NH_2OH\cdot HCl \; (HO\overset{+}{N}H_3\, Cl^-) \xrightarrow[\text{(sodium acetate)}]{CH_3CO_2Na} H_3C-C(CH_3)=NOH$$

Acetone | **Hydroxylamine hydrochloride** | **Acetoxime (acetone oxime)**

The reaction occurs in two steps. Hydroxylamine hydrochloride reacts with sodium acetate to release free hydroxylamine which then reacts with acetone to form the oxime.

$$HO\overset{+}{N}H_3\, Cl^- \xrightarrow{CH_3CO_2^-\, Na^+} HONH_2 + CH_3CO_2H + Na^+Cl^-$$

Hydroxylamine hydrochloride → **Hydroxylamine**

$$HONH_2 \xrightarrow[-H_2O]{(CH_3)_2C=O} (CH_3)_2C=NOH$$

Acetoxime

With other reagents such as hydrazine hydrochloride, phenylhydrazine hydrochloride or semicarbazide hydrochloride, the corresponding reactions occur in the similar fashion.

Note: Oximes, hydrazones, phenylhydrazones, 2,4-dinitrophenylhydrazones and semicarbazones are often stable crystalline solids with characteristic melting points. Oximes, phenylhydrazones and semicarbazones are usually colourless, whereas 2,4-dinitrophenylhydrazones are usually yellow or orange. They are used to characterize aldehydes and ketones.

Qualitative Test for Carbonyl Group in Aldehydes and Ketones

Aldehydes and ketones contain the carbonyl group (C=O). The reaction with **2,4-dinitrophenylhydrazine (2,4-DNP)** is commonly used as a qualitative test for the carbonyl group. 2,4-DNP is generally used as its solution in a mixture of methanol and sulphuric acid, known as **Brady's reagent.**

To perform the test, a solution of the suspected aldehyde or ketone in methanol or ethanol is mixed with the Brady's reagent and shaken well. The formation of a crystalline **yellow or orange precipitate** indicates the presence of a carbonyl group in aldehyde or ketone.

$$R-C(R')=O + H_2NNH-C_6H_3(NO_2)_2 \xrightarrow{-H_2O} R-C(R')=N-NH-C_6H_3(NO_2)_2\downarrow$$

Aldehyde or ketone | **2,4-DNP (Brady's reagent)** | ***Yellow or orange precipitate***

9.5.3 Enamines

Secondary amines react with aldehydes or ketones to give products, called **enamines**. An enamine is an alkene-amine (ene + amine = enamine). The initial addition product cannot eliminate water to form an imine because there is *no hydrogen atom on the nitrogen atom.* Instead, it eliminates water to form an alkene double bond (C=C) through the **loss of α-hydrogen atom** of aldehyde or ketone. The reaction is acid-catalysed.

Aldehyde or ketone having α-H atom + 2° Amine (R_2NH) $\xrightarrow{H^+}$ [Initial addition product] $\xrightarrow{-H_2O}$ Enamine

Enamines are highly useful as synthetic intermediates. Most commonly, *cyclic* secondary amines are used to prepare enamines.

For example, **cyclohexanone** reacts with **pyrrolidine** (a cyclic secondary amine) to give an **enamine**.

Cyclohexanone + Pyrrolidine $\xrightarrow{H^+}$ [ ] $\xrightarrow{-H_2O}$ An enamine

9.6 ADDITION OF HYDRIDE ION: REDUCTION TO ALCOHOLS

A **hydride ion** is a negative charged hydrogen atom with a pair of electrons (H:$^-$) that can act as a nucleophile. Complex metal hydrides are an important source of the hydride ion. Two such metal hydrides commonly used are **sodium borohydride** ($NaBH_4$) and **lithium aluminium hydride** ($LiAlH_4$).

The borohydride ion (BH_4^-) from $NaBH_4$ or aluminium hydride ion (AlH_4^-) from $LiAlH_4$ acts as a σ bond nucleophile (see Section 7.4.1). **Addition of hydride ion** from the metal (B or Al) to the carbonyl group of an aldehyde or a ketone accomplishes **reduction** of the carbonyl group.

Then treatment of the reaction mixture with water (known as aqueous work-up) produces an alcohol as the product.

Aldehyde or ketone + ($NaBH_4$) $\xrightarrow[\text{Reduction}]{\text{Addition of hydride ion}}$ Alkoxide ion $\xrightarrow[\text{(work-up)}]{H_2O}$ Alcohol

$$-\overset{|}{\underset{|}{C}}=O + H-\overset{H}{\underset{H}{\overset{|}{\underset{|}{Al^{-}}}}}-H \xrightarrow[\textit{Reduction}]{\text{Addition of hydride ion}} -\overset{H}{\underset{|}{\overset{|}{C}}}-O^{-} \xrightarrow[\text{(work-up)}]{H_2O} -\overset{H}{\underset{|}{\overset{|}{C}}}-OH$$

Aldehyde or ketone (LiAlH4) Alkoxide ion Alcohol

NaBH$_4$ reductions can be carried out in **water**, but reductions with **LiAlH$_4$** are performed under **anhydrous conditions**, usually in *anhydrous ether* such as **diethyl ether** [$(C_2H_5)_2O$ or Et_2O] or **tetrahydrofuran** (THF), because $LiAlH_4$ reacts violently with water to liberate hydrogen.

$$LiAlH_4 + 4\,H_2O \longrightarrow 4\,H_2 + LiOH + Al(OH)_3$$

Examples

Reduction of **aldehydes** with $NaBH_4$ or $LiAlH_4$ gives **primary alcohols**, whereas **ketones** are reduced to **secondary alcohols**. The following are two examples of $NaBH_4$ and $LiAlH_4$ reductions.

$$CH_3CH_2\overset{O}{\overset{\|}{C}}H \xrightarrow[H_2O]{NaBH_4} CH_3CH_2CH_2OH$$

Propanal → 1-Propanol

Aldehyde → *1° Alcohol*

$$\text{Cyclohexanone} \xrightarrow[2.\ H_2O]{1.\ LiAlH_4/Et_2O} \text{Cyclohexanol}$$

Cyclohexanone (*Ketone*) → Cyclohexanol (*2° Alcohol*)

Note: $NaBH_4$ or $LiAlH_4$ can reduce both aldehydes and ketones. However, $NaBH_4$ is a less powerful reducing agent than $LiAlH_4$. Thus $NaBH_4$ does not reduce carboxylic acids and esters but $LiAlH_4$ does (see later Section 12.8).

Mechanism of Reduction

The mechanism of reduction of aldehydes or ketones with $NaBH_4$ and $LiAlH_4$ is similar. Figure 9.3 shows the mechanism of reduction of a ketone with sodium borohydride ($NaBH_4$).

The reaction proceeds by the transfer of a hydride ion from B to the carbonyl carbon, and this step is **repeated** until all the hydrogen atoms attached to boron is transferred. Then hydrolysis with water produces the alcohol.

Thus, *one mole* of $NaBH_4$ is capable of reducing *four moles* of aldehyde or ketone.

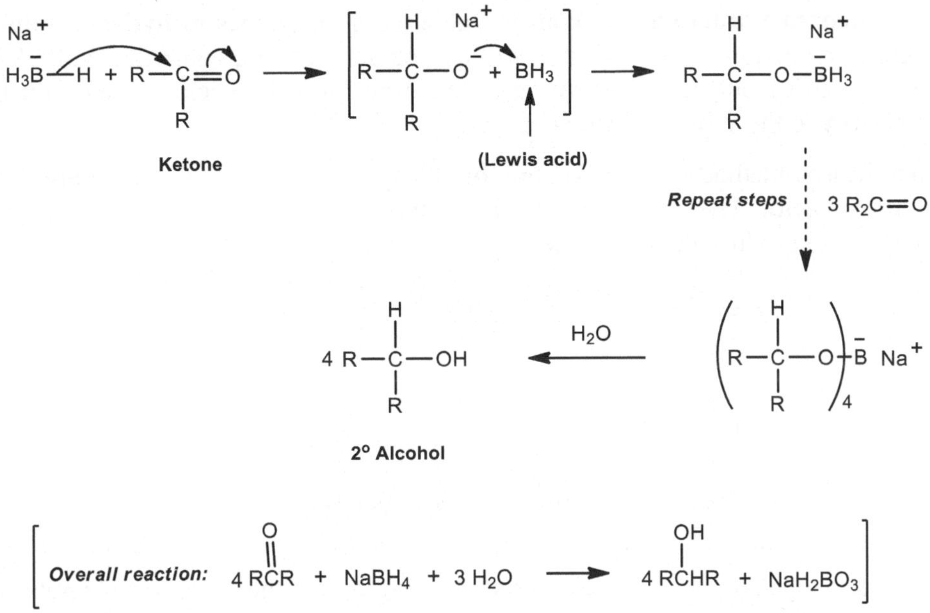

Ketone

(Lewis acid)

Repeat steps $3\ R_2C{=}O$

H_2O

2° Alcohol

Overall reaction: $4\ RCOR + NaBH_4 + 3\ H_2O \longrightarrow 4\ RCH(OH)R + NaH_2BO_3$

Figure 9.3: Mechanism of reduction of a ketone with sodium borohydride.

9.7 WOLFF-KISHNER AND CLEMMENSEN REDUCTIONS

The **Wolff-Kishner reduction** is a useful method for reducing the **carbonyl group (C=O)** of an aldehyde or a ketone to a **methylene (CH_2) group** via a hydrazone. The method involves heating of aldehydes or ketones with **hydrazine** and a strong **base**.

The reaction of the aldehyde or ketone with hydrazine forms a hydrazone, which then eliminates **nitrogen gas** in the presence of base to produce a **hydrocarbon**. The elimination of a stable nitrogen gas provides the driving force for the reaction.

The Wolff-Kishner reduction of ethyl phenyl ketone, for example, gives propylbenzene.

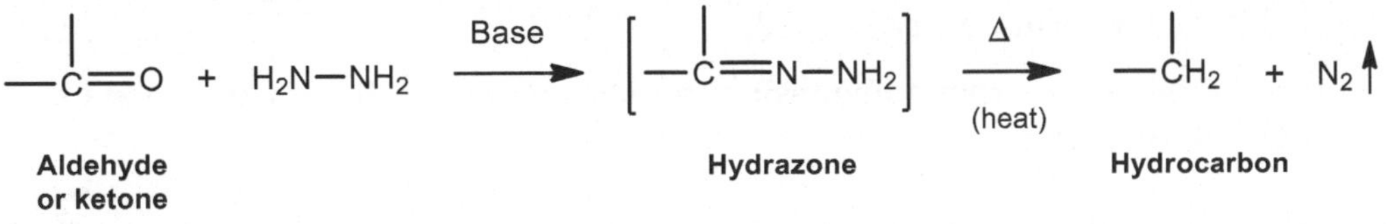

$$>C{=}O + H_2N{-}NH_2 \xrightarrow{\text{Base}} [>C{=}N{-}NH_2] \xrightarrow[\text{(heat)}]{\Delta} >CH_2 + N_2\uparrow$$

Aldehyde or ketone **Hydrazone** **Hydrocarbon**

Example:

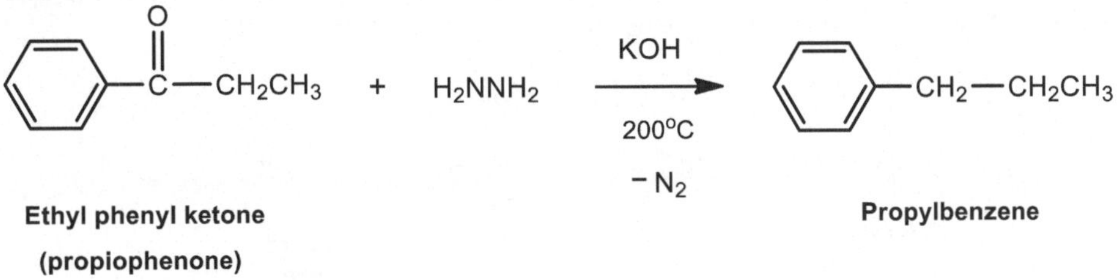

$$C_6H_5{-}CO{-}CH_2CH_3 + H_2NNH_2 \xrightarrow[200^{\circ}C,\ -N_2]{KOH} C_6H_5{-}CH_2{-}CH_2CH_3$$

Ethyl phenyl ketone (propiophenone) **Propylbenzene**

The **Clemmensen reduction** also converts aldehydes or ketones to **hydrocarbons**. The Clemmensen reduction takes place in a strongly acidic medium, while the Wolff-Kishner reduction is carried out in a strongly basic medium. Thus the Clemmensen reduction is *complementary* to the Wolff-Kishner reduction.

The Clemmensen method involves heating of aldehydes or ketones with **zinc amalgam in hydrochloric acid**. The following is an example of the Clemmensen reduction of acetophenone to produce ethylbenzene.

$$C_6H_5-\overset{O}{\overset{\|}{C}}-CH_3 \xrightarrow[\text{HCl}, \Delta]{\text{Zn(Hg)}} C_6H_5-CH_2-CH_3$$

Acetophenone **Ethylbenzene**

9.8 ADDITION OF ORGANOMETALLIC COMPOUNDS

Organometallic compounds contain a **carbon-metal bond (C–M)**. Since the carbon atom is more electronegative than the metal atom (e.g. Mg or Li), the C–M bond is polar with a partial negative charge on the carbon atom and a partial positive charge on the metal atom.

$$-\overset{|}{\underset{|}{C}}^{\delta-}-M^{\delta+} \qquad (M = Mg \text{ or } Li)$$

Thus, organometallic compounds can act as a **carbon nucleophile** in their addition to aldehydes and ketones. Here the carbon nucleophile is a σ bond nucleophile (see Section 7.4.1).

9.8.1 Grignard Reagents: Preparation and Reactions with Water, Alcohols and Terminal Alkynes

An important class of organometallic compounds was discovered by the French chemist Victor Grignard in 1900 which are known as **Grignard reagents**. Grignard was awarded the Nobel Prize for his discovery in 1912.

Grignard reagents are **organomagnesium halides**, written as **RMgX** (R = alkyl or aryl group; X = Cl, Br or I).

Preparation of Grignard Reagents

Grignard reagents are usually prepared by the reaction of magnesium turnings with an alkyl or aryl halide in *anhydrous* ether such as diethyl ether [$(C_2H_5)_2O$ or Et_2O]. The order of reactivity of organic halides (RX) toward Mg is: RI > RBr > RCl.

$$R-X + Mg \xrightarrow[(Et_2O)]{\text{Anhydrous ether}} R-MgX$$

Alkyl or aryl halide **Grignard reagent**

Examples

$$CH_3{-}I + Mg \xrightarrow{Et_2O} CH_3{-}MgI$$

Methyl iodide → **Methylmagnesium iodide**

$$C_6H_5{-}Br + Mg \xrightarrow{Et_2O} C_6H_5{-}MgBr$$

Phenyl bromide (bromobenzene) → **Phenylmagnesium bromide**

Reactions of Grignard Reagents with Water and Alcohols

Grignard reagents are prepared in anhydrous solutions because they react with water; they also react with other protic solvents such as alcohols. This is because Grignard reagents can act not only as a *nucleophile* but also as a *strong base*. As a **strong base**, it can abstract a proton from any compound containing an acidic hydrogen atom (a hydrogen atom attached to an electronegative oxygen or nitrogen) such as water and alcohols.

The reaction of a Grignard reagent with water or an alcohol gives a **hydrocarbon** corresponding to the alkyl or aryl part of the reagent.

$$IMg{-}CH_3 + H{-}OH \longrightarrow CH_4 + OH^- + Mg^{2+} + I^-$$

Methylmagnesium iodide + **Water** → **Methane**

$$BrMg{-}CH_2CH_3 + H{-}OC_2H_5 \longrightarrow CH_3CH_3 + C_2H_5O^- + Mg^{2+} + Br^-$$

Ethylmagnesium bromide + **Ethanol** → **Ethane**

Reactions of Grignard Reagents with Terminal Alkynes

In **terminal alkynes** (–C≡CH), a hydrogen atom attached to the sp-hybridized carbon atom is sufficiently acidic so that it can be abstracted by a Grignard reagent. The reaction of the terminal alkyne with an alkyl Grignard reagent produces a new **acetylenic Grignard reagent**.

$$CH_3C{\equiv}C{-}H + CH_3CH_2CH_2CH_2{-}MgBr \longrightarrow CH_3C{\equiv}C{-}MgBr + CH_3CH_2CH_2CH_3$$

Propyne + **Butylmagnesium bromide** (*Alkyl Grignard reagent*) → **Propynylmagnesium bromide** (*Acetylenic Grignard reagent*) + **Butane**

Reactions with D_2O

An incorporation of deuterium in organic molecules is possible by a **Grignard reaction with D_2O** (heavy water). For example, **benzene** can be converted to **deuterobenzene** (C_6H_5D) by the following sequence of reactions:

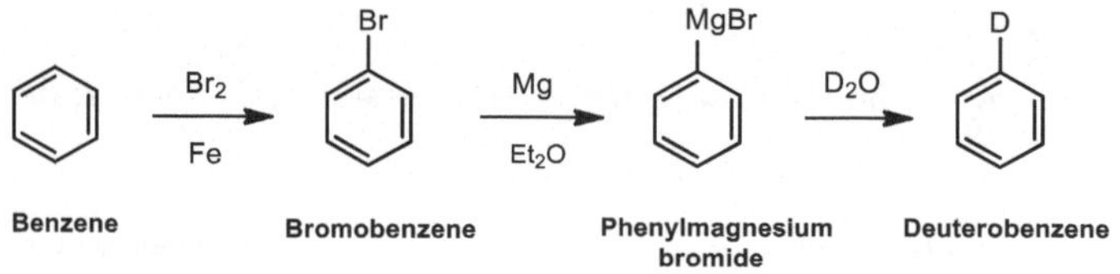

Benzene —(Br_2, Fe)→ Bromobenzene —(Mg, Et_2O)→ Phenylmagnesium bromide —(D_2O)→ Deuterobenzene

9.8.2 Grignard reagents: Reactions with Aldehydes, Ketones and Carbon Dioxide

To carry out Grignard reactions, Grignard reagents are *not* isolated but are used *directly* in the reactions with aldehydes or ketones in anhydrous ether solutions.

In a typical Grignard reaction, an alkyl or aryl halide is slowly added to magnesium metal suspended in dry ether. The Grignard reagent thus produced is utilized immediately by slowly adding to it an ether solution of the aldehyde or ketone. After the reaction is over, the resulting mixture is treated with aqueous acid (such as dilute HCl or dilute H_2SO_4) to isolate the product.

Figure 9.4 shows a general Grignard reaction with aldehydes and ketones. The product is an **alcohol**.

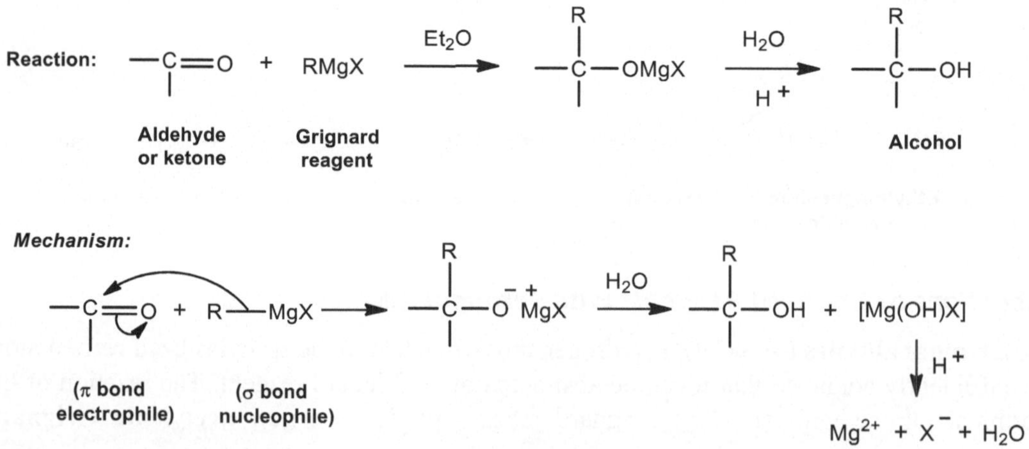

Reaction: $-\overset{|}{\underset{|}{C}}=O$ (Aldehyde or ketone) + RMgX (Grignard reagent) $\xrightarrow{Et_2O}$ $R-\overset{|}{\underset{|}{C}}-OMgX$ $\xrightarrow[H^+]{H_2O}$ $R-\overset{|}{\underset{|}{C}}-OH$ (Alcohol)

Mechanism:

$-\overset{|}{C}=O$ (π bond electrophile) + R—MgX (σ bond nucleophile) ⟶ $R-\overset{|}{\underset{|}{C}}-\overset{-}{O}\ \overset{+}{Mg}X$ $\xrightarrow{H_2O}$ $R-\overset{|}{\underset{|}{C}}-OH$ + [Mg(OH)X] $\xrightarrow{H^+}$ $Mg^{2+} + X^- + H_2O$

Figure 9.4: A general Grignard reaction with aldehydes or ketones and its mechanism.

Reactions with aldehydes

Formaldehyde reacts with Grignard reagents to give **primary alcohols**, whereas **other aldehydes** react with Grignard reagents to produce **secondary alcohols**.

$$H{-}\overset{\overset{H}{|}}{C}{=}O + R{-}MgX \xrightarrow{Et_2O} H{-}\underset{H}{\overset{R}{\overset{|}{\underset{|}{C}}}}{-}OMgX \xrightarrow{H_3O^+} H{-}\underset{H}{\overset{R}{\overset{|}{\underset{|}{C}}}}{-}OH$$

Formaldehyde **Grignard reagent** **1° Alcohol**

Example

$$H_2C{=}O + C_6H_5MgBr \xrightarrow[2.\ H_3O^+]{1.\ Et_2O} C_6H_5CH_2OH$$

(C_6H_5MgBr)

Formaldehyde **Phenylmagnesium bromide** **Benzyl alcohol**

$$R'{-}\underset{H}{\underset{|}{C}}{=}O + R{-}MgX \xrightarrow{Et_2O} R'{-}\underset{H}{\overset{R}{\overset{|}{\underset{|}{C}}}}{-}OMgX \xrightarrow{H_3O^+} R'{-}\underset{H}{\overset{R}{\overset{|}{\underset{|}{C}}}}{-}OH$$

Higher aldehyde **Grignard reagent** **2° Alcohol**

Example

$$CH_3CH{=}O + CH_3CH_2MgBr \xrightarrow[2.\ H_3O^+]{1.\ Et_2O} CH_3\overset{\overset{OH}{|}}{C}HCH_2CH_3$$

Acetaldehyde **Ethylmagnesium bromide** **2-Butanol**

Reactions with ketones

Ketones react with Grignard reagents to produce **tertiary alcohols**. Since tertiary alcohols are susceptible to acid-catalysed *dehydration*, aqueous NH_4Cl solution (instead of aqueous HCl or H_2SO_4) is often used. Aqueous NH_4Cl solution is mildly acidic and gives the alcohol product without causing dehydration.

$$R'{-}\underset{R''}{\underset{|}{C}}{=}O + R{-}MgX \xrightarrow{Et_2O} R'{-}\underset{R''}{\overset{R}{\overset{|}{\underset{|}{C}}}}{-}OMgX \xrightarrow{H_3O^+} R'{-}\underset{R''}{\overset{R}{\overset{|}{\underset{|}{C}}}}{-}OH$$

Ketone **Grignard reagent** **3° Alcohol**

Example

$$CH_3\underset{CH_3}{\underset{|}{C}}{=}O + CH_3MgI \xrightarrow[2.\ H_3O^+]{1.\ Et_2O} CH_3\underset{CH_3}{\overset{CH_3}{\overset{|}{\underset{|}{C}}}}{-}OH$$

Acetone **Methylmagnesium iodide** ***t*-Butanol**

Reactions with carbon dioxide (O=C=O)

Carbon dioxide (CO_2) has C=O groups which can react with a Grignard reagent in a similar manner as the carbonyl group (C=O) of aldehydes and ketones.

In the Grignard reaction with CO_2, one of the C=O groups reacts to form a magnesium carboxylate, which on treatment with aqueous acid yields a **carboxylic acid**.

For example, the reaction of *t*-butylmagnesium chloride (a Grignard reagent) with CO_2 followed by acidification produces pivalic acid (trimethylacetic acid).

$$O{=}C{=}O + R{-}MgX \xrightarrow{Et_2O} R{-}\overset{O}{\overset{\|}{C}}{-}OMgX \xrightarrow{H_3O^+} R{-}\overset{O}{\overset{\|}{C}}{-}OH$$

Carbon dioxide **Grignard reagent** **Carboxylic acid**

Example

$$(CH_3)_3C{-}Cl \xrightarrow[Et_2O]{Mg} (CH_3)_3C{-}MgCl \xrightarrow{CO_2} (CH_3)_3C{-}\overset{O}{\overset{\|}{C}}{-}OMgCl \xrightarrow{H_3O^+} (CH_3)_3C{-}CO_2H$$

t*-Butyl chloride** ***t*-Butylmagnesium chloride** ***Grignard reagent **Pivalic acid (trimethylacetic acid)**

For organic syntheses using Grignard reactions, see Chapter 18.

PROBLEM 9.4: Give the structures of **A–D** in the following scheme:

$$C_6H_5{-}C{\equiv}CH + C_2H_5MgBr \longrightarrow \textbf{A} + \textbf{B}$$

$$\textbf{A} \xrightarrow{CH_3CHO} \textbf{C} \xrightarrow{H_3O^+} \textbf{D}$$

Phenylacetylene

Answer

Phenylacetylene is a terminal alkyne with an acidic hydrogen atom which is abstracted by the Grignard reagent (ethylmagnesium bromide) to form an acetylenic Grignard reagent (**A**) and ethane (**B**). The new Grignard reagent (**A**) adds to the carbonyl group of acetaldehyde to form a bromomagnesium alkoxide (**C**), which on treatment with aqueous acid produces an acetylenic secondary alcohol (**D**).

$$C_6H_5-C\equiv C-H + C_2H_5-MgBr \longrightarrow \underset{\textbf{(A)}}{C_6H_5-C\equiv C-MgBr} + \underset{\textbf{(B)}}{C_2H_6}$$

$$\textbf{(A)} \xrightarrow{CH_3CH=O} \underset{\textbf{(C)}}{C_6H_5-C\equiv C-CH(OMgBr)CH_3} \xrightarrow{H_3O^+} \underset{\textbf{(D)}}{C_6H_5-C\equiv C-CH(OH)CH_3}$$

9.8.3 Organolithium Reagents: Preparation and Reactions with Aldehydes and Ketones

Organolithium reagents contain a *polar* carbon-lithium bond (C–Li), and like Grignard reagents, act as a *carbon nucleophile.* Organolithium reagents are usually **alkyllithiums or aryllithiums** (RLi, R = alkyl or aryl group).

Preparation

Alkyllithiums or aryllithiums can be prepared by reduction of **alkyl or aryl halides** with **lithium metal** in *anhydrous* ether such as diethyl ether (Et_2O) or tetrahydrofuran (THF).

The reaction requires anhydrous conditions because organolithium reagents are strong bases and react with water.

$$\underset{\textbf{Alkyl or aryl halide}}{R-X} + \underset{\textbf{Lithium metal}}{2\,Li} \xrightarrow[\text{(Et}_2\text{O or THF)}]{\textit{Anhydrous}\ \text{ether}} \underset{\textbf{Alkyl- or aryllithium}}{R-Li} + LiX$$

Examples

$$\underset{\textbf{Butyl bromide}}{CH_3CH_2CH_2CH_2Br} + 2\,Li \xrightarrow{Et_2O} \underset{\textbf{Butyllithium}}{CH_3CH_2CH_2CH_2Li} + LiBr$$

$$\underset{\textbf{Phenyl bromide}}{C_6H_5-Br} + 2\,Li \xrightarrow{THF} \underset{\textbf{Phenyllithium}}{C_6H_5-Li} + LiBr$$

Reactions with Aldehydes and Ketones

Organolithium reagents react with aldehydes and ketones to produce alcohols in the same way as Grignard reagents (see Section 9.8.2).

But organolithium reagents are **more reactive** than Grignard reagents because lithium is more electropositive than magnesium and the carbon in the C–Li bond has more negative charge (see Section 9.8).

$$\text{—C=O} + \overset{\delta-}{R}\text{—}\overset{\delta+}{Li} \xrightarrow{Et_2O} \text{—C(R)—}O^- Li^+ \xrightarrow{H_3O^+} \text{—C(R)—OH}$$

Aldehyde or ketone **Alkyl- or aryllithium** **Lithium alkoxide** **Alcohol**

Example

$$\text{Cyclohexanone} + CH_3Li \xrightarrow[2.\ H_3O^+]{1.\ THF} \text{1-Methylcyclohexanol}$$

Cyclohexanone *Ketone* **Methyllithium** **1-Methylcyclohexanol** *3° Alcohol*

9.9 ADDITION OF YLIDES: THE WITTIG REACTION

Ylides are compounds in which a negatively charged carbon atom is *adjacent* to a positively charged heteroatom (nitrogen, phosphorus or sulphur). They are electrically neutral.

Phosphorus ylides are neutral molecules having a negatively charged carbon atom adjacent to a positively charged phosphorus atom.

$$(C_6H_5)_3\overset{+}{P}\text{—}\overset{-}{C}R_2 \equiv Ph_3\overset{+}{P}\text{—}\overset{-}{C}(R)\text{—}R$$

Phosphorus ylide (Ph = C_6H_5)

Phosphorus ylides are stabilized by **resonance**; one resonance structure represents an *ylide* structure while the other resonance structure denotes a *phosphorane* structure. Thus **phosphorus ylides are also called phosphoranes**. They are extremely useful in synthesis.

$$Ph_3\overset{+}{P}\text{—}\overset{-}{C}(R)\text{—}R \longleftrightarrow Ph_3P\text{=}C(R)\text{—}R$$

(ylide structure) **(phosphorane structure)**

Phosphorus ylide or phosphorane

Preparation of Phosphorus Ylides

Phosphorus ylides are usually prepared from **triphenylphosphine** and **alkyl halides**. The reaction of triphenylphosphine (Ph_3P) with an alkyl halide (RCH_2X) forms a **phosphonium**

salt, which on treatment with a *strong base* such as butyllithium ($CH_3CH_2CH_2CH_2Li$ or C_4H_9Li or BuLi) or sodium hydride (NaH) gives the **ylide**.

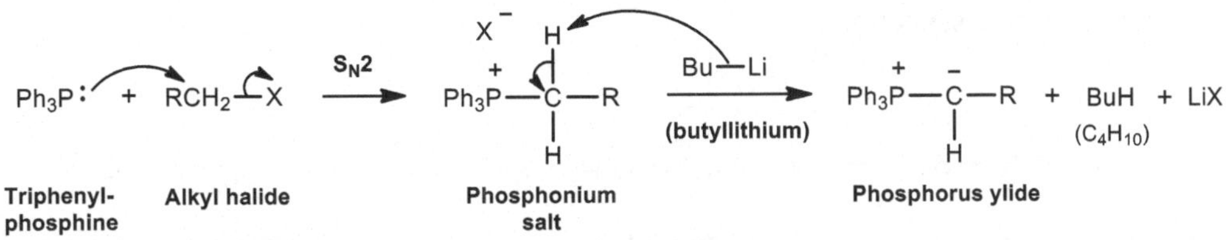

$$Ph_3P: + RCH_2-X \xrightarrow{S_N2} Ph_3\overset{+}{P}-CH_2R\ X^- \xrightarrow[\text{(butyllithium)}]{Bu-Li} Ph_3\overset{+}{P}-\overset{-}{C}HR + BuH\ (C_4H_{10}) + LiX$$

Triphenylphosphine + **Alkyl halide** → **Phosphonium salt** → **Phosphorus ylide**

Triphenylphosphine (Ph_3P) is an excellent phosphorus nucleophile which readily displaces halide ion from the alkyl halide (RCH_2X) in an S_N2 reaction (see later Section 11.3).

The Wittig Reaction

A phosphorus ylide can react with an aldehyde or a ketone to give an **alkene** and a phosphine oxide. This reaction was discovered by Georg Wittig in 1954 and is called the **Wittig reaction**. Wittig was awarded the Nobel Prize for his discovery in 1979.

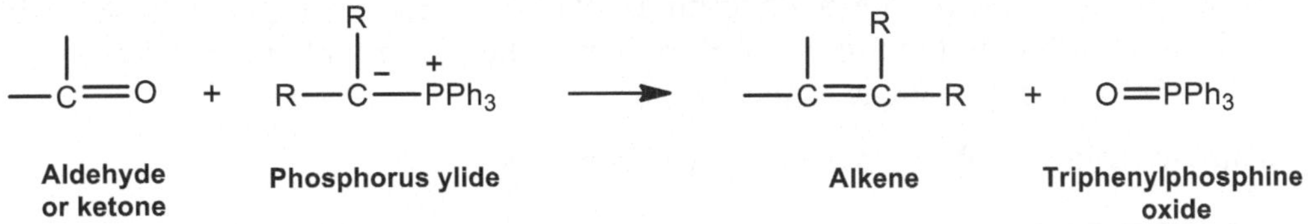

$$>C=O + R_2\overset{-}{C}-\overset{+}{P}Ph_3 \longrightarrow >C=CR_2 + O=PPh_3$$

Aldehyde or ketone + **Phosphorus ylide** → **Alkene** + **Triphenylphosphine oxide**

The important consequence of the Wittig reaction is the **transformation of a C=O bond to a C=C bond.**

Mechanism

The Wittig reaction is thought to take place by a **pericyclic mechanism** rather than an ionic mechanism as described for carbonyl additions so far.

Figure 9.5 shows a mechanism for the Wittig reaction, which involves two pericyclic steps. **Step 1** is a **cycloaddition** of the ylide with the carbonyl group (C=O) of aldehydes or ketones to form a *four-membered ring* intermediate, called an **oxaphosphetane**.

In **step 2**, the *unstable* oxaphosphetane intermediate fragments by a **retro-cycloaddition** (the reverse of a cycloaddition) to produce an **alkene** and a **phosphine oxide**. The driving force for the reaction is the formation of very strong P=O bond in triphenylphosphine oxide.

Aldehyde or ketone + Phosphorus ylide —Cycloaddition (*Step 1*, *Pericyclic*)→ Oxaphosphetane —Retro-cycloaddition (*Step 2*, *Pericyclic*)→ Alkene + Triphenylphosphine oxide ($O=PPh_3$)

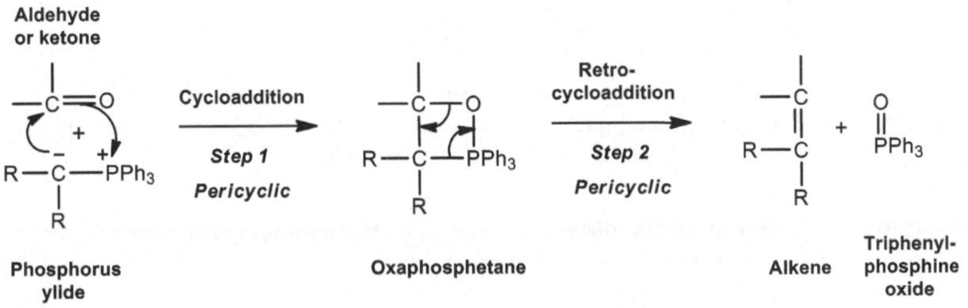

Figure 9.5: A mechanism for the Wittig reaction.

Example

The following is an example of the Wittig reaction starting from **benzaldehyde** and **methyl bromide** to produce **styrene**.

$$\underset{\text{Triphenylphosphine}}{Ph_3P{:}} + \underset{\text{Methyl bromide}}{CH_3{-}Br} \longrightarrow \underset{\text{Phosphonium salt}}{Ph_3\overset{+}{P}{-}CH_2{-}H\ \ Br^-} \xrightarrow{BuLi} \underset{\text{Phosphorus ylide}}{Ph_3\overset{+}{P}{-}\overset{-}{C}H_2}$$

$$\underset{\text{Benzaldehyde}}{C_6H_5{-}CH{=}O} + Ph_3\overset{+}{P}{-}\overset{-}{C}H_2 \xrightarrow{\textit{Wittig}} \underset{\text{Styrene}}{C_6H_5{-}CH{=}CH_2} + Ph_3P{=}O$$

Methyl bromide reacts with triphenylphosphine (Ph_3P) to form a phosphonium salt, which on treatment with butyllithium gives a phosphorus ylide. The ylide is then reacted with benzaldehyde to produce styrene as the alkene.

For syntheses using the Wittig reaction, see Chapter 18.

PROBLEM 9.5: Accomplish the following transformation using the Wittig reaction:

$$\underset{\text{Cyclohexanone}}{C_6H_{10}{=}O} \longrightarrow \underset{\text{Methylenecyclohexane}}{C_6H_{10}{=}CH_2}$$

Answer

The required phosphorus ylide in the Wittig reaction is made from triphenylphosphine [$(C_6H_5)_3P$] and methyl bromide (CH_3Br). Treatment of cyclohexanone with the ylide gives methylenecyclohexane.

$$\underset{\text{Triphenylphosphine}}{(C_6H_5)_3P} + \underset{\text{Methyl bromide}}{CH_3Br} \longrightarrow (C_6H_5)_3\overset{+}{P}{-}CH_3\ \ Br^- \xrightarrow[\text{(base)}]{NaH} \underset{\text{Phosphorus ylide}}{(C_6H_5)_3\overset{+}{P}{-}\overset{-}{C}H_2}$$

$$\underset{\text{Cyclohexanone}}{C_6H_{10}{=}O} + \underset{\text{Phosphorus ylide}}{(C_6H_5)_3\overset{+}{P}{-}\overset{-}{C}H_2} \xrightarrow{\textit{Wittig}} \underset{\text{Methylenecyclohexane}}{C_6H_{10}{=}CH_2} + (C_6H_5)_3P{=}O$$

9.10 OXIDATION OF ALDEHYDES AND KETONES

Aldehydes are easily oxidized to **carboxylic acids** by potassium permanganate ($KMnO_4$) in basic solution or potassium dichromate ($K_2Cr_2O_7$) in acidic medium. But in basic solution, oxidation produces a *carboxylate anion* as a salt of carboxylic acid. The free acid is then liberated by acidification of the mixture.

$$\underset{\textbf{Aldehyde}}{R{-}\overset{\overset{\displaystyle O}{\|}}{C}{-}H} \xrightarrow[KMnO_4/OH^-]{[O]} \underset{\textbf{Carboxylate ion}}{R{-}\overset{\overset{\displaystyle O}{\|}}{C}{-}O^-} \xrightarrow{H_3O^+} \underset{\textbf{Carboxylic acid}}{R{-}\overset{\overset{\displaystyle O}{\|}}{C}{-}OH}$$

$$\underset{\textbf{Aldehyde}}{R{-}\overset{\overset{\displaystyle O}{\|}}{C}{-}H} \xrightarrow[K_2Cr_2O_7/H^+]{[O]} \underset{\textbf{Carboxylic acid}}{R{-}\overset{\overset{\displaystyle O}{\|}}{C}{-}OH}$$

The *hydrogen atom* attached to the carbonyl carbon atom in aldehydes is *lost* in the oxidation reaction. But ketones have *no* hydrogen atom attached to the carbonyl group, so they are more resistant to oxidation.

However, ketones can be oxidized with a **cleavage of a C–C bond** between the carbonyl carbon and the adjacent carbon atom. This can be done by strong oxidizing agents such as acid dichromate or nitric acid.

$$\underset{\textbf{Ketone}}{R{-}\overset{\overset{\displaystyle O}{\|}}{C}{\vdots}CH_2CH_3} \xrightarrow[K_2Cr_2O_7/H^+]{[O]} \underset{\textbf{Carboxylic acid}}{R{-}\overset{\overset{\displaystyle O}{\|}}{C}{-}OH} + \underset{\textbf{Acetic acid}}{HO{-}\overset{\overset{\displaystyle O}{\|}}{C}{-}CH_3}$$

$$\underset{\substack{\textbf{Cyclohexanone}\\ \textbf{(cyclic ketone)}}}{C_6H_{10}O} \xrightarrow[HNO_3]{[O]} \underset{\textbf{Adipic acid}}{HO(O)C{-}(CH_2)_4{-}CO_2H} \equiv HO_2CCH_2CH_2CH_2CH_2CO_2H$$

Tollens' Test to Distinguish between Aldehydes and Ketones

Both aldehydes and ketones possess a carbonyl group (C=O). The test for the C=O group in aldehydes and ketones was described earlier using 2,4-DNP reagent or Brady's reagent (see Section 9.5.2). A qualitative test that can distinguish between aldehydes and ketones is known as the **Tollens' test**.

Tollens' reagent contains the **complex diamminesilver(I) ion** [$Ag(NH_3)_2^+$]. It is a *mild* **oxidizing agent** which can oxidize aldehydes but *not* ketones. The reagent is freshly prepared by adding a drop of sodium hydroxide solution to silver nitrate ($AgNO_3$) solution to give a precipitate of silver(I) oxide (Ag_2O). Dilute ammonia solution is then added to just dissolve the precipitate. The resulting Tollens' reagent is a colourless solution.

The test is performed by adding a small amount of the aldehyde or ketone to the freshly prepared Tollens' reagent and warming it in a hot water bath for a few minutes.

In the oxidation of an aldehyde to a carboxylate ion by the Tollens' reagent, Ag^+ is reduced to *metallic silver* which is deposited on the walls of the test tube as a **silver mirror** (under the proper conditions) or as a **grey to black precipitate**. Thus the appearance of a silver mirror or grey precipitate indicates the presence of an **aldehyde**. If silver mirror or grey precipitate is not formed, the carbonyl compound is a **ketone**. The Tollens' test is sometimes called the silver mirror test.

$$\underset{\text{Aldehyde}}{R-\overset{\overset{O}{\|}}{C}-H} + \underset{\substack{\text{Tollens' reagent}\\ \textit{(colourless solution)}}}{Ag(NH_3)_2^+} \longrightarrow \underset{\text{Carboxylate ion}}{R-\overset{\overset{O}{\|}}{C}-O^-} + \underset{\substack{\textit{Siver mirror or}\\ \textit{grey precipitate}}}{Ag\downarrow}$$

$$\underset{\text{Ketone}}{R-\overset{\overset{O}{\|}}{C}-R'} \xrightarrow[\times]{Ag(NH_3)_2^+} \text{No silver mirror or grey precipitate}$$

9.10.1 The Baeyer-Villiger Oxidation

Ketones (RCOR) can be oxidized to esters (RCO_2R) by peroxy acids ($R'CO_3H$). The reaction was discovered by two German chemists, A. Baeyer and V. Villiger in 1899 and is called the **Baeyer-Villiger oxidation**.

In the Baeyer-Villiger oxidation, an oxygen atom is **inserted** next to the carbonyl group of the ketone.

oxygen atom inserted

$$\underset{\text{Ketone}}{R-\underset{\underset{O}{\|}}{C}-R} \xrightarrow[\text{(peroxy acid)}]{R'CO_3H} \underset{\text{Carboxylic ester}}{R-\underset{\underset{O}{\|}}{C}-O-R} + R'CO_2H$$

For **unsymmetrical ketones** (when the two groups attached to the carbonyl carbon are different), **two ester products are possible**. However, *one* of them is often the **major or exclusive product**.

For example, the Baeyer-Villiger oxidation of **acetophenone** with peroxytrifluoroacetic acid (CF_3CO_3H) produces **phenyl acetate** as the ester product. In this oxidation, an oxygen atom is inserted preferentially between the carbonyl group and the phenyl group.

O inserted

$$C_6H_5{-}C({=}O){-}CH_3 \xrightarrow{CF_3CO_3H} C_6H_5{-}O{-}C({=}O){-}CH_3$$

Acetophenone **Phenyl acetate**

The preference for the insertion of oxygen toward a substituent group of ketones generally follows the order:

3° alkyl > 2° alkyl > phenyl > 1° alkyl > methyl

Therefore, the Baeyer-Villiger oxidation of ***t*-butyl phenyl ketone** with CF_3CO_3H yields ***t*-butyl benzoate** as the major product.

(*t*-butyl > phenyl)

O inserted

$$C_6H_5{-}C({=}O){-}C(CH_3)_3 \xrightarrow{CF_3CO_3H} C_6H_5{-}C({=}O){-}O{-}C(CH_3)_3$$

***t*-Butyl phenyl ketone** ***t*-Butyl benzoate**

The **Baeyer-Villiger oxidation of aldehydes** with a peroxy acid usually gives a **carboxylic acid** (instead of ester) because an oxygen atom is inserted preferentially between the carbonyl group and the hydrogen atom.

For example, treatment of **benzaldehyde** with CF_3CO_3H converts it to **benzoic acid**.

O inserted

$$C_6H_5{-}C({=}O){-}H \xrightarrow{CF_3CO_3H} C_6H_5{-}C({=}O){-}O{-}H$$

Benzaldehyde **Benzoic acid**

For the mechanism of the Baeyer-Villiger oxidation, see later Section 16.4.1.

PROBLEM 9.6: Predict the structure of the product in the following Baeyer-Villiger oxidation:

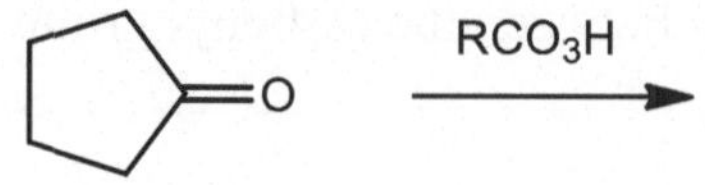

Cyclopentanone $\xrightarrow{RCO_3H}$

Cyclopentanone

Answer

The cyclic ketone (cyclopentanone) is symmetrical, and hence gives a single cyclic ester (called a lactone). Because of the insertion of an oxygen atom in the five-membered ring, the product (lactone) has a six-membered ring.

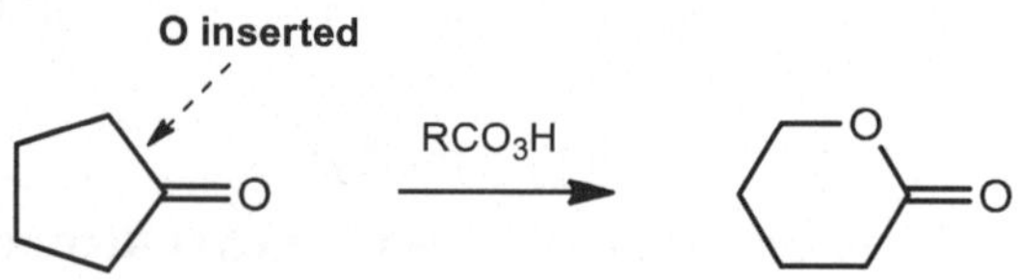

O inserted

RCO_3H

Cyclopentanone

Cyclic ester (lactone)

9.11 ADDITIONS TO CONJUGATED ALDEHYDES AND KETONES

A conjugated aldehyde or ketone is an **α,β-unsaturated aldehyde or ketone** in which the double bond (C=C) is between the α-carbon (adjacent to carbonyl carbon) and the β-carbon (next to α-carbon).

The resonance structures for an α,β-unsaturated aldehyde or ketone show that there are **two electrophilic sites (carbonyl carbon and β-carbon)** that can be attacked by a nucleophile.

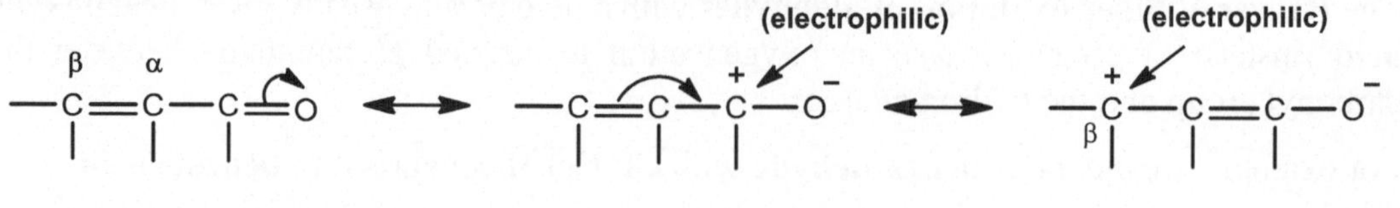

(electrophilic) (electrophilic)

β α

—C=C—C=O ⟷ —C=C—C⁺—O⁻ ⟷ —C⁺—C=C—O⁻

β

α,β-Unsaturated aldehyde or ketone

Three resonance strutures

The **nucleophilic addition** to an α,β-unsaturated aldehyde or ketone is shown in Figure 9.6. There are *two* possible pathways: 1,2-addition and 1,4-addition, similar to those for conjugated dienes (see Section 8.16).

1,2-Addition is the nucleophilic addition to the carbonyl group (similar to that for simple aldehydes and ketones). It is also called a **direct addition**.

1,4-Addition involves the nucleophilic attack at the β-carbon leading to an **enol** form. But the enol form rapidly tautomerizes to a **keto** form. Thus, **the net result of 1,4-addition** to an α,β-unsaturated aldehyde or ketone is the **addition to the carbon-carbon double bond (C=C)**. The 1,4-addition is frequently called the **conjugate addition**.

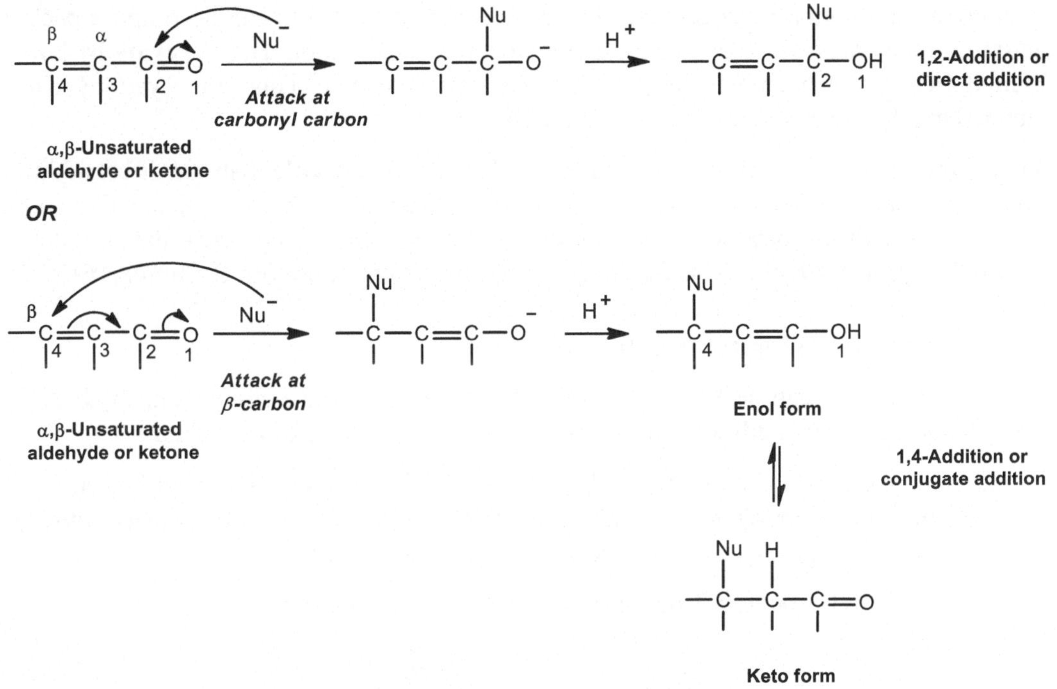

Figure 9.6: Nucleophilic addition to α,β-unsaturated aldehyde or ketone: 1,2-Addition (direct addition) and 1,4-addition (conjugate addition). The conjugate addition leads to the net addition of the nucleophilic reagent (HNu) to the C=C of the conjugated aldehyde or ketone.

9.11.1 Addition of Hydrogen Cyanide

Consider, for example, the addition of HCN to **methyl vinyl ketone** which is an α,β-unsaturated ketone.

At a **low temperature** (5–10°C), the reaction gives a **cyanohydrin** as a result of 1,2- or direct addition, but at a **higher temperature** (80°C), the reaction gives the **product of conjugate addition**. It should be noted that conjugate addition finally leads to a product for addition to carbon-carbon double bond.

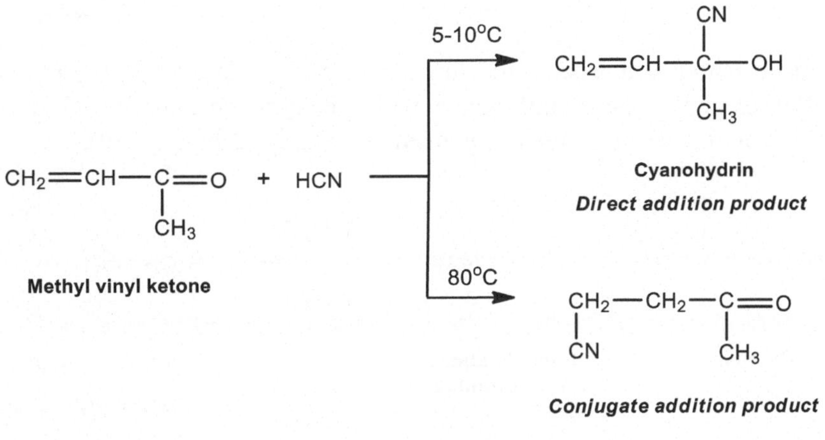

Cyanohydrin formation is reversible. Therefore, some cyanohydrin can decompose back to the starting materials which can then undergo a conjugate addition. But the **cyanohydrin** is **formed faster** and becomes the product of **kinetic control**. Thus, the cyanohydrin is obtained at a low temperature (cf. Section 8.16).

On the other hand, the **conjugate addition product** is **more stable** than the direct addition product because a C=O bond in the conjugate addition product is stronger than a C=C bond in the direct addition product. The conjugate addition product becomes the product of **thermodynamic control**, and is favoured at a higher temperature (cf. Section 8.16).

9.11.2 Addition of Organometallic Reagents

Grignard reagents can add to α,β-unsaturated aldehydes and ketones, and both direct addition and conjugate addition can occur resulting in a mixture of products.

For example, the **Grignard reaction** of 3-penten-2-one (an **α,β-unsaturated ketone**) with methylmagnesium bromide gives a mixture of 2-methyl-3-penten-2-ol (**direct addition product**) and 4-methyl-2-pentan**one** (**conjugate addition product**).

However, 2-methyl-3-penten-2-ol is formed as the **major product**.

$$CH_3CH{=}CH{-}\overset{\overset{\displaystyle O}{\|}}{C}{-}CH_3 + CH_3{-}MgBr \xrightarrow[2.\ H_3O^+]{1.\ Et_2O}$$

3-Penten-2-one; Methylmagnesium bromide

Addition to C=O (*Direct addition*):

$$CH_3CH{=}CH{-}\underset{OH}{\overset{CH_3}{C}}{-}CH_3$$

2-Methyl-3-penten-2-ol
Major product

Net addition to C=C (*Conjugate addition*):

$$CH_3\underset{CH_3}{CH}{-}CH_2{-}\overset{O}{\overset{\|}{C}}{-}CH_3$$

4-Methyl-2-pentanone
Minor product

With **α,β-unsaturated aldehydes**, the direct addition of Grignard reagents is generally favoured. For example, **2-butenal** reacts with phenylmagnesium bromide to produce almost exclusively the **direct addition product** (1-phenyl-2-buten-1-ol).

$$CH_3CH{=}CH{-}\overset{O}{\overset{\|}{C}}{-}H + C_6H_5{-}MgBr \xrightarrow[2.\ H_3O^+]{1.\ Et_2O} CH_3CH{=}CH{-}\underset{H}{\overset{OH}{C}}{-}C_6H_5$$

2-Butenal; Phenylmagnesium bromide; **1-Phenyl-2-buten-1-ol** ***Direct addition product***

Choice of Organometallic Reagent

Depending on the choice of an organometallic reagent, a particular mode of addition can be favoured.

To illustrate, consider an addition to **1,3-diphenyl-2-propen-1-one** (an α,β-unsaturated ketone) in which the carbonyl group is somewhat less reactive by electronic and steric factors. Since Grignard reagents are *less reactive* than organolithium reagents, the Grignard reaction with **phenylmagnesium bromide** leads to **conjugate addition**, whereas the reaction of the same compound with **phenyllithium** favours **direct addition**.

$$C_6H_5CH{=}CH{-}C({=}O){-}C_6H_5 + C_6H_5{-}MgBr \xrightarrow[\text{Conjugate addition}]{1.\ Et_2O\quad 2.\ H_3O^+} C_6H_5CH(C_6H_5){-}CH_2{-}C({=}O){-}C_6H_5$$

1,3-Diphenyl-2-propen-1-one **Phenylmagnesium bromide** **1,3,3-Triphenyl-1-propanone**

$$C_6H_5CH{=}CH{-}C({=}O){-}C_6H_5 + C_6H_5{-}Li \xrightarrow[\text{Direct addition}]{1.\ THF\quad 2.\ H_3O^+} C_6H_5CH{=}CH{-}C(OH)(C_6H_5){-}C_6H_5$$

1,3-Diphenyl-2-propen-1-one **Phenyllithium** **1,1,3-Triphenyl-2-propen-1-ol**

PROBLEM 9.7: Predict the product of the following reaction:

3-Methyl-2-cyclohexenone $\xrightarrow[2.\ H_3O^+]{1.\ CH_3MgI/CuCl,\ THF}$?

3-Methyl-2-cyclohexenone

Answer

The use of a catalytic quantity of cuprous salt (CuCl) in the Grignard reaction promotes conjugate addition. Therefore, the reaction proceeds through conjugate addition to give the product (3,3-dimethylcyclohexanone).

It is pertinent to mention here that organocopper reagents such as alkylcopper (RCu) and lithium dialkylcuprate (R_2CuLi) also undergo conjugate addition to α,β-unsaturated ketones.

3-Methyl-2-cyclohexenone $\xrightarrow[2.\ H_3O^+]{1.\ CH_3MgI/CuCl}$ 3,3-Dimethylcyclohexanone

Conjugate addition

3-Methyl-2-cyclohexenone **3,3-Dimethylcyclohexanone**

SUPPLEMENTARY PROBLEMS

9.8 Explain the following:

(a) When acetone (CH_3COCH_3) is dissolved in $H_2{}^{18}O$ (water containing ^{18}O isotope) instead of ordinary H_2O, the recovered acetone from the solution contains acetone containing ^{18}O ($CH_3C^{18}OCH_3$).

(b) The hydrate of trichloroacetaldehyde (CCl_3CHO) is stable and isolated as a colourless solid. The hydrate is known as chloral hydrate which is a hypnotic drug.

[*Hints:* (a) hydration of acetone is reversible; (b) repulsion between adjacent positive charges in trichloroacetaldehyde destabilizes the molecule]

9.9 The reaction of benzaldehyde with 2,4-dinitrophenylhydrazine gives an orange precipitate. Write the chemical equation for the reaction.

9.10 Outline two approaches to the synthesis of the following alkene using the Wittig reaction:

$C_6H_5-CH{=}C(CH_3)-CH_3$

2-Methyl-1-phenyl-1-propene

[*Hints:* work backward to derive the possible ylides and aldehyde or ketone]

(ylide) (carbonyl)

$C_6H_5-CH{=}C(CH_3)CH_3 \Rightarrow C_6H_5-\overset{-}{C}H-\overset{+}{P}Ph_3 + O{=}C(CH_3)CH_3$

(carbonyl) (ylide)

$C_6H_5-CH{=}C(CH_3)CH_3 \Rightarrow C_6H_5-CH{=}O + Ph_3\overset{+}{P}-\overset{-}{C}(CH_3)CH_3$

9.11 Suggest the major product that would result from the Baeyer-Villiger oxidation of 3-methyl-2-butanone.

[*Ans:* isopropyl acetate]

9.12 How would you accomplish the following transformations?

(a) $C_6H_5-Br \longrightarrow C_6H_5-D$

(b) $C_6H_5-Br \longrightarrow C_6H_5-CO_2H$

9.13 Predict the product in each of the following reactions. Draw structures of products in bond-line formulas.

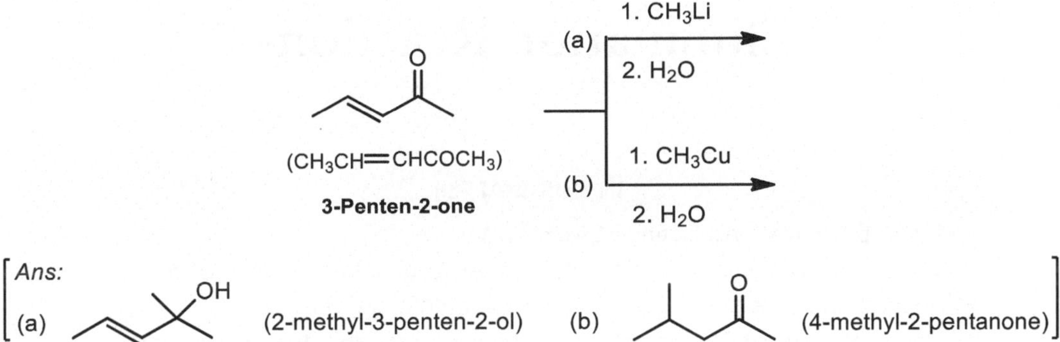

CHAPTER 10

Alkyl Halides and Alcohols: Elimination Reactions

IN THIS CHAPTER

- 1,2- or β-Eliminations: E2 and E1 Mechanisms
- Dehydrohalogenation of alkyl halides and dehalogenation of dihaloalkanes
- Orientation and regioselectivity: Saytzeff and Hofmann eliminations
- Stereochemistry of E2 eliminations: Stereoselectivity and stereospecificity
- Dehydration of alcohols: The E1 Mechanism
- Pyrolytic syn eliminations
- Formation of alkynes by 1,2-eliminations
- Dehydrogenation of alcohols and dehydration of amides

10.1 1,2- OR β-ELIMINATION

Elimination reactions are the reverse of addition reactions. The most common type of elimination is the **1,2-elimination or β-elimination**.

A 1,2-elimination involves the loss of atoms or groups from two *adjacent* atoms of the reactant molecule, with consequent formation of a *double bond* between the two atoms. The **relative positions** of the two adjacent atoms (usually carbons) are numbered **1, 2**. The product of the 1,2-elimination is an **alkene**.

$$\underset{\text{(X, Y = atoms or groups to be lost)}}{-\overset{|}{\underset{\substack{|\\Y}}{\overset{2}{C}}}-\overset{|}{\underset{\substack{|\\X}}{\overset{1}{C}}}-} \xrightarrow[\text{- XY}]{\text{1,2-Elimination}} \underset{\text{Alkene}}{-\overset{|}{C}=\overset{|}{C}-}$$

More often, a **hydrogen atom** (H) is *one* of the two groups eliminated. The other group that is lost is called a **leaving group** (L). The carbon atom that bears the leaving group is usually designated as the **alpha (α) carbon atom** and *any* carbon atom adjacent to it is called a **beta (β) carbon atom**. A hydrogen atom attached to the β carbon is referred to as a **β hydrogen atom**. In the 1,2-elimination, a β hydrogen atom is lost, and the 1,2-elimination is also termed a **β-elimination**.

(β hydrogen atom)

$$\overset{\beta}{\underset{2}{-\overset{|}{\underset{|}{C}}H}}-\overset{\alpha}{\underset{1}{\overset{|}{\underset{|}{C}}}}- \xrightarrow[-\ HL]{1,2\text{-or }\beta\text{-Elimination}} -\overset{|}{C}=\underset{|}{C}-$$

(L = leaving group) Alkene

10.2 DEHYDROHALOGENATION AND DEHALOGENATION

Dehydrohalogenation and dehalogenation are two common 1,2-elimination reactions. Dehydrohalogenation involves the elimination of a molecule of hydrogen halide (HCl, HBr or HI), while dehalogenation involves the elimination of a molecule of halogen (Cl_2, Br_2 or I_2).

10.2.1 Dehydrohalogenation

The **elimination of hydrogen halide** (HX, X = Cl, Br, I) from **alkyl halides** is called the **dehydrohalogenation**, which is *opposite* to hydrohalogenation (addition of hydrogen halide to alkenes).

Dehydrohalogenation reactions are usually carried out by heating an alkyl halide with a **strong base**, and are widely used for synthesizing alkenes.

Various bases have been employed for dehydrohalogenation reactions. Alcoholic potassium hydroxide (KOH dissolved in ethanol) has been used, but sodium or potassium alkoxides are better. A common alkoxide base is **sodium ethoxide** (C_2H_5ONa), which can be prepared by treating an excess of ethanol with sodium metal.

$$2\ C_2H_5OH \quad + \quad 2\ Na \longrightarrow 2\ C_2H_5ONa \quad + \quad H_2$$

Ethanol   Sodium ethoxide

With sodium ethoxide, ethanol is used as a *solvent* for the dehydrohalogenation reaction. For other alkoxides, the corresponding solvents are used. For example, with **potassium *t*-butoxide**, *t*-butanol is used as the solvent.

Dehydrohalogenation of secondary and tertiary alkyl halides

Secondary (2°) and tertiary (3°) alkyl halides undergo easy dehydrohalogenation with sodium ethoxide in ethanol to produce **alkenes.**

Note that under basic conditions, NaBr as a salt (not free HBr) is liberated in these reactions.

$$CH_3-\underset{Br}{\underset{|}{CH}}-\underset{H}{\underset{|}{CH_2}} \xrightarrow[C_2H_5OH,\ 55^oC]{C_2H_5ONa} CH_3-CH=CH_2 \quad + \quad NaBr$$

2-Bromopropane
2° Alkyl bromide

Propene

$$CH_3-\underset{Br}{\overset{CH_3}{C}}-\underset{H}{CH_2} \xrightarrow[C_2H_5OH,\ 25^oC]{C_2H_5ONa} CH_3-\overset{CH_3}{C}=CH_2 + NaBr$$

2-Bromo-2-methylpropane — ***3° Alkyl bromide*** → **2-Methylpropene**

Dehydrohalogenation of primary alkyl halides

If the substrate is a **primary (1°) alkyl halide**, the elimination is *not* favoured with sodium ethoxide as a base. For example, when ethyl bromide is treated with sodium ethoxide in ethanol, the **elimination product** (ethylene) is a **minor** product. The **major** product is diethyl ether which is formed by a substitution reaction (see later Section 11.6).

$$CH_3-CH_2-Br + CH_3CH_2O^-Na^+ \xrightarrow[55^oC,\ (-NaBr)]{C_2H_5OH} CH_2=CH_2 + CH_3CH_2OCH_2CH_3$$

Ethyl bromide (***1° Alkyl bromide***) + **Sodium ethoxide** (C_2H_5ONa) → **Ethylene** (***Minor product***, **(10%)**) + **Diethyl ether** (***Major product***, **(90%)**)

The elimination reaction of a primary alkyl halide becomes favourable when a **strong and bulky base** such as **potassium *t*-butoxide** [$(CH_3)_3COK$] is used. For example, when **propyl bromide** (1-bromopropane) is treated with potassium *t*-butoxide in *t*-butanol, the major product is **propene**.

$$CH_3-\underset{H}{CH}-\underset{Br}{CH_2} + (CH_3)_3CO^-K^+ \xrightarrow[50^oC,\ (-NaBr)]{(CH_3)_3COH} CH_3-CH=CH_2$$

Propyl bromide (1-bromopropane) (***1° Alkyl bromide***) + **Potassium *t*-butoxide** → **Propene** (***Major product***)

This is because the approach of the bulky *t*-butoxide ion (with three methyl groups) to the *internal* carbon atom of the primary halide for a substitution reaction is **sterically hindered**, but the *t*-butoxide ion can easily abstract an exposed *peripheral* hydrogen atom from the alkyl halide molecule for the **elimination to take place** (see also Section 11.6).

The bulky *t*-butoxide ion is generally referred to as a **sterically hindered base**, while the ethoxide ion is a relatively **unhindered base**.

10.2.2 Dehalogenation

The **elimination of halogen** (X_2, X = Cl, Br, I) from **1,2-dihaloalkanes** (or ***vic*-dihalides**) is called the **dehalogenation**. This is the reverse of halogenation (addition of halogen to alkenes).

The following is a general case of **debromination** of a 1,2-dibromoalkane which occurs in the presence of **iodide ion** or **zinc metal**.

$$\underset{\substack{\textbf{1,2-Dibromoalkane}\\(\textbf{\textit{vic}-dibromide})}}{-\overset{|}{\underset{\underset{Br}{|}}{C}}-\overset{|}{\underset{\underset{Br}{|}}{C}}-} \xrightarrow{I^-} \underset{\textbf{Alkene}}{-\overset{|}{C}=\overset{|}{C}-} + [IBr] + Br^-$$

$$[IBr] \xrightarrow{I^-} I_2 + Br^-$$

$$\underset{\substack{\textbf{1,2-Dibromoalkane}\\(\textbf{\textit{vic}-dibromide})}}{-\overset{|}{\underset{\underset{Br}{|}}{C}}-\overset{|}{\underset{\underset{Br}{|}}{C}}-} \xrightarrow{Zn} \underset{\textbf{Alkene}}{-\overset{|}{C}=\overset{|}{C}-} + ZnBr_2$$

The debromination reaction is carried out with **potassium iodide** (KI) in acetone or **zinc dust** in acetic acid (or ethanol). No abstraction of hydrogen atom is involved in the reaction, and the reagents are relatively nonbasic. Two examples are the following:

$$\underset{\textbf{2,3-Dibromobutane}}{CH_3\underset{\underset{Br}{|}}{C}H-\underset{\underset{Br}{|}}{C}HCH_3} \xrightarrow[\text{Acetone}]{\text{KI}} \underset{\textbf{2-Butene}}{CH_3CH=CHCH_3}$$

$$\underset{\textbf{1,2-Dibromobutane}}{CH_3CH_2\underset{\underset{Br}{|}}{C}H-CH_2Br} \xrightarrow[\text{Ethanol}]{\text{Zn}} \underset{\textbf{1-Butene}}{CH_3CH_2CH=CH_2}$$

PROBLEM 10.1: Predict the product in each of the following reactions:

(a) 1,2-Dibromocyclohexane $\xrightarrow[(CH_3)_2CHOH,\ 110^{\circ}C]{(CH_3)_2CHO^-K^+}$

(b) 1,2-Dibromocyclohexane $\xrightarrow[\text{Acetone}]{\text{NaI}}$

(c) $H_2C{=}C(Cl)CH_2Cl$ $\xrightarrow[CH_3CO_2H]{Zn}$

2,3-Dichloropropene

Answer

(a) Treatment of 1,2-dibromocyclohexane with an alkoxide base (potassium isopropoxide in isopropanol) at an elevated temperature leads to double dehydrohalogenation with elimination of two molecules of hydrogen bromide, and gives 1,3-cyclohexadiene.

1,2-Dibromocyclohexane $\xrightarrow[-\ HBr]{(CH_3)_2CHO^-}$ [3-bromocyclohexene intermediate] $\xrightarrow[-\ HBr]{(CH_3)_2CHO^-}$ 1,3-Cyclohexadiene

(b) However, when the same compound (1,2-dibromocyclohexane) is treated with a relatively nonbasic iodide ion, debromination occurs to form cyclohexene.

1,2-Dibromocyclohexane $\xrightarrow[-\ Br_2]{I^-}$ Cyclohexene

(c) This is a dehalogenation reaction. Dechlorination of 2,3-dichloropropene with zinc produces allene.

$H_2C{=}C(Cl){-}CH_2Cl$ $\xrightarrow{Zn}$ $H_2C{=}C{=}CH_2$

2,3-Dichloropropene Allene

10.3 THE ELIMINATION MECHANISM

Most 1,2-eliminations involve **ionic mechanisms**. Two ionic mechanisms are especially important. One mechanism is called the **E2 (elimination, bimolecular) mechanism**, and the other mechanism is called the **E1 (elimination, unimolecular) mechanism**.

10.3.1 The E2 Mechanism

The **E2 elimination** is a one-step, concerted process. Both substrate and nucleophile/base participate in the *single* step. The nucleophile (Nu^-) removes an electrophilic group (E) from the substrate at the same time the leaving group (L) on the adjacent carbon departs.

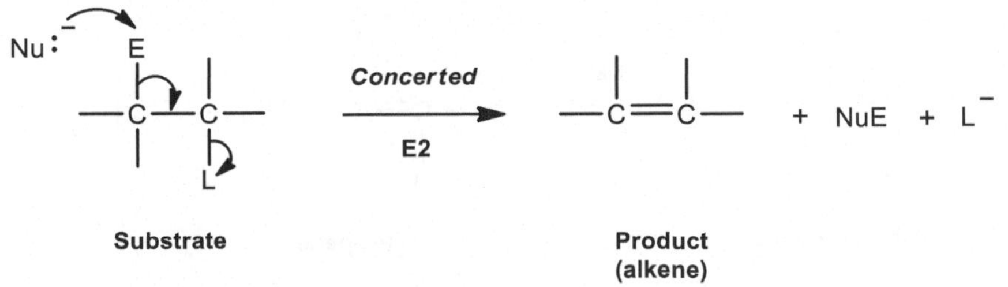

Nu: E — C — C — L
Concerted
E2
— C=C — + NuE + L^-

Substrate Product (alkene)

Since two species (substrate and nucleophile) are involved in the single, r/d step, the mechanism is designated **bimolecular (E2)**.

Dehydrohalogenation of alkyl halides

In **dehydrohalogenation**, the nucleophile is a **base** which donates an electron pair to an electrophilic **β-hydrogen atom** of the alkyl halide.

In the **E2 mechanism**, the abstraction of β-hydrogen atom by the base (B^-) and the departure of the leaving group (X = Cl, Br or I) are **concerted or simultaneous** (Figure 10.1).

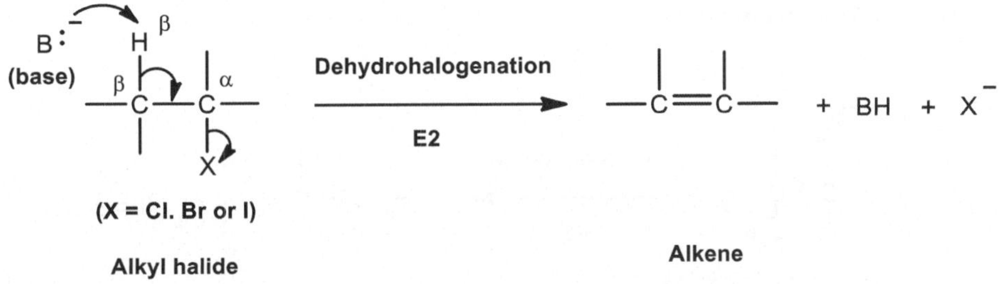

B: (base) H β β α — C — C — X
Dehydrohalogenation
E2
— C=C — + BH + X^-

(X = Cl. Br or I)

Alkyl halide Alkene

Figure 10.1: The E2 mechanism for dehydrohalogenation of alkyl halides with a base.

Example

The reaction of **2-bromopropane** with ethoxide ion gives **propene**.

$$CH_3CH(Br)CH_3 + C_2H_5O^- \longrightarrow CH_3CH{=}CH_2 + C_2H_5OH + Br^-$$

2-Bromopropane Ethoxide ion (base) Propene

It has been observed *experimentally* that the rate of the reaction depends on the concentration of 2-bromopropane and on the concentration of ethoxide ion.

Rate = $k\ [CH_3CHBrCH_3]\ [C_2H_5O^-]$

Thus the reaction is **second order** overall. The rate equation is consistent with the **E2 mechanism** involving both the alkyl halide (2-bromopropane) and the ethoxide ion in the single (r/d) step.

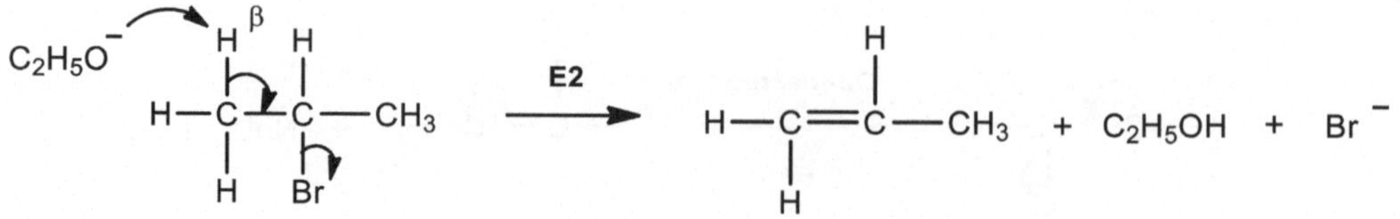

$C_2H_5O^-$ + $H-CH_2-CH(Br)-CH_3$ $\xrightarrow{E2}$ $H_2C=CH-CH_3$ + C_2H_5OH + Br^-

2-Bromopropane Propene

Dehalogenation of dihaloalkanes

Debromination of 1,2-dibromoalkanes in the presence of iodide ion or zinc metal (see section 10.2.2) involves an **E2 mechanism**, as shown in Figure 10.2. The nucleophile (I^- or Zn:) attacks the electrophilic bromine atom as the leaving group (Br) on the adjacent carbon departs.

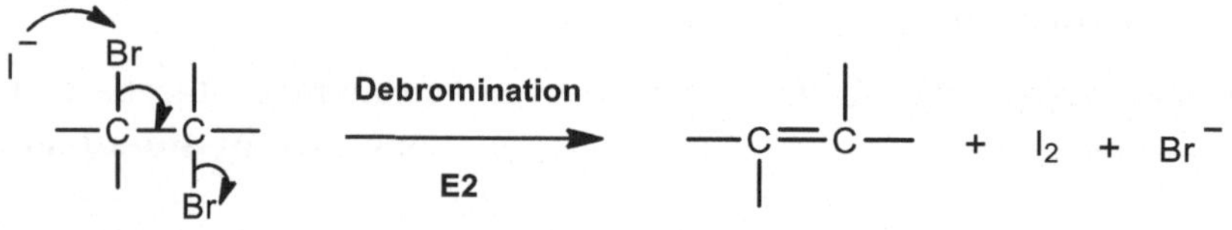

I^- + 1,2-Dibromoalkane $\xrightarrow[E2]{Debromination}$ Alkene + I_2 + Br^-

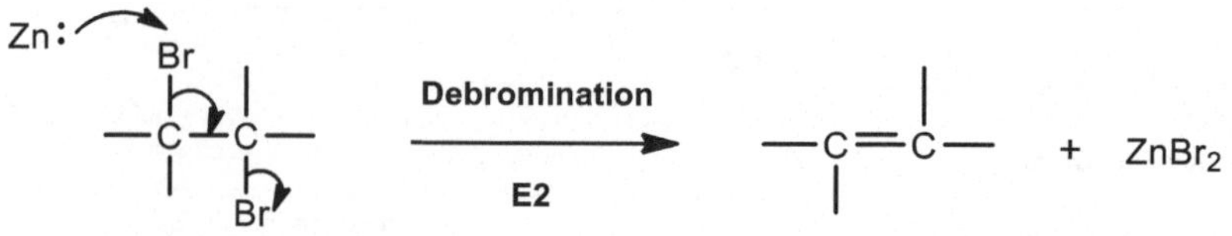

Zn: + 1,2-Dibromoalkane $\xrightarrow[E2]{Debromination}$ Alkene + $ZnBr_2$

Figure 10.2: The E2 mechanism for debromination of 1,2-dibromoalkanes in the presence of iodide ion or zinc metal.

10.3.2 The E1 Mechanism

The **E1 elimination** for dehydrohalogenation of alkyl halides involves a **stepwise ionic mechanism**, as shown in Figure 10.3.

In **step 1**, the leaving group (X) departs when a **carbocation** is formed as an intermediate. In **step 2**, the base/nucleophile abstracts a β-proton from the carbocation to produce an alkene.

The **step 1 is the slow, r/d step** which involves only the substrate (alkyl halide) and does not involve the base/nucleophile. Thus only one species (alkyl halide) is involved in the r/d step, so the mechanism is designated **unimolecular (E1)**.

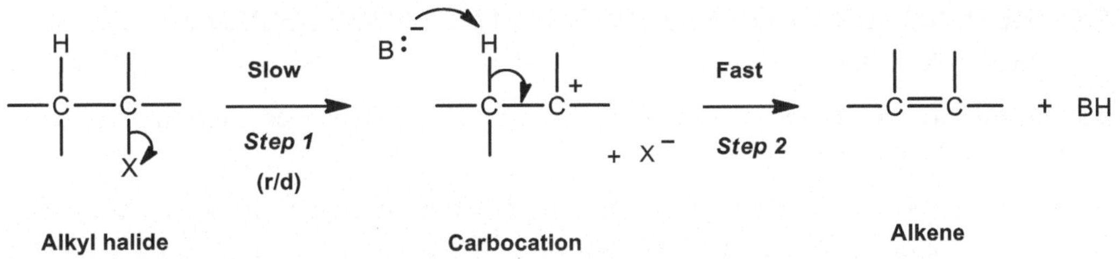

Alkyl halide —(Slow, Step 1 (r/d))→ Carbocation + X^- —(Fast, Step 2)→ Alkene + BH

Figure 10.3: The E1 mechanism for dehydrohalogenation of alkyl halides.

A carbocation can form when it is relatively stable. Thus **tertiary alkyl halides** can undergo E1 eliminations since a stable tertiary carbocation is formed by the departure of the halide ion. However, E1 reactions almost always accompany substitution reactions.

Example

The reaction of ***t*-butyl chloride** with aqueous ethanol (C_2H_5OH/H_2O) gives **isobutene** along with substitution products (*t*-butyl alcohol and *t*-butyl ethyl ether). In this reaction, **the solvent (water or alcohol) acts as a weak base or nucleophile.**

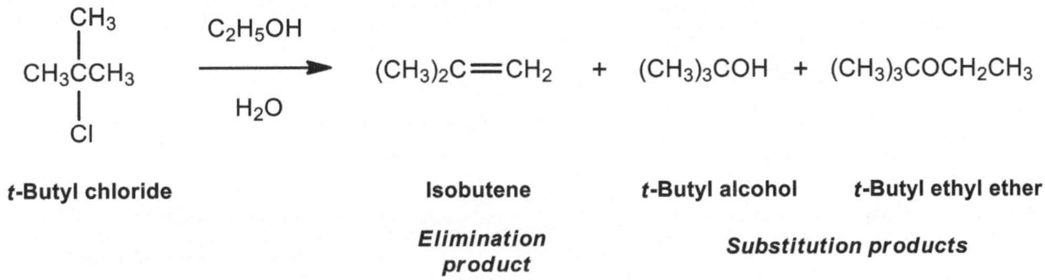

$$CH_3CCl(CH_3)CH_3 \xrightarrow[H_2O]{C_2H_5OH} (CH_3)_2C{=}CH_2 + (CH_3)_3COH + (CH_3)_3COCH_2CH_3$$

t*-Butyl chloride** → **Isobutene** (Elimination product***) + ***t*-Butyl alcohol** + ***t*-Butyl ethyl ether** (***Substitution products***)

The rate of the reaction is found to depend only on the concentration of the substrate (*t*-butyl chloride) and the reaction is **first order**.

$$\text{Rate} = k\,[(CH_3)_3CCl]$$

The rate equation is consistent with the **E1 mechanism** involving only the alkyl halide (*t*-butyl chloride) in the r/d step, as shown in Figure 10.4. The solvent (water or ethanol) serves as a weak base to remove a proton from the intermediate carbocation to produce isobutene.

For the formation of substitution products, see later Section 11.4.1.

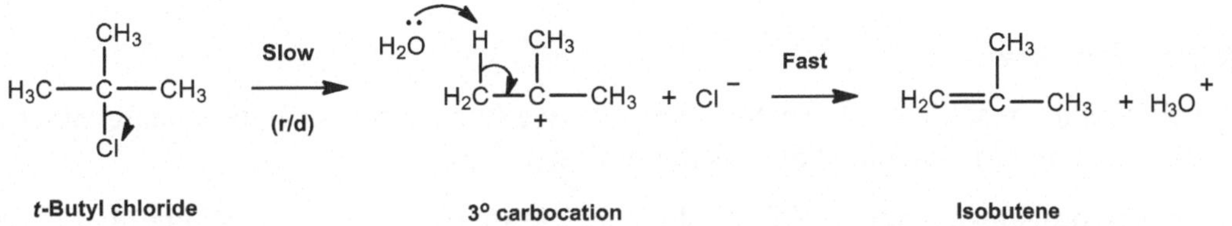

$$H_3C{-}C(CH_3)(Cl){-}CH_3 \xrightarrow[\text{(r/d)}]{\text{Slow}} H_2C(H){-}\overset{+}{C}(CH_3){-}CH_3 + Cl^- \xrightarrow{\text{Fast}} H_2C{=}C(CH_3){-}CH_3 + H_3O^+$$

***t*-Butyl chloride** → **3° carbocation** (with H_2O) → **Isobutene**

Figure 10.4: The E1 mechanism for dehydrohalogenation of *t*-butyl chloride in aqueous ethanol.

Note: In the presence of a strong base such as ethoxide ion ($C_2H_5O^-$) or ethanolic potassium hydroxide (C_2H_5OH/KOH), *t*-butyl chloride undergoes the E2 elimination.

10.4 THE DIRECTION OF ELIMINATION: SAYTZEFF AND HOFMANN ORIENTATIONS

Unsymmetrical substrates have more than one kind of β-hydrogen atom that can be lost in elimination reactions.

For example, consider dehydrohalogenation of **2-bromobutane** with sodium ethoxide in ethanol. In the elimination of HBr, the removal of a β-hydrogen atom from the methylene group (CH_2) gives **2-butene**, whereas the loss of a β-hydrogen atom from the methyl group (CH_3) forms **1-butene**, Therefore, two alkene products (2-butene and 1-butene) are possible.

$C_2H_5O^-$ H β

$CH_3CH{-}CHCH_3$ (Br) → **Removal of β-H from CH_2 group** → $CH_3CH{=}CHCH_3$

2-Bromobutane

2-Butene
(more substituted alkene)
Saytzeff product

β H $^-OC_2H_5$

$CH_3CH_2CH{-}CH_2$ (Br) → **Removal of β-H from CH_3 group** → $CH_3CH_2CH{=}CH_2$

2-Bromobutane

1-Butene
(less substituted alkene)
Hofmann product

The **direction or orientation** of elimination is commonly described in terms of the *degree of substitution* on the carbon-carbon double bond. The direction which gives the **more substituted alkene** is called the **Saytzeff orientation**, while the direction which leads to the **less substituted alkene** is called the **Hofmann orientation**.

In the above reaction, 2-Butene is the more substituted alkene (two hydrogen atoms at C=C are substituted by two methyl groups) while 1-butene is the less substituted alkene (only one hydrogen atom at C=C is substituted by ethyl group). Therefore, **2-butene is the Saytzeff product and 1-butene is the Hofmann product**.

Regioselectivity

The elimination reactions are usually **regioselective** (see Section 8.2), producing either the Saytzeff alkene or Hofmann alkene as the **major product**.

In the dehydrohalogenation of 2-bromobutane, **2-butene** (the Saytzeff alkene) is found to be the **major product** (81%). The *more substituted alkene is usually the more stable alkene*. Thus, the elimination reaction produces predominantly the **Saytzeff product** as a **more stable alkene**.

In general, there are several factors that may lead to the formation of the Saytzeff alkene or the Hofmann alkene as the predominant product. An important factor is the **choice of base** in dehydrohalogenation reactions.

To illustrate, consider the dehydrohalogenation of **2-bromo-2-methylbutane.**

When dehydrohalogenation of 2-bromo-2-methylbutane is carried out with **sodium ethoxide** as the base, the **major product** is the more substituted or more stable alkene (2-methyl-2-butene) as the **Saytzeff product.**

$$CH_3\overset{\beta\,H}{\underset{|}{C}}H-\overset{CH_2-H\,\beta}{\underset{Br}{C}}CH_3 \xrightarrow[C_2H_5OH,\ 70^\circ C]{C_2H_5ONa} CH_3CH=\overset{CH_3}{C}CH_3 + CH_3CH_2-\overset{CH_2}{\overset{\|}{C}}CH_3$$

2-Bromo-2-methyl-butane

2-Methyl-2-butene (more stable) *Saytzeff product* **Major (69%)**

2-Methyl-1-butene (less stable) *Hofmann product* **Minor (31%)**

But when dehydrohalogenation of 2-bromo-2-methylbutane is carried with **potassium *t*-butoxide** as the base, the **major product** is the less substituted alkene (2-methyl-1-butene) as the **Hofmann product.**

(more crowded) (more exposed)

$$CH_3\overset{\beta\,H}{\underset{|}{C}}H-\overset{CH_2-H\,\beta}{\underset{Br}{C}}CH_3 \xrightarrow[(CH_3)_3COH,\ 75^\circ C]{(CH_3)_3COK} CH_3CH=\overset{CH_3}{C}CH_3 + CH_3CH_2-\overset{CH_2}{\overset{\|}{C}}CH_3$$

2-Bromo-2-methyl-butane

2-Methyl-2-butene (more substituted) *Saytzeff product* **Minor (28%)**

2-Methyl-1-butene (less substituted) *Hofmann product* **Major (72%)**

Here the Hofmann elimination is favoured mainly due to **steric effects**. The approach of a base generally increases crowding within the substrate molecule in E2 eliminations. Since the *t*-butoxide ion [$(CH_3)_3CO^-$] is a **bulky and hindered base**, it will experience more difficulty in removing a β-hydrogen atom from the more substituted or more crowded methylene carbon atom. Therefore, it will remove a more exposed or more accessible β-hydrogen atom from the less substituted methyl carbon atom, and produce the less substituted alkene (cf. Section 11.6).

For more discussion of the Hofmann elimination, see Chapter 14 (Section 14.7).

PROBLEM 10.2: Give the structures of alkenes that would be formed in each of the following reactions. Suggest the major product in each case.

(a) $CH_3CH_2CH_2CH(Br)CH_3 \xrightarrow[C_2H_5OH]{C_2H_5ONa}$

2-Bromopentane

(b) 1-Chloro-1-methylcyclohexane $\xrightarrow[C_2H_5OH]{C_2H_5ONa}$

1-Chloro-1-methylcyclohexane

Answer

Each of the elimination reactions produces a mixture of two alkenes. In the presence of a relatively unhindered base (ethoxide ion), the reactions proceed preferentially by **Saytzeff orientation** to give the more substituted or more stable alkene as the major product.

(a) The products are 1-pentene and 2-pentene. The major product is the more substituted alkene (2-pentene).

$CH_3CH_2CH(H_{(a)})—CH(Br)—CH_2(H_{(b)}) \xrightarrow[(-HBr)]{C_2H_5O^-}$

Loss of H(*a*): $CH_3CH_2CH{=}CHCH_3$ **2-Pentene** ***Major product***

Loss of H(*b*): $CH_3CH_2CH_2CH{=}CH_2$ **1-Pentene** ***Minor product***

(b) The products are 1-methylcyclohexene and methylenecyclohexane. The more substituted alkene (1-methylcyclohexene) is the major product.

1-Chloro-1-methylcyclohexane (H(*b*) on CH_2 of methyl, H(*a*) on ring carbon) $\xrightarrow[(-HBr)]{C_2H_5O^-}$

Loss of H(*a*): **1-Methylcyclohexene** ***Major product***

Loss of H(*b*): **Methylenecyclohexane** ***Minor product***

10.5 THE STEREOCHEMISTRY OF E2 ELIMINATION

Consider the following general molecule for E2 eliminations where E and L represent electrophilic and leaving groups (see also Section 10.3.1).

(E and L are anti)

Antiperiplanar conformation

The most favourable pathway for the E2 elimination requires that the E–C–C–L unit is *planar* and the two departing groups (E and L) are *anti* to each other. This spatial arrangement or conformation of the molecule is called the **antiperiplanar conformation.**

The E2 elimination is then called an **anti elimination** which determines the stereochemistry or stereochemical outcome of E2 reactions.

10.5.1 Stereoselectivity

Consider, for example, the E2 dehydrohalogenation of **2-bromobutane** with ethoxide ion. Two antiperiplanar conformations of 2-bromobutane are possible. One conformation undergoes anti elimination to give ***cis*-2-butene**, while the other conformation undergoes anti elimination to give ***trans*-2-butene**, as shown in Figure 10.5.

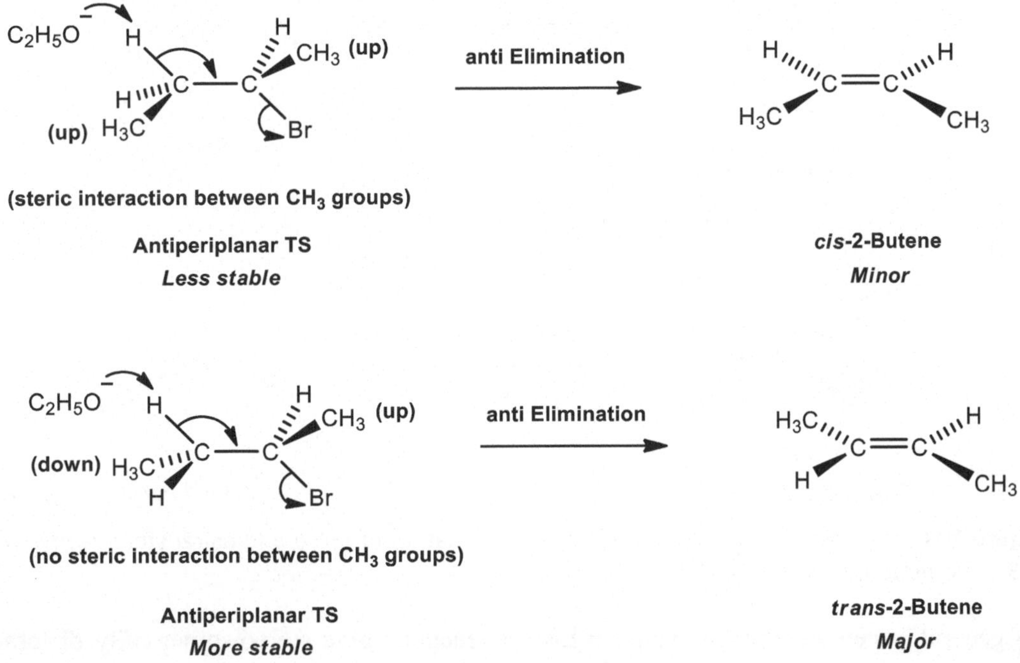

Figure 10.5: A stereoselective E2 elimination: Dehydrohalogenation of 2-bromobutane with ethoxide ion.

The concerted E2 elimination involves a single transition structure (TS), which includes the base ($C_2H_5O^-$). The antiperiplanar TS leading to ***cis*-2-butene** suffers from **steric interactions** between two methyl groups that are on the same side. This is **less stable** than the antiperiplanar TS leading to ***trans*-2-butene** in which there is **no steric interactions** between the two methyl groups on opposite sides.

As a result, the two stereoisomeric products (*cis* and *trans*) are formed in unequal amounts, and ***trans*-2-butene** arising from the more stable antiperiplanar TS becomes the **major product**.

In general, when a mixture of stereoisomeric products is formed in unequal quantities, the reaction is said to be **stereoselective**. Therefore, the **E2 elimination of 2-bromobutane is a stereoselective reaction.**

10.5.2 Stereospecificity

Consider, for example, the E2 dehalogenation of *two* different stereoisomers (chiral and *meso*) of **2,3-dibromobutane** with iodide ion, as shown in Figure 10.6.

The **chiral diastereomer** of 2,3-dibromobutane (as one of the enantiomers or a racemate) undergoes anti elimination of bromine to produce ***cis*-2-butene**. In contrast, the ***meso* diastereomer** of 2,3-dibromobutane (which is achiral with a centre of symmetry) undergoes anti elimination of bromine to give ***trans*-2-butene**.

I⁻ Br H CH₃ H C C H₃C Br — anti Elimination, $-Br_2$ → H H C=C H₃C CH₃

Chiral

2,3-Dibromobutane (one enantiomer) — ***cis*-2-Butene**

I⁻ Br H CH₃ H₃C C C H Br — anti Elimination, $-Br_2$ → H₃C H C=C H CH₃

Achiral

***meso*-2,3-Dibromobutane** — ***trans*-2-Butene**

Figure 10.6: Stereospecific E2 eliminations: Debromination of *meso* and chiral stereoisomers of 2,3-dibromobutane with iodide ion.

In general, when stereoisomerically different reactants give stereoisomerically different products, the reactions are said to be **stereospecific** (see also Section 8.6.2). Therefore, the above E2 eliminations are **stereospecific reactions**.

10.6 DEHYDRATION OF ALCOHOLS

The **1,2-elimination of water** from **alcohols** to form alkenes is called the **dehydration.** This is opposite to the hydration (addition of water) to alkenes.

Dehydration of alcohols is usually carried out by heating alcohols with an acid catalyst such as sulphuric acid or phosphoric acid.

$$\underset{\textbf{Alcohol}}{-\overset{|}{\underset{\underset{H}{|}}{C}}^{2}-\overset{|}{\underset{\underset{OH}{|}}{C}}^{1}-} \xrightarrow[\text{Heat }(\Delta)]{H^+} \underset{\textbf{Alkene}}{-\overset{|}{C}=\overset{|}{C}-} + H_2O$$

Tertiary alcohols dehydrate most easily and primary alcohols are the most difficult to dehydrate. The order of reactivity of alcohols for dehydration is: **tertiary > secondary > primary**.

Dehydration of tertiary and secondary alcohols

A tertiary alcohol such as ***t*-butyl alcohol** undergoes dehydration in 20% H_2SO_4 at 85°C, whereas **isopropyl alcohol** (a secondary alcohol) dehydrates in 75% H_2SO_4 at 100°C. This shows that a 3° alcohol dehydrates under milder conditions than a 2° alcohol does.

$$\underset{\substack{\textbf{\textit{t}-Butyl alcohol}\\ \textbf{\textit{3° Alcohol}}}}{H_3C-\overset{\overset{CH_3}{|}}{\underset{\underset{OH}{|}}{C}}-\underset{\underset{H}{|}}{CH_2}} \xrightarrow[\substack{85°C\\ -H_2O}]{20\%\ H_2SO_4} \underset{\textbf{Isobutene}}{H_3C-\overset{\overset{CH_3}{|}}{C}=CH_2}$$

$$\underset{\substack{\textbf{Isopropyl alcohol}\\ \textbf{\textit{2° Alcohol}}}}{H_3C-\overset{\overset{H}{|}}{\underset{\underset{OH}{|}}{C}}-\underset{\underset{H}{|}}{CH_2}} \xrightarrow[\substack{100°C\\ -H_2O}]{75\%\ H_2SO_4} \underset{\textbf{Propene}}{H_3C-CH=CH_2}$$

Dehydration of primary alcohols

Dehydration of primary alcohols generally takes place under much harsher experimental conditions. For example, dehydration of **ethyl alcohol** to **ethylene** requires concentrated H_2SO_4 and a high temperature of **180°C**.

Interestingly, at a lower temperature near **140°C**, the reaction produces **diethyl ether** by elimination of water molecule from *two* molecules of ethyl alcohol.

$$H_2C{-}CH_2 \text{ (H, OH)} \xrightarrow[180^\circ C,\ -H_2O]{\text{Conc. } H_2SO_4} CH_2{=}CH_2$$

Ethyl alcohol **Ethylene**

$$CH_3CH_2{-}OH + H{-}O{-}CH_2CH_3 \xrightarrow[140^\circ C,\ -H_2O]{\text{Conc. } H_2SO_4} CH_3CH_2{-}O{-}CH_2CH_3$$

Two molecules of ethyl alcohol **Diethyl ether**

Gas-phase dehydration

Dehydration of alcohols can be accomplished using **Lewis acid catalysts** such as alumina (Al_2O_3). This method is often used in gas-phase dehydration of alcohols, particularly in industrial applications.

For example, when the vapour of 1-propanol is passed over hot alumina at 350°C, propene is produced.

$$CH_3CH(H)CH_2OH \xrightarrow[350^\circ C,\ -H_2O]{Al_2O_3} CH_3CH{=}CH_2$$

1-Propanol **Propene**

10.6.1 Acid-catalysed Dehydration: The E1 Mechanism

In an elimination reaction, the leaving group (L) departs with the electron pair of the C–L bond, i.e. as L^- (when L is uncharged). But in the dehydration of alcohols, OH (to be lost as OH^-) is *not* a good leaving group.

However, the alcohol can accept a proton from an acid to form a **protonated alcohol** in which the leaving group is a protonated OH (H_2O^+), which can be easily lost as a neutral water molecule (H_2O). In effect, **protonation converts a poor leaving group to a good leaving group**.

The mechanism for acid-catalysed dehydration of alcohols is shown in Figure 10.7, taking ***t*-butyl alcohol** as an example.

$$(CH_3)_3C{-}\ddot{O}H \xrightarrow[\text{Fast, Step 1}]{H^+} (CH_3)_3C{-}\overset{+}{O}H_2 \xrightarrow[\text{Slow (r/d), Step 2}]{-H_2O} (CH_3)_2\overset{+}{C}{-}CH_2{-}H$$

***t*-Butyl alcohol** **Protonated alcohol** **Carbocation**

Fast *Step 3* (H_2O:)

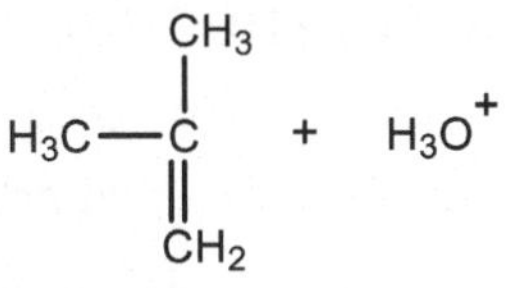

Figure 10.7: The E1 mechanism for the acid-catalysed dehydration of *t*-butyl alcohol.

Proton transfer reactions are generally fast; so the **step 1** for the protonation of the alcohol is **fast**. In **step 2**, the protonated alcohol loses a water molecule to form a **carbocation,** which is **slow and rate-determining**. Finally in **step 3**, the carbocation (*t*-butyl cation) loses a proton to a water molecule acting as a base, and produces **isobutene** as the alkene product.

The mechanism is described as the **E1 mechanism** since the leaving group is lost from the protonated alcohol acting as the substrate.

PROBLEM 10.3: Predict the product(s) in each of the following reactions. Indicate the major product, wherever applicable.

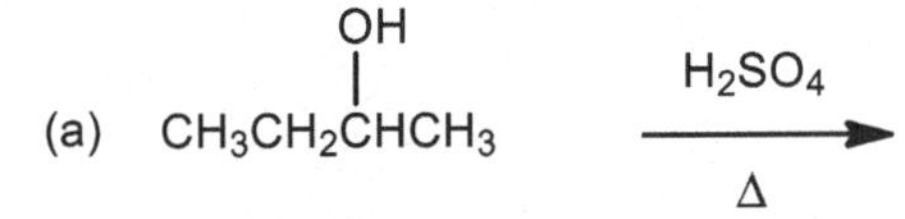

(a) $CH_3CH_2CH(OH)CH_3 \xrightarrow[\Delta]{H_2SO_4}$
2-Butanol

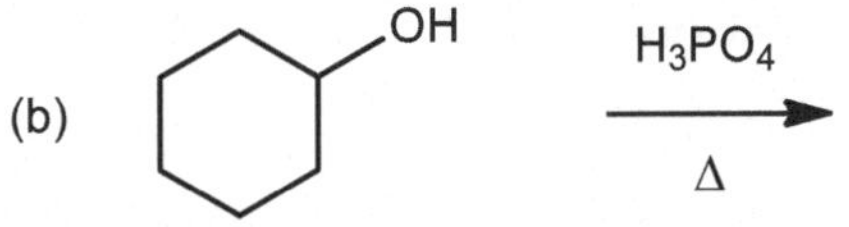

(b) Cyclohexanol $\xrightarrow[\Delta]{H_3PO_4}$
Cyclohexanol

Answer

(a) Two products are possible (excluding stereoisomerism) because the alcohol has two kinds of β-hydrogen atoms that can be lost. The dehydration follows the Saytzeff rule to give the more substituted alkene (2-butene) as the major product.

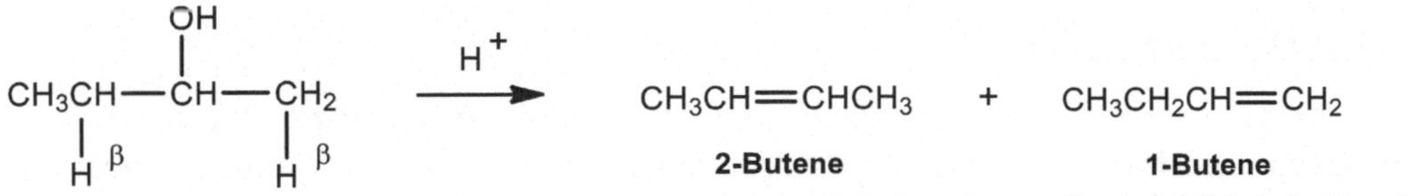

$CH_3CH(H)—CH(OH)—CH_2(H) \xrightarrow{H^+} CH_3CH{=}CHCH_3 + CH_3CH_2CH{=}CH_2$

2-Butanol	2-Butene (more substituted alkene) *Major product*	1-Butene (less substituted alkene) *Minor product*

(b) The dehydration of cyclohexanol gives a single product (cyclohexene) because the alcohol has only one kind of β-hydrogen atom.

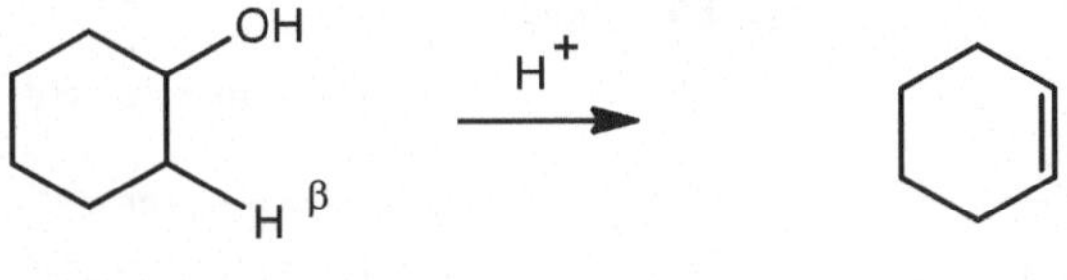

Cyclohexanol (OH, H β) $\xrightarrow{H^+}$ Cyclohexene

Cyclohexanol **Cyclohexene**

10.7 PYROLYTIC ELIMINATIONS

1,2- or β-Elimination of certain compounds is possible without the use of external reagents (e.g. acids or bases) that promote elimination.

For these compounds (such as esters), **heating** alone can provide sufficient energy for the leaving groups (H and L) to depart, forming an alkene. Such eliminations involving the thermal cleavage of a molecule is called the **pyrolytic elimination.**

Esters containing a β-hydrogen atom on the *alcohol portion* of the molecule are the common substrates. When heated at 300–600°C, they can undergo a pyrolytic elimination with the departure of a β-H atom and a carboxy group (OCOR). The product is an **alkene** with the loss of a carboxylic acid.

$$\underset{\text{Ester}}{-\overset{|}{\underset{\underset{\beta\,H}{|}}{C}}{}^{\beta}-\overset{|}{\underset{\underbrace{O-\underset{\underset{O}{\|}}{C}-R}_{(L)}}{\overset{|}{C}}}{}^{\alpha}-} \xrightarrow[300\text{-}600^{\circ}C]{\Delta} \underset{\text{Alkene}}{-\overset{|}{C}=\overset{|}{C}-} + \underset{\text{Carboxylic acid}}{HO-\underset{\underset{O}{\|}}{C}-R}$$

The esters employed for pyrolytic eliminations are usually **acetates**. For example, **butyl acetate** when heated at 500°C produces **1-butene** with the loss of acetic acid.

$$\underset{\substack{(CH_3CO_2CH_2CH_2CH_2CH_3\\ \textbf{Butyl acetate}}}{CH_3\overset{\overset{O}{\|}}{C}O-\overset{\alpha}{C}H_2-\overset{\overset{H^{\beta}}{|}}{{}^{\beta}C}HCH_2CH_3} \xrightarrow{500^{\circ}C} \underset{\textbf{1-Butene}}{CH_2=CHCH_2CH_3} + \underset{\text{Acetic acid}}{CH_3\overset{\overset{O}{\|}}{C}OH}$$

Pericyclic mechanism

The pyrolytic elimination does not involve an ionic mechanism, but it occurs by a **pericyclic mechanism** with the cyclic movement of electrons in a single, concerted step, as shown in Figure 10.8.

$$\underset{\textbf{Butyl acetate}}{\text{[cyclic TS: } H_3C-C(=O)-O-CH_2-CHCH_2CH_3,\ H\text{]}} \xrightarrow[\textbf{syn Elimination}]{\textbf{\textit{Pericyclic mechanism}}} \underset{\textbf{(acetic acid)}}{H_3C-C(=O)-OH} + \underset{\textbf{1-Butene}}{CH_2=CHCH_2CH_3}$$

Figure 10.8: The pericyclic mechanism for the pyrolytic elimination of butyl acetate.

In contrast to anti elimination in the E2 mechanism, the two departing groups (H and $OCOCH_3$) in the pyrolytic elimination are lost from the same side of the molecule. Thus the **pyrolytic elimination is a syn elimination**. The mechanism is sometimes designated as **E*i* (elimination, internal)**.

For more examples of pyrolytic eliminations, see Sections 14.8 and 15.8.1.

10.8 FORMATION OF ALKYNES BY 1,2-ELIMINATIONS

1,2-or β-Eliminations can provide suitable routes to the formation of alkynes.

Dehydrohalogenation of haloalkenes with a base gives an alkyne.

Double dehydrohalogenation of dihaloalkanes also gives alkynes. The double dehydrohalogenation reaction proceeds through a haloalkene intermediate. Generally, a strong base such as sodium amide ($NaNH_2$) in liquid ammonia is used to carry out double dehydrohalogenation.

$$\underset{\substack{\textbf{Chloroethene}\\\textbf{(vinyl chloride)}\\\textit{\textbf{Haloalkene}}}}{H_2C{=}CH{-}Cl} \xrightarrow[\Delta,\ -HCl]{NaOH} \underset{\substack{\textbf{Ethyne}\\\textbf{(acetylene)}\\\textit{\textbf{Alkyne}}}}{HC{\equiv}CH}$$

$$\underset{\substack{\textbf{2,3-Dibromobutane}\\\textit{\textbf{Dihaloalkane}}}}{CH_3CH(Br){-}CH(Br)CH_3} \xrightarrow[\text{liq. }NH_3,\ -HBr]{NaNH_2} \underset{\substack{\textit{\textbf{Haloalkene}}\\\textit{\textbf{intermediate}}}}{\left[CH_3C(Br){=}CHCH_3\right]} \xrightarrow[\text{liq. }NH_3,\ -HBr]{NaNH_2} \underset{\substack{\textbf{2-Butyne}\\\textit{\textbf{Alkyne}}}}{CH_3C{\equiv}CCH_3}$$

Alkynes from alkenes

Alkynes can be prepared from alkenes. First, an **alkene** is treated with bromine to form a **dibromoalkane** (*vic*-dibromide); the dibromoalkane is then subjected to double dehydrohalogenation with $NaNH_2$ in liquid NH_3 to produce an **alkyne**.

$$\underset{\textbf{Alkene}}{RCH{=}CHR} \xrightarrow[\textit{\textbf{Bromination}}]{Br_2} \underset{\substack{\textbf{Dibromoalkane}\\(\textit{\textbf{vic}}\textbf{-Dibromide})}}{RCH(Br){-}CH(Br)R} \xrightarrow[\substack{\text{liq. }NH_3,\ -2\,HBr\\\textit{\textbf{Double dehydrohalogenation}}}]{2\ NaNH_2} \underset{\textbf{Alkyne}}{RC{\equiv}CR}$$

When the above method produces a **terminal alkyne** (RC≡CH), the acidic hydrogen atom is abstracted by the strong base ($NaNH_2$) to form a salt, **sodium alkynide** ($RC{\equiv}C^-Na^+$).

$$RC\equiv C{-}H + Na^+\,^-NH_2 \longrightarrow RC\equiv C^-\,Na^+ + NH_3$$

Terminal alkyne → Sodium alkynide

As a result, *three* moles of $NaNH_2$ are consumed per mole of the dibromoalkane, thereby forming a sodium alkynide. The terminal alkyne is regenerated by treating the sodium alkynide with aqueous NH_4Cl under mildly acidic condition.

The method is illustrated with the preparation of **propyne** from **propene**.

$$CH_3CH{=}CH_2 \xrightarrow{Br_2} CH_3CH(Br){-}CH_2Br \xrightarrow[\text{liq. } NH_3,\ -HBr]{NaNH_2} [CH_3CH{=}CHBr + CH_3CBr{=}CH_2]$$

Propene; 1,2-Dibromopropane; A mixture of bromopropenes

$$\xrightarrow[-HBr]{NaNH_2,\ \text{liq. } NH_3} [CH_3C\equiv CH] \xrightarrow[-NH_3]{NaNH_2} CH_3C\equiv C^-\,Na^+ \xrightarrow[H_2O]{NH_4Cl} CH_3C\equiv CH$$

A terminal alkyne; Sodium propynide; Propyne

Alkynes from ketones

Ketones can be converted to **dichloroalkanes** (*gem*-dichlorides) by their reactions with PCl_5. A *gem*-dichloride has two Cl atoms attached to the same carbon atom. The ***gem*-dichloride** undergoes **double dehydrohalogenation** in the presence of $NaNH_2$ in liquid ammonia (or $NaNH_2$ in mineral oil) to produce an alkyne.

The method is illustrated with the synthesis of **phenylacetylene** from **acetophenone**.

$$C_6H_5C(=O)CH_3 \xrightarrow[-POCl_3]{PCl_5} C_6H_5CCl_2CH_3 \xrightarrow[\text{Mineral oil},\ \Delta,\ -HCl]{NaNH_2} [C_6H_5CCl{=}CH_2]$$

Acetophenone; A *gem*-dichloride; A chloroalkene

$$\xrightarrow[-HCl]{NaNH_2,\ \text{Mineral oil},\ \Delta} [C_6H_5C\equiv CH] \xrightarrow{NaNH_2} C_6H_5C\equiv C^-\,Na^+ \xrightarrow[H_2O]{H^+} C_6H_5C\equiv CH$$

A terminal alkyne; A sodium alkynide; Phenylacetylene

Acetophenone reacts with PCl_5 to form a *gem*-dichloride which then undergoes double dehydrogenation on heating with $NaNH_2$ in mineral oil to yield phenylacetylene. Since phenylacetylene is a terminal alkyne, it further reacts with $NaNH_2$ to form a salt (sodium alkynide) and the alkyne is regenerated with aqueous acid.

Preparation of acetylene

Acetylene (the simplest alkyne) is an important industrial gas which is prepared by catalytic dehydrogenation of hydrocarbons in petroleum 'cracking' units at high temperatures (>1200°C).

Historically, acetylene was manufactured by the hydrolysis of calcium carbide. Calcium carbide was obtained by heating coke (carbon) with lime (calcium oxide).

$$\underset{\text{Coke}}{3\,C} + \underset{\text{Lime}}{CaO} \xrightarrow{\Delta} \underset{\text{Calcium carbide}}{CaC_2} + CO$$

$$\underset{\text{Calcium carbide}}{CaC_2} \xrightarrow{2\,H_2O} \underset{\text{Acetylene}}{HC\equiv CH} + Ca(OH)_2$$

PROBLEM 10.4: Outline a synthesis of 1-butyne from 1-butene. Draw the structures in bond-line formulas.

Answer

$$\underset{\text{1-Butene}}{CH_3CH_2CH=CH_2} \xrightarrow[CCl_4]{Br_2} \underset{\text{1,2-Dibromobutane}}{CH_3CH_2CHBrCH_2Br} \xrightarrow[\text{Mineral oil, }\Delta]{3\,NaNH_2} CH_3CH_2C\equiv C^-\,Na^+ \xrightarrow{NH_4Cl,\ H_2O} \underset{\text{1-Butyne}}{CH_3CH_2C\equiv CH}$$

10.9 DEHYDROGENATION OF ALCOHOLS

A very useful elimination reaction is the **dehydrogenation** of alcohols, which produces aldehydes and ketones. Dehydrogenation is the **loss of hydrogen molecule** from alcohols; thus dehydrogenation indicates the **oxidation** of alcohols.

Oxidation (or dehydrogenation) of primary alcohols

Oxidation or dehydrogenation of **primary alcohols** gives **aldehydes**.

$$R-\underset{H}{\overset{H}{C}}-\underset{H}{O} \xrightarrow[-H_2]{[O]} R-\overset{H}{C}=O$$

1° Alcohol **Aldehyde**

The most commonly used **oxidizing agent** is **chromic acid (H_2CrO_4)**. Chromic acid can be prepared by dissolving chromium (VI) oxide (CrO_3) in aqueous sulphuric acid or by dissolving sodium or potassium dichromate ($Na_2Cr_2O_7$ or $K_2Cr_2O_7$) in aqueous sulphuric acid.

However, aldehydes can be further oxidized by chromic acid to **carboxylic acids** and it may be difficult to stop the oxidation at the aldehyde stage.

$$R-\underset{H}{\overset{H}{C}}-OH \xrightarrow[H_2CrO_4]{[O]} R-\overset{H}{C}=O \xrightarrow[H_2CrO_4]{[O]} R-\overset{OH}{C}=O$$

1° Alcohol **Aldehyde** **Carboxylic acid**

This problem can be avoided by distilling off the aldehyde from the reaction mixture as it is formed. This is possible because an aldehyde has a lower boiling point than an alcohol.

For example, **1-propanol** can be converted to **propanal** by oxidation with chromic acid.

$$CH_3CH_2-CH_2OH \xrightarrow[H_2SO_4]{K_2Cr_2O_7} CH_3CH_2-\overset{H}{C}=O$$

1-Propanol (bp, 97°C) **Propanal (bp, 50°C)**

But the method is not suitable for high molecular weight aldehydes that boil above 100°C.

Nowadays, a variety of **special oxidizing agents** has been employed to oxidize a primary alcohol to an aldehyde when the reaction stops at the aldehyde stage. An excellent reagent is **pyridinium chlorochromate (PCC)**.

The **PCC reagent** is prepared by adding CrO_3 to hydrochloric acid and then treating the resulting solution with pyridine.

$$CrO_3 + HCl \xrightarrow{\text{(pyridine)}} C_5H_5\overset{+}{N}-H \quad CrO_3Cl^-$$

Pyridinium chlorochromate (PCC)

The **PCC oxidation** is usually carried out by treating a primary alcohol with PCC reagent dissolved in a **nonaqueous solvent** such as dichloromethane (CH_2Cl_2).

For example, oxidation of **2-methyl-1-propanol** with PCC in dichloromethane stops at the aldehyde stage and yields **2-methylpropanal**.

$$\underset{\textbf{2-Methyl-1-propanol}}{CH_3CH(CH_3)-CH_2OH} \xrightarrow[CH_2Cl_2,\ 25^{\circ}C]{PCC} \underset{\textbf{2-Methylpropanal}}{CH_3CH(CH_3)-CH{=}O}$$

The PCC oxidation in a **nonaqueous** solvent (CH_2Cl_2) is in contrast to the oxidation of primary alcohols by chromic acid in **aqueous** medium. With **chromic acid**, aldehydes are further oxidized to carboxylic acids via the formation of their hydrates by addition of water.

$$\underset{\textbf{1° Alcohol}}{R-CH_2OH} \xrightarrow{H_2CrO_4} \underset{\textbf{Aldehyde}}{R-CH{=}O} \overset{H_2O}{\rightleftharpoons} \underset{\textbf{Aldehyde hydrate}}{R-CH(OH)-OH} \xrightarrow{H_2CrO_4} \underset{\textbf{Carboxylic acid}}{R-C(OH){=}O}$$

Oxidation of secondary alcohols

Oxidation of **secondary alcohols** produces **ketones**. With chromic acid, the reaction usually stops at the ketone stage, without further oxidation involving the cleavage of C–C bond.

$$\underset{\textbf{2° Alcohol}}{R_2C(H)-O-H} \xrightarrow[-\ H_2]{[O]} \underset{\textbf{Ketone}}{R_2C{=}O}$$

Examples

$$\underset{\textbf{2-Butanol}}{CH_3CH_2-CH(OH)-CH_3} \xrightarrow[H_2SO_4]{CrO_3} \underset{\textbf{Butanone}}{CH_3CH_2-C(=O)-CH_3}$$

$$\textbf{Cyclohexanol} \xrightarrow[H_2SO_4]{Na_2Cr_2O_7} \textbf{Cyclohexanone}$$

Qualitative test for primary and secondary alcohols

A primary alcohol (RCH_2OH) or a secondary alcohol (R_2CHOH) is *colourless* and the chromic acid solution is *orange* in colour. In chromic acid (H_2CrO_4) oxidation, the chromium is Cr(VI) which itself is reduced to Cr^{3+}. The colour of the solution containing Cr^{3+} is *green* and the mixture containing Cr^{3+} and oxidation products gives an opaque greenish solution.

A change in colour from **orange to green** is used as a **qualitative test** for primary and secondary alcohols.

$$\underset{\substack{\textbf{1° or 2° Alcohol} \\ \textit{Colourless}}}{RCH_2OH \text{ or } R_2CHOH} + \underset{\substack{(CrO_3,\ H_2SO_4,\ H_2O) \\ \textbf{Chromic acid} \\ \textit{Orange}}}{H_2CrO_4} \longrightarrow \underset{\substack{\textbf{Opaque solution} \\ \textit{Green}}}{Cr^{3+} + \text{oxidation products}}$$

Catalytic dehydrogenation of alcohols

Catalytic dehydrogenation of primary and secondary alcohols is used industrially for the preparation of aldehydes and ketones.

The **catalytic dehydrogenation** is usually carried out by passing the alcohol vapour over copper catalyst at high temperatures. The following are two examples:

$$\underset{\substack{\textbf{Ethyl alcohol} \\ \textit{1° Alcohol}}}{CH_3CH_2OH} \xrightarrow[300^{\circ}C]{Cu} \underset{\textbf{Acetaldehyde}}{CH_3CH{=}O} + H_2$$

$$\underset{\substack{\textbf{Isopropyl alcohol} \\ \textit{2° Alcohol}}}{CH_3CH(OH)CH_3} \xrightarrow[300^{\circ}C]{Cu} \underset{\textbf{Acetone}}{CH_3C(=O)CH_3} + H_2$$

10.10 DEHYDRATION OF AMIDES

Dehydration of amides involves **elimination of water** to produce **nitriles or cyanides**. The most common dehydrating agent is phosphorus pentoxide (P_2O_5 or P_4O_{10}). In the dehydration reaction, phosphorus pentoxide (P_4O_{10}) itself is converted to phosphoric acid (H_3PO_4) by reaction with water.

$$R{-}\overset{O}{\overset{\|}{C}}{-}NH_2 \xrightarrow[\Delta]{P_4O_{10}} R{-}C{\equiv}N + H_3PO_4$$

Amide **Nitrile (cyanide)**

Examples

$$CH_3{-}\overset{O}{\overset{\|}{C}}{-}NH_2 \xrightarrow[-\,H_2O]{P_4O_{10}} CH_3{-}C{\equiv}N$$

Acetamide **Acetonitrile (methyl cyanide)**

$$C_6H_5{-}\overset{O}{\overset{\|}{C}}{-}NH_2 \xrightarrow[-\,H_2O]{P_4O_{10}} C_6H_5{-}C{\equiv}N$$

Benzamide **Benzonitrile (phenyl cyanide)**

SUPPLEMENTARY PROBLEMS

10.5 Write the product(s) in each of the following reactions. If more than one product is formed, indicate the major product.

(a) $CH_3CH_2CH_2CH_2CH_2Br \xrightarrow[t\text{-BuOH}]{t\text{-BuO}^-}$

1-Bromopentane

(b) $CH_3CH_2CH_2\underset{Br}{\underset{|}{C}}HCH_3 \xrightarrow[t\text{-BuOH}]{t\text{-BuO}^-}$

2-Bromopentane

(c) 1,4-dichlorocyclohexane (Cl at C-1 and C-4 of the cyclohexane ring) $\xrightarrow[EtOH]{EtO^-}$

1,4-Dichlorocyclohexane

[*Ans:* (a) 1-pentene; (b) 1-pentene (Hofmann, major) + 2-pentene; (c) 1,3-cyclohexadiene (conjugated, major) + 1,4-cyclohexadiene]

10.6 Give the structure of the product that you would expect to be formed in each of the following reactions:

(a) $H_3C-C(Br)(CH_3)-C(Br)=O$ $\xrightarrow[\text{Acetone}]{\text{Zn}}$ A ketene

2-Bromo-2-methylpropanoyl bromide

(b) $H_3C-C(CH_3)(OH)-C(CH_3)(OH)-CH_3$ $\xrightarrow[\Delta]{Al_2O_3}$ A diene

2,3-Dimethyl-2,3-butanediol

10.7 The Chugaev reaction is the pyrolytic elimination of a xanthate ester which can be prepared as shown in the following scheme. Predict the alkene (**X**) that would be formed when the xanthate ester is heated at 200°C.

$CH_3CH_2CH_2OH$ $\xrightarrow[\text{(base)}, -H_2]{\text{NaH}}$ $CH_3CH_2CH_2O^- Na^+$ $\xrightarrow[(CS_2)]{S=C=S}$ $CH_3CH_2CH_2O-C(=S)-S^- Na^+$

$\xrightarrow[-\text{NaI}]{CH_3I}$ $CH_3CH_2CH_2O-C(=S)-SCH_3$ (**Xanthate ester**) $\xrightarrow[\textit{Chugaev reaction}]{200°C}$ **X** (alkene)

[*Ans:* X is propene; the other product decomposes into methanethiol (CH_3SH) and carbonyl sulphide (COS)]

10.8 How would you carry out the following transformation?

$C_6H_5-CH=CH_2$ (**Styrene**) $\longrightarrow$ $C_6H_5-C\equiv CH$ (**Phenylacetylene**)

CHAPTER 11

Alkanes, Alkyl Halides, Alcohols and Ethers: Substitution Reactions

IN THIS CHAPTER

- Halogenation of alkanes and allylic halogenation: Radical substitution
- Nucleophilic substitution of alkyl halides: S_N2 and S_N1 Mechanisms
- Stereochemistry of S_N2 and S_N1 reactions
- Nucleophilic substitution versus elimination
- Nucleophilic substitution of alcohols via protonation and via sulphonates
- The Williamson synthesis of ethers
- Nucleophilic substitution of ethers and ring opening of epoxides

The characteristic reactions of alkanes, alkyl halides, alcohols and ethers involve **substitution** at a **saturated or sp^3-hybridized** carbon atom.

The substitution reactions of **alkanes** generally involve a **radical mechanism**, while substitution reactions of **alkyl halides, alcohols** and **ethers** proceed by **ionic mechanisms**.

11.1 Reactions of Alkanes with Halogens: Radical Substitution

The reaction of an alkane with a halogen (chlorine or bromine) is called the **halogenation**. The reaction is promoted by **heat (Δ)** or **light (*hν*)**.

The halogenation reaction is a **substitution reaction** which involves the replacement of a hydrogen atom of alkane with a halogen atom. The product is an **alkyl halide** (or haloalkane).

$$\text{—C—H (Alkane, sp}^3\text{)} + X_2 \ (X = Cl, Br) \xrightarrow[\textit{Substitution reaction}]{\text{Heat }(\Delta)\text{ or Light }(h\nu)} \text{—C—X (Alkyl halide (haloalkane))} + HX$$

11.1.1 Chlorination of Methane

The simplest halogenation of alkanes is **chlorination** of methane, which gives methyl chloride (chloromethane) and hydrogen chloride.

$$H_3C—H + Cl_2 \xrightarrow[\text{or } h\nu]{\Delta} H_3C—Cl + HCl$$

Methane → **Methyl chloride (chloromethane)**

However, a problem of the reaction is that *multiple substitutions* almost always occur. A *second* substitution gives methylene chloride (dichloromethane) by reaction of methyl chloride with chlorine. A *third* substitution gives chloroform (trichloromethane), and a *fourth* substitution gives carbon tetrachloride (tetrachloromethane).

$$CH_3Cl + Cl_2 \longrightarrow CH_2Cl_2 + HCl$$

Methylene chloride (dichloromethane)

$$CH_2Cl_2 + Cl_2 \longrightarrow CHCl_3 + HCl$$

Chloroform (trichloromethane)

$$CHCl_3 + Cl_2 \longrightarrow CCl_4 + HCl$$

Carbon tetrachloride (tetrachloromethane)

The problem of multiple substitutions can be minimized by using a *large excess* of methane in the reaction mixture. This ensures that the probability of attack by chlorine on methane molecules is a maximum because the concentration of methane in the mixture will always be quite large, and methyl chloride (CH_3Cl) will then be obtained in high yield.

$$\underset{\textbf{1.0 mol}}{CH_4} + \underset{\textbf{0.6 mol}}{Cl_2} \xrightarrow{400^\circ C} \underset{\textbf{59\%}}{CH_3Cl} + \underset{\textbf{29\%}}{CH_2Cl_2} + \underset{\textbf{10\%}}{CHCl_3} + \underset{\textbf{2\%}}{CCl_4}$$

11.1.2 Radical Mechanism

The **chlorination of methane** takes place by a **radical chain mechanism**, as shown in Figure 11.1.

In **step 1**, the homolysis of Cl–Cl bond under the influence of heat or light generates two **chlorine atoms or radicals**. This is called **chain initiation**.

Steps 2 and 3 denote **chain propagation**. In **step 2**, the chlorine radical (Cl·) abstracts a hydrogen atom from a methane molecule to form **methyl radical** ($CH_3\cdot$) and **HCl**. In **step 3**, the highly reactive methyl radical abstracts a chlorine atom from a chlorine molecule to form **methyl chloride** and **chlorine radical**. The regenerated chlorine radical propagates the chain; steps 2 and 3 are repeated, and so on.

Chain Initiation

$$Cl{-}Cl \xrightarrow[\textit{Step 1}]{\Delta \text{ or } h\nu} Cl^{\bullet} + Cl^{\bullet}$$

Chlorine radicals

Chain Propagation

$$Cl^{\bullet} + H{-}CH_3 \xrightarrow{\textit{Step 2}} H{-}Cl + H_3C^{\bullet}$$

Methyl radical

$$H_3C^{\bullet} + Cl{-}Cl \xrightarrow{\textit{Step 3}} H_3C{-}Cl + Cl^{\bullet}$$

Methyl chloride

Figure 11.1: A radical chain mechanism for chlorination of methane.

The chain reaction indicates that a few chlorine atoms at any given moment lead to the formation of thousands of methyl chloride molecules.

However, chains can terminate and make the reaction slow or stop. **Chain termination** can occur by the combination of radicals present in the reaction mixture.

Chain Termination

$$Cl^{\bullet} + {}^{\bullet}Cl \longrightarrow Cl{-}Cl \qquad CH_3^{\bullet} + {}^{\bullet}Cl \longrightarrow CH_3{-}Cl$$

$$CH_3^{\bullet} + {}^{\bullet}CH_3 \longrightarrow CH_3{-}CH_3$$

Because of the existence of chain termination which uses up the radicals, the reaction requires the regeneration of radicals by keeping the reaction mixture continuously under the action of heat or light.

11.1.3 Halogenation of Higher Alkanes

Chlorination of higher alkanes with two or more carbon atoms occurs by the similar radical chain mechanism as described for methane. We will consider here only **monohalogenation** which involves the replacement of only one of the hydrogen atoms of the alkane by a halogen atom.

Ethane

Monochlorination of ethane gives **only one** product (ethyl chloride) because all six primary hydrogen atoms of ethane are equivalent.

$$\overset{1^\circ}{CH_3}{-}\overset{1^\circ}{CH_3} \xrightarrow[h\nu]{Cl_2} CH_3{-}CH_2{-}Cl$$

Ethane — ***Monochlorination*** — **Ethyl chloride** *Only one product*

Propane, butane and isobutane

Monochlorination of propane, butane or isobutane produces a mixture of isomeric products, as shown below.

$$\underset{\textbf{Propane}}{\overset{1^\circ\ 2^\circ\ 1^\circ}{CH_3CH_2CH_3}} \xrightarrow[h\nu]{Cl_2} \underset{\textbf{Propyl chloride}}{CH_3CH_2CH_2Cl} + \underset{\textbf{Isopropyl chloride}}{CH_3\overset{Cl}{\overset{|}{C}}HCH_3}$$

$$\underset{\textbf{Butane}}{\overset{1^\circ\ 2^\circ\ 2^\circ\ 1^\circ}{CH_3CH_2CH_2CH_3}} \xrightarrow[h\nu]{Cl_2} \underset{\textbf{Butyl chloride}}{CH_3CH_2CH_2CH_2Cl} + \underset{\textbf{\textit{sec}-Butyl chloride}}{CH_3\overset{Cl}{\overset{|}{C}}HCH_2CH_3}$$

$$\underset{\textbf{Isobutane}}{\overset{1^\circ\ 3^\circ\ 1^\circ}{CH_3\underset{\underset{1^\circ}{CH_3}}{\underset{|}{C}}HCH_3}} \xrightarrow[h\nu]{Cl_2} \underset{\textbf{Isobutyl chloride}}{CH_3\underset{CH_3}{\underset{|}{C}}HCH_2Cl} + \underset{\textbf{\textit{tert}-Butyl chloride}}{CH_3\overset{Cl}{\overset{|}{\underset{CH_3}{\underset{|}{C}}}}CH_3}$$

All hydrogen atoms that are equivalent lead to one monochloro product. Propane has six equivalent primary hydrogens which give **propyl chloride**. It has also two equivalent secondary hydrogens which give **isopropyl chloride**. Thus propane gives *two* monochloro products (propyl chloride and isopropyl chloride).

Similarly, monochlorination of **butane** gives two products: **butyl chloride** (from six equivalent primary hydrogens) and ***sec*-butyl chloride** (from four equivalent secondary hydrogens).

Monochlorination of **isobutane** also gives two products: **isobutyl chloride** (from nine equivalent primary hydrogens) and ***tert*-butyl chloride** (from the single tertiary hydrogen).

Reactivity of primary (1°), secondary (2°) and tertiary (3°) hydrogen atoms

The **relative order of reactivity** of primary, secondary and tertiary hydrogen atoms towards halogenation is: **tertiary > secondary > primary**.

This follows the **stability order** of the corresponding **radicals: 3° > 2° > 1°**. In the radical chain mechanism, an alkyl radical is formed by the homolytic cleavage of a C–H bond of alkane (cf. step 2 in Figure 11.1).

The cleavage of a tertiary C–H bond gives the **most stable** tertiary radical; therefore this bond breaks most easily, and *tertiary hydrogen atoms* become **most reactive**. For similar reasons, *secondary hydrogen atoms* would be the **next most reactive** and *primary hydrogen atoms* would be the **least reactive**.

Cleavage of a tertiary C—H bond

3° Radical

Most stable

Cleavage of a secondary C—H bond

2° Radical

Cleavage of a primary C—H bond

1° Radical

Least stable

Reactivity order of hydrogens: 3° > 2° > 1°

The relative reactivity of hydrogen atoms to **chlorination** is estimated to be

1° : 2° : 3° = 1 : 3.8 : 5

whereas, the relative reactivity of hydrogen atoms to **bromination** is:

1° : 2° : 3° = 1 : 82 : 1600.

Determination of relative amounts of monochloro products in a mixture

The relative amounts of monochloro products in a mixture can be estimated by **multiplying** the **relative reactivity** of the hydrogen atom (1°, 2° or 3°) with the corresponding **number of equivalent hydrogen atoms**.

The **percentage** of each product is then obtained by dividing the relative amount of the product by the sum of relative amounts and multiplying by 100.

This is illustrated in Figure 11.2 with monochlorination of propane. Note that propane has 6 equivalent 1° hydrogens that give propyl chloride, and 2 equivalent 2° hydrogens that give isopropyl chloride. The relative reactivity of hydrogen atoms to chlorination is: 1° : 2° = 1 : 3.8.

$$\underset{\text{(6 primary and 2 secondary H's)}}{\overset{1^\circ\ 2^\circ\ 1^\circ}{CH_3CH_2CH_3}} \xrightarrow{Cl_2} \underset{\textbf{Propyl chloride}}{CH_3CH_2CH_2Cl} + \underset{\textbf{Isopropyl chloride}}{CH_3\overset{Cl}{\overset{|}{C}}HCH_3}$$

Reactivity 1° : 2° = 1 : 3.8

	Relative amount (reactivity x number of equivalent H's)	Percentage [(relative amount/sum) x 100]
Propyl chloride (1° chloride)	1 x 6 = 6	(6/13.6) x 100 = 44
Isopropyl chloride (2° chloride)	3.8 x 2 = 7.6	(7.6/13.6) x 100 = 56
	Sum = 13.6	

Figure 11.2: Estimation of relative amounts of propyl chloride and isopropyl chloride and their percentages in the product mixture obtained from monochlorination of propane.

The estimated yields of propyl chloride (44%) and isopropyl chloride (56%) agree well with their observed yields of 45% and 55% respectively.

Problem 11.1: Write the structures of all monochloro products obtained from the chlorination of 2-methylbutane. Calculate their percentages in the product mixture. The relative reactivity of hydrogen atoms to chlorination is 1° : 2° : 3° = 1 : 3.8 : 5.

Answer

2-Methylbutane has four different sets of equivalent H atoms: six equivalent 1° H atoms (labelled *a*), three equivalent 1° H atoms (labelled *b*), two equivalent 2° H atoms (labelled *c*), and one 3° H atom (labelled *d*). Therefore, 2-methylbutane gives four different monochloro products. The structures of the products and the calculation of their percentages in the mixture are shown below.

Starting material: **2-Methylbutane**, (a)(d)(c)(b) $CH_3CHCH_2CH_3$ with CH_3 (a) on C-2; reagents Cl_2, $h\nu$

From	Product	Relative amount	%
From (*a*)	$ClCH_2CH(CH_3)CH_2CH_3$ **1-Chloro-2-methylbutane**	1 x 6 = 6	(6/21.6) x 100 = 28
From (*b*)	$CH_3CH(CH_3)CH_2CH_2Cl$ **1-Chloro-3-methylbutane**	1 x 3 = 3	(3/21.6) x 100 = 14
From (*c*)	$CH_3CHClCH(CH_3)CH_3$ **2-Chloro-3-methylbutane**	3.8 x 2 = 7.6	(7.6/21.6) x 100 = 35
From (*d*)	$CH_3CCl(CH_3)CH_2CH_3$ **2-Chloro-2-methylbutane**	5 x 1 = 5	(5/21.6) x 100 = 23
		Sum = 21.6	

Bromination versus chlorination

Bromine is *less reactive* toward alkanes than chlorine, but the *difference in reactivity* of bromine for different types of hydrogen atoms (1°, 2° or 3°) is far greater when it reacts.

The relative reactivity of hydrogen atoms to bromination is 1° : 2° : 3° = **1 : 82 : 1600**. This implies that **bromine is more selective** than chlorine in attacking the different types of hydrogen atoms.

For example, consider **isobutane**. It has one 3° H atom and nine 1° H atoms but monobromination of isobutane gives practically one product (***t*-butyl bromide**).

This is in sharp contrast to monochlorination of isobutane which yields ***t*-butyl chloride** as the minor product (37%) and **isobutyl chloride** as the major product (63%).

$$\underset{\textbf{Isobutane}}{\overset{1° \quad 3° \quad 1°}{CH_3CHCH_3} \atop {| \atop {CH_3 \atop 1°}}} \xrightarrow[h\nu]{Br_2} \underset{\textbf{\textit{tert}-Butyl bromide (> 99\%)}}{CH_3C(Br)(CH_3)CH_3}$$

$$\underset{\textbf{Isobutane}}{\overset{1° \quad 3° \quad 1°}{CH_3CHCH_3} \atop {| \atop {CH_3 \atop 1°}}} \xrightarrow[h\nu]{Cl_2} \underset{\substack{\textbf{\textit{tert}-Butyl chloride} \\ \textbf{(37\%)} \\ \textbf{\textit{Minor product}}}}{CH_3C(Cl)(CH_3)CH_3} + \underset{\substack{\textbf{Isobutyl chloride} \\ \textbf{(63\%)} \\ \textbf{\textit{Major product}}}}{CH_3CH(CH_3)CH_2Cl}$$

PROBLEM 11.2: An alkane (C_5H_{12}) reacts with chlorine to give a single monochloro product. Predict the structure of alkane.

Answer

Formation of a single monochloro product indicates that twelve H atoms of the alkane are equivalent. This is possible if all H atoms are primary in the five-carbon alkane, which can be accounted for by the presence of four methyl (CH_3) groups. Therefore, the alkane is neopentane (2,2-dimethylpropane).

$$\underset{\substack{\textbf{Neopentane} \\ \textbf{(2,2-Dimethylpropane)}}}{\overset{1°}{CH_3}\text{-}\overset{1°\;\;1°}{C(CH_3)_2}\text{-}\underset{1°}{CH_3}} \xrightarrow[h\nu]{Cl_2} \underset{\textbf{Single monochloro product}}{CH_3C(CH_3)_2CH_2Cl}$$

11.1.4 Allylic Halogenation

A saturated (sp^3) carbon atom *adjacent* to a carbon-carbon double bond (C=C) is called an **allylic carbon atom** and the hydrogen atoms attached to the allylic carbon atom are called **allylic hydrogen atoms**.

For example, methyl carbon in propene is an allylic carbon atom and three methyl hydrogen atoms are allylic hydrogen atoms.

$$—C{=}C—C—$$

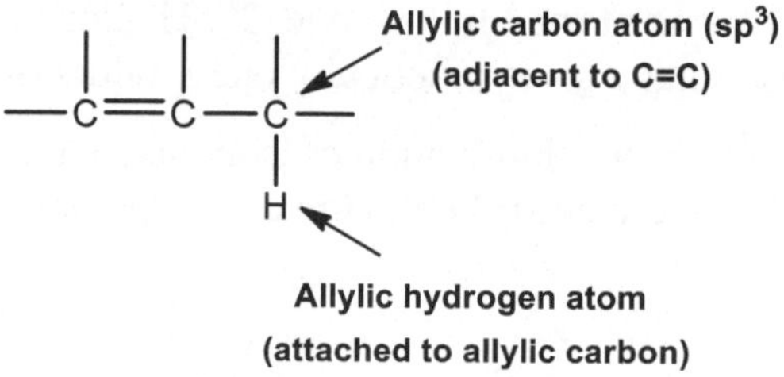

Allylic carbon atom (sp^3)
(adjacent to C=C)

H

Allylic hydrogen atom
(attached to allylic carbon)

Example

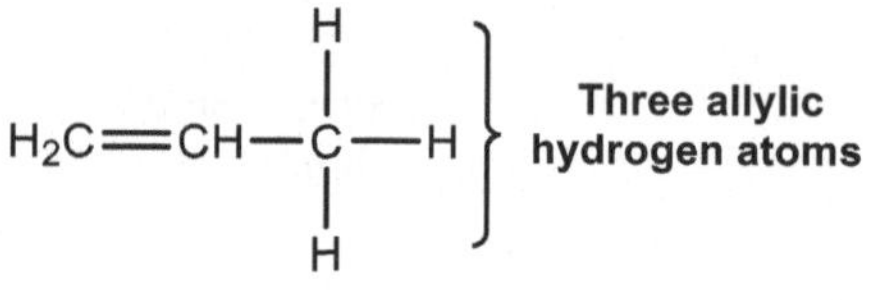

$$H_2C{=}CH—CH_3$$

Three allylic hydrogen atoms

Propene

The substitution of an allylic hydrogen atom by a halogen atom (Cl or Br) is called the **allylic halogenation**.

Allylic chlorination

Allylic chlorination can be carried out with chlorine at a high temperature. For example, propene and chlorine react in the gas phase at 400°C to produce **allyl chloride** (3-chloropropene).

$$H_2C{=}CH—CH_3 + Cl_2 \xrightarrow{400^\circ C} H_2C{=}CH—CH_2Cl + HCl$$

Propene **Allyl chloride (3-chloropropene)**

The **mechanism** for allylic chlorination of propene is similar to the radical chain mechanism for chlorination of methane, and is shown below.

Chain Initiation

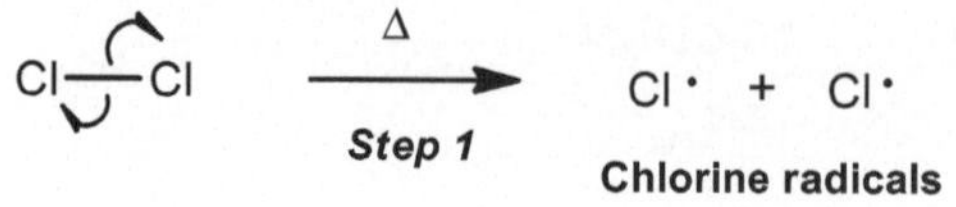

$$Cl—Cl \xrightarrow[\text{Step 1}]{\Delta} Cl^\bullet + Cl^\bullet$$

Chlorine radicals

Chain Propagation

$Cl\cdot + H{-}CH_2{-}CH{=}CH_2 \xrightarrow{\text{Step 2}} HCl + [H_2\dot{C}{-}CH{=}CH_2 \leftrightarrow H_2C{=}CH{-}\dot{C}H_2]$

Propene

Allyl radical

Stabilized by resonance

$H_2C{=}CH{-}\dot{C}H_2 + Cl{-}Cl \xrightarrow{\text{Step 3.}} H_2C{=}CH{-}CH_2Cl + Cl\cdot$

Allyl chloride

Step 1 is the chain initiation, and **steps 2 and 3** denote the chain propagation (cf. Figure 11.1). In **step 2**, the carbon radical generated in chain propagation is an **allyl radical** which is stabilized by **resonance**. Two resonance structures of allyl radical are equivalent.

Allylic bromination

Allylic bromination is usually carried out under conditions in which the *concentration of bromine is very small*; otherwise, bromine addition to the double bond (C=C) will compete.

A useful reagent for allylic bromination is ***N*-Bromosuccinimide (NBS)**.

Succinimide

N-Bromosuccinimide (NBS)

To illustrate, consider propene and cyclohexene. **Propene** reacts with NBS in CCl_4 in the presence of peroxides or light to produce **allyl bromide** (3-bromopropene). Similarly, **cyclohexene** undergoes allylic bromination with NBS to give **3-bromocyclohexene**.

$H_2C{=}CH{-}CH_3 + \text{NBS} \xrightarrow[CCl_4]{\text{Peroxide or } h\nu} H_2C{=}CH{-}CH_2Br$

Propene

NBS

Allyl bromide (3-bromopropene)

Cyclohexene + NBS —(Peroxide, CCl_4)→ 3-Bromocyclohexene

The allylic bromination is a **radical chain reaction** (cf. mechanism of allylic chlorination) in which bromine (Br_2), *not* NBS, is involved in a chain propagation step to produce an allylic bromide.

In **step 1**, chain initiation involves the homolytic cleavage of the N–Br bond in NBS to generate the bromine radical.

In **step 2**, the Br radical abstracts an allylic hydrogen atom to form the corresponding allylic radical and HBr.

The HBr formed then reacts with NBS in **step 3** to give Br_2 in low concentration. Thus **NBS provides a constant but very low concentration of bromine** in the reaction mixture. Notably, this step occurs by an *ionic mechanism.*

In **step 4**, the allylic radical abstracts a Br atom from Br_2 molecule to produce the allylic bromide.

Chain Initiation

NBS —(*hν*, *Step 1*)→ succinimidyl radical (N·) + Br· (**Bromine radical**)

Chain Propagation

Br· + H–CH_2–CH=CH_2 (**Propene**) —(*Step 2*)→ HBr + $H_2\dot{C}$–CH=CH_2 (**Allyl radical**)

NBS (N–Br) + H–Br —(*Step 3*)→ succinimide (N–H) + Br_2

Br–Br + $H_2\dot{C}$–CH=CH_2 —(*Step 4*)→ Br· + $H_2C(Br)$–CH=CH_2 (**Allyl bromide**)

11.2 REACTIONS OF ALKYL HALIDES: NUCLEOPHILIC SUBSTITUTION

The most important reactions of alkyl halides involve a **nucleophilic substitution at a saturated (sp^3) carbon**.

In this type of substitution, a **nucleophile** reacts with an alkyl halide (the **substrate**) by replacing the halogen substituent (the **leaving group**). Since the substitution reaction is initiated by a nucleophile, it is called a nucleophilic substitution.

The nucleophile is usually a species with at least one lone pair of electrons on an atom such as carbon, nitrogen, oxygen or halogen atom. In the substitution reaction of alkyl halides, the leaving group departs as a halide ion, and the *departing anion is often indicated as the leaving group*.

$$\text{—C—X} + \text{Nu:}^- \longrightarrow \text{—C—Nu} + \text{X:}^-$$

sp^3 (X = Cl, Br, I)

Alkyl halide (substrate) **Nucleophile** **Product** **Halide ion (leaving group)**

Examples

1. **Oxygen nucleophiles** with lone pairs on the oxygen atom are quite common in nucleophilic substitution of alkyl halides. The following are three examples with HO^- (hydroxide ion), CH_3O^- (methoxide ion) and CH_3COO^- (acetate ion) as nucleophiles:

$$CH_3CH_2\text{—}Cl + {}^-OH \xrightarrow{H_2O} CH_3CH_2\text{—}OH + Cl^-$$

Ethyl chloride (NaOH) **Ethyl alcohol**

$$CH_3CH_2CH_2\text{—}Br + CH_3O^- \xrightarrow{CH_3OH} CH_3CH_2CH_2\text{—}OCH_3 + Br^-$$

Propyl bromide (CH_3ONa) **Methyl propyl ether**

$$CH_3CH_2CH_2CH_2\text{—}Br + CH_3\text{—}\overset{O}{\overset{\|}{C}}\text{—}O^- \xrightarrow{DMF} CH_3CH_2CH_2CH_2\text{—}O\text{—}\overset{O}{\overset{\|}{C}}\text{—}CH_3 + Br^-$$

Butyl bromide (CH_3CO_2K) **(dimethylformamide)** **Butyl acetate**

Note that three lone pairs on the anionic oxygen atom are not shown (see Section 7.4.1).

2. **Halide ions** also act as nucleophiles. For example, treatment of butyl chloride with I^- gives butyl iodide.

$$CH_3CH_2CH_2CH_2{-}Cl + \underset{(KI)}{I^-} \xrightarrow{\text{Acetone}} CH_3CH_2CH_2CH_2{-}I + Cl^-$$

Butyl chloride → **Butyl iodide**

3. A useful **carbon nucleophile** is cyanide ion (CN^-) with a lone pair on carbon atom. For example, the reaction of propyl bromide with cyanide ion gives propyl cyanide (butyronitrile) as the substitution product.

   Cyanides or nitriles are useful intermediates in synthesis. They can be easily converted to **amines** by reduction or **carboxylic acids** by hydrolysis.

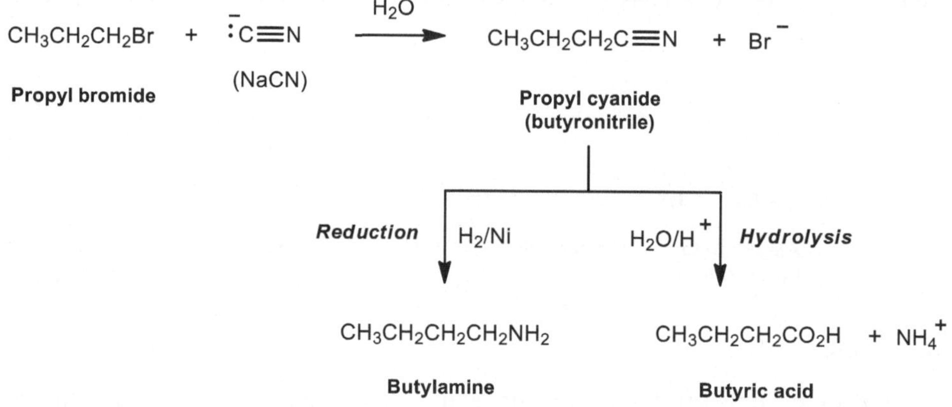

4. Acetylenic hydrogen atoms are sufficiently acidic so that they can be abstracted by a strong base. Treatment of acetylene with sodium amide ($NaNH_2$) in liquid ammonia gives sodium acetylide ($HC{\equiv}C^-\,Na^+$).

   Acetylide ion ($HC{\equiv}C^-$) can act as a **carbon nucleophile** in nucleophilic substitution. For example, reaction of methyl bromide with acetylide ion yields propyne.

$$HC{\equiv}C{-}H \xrightarrow[\text{liq. } NH_3]{NaNH_2} HC{\equiv}C^-\,Na^+ + NH_3$$

Acetylene → **Sodium acetylide**

$$CH_3{-}Br + {:}\bar{C}{\equiv}CH \xrightarrow{\text{liq. } NH_3} H_3C{-}C{\equiv}CH + Br^-$$

Methyl bromide → **Propyne**

In general, this reaction represents *alkylation of alkynes* (substitution of the acetylenic hydrogen atom of alkynes by alkyl groups).

Mechanisms

Nucleophilic substitution of alkyl halides can proceed by **ionic mechanisms**, of which two are most important.

One mechanism is called the **S$_N$2** (substitution, nucleophilic, bimolecular) mechanism, and the other is called the **S$_N$1** (substitution, nucleophilic, unimolecular) mechanism. These are described in the following sections.

11.3 S_N2 (SUBSTITUTION, NUCLEOPHILIC, BIMOLECULAR) REACTIONS

Consider the following nucleophilic substitution reaction of methyl chloride with hydroxide ion in aqueous solution:

$$\underset{\text{Methyl chloride}}{CH_3Cl} + OH^- \xrightarrow[60^oC]{H_2O} \underset{\text{Methyl alcohol}}{CH_3OH} + Cl^-$$

The rate of the reaction is found to depend on both the concentration of CH_3Cl and the concentration of OH^-.

$$\text{Rate} = k\,[CH_3Cl]\,[OH^-]$$

Therefore, the reaction is **second order** overall. In consistent with the second order kinetics, an ionic mechanism can be proposed for the reaction that involves a single, concerted step with a single transition structure (TS) (see also Section 7.8.2).

$$HO^- + H_3C\text{—}Cl \longrightarrow \underset{\text{Transition structure (TS)}}{\left[\overset{\delta-}{HO}\text{---}CH_3\text{---}\overset{\delta-}{Cl}\right]^{\ddagger}} \longrightarrow HO\text{—}CH_3 + Cl^-$$

The reaction involves the breaking of C–Cl bond and the making of C–OH bond. In the TS, the C–Cl bond is partially broken and the C–OH bond is partially formed. Since **two** species (CH_3Cl and HO^-) are involved in the TS, the mechanism is **bimolecular**. The mechanism is fully designated as **S_N2 (substitution, nucleophilic, bimolecular)**.

A **free energy diagram** for the one-step reaction between methyl chloride and hydroxide ion has been described earlier (see Figure 7.2).

Reactions which take place by the S_N2 mechanism are called **S_N2 reactions**. Some S_N2 reactions may be *fast*, while some others may be quite *slow*.

Several factors affect the **relative rates** of S_N2 reactions. The most important factors are: the structure of the **substrate** (alkyl halide), the strength of the **nucleophile**, the nature of the **leaving group**, and the effect of the **solvent**.

11.3.1 Substrate Structures

An alkyl halide (substrate) may be a **methyl halide** or it may be a **primary, secondary or tertiary halide**. The order of reactivity of simple alkyl halides in S_N2 reactions generally follows the order:

$$\underset{\text{Methyl halide}}{H\text{—}CH_2\text{—}X} > \underset{1^o\text{ Alkyl halide}}{R\text{—}CH_2\text{—}X} > \underset{2^o\text{ Alkyl halide}}{R\text{—}CHR\text{—}X} \gg \underset{3^o\text{ Alkyl halide}}{R\text{—}CR_2\text{—}X}$$

This order of reactivity is mainly attributed to **steric effects**. The approach of the nucleophile to the tetrahedral carbon bearing the leaving group (X) may be hindered by the substituents (R's) present at the reacting site. If the substituent groups are bulky, the **steric hindrance** to the approaching nucleophile will be large. As a result of the steric hindrance, the S_N2 reaction will get slower.

Methyl halides (CH_3X) react most rapidly because three small hydrogen atoms interfere minimally with the approaching nucleophile. On the other hand, tertiary halides (R_3X) react extremely slowly because of a severe steric hindrance from the three large alkyl groups. In practice, tertiary halides *do not react* by the S_N2 mechanism.

Nu⁻ (approach is sterically hindered) → R_3C—X —S_N2 (✕)→

3° Alkyl halide

The S_N2 reactions of methyl halides and 1° alkyl halides are highly favourable. For example, all reactions described earlier in Section 11.2 proceed by the S_N2 mechanism.

Allyl halides and benzyl halides

If a primary (1°) alkyl halide has a π system (e.g. C=C or benzene ring) *adjacent* to the reacting primary carbon, the rate of the S_N2 reaction is greatly increased.

For example, **allyl chloride** with a C=C adjacent to the reacting carbon undergoes an S_N2 reaction with iodide ion 80 times **faster** than ethyl chloride. Similarly, **benzyl chloride** with a benzene ring adjacent to the reacting carbon reacts with iodide ion 200 times **faster** than ethyl chloride.

Reaction (Acetone, S_N2)	Relative rate
$CH_3CH_2{-}Cl$ (1°) + I^- → $CH_3CH_2{-}I$ + Cl^- **Ethyl chloride** → **Ethyl iodide**	**1**
$CH_2{=}CH{-}CH_2{-}Cl$ (1°) + I^- → $CH_2{=}CH{-}CH_2{-}I$ + Cl^- **Allyl chloride** → **Allyl iodide**	**80**
$C_6H_5{-}CH_2{-}Cl$ (1°) + I^- → $C_6H_5{-}CH_2{-}I$ + Cl^- **Benzyl chloride** → **Benzyl iodide**	**200**

The increased reactivity of allyl chloride or benzyl chloride in S_N2 reactions is attributed to **electronic effects**. In an S_N2 TS, the hybridization of the reacting carbon becomes sp^2 when the two lobes of the p orbital are involved in the formation of partial bonds with the nucleophile and the leaving group. The π system of the adjacent double bond or aromatic ring can **stabilize the transition structure by conjugation with the p orbital** at the reacting carbon. This leads to the *lowering* of the free energy barrier ($\Delta G^{\ddagger}$) (cf. Figure 7.2), and hence an *increase* in the rate of the reaction.

The following is an orbital picture for the stabilization of the TS for the S_N2 reaction of an allyl halide.

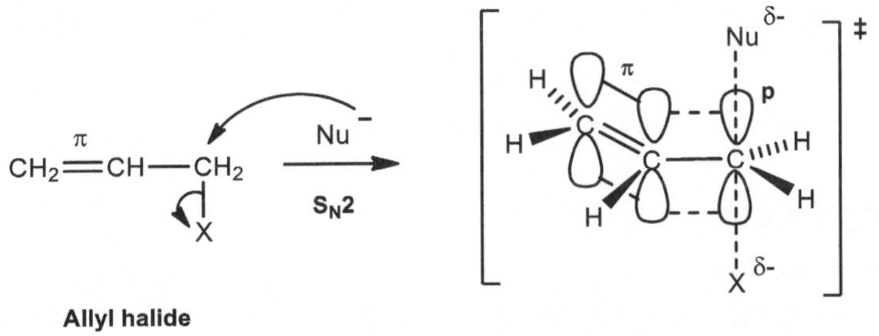

Allyl halide

Transition structure (TS)

Stabilized by π-p conjugation

PROBLEM 11.3: Neopentyl bromide is a primary alkyl bromide but it does not undergo an S_N2 reaction. Explain.

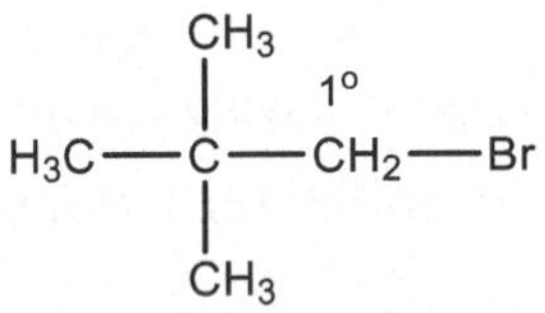

Neopentyl bromide

Answer

In neopentyl bromide, the primary carbon bearing Br (leaving group) has a very bulky *t*-butyl substituent [$(CH_3)_3C$] which gives a severe steric hindrance to the approaching nucleophile. Hence neopentyl bromide is unreactive in the S_N2 reaction.

This is an example where a *single* bulky substituent at the reactive site inhibits an S_N2 reaction.

11.3.2 Nucleophiles

An S_N2 reaction involves the **nucleophile** in the single and **rate-determining (r/d) step** of the reaction. Thus, the stronger or more reactive the nucleophile is, the greater is the rate of the reaction.

In general, **a negatively charged nucleophile is more reactive than the analogous uncharged nucleophile**. Thus, HO^- (hydroxide ion) is a better nucleophile (more reactive) than H_2O. Similarly, RO^- (alkoxide ion) is a better nucleophile than ROH. The ability of a species to act as a nucleophile is called its nucleophilicity.

The nucleophilicity order is

$HO^- > H_2O$; $RO^- > ROH$

In a group of nucleophiles in which the nucleophilic atom is the *same*, **nucleophilicity parallels basicity**.

For example, consider the three oxygen nucleophiles: RO^-, HO^- and RCO_2^-. Since the order of basicity is $RO^- > HO^- > RCO_2^-$ (see Section 5.6), RO^- is more nucleophilic than HO^-, which in turn is more nucleophilic than RCO_2^- (a carboxylate ion).

Basicity: $RO^- > HO^- > RCO_2^-$

Nucleophilicity: $RO^- > HO^- > RCO_2^-$

In a group of nucleophiles in which the nucleophilic atom is *not* the same, nucleophilicity does not always parallel basicity (see Section 11.3.4).

11.3.3 Leaving Groups

In the nucleophilic substitution of alkyl halides, the leaving group departs as a negative ion (halide ion). As such, the departing anion is also indicated as the **leaving group**.

A good leaving group is an anion of a strong acid. The stronger the acid, the weaker is the conjugate base. Thus, a good leaving group is a **weak base**.

For example, consider the halide ions (X^-) as leaving groups. The acidity order of HX is: HF < HCl < HBr < HI (see Problem 5.5). Since iodide ion (I^-) is the anion of the strongest acid (HI), it is the best leaving group. On the other hand, the fluoride ion (F^-) is the anion of the weakest acid (HF), so it is a very poor leaving group.

Thus, the order of the **leaving group ability (or leavability)** of halide ions is:

$I^- > Br^- > Cl^- > F^-$.

To illustrate, consider the relative rates of the following substitution reactions of ethyl halides:

$$CH_3CH_2X \xrightarrow{OH^-} CH_3CH_2OH + X^-$$

(X = F, Cl, Br. I)

Ethyl halides → **Ethyl alcohol**

X	Relative rate
F	**Very slow**
Cl	**Moderate**
Br	**Fast**
I	**Very fast**

The reaction is very fast with ethyl iodide (CH_3CH_2I) because I^- is the best leaving group, whereas the reaction of ethyl fluoride (CH_3CH_2F) is very slow as F^- is the poorest leaving group.

11.3.4 Solvent Effects

Polar solvents which dissolve ionic compounds are normally used to carry out a nucleophilic substitution reaction. Polar solvents are of two types, called **polar protic solvents** and **polar aprotic solvents** (see Section 5.2.3).

We will describe here the effect of solvents on the relative nucleophilicitiy of halide ions.

Polar protic solvents

The **relative nucleophilicity** of halide ions in a **polar protic solvent** such as water or an alcohol is:

$I^- > Br^- > Cl^- > F^-$

This order of nucleophilicity in a polar protic solvent is *opposite* to the order of basicity: $I^- < Br^- < Cl^- < F^-$. Here the **nucleophilicity does not parallel basicity**. The weakest base (I^-) is the strongest nucleophile, while the strongest base (F^-) is the weakest nucleophile.

The nucleophilicity order can be explained on the basis of **differential solvation** of halide ions in a polar protic solvent. A polar protic solvent (such as water) has a hydrogen atom attached to an electronegative oxygen atom and is therefore capable of solvating the nucleophile (halide ion) through hydrogen bonding.

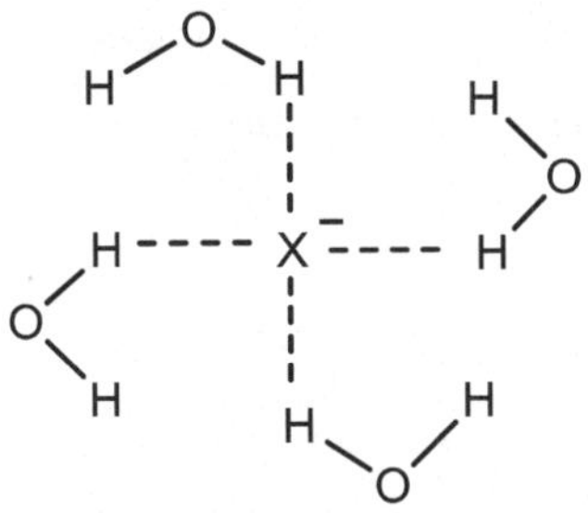

Solvation of halide ion (X^-) by hydrogen bonding in a polar protic solvent (water)

(X = F, Cl, Br, I)

The differential solvation arises due to the different sizes of halide ions. **The size increases from F^- → Cl^- → Br^- → I^-.** In a *smaller* halide ion, the negative charge is more concentrated, and so *hydrogen bonds* to a smaller ion are *stronger* than those to a larger ion. Therefore, **F^- is most strongly solvated and I^- is most weakly solvated through hydrogen bonding**.

For a halide ion to react as a nucleophile, it must shed at least some of the solvent molecules to donate an electron pair to the carbon bearing the leaving group. Since **I^-** is solvated most weakly, it can shed the solvent molecules most easily and become **most nucleophilic**. On the other hand, it is most difficult for the highly solvated **F^-** to shed the solvent molecules, and it will be **least nucleophilic**.

Another factor responsible for the nucleophilicity order is the **polarizability of the halide ions** (see Section 5.1). A lone pair on a larger halide ion is less tightly held by the nucleus and is more polarizable than a smaller one whose electrons are more tightly held. Thus a larger ion can donate an electron pair more easily and so it will be more nucleophilic.

Polar aprotic solvents

Interestingly, in a **polar aprotic solvent** such as acetone, dimethyl sulphoxide (DMSO) or *N,N*-dimethylformamide (DMF), the nucleophilicity order of the halide ions is reversed:

$$F^- > Cl^- > Br^- > I^-$$

A **polar aprotic solvent** has *no* hydrogen atom attached to electronegative oxygen or nitrogen atom and therefore it **cannot solvate the halide ions (X^-) by hydrogen bonding**. But it can solvate the counter ions (metal cations, M^+) of the nucleophilic reagents (MX) by donating an unshared electron pair to the metal. Thus, the **halide ions** are much less solvated in a polar aprotic solvent and therefore **more free** to react.

Polar aprotic solvents:

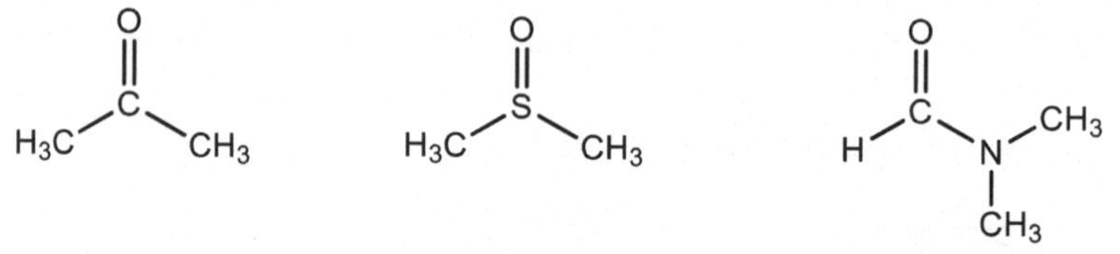

Acetone

Dimethyl sulphoxide (DMSO)

***N,N*-Dimethylformamide (DMF)**

Solvation:

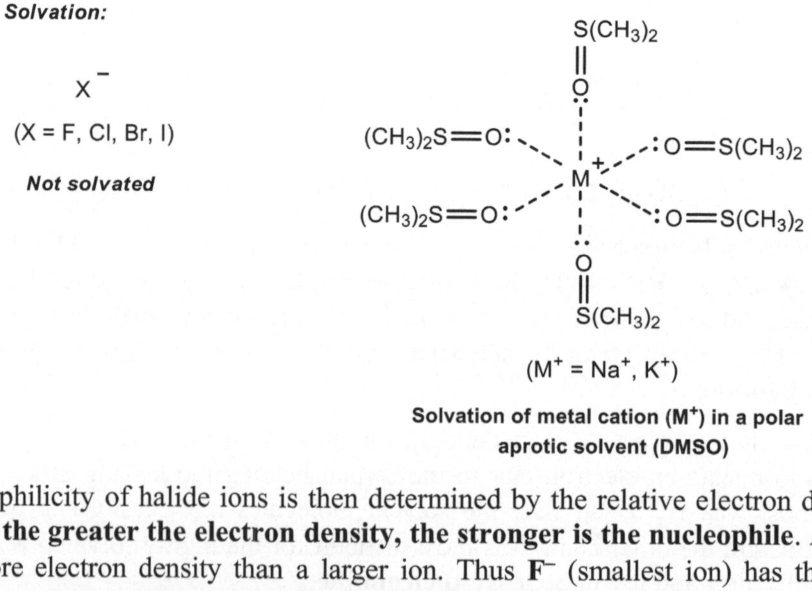

Solvation of metal cation (M^+) in a polar aprotic solvent (DMSO)

The nucleophilicity of halide ions is then determined by the relative electron density on the anions: **the greater the electron density, the stronger is the nucleophile**. A smaller ion has more electron density than a larger ion. Thus $\mathbf{F^-}$ (smallest ion) has the highest electron density and is **most nucleophilic**, whereas $\mathbf{I^-}$ (largest ion) has the lowest electron density and is **least nucleophilic**. Note that in this case, nucleophilicity parallels basicity: $F^- > Cl^- > Br^- > I^-$.

In general, the rates of S_N2 reactions are increased in polar aprotic solvents since the 'naked' or 'free' nucleophiles in these solvents are highly reactive.

11.4 S_N1 (SUBSTITUTION, NUCLEOPHILIC, UNIMOLECULAR) REACTIONS

Consider the following nucleophilic substitution reaction of ***t*-butyl chloride** with dilute sodium hydroxide solution (0.05M) in water:

$$(CH_3)_3C{-}Cl + OH^- \xrightarrow{H_2O} (CH_3)_3C{-}OH + Cl^-$$

***t*-Butyl chloride** ***t*-Butyl alcohol**

It has been observed that the rate of the reaction depends only on the concentration of $(CH_3)_3CCl$, and is *independent* of the concentration of OH^-.

Rate = k [$(CH_3)_3CCl$]

Thus the reaction is **first order** overall. A stepwise ionic mechanism is proposed for the reaction in which the **rate-determining (r/d) step** does *not* involve the hydroxide ion (OH^-) consistent with the rate equation.

$$(CH_3)_3C{-}Cl \xrightarrow[\text{(r/d) Step 1}]{\text{Slow}} (CH_3)_3C^+ + Cl^- \xrightarrow[\text{Fast Step 2}]{^-OH} (CH_3)_3C{-}OH$$

***t*-Butyl chloride** **Carbocation (*t*-butyl cation)** ***t*-Butyl alcohol**

The mechanism involves two steps. **Step 1** is the ionization of *t*-butyl chloride to a **carbocation** (*t*-butyl cation) and a chloride ion. This step is the **slow, r/d step**. In **step 2**, the intermediate carbocation reacts rapidly with hydroxide ion to produce *t*-butyl alcohol.

Since only one species (*t*-butyl chloride) is involved in the r/d step, the mechanism is **unimolecular**. The mechanism is fully designated as **S_N1 (substitution, nucleophilic, unimolecular)**.

A **free energy diagram** for this two-step S_N1 reaction is similar to the free energy diagram of a two-step reaction described previously (see Figure 7.3). Try yourself to draw the free energy diagram for the S_N1 reaction, where the reactant is [$(CH_3)_3CCl$ + OH^-], the intermediate is [$(CH_3)_3C^+ + Cl^- + OH^-$] and the product is [$(CH_3)_3COH + Cl^-$].

Reactions which take place by the S_N1 mechanism are called **S_N1 reactions**. The most familiar S_N1 reactions are **solvolysis** of alkyl halides.

11.4.1 Solvolysis: Solvent as the Nucleophile

In an S_N1 reaction, the nature or strength of the nucleophile is not important because it does not participate in the r/d step of the reaction.

Solvents are usually poor nucleophiles but they can react as a **nucleophile** in S_N1 reactions. The reaction is called **solvolysis** (cleavage by the solvent). Solvolysis is most common for S_N1 reactions, where **solvation can stabilize a forming carbocation intermediate**.

If the solvent is water, the solvolysis is called **hydrolysis**. Similarly, if the S_N1 reaction is carried out in methanol (CH_3OH), formic acid (HCO_2H) or acetic acid (CH_3CO_2H) as a solvent, the solvolysis is known as **methanolysis**, **formolysis** and **acetolysis**, respectively.

The following are some examples of S_N1 solvolysis:

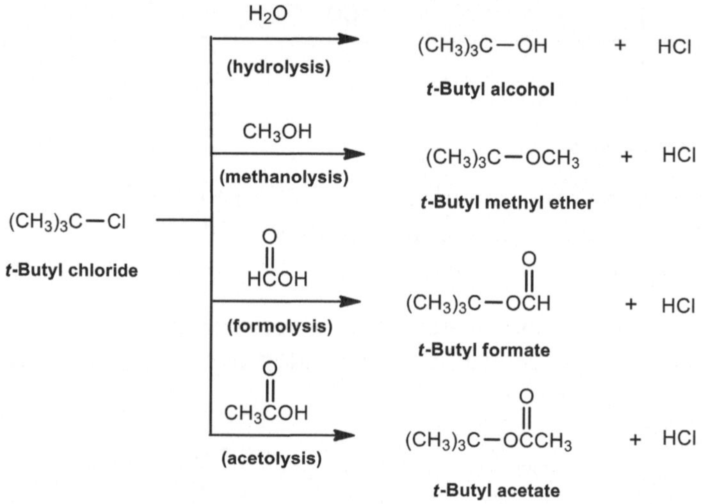

$(CH_3)_3C{-}Cl$ (*t*-Butyl chloride) $\xrightarrow[\text{(hydrolysis)}]{H_2O}$ $(CH_3)_3C{-}OH$ (*t*-Butyl alcohol) + HCl

$(CH_3)_3C{-}Cl$ $\xrightarrow[\text{(methanolysis)}]{CH_3OH}$ $(CH_3)_3C{-}OCH_3$ (*t*-Butyl methyl ether) + HCl

$(CH_3)_3C{-}Cl$ $\xrightarrow[\text{(formolysis)}]{HCOOH}$ $(CH_3)_3C{-}OCH{=}O$ (*t*-Butyl formate) + HCl

$(CH_3)_3C{-}Cl$ $\xrightarrow[\text{(acetolysis)}]{CH_3COOH}$ $(CH_3)_3C{-}OC(=O)CH_3$ (*t*-Butyl acetate) + HCl

Mechanism

A mechanism for the S_N1 solvolysis (hydrolysis) of ***t*-butyl chloride** in water is shown in Figure 11.3.

The mechanism involves *three* steps. **Step 1** involves the formation of a **carbocation** (*t*-butyl cation), which reacts with a solvent molecule (water) as a nucleophile in **step 2** to form a **protonated alcohol** (protonated *t*-butyl alcohol). In **step 3**, the removal of a proton from the protonated alcohol by the solvent molecule (water) acting as a base produces ***t*-butyl alcohol**.

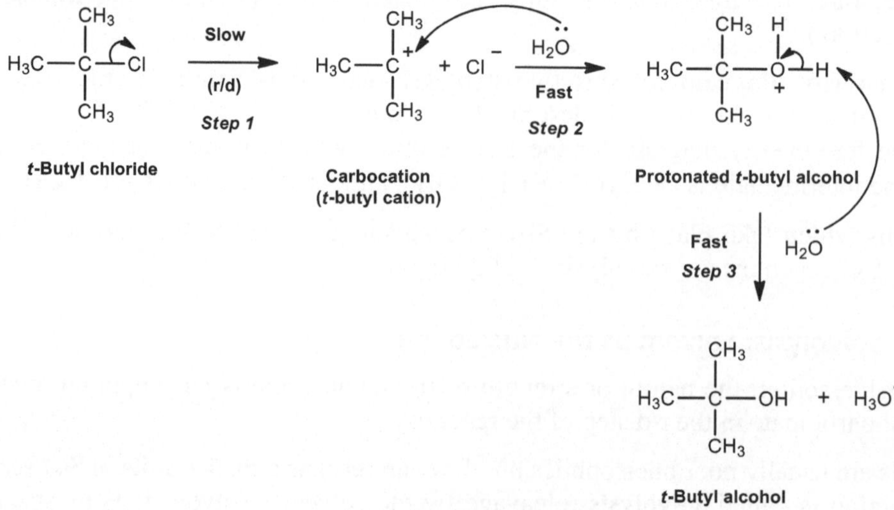

Step 1 (Slow, r/d): $(CH_3)_3C{-}Cl$ (*t*-Butyl chloride) → $(CH_3)_3C^+$ (Carbocation (*t*-butyl cation)) + Cl^-

Step 2 (Fast): $(CH_3)_3C^+ + H_2O$ → $(CH_3)_3C{-}O^+H_2$ (Protonated *t*-butyl alcohol)

Step 3 (Fast): $(CH_3)_3C{-}O^+H_2 + H_2O$ → $(CH_3)_3C{-}OH$ (*t*-Butyl alcohol) + H_3O^+

Figure 11.3: An S_N1 mechanism for the solvolysis (hydrolysis) of *t*-butyl chloride in water.

In the S_N1 reaction, the ionization of an alkyl halide to a carbocation and a halide ion is the r/d step. A polar protic solvent such as water can solvate and stabilize the carbocation by ion-dipole interactions (see Section 5.1) and thereby facilitates its formation. The halide ion formed by ionization is also solvated by water through hydrogen bonding.

Therefore, S_N1 reactions are promoted in polar protic solvents. In contrast, S_N2 reactions are favoured in polar aprotic solvents (see Section 11.3.4).

11.4.2 Substrate Structures and Leaving Groups

Alkyl halides as substrates form a carbocation in the r/d step of S_N1 reactions. Thus, substrates that can form relatively **stable carbocations** are capable of undergoing substitution reactions by the S_N1 mechanism.

The order of stability of carbocations is:

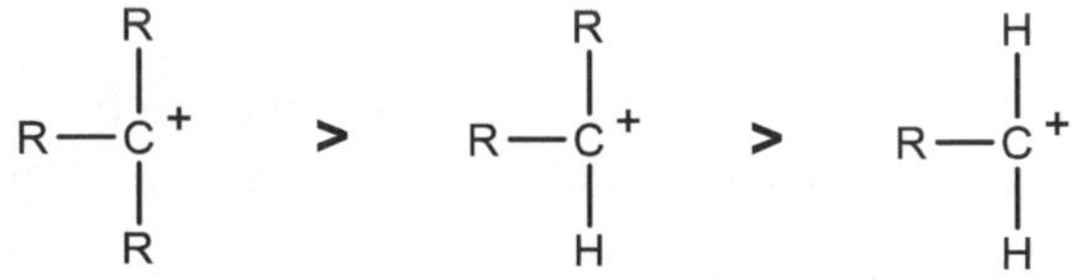

R_3C^+ > R_2CH^+ > RCH_2^+

3° Carbocation **2° Carbocation** **1° Carbocation**

In practice, mainly **tertiary alkyl halides** (e.g. *t*-butyl chloride) that can form a stable tertiary carbocation react by the S_N1 mechanism (see Figure 11.3).

Allyl halides and benzyl halides

Allylic halides such as allyl chloride and **benzylic halides** such as benzyl chloride can also undergo S_N1 reactions.

Allyl chloride ($CH_2{=}CHCH_2Cl$) and benzyl chloride ($C_6H_5CH_2Cl$) are primary alkyl halides but they can form **carbocations** (allyl carbocation and benzyl carbocation) that are **stabilized by resonance**.

$CH_2{=}CH{-}\overset{+}{C}H_2 \longleftrightarrow \overset{+}{C}H_2{-}CH{=}CH_2$ ***Stabilzed by resonance***

Allyl carbocation

$C_6H_5{-}\overset{+}{C}H_2 \longleftrightarrow \; =CH_2 \longleftrightarrow \; =CH_2 \longleftrightarrow \; =CH_2$

Benzyl carbocation

Stabilized by resonance

Since allyl chloride and benzyl chloride form relatively stable carbocations, they can undergo S_N1 solvolysis (hydrolysis) in water.

$$CH_2{=}CHCH_2Cl \xrightarrow[S_N1]{H_2O} CH_2{=}CHCH_2OH + HCl$$

Allyl chloride → **Allyl alcohol**

$$C_6H_5CH_2Cl \xrightarrow[S_N1]{H_2O} C_6H_5CH_2OH + HCl$$

Benzyl chloride → **Benzyl alcohol**

Remember that allyl chloride and benzyl chloride can also react by the S_N2 mechanism (see Section 11.3.1).

Leaving groups

In both S_N1 and S_N2 reactions, the leaving group is involved in the r/d step. So the effect of the leaving group in S_N1 reactions is similar to that for S_N2 reactions.

Since I^- is a better leaving group than Br^- which, in turn, is a better leaving group than Cl^-, the order of **reactivity of alkyl halides** in S_N1 reactions is:

R–I > R–Br > R–Cl

PROBLEM 11.4: The following S_N1 solvolysis (methanolysis) proceeds rapidly, although the substrate is a primary halide. Explain and give the structure of the product.

$$CH_3OCH_2Cl \xrightarrow[S_N1]{CH_3OH}$$

Chloromethyl methyl ether

Answer

The S_N1 reaction proceeds well because the substrate forms a relatively stable carbocation by the departure of chloride ion (leaving group). The intermediate carbocation is stabilized by resonance with an adjacent lone pair on oxygen atom. The product is methoxymethyl methyl ether.

$$CH_3O{-}CH_2{-}Cl \xrightarrow{-Cl^-} \left[CH_3\ddot{O}{-}\overset{+}{C}H_2 \leftrightarrow CH_3\overset{+}{O}{=}CH_2\right] \xrightarrow{CH_3\ddot{O}H} CH_3O{-}CH_2{-}\overset{+}{O}(H)CH_3 \xrightarrow{-H^+} CH_3OCH_2{-}OCH_3$$

Stabilized by resonance

Methoxymethyl methyl ether

11.5 STEREOCHEMISTRY OF NUCLEOPHILIC SUBSTITUTION

For stereochemistry, we generally consider a **substrate with a chiral centre** that undergoes a nucleophilic substitution at the chiral centre. Then we determine what **change in configuration** at the chiral centre (chiral carbon) occurs as a result of nucleophilic substitution. (Remember that the configuration at a chiral carbon denotes a particular arrangement of groups or ligands in space around that carbon atom; see Section 4.7).

There are three possible stereochemical results, namely, **retention**, **inversion** or **racemization** for the nucleophilic substitution at a chiral centre:

(a) **Retention of configuration** implies that the nucleophile assumes the *original* position of the leaving group, i.e. when the configuration of the product resembles the configuration of the substrate.

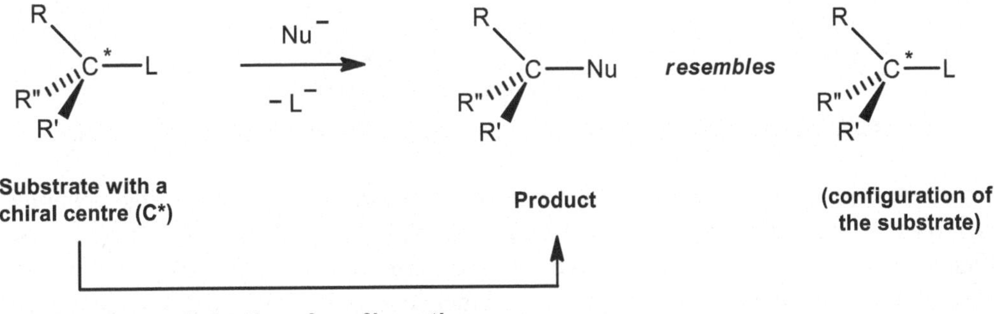

Retention of configuration

(b) **Inversion of configuration** implies that the nucleophile assumes a position on the side of the chiral carbon *opposite* to the original position of the leaving group, i.e. when the configuration of the product resembles the mirror image configuration of the substrate.

Nu⁻

– L⁻

resembles

Substrate with a chiral centre (C*)

Product

(mirror image configuration of the substrate)

Inversion of configuration

(c) **Racemization** implies equal amounts of retention and inversion.

Nu⁻

– L⁻

(1 : 1)

Substrate with a chiral centre (C*)

(retention)

(inversion)

Racemization

The S_N2 and S_N1 reactions of chiral substrates can lead to different stereochemical outcomes.

11.5.1 The Stereochemistry of S_N2 Reactions

The most favourable pathway for an S_N2 reaction requires that the **nucleophile approaches from the backside**, i.e. from the side directly opposite to the leaving group, as shown in Figure 11.4.

Nu⁻ (approaches from backside) + C–X (sp³) → [Nu---C---X]‡ (δ-, δ-, sp², p) → Nu—C (sp³) + X⁻

Transition structure (TS)

Trigonal bipyramidal

Figure 11.4: An S_N2 mechanism involving backside attack of the nucleophile.

As the reaction progresses, the hybridization of the carbon atom under attack changes from **sp^3 to sp^2** in the transition structure (TS). In the TS, the central carbon atom has a **p orbital.** One lobe of the p orbital is involved in the formation of a *partial bond* between the carbon atom and the nucleophile, and the other lobe is involved for developing a *partial bond* between the carbon atom and the leaving group. The other three atoms or groups on the central carbon atom lie on a *perpendicular plane*. Overall, **the geometry of the five-coordinate TS is trigonal bipyramidal.**

In the product, the hybridization of the reacting carbon atom is re-established and becomes **sp^3** with the loss of the leaving group.

An analogy for the whole process is like this: **the groups on the central carbon atom end up turning inside out, just like an umbrella inverts or turns inside out in a high wind.**

Chiral substrates with a chiral centre

If the substrate (alkyl halide) is chiral and the nucleophile attacks **a chiral centre (chiral carbon)**, the S_N2 mechanism will lead to **inversion of configuration.**

HO⁻ + R, H, R' C*–Br → (S_N2) → HO–C*(R)(H)(R') + Br⁻

(* = chiral centre)

Chiral substrate (chiral alkyl halide)

Chiral product (chiral alcohol)

Inversion of configuration

An example of a chiral alkyl halide is **2-chlorobutane**. It contains a **chiral centre (C2)** and has two enantiomers.

Consider an S_N2 reaction of *one* of the enantiomers of 2-chlorobutane with sodium hydroxide. The reaction leads to **inversion of configuration** and produces a **single enantiomer** of chiral **2-butanol**.

$CH_3CH_2\overset{2}{C}H(Cl)\overset{1}{C}H_3$ (chiral centre: C2)

2-Chlorobutane

Cl—C(C_2H_5)(H)(CH_3) $\xrightarrow[S_N2]{HO^-}$ (C_2H_5)(H)(H_3C)C—OH

2-Chlorobutane (one enantiomer) = (*R*)-2-Chlorobutane $[\alpha]_D^{25} = -36.0$

2-Butanol (single enantiomer) = (*S*)-2-Butanol $[\alpha]_D^{25} = +13.5$

In the above reaction, the enantiomer of 2-chlorobutane has *R* configuration, while the enantiomer of 2-butanol has *S* configuration. Thus, **(*R*)-2-chlorobutane reacts with NaOH to give (*S*)-2-butanol.**

It may be mentioned that (*R*)-2-chlorobutane is levorotatory with a specific rotation $[\alpha]_D^{25} = -36.0$, whereas (S)-2-butanol is dextrorotatory with $[\alpha]_D^{25} = +13.5$.

If the **other enantiomer** of 2-chlorobutane, i.e. **(*S*)-2-chlorobutane** is reacted with NaOII, the product will be **(*R*)-2-butanol** (draw the scheme).

Note: The chiral product (2-butanol) has two possible enantiomers. Thus the formation of a single enantiomer of 2-butanol from a particular enantiomer of 2-chlorobutane suggests that the S_N2 reaction occurs with *complete* (100%) inversion of configuration.

PROBLEM 11.5: Predict the stereochemistry of the product in each of the following reactions:

(a) (*R*)-2-Bromobutane $\xrightarrow{KI}$

(b) *cis*-1-Chloro-4-methylcyclohexane $\xrightarrow{NaOH}$

Answer

Each reaction proceeds by an S_N2 mechanism with inversion of configuration.

(a) (*R*)-2-bromobutane on treatment with I^- gives (*S*)-2-iodobutane.

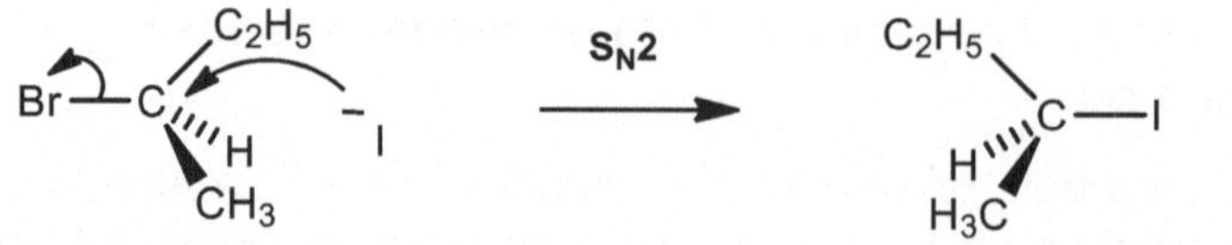

S_N2

(R)-2-Bromobutane **(S)-2-Iodobutane**

(b) Backside attack by the nucleophile (HO^-) on the carbon bearing chlorine atom leads to *trans*-4-methylcyclohexanol.

S_N2

Attack from the side opposite to Cl atom

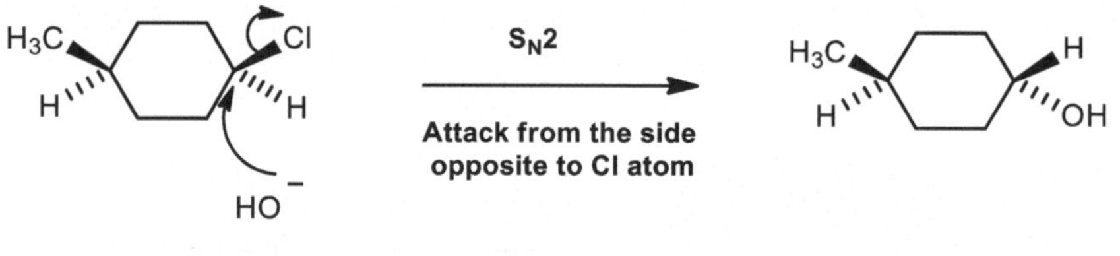

***cis*-1-Chloro-4-methylcyclohexane** ***trans*-4-methylcyclohexanol**

11.5.2 The Stereochemistry of S_N1 Reactions

An alkyl halide with a **chiral centre** has two optically active enantiomers. The S_N1 reaction of an **optically active** alkyl halide (one enantiomer) often gives an **optically inactive** racemic product (a mixture of equal amounts of two enantiomers of the product). Thus **an S_N1 reaction can lead to racemization.**

Consider, for example, the S_N1 hydrolysis of **optically active 3-bromo-3-methylhexane** (one enantiomer). The product is an optically inactive, **racemic 3-methyl-3-hexanol.**

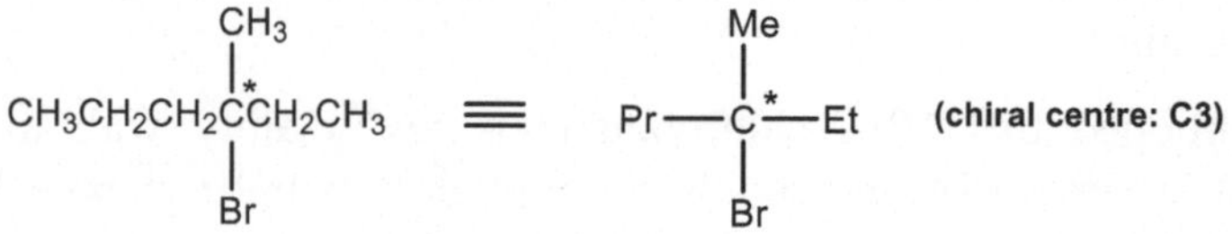

$CH_3CH_2CH_2\overset{*}{C}(CH_3)(Br)CH_2CH_3 \equiv Pr{-}\overset{*}{C}(Me)(Br){-}Et$ **(chiral centre: C3)**

3-Bromo-3-methylhexane

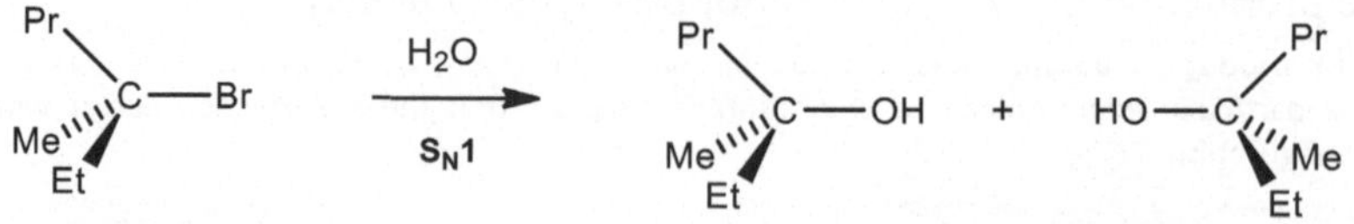

$\xrightarrow[S_N1]{H_2O}$ +

3-Bromo-3-methylhexane
(one enantiomer)
Optically active

3-Methyl-3-hexanol
(racemic form)
Optically inactive

An S_N1 reaction involves a **carbocation** intermediate which is **trigonal planar**. A nucleophile can then react from either the **frontside** or the **backside** of the planar carbocation.

In the solvolysis of optically active 3-bromo-3-methylhexane, the frontside attack and the backside attack of the nucleophile (water) can take place with equal probability, leading to a product with equal amounts of enantiomers, i.e. racemic form, as shown in Figure 11.5.

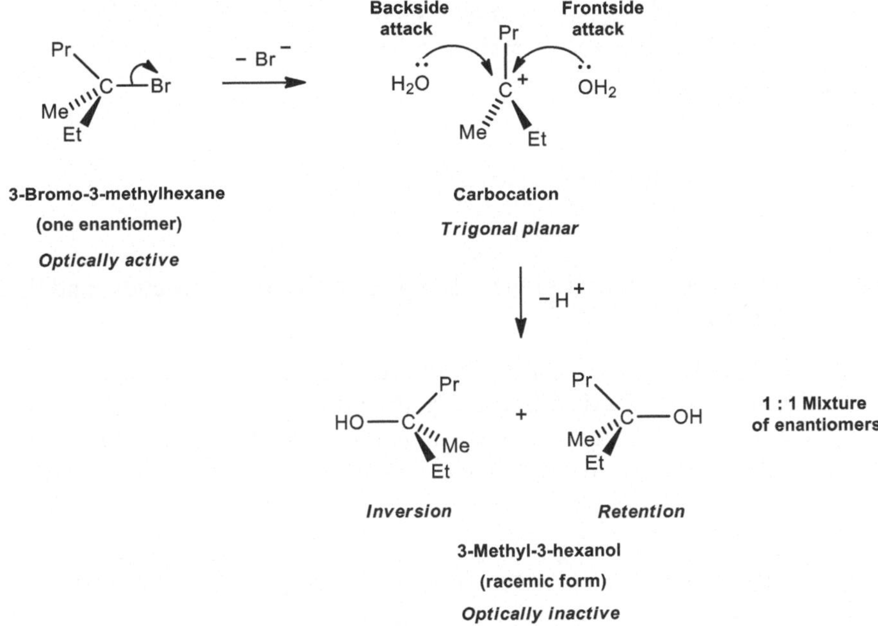

Figure 11.5: A mechanism for the S_N1 hydrolysis of optically active 3-bromo-3-methylhexane that occurs with racemization.

The frontside attack leads to retention of configuration while the backside attack leads to inversion of configuration. Racemization results from equal amounts of retention and inversion.

Note: The above S_N1 reaction occurs with *complete* racemization (50% retention + 50% inversion), and the product is *optically inactive*. However, all S_N1 reactions do not proceed in the similar fashion; some S_N1 reactions can proceed with *partial* racemization when retention and inversion are unequal. The product is then *optically active* due to the presence of an excess of an enantiomer over the other, i.e. there is partial loss of optical activity of the product relative to its pure enantiomer.

11.6 SUBSTITUTION VERSUS ELIMINATION

Nucleophiles and **bases** are both electron pair donors. So all nucleophiles are potential bases and all bases are potential nucleophiles. Thus, **substitution reactions** and **elimination reactions** with nucleophiles/bases often **compete** with each other.

A **nucleophile** donates an electron pair to a **carbon atom** at the *internal* position of a molecule, whereas a **base** donates an electron pair to a **hydrogen atom** on the *periphery* of the molecule. When the nucleophile/base attacks the carbon atom of the substrate molecule, **substitution** takes place, but when the nucleophile/base attacks a β-hydrogen atom, **elimination** occurs.

Y⁻ (nucleophile/base) — β H — (internal C) — X —*Substitution*→ —C(H)—C—Y + X⁻

Y^- (nucleophile/base) + $-\overset{|}{\underset{\underset{H\,\beta}{|}}{C}}-\overset{|}{\underset{|}{C}}-X$ $\xrightarrow{\textit{Elimination}}$ $-\overset{|}{C}=\overset{|}{C}-$ + HY + X^-

(peripheral H)

S_N2 versus E2

A nucleophile or a base is involved in the r/d step of an S_N2 or E2 reaction, and S_N2 and E2 reactions can compete.

Ethoxide ion ($C_2H_5O^-$) is a relatively *unhindered* nucleophile/base. If the substrate is a **primary alkyl halide, substitution is usually favoured** in the presence of ethoxide ion. Here, the ethoxide ion as a nucleophile can readily attack the less hindered primary carbon atom attached to only *one* alkyl group.

$CH_3CH(H^\beta)-C(H)(H)(Br)$ (1°) $\xrightarrow[C_2H_5OH]{^-OC_2H_5}$ $CH_3CH_2CH_2OC_2H_5$ (*Substitution*) + $CH_3CH{=}CH_2$ (*Elimination*)

Propyl bromide — ***1° Alkyl halide*** | **Ethyl propyl ether (major product)** | **Propene** (minor product)

But with a **secondary alkyl halide** in which the secondary carbon atom attached to *two* alkyl groups are more hindered, **elimination is usually favoured** with ethoxide ion (sodium ethoxide) as a base.

With a **tertiary alkyl halide**, the S_N2 reaction does not occur and the E2 elimination takes place almost exclusively.

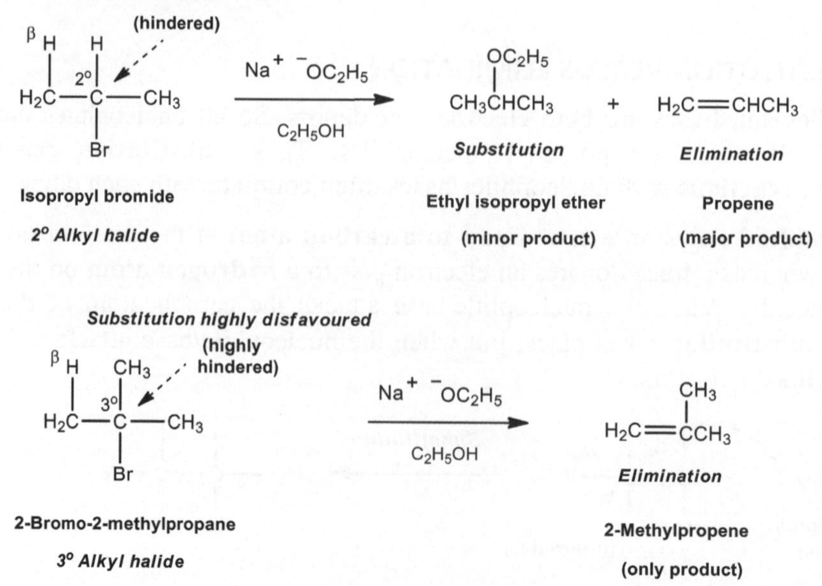

When a *sterically hindered* **t-butoxide ion** [$(CH_3)_3CO^-$] is used, **elimination is favoured even with a primary alkyl halide**. In this case, the bulky *t*-butoxide ion in potassium *t*-butoxide reagent encounters a severe steric hindrance for the attack on an internal carbon atom of the molecule for a substitution reaction, but it can easily abstract an exposed peripheral hydrogen atom for the elimination reaction.

Elimination favoured **(no steric hindrance)**

$(CH_3)_3C{-}O^-$ + $CH_3CH(H^\beta){-}CH_2{-}Br$ (1°) $\xrightarrow[(CH_3)_3COH]{(CH_3)_3CO^-\ K^+}$ $CH_3CH{=}CH_2$ + $CH_3CH_2CH_2OC(CH_3)_3$

(severe steric hindrance) ***Substitution disfavoured***

t-Butoxide ion	Propyl bromide *1° Alkyl halide*	Propene (major product) *Elimination*	*t*-Butyl propyl ether (minor product) *Substitution*

PROBLEM 11.6: Predict the major product of the following reaction. Draw structures in bond-line formulas.

$$CH_3CH(CH_3){-}Br \xrightarrow{CH_3CO_2^-}$$

Isopropyl bromide

Answer

The substrate is a 2° alkyl halide but acetate ion is a *weak base*. Therefore, elimination reaction is less likely and substitution reaction gives the major product (isopropyl acetate).

$$(CH_3)_2CH{-}Br + {}^-O{-}C(=O){-}CH_3 \xrightarrow[-\ Br^-]{\textit{Substitution}} (CH_3)_2CH{-}O{-}C(=O){-}CH_3$$

Isopropyl bromide **Acetate ion** **Isopropyl acetate**

11.7 REACTIONS OF ALCOHOLS: NUCLEOPHILIC SUBSTITUTION

Nucleophilic substitution of alcohols involves the replacement of the hydroxyl group by a nucleophile. However, OH^- is a *poor leaving group* since it is a strong base. A good leaving group is a weak base. Therefore, **nucleophilic substitution of alcohols requires the conversion of the OH group into a good leaving group**.

11.7.1 Halide as the Nucleophile: Formation of Alkyl Halides

Halide ions are readily available from inorganic acids or their salts. But the halide ion itself (e.g. Br^- from NaBr) cannot displace OH^-, a poor leaving group.

$$Br^- \; R{-}OH \not\longrightarrow Br{-}R + {}^-OH$$

(NaBr) — Poor leaving group

Protonation of the alcohol converts the poor leaving group (OH^-) into a good leaving group (H_2O). Thus **alcohols** react with strongly acidic hydrogen halides (HCl, HBr and HI) to form **alkyl halides** as substitution products. The order of reactivity is: HI > HBr > HCl.

$$-\overset{|}{\underset{|}{C}}-OH + H{-}X \longrightarrow -\overset{|}{\underset{|}{C}}-X + H_2O$$

Alcohol; (X = Cl, Br, I); Alkyl halide; (good leaving group)

Reactions of tertiary alcohols

Tertiary alcohols react with hydrogen halides by the **S_N1** mechanism because the reaction involves the formation of a stable tertiary carbocation intermediate.

For example, consider the reaction of ***t*-butyl alcohol** with HCl to form ***t*-butyl chloride**. The mechanism involves three steps. The addition of a proton to the alcohol forms a protonated alcohol, which gives a stable tertiary carbocation after the departure of a good leaving group (water). Finally, the carbocation is attacked by chloride ion as a nucleophile to produce *t*-butyl chloride.

$$(CH_3)_3COH \xrightarrow{HCl} (CH_3)_3CCl + H_2O$$

t-Butyl alcohol; *t*-Butyl chloride

Mechanism:

$$(CH_3)_3C{-}\ddot{O}H \xrightarrow{H^+} (CH_3)_3C{-}\overset{+}{O}H_2 \xrightarrow[S_N1]{-H_2O} (CH_3)_3C^+ \xrightarrow{Cl^-} (CH_3)_3C{-}Cl$$

t-Butyl alcohol; Protonated alcohol; Carbocation (*t*-butyl cation); *t*-Butyl chloride

Reactions of primary alcohols

A **primary alcohol**, (e.g. propyl alcohol) reacts with **HBr** by the **S_N2 mechanism.** Protonation of the alcohol forms a protonated alcohol which can undergo nucleophilic substitution by **bromide ion** as the departing water molecule is a good leaving group. The product is propyl bromide.

$$CH_3CH_2CH_2{-}OH \xrightarrow{H^+} CH_3CH_2CH_2{-}\overset{+}{O}H_2 \xrightarrow[S_N2]{Br^-} CH_3CH_2CH_2Br + H_2O$$

Propyl alcohol **Protonated alcohol** **Propyl bromide**

HCl does not react with primary or secondary alcohols because chloride ion is a weaker nucleophile than bromide ion or iodide ion. However, the reaction can take place when a Lewis acid such as zinc chloride is present.

Consider the following reaction between a **primary alcohol and HCl in the presence of $ZnCl_2$.** The mechanism involves the formation of a complex of the alcohol (Lewis base) with $ZnCl_2$ (Lewis acid). This complex provides a better leaving group than water. As a result, a relatively weaker nucleophile (Cl^-) can displace a group which is more easily lost.

$$RCH_2OH + HCl \xrightarrow{ZnCl_2} RCH_2Cl + H_2O$$

1° Alcohol **Alkyl chloride**

Mechanism:

$$RCH_2{-}\ddot{O}H \xrightarrow[\text{(Lewis acid)}]{ZnCl_2} RCH_2{-}\overset{+}{O}(H){-}\overset{-}{Zn}Cl_2 \xrightarrow{^-Cl} RCH_2{-}Cl + [HO{-}\overset{-}{Zn}Cl_2] \xrightarrow{H^+} ZnCl_2 + H_2O$$

1° Alcohol **Lewis acid-Lewis base complex** **Alkyl chloride**

Use of phosphorus tribromide and thionyl chloride

Phosphorus tribromide (PBr_3) and **thionyl chloride** ($SOCl_2$) are also used to prepare alkyl halides from alcohols. The following are two examples:

$$3\ C_6H_5{-}CH_2OH \xrightarrow{PBr_3} 3\ C_6H_5{-}CH_2Br + H_3PO_3$$

Benzyl alcohol **Benzyl bromide**

$$C_6H_{11}{-}OH \xrightarrow{SOCl_2} C_6H_{11}{-}Cl + SO_2 + HCl$$

Cyclohexanol **Cyclohexyl chloride**

11.7.2 Nucleophilic Substitution via Sulphonates (mesylates and tosylates)

Alcohols can be converted into **alkyl sulphonates** in which the sulphonate group can act a very good leaving group. Thus nucleophilic substitution of alcohols is possible via their sulphonates.

Methanesulphonate ion (mesylate ion) and p-toluenesulphonate ion (tosylate ion)

Sulphuric acid (H_2SO_4) is a very strong acid. Replacement of one OH group by an alkyl group gives an **alkanesulphonic acid** (RSO_3H), which is a strong organic acid. Therefore, the corresponding anion, an **alkanesulphonate ion** (RSO_3^-) is a much better leaving group than water.

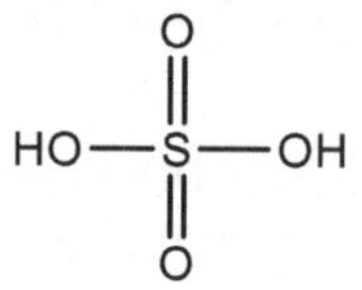

(H_2SO_4) **Sulphuric acid**

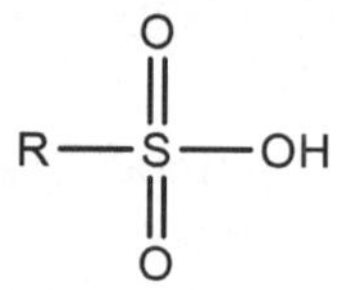

(RSO_3H) **Alkanesulphonic acid**

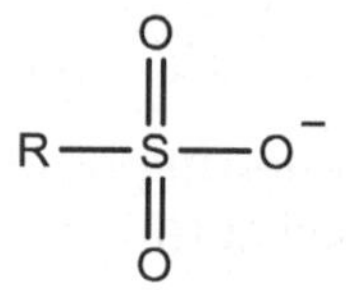

(RSO_3^-) **Alkanesulphonate ion**

A very good leaving group

If R = CH_3 (methyl), the alkanesulphonate ion ($CH_3SO_3^-$) is called **methanesulphonate ion or mesylate ion**. In short, the mesylate ion ($CH_3SO_3^-$) is written as **MsO**$^-$ where Ms = mesyl = methanesulphonyl = CH_3SO_2.

Similarly, if R = *p*-$CH_3C_6H_4$ (*p*-tolyl), the sulphonate ion (*p*-$CH_3C_6H_4SO_3^-$) is called ***p*-toluenesulphonate ion or tosylate ion**. In short, the tosylate ion (*p*-$CH_3C_6H_4SO_3^-$) is written as **TsO**$^-$ where Ts = tosyl = *p*-toluenesulphonyl = *p*-$CH_3C_6H_4SO_2$.

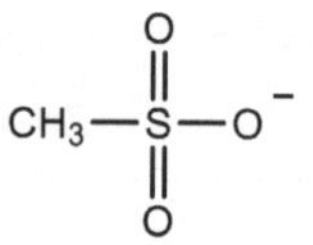

($CH_3SO_3^-$) **Methanesulphonate ion or mesylate ion (MsO$^-$)**

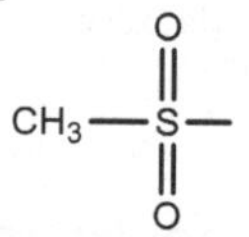

(CH_3SO_2) **= methanesulphonyl or mesyl (Ms) group**

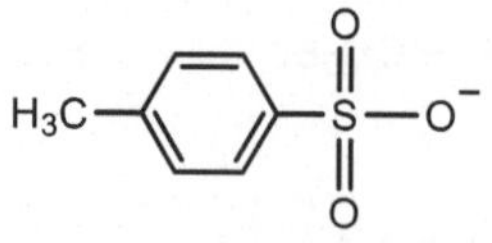

(*p*-$CH_3C_6H_4SO_3^-$) ***p*-Toluenesulphonate ion or tosylate ion (TsO$^-$)**

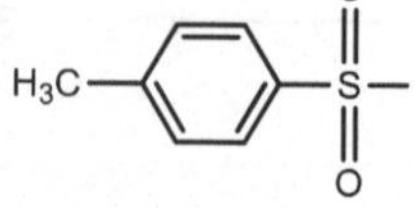

(*p*-$CH_3C_6H_4SO_2$) **= *p*-toluenesulphonyl or tosyl (Ts) group**

Mesylate ion (MsO$^-$) and tosylate ion (TsO$^-$) are excellent leaving groups.

Alkyl sulphonates: Alkyl mesylates and alkyl tosylates

Alcohols can be easily converted into alkyl mesylates or tosylates. The reaction of an **alcohol** with **methanesulphonyl chloride** (CH_3SO_2Cl or MsCl) in the presence of a base (pyridine) gives a **mesylate**.

Similarly, the reaction of an **alcohol** with ***p*-toluenesulphonyl chloride** (*p*-$CH_3C_6H_4SO_2Cl$ or TsCl) and pyridine produces a **tosylate**.

$R{-}OH + CH_3{-}S(=O)_2{-}Cl \xrightarrow[-\ HCl]{\text{Pyridine (base)}} R{-}O{-}S(=O)_2{-}CH_3$

Alcohol **Methanesulphonyl chloride (MsCl)** ($ROSO_2CH_3$ or ROMs) **Alkyl mesylate**

$R{-}OH + H_3C{-}C_6H_4{-}S(=O)_2{-}Cl \xrightarrow[-\ HCl]{\text{Pyridine (base)}} R{-}O{-}S(=O)_2{-}C_6H_4{-}CH_3$

Alcohol ***p*-Toluenesulphonyl chloride (TsCl)** ($ROSO_2C_6H_4CH_3$-*p* or ROTs) **Alkyl tosylate**

Nucleophilic substitution

An alkyl mesylate (or tosylate) undergoes nucleophilic substitution similar to an alkyl halide by the S_N2 mechanism.

The alcohol is first converted into a mesylate (or tosylate) which then reacts with a nucleophile to give the substitution product. This is illustrated with the nucleophilic substitution of an alcohol via its tosylate in a general case.

$RCH_2{-}OH + H_3C{-}C_6H_4{-}SO_2Cl \xrightarrow[-\ HCl]{\text{Pyridine}} RCH_2{-}OTs$ **Alkyl tosylate**

Alcohol ***p*-Toluenesulphonyl chloride (tosyl chloride or TsCl)**

Nu^- S_N2

$RCH_2{-}Nu + {}^-OTs$

Substitution product **Tosylate ion** ***An excellent leaving group***

11.7.3 Stereochemistry

Consider an alcohol with a **chiral centre** that bears the OH group. The conversion of the **alcohol to an alkyl tosylate** proceeds with **retention of configuration** because no bond to the chiral carbon is broken.

An S_N2 reaction of the alkyl tosylate with a nucleophile would then proceed with inversion of configuration.

The net result is the nucleophilic substitution of the chiral alcohol with inversion of configuration (because retention + inversion = inversion).

This is illustrated with the conversion of (*S*)-2-butanol to (*R*)-2-butanol, as shown in Figure 11.6.

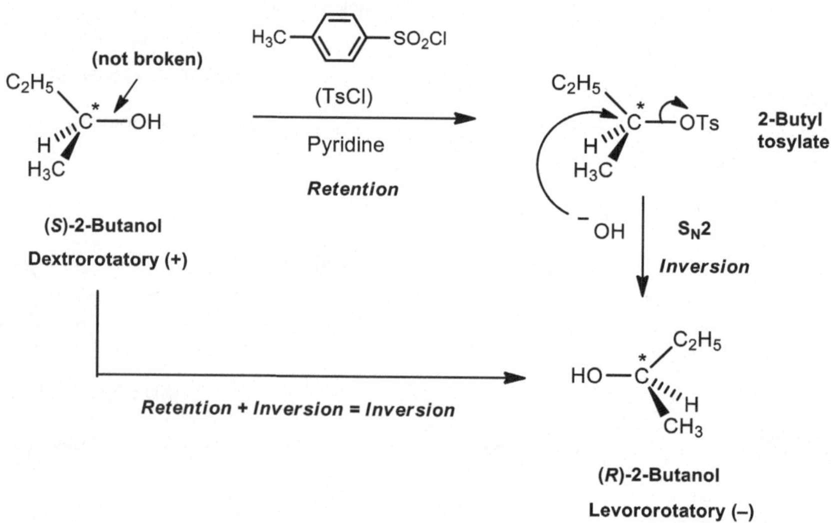

Figure 11.6: A scheme for the conversion of (*S*)-2-butanol to (*R*)-2-butanol.

(*S*)-2-Butanol is found to be dextrorotatory (+); therefore its enantiomer (*R*)-2-butanol must be levorotatory (–). Thus the scheme demonstrates how one enantiomer of an alcohol [*S*-(+)-2-butanol] can be transformed into the other enantiomer of the same alcohol [*R*-(–)-2-butanol].

11.8 SYNTHESIS OF ETHERS USING NUCLEOPHILIC SUBSTITUTION

Primary alcohols can *dehydrate* to form ethers. For example, dehydration of ethanol with H_2SO_4 at 140°C gives mainly diethyl ether (see also Section 10.6).

$$\underset{\text{Ethanol (2 molecules)}}{CH_3CH_2{-}OH \;+\; HO{-}CH_2CH_3} \xrightarrow[140^\circ C,\ -H_2O]{H_2SO_4} \underset{\text{Diethyl ether}}{CH_3CH_2{-}O{-}CH_2CH_3}$$

This is an *intermolecular* dehydration which involves a **nucleophilic substitution**. One molecule of ethanol is protonated to form a protonated ethanol which is then attacked by another molecule of ethanol acting as a nucleophile in an S_N2 reaction. A protonated ether is formed which then loses a proton to produce diethyl ether.

$$CH_3CH_2{-}\ddot{O}H \xrightarrow{H^+} CH_3CH_2{-}\overset{+}{O}H_2 \xrightarrow[S_N2]{HO{-}CH_2CH_3} CH_3CH_2{-}\overset{+}{\underset{|}{O}}(H){-}CH_2CH_3 + H_2O$$

Ethanol — **Protonated ethanol** — **Protonated ether**

$$\xrightarrow{-H^+} CH_3CH_2{-}O{-}CH_2CH_3$$

Diethyl ether

11.8.1 The Williamson Synthesis of Ethers

The **Williamson synthesis** provides an important route to the synthesis of ethers. The synthesis involves an S_N2 reaction of an **alkoxide ion** with an **alkyl halide** or an **alkyl sulphonate** (e.g. alkyl tosylate).

$$RO^- + R'{-}X \xrightarrow{S_N2} RO{-}R' + X^-$$

(X = Cl, Br, I)

Alkoxide ion	**Alkyl halide**	**Ether**
Reagent	***Substrate***	***Product***

$$RO^- + R'{-}OTs \xrightarrow{S_N2} RO{-}R' + {}^-OTs$$

Alkoxide ion	**Alkyl tosylate**	**Ether**
Reagent	***Substrate***	***Product***

As for the S_N2 reaction, the synthesis proceeds successfully when the **substrate is a methyl, primary or secondary alkyl halide or sulphonate**. But with a *tertiary* substrate, the synthesis fails because instead of substitution, elimination occurs exclusively.

Example

Sodium or potassium alkoxide can be prepared by the reaction of an alcohol with sodium or potassium metal or with metal hydride such as sodium hydride (NaH).

The following is an example of the Williamson synthesis of an ether (**methyl propyl ether**) from an alkoxide (sodium propoxide) and methyl halide (methyl iodide).

The alkoxide is prepared from the corresponding alcohol (propyl alcohol).

$$CH_3CH_2CH_2OH \xrightarrow[-H_2]{NaH} CH_3CH_2CH_2O^-\,Na^+ \xrightarrow[-NaI]{CH_3{-}I} CH_3CH_2CH_2OCH_3$$

Propyl alcohol — **Sodium propoxide** — **Methyl propyl ether**

Stereochemistry

If the substrate (alkyl halide or tosylate) has a **chiral centre**, the S_N2 reaction involved in the Williamson synthesis will proceed with **inversion of configuration**.

This is illustrated with the Williamson synthesis of an optically active ether **[(*S*)-2-butyl ethyl ether]** starting from an optically active alcohol [(*R*)-2-butanol]. The alcohol is treated with *p*-toluenesulphonyl chloride (tosyl chloride or TsCl) to form an alkyl tosylate which then reacts with ethoxide ion (sodium ethoxide) to give the product.

$HO-C^*(C_2H_5)(H)(CH_3)$ —TsCl, Pyridine→ $TsO-C^*(C_2H_5)(H)(CH_3)$ —$^-OC_2H_5$, S_N2, *Inversion*→ $C_2H_5C^*(H)(H_3C)-OC_2H_5$

(*R*)-2-Butanol *Optically active* — **(*R*)-2-Butyl tosylate** *Optically active* — **(*S*)-2-Butyl ethyl ether** *Optically active*

Synthesis of cyclic ethers

Cyclic ethers such as **epoxides** can be prepared by an **intramolecular Williamson synthesis**.

Consider, for example, the synthesis of the simplest epoxide (**ethylene oxide**) from ethylene chlorohydrin. Treatment of ethylene chlorohydrin with aqueous sodium hydroxide forms an alkoxide ion which then undergoes an intramolecular S_N2 reaction to form the epoxide.

ClH_2C-CH_2-O-H —^-OH, $-H_2O$→ $ClH_2C-CH_2-O^-$ —*Intramolecular* S_N2, $-Cl^-$→ H_2C-CH_2 (O bridged)

Ethylene chlorohydrin — **An alkoxide ion** — **Ethylene oxide (an epoxide)**

Ethylene chlorohydrin can be obtained by treating ethylene with chlorine in the presence of water (see Section 8.6.3).

PROBLEM 11.7: Which of the following reactions would you use to prepare *t*-butyl methyl ether? Give your reasoning.

(a) $CH_3O^- K^+ + (CH_3)_3CBr \longrightarrow$

(b) $(CH_3)_3CO^- K^+ + CH_3Br \longrightarrow$

Answer

Reaction (a) involves a *tertiary* alkyl halide as the substrate. Therefore, no S_N2 substitution but E2 elimination of *t*-butyl bromide will take place in the presence of methoxide ion as a *base* to give an alkene (isobutene).

On the other hand, reaction (b) involves a *methyl* halide. With methyl bromide, only substitution (S_N2) is possible, and the reaction with *t*-butoxide ion as a *nucleophile* yields exclusively an ether (*t*-butyl methyl ether). Hence reaction (b) will be used to prepare the ether.

(a) $H_3C—C(CH_3)(Br)—CH_2H \xrightarrow[E2]{^{-}OCH_3} H_3C—C(CH_3)=CH_2$

t-Butyl bromide → Isobutene

S_N2 ✗ $^{-}OCH_3$ ↓

Ether not formed

(b) $(CH_3)_3CO^{-} + CH_3—Br \xrightarrow{S_N2} (CH_3)_3C—O—CH_3$

Methyl bromide → *t*-Butyl methyl ether

Only product

11.9 REACTIONS OF ETHERS: NUCLEOPHILIC SUBSTITUTION

Ethers are relatively unreactive but they can solvate cations by donating an electron pair from the oxygen atom. Because of their lack of reactivity and their ability to solvate cations, ethers are widely used as **solvents** for many reactions.

However, **ethers** can be **cleaved** by very strong acids such as **hydroiodic acid** (HI) or **hydrobromic acid** (HBr). In this reaction, the C–O bond of ethers is broken, and the products are an **alkyl halide** and an **alcohol**.

$$R—O—R + H—X \longrightarrow R—OH + R—X$$

(X = Br, I)

Ether → Alcohol + Alkyl bromide or iodide

Examples

Diethyl ether reacts with cold concentrated **HI** (*one* mole) to form ethyl iodide and ethyl alcohol. But with *two* moles of HI, only ethyl iodide is the product because ethyl alcohol just formed reacts further with a second mole of HI to form the second mole of ethyl iodide.

$$CH_3CH_2O{-}CH_2CH_3 + \underset{(1\ mol)}{HI} \longrightarrow CH_3CH_2OH + CH_3CH_2I$$

Diethyl ether **Ethyl alcohol** **Ethyl iodide**

$$CH_3CH_2O{-}CH_2CH_3 + \underset{(2\ mol)}{HI} \longrightarrow \underset{(2\ mol)}{CH_3CH_2I} + H_2O$$

Diethyl ether **Ethyl iodide**

$$\left[CH_3CH_2OCH_2CH_3 \xrightarrow{HI} CH_3CH_2OH + CH_3CH_2I \xrightarrow{HI} CH_3CH_2I + H_2O \right]$$

Similarly, **diethyl ether** reacts with hot concentrated **HBr** (2 moles) to form two moles of ethyl bromide.

$$CH_3CH_2OCH_2CH_3 + 2\ HBr \longrightarrow 2\ CH_3CH_2Br + H_2O$$

Diethyl ether **Ethyl bromide**

Mechanism

The mechanism for the cleavage of diethyl ether with HI (or HBr) involves protonation of the ether followed by the attack of the iodide ion (or bromide ion) as a nucleophile in an S_N2 process. The **protonated ether** loses an alcohol as a good leaving group (cf. the loss of water from a protonated alcohol, see Section 11.7.1).

$$CH_3CH_2\ddot{O}{-}CH_2CH_3 \xrightarrow{H-I} CH_3CH_2\overset{+}{\underset{H}{O}}{-}CH_2CH_3 + I^- \xrightarrow{S_N2} CH_3CH_2I + CH_3CH_2OH$$

Diethyl ether **Protonated ether** **Ethyl iodide** **Ethyl alcohol**

A good leaving group

Reactions of unsymmetrical ethers

With unsymmetrical or mixed ethers (in which two alkyl groups linked to the oxygen atom are not the same), the alkyl halide and alcohol that are formed depend on the *nature* of the alkyl groups. This is illustrated with two examples:

(a) When **methyl phenyl ether (anisole)** reacts with cold concentrated HI, the products are methyl iodide and phenol. This is an **S_N2** reaction involving a favourable attack of the iodide ion on the methyl carbon of the protonated ether molecule.

$$C_6H_5{-}OCH_3 \xrightarrow{HI} C_6H_5{-}OH + CH_3I$$

Methyl phenyl ether (anisole) **Phenol** **Methyl iodide**

Mechanism:

$C_6H_5\ddot{O}-CH_3 \xrightarrow{H^+} C_6H_5\overset{+}{O}(H)-CH_3 \xrightarrow[S_N2]{I^-} C_6H_5-OH + CH_3I$

(b) When ***t*-butyl methyl ether** is cleaved with HI, the products are *t*-butyl iodide and methanol. Here the reaction proceeds by an **S_N1** mechanism because the protonated ether can readily form a stable 3° carbocation (*t*-butyl cation).

$$(CH_3)_3C-O-CH_3 \xrightarrow{HI} (CH_3)_3C-I + CH_3OH$$

***t*-Butyl methyl ether** — ***t*-Butyl iodide** — **Methanol**

Mechanism:

$$(CH_3)_3C-\ddot{O}-CH_3 \xrightarrow{H^+} (CH_3)_3C-\overset{+}{O}(H)-CH_3 \xrightarrow{S_N1} (CH_3)_3C^+ + CH_3OH \xrightarrow{I^-} (CH_3)_3C-I$$

3° Carbocation

11.9.1 Reactions of Cyclic Ethers (epoxides)

Epoxides or oxiranes are cyclic ethers. The three-membered ring in epoxides is highly strained and can be opened easily by the attack of a nucleophile.

Acid-catalysed ring opening

An acid-catalysed hydrolysis of epoxides by ring opening to produce a 1,2-diol has been described previously (see Section 8.8.1). In general, the acid-catalysed ring opening for epoxides takes place by the **attack of the nucleophile** mainly at the **more substituted carbon atom.**

For example, consider the reaction of **2,2-dimethyloxirane** with **methanol** in the presence of an acid catalyst. The product is **2-methoxy-2-methyl-1-propanol.**

$$H_3C-C(CH_3)(-O-)CH_2 + CH_3OH \xrightarrow{H^+} H_3C-C(CH_3)(OCH_3)-CH_2-OH$$

2,2-Dimethyloxirane — **2-Methoxy-2-methyl-1-propanol (product)**

Mechanism:

$$H_3C-C(CH_3)(-O-)CH_2 \xrightarrow{H^+} H_3C-C(CH_3)(-\overset{+}{O}H-)CH_2 \xrightarrow{S_N1} H_3C-\overset{+}{C}(CH_3)-CH_2OH \xrightarrow{CH_3\ddot{O}H} H_3C-C(CH_3)(\overset{+}{O}HCH_3)-CH_2OH \xrightarrow{-H^+} \text{Product}$$

2,2-Dimethyloxirane — **Protonated epoxide** — **3° Carbocation** — **Product**

OR

$$H_3C{-}C(CH_3)(O)CH_2 \xrightarrow{H^+} H_3C{-}\overset{\delta+}{C}(CH_3)(\overset{\delta+}{O}H)CH_2 \xrightarrow{CH_3\ddot{O}H} H_3C{-}C(CH_3)(\overset{+}{O}(H)CH_3){-}CH_2OH \xrightarrow{-H^+} \text{Product}$$

2,2-Dimethyloxirane **Protonated epoxide**

Methanol is a weak nucleophile. Thus, protonation and the consequent ring opening of the protonated epoxide can proceed by an S_N1 mechanism with the formation of a relatively stable tertiary carbocation.

It is also possible that the nucleophile (methanol) can attack the tertiary carbon atom with a partial positive charge when the carbocation is about to form but not formed yet.

Base-catalysed ring opening

The base-catalysed ring opening of an epoxide can take place in the presence of hydroxide ion (HO^-) or an alkoxide ion (RO^-) as a strong base/nucleophile. With *unsymmetrical* epoxides, the **attack of the nucleophile** occurs primarily at the **less substituted carbon atom**.

When the same epoxide (**2,2-dimethyloxirane**) is treated with **sodium methoxide** in methanol, the product is **1-methoxy-2-methyl-2-propanol**.

Here the nucleophilic attack occurs at the less hindered primary carbon atom in an S_N2 process.

$$H_3C{-}C(CH_3)(O)CH_2 + CH_3O^- Na^+ \xrightarrow{CH_3OH} H_3C{-}C(CH_3)(OH){-}CH_2{-}OCH_3$$

2,2-Dimethyloxirane **1-Methoxy-2-methyl-2-propanol**

Mechanism:

$$H_3C{-}C(CH_3)(O)\underset{1^\circ}{CH_2} \xrightarrow[S_N2]{^-OCH_3} H_3C{-}C(CH_3)(O^-){-}CH_2{-}OCH_3 \xrightarrow{CH_3OH} H_3C{-}C(CH_3)(OH){-}CH_2OCH_3 + CH_3O^-$$

2,2-Dimethyloxirane **1-Methoxy-2-methyl-2-propanol**

The Grignard reaction

A Grignard reagent (RMgX) is a carbon nucleophile. The nucleophilic alkyl or aryl group of the Grignard reagent can attack the ring carbon of the epoxide and lead to ring opening of the strained epoxide ring.

For example, the simplest epoxide (**ethylene oxide**) reacts with **ethylmagnesium bromide** by ring opening in an **S_N2** process. Subsequent acidification of the reaction mixture produces **1-butanol** (a primary alcohol).

$$H_2C{-}CH_2\ (\text{epoxide}) + C_2H_5{-}MgBr \xrightarrow[S_N2]{Et_2O} C_2H_5CH_2CH_2OMgBr \xrightarrow[H_2O]{H^+} CH_3CH_2CH_2CH_2OH$$

Ethylene oxide (an epoxide) **Ethylmagnesium bromide** **1-Butanol (a 1° alcohol)**

SUPPLEMENTARY PROBLEMS

11.8 Each of the following hydrocarbons reacts with chlorine to give only one monochloro product. Identify the hydrocarbon and give the structure of product in each case.

(a) C_5H_{10} (b) C_8H_{18}

[*Hints:* (a) the hydrocarbon is a cycloalkane; (b) the hydrocarbon is a substituted butane]

1.9 Arrange the following alkyl halides in order of decreasing reactivity in S_N2 reactions with methoxide ion (CH_3O^-):

CH_3Br, CH_3CH_2Br, $(CH_3)_2CHBr$, $(CH_3)_3CBr$

11.10 Which of the following reactions would occur faster? Give reasons.

(a) $CH_3CH_2CH_2Br + {}^-CN \xrightarrow{DMF} CH_3CH_2CH_2CN + Br^-$

(b) $CH_3CH_2CH_2Br + {}^-CN \xrightarrow{C_2H_5OH} CH_3CH_2CH_2CN + Br^-$

11.11 Which of the following compounds would you expect to react by the S_N1 or S_N2 mechanism? Explain.

$H_3CO{-}C_6H_4{-}CH_2Cl$ $O_2N{-}C_6H_4{-}CH_2Cl$

***p*-Methoxybenzyl chloride** ***p*-Nitrobenzyl chloride**

[*Hints: p*-methoxybenzyl cation is highly stablized but *p*-nitrobenzyl cation is destabilized]

11.12 Predict the major product in each of the following reactions:

(a) $(CH_3)_3CBr \xrightarrow[MeOH]{MeONa}$

***t*-Butyl bromide**

(b) $(CH_3)_3CBr \xrightarrow{MeOH}$

***t*-Butyl bromide**

11.13 Write the structures of the compounds (**A–D**) in the following scheme:

$$H_3C-C_6H_4-SO_2-OH \xrightarrow{PCl_5} \mathbf{A} \xrightarrow{(CH_3)_2CHOH} \mathbf{B} \xrightarrow{CH_3CO_2Na} \mathbf{C} \xrightarrow[H_2O]{OH^-} \mathbf{D}$$

p-Toluenesulphonic acid

[*Ans:* **A** = *p*-toluenesulphonyl chloride; **B** = isopropyl tosylate; **C** = isopropyl acetate; **D** = isopropanol]

11.14 Identify the compounds **E** and **F** in the following reactions:

$$\text{Cyclohexene} \xrightarrow[H_2O]{Cl_2} \mathbf{E} \xrightarrow{OH^-} \mathbf{F}$$

Cyclohexene

[*Ans:* **E** = *trans*-2-chlorocyclohexanol; **F** = cyclohexene oxide (an epoxide)]

11.15 Predict the major product of the following reaction:

$$H_2C\overset{O}{—}CHCH_3 + C_6H_5MgBr \xrightarrow[2.\ H_3O^+]{1.\ Et_2O}$$

2-Methyloxirane

[*Ans:* 3-phenyl-2-propanol]

CHAPTER 12

Carboxylic Acids and Their Derivatives: Substitution Reactions

IN THIS CHAPTER

- Nucleophilic substitution at the carbonyl group: The addition-elimination mechanism
- Relative reactivity of esters, amides, acyl chlorides and acid anhydrides
- Substitution by alcohols: Formation of esters and lactones
- Substitution by amines: Formation of amides and lactams
- Acid chlorides and acid anhydrides: Acyl transfer reactions
- Substitution by water: Hydrolysis of esters and amides
- Substitution by hydride reagents: Reduction of esters, carboxylic acids and amides
- Substitution by organometallic reagents: Reactions with Grignard reagents, organolithium and organocadmium reagents, and the Reformatsky reaction
- Nitriles: Hydrolysis, reduction, and Grignard reactions

Like aldehydes and ketones, **carboxylic acids and their derivatives** (which constitute the carboxylic acid family) also contain a **carbonyl group (C=O)**.

But there is a *difference*. The **C=O of carboxylic acids or their derivatives is bonded to a heteroatom** (oxygen, nitrogen or halogen atom), whereas the C=O of aldehydes or ketones is attached to a carbon or hydrogen atom.

The general formulas of carboxylic acids and their derivatives are given below.

$R-\overset{O}{\overset{\|}{C}}-OH$

Carboxylic acid

$R-\overset{O}{\overset{\|}{C}}-X$

Carboxylic acid derivatives

(X = heteroatom-containing substituent)

$R-\overset{O}{\overset{\|}{C}}-OR'$

[X = OR' (alkoxy)]

Ester

$R-\overset{O}{\overset{\|}{C}}-NH_2$

[X = NH_2 (amino)]

Amide

$R-\overset{O}{\overset{\|}{C}}-Cl$

[X = Cl (or Br)]

Acid chloride (or acyl chloride)

$R-\overset{O}{\overset{\|}{C}}-O-\overset{O}{\overset{\|}{C}}-R'$

[X = OCOR' (carboxylate)]

Acid anhydride

Esters usually have pleasant odours, and they contribute to the flavours and fragrances of fruits and flowers. **Amides** are quite often found in nature, and they provide the structural units (amide bonds) that hold amino acids together in proteins. **Acyl halides** and **acid**

anhydrides are reactive compounds which are used as synthetic reagents. They are also generated in many reactions as synthetic intermediates.

In carboxylic acids and their derivatives, the **heteroatom-containing substituent** attached to the carbonyl group can act as a **leaving group**. Thus, carboxylic acids and their derivatives can undergo **substitution reactions at the unsaturated (sp^2) carbonyl carbon.**

These substitution reactions at the electrophilic carbonyl group are initiated by nucleophiles, and are called **nucleophilic substitution at the carbonyl group**.

12.1 NUCLEOPHILIC SUBSTITUTION AT THE CARBONYL CARBON

In Chapter 9, we have learned that a nucleophilic attack at the carbonyl group of aldehydes and ketones leads to addition.

In contrast, the nucleophilic attack at the carbonyl group of carboxylic acids and their derivatives leads to **substitution**.

$$\underbrace{R-\overset{O}{\overset{\|}{C}}}_{\text{(acyl group)}}-X + Nu^- \xrightarrow{\text{Substitution}} \underbrace{R-\overset{O}{\overset{\|}{C}}}_{\text{(acyl group)}}-Nu + X^-$$

Carboxylic acid or derivatives — *Acyl compounds*; **Nucleophile**; **Product** — *Acylated nucleophile*; **Leaving group**

The **RC=O group** is called an **acyl group**. Thus, carboxylic acids and their derivatives represent **acyl compounds**, and the products are **acylated nucleophiles**.

Mechanism

The mechanism of nucleophilic substitution at the carbonyl group involves *two* steps, as shown in Figure 12.1. In **step 1**, the nucleophile (Nu^-) adds to the carbonyl group and forms a **tetrahedral intermediate**. The intermediate is unstable and eliminates a leaving group (X^-) in **step 2**. As a result, the C=O bond is regenerated, thereby forming the **substitution product**.

Since step 1 is a nucleophilic addition and step 2 involves elimination, this mechanism is called a **nucleophilic addition–elimination mechanism**.

$$R-\underset{X}{\underset{|}{C}}{=}O \xrightarrow[\text{Step 1: Nucleophilic addition}]{Nu^-} R-\overset{Nu}{\overset{|}{\underset{X}{\underset{|}{C}}}}-O^- \xrightarrow[\text{Step 2: Elimination}]{} R-\overset{Nu}{\overset{|}{C}}{=}O + X^-$$

Carboxylic acid or derivatives — **Tetrahedral intermediate** — **Substitution product**

Figure 12.1: A nucleophilic addition-elimination mechanism for nucleophilic substitution at the carbonyl group of carboxylic acids or their derivatives.

Note: For the nucleophilic addition to **aldehydes** or **ketones,** the tetrahedral intermediate *cannot* eliminate a hydride ion (H^-) or a carbanion (R^-) because they are very poor leaving groups. Instead, the intermediate accepts a proton to form a stable addition product (see Figure 9.1).

R—C(=O)—R' (R' = H or alkyl) **Aldehyde or ketone** —Nu^-→ R—C(O^-)(Nu)—R' **Tetrahedral intermediate** —✕→ R—C(=O)—Nu ***Does not form*** + R'^- **(hydride or carbanion)** ***Very poor leaving group***

12.2 RELATIVE REACTIVITY OF CARBOXYLIC ACID DERIVATIVES

Carboxylic acid derivatives are *not* equally reactive toward nucleophilic substitution. Acyl chlorides are the **most reactive** and amides are the **least reactive**.

The general order of reactivity is:

RCOCl (**Acyl chloride**) > RCOOCOR' (**Acid anhydride**) > RCOOR' (**Ester**) > $RCONH_2$ (**Amide**)

The reactivity order can be explained by considering the electrophilicity of the carbonyl group and the leaving group ability of the departing groups.

Electrophilicity of the carbonyl group

If the heteroatom-containing substituent is more electron-donating, the carbonyl group will be less electrophilic, and hence it will be less reactive toward an attack by a nucleophile. The electron-donating +***M* effect** of the substituents increases in the order:

Cl < OCOR′ < OR′ < NH_2

Since Cl is *least* electron-donating by + *M* effect, the carbonyl group in an **acyl chloride** is **most electrophilic**, and *acyl chlorides are most reactive.* On the other extreme, the NH_2 is is *most* electron-donating by + *M* effect, and the carbonyl group in an **amide** is **least electrophilic**, and hence *amides are least reactive.*

R—C(=O)—Cl **Acid chloride** | R—C(=O)—O—C(=O)—R' **Acid anhydride** | R—C(=O)—OR' **Ester** | R—C(=O)—NH_2 **Amide**

+*M* effect: Cl < OCOR' < OR' < NH_2

→ ***Electrophilicity of the carbonyl group (C=O) decreases; reactivity decreases***

Note that the substituent OCOR′ is less electron donating than OR′ because the lone pair on the oxygen atom in :O–COR′ is also *partly* involved in resonance with its *own* carbonyl group (C=O).

Leavability of the departing groups

A good leaving group is a weak base (anion of a strong acid). **The weaker the base, the better is the leaving group.**

For the substitution reaction of an acid chloride, acid anhydride, ester and amide, the **leaving group** is chloride ion (Cl^-), carboxylate ion ($R'CO_2^-$), alkoxide ion ($R'O^-$) and amide ion (NH_2^-) respectively.

The order of basicity of the leaving groups is:

$$Cl^- < R'CO_2^- < R'O^- < NH_2^-$$

Since **Cl^-** (the weakest base) is the **best leaving group**, *acyl chlorides are the most reactive.* On the other extreme, **NH_2^-** (the strongest base) has the **least leaving group ability**, and *amides are the least reactive.*

Relative reactivity

Both factors (the electrophilicity of the carbonyl group and the leaving group ability of the departing groups) lead to the *same* relative reactivity of carboxylic acid derivatives.

Therefore, the **reactivity order** of carboxylic acid derivatives is:

acyl chloride (RCOCl) > **acid anhydride** (RCOOCOR′ > **ester** (RCOOR′ > **amide** ($RCONH_2$).

12.3 SUBSTITUTION BY ALCOHOLS: FORMATION OF ESTERS

Carboxylic acids react with alcohols to form **esters. Esterification** of a carboxylic acid involves *substitution* of hydroxyl group at the carbonyl carbon by the alkoxy group of an alcohol.

Esterification reactions are **reversible** and occur in the presence of an acid catalyst (usually HCl or H_2SO_4). The reaction is known as the **Fischer esterification.**

$$\underset{\text{Carboxylic acid}}{R{-}\overset{\overset{\displaystyle O}{\|}}{C}{-}OH} + \underset{\text{Alcohol}}{H{-}OR'} \underset{}{\overset{H^+}{\rightleftharpoons}} \underset{\text{Ester}}{R{-}\overset{\overset{\displaystyle O}{\|}}{C}{-}OR'} + H_2O$$

The yield of the ester depends on the position of equilibrium. The use of an *excess* of one of the reactants (usually alcohol) pushes the equilibrium to the product side. The yield of ester can also be increased by *removing water* from the reaction mixture as it is formed.

The following are two examples of Fischer esterification:

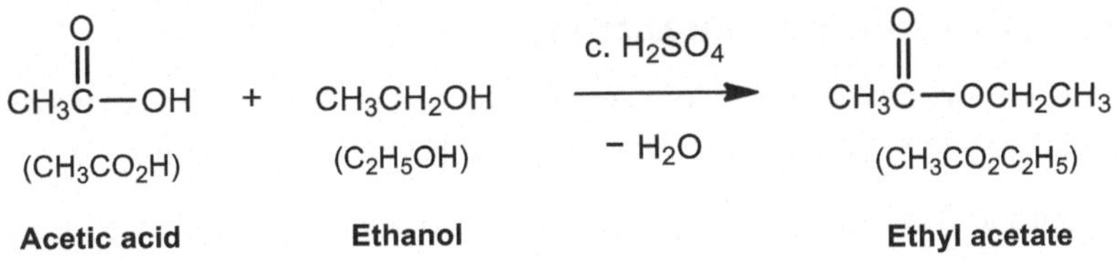

$$CH_3COOH\ (CH_3CO_2H) + CH_3CH_2OH\ (C_2H_5OH) \xrightarrow[-\ H_2O]{c.\ H_2SO_4} CH_3C(=O){-}OCH_2CH_3\ (CH_3CO_2C_2H_5)$$

Acetic acid **Ethanol** **Ethyl acetate**

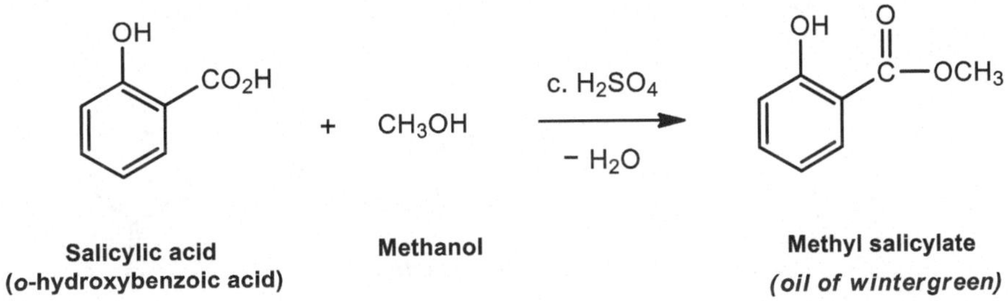

$$\text{Salicylic acid} + CH_3OH \xrightarrow[-\ H_2O]{c.\ H_2SO_4} \text{Methyl salicylate}$$

Salicylic acid (o-hydroxybenzoic acid) **Methanol** **Methyl salicylate** ***(oil of wintergreen)***

In the first example, **acetic acid** reacts with ethanol in the presence of concentrated sulphuric acid (c. H_2SO_4) to produce **ethyl acetate**, which has a characteristic fruity smell. In the second example, **salicylic acid** (*o*-hydroxybenzoic acid) reacts with methanol to form **methyl salicylate**. Methyl salicylate occurs in nature and is produced by wintergreen plants. It is found in *oil of wintergreen* and is used as an external analgesic.

Mechanism

A mechanism for the **acid-catalysed esterification** is shown in Figure 12.2.

Addition of a proton to carboxylic acid forms a protonated carboxylic acid in **step 1**. The protonation increases the electrophilicity of the carbonyl group; thus an alcohol as a nucleophile can easily attack the protonated carbonyl group to form a tetrahedral intermediate (**step 2**). In **step 3**, a *proton exchange* converts OH to a better leaving group (H_2O^+). Subsequent loss of water in **step 4** forms a protonated ester, which then loses a proton to give the ester product (**step 5**).

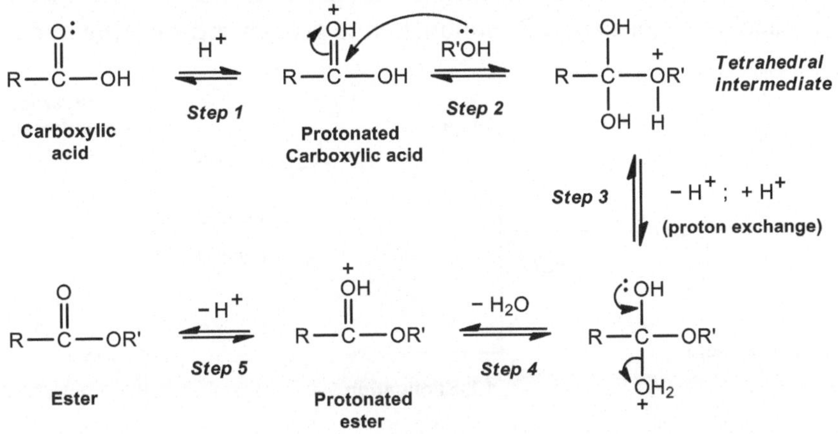

Carboxylic acid ⇌ (H^+, Step 1) Protonated Carboxylic acid ⇌ (R'OH, Step 2) Tetrahedral intermediate

Step 3 ⇅ $-H^+$; $+H^+$ (proton exchange)

Ester ⇌ ($-H^+$, Step 5) Protonated ester ⇌ ($-H_2O$, Step 4)

Figure 12.2: A mechanism for an acid-catalysed esterification reaction.

12.3.1 Intramolecular Esterification: Lactones

When OH and CO_2H groups are present in the *same* molecule, an **intramolecular esterification** can take place. The product is a cyclic ester called **lactone.**

Lactonization (or lactone formation) is acid-catalysed and occurs readily with a **γ-hydroxy acid** or a **δ-hydroxy acid** to form *unstrained* five- or six-membered ring.

A five-membered ring lactone is called a γ-lactone, and a six-membered ring lactone is called a δ-lactone.

γ β α
$CH_2CH_2CH_2$
H—O C=O
OH

H^+, $-H_2O$

$CH_2CH_2CH_2$
O——C=O
Cyclic ester
≡

γ-Hydroxy acid

γ-Lactone
(5-membered ring lactone)

δ γ β α
$CH_2CH_2CH_2CH_2$
H—O C=O
OH

H^+, $-H_2O$

$CH_2CH_2CH_2CH_2$
O——C=O
Cyclic ester
≡

δ-Hydroxy acid

δ-Lactone
(6-membered ring lactone)

It may be mentioned that many naturally occurring lactones are γ- or δ-lactones. For example, **vitamin C is a γ-lactone.**

A **β-hydroxy acid** does not form a β-lactone which has a *strained* four-membered ring. Instead, the reaction proceeds by 1,2- elimination to give an **α,β-unsaturated acid.**

β α
$CH_2—CH_2—C=O$
OH OH

β-Hydroxy acid

H^+, $-H_2O$

H_2C——CH_2
O——C=O
Strained 4-membered ring

β-Lactone
Not formed

1,2-Elimination

β α
$H_2C=CH—C—OH$
O

α,β-Unsaturated acid

12.4 SUBSTITUTION BY AMINES: FORMATION OF AMIDES

Carboxylic acids and their derivatives can react with ammonia (NH_3), primary amine (RNH_2) or secondary amine (R_2NH) to produce **amides**. The reactions involve *nucleophilic substitution* at an acyl carbon. Ammonia, primary amine or secondary amine acts as a nitrogen nucleophile.

Carboxylic acids react with aqueous NH_3 to form ammonium salts. This is an acid-base reaction where ammonia acts as a base. On heating the *dry* salt, dehydration (elimination of water) takes place and an amide is formed. The following is an example:

$$\underset{\text{Acetic acid}}{CH_3C(=O){-}OH} + \underset{\text{Ammonia}}{NH_3} \xrightarrow{H_2O} \underset{\text{Ammonium acetate (aqueous solution)}}{CH_3C(=O){-}O^- \, NH_4^+} \xrightarrow[-\,H_2O]{\text{Evaporate water and heat}} \underset{\text{Acetamide}}{CH_3C(=O){-}NH_2}$$

However, this is not a good method for preparing amides.

Formation of amides via acid chlorides

A much better method for the formation of amides involves the **conversion of a carboxylic acid** to an **acid chloride** using phosphorus pentachloride (PCl_5) or thionyl chloride ($SOCl_2$) (see Section 12.5) and then treating the acid chloride with **ammonia**, a **primary amine** or a **secondary amine**.

The products are an amide or a substituted amide. When one hydrogen atom of the NH_2 group of an amide is replaced by an alkyl or aryl group, the product is an ***N*-substituted amide**. If both hydrogen atoms of the NH_2 group are replaced by alkyl or aryl groups, the product is an ***N,N*-disubstituted amide**.

The HCl liberated in the reaction forms a salt with the base, i.e. $NH_4^+ \, Cl^-$ (from HCl and NH_3), $R'NH_3^+ \, Cl^-$ (from HCl and $R'NH_2$) and $R'R''NH_2^+ \, Cl^-$ (from HCl and $R'R''NH$).

$$\underset{\text{Carboxylic acid}}{R{-}C(=O){-}OH} \xrightarrow[SOCl_2]{PCl_5 \text{ or}} \underset{\text{Acid chloride}}{R{-}C(=O){-}Cl}$$

$$R{-}C(=O){-}Cl \xrightarrow[\text{(ammonia)}]{NH_2{-}H} \underset{\text{Amide}}{R{-}C(=O){-}NH_2} + NH_4^+ \, Cl^-$$

$$R{-}C(=O){-}Cl \xrightarrow[(1^\circ \text{ amine})]{R'NH{-}H} \underset{N\text{-Substituted amide}}{R{-}C(=O){-}NH{-}R'} + R'NH_3^+ \, Cl^-$$

$$R{-}C(=O){-}Cl \xrightarrow[(2^\circ \text{ amine})]{R'R''N{-}H} \underset{N,N\text{-Disubstituted amide}}{R{-}C(=O){-}N(R''){-}R'} + R'R''NH_2^+ \, Cl^-$$

The following scheme illustrates the **preparation of benzamide** from benzoic acid via its acid chloride. Treatment of benzoic acid with PCl_5 forms benzoyl chloride which is then reacted with aqueous ammonia to produce benzamide.

$$C_6H_5\text{—}\overset{O}{\overset{\|}{C}}\text{—}OH \xrightarrow{PCl_5} C_6H_5\text{—}\overset{O}{\overset{\|}{C}}\text{—}Cl \xrightarrow{NH_3} C_6H_5\text{—}\overset{O}{\overset{\|}{C}}\text{—}NH_2$$

Benzoic acid **Benzoyl chloride** **Benzamide**

Amides from esters

Esters can react with ammonia or an amine to form **amides**. Since esters are less reactive than acid chlorides, the reactions with esters are slower than those of acid chlorides.

$$R\text{—}\overset{O}{\overset{\|}{C}}\text{—}OR' + NH_3 \xrightarrow{H_2O} R\text{—}\overset{O}{\overset{\|}{C}}\text{—}NH_2 + R'OH$$

Ester **Amide**

Example

$$ClCH_2\overset{O}{\overset{\|}{C}}\text{—}OC_2H_5 \xrightarrow{NH_3} ClCH_2\overset{O}{\overset{\|}{C}}\text{—}NH_2 + C_2H_5OH$$

Ethyl chloroacetate **Chloroacetamide**

Imides

Imides are the nitrogen analogues of acid anhydrides in which the anhydride oxygen is replaced by NH group.

$$R\text{—}\overset{O}{\overset{\|}{C}}\text{—}O\text{—}\overset{O}{\overset{\|}{C}}\text{—}R \qquad R\text{—}\overset{O}{\overset{\|}{C}}\text{—}\overset{H}{\overset{|}{N}}\text{—}\overset{O}{\overset{\|}{C}}\text{—}R$$

Acid anhydride **Imide**

Cyclic imides are particularly useful in synthesis. An example is phthalimide in which NH replaces the anhydride oxygen of phthalic anhydride (see Section 12.5).

Phthalimide can be prepared by heating phthalic anhydride with aqueous ammonia.

$$\text{Phthalic anhydride} + NH_3 \xrightarrow[\Delta]{H_2O} \text{Phthalimide}$$

Phthalic anhydride **Phthalimide**

12.4.1 Cyclic Amides: Lactams

Lactams are cyclic amides. They are the **nitrogen analogues of lactones** (the alkoxy oxygen of lactones is replaced by NH group). A **five-membered ring lactam** is called a **γ-lactam** and a **six-membered ring lactam** is called a **δ-lactam**. The five- and six-membered rings are *unstrained* and *stable*.

In contrast, **β-lactams** have a *strained* four-membered ring and are highly reactive. The famous antibiotic **penicillin** contains a β-lactam ring, which readily reacts with enzymes responsible for building the cell walls of bacteria, and thereby leads to bacterial death.

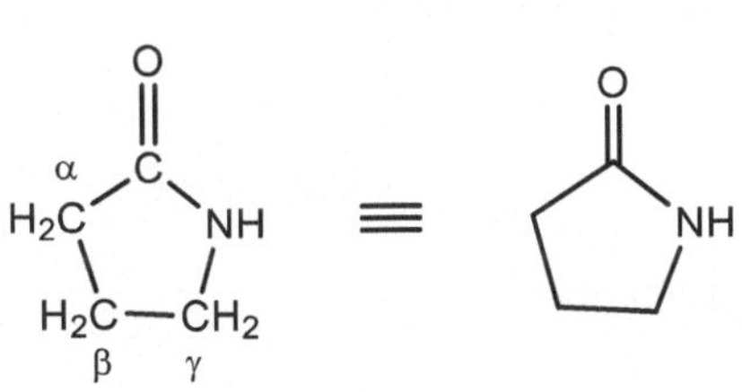

γ-Lactam

(5-membered ring lactam)

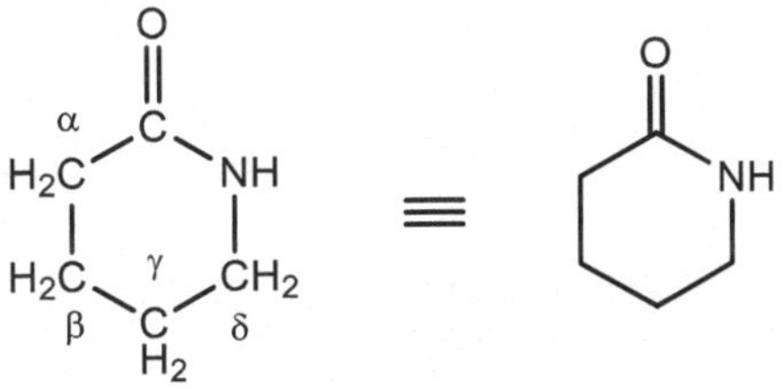

δ-Lactam

(6-membered ring lactam)

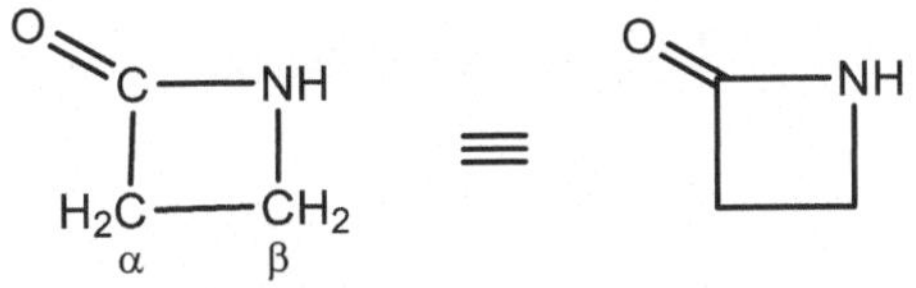

β-Lactam

(4-membered ring lactam)

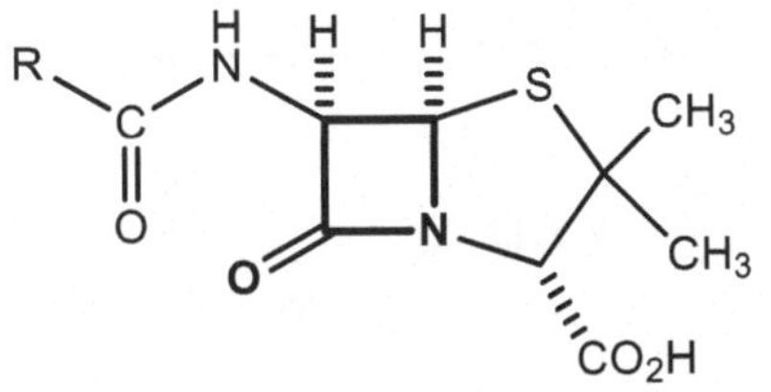

Penicillin antibiotic with a β-lactam ring

γ-Lactams can be prepared by heating **γ-amino acids**. For example, heating γ-aminobutyric acid produces a γ-lactam (γ-butyrolactam). An *intramolecular* amide formation between NH_2 and CO_2H occurs, which is *much more favourable* than an intermolecular amide formation.

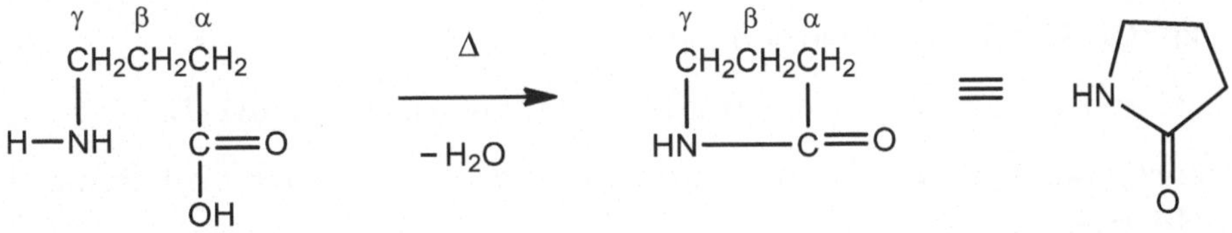

γ-Aminobutyric acid

A γ-amino acid

γ-Butyrolactam

A γ-lactam

Similarly, **δ-lactams** can be prepared by heating δ-amino acids.

12.5 ACID CHLORIDES AND ACID ANHYDRIDES

Acid chlorides (or acyl chlorides) are usually prepared from carboxylic acids by treatment with thionyl chloride ($SOCl_2$) or phosphorus pentachloride (PCl_5). The reaction involves the substitution of OH by Cl.

$$\underset{\textbf{Carboxylic acid}}{R{-}\overset{O}{\overset{\|}{C}}{-}OH} + SOCl_2 \longrightarrow \underset{\textbf{Acyl chloride}}{R{-}\overset{O}{\overset{\|}{C}}{-}Cl} + SO_2 + HCl$$

$$\underset{\textbf{Carboxylic acid}}{R{-}\overset{O}{\overset{\|}{C}}{-}OH} + PCl_5 \longrightarrow \underset{\textbf{Acyl chloride}}{R{-}\overset{O}{\overset{\|}{C}}{-}Cl} + POCl_3 + HCl$$

Examples

$$\underset{\substack{(CH_3CO_2H)\\ \textbf{Acetic acid}}}{CH_3\overset{O}{\overset{\|}{C}}{-}OH} \xrightarrow{SOCl_2} \underset{\substack{(CH_3COCl)\\ \textbf{Acetyl chloride}}}{CH_3\overset{O}{\overset{\|}{C}}{-}Cl} \qquad [CH_3CO = \textbf{Acetyl (Ac)}]$$

$$\underset{\substack{(C_6H_5CO_2H)\\ \textbf{Benzoic acid}}}{C_6H_5{-}\overset{O}{\overset{\|}{C}}{-}OH} \xrightarrow{PCl_5} \underset{\substack{(C_6H_5COCl)\\ \textbf{Benzoyl chloride}}}{C_6H_5{-}\overset{O}{\overset{\|}{C}}{-}Cl} \qquad [C_6H_5CO = \textbf{Benzoyl (Bz)}]$$

The RC=O group is called an acyl group (see Section 12.1). When R = CH_3 (methyl), the acyl group (CH_3CO–) is called **acetyl (Ac)**.

If R = C_6H_5 (phenyl), the acyl group (C_6H_5CO–) is known as **benzoyl (Bz)**.

The benzoyl group (Bz) must not be confused with the benzyl group (Bn). Remember that Bn = $C_6H_5CH_2$–.

Acid anhydrides

Acid anhydrides are regarded as the dehydration products of carboxylic acids. The elimination of water from *two* molecules of carboxylic acids (either identical or different) gives an acid anhydride.

$$R{-}\overset{O}{\overset{\|}{C}}{-}OH + H{-}O{-}\overset{O}{\overset{\|}{C}}{-}R \xrightarrow{-H_2O} R{-}\overset{O}{\overset{\|}{C}}{-}O{-}\overset{O}{\overset{\|}{C}}{-}R$$

Two molecules of carboxylic acid **Acid anhydride**

Both *acyclic* anhydrides and *cyclic* anhydrides are useful reagents in synthesis.

Acyclic anhydrides are commonly prepared from the reaction of carboxylic acids with acyl chlorides in the presence of pyridine. Pyridine functions as a base to neutralize HCl liberated in the reaction by forming a salt (pyridinium hydrochloride).

$$R{-}\overset{O}{\overset{\|}{C}}{-}OH + Cl{-}\overset{O}{\overset{\|}{C}}{-}R' \xrightarrow[\text{(pyridine)}]{} R{-}\overset{O}{\overset{\|}{C}}{-}O{-}\overset{O}{\overset{\|}{C}}{-}R' + C_5H_5\overset{+}{N}H\ Cl^-$$

Carboxylic acid **Acyl chloride** **Acid anhydride** **Pyridinium hydrochloride**

Example

$$CH_3\overset{O}{\overset{\|}{C}}{-}OH + CH_3\overset{O}{\overset{\|}{C}}{-}Cl \xrightarrow{\text{Pyridine}} CH_3\overset{O}{\overset{\|}{C}}{-}O{-}\overset{O}{\overset{\|}{C}}CH_3 = (CH_3CO)_2O = Ac_2O$$

Acetic acid **Acetyl chloride** **Acetic anhydride**

Cyclic anhydrides can be prepared directly by heating a dicarboxylic acid when intramolecular dehydration occurs. This method is successful when anhydride formation leads to a five- or six-membered ring.

$$H_2C(COOH){-}H_2C(COOH) \xrightarrow[-H_2O]{300^\circ C} \text{succinic anhydride}$$

Succinic acid **Succinic anhydride**

$$C_6H_4(COOH)_2 \xrightarrow[-H_2O]{230^\circ C} \text{phthalic anhydride}$$

Phthalic acid **Phthalic anhydride**

12.6 ACYL TRANSFER REACTIONS

The transfer of an acyl group (RCO–) from one molecule to another is called an **acyl transfer reaction**. Two common acyl transfer reactions are **acetylation** and **benzoylation**. Acyl transfer reactions are common in living organisms.

12.6.1 Acetylation

Acetylation reaction usually involves the transfer of an **acetyl group** (CH_3CO–) to the *oxygen atom* of a hydroxyl group (OH) or the *nitrogen atom* of an amino group (NH_2). The two common reagents for acetylation reactions are **acetyl chloride** (CH_3COCl) and **acetic anhydride** [$(CH_3CO)_2O$].

Acetylation of OH group

Acetylation of the OH group of **alcohols** or **phenols** gives an **ester**. This is illustrated with the following examples:

$$\underset{\substack{(C_2H_5OH)\\ \textbf{Ethyl alcohol}}}{CH_3CH_2{-}OH} + \underset{\textbf{Acetyl chloride}}{CH_3\overset{O}{\overset{\|}{C}}Cl} \xrightarrow{\textit{Acetylation}} \underset{\substack{(CH_3CO_2C_2H_5)\\ \textbf{Ethyl acetate}}}{CH_3CH_2{-}O{-}\overset{O}{\overset{\|}{C}}{-}CH_3} + HCl$$

$$\underset{\substack{(C_6H_5OH)\\ \textbf{Phenol}}}{C_6H_5{-}OH} + \underset{\textbf{Acetic anhydride}}{CH_3\overset{O}{\overset{\|}{C}}O\overset{O}{\overset{\|}{C}}CH_3} \xrightarrow{\textit{Acetylation}} \underset{\substack{(CH_3CO_2C_6H_5)\\ \textbf{Phenyl acetate}}}{C_6H_5{-}O{-}\overset{O}{\overset{\|}{C}}{-}CH_3} + CH_3\overset{O}{\overset{\|}{C}}OH$$

Acetylation of NH_2 group

Acetylation of the NH_2 group of **aliphatic amines** produces an ***N*-alkyl substituted amide**. For example, acetylation of methylamine with acetyl chloride gives *N*-methylacetamide.

On the other hand, **acetylation** of **aromatic amines** gives an ***N*-aryl substituted amide**, which is called an **anilide**. For example, acetylation of aniline with acetic anhydride produces **acetanilide** (*N*-phenylacetamide).

Note that *N*-phenylacetamide is an anilide because phenyl substituent on nitrogen is an aryl group but *N*-methylacetanilide is *not* an anilide because methyl substituent on nitrogen is an alkyl group, not an aryl group.

$$CH_3NH_2 + CH_3COCl \xrightarrow[-\ HCl]{\textit{Acetylation}} CH_3NHCCH_3\ (CH_3CONHCH_3)$$

Methylamine **Acetyl chloride** ***N*-Methylacetamide**

Not an anilide

$$C_6H_5NH_2 + (CH_3CO)_2O \xrightarrow[-\ CH_3CO_2H]{\textit{Acetylation}} C_6H_5NHCCH_3\ (CH_3CONHC_6H_5)$$

$(C_6H_5NH_2)$

Aniline **Acetic anhydride** ***N*-Phenylacetamide (acetanilide)**

An anilide

PROBLEM 12.1: Suggest a mechanism for each of the following acetylation reactions:

(a) $ROH + CH_3COCl \longrightarrow ROCOCH_3 + HCl$

(b) $RNH_2 + (CH_3CO)_2O \longrightarrow RNHCOCH_3 + CH_3CO_2H$

Answer

Each reaction involves a nucleophilic substitution at the carbonyl group (C=O). In reaction (a), the nucleophile is ROH (an alcohol), while RNH_2 (an amine) is the nucleophile in reaction (b).

(a) $CH_3C(=O){-}Cl \xrightarrow{ROH} CH_3C(O^-)(\overset{+}{R}O{-}H){-}Cl \longrightarrow CH_3C(=O){-}\overset{+}{R}O{-}H + Cl^- \longrightarrow CH_3C(=O){-}OR + HCl$

(b) $CH_3C(=O){-}O{-}C(=O)CH_3 \xrightarrow{RNH_2} CH_3C(=O){-}O{-}C(O^-)(H{-}\overset{+}{N}HR)CH_3 \longrightarrow CH_3CO_2^- + CH_3C(=O){-}\overset{+}{N}HR{-}H \longrightarrow CH_3CO_2H + CH_3C(=O){-}NHR$

Drugs: Aspirin and paracetamol

Aspirin and paracetamol are two common drugs that are used as a pain reliever and for reducing fever.

Aspirin can be prepared by acetylation of **salicylic acid** (*o*-hydroxybenzoic acid) with acetic anhydride in the presence of an acid catalyst (H_2SO_4).

OH, CO_2H + CH_3COCCH_3 (O O) $\xrightarrow{H^+}$ $OCCH_3$ (O), CO_2H

Salicylic acid | Acetic anhydride | Aspirin (acetylsalicylic acid)

Paracetamol (also known as acetaminophen) is *p*-hydroxyacetanilide which can be prepared by acetylation of ***p*-aminophenol** with acetic anhydride.

NH_2, OH + $(CH_3CO)_2O$ (Ac_2O) $\longrightarrow$ $NHCCH_3$ (O), OH

p-Aminophenol | Acetic anhydride | Paracetamol (*p*-hydroxyacetanilide)

In this acetylation reaction, the NH_2 (which is more nucleophilic than the OH) reacts selectively. This type of selectivity is called **chemoselectivity**. Chemoselectivity refers to *which* functional group reacts when there are two separate functional groups that are capable of reacting with the reagent.

(The retrosynthetic analysis and synthesis of these drugs have also been described later; see Section 18.14)

12.6.2 Benzoylation

Benzoylation is also an acylation reaction that involves the transfer of a **benzoyl group** (COC_6H_5) to NH_2 or OH groups.

Benzoylation can be carried out by treating **amines** or **phenols** with benzoyl chloride (C_6H_5COCl) in aqueous sodium hydroxide. This method of acylation with acid chlorides in aqueous alkaline solution is known as the **Schotten-Baumann reaction**.

For example, benzoylation of **aniline** gives *N*-phenylbenzamide, which is an *N*-aryl substituted amide. Thus *N*-phenylbenzamide is an anilide called **benzanilide**.

Another example is benzoylation of **phenol** which gives phenyl benzoate as an **ester**.

NH_2 + C(=O)—Cl $\xrightarrow[H_2O]{NaOH}$ N(H)—C(=O)

($C_6H_5NH_2$) | (C_6H_5COCl) | ***Benzoylation*** | ($C_6H_5CONHC_6H_5$)

Aniline **Benzoyl chloride** ***N*-Phenylbenzamide (benzanilide)**

$$C_6H_5OH + C_6H_5COCl \xrightarrow[H_2O]{NaOH} C_6H_5CO_2C_6H_5$$

(C_6H_5OH) **Phenol** + (C_6H_5COCl) **Benzoyl chloride** → ***Benzoylation*** → ($C_6H_5CO_2C_6H_5$) **Phenyl benzoate**

12.7 SUBSTITUTION BY WATER: FORMATION OF CARBOXYLIC ACIDS

Esters are hydrolysed to **carboxylic acids** by their reaction with water. Hydrolysis involves a substitution of the alkoxy group (OR) of esters with a hydroxyl group (OH).

The hydrolysis can occur in the presence of an **acid catalyst** or can be *promoted* by a **base**. Since the acid is a catalyst, it is regenerated after the reaction, but the base is a reactant which is consumed in the reaction.

12.7.1 Acid-catalysed ester hydrolysis

Acid-catalysed hydrolysis of esters is the reverse of acid-catalysed esterification. Thus the mechanism of hydrolysis follows the reverse steps of the mechanism of esterification shown in Figure 12.2. *Try yourself to draw the hydrolysis mechanism.*

Acid-catalysed esterification and hydrolysis are **equilibrium reactions**. While esterification is favoured using an excess of the alcohol, hydrolysis occurs in the presence of a large excess of water. In practice, **esters are hydrolysed by heating with dilute aqueous acid (HCl or H_2SO_4) under reflux.**

For example, **methyl benzoate** on heating with aqueous HCl produces **benzoic acid** and **methanol**.

$$R{-}C(=O){-}OR' + H{-}OH \xrightleftharpoons{H^+} R{-}C(=O){-}OH + R'OH$$

Ester **Carboxylic acid** **Alcohol**

Example

$$C_6H_5{-}C(=O){-}OCH_3 + H_2O \xrightarrow[\Delta]{HCl} C_6H_5{-}C(=O){-}OH + CH_3OH$$

Methyl benzoate **Benzoic acid** **Methanol**

12.7.2 Base-promoted Ester Hydrolysis: Saponification

Esters undergo hydrolysis with a base (usually aqueous NaOH or KOH) to produce **sodium or potassium salt of a carboxylic acid** and an **alcohol.**

In contrast to acid-catalysed hydrolysis, the basic hydrolysis is essentially an **irreversible** reaction.

$$R-\overset{O}{\overset{\|}{C}}-OR' \;+\; NaOH \xrightarrow[\Delta]{H_2O} R-\overset{O}{\overset{\|}{C}}-\overset{-}{O}\,\overset{+}{Na} \;+\; R'OH$$

Ester **Sodium salt of carboxylic acid** **Alcohol**

Since the carboxylic acid is not formed under basic conditions, **acidification** of the reaction mixture (acid work-up) is needed to generate the carboxylic acid.

For example, consider the basic hydrolysis of **ethyl benzoate**. Refluxing ethyl benzoate with aqueous NaOH produces sodium benzoate and ethanol. Acidification of the reaction mixture with aqueous HCl generates benzoic acid as the product.

$$C_6H_5-\overset{O}{\overset{\|}{C}}-OCH_2CH_3 \xrightarrow[H_2O,\ \Delta]{NaOH} C_6H_5-\overset{O}{\overset{\|}{C}}-\overset{-}{O}\,\overset{+}{Na} \;+\; CH_3CH_2OH$$

Ethyl benzoate **Sodium benzoate** **Ethanol**

(work-up) HCl, H_2O

$$C_6H_5-\overset{O}{\overset{\|}{C}}-OH \;+\; NaCl$$

Benzoic acid

Mechanism

A mechanism for the base-promoted hydrolysis of an ester is shown in Figure 12.3.

Nucleophilic attack at the carbonyl group by hydroxide ion forms a **tetrahedral intermediate** which then eliminates an alkoxide ion. The **carboxylic acid** so formed immediately *transfers* its proton to a strongly basic species (alkoxide ion or hydroxide ion) present in the basic reaction medium, thereby forming a **carboxylate ion**. The *proton transfer step is irreversible*, which drives the reaction to the product side.

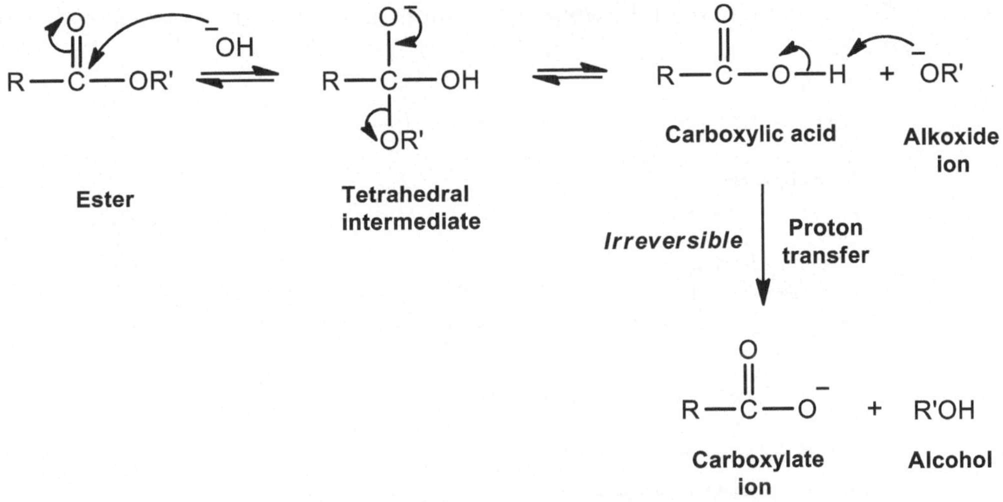

R—C(=O)—OR' + $^-$OH ⇌ R—C(O$^-$)(OH)—OR' ⇌ R—C(=O)—O—H + $^-$OR'

Ester

Tetrahedral intermediate

Carboxylic acid

Alkoxide ion

Irreversible

Proton transfer

R—C(=O)—O$^-$ + R'OH

Carboxylate ion

Alcohol

Figure 12.3: A mechanism for the base-promoted hydrolysis of esters.

Acyl–oxygen bond cleavage versus alkyl–oxygen bond cleavage

The hydrolysis of esters is possible either by the **cleavage of acyl–oxygen bond** or by the **cleavage of alkyl–oxygen bond**, as shown below. The acyl–oxygen bond cleavage involves a nucleophilic substitution at acyl or carbonyl carbon, while the alkyl–oxygen bond cleavage involves a nucleophilic substitution at alkyl carbon.

R—C(=O)┆OR' $\xrightarrow{OH^-}$ RCO_2^- + R'OH

Acyl-oxygen bond cleavage

Nucleophilic substitution at acyl carbon

R—C(=O)—O┆R' $\xrightarrow{OH^-}$ RCO_2^- + R'OH

Alkyl-oxygen bond cleavage

Nucleophilic substitution at alkyl carbon

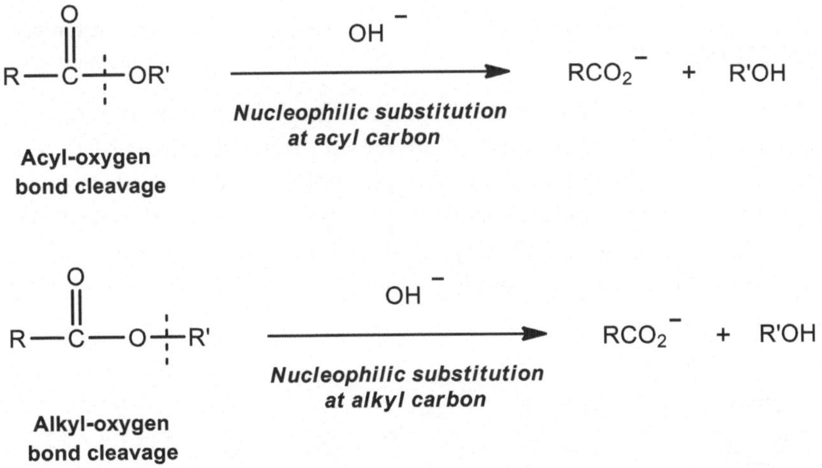

However, most esters undergo hydrolysis by **acyl–oxygen bond cleavage** through nucleophilic substitution at the acyl carbon (see the mechanism in Figure 12.3).

The **alkyl–oxygen bond cleavage** by nucleophilic substitution at the alkyl carbon is not generally observed, although it might occur in some special cases (see Problem 12.2b).

PROBLEM 12.2: Explain the following observations in the hydrolysis of esters labelled with ^{18}O isotope:

(a) $$CH_3CH_2\overset{O}{\overset{\|}{C}}-{}^{18}OCH_2CH_3 \xrightarrow{OH^-} CH_3CH_2\overset{O}{\overset{\|}{C}}-O^- + CH_2CH_2{}^{18}OH$$

Ethyl propionate

(b) $$H_3C-C_6H_2(CH_3)_2-\overset{O}{\overset{\|}{C}}-{}^{18}OCH_3 \xrightarrow{OH^-} H_3C-C_6H_2(CH_3)_2-\overset{O}{\overset{\|}{C}}-{}^{18}O^- + CH_3OH$$

Methyl mesitoate
(methyl 2,4,6-trimethylbenzoate)

Answer

(a) The hydrolysis involves a nucleophilic substitution at the acyl carbon of ethyl propionate with **acyl–oxygen bond cleavage**. As a result, the product alcohol (ethanol) contains the ^{18}O isotope label.

$$CH_3CH_2\overset{O}{\overset{\|}{C}} \,\vdots\, {}^{18}OCH_2CH_3 \xrightarrow{OH^-} CH_3CH_2\overset{O}{\overset{\|}{C}}-O^- + CH_2CH_2{}^{18}OH$$

Acyl-oxygen cleavage

Ethanol
(labelled with ^{18}O)

(b) In the hydrolysis of methyl mesitoate, the nucleophilic attack at the acyl or carbonyl carbon is highly hindered because of severe steric interferences of two *ortho* methyl substituents on the aromatic ring. An alternative pathway of hydrolysis is then possible with a nucleophilic attack at the unhindered methyl carbon in an S_N2 process, which leads to **alkyl–oxygen bond cleavage**. In this S_N2 reaction, the carboxylate ion acts as a good leaving group. Thus, the isotopic label (^{18}O) resides in the mesitoate ion.

$$H_3C-C_6H_2(CH_3)_2-\overset{O}{\overset{\|}{C}}-{}^{18}O \,\vdots\, CH_3 \xrightarrow[S_N2]{OH^-} H_3C-C_6H_2(CH_3)_2-\overset{O}{\overset{\|}{C}}-{}^{18}O^- + CH_3OH$$

Alkyl-oxygen cleavage

Mesitoate ion
(labelled with ^{18}O)

Saponification

The **basic hydrolysis of esters** is sometimes called **saponification** (from the Latin *sapo* meaning soap) because a **soap** is the sodium or potassium salt of a long chain carboxylic acid (RCO_2H; R = C_{13}–C_{19}).

Animal **fats** are triesters of glycerol (a trihydric alcohol), called **triglycerides** (see later Section 20.2.2). Hydrolysis of fats by heating in aqueous NaOH (caustic soda) gives soaps and glycerol.

For example, **sodium palmitate** (RCO_2Na; R = $C_{15}H_{31}$) is a **soap** which is formed by the hydrolysis of the corresponding triglyceride with NaOH.

Acidification of sodium palmitate gives **palmitic acid**. Since carboxylic acids derived from natural fats are called **fatty acids**, palmitic acid is a **fatty acid**.

$$\begin{array}{l}CH_2\text{—}OCOR\\ |\\ CH\text{—}OCOR\\ |\\ CH_2\text{—}OCOR\end{array} + 3\ NaOH \xrightarrow[\Delta]{H_2O} \begin{array}{l}CH_2\text{—}OH\\ |\\ CH\text{—}OH\\ |\\ CH_2\text{—}OH\end{array} + 3\ R\overset{\overset{O}{\|}}{C}O^-\ Na^+$$

(R = $C_{15}H_{31}$) — **Triglyceride**; **Glycerol**; (R = $C_{15}H_{31}$) **Soap (sodium palmitate)**

$$CH_3(CH_2)_{14}CO_2^-\ Na^+ \xrightarrow[H_2O]{H^+} CH_3(CH_2)_{14}CO_2H$$

Sodium palmitate → **Palmitic acid** ***A fatty acid***

12.7.3 Hydrolysis of Amides, Acyl Halides and Acid Anhydrides

Amides are less easily hydrolysed than esters because amides are less reactive toward nucleophilic substitution.

Hydrolysis of amides can be carried out by heating in aqueous acid (e.g. aqueous HCl) or aqueous base (such as aqueous NaOH or KOH). The acidic hydrolysis gives directly the carboxylic acid, whereas the basic hydrolysis gives sodium or potassium salt of the carboxylic acid. Acidification of the basic reaction mixture then produces the carboxylic acid.

The hydrolysis reaction involves **acyl–nitrogen bond cleavage**.

The following are two examples:

$$CH_3\overset{\overset{O}{\|}}{C}\text{—}NH_2 \xrightarrow[H_2O,\ \Delta]{HCl} CH_3\overset{\overset{O}{\|}}{C}\text{—}OH + NH_4Cl$$

Acetamide → **Acetic acid**

$$C_6H_5-\overset{O}{\overset{||}{C}}-NH_2 \xrightarrow[H_2O,\ \Delta]{KOH} C_6H_5-\overset{O}{\overset{||}{C}}-O^- K^+ + NH_3$$

Benzamide — Potassium benzoate

(work-up) H_3O^+

$$C_6H_5-\overset{O}{\overset{||}{C}}-OH$$

Benzoic acid

Qualitative test for amides

When an amide is heated with NaOH, ammonia gas is evolved which can be easily detected by its characteristic smell. Ammonia gas also turns moist red litmus paper blue. This reaction is used as a **test for amides**.

$$R-\overset{O}{\overset{||}{C}}-NH_2 \xrightarrow{NaOH} R-\overset{O}{\overset{||}{C}}-ONa + NH_3\uparrow$$

Amide — Ammonia

Detected by its characteristic smell

Hydrolysis of acyl halides

Acyl halides are *most easily* hydrolysed. For example, **acetyl chloride** is so reactive that it reacts vigorously with cold water to produce acetic acid. In this reaction, steamy acidic fumes of HCl are evolved.

In comparison, **benzoyl chloride** is somewhat less reactive and it can undergo hydrolysis in aqueous base.

$$CH_3\overset{O}{\overset{||}{C}}-Cl \xrightarrow{H_2O} CH_3\overset{O}{\overset{||}{C}}-OH + HCl$$

Acetyl chloride — Acetic acid

$$C_6H_5-\overset{O}{\overset{||}{C}}-Cl \xrightarrow[H_2O]{OH^-} C_6H_5-\overset{O}{\overset{||}{C}}-O^- + Cl^-$$

Benzoyl chloride — Benzoate ion

Hydrolysis of acid anhydrides

Acid anhydrides also undergo hydrolysis but they are *less reactive* than acyl halides. For example, acetic anhydride is hydrolysed to two molecules of acetic acid, but it occurs slowly at room temperature.

$$CH_3\overset{O}{\overset{\|}{C}}-O-\overset{O}{\overset{\|}{C}}CH_3 \xrightarrow{H_2O} 2\ CH_3\overset{O}{\overset{\|}{C}}-OH$$

Acetic anhydride **Acetic acid**

12.8 SUBSTITUTION BY HYDRIDE NUCLEOPHILE: REDUCTION

In Chapter 9, we have learned that complex metal hydrides such as **lithium aluminium hydride** ($LiAlH_4$) and **sodium borohydride** ($NaBH_4$) can act as hydride donors and reduce aldehydes and ketones to alcohols (see Section 9.6).

However, $LiAlH_4$ is a more powerful reducing agent than $NaBH_4$. Thus $LiAlH_4$ can reduce carboxylic acids, esters and amides but $NaBH_4$ cannot.

12.8.1 Reduction of Esters

$LiAlH_4$ reduces esters to **primary alcohols**. The substitution of the alkoxy group of esters by hydride converts esters to **aldehydes**. However, aldehydes cannot be isolated since they are much more easily reduced than esters. As a result, the reduction product of an ester is a primary alcohol.

For example, **ethyl benzoate** is reduced by $LiAlH_4$ to **benzyl alcohol** (a primary alcohol).

$$R-\overset{O}{\overset{\|}{C}}-OR' \xrightarrow[2.\ H_2O]{1.\ LiALH_4/Et_2O} R-CH_2OH \quad (+\ R'OH)$$

Ester **1° Alcohol**

$$R-\overset{O}{\overset{\|}{C}}-OR' \xrightarrow[Et_2O]{LiAlH_4} \left[R-\overset{O}{\overset{\|}{C}}-H\right] \xrightarrow[2.\ H_2O]{1.\ LiAlH_4} R-CH_2OH$$

(aldehyde)

Example

$$C_6H_5-\overset{O}{\overset{\|}{C}}-OC_2H_5 \xrightarrow[2.\ H_2O]{1.\ LiALH_4/Et_2O} C_6H_5-CH_2OH$$

Ethyl benzoate **Benzyl alcohol**

Selective reduction of ester group in the presence of keto group

Ketones are more easily reduced than esters. Therefore, if a compound contains both keto and ester groups, any reducing agent (e.g. $LiAlH_4$) that can reduce the ester group will also reduce the keto group. However, a **selective reduction of the ester group** in the presence of a keto group is possible using a **protection-deprotection** protocol.

The **keto group** is first **protected** by **acetal formation** (see Section 9.4). Acetals are stable and not hydrolysed back to ketones in the *basic medium* of reduction. Thus the ester group can only be reduced with $LiAlH_4$ with the keto group being protected as an acetal.

After the ester reduction in the basic solution, the **deprotection of the keto group** is accomplished by **hydrolysis of the acetal** with aqueous acid.

The acetal protection-deprotection strategy is illustrated here with the selective reduction of the ester group of a keto-ester (ethyl 4-oxopentanoate).

$CH_3-C(=O)-CH_2CH_2-C(=O)-OC_2H_5$ —[$HOCH_2CH_2OH$, H^+; *Acetal protection of the keto group*]→ cyclic acetal: $CH_3-C(OCH_2CH_2O)-CH_2CH_2-C(=O)-OC_2H_5$

Ethyl 4-oxopentanoate (a keto-ester)

—[*Reduction of the ester group*: 1. $LiALH_4/Et_2O$; 2. H_2O]→ $CH_3-C(OCH_2CH_2O)-CH_2CH_2-CH_2OH$

—[H_2O, H^+; *Deprotection (acetal hydrolysis)*]→ $CH_3-C(=O)-CH_2CH_2-CH_2OH$

4-Oxo-1-pentanol

PROBLEM 12.3: How would you carry out the following conversions?

(a) Cyclohexyl$-CO_2CH_3$ ⟶ Cyclohexyl$-CH_2OH$

(b) $OHC-$cyclohexane$-CO_2CH_3$ ⟶ HOH_2C-cyclohexane$-CO_2CH_3$

Answer

(a) $LiAlH_4$ reduces the ester group to a primary alcohol.

$$\text{C}_6\text{H}_{11}\text{–CO}_2\text{CH}_3 \xrightarrow[\text{2. H}_2\text{O}]{\text{1. LiAlH}_4/\text{Et}_2\text{O}} \text{C}_6\text{H}_{11}\text{–CH}_2\text{OH}$$

(b) $NaBH_4$ reduces the aldehyde group to a primary alcohol but cannot reduce the ester group.

$$\text{OHC–C}_6\text{H}_{10}\text{–CO}_2\text{CH}_3 \xrightarrow[\text{H}_2\text{O}]{\text{NaBH}_4} \text{HOH}_2\text{C–C}_6\text{H}_{10}\text{–CO}_2\text{CH}_3$$

12.8.2 Reduction of Carboxylic Acids

Reduction of **carboxylic acids** is generally difficult; however, they can be reduced by **$LiAlH_4$** to **primary alcohols**. $LiAlH_4$ first reacts with OH of the carboxyl group to liberate hydrogen, and reduction takes place thereafter.

$$\underset{\textbf{Carboxylic acid}}{\text{R–C(=O)–OH}} \xrightarrow[\text{2. H}_2\text{O}]{\text{1. LiALH}_4/\text{Et}_2\text{O}} \underset{\textbf{1° Alcohol}}{\text{R–CH}_2\text{OH}}$$

$$\left[\text{R–C(=O)–OH} \xrightarrow{\text{LiAlH}_4} \text{H}_2 + \text{R–C(=O)–O}\overset{-}{\text{Al}}\text{H}_3\ \text{Li}^+ \xrightarrow[\text{2. H}_2\text{O}]{\text{1. LiALH}_4} \text{RCH}_2\text{OH}\right]$$

Example

$$\underset{\textbf{2,2-Dimethylpropanoic acid}}{\text{H}_3\text{C–C(CH}_3)_2\text{–CO}_2\text{H}} \xrightarrow[\text{2. H}_2\text{O}]{\text{1. LiALH}_4/\text{Et}_2\text{O}} \underset{\textbf{2,2-Dimethyl-1-propanol (neopentyl alcohol)}}{\text{H}_3\text{C–C(CH}_3)_2\text{–CH}_2\text{OH}}$$

Diborane [B_2H_6 or $(BH_3)_2$] can act as a **hydride source** to reduce not only aldehydes and ketones but also carboxylic acids. Diborane reduces carboxylic acids to **primary alcohols**.

For example, *p*-chlorobenzoic acid is reduced by diborane to *p*-chlorobenzyl alcohol.

$$\underset{\textbf{\textit{p}-Chlorobenzoic acid}}{\text{Cl–C}_6\text{H}_4\text{–CO}_2\text{H}} + \underset{\textbf{Diborane}}{(\text{BH}_3)_2} \xrightarrow[\text{2. H}_2\text{O}]{\text{1. THF}} \underset{\textbf{\textit{p}-Chlorobenzyl alcohol}}{\text{Cl–C}_6\text{H}_4\text{–CH}_2\text{OH}}$$

12.8.3 Reduction of Amides

Reduction of **amides** with $LiAlH_4$ produces **amines**. Here the **carbonyl group (C=O) is reduced to a methylene group (CH_2)**. Thus the reduction of amides differs from the reduction of esters and carboxylic acids by $LiAlH_4$.

For *N*-substituted amides or *N,N*-disubstituted amides, the reduction by $LiAlH_4$ proceeds similarly.

$$R-\overset{O}{\overset{\|}{C}}-NH_2 \xrightarrow[2.\ H_2O]{1.\ LiALH_4/Et_2O} R-CH_2-NH_2$$

Amide **Amine**

Examples

$$CH_3-\overset{O}{\overset{\|}{C}}-NH_2 \xrightarrow[2.\ H_2O]{1.\ LiALH_4/Et_2O} CH_3-CH_2-NH_2$$

Acetamide **Ethylamine**

$$C_6H_5-\underset{CH_3}{\underset{|}{N}}-\overset{O}{\overset{\|}{C}}CH_3 \xrightarrow[2.\ H_2O]{1.\ LiALH_4/Et_2O} C_6H_5-\underset{CH_3}{\underset{|}{N}}-CH_2CH_3$$

***N*-Methylacetanilide** ***N*-Ethyl-*N*-methylaniline**

12.8.4 Reduction of Acyl Chlorides

Acyl chlorides are readily reduced by $LiAlH_4$ to **primary alcohols** via aldehydes.

$$R-\overset{O}{\overset{\|}{C}}-Cl \xrightarrow[2.\ H_2O]{1.\ LiALH_4/Et_2O} R-CH_2OH$$

Acyl chloride **1° Alcohol**

$$R-\overset{O}{\overset{\|}{C}}-Cl \xrightarrow{LiAlH_4} \left[R-\overset{O}{\overset{\|}{C}}-H\right] \xrightarrow[2.\ H_2O]{1.\ LiAlH_4} R-CH_2OH$$

Aldehyde

However, it is possible to **stop** the reduction at the **aldehyde** stage by using a *sterically hindered* derivative of $LiAlH_4$. One such derivative is **lithium tri-*t*-butoxyaluminium hydride** $LiAlH[OC(CH_3)_3]_3$ in which three hydrogen atoms of $LiAlH_4$ are replaced by three bulky *t*-butoxy group $[(CH_3)_3CO]$.

Thus, **benzoic acid** can be converted to **benzaldehyde** as follows:

$$C_6H_5-\overset{O}{\overset{\|}{C}}-OH \xrightarrow{PCl_5} C_6H_5-\overset{O}{\overset{\|}{C}}-Cl \xrightarrow[2.\ H_2O]{1.\ LiAlH[OC(CH_3)_3]_3} C_6H_5-\overset{O}{\overset{\|}{C}}-H$$

Benzoic acid **Benzoyl chloride** **Benzaldehyde**

An older method of reducing an acyl chloride to an aldehyde is known as the **Rosenmund reduction**. The reduction involves catalytic hydrogenation using a palladium catalyst supported on barium sulphate, where the catalyst is partially poisoned with quinoline and sulphur.

$$C_6H_5-\overset{O}{\overset{\|}{C}}-Cl \xrightarrow[\text{quinoline, sulphur}]{H_2,\ Pd/BaSO_4} C_6H_5-\overset{O}{\overset{\|}{C}}-H$$

Benzoyl chloride ***Rosenmund reduction*** **Benzaldehyde**

12.9 SUBSTITUTION BY CARBON NUCLEOPHILES: REACTIONS WITH ORGANOMETALLIC REAGENTS

We have described earlier that organometallic reagents such as Grignard reagents and organolithium reagents can act as carbon nucleophiles in the addition reactions of aldehydes and ketones (see Section 9.8).

In this section, we will consider the reactions of organometallic reagents with carboxylic acids and their derivatives.

12.9.1 Grignard Reagents: Reactions with Esters

A **Grignard reagent** (RMgX) can react with an **ester** ($R'CO_2R''$) with *substitution* of the alkoxy group (OR″) of the ester by the alkyl group (R) of the Grignard reagent to form a **ketone** (RCOR′). However, the ketone is not isolated as the product. This is because a ketone is *more reactive* toward Grignard reagents than the parent ester. Therefore, as soon as the ketone is formed, it reacts rapidly with a *second molecule* of the Grignard reagent to produce a **tertiary alcohol** as the product.

It should be noted that the **tertiary alcohol** possesses **two identical alkyl groups** (R, R) since *two molecules of the same Grignard reagent* are involved in the reaction.

$R'-C(=O)-OR''$ + $R-MgX$ $\xrightarrow{Et_2O}$ $R'-C(R)(OR'')-O^- \; \overset{+}{MgX}$ ⟶ [$R'-C(R)=O$] + $R''OMgX$

Ester — Grignard reagent — (ketone)

[ketone] $\xrightarrow{R-MgX}$ $R'-C(R)(R)-OMgX$ $\xrightarrow{H_3O^+}$ $R'-C(R)(R)-OH$

3° Alcohol

Two groups are identical (R, R)

Example

$CH_3-C(=O)-OC_2H_5$ + C_2H_5MgBr $\xrightarrow{Et_2O}$ [$CH_3-C(=O)-C_2H_5$] $\xrightarrow[2.\ H_3O^+]{1.\ C_2H_5MgBr}$ $CH_3-C(OH)(CH_2CH_3)-CH_2CH_3$

Ethyl acetate — **Ethylmagnesium bromide** — **3-Methyl-3-pentanol**

There is however **one exception**; the **esters of formic acid** lead to a **secondary alcohol** with two identical alkyl groups. For example, ethyl formate reacts with ethylmagnesium bromide to give 2-pentanol (a 2° alcohol).

$H-C(=O)-OC_2H_5$ + C_2H_5MgBr $\xrightarrow{Et_2O}$ [$H-C(=O)-C_2H_5$] $\xrightarrow[2.\ H_3O^+]{1.\ C_2H_5MgBr}$ $H-C(OH)(CH_2CH_3)-CH_2CH_3$

Ethyl formate — **Ethylmagnesium bromide** — **3-Pentanol (a 2° alcohol)**

PROBLEM 12.4: How would you carry out the following transformation?

3-(ethoxycarbonyl)cyclohexanone ($CO_2C_2H_5$) ⟶ 3-[$C(OH)(CH_3)_2$]cyclohexanone

Answer

In this transformation, the ester unit can be converted to a tertiary alcohol by a Grignard reaction. But a keto group present in the substrate is more reactive than the ester group, so the keto group will undergo the Grignard reaction more easily. With excess of Grignard reagents, both keto and ester groups will react.

However, as the keto group is retained in the product, it is necessary to protect the keto group by acetal formation. Thus an acetal protection-deprotection strategy could accomplish the desired transformation using a Grignard reaction (cf. selective reduction of ester group in Section 12.8.1).

CH_2CH_2 / OH OH, H^+ — ***Acetal protection***

1. CH_3MgBr/Et_2O
2. H_3O^+

HCl, H_2O — ***Deprotection***

$CO_2C_2H_5$; C—OC_2H_5; OH, C—CH_3, CH_3; OH, $C(CH_3)_2$

12.9.2 Organolithium Reagents: Reactions with Esters and Carboxylic Acids

Organolithium reagents also react with **esters** in the same fashion as Grignard reagents to produce a **tertiary alcohol**.

$$CH_3-\overset{O}{\overset{\|}{C}}-OC_2H_5 + CH_3Li \xrightarrow{THF} \left[CH_3-\overset{O}{\overset{\|}{C}}-CH_3\right] \xrightarrow[2.\ H_3O^+]{1.\ CH_3Li} CH_3-\overset{OH}{\underset{CH_3}{C}}-CH_3$$

Ethyl acetate — **Methyllithium** — **(ketone)** — ***t*-Butanol**

Reactions with carboxylic acids

With carboxylic acids, Grignard reagents and organolithium reagents readily form carboxylate salts in an acid-base reaction (cf. Section 9.8.1).

The **carboxylate ion** is much less reactive toward nucleophilic attack and no further reaction takes place with Grignard reagents. But an organolithium reagent is a powerful nucleophilic reagent and further reaction can take place.

$$R'-\overset{O}{\overset{\|}{C}}-OH + RMgX \longrightarrow R'-\overset{O}{\overset{\|}{C}}-\overset{-}{O}\ \overset{+}{MgX} + RH$$

Carboxylic acid (***Acid***) — **Grignard reagent** (***Base***) — **Carboxylate salt**

Carboxylate salt $\xrightarrow{RMgX}$ (crossed) **No further reaction**

$$R'-C(=O)-OH + RLi \longrightarrow R'-C(=O)-O^{-}Li^{+} + RH$$

Carboxylic acid (Acid) + Organolithium reagent (Base) → Carboxylate salt; Carboxylate salt + RLi → Further reaction can take place

After **lithium carboxylate** is formed from the initial reaction between the carboxylic acid and organolithium reagent, a *second molecule* of alkyllithium (or aryllithium) adds to the lithium carboxylate. The resulting adduct is then hydrolysed to a **ketone** via the formation of an unstable *gem*-diol.

Thus a direct conversion of a **carboxylic acid to a ketone** is possible using organolithium reagents.

$$R'-C(=O)-OH + RLi \longrightarrow R'-C(=O)-O^{-}Li^{+} \xrightarrow{RLi} R'-C(R)(O^{-}Li^{+})-O^{-}Li^{+} \xrightarrow{H_3O^+} [R'-C(R)(OH)-OH] \xrightarrow{-H_2O} R'-C(=O)-R$$

Carboxylic acid + Organolithium reagent → Lithium carboxylate → … → *gem*-Diol → Ketone

Example

$$C_6H_5-C(=O)-OH + 2\,CH_3Li \xrightarrow[2.\ H_3O^+]{1.\ Et_2O} C_6H_5-C(=O)-CH_3$$

Benzoic acid + Methyllithium → Acetophenone

12.9.3 Organocadmium and Organocopper Reagents: Reactions with Acid Chlorides

Acid chlorides react rapidly with Grignard reagents or organolithium reagents to form tertiary alcohols, but the method is not quite useful because side reactions often predominate.

In contrast, **organocadmium reagents** and **organocopper reagents** are much less reactive and they can react with *highly reactive* **acid chlorides**, but do not react with aldehydes and ketones.

Organocadmium reagents

Organocadmium reagents are usually prepared by the reaction between cadmium chloride and Grignard reagents. For example, **ethylmagnesium chloride** (two moles) react with $CdCl_2$ (one mole) to produce **diethylcadmium**.

$$2\ CH_3CH_2MgCl \quad + \quad CdCl_2 \longrightarrow (CH_3CH_2)_2Cd \quad + \quad 2\ MgCl_2$$

Ethylmagnesium chloride **Diethylcadmium**

In general, the reaction of a **dialkylcadmium** (R_2Cd) with an **acid chloride** causes substitution of the chlorine atom by the nucleophilic alkyl group and produces a **ketone**. The ketone can be isolated because it does not react further with the dialkylcadmium.

The following is an example:

$$C_6H_5-\overset{O}{\overset{\|}{C}}-Cl \quad + \quad (CH_3CH_2)_2Cd \xrightarrow{\text{Benzene}} C_6H_5-\overset{O}{\overset{\|}{C}}-CH_2CH_3$$

Benzoyl chloride **Diethylcadmium** **1-Phenyl-1-propanone**

Organocopper reagents

Organocopper reagents such as **lithium dialkylcuprate** (R_2CuLi) can be prepared from the reaction of *two moles* of **alkyllithium** (RLi) with *one mole* of **cuprous iodide** (CuI).

Similar to organocadmium reagents, **R_2CuLi** reacts with **acid chlorides** to produce **ketones**. An example is the reaction of propanoyl chloride with lithium dimethylcuprate to form butanone.

$$2\ RLi \quad + \quad CuI \longrightarrow R_2CuLi \quad + \quad LiI$$

Alkyllithium **Lithium dialkylcuprate**

$$CH_3CH_2-\overset{O}{\overset{\|}{C}}-Cl \quad + \quad (CH_3)_2CuLi \xrightarrow{\text{THF}} CH_3CH_2-\overset{O}{\overset{\|}{C}}-CH_3$$

Propanoyl chloride **Lithium dimethylcuprate** **Butanone**

12.9.4 Organozinc Reagents: The Reformatsky Reaction

Organozinc reagents do not react with esters. So an organozinc reagent can be derived from an ester containing a halogen atom (i.e. α-halo ester).

An **organozinc reagent** is usually prepared *in situ* from an **α-halo ester** and **zinc metal**. The reaction of the organozinc reagent with an aldehyde or ketone is known as the **Reformatsky reaction**. The product is a **β-hydroxy ester**.

The Reformatsky reaction is performed by treating an aldehyde or ketone with an α-bromo ester in the presence of zinc metal in a suitable solvent such as benzene. The reaction of the organozinc compound so formed with the carbonyl group of aldehyde or ketone gives a zinc alkoxide, which is then hydrolysed to yield a β-hydroxy ester as the product.

α-Bromo ester $\xrightarrow[\text{Benzene}]{\text{Zn}}$ Organozinc compound $\xrightarrow{\text{C=O (aldehyde or ketone)}}$ Zinc alkoxide $\xrightarrow{H_3O^+}$ β-Hydroxy ester

Example

$$CH_3CHO + BrCH_2CO_2C_2H_5 \xrightarrow[2.\ H_3O^+]{1.\ Zn} CH_3CH(OH)CH_2CO_2C_2H_5$$

Acetaldehyde, Ethyl bromoacetate, Ethyl 3-hydroxybutanoate

The **β-hydroxy ester** obtained from the Reformatsky reaction can be easily **dehydrated** to produce an **α,β-unsaturated ester**. For example, ethyl 3-hydroxybutanoate undergoes acid-catalysed dehydration to produce ethyl crotonate (ethyl 2-butenoate).

$$CH_3CH(OH)\text{—}CH_2\text{—}C(=O)\text{—}OC_2H_5 \xrightarrow[\Delta]{H_3O^+} CH_3CH\text{=}CH\text{—}C(=O)\text{—}OC_2H_5$$

Ethyl 3-hydroxybutanoate — *A β-hydroxy ester*

Ethyl crotonate (ethyl 2-butenoate) — *An α,β-unsaturated ester*

12.10 NITRILES AND THEIR REACTIONS

Nitriles or cyanides can be prepared by nucleophilic substitution of **alkyl halides** with cyanide ion (see Section 11.2). They can also be prepared from **amides** by dehydration with phosphorus pentoxide (see Section 10.10).

PROBLEM 12.5: Which of the following reactions (a or b) would you use to prepare the nitrile $(CH_3)_3CCN$? Explain.

(a) $(CH_3)_3C\text{—}Br \xrightarrow{CN^-}$

(b) $(CH_3)_3C\text{—}C(=O)\text{—}NH_2 \xrightarrow{P_4O_{10}}$

Answer

The cyanide ion (CN^-) is a good nucleophile as well as a strong base. In reaction (a), the substrate is a tertiary bromide; thus the reaction with cyanide ion would favour elimination (E2) rather than substitution (S_N2). The major product is therefore isobutene (from elimination), *not* the nitrile (from substitution).

On the other hand, in reaction (b), the dehydration of the amide with phosphorus pentoxide would produce the desired nitrile. Hence the reaction (b) will be used for the synthesis of the nitrile.

(a) $(CH_3)_3C{-}Br \xrightarrow[E2]{CN^-} H_3C{-}C(CH_3){=}CH_2$

t-Butyl bromide → **Isobutene**

(b) $(CH_3)_3C{-}C(=O){-}NH_2 \xrightarrow[-H_2O]{P_4O_{10}} (CH_3)_3C{-}C{\equiv}N$

2,2-Dimethylpropanamide → **2,2-Dimethylpropanenitrile**

12.10.1 Hydrolysis of Nitriles

Nitriles are hydrolysed to **carboxylic acids** under acidic or basic conditions. The addition of water to a nitrile forms an intermediate *amide* which is then hydrolysed to a carboxylic acid.

The complete hydrolysis is carried out by heating nitriles with aqueous acids or bases without isolating the amide. However, under appropriate conditions, amides can be isolated.

$R{-}C{\equiv}N \xrightarrow[\Delta]{H_2O/H^+} R{-}C(=O){-}OH + NH_4^+$ (**Carboxylic acid**)

$R{-}C{\equiv}N \xrightarrow[\Delta]{OH^-/H_2O} R{-}C(=O){-}O^- + NH_3$ (**Carboxylate ion**)

Nitrile

$\left[R{-}C{\equiv}N \xrightarrow{H_2O} R{-}C(=O){-}NH_2 \xrightarrow[\text{or } OH^-]{H_3O^+} RCO_2H \text{ or } RCO_2^- \right]$

Amide
Not isolated

Example

$$CH_3CH_2CH_2Br \xrightarrow{CN^-} CH_3CH_2CH_2CN \xrightarrow[H^+]{H_2O} CH_3CH_2CH_2CO_2H$$

1-Bromopropane **Butanenitrile** **Butanoic acid**

12.10.2 Reduction of Nitriles

Nitriles can be reduced to **primary amines** using catalytic hydrogenation or with $LiAlH_4$. The following are two examples:

$$CH_3CH_2CH_2C{\equiv}N + 2\ H_2 \xrightarrow[\Delta]{\text{Raney Ni}} CH_3CH_2CH_2CH_2NH_2$$

Butanenitrile **Butylamine**

$$C_6H_5{-}C{\equiv}N \xrightarrow[2.\ H_2O]{1.\ LiAlH_4/Et_2O} C_6H_5{-}CH_2NH_2$$

Benzonitrile **Benzylamine**

12.10.3 Reactions of Nitriles with Organometallic Reagents

Addition of a **Grignard reagent** to a **nitrile** followed by hydrolysis produces a **ketone**. The reaction provides a useful approach to the synthesis of ketones.

$$R'{-}C{\equiv}N + R{-}MgX \longrightarrow R'{-}C(R){=}N^- \ \overset{+}{Mg}X \xrightarrow{H_3O^+} R'{-}C(R){=}O + NH_4^+ + Mg^{2+} + X^-$$

Nitrile **Grignard reagent** **Ketone**

Example

$$CH_3C{\equiv}N + C_6H_5{-}MgBr \xrightarrow[2.\ H_3O^+]{1.\ Et_2O} CH_3C(=O){-}C_6H_5$$

Acetonitrile **Phenylmagnesium bromide** **Acetophenone**

An **organolithium reagent** reacts with a **nitrile** in a similar manner as the Grignard reagent to yield a **ketone**.

$$R'{-}C{\equiv}N + R{-}Li \longrightarrow R'{-}C(R){=}N^- \ Li^+ \xrightarrow{H_3O^+} R'{-}C(R){=}O + NH_4^+ + Li^+$$

Nitrile **Organolithium reagent** **Ketone**

Example

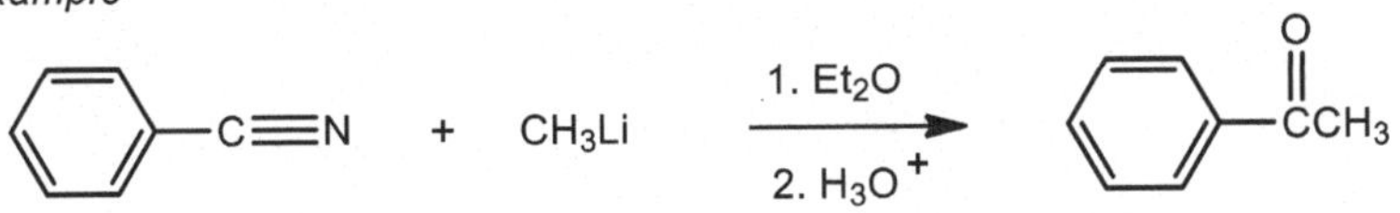

Benzonitrile (C_6H_5–C≡N) + CH_3Li (Methyllithium) —1. Et_2O; 2. H_3O^+→ Acetophenone (C_6H_5–C(=O)CH_3)

Benzonitrile **Methyllithium** **Acetophenone**

SUPPLEMENTARY PROBLEMS

12.6 Predict the products of the following reactions. Draw their structures in bond-line formulas.

(a) Benzyl alcohol + Acetic anhydride →

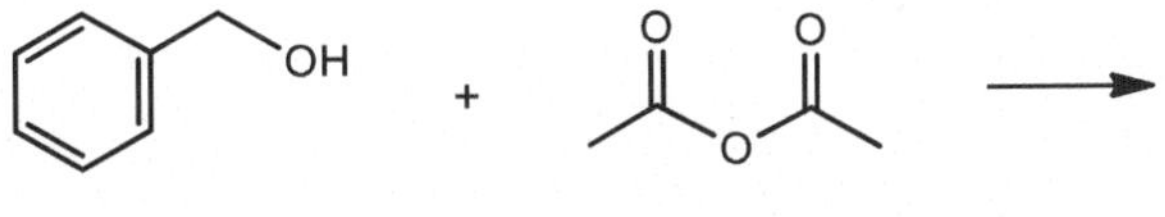

Benzyl alcohol **Acetic anhydride**

(b) Benzoyl chloride + Ethanol —Pyridine→

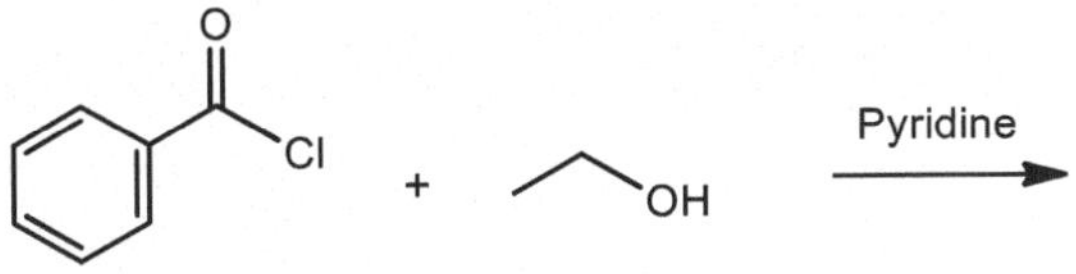

Benzoyl chloride **Ethanol**

12.7 In an acid-catalysed hydrolysis of methyl benzoate in ^{18}O-labelled water ($H_2{}^{18}O$), predict where would you expect to find the labelled oxygen (^{18}O) in the products.

12.8 Draw stereochemical formulas and give *R* or *S* configuration for compounds **A** and **B** in the following scheme:

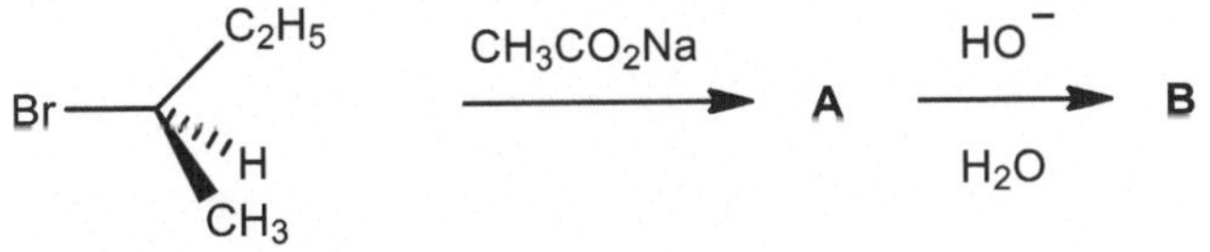

(*R*)-2-Bromobutane (Br, C_2H_5, H, CH_3) —CH_3CO_2Na→ **A** —HO^-, H_2O→ **B**

(*R*)-2-Bromobutane

[*Ans:* **A** = (*S*)-2-butyl acetate, **B** = (*S*)-2-butanol]

12.9 Predict the product(s) in each of the following reactions:

(a) Phthalic anhydride —MeOH→

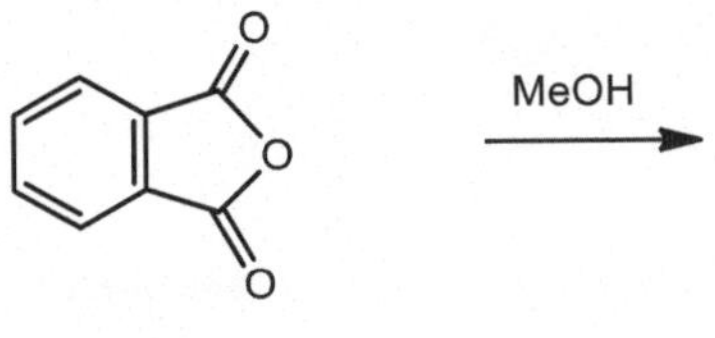

Phthalic anhydride

(b) N,N-Dimethylbenzamide ($C_6H_5CONMe_2$) $\xrightarrow[H_2O]{OH^-}$

N,N-Dimethylbenzamide

[*Ans:* (a) methyl hydrogen phthalate (a half-ester); (b) benzoate ion and dimethylamine]

12.10 Write structures for compounds **A–F** in the following scheme:

$$C_6H_5CO_2H \xrightarrow{SOCl_2} \textbf{A} \xrightarrow{NH_3} \textbf{B} \xrightarrow{P_4O_{10}} \textbf{C} \xrightarrow[2.\ H_3O^+]{1.\ C_6H_5Li} \textbf{D}$$

$$\textbf{A} \xrightarrow{C_2H_5OH} \textbf{E} \xrightarrow[2.\ H_3O^+]{1.\ C_6H_5MgBr} \textbf{F}$$

CHAPTER 13

Aromatic Compounds: Substitution Reactions

IN THIS CHAPTER

- Electrophilic aromatic substitution: The addition–elimination mechanism
- Substituent effects: Reactivity and orientation (*ortho/para* and *meta*)
- Halogenation of benzene and substituted benzenes
- Nitration of benzene and substituted benzenes
- Sulphonation, desulphonation and chlorosulphonation
- Friedel-Crafts reactions: Alkylation and acylation
- Nucleophilic aromatic substitution: Addition–elimination (S_NAr) and elimination–addition (benzyne) mechanisms
- Radical substitution of the aromatic side chain
- Oxidation of aromatic compounds
- Reduction of aromatic nitro compounds and the Birch reduction

13.1 ELECTROPHILIC AROMATIC SUBSTITUTION: THE ADDITION–ELIMINATION MECHANISM

The most characteristic reactions of aromatic compounds are **substitution reactions.** The *parent* aromatic hydrocarbon is **benzene**. A substitution reaction of benzene involves the replacement of one of its hydrogen atoms by another atom or group.

The exposed π electrons of the benzene ring, like alkenes, are susceptible to electrophilic attack. Thus most substitution reactions of benzene and its derivatives are initiated by an electrophile, and they are called electrophilic aromatic substitutions.

$$C_6H_5\text{–}H + E^+ \longrightarrow C_6H_5\text{–}E + H^+$$

Benzene **Electrophile** **Substitution product**

Mechanism

A mechanism for the electrophilic aromatic substitution of benzene is shown in Figure 13.1.

The mechanism involves *two* steps. In **step 1**, the electrophile (E^+) attacks the π system of the benzene ring to form a σ bond to one of the carbon atoms of the ring. This **addition** of the electrophile to benzene result in the formation of an *intermediate carbocation*, called an **arenium ion** (or **σ complex**). In **step 2**, the arenium ion **eliminates** a proton from the carbon atom that bears the electrophile to give the substitution product.

Since step 1 involves an addition and step 2 involves an elimination, the mechanism is called an **addition-elimination mechanism**.

Benzene + E^+ → (Slow (r/d), *Step 1*, *Addition*) → Arenium ion (σ complex) → (Fast, *Step 2*, *Elimination*) → Substitution product + H^+

Resonance stabilization of the arenium ion:

(*ortho*) ⟷ (*para*) ⟷ (*ortho*)

Three resonance structures

≡ (δ+, δ+, δ+) or

Resonance hybrid

Figure 13.1: An addition-elimination mechanism for the electrophilic aromatic substitution.

Step 1 is **slow and rate-determining (r/d)** because the arenium ion intermediate is *nonaromatic* and its formation involves the **loss of aromaticity** of the benzene ring. **Step 2** is **fast** because the loss of proton **restores the aromaticity** of the benzene ring.

Notably, the **arenium ion** is stabilized somewhat by **resonance**. Each of the three resonance structures has a positive charge at the *ortho* or *para* position with respect to the electrophile. These resonance structures give a **resonance hybrid** that can be drawn as a single delocalized structure with partial (dotted) bonds or with a partial (one-third) positive charge at the two *ortho* positions and the *para* position.

Note: The arenium ion is a carbocation that *does not* add to a nucleophile like the carbocation formed from an alkene (see Section 8.1), but loses a proton to give the substitution product.

13.2 EFFECT OF SUBSTITUENTS IN ELECTROPHILIC AROMATIC SUBSTITUTION: REACTIVITY AND ORIENTATION

In electrophilic aromatic substitution of substituted benzenes, a **substituent** already on the ring affects the **rate of the reaction** and also directs the **position of subsequent**

substitution. Thus, substituents influence both **reactivity** and **orientation** in electrophilic aromatic substitution.

13.2.1 Reactivity: Activating and Deactivating Substituents

Substituents are classified into two categories depending on their influence on the **reactivity** of the benzene ring toward electrophilic aromatic substitution.

Substituents that cause the ring to be **more reactive** than benzene are called **activating substituents**. On the other hand, substituents that cause the ring to be **less reactive** than benzene are called **deactivating substituents**.

Activating substituents

Electron-donating substituents such as nitrogen substituents (e.g. NH_2), oxygen substituents (e.g. OH) and alkyl or aryl groups are **activating substituents**. They *increase* the rate of the reaction relative to benzene itself.

Activating substituents are further divided into *strongly activating*, *moderately activating* and *weakly activating* substituents. The following are some examples:

Activating substituents				
Strongly activating	$-\ddot{N}H_2$	$-\ddot{N}R_2$	$-\ddot{O}H$	$-\ddot{O}^-$
Moderately activating	$-\ddot{N}HCOCH_3$	$-\ddot{O}CH_3$	$-\ddot{O}R$	
Weakly activating	$-CH_3$	$-R$	$-C_6H_5$	

Deactivating substituents

Electron-withdrawing substituents are **deactivating substituents**. They *decrease* the rate of the reaction relative to benzene itself.

Deactivating substituents are also divided into *strongly deactivating*, *moderately deactivating*, and *weakly deactivating* substituents. Some examples are the following:

Deactivating substituents				
Strongly deactivating	$-NO_2$	$-\overset{+}{N}H_3$	$-CF_3$	
Moderately deactivating	$-CHO$	$-COR$	$-CO_2H$	
	$-CO_2R$	$-SO_3H$	$-CN$	
Weakly deactivating	$-F$	$-Cl$	$-Br$	$-I$

Reactivity based on the relative stability of arenium ions

The rate-determining (r/d) step of an electrophilic aromatic substitution involves the formation of an intermediate carbocation or **arenium ion**. Therefore, the relative stabilities of arenium ions formed from substituted benzenes would determine their relative rates of substitution. **The more stable the arenium ion the faster is the rate or more reactive is the substituted benzene.**

An **activating substituent** is an *electron-donating group* that **stabilizes** the arenium ion intermediate by dispersal of its positive charge, and the reaction is **faster**.

In contrast, a **deactivating substituent** is an *electron-withdrawing group* that **destabilizes** the arenium ion intermediate by concentrating its positive charge, and the reaction is **slower**.

A — (benzene ring) — H →(E^+) A → (arenium ion, +) E, H ⟹ ***Reaction is faster***

(A = activating substituent at any position of the benzene ring)

Electron donation by A stabilizes the arenium ion by dispersal of positive charge

D — (benzene ring) — H →(E^+) D ← (arenium ion, +) E, H ⟹ ***Reaction is slower***

(D = deactivating substituent at any position of the benzene ring)

Electron withdrawal by D destabilizes the arenium ion by concentration of positive charge

13.2.2 Orientation: *ortho, meta* and *para* Directing Substituents

An electrophilic aromatic substitution is **regioselective**, leading to *ortho*, *meta* or *para* substitution depending on the *nature of the substituent* already present on the benzene ring.

All activating substituents tend to orient or direct the incoming electrophile mainly to the positions *ortho* and *para* to the substituent. An **activating substituent** is therefore an ***ortho/para* director**.

On the other hand, **most deactivating substituents** tend to direct the incoming electrophile mainly to the *meta* position, and are called ***meta* directors**.

Notably, halogen substituents, although *deactivating*, are *ortho/para* directors.

The *ortho/para* or *meta* orientation of a substituent can be explained by considering the relative stabilities of the intermediate arenium ions.

Activating substituents: ortho/para Directors

An activating nitrogen substituent (e.g. NH_2) or oxygen substituent (e.g. OH) has a lone pair of electrons on the nitrogen or oxygen atom, and is electron-donating through its dominant $+M$ effect ($+M > -I$).

For example, consider a strongly activating **amino substituent** (NH_2) on the benzene ring. Figure 13.2 shows the relative stabilities of arenium ions (σ complexes) for *ortho* attack, *para* attack and *meta* attack of the electrophile.

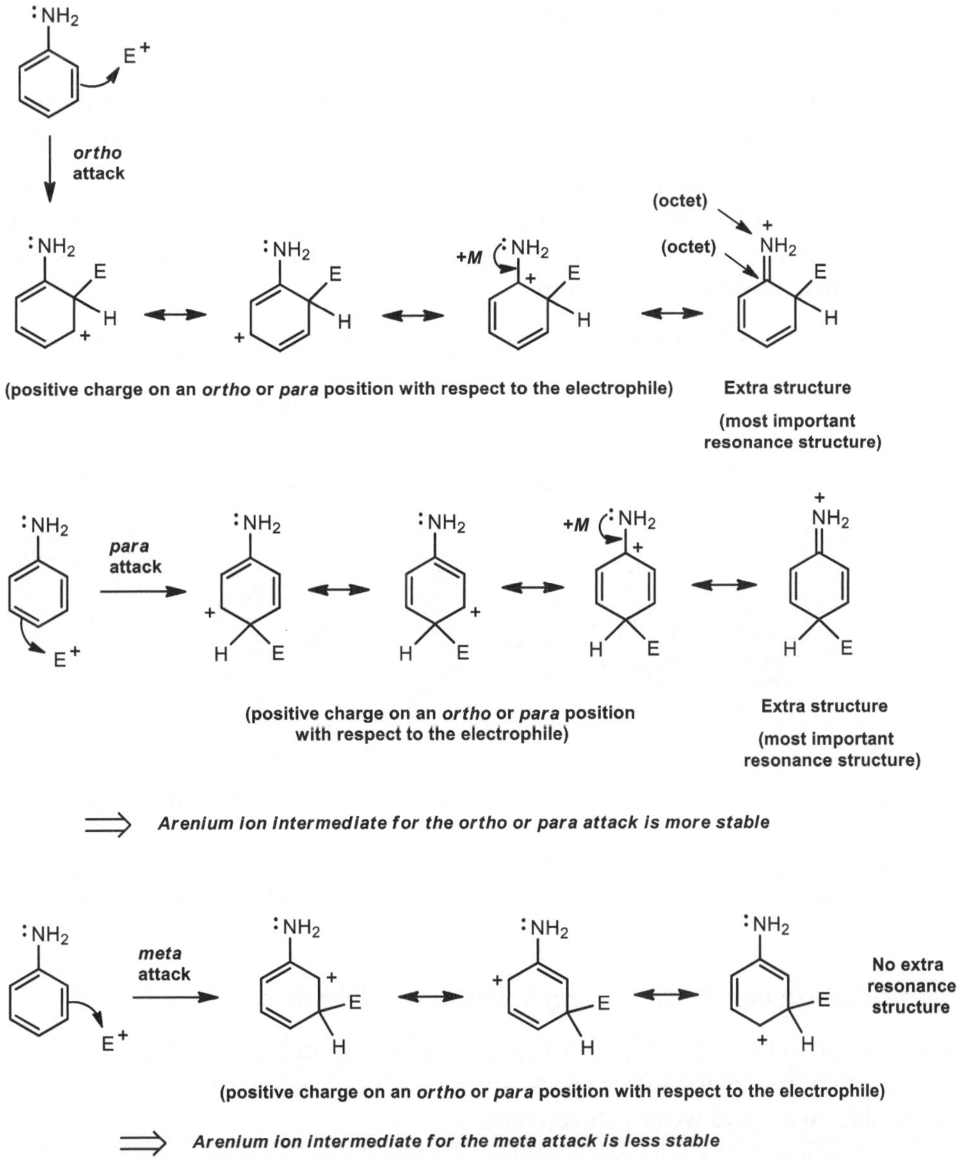

Figure 13.2: The *ortho/para* orientation of an activating substituent (NH_2) in electrophilic aromatic substitution.

We know that three resonance structures for an arenium ion bear a positive charge at either an *ortho* position or the *para* position with respect to the electrophile (see Figure 13.1).

Therefore, an NH_2 group can exert its $+M$ effect to stabilize the arenium ion intermediate when the positive charge is adjacent to the NH_2 group, i.e. when the NH_2 and the electrophile (E) are in the *ortho* or *para* relationship. The $+M$ effect leads to an *extra resonance structure* which is also the most important contributor as both carbon and nitrogen atoms have an octet of electrons.

But an arenium ion cannot be stabilized by NH_2 for a *meta* relationship because the positive charge is *not* adjacent to the NH_2 group for its $+M$ effect to operate. As such, there is *no extra resonance structure* for the *meta* attack.

Thus, the **arenium ion** intermediate for the ***ortho*/*para* attack** is **more stable** than that for the *meta* attack, and the activating NH_2 substituent acts as an ***ortho*/*para* director**.

Next, consider a weakly activating **alkyl group** (e.g. CH_3), which also directs the incoming electrophile to the ***ortho* or *para*** position. This is because the positive charge in the arenium ion can end up on the carbon bearing the methyl group. The **carbocation centre is then tertiary and becomes more stable**.

If the electrophile attacks at the *meta* position, the arenium ion becomes less stable because the carbocation is *secondary*.

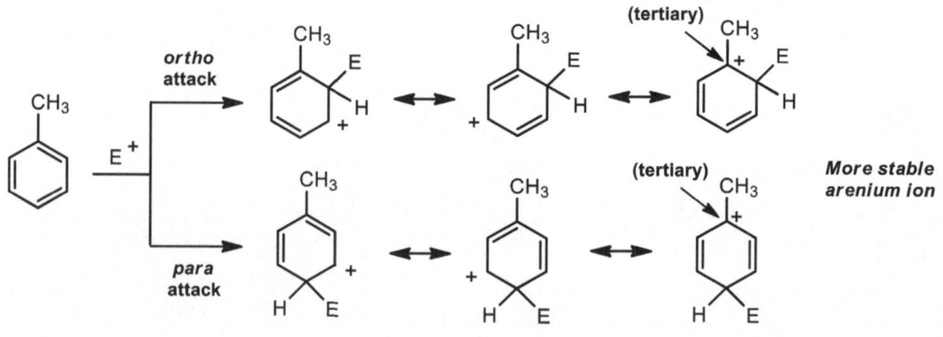

CH3, E+, *ortho* attack, CH3, E, H, +, CH3, E, H, +, (tertiary), CH3, +, E, H, *para* attack, CH3, +, H, E, CH3, +, H, E, (tertiary), CH3, +, H, E, *More stable arenium ion*

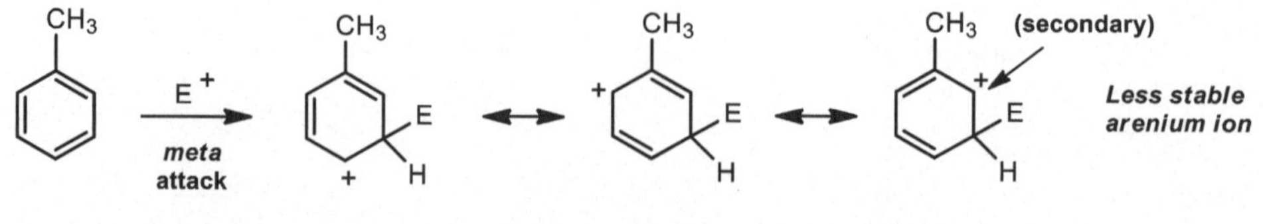

CH3, E+, *meta* attack, CH3, E, +, H, CH3, +, E, H, CH3, (secondary), +, E, H, *Less stable arenium ion*

Halogen substituents: Deactivating ortho/para directors

A halogen substituent (e.g. Cl) has electron-withdrawing inductive effect ($-I$) as well as electron-donating mesomeric effect ($+M$). Since $-I > +M$, the halogen substituent is mildly electron-withdrawing and weakly deactivating.

However, the ***+M* effect** of Cl can **stabilize** the arenium ion intermediate resulting from the ***ortho* or *para*** attack because the positive charge is adjacent to Cl. No such stabilization is possible for the *meta* attack since the positive charge is not adjacent to Cl.

Therefore, the weakly deactivating Cl substituent acts as an ***ortho/para* director**.

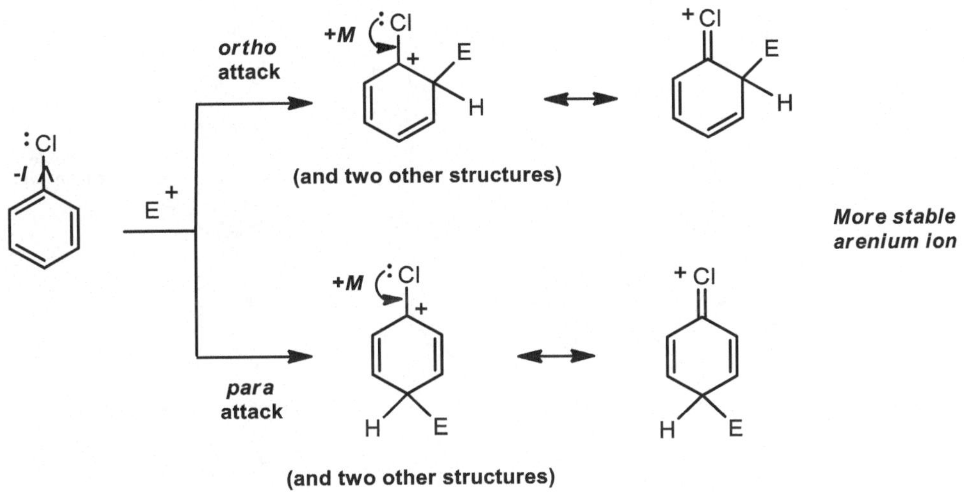

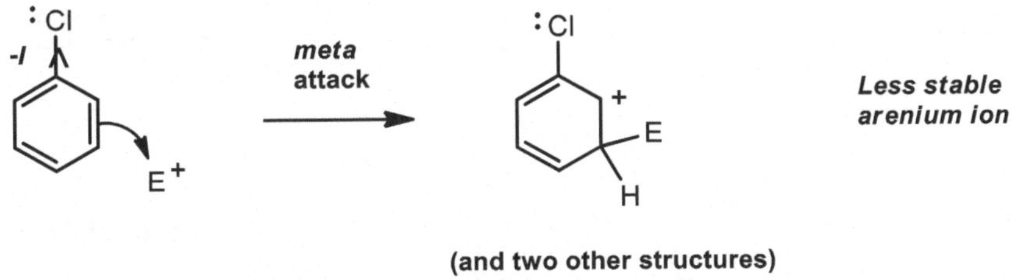

ortho attack
+M
(and two other structures)
More stable arenium ion
para attack
(and two other structures)
meta attack
Less stable arenium ion
(and two other structures)

Deactivating substituents: meta Directors

The *meta* orientation of a **strongly or moderately deactivating** substituent such as NO_2, CF_3 or CHO can be explained by considering the relative stabilities of arenium ions from the *ortho*, *para* or *meta* attack.

Consider, for example, the **nitro substituent** (NO_2). Figure 13.3 shows the relative stabilities of arenium ions (σ complexes) for *ortho* attack, *para* attack and *meta* attack of the electrophile.

The NO_2 group is *strongly* electron-withdrawing (*–I*, *–M*). The ***ortho* or *para* attack** gives a positive charge on the carbon bearing the nitro group in one of the three resonance structures. This resonance structure is **highly unstable** due to the severe electrostatic repulsion between two adjacent positive charges, and contributes to the **destabilization of the arenium ion**.

But for the ***meta* attack**, the positive charge is *not* adjacent to the nitro group in any of the three resonance structures. As such, there is *no* highly unstable resonance structure.

Thus, the **arenium ion** from the *meta* attack is relatively **more stable** than that from the *ortho* or *para* attack, and the NO_2 substituent acts as a ***meta* director**.

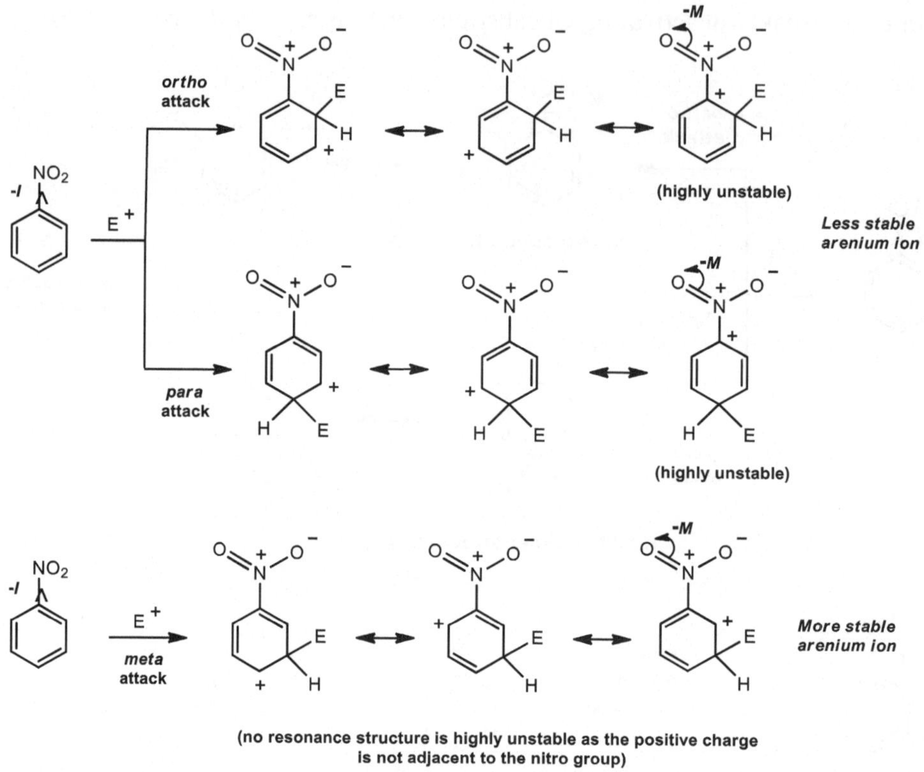

Figure 13.3: The *meta* orientation of a deactivating substituent (NO_2) in electrophilic aromatic substitution.

13.3 HALOGENATION

Benzene reacts with chlorine or bromine in the presence of a Lewis acid catalyst ($FeCl_3$ or $FeBr_3$) to produce chlorobenzene or bromobenzene.

Typically, **metallic iron (Fe)** is used in the halogenation reaction instead of $FeCl_3$ or $FeBr_3$ reagent. But iron (Fe) itself is *not* a catalyst; it is converted to the Lewis acid catalyst $FeCl_3$ or $FeBr_3$ in the presence of Cl_2 or Br_2.

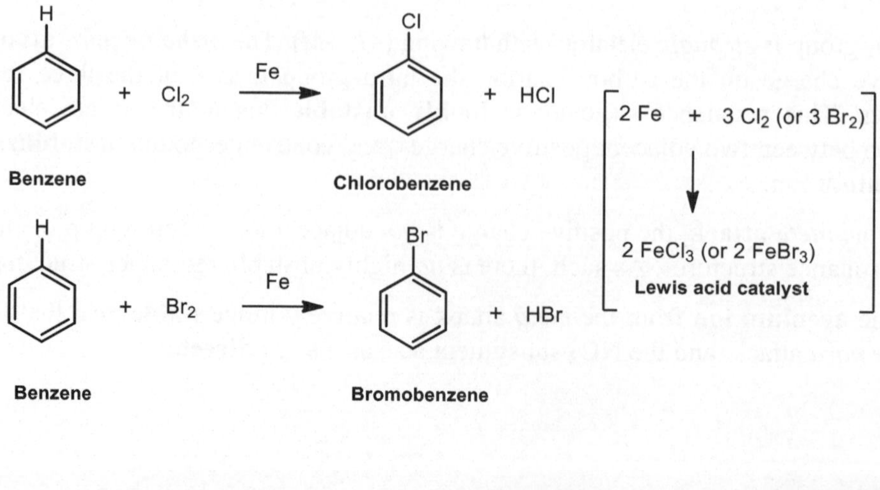

Mechanism

A mechanism for the **chlorination** of benzene in the presence of $FeCl_3$ is shown in Figure 13.4.

Cl_2 combines with $FeCl_3$ to form a Lewis base-Lewis acid complex, which can dissociate to form a chlorine cation (Cl^+) and $FeCl_4^-$. The $\mathbf{Cl^+}$ or the **complex** itself can act as the **electrophile.**

The electrophilic substitution involves two steps as described for a general case (see Figure 13.1). In **step 1**, the electrophile attacks benzene to form an arenium ion intermediate. The arenium ion then loses a proton to $FeCl_4^-$ in **step 2** to produce chlorobenzene. The catalyst ($FeCl_3$) is also regenerated in this step.

Generation of the electrophile:

$Cl—Cl: + FeCl_3 \rightleftharpoons Cl—\overset{+}{Cl}—\overset{-}{Fe}Cl_3 \rightleftharpoons Cl^+ + Cl—\overset{-}{Fe}Cl_3$

Lewis base **Lewis acid** **Complex** ($FeCl_4^-$)

Electrophilic substitution (chlorination):

\+ Cl^+ **(electrophile)**

OR

\+ $Cl—\overset{+}{Cl}—\overset{-}{Fe}Cl_3$ **(electrophile)**

Step 1

An arenium ion

$\overset{-}{Cl—FeCl_3}$

Step 2

\+ HCl + $FeCl_3$

(catalyst regenerated)

Arenium ion **Chlorobenzene**

Figure 13.4: A mechanism for chlorination of benzene.

Note: The mechanism of the **bromination** of benzene in the presence of $FeBr_3$ is similar to that for the chlorination.

13.3.1 Halogenation of Substituted Benzenes

The substituted benzenes may contain an activating or a deactivating substituent, and the substituent may be an *ortho*/*para* director or a *meta* director.

Halogenation of toluene and nitrobenzene

Toluene (methylbenzene) undergoes chlorination *faster* than benzene since methyl group (CH_3) is an activating substituent, whereas nitrobenzene reacts *slower* than benzene because nitro group (NO_2) is a deactivating substituent.

Toluene produces a mixture of ***o*-chlorotoluene and *p*-chlorotoluene** because CH_3 is an *ortho/para* director.

In contrast, **nitrobenzene** gives ***m*-chloronitrobenzene** since NO_2 is a *meta* director.

(*o*-/*p*-directing)

$$C_6H_5CH_3 + Cl_2 \xrightarrow{Fe} o\text{-}ClC_6H_4CH_3 + p\text{-}ClC_6H_4CH_3 + HCl$$

Toluene → ***o*-Chlorotoluene** + ***p*-Chlorotoluene**

(*m*-directing)

$$C_6H_5NO_2 + Cl_2 \xrightarrow{Fe} m\text{-}ClC_6H_4NO_2 + HCl$$

Nitrobenzene → ***m*-Chloronitrobenzene**

Halogenation of phenol

The OH group in phenol is a strongly activating group and an *ortho/para* director. Thus phenol reacts *very rapidly* with bromine in aqueous solution.

Notably, no Lewis acid catalyst is needed for the bromination of the highly activated phenol ring. In *aqueous medium*, **polybromination** (tribromination) occurs with substitution occurring at the *para* position and both *ortho* positions. The product is almost exclusively **2,4,6-tribromophenol.**

(*o*-/*p*-directing)

$$C_6H_5OH + 3\,Br_2 \xrightarrow[\text{Polybromination}]{H_2O} 2,4,6\text{-}Br_3C_6H_2OH + 3\,HBr$$

Phenol → **2,4,6-Tribromophenol**

However, it is possible to achieve **monobromination** of phenol by performing the reaction in carbon disulphide (CS_2) solvent at a low temperature when the reactivity of the aromatic ring and also bromine are reduced. The major product is ***p*-bromophenol.**

Phenol + Br_2 —(CS_2, 5°C, *Monobromination*)→ *p*-Bromophenol + HBr

Halogenation of aniline

Like phenol, aniline also has a highly activated aromatic ring since NH_2 is a strongly activating substituent.

Aniline reacts with bromine in *water* to undergo **polybromination**, thereby producing **2,4,6-tribromoaniline** in almost quantitative yield.

Aniline (*o*-/*p*-directing) + 3 Br_2 —(H_2O, *Polybromination*)→ 2,4,6-Tribromoaniline + 3 HBr

It is also possible to achieve **monobromination** of aniline via the conversion of the strongly activating amino substituent (–NH_2) to a moderately activating acetamido substituent (–$NHCOCH_3$), which is also an *ortho*/*para* director.

The –$NHCOCH_3$ group is less activating than –NH_2 since the acetamido group is a less powerful electron donor by its +*M* effect. This is because the lone pair on the nitrogen atom in the acetamido group is partly involved in resonance with the carbonyl group.

+*M* —NH_2 — **Amino group**, *Strongly activating*

+*M* —NH—C(=O)—CH_3 or —NH—C(=O)—CH_3 (reduces +*M*) — **Acetamido group**, *Moderately activating (less activating than NH_2)*

The following scheme provides a convenient method to convert aniline to ***p*-bromoaniline** as a monobromo derivative.

Acetylation of **aniline** with acetic anhydride gives **acetanilide** which reacts with bromine in acetic acid to form ***p*-bromoacetanilide** almost exclusively because the steric influence of the large acetamido group hinders the *ortho* attack. Hydrolysis of *p*-bromoacetanilide in aqueous acid yields ***p*-bromoaniline**.

Aniline (NH_2) —[$(CH_3CO)_2O$, *Acetylation*]→ Acetanilide ($NHCOCH_3$) —[Br_2, CH_3CO_2H, *Bromination*]→ *p*-Bromoacetanilide (Br) —[H^+, H_2O, *Hydrolysis*]→ *p*-Bromoaniline (NH_2, Br)

PROBLEM 13.1: Predict the major product in each of the following reactions:

(a) CH_3 / NO_2 (para-substituted benzene) —[Br_2, Fe]→

(b) CH_3 / $NHCOCH_3$ (para-substituted benzene) —[Cl_2, CH_3CO_2H]→

Answer

(a) CH_3 is an activating substituent while NO_2 is a deactivating substituent. In this situation, the activating substituent (CH_3) will determine the outcome of the electrophilic aromatic substitution. CH_3 is an *ortho*/*para* director. But since the *para* position is blocked (as there is already a NO_2 substituent), the bromination occurs at the position *ortho* to the CH_3 group.

p-Nitrotoluene (CH_3, NO_2) —[Br_2, Fe]→ 2-Bromo-4-nitrotoluene (CH_3, Br, NO_2)

(b) Both CH_3 and $NHCOCH_3$ are activating groups. However, since $NHCOCH_3$ is a stronger activating group than CH_3, $NHCOCH_3$ (an *ortho/para* director) will determine the outcome of the reaction. Chlorination occurs mainly at the position *ortho* to $NHCOCH_3$ group as the *para* position is blocked.

CH_3, $NHCOCH_3$ — Cl_2 / CH_3CO_2H → CH_3, Cl, $NHCOCH_3$

p-Methylacetanilide **2-Chloro-4-methylacetanilide**

13.4 NITRATION

Benzene reacts with a mixture of concentrated nitric acid (c. HNO_3) and concentrated sulphuric acid (c. H_2SO_4) to yield nitrobenzene.

H — c. HNO_3 / c. H_2SO_4 / 40-50°C → NO_2

Benzene **Nitrobenzene**

Mechanism

A mechanism for the nitration of benzene is shown in Figure 13.5. The **electrophile** is **nitronium ion** (NO_2^+) which is generated in the reaction mixture.

In the presence of a stronger acid (sulphuric acid), *nitric acid* acts as a base and is *protonated.* The protonated nitric acid then dissociates to form nitronium ion as the electrophile. Notably, the electrophile (NO_2^+) is derived from nitric acid.

The electrophilic substitution (nitration) involves two steps as usual. In **step 1** the electrophile (NO_2^+) attacks benzene to form an arenium ion intermediate, which then loses a proton in **step 2** to form nitrobenzene.

Generation of the electrophile:

O, $\overset{+}{N}$—OH, ^-O (HNO_3) — H^+ / (H_2SO_4) → O, $\overset{+}{N}$—$\overset{+}{O}H_2$, ^-O — $- H_2O$ → $O{=}\overset{+}{N}{=}O$ (NO_2^+)

Nitric acid **Protonated nitric acid** **Nitronium ion** ***Electrophile***

Electrophilic substitution (nitration):

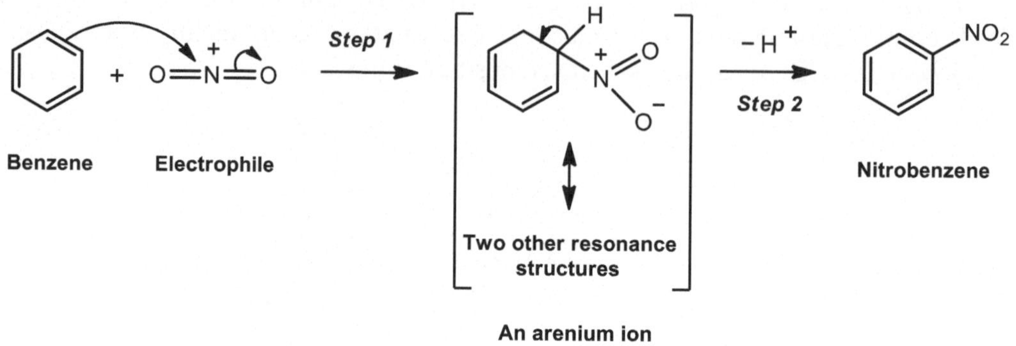

Benzene + Electrophile —Step 1→ [An arenium ion ↔ Two other resonance structures] —$-H^+$, Step 2→ Nitrobenzene

Figure 13.5: A mechanism for the nitration of benzene.

13.4.1 Nitration of Substituted Benzenes

As substituted benzenes, consider toluene, nitrobenzene and phenol for nitration reactions.

Nitration of toluene

Toluene possesses an activating CH_3 group and therefore reacts *faster* than benzene. Thus nitration of toluene can occur at room temperature (25°C).

Since the methyl group is an *ortho/para* director, **toluene** undergoes nitration primarily at the *ortho* or *para* position to produce a mixture of ***o*-nitrotoluene** and ***p*-nitrotoluene**.

Toluene —HNO_3, H_2SO_4, 25°C→ *o*-Nitrotoluene + *p*-Nitrotoluene

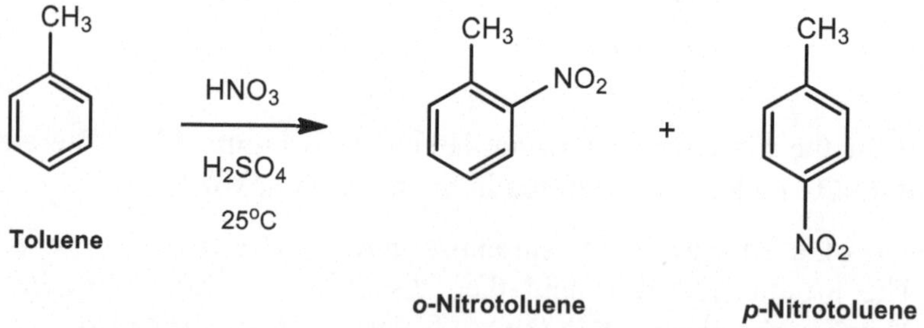

Mononitrotoluenes (*o*-nitrotoluene and *p*-nitrotoluene) can undergo further nitration. But since NO_2 is a deactivating group, mononitoluenes react *slower* than toluene. Therefore, further nitration (i.e. dinitration) requires a higher temperature (90°C).

Nitration of *o*-nitrotoluene gives **2,4-dinitrotoluene and 2,6-dinitrotoluene** while *p*-nitrotoluene gives only **2,4-dinitrotoluene**, as shown below.

Further nitration of dinitrotoluenes (2,4-dinitrotoluene and 2,6-dinitrotoluene) is also possible under more vigorous conditions using a mixture of fuming nitric acid and fuming sulphuric acid (or oleum). The final product is **2,4,6-trinitrotoluene (TNT)**, which is used as an explosive.

(b) (a) (c)

CH_3, NO_2 + CH_3, NO_2 $\xrightarrow[H_2SO_4\ 90°C]{HNO_3}$ CH_3, NO_2, NO_2 + CH_3, O_2N, NO_2

o-Nitrotoluene **p-Nitrotoluene**

[same product from (a) and (c)]
2,4-Dinitrotoluene

[from (b)]
2,6-Dinitrotoluene

fuming HNO_3, fuming H_2SO_4

CH_3, O_2N, NO_2, NO_2

2,4,6-Trinitrotoluene (TNT)

Nitration of nitrobenzene

Nitrobenzene has a strongly deactivating nitro group (NO_2) and hence is much less reactive than benzene.

Nitrobenzene undergoes nitration with the mixed acid reagent (c. HNO_3/c. H_2SO_4) at a high temperature (~100°C). The major product is ***m*-dinitrobenzene** because NO_2 group is a *meta* director.

NO_2 $\xrightarrow[H_2SO_4\ \sim100°C]{HNO_3}$ NO_2, NO_2

Nitrobenzene ***m*-Dinitrobenzene**

Nitration of phenol

Phenol has a strongly activating hydroxyl group (OH); so it is highly reactive toward electrophilic substitution.

Nitration of **phenol** can occur in *dilute* nitric acid and a mixture of ***o*-nitrophenol** and ***p*-nitrophenol** is formed because the OH group is an *ortho*/*para* director.

OH

dilute HNO_3 / 25°C

Phenol → o-Nitrophenol + p-Nitrophenol

Note: Since nitric acid is a strong oxidising agent, *concentrated* HNO_3 will oxidize phenol and nitration does not take place. But in *dilute* nitric acid, the oxidation is much less and nitration of phenol can occur.

For aniline, direct nitration with nitric acid is not a suitable method to prepare *o*-nitroaniline and *p*-nitroaniline because nitric acid causes considerable destruction of the aromatic ring by oxidation.

PROBLEM 13.2: How would you carry out the following transformation?

NH_2 (aniline) → NH_2 / NO_2 (p-nitroaniline)

Answer

Acetylation of aniline with acetic anhydride gives acetanilide. The acetamido group ($NHCOCH_3$) in acetanilide is moderately activating and is also not susceptible to oxidation by nitric acid. The acetyl group ($COCH_3$) thus serves as a **protecting group** for NH_2 from oxidation.

Acetanilide reacts with mixed acid (c. HNO_3 and c. H_2SO_4) to form *p*-nitroacetanilide almost exclusively because the steric influence of the large acetamido group hinders the *ortho* attack. Hydrolysis of *p*-nitroacetanilide with aqueous acid causes removal of the acetyl group (i.e. **deprotection**) and produce *p*-nitroaniline.

It should be noted that hydrolysis in acidic medium actually forms a protonated amine, which on treatment with aqueous base (work-up) generates the free amine (*p*-nitroaniline).

Aniline —($(CH_3CO)_2O$; Acetylation, *Protection*)→ Acetanilide ($NHCOCH_3$) —(HNO_3, H_2SO_4; Nitration)→ p-Nitroacetanilide —(1. H^+/H_2O; 2. OH^- (work-up); Hydrolysis, *Deprotection*)→ p-Nitroaniline

---

13.5 SULPHONATION

Benzene reacts with **fuming sulphuric acid (or oleum)** to produce **benzenesulphonic acid**. Fuming sulphuric acid contains added sulphur trioxide (SO_3).

Unlike halogenation or nitration, the **sulphonation reaction is reversible**. However, the equilibrium for the sulphonation reaction of benzene with oleum ($H_2SO_4 + SO_3$) lies mostly to the right or product side; thus the product is obtained in good yield. *Concentrated* sulphuric acid can also be used, but the reaction then occurs slowly.

Sulphonation: Benzene + H_2SO_4 $\underset{25^\circ C}{\overset{SO_3}{\rightleftharpoons}}$ Benzenesulphonic acid + H_2O

Desulphonation is the reverse of sulphonation, which can be carried out under appropriate conditions. *Dilute* sulphuric acid ($H_2O + H_2SO_4$) is usually employed to remove the sulphonic acid group (SO_3H) from the benzene ring. Under this condition with a high concentration of water, the equilibrium favours desulphonation.

Desulphonation: Benzenesulphonic acid + H_2O $\underset{\Delta}{\overset{H_2SO_4}{\rightleftharpoons}}$ Benzene + H_2SO_4

Mechanism for sulphonation

A mechanism for the sulphonation of benzene is shown in Figure 13.6. The **electrophile** is **sulphur trioxide** (SO_3), which is neutral and a π bond electrophile. The electrophile (SO_3) itself is present in fuming H_2SO_4. Concentrated sulphuric acid can also generate SO_3 by dissociation.

The sulphonation reaction is an electrophilic aromatic substitution. The electrophilic attack by SO_3 on benzene forms an arenium ion, which then loses a proton to form benzenesulphonate ion. Finally, the benzenesulphonate ion accepts a proton to produce benzenesulphonic acid.

Electrophile (sulphur trioxide):

SO_3 (present in fuming H_2SO_4) or $2\,H_2SO_4$ (conc. H_2SO_4) $\rightleftharpoons$ SO_3 + H_3O^+ + HSO_4^-

Electrophilic substitution (sulphonation):

Benzene + Electrophile (SO_3) $\rightleftharpoons$ [An arenium ion ↔ Two other resonance structures] $\overset{-H^+}{\rightleftharpoons}$ Benzenesulphonate ion (SO_3^-) $\overset{H_3O^+}{\rightleftharpoons}$ Benzenesulphonic acid (SO_3H) + H_2O

Figure 13.6: A mechanism for the sulphonation of benzene.

13.5.1 Sulphonation of Substituted Benzenes

Toluene undergoes sulphonation *faster* than benzene. **Toluene** reacts with concentrated sulphuric acid (c. H_2SO_4) to produce a mixture of ***o*-toluenesulphonic acid** and ***p*-toluenesulphonic acid**. (Remember that the methyl group in toluene is an activating *ortho*/*para* director.)

CH3 | c. H_2SO_4, 0°C | CH3, SO3H + CH3, SO3H

Toluene

o-Toluenesulphonic acid **p-Toluenesulphonic acid**

Phenol reacts with concentrated H_2SO_4 to give mainly an ***ortho*-product** (*o*-hydroxy-benzenesulphonic acid) at a **low temperature** (25°C). But at a **higher temperature** (100°C), the reaction gives mainly a ***para*-product** (*p*-hydroxybenzenesulphonic acid).

This is another case of **kinetic control versus thermodynamic control** of a reversible reaction (see Sections 8.16 and 9.11.1). The *ortho*-product is formed under kinetic control and hence is the major product at a low temperature (25°C). On the other hand, the *para*-product (which is more stable than the *ortho*-product) is obtained under thermodynamic control and is therefore the major product at a higher temperature (100°C).

OH | c. H_2SO_4

25°C *Kinetic control* → OH, SO3H *Major product*

o-Hydroxybenzenesulphonic acid

100°C *Thermodynamic control* → OH, SO3H *Major product*

p-Hydroxybenzenesulphonic acid

13.5.2 Sulphonation–Desulphonation Strategy in Synthesis

In the synthesis of aromatic compounds, **sulphonation** is sometimes used to introduce a sulphonic acid group (SO_3H) into a benzene ring **to determine the course of some subsequent reaction**. Thereafter, the SO_3H group is removed by **desulphonation**. This is known as the **sulphonation-desulphonation strategy** in synthesis.

To illustrate, consider the following sequence of reactions in the synthesis of ***o*-nitroaniline** from aniline.

Aniline → $(CH_3CO)_2O$ (**Acetylation**) → $NHCOCH_3$-benzene → c. H_2SO_4 (***Sulphonation***) → *p*-SO_3H acetanilide → HNO_3 (**Nitration**) → 2-NO_2, 4-SO_3H acetanilide → H_2O/H_2SO_4 (***Desulphonation***) → [*o*-NO_2 acetanilide] → (H_2O/H_2SO_4) (**Hydrolysis**) → *o*-NO_2-$\overset{+}{N}H_3$ benzene → OH^- (**(work-up)**) → ***o*-Nitroaniline**

Acetylation of aniline with acetic anhydride gives acetanilide. Since the acetamido group ($NHCOCH_3$) is large, the subsequent sulphonation can occur exclusively at the *para* position. (Remember that $NHCOCH_3$ is a moderately activating *ortho*/*para* director). Thus, a **SO_3H group can be used to 'block' the *para* position of the benzene ring**.

The subsequent **nitration** then takes place exclusively at the position that is ***ortho* to $NHCOCH_3$**.

After nitration, the SO_3H group is removed by **desulphonation** with dilute H_2SO_4. This treatment with aqueous acid (dilute H_2SO_4) also leads to the **hydrolysis** of the acetamido group to produce ***o*-nitroaniline**.

It is of note that hydrolysis in acidic medium actually forms a protonated amine, which on treatment with aqueous base (work-up) generates the free amine (*o*-nitroaniline).

13.5.3 Chlorosulphonation

Chlorosulphonation is an electrophilic aromatic substitution that introduces a **chlorosulphonyl group (SO_2Cl)** into a benzene ring.

The reagent generally used for chlorosulphonation is **chlorosulphonic acid** ($ClSO_3H$ or $HOSO_2Cl$), which is a derivative of sulphuric acid (H_2SO_4 or $HOSO_2OH$).

Benzene reacts with chlorosulphonic acid ($HOSO_2Cl$) to produce **benzenesulphonyl chloride**.

Benzene + Chlorosulphonic acid $(HOSO_2Cl = ClSO_3H)$ → Benzene sulphonyl chloride $(C_6H_5SO_2Cl)$ + H_2O

Toluene undergoes chlorosulphonation with chlorosulphonic acid to produce a mixture of ***o*-toluenesulphonyl chloride** and ***p*-toluenesulphonyl chloride**. The mixture can be separated by distillation.

Toluene $\xrightarrow{ClSO_3H}$ *o*-Toluene sulphonyl chloride + *p*-Toluene sulphonyl chloride

Separated by distillation

13.6 FRIEDEL-CRAFTS REACTIONS

Benzene (C_6H_6) reacts with an alkyl halide (RX) or acyl halide (RCOX) in the presence of a *Lewis acid catalyst* such as aluminium chloride ($AlCl_3$) to produce **alkylbenzenes** (C_6H_5–R) or **acylbenzenes** (C_6H_5–COR).

These reactions were discovered jointly by the French chemist Charles Friedel and the American chemist James F. Crafts in 1877 and are known as **Friedel-Crafts reactions.**

The substitution of a hydrogen atom on the benzene ring by an alkyl group (R) is called the **Friedel-Crafts alkylation**, whereas the substitution of a hydrogen atom on the benzene ring by an acyl group (COR) is called the **Friedel-Crafts acylation.**

Importantly, a Friedel-Crafts reaction leads to the formation of a new C–C bond.

Limitations of Friedel-Crafts reactions

One limitation of the Friedel-Crafts alkylation or acylation is that the reaction does not take place with *strongly deactivated* aromatic substrates. When the benzene ring bears a strong electron-withdrawing group, the ring becomes too electron-deficient or deactivated to undergo a Friedel-Crafts reaction.

For example, **nitrobenzene** that contains a strong electron-withdrawing nitro group (NO_2) does not undergo the Friedel-Crafts alkylation or acylation. This is why nitrobenzene can be used as a *solvent* in Friedel-Crafts reactions.

NO_2

+ R—X $\xrightarrow{AlCl_3}$ (crossed arrow) ***No Friedel-Crafts reaction***

Nitrobenzene **Alkyl halide**

Interestingly, aromatic amines which are strongly activated due to the presence of a powerful *electron-donating* amino group (NH_2) do not undergo the Friedel-crafts alkylation or acylation.

For example, **aniline** does not undergo the Friedel-Crafts alkylation or acylation. This is because the NH_2 group (a Lewis base) forms a complex with the Lewis acid catalyst ($AlCl_3$). The complex group with a positive charge on the nitrogen atom is *strongly electron-withdrawing*. Thus the aniline–Lewis acid complex becomes strongly deactivated and fails to undergo the Friedel-Crafts reaction.

(strongly electron-withdrawing group)

(Lewis base) $\ddot{N}H_2$ $AlCl_3$ **(Lewis acid)** → $H_2\overset{+}{N}—\overset{-}{AlCl_3}$ $\xrightarrow[AlCl_3]{RX}$ (crossed arrow) ***No Friedel-Crafts reaction***

Aniline **Aniline-Lewis acid complex**

13.6.1 The Friedel-Crafts Alkylation

Treating **benzene** with an alkyl halide in the presence of a Lewis acid catalyst converts benzene to an **alkylbenzene**. The most common Lewis acid catalyst is ***anhydrous* aluminium chloride ($AlCl_3$)**. The anhydrous conditions are required because $AlCl_3$ reacts with water to produce HCl.

An example of the Friedel-Crafts alkylation is the reaction of **benzene** with ***t*-butyl chloride** in the presence of anhydrous $AlCl_3$ to produce ***t*-butylbenzene**.

H + R—X $\xrightarrow{\text{Anhydrous } AlCl_3}$ R + HX

Benzene **Alkyl halide** **Alkylbenzene**

Example

Benzene + $(CH_3)_3C—Cl$ —Anhydrous $AlCl_3$→ $C(CH_3)_3$-benzene

Benzene **t-Butyl chloride** **t-Butylbenzene**

Mechanism

The Friedel-Crafts alkylation is reversible. A mechanism for the Friedel-Crafts alkylation of benzene is shown in Figure 13.7.

The **electrophile** is usually a **carbocation**. An alkyl chloride (a Lewis base) forms a complex with the Lewis acid catalyst ($AlCl_3$) and the carbocation is generated by dissociation of the complex formed.

The Friedel Crafts alkylation is an electrophilic aromatic substitution that involves two steps as usual. The carbocation as an electrophile attacks the benzene ring to form an arenium ion which then loses a proton to $AlCl_4^-$ to produce an alkylbenzene. The catalyst is also regenerated with the liberation of HCl.

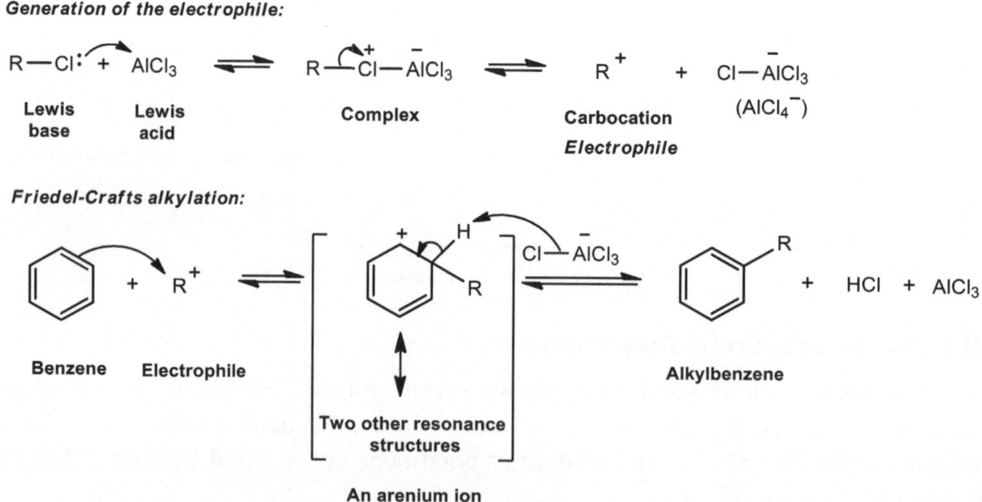

Figure 13.7: A mechanism for the Friedel-Crafts alkylation of benzene.

Note: If the alkyl halide (RCl) is a *methyl* halide (CH_3Cl) or *primary* halide, a relatively unstable methyl carbocation or primary carbocation is not formed easily; the complex itself then can react as an electrophile (cf. halogenation in Figure 13.4).

Benzene + $H_3C—\overset{+}{Cl}—\overset{-}{AlCl_3}$ ⇌ arenium ion (H, CH_3) + $AlCl_4^-$ ⇌ CH_3-benzene + HCl + $AlCl_3$

Complex
Electrophile

Alkenes as alkylating reagents

Besides alkyl halides, **alkenes** can also be used as **alkylating reagents**. A carbocation electrophile in the Friedel-Crafts alkylation can be generated from an alkene in the presence of an acid catalyst.

For example, **benzene** reacts with **isobutene** in the presence of HF to produce ***t*-butylbenzene**. In this reaction, the electrophile is *t*-butyl cation which is generated by the addition of proton to the double bond of isobutene.

$$\text{Benzene} + (CH_3)_2C{=}CH_2 \xrightarrow{HF} C_6H_5C(CH_3)_3$$

Benzene **Isobutene** ***t*-Butylbenzene**

Mechanism:

$$CH_3C(CH_3){=}CH_2 \xrightarrow{H{-}F} CH_3\overset{+}{C}(CH_3){-}CH_3 + F^-$$

Isobutene ***t*-Butyl cation (electrophile)**

$$\text{Benzene} + \overset{+}{C}(CH_3)_3 \longrightarrow \text{arenium ion} \xrightarrow{F^-} C_6H_5C(CH_3)_3 + HF$$

Benzene **Electrophile** **An arenium ion** ***t*-Butylbenzene**

Alcohols as alkylating reagents

It is also possible to carry out Friedel-Crafts alkylations using **alcohols** as **alkylating reagents.**

For example, **benzene** reacts with **isopropanol** in the presence of BF_3 as a Lewis acid catalyst to produce **isopropylbenzene (or cumene)**.

Here the participating electrophile is isopropyl cation which is generated from the dissociation of a complex between the alcohol (a Lewis base) and BF_3 (a Lewis acid).

$$\text{Benzene} + CH_3CH(OH)CH_3 \xrightarrow{BF_3} C_6H_5CH(CH_3)_2$$

Benzene **Isopropanol** **Isopropylbenzene (cumene)**

Mechanism:

$CH_3CH(CH_3)-\ddot{O}H$ + BF_3 ⇌ $CH_3CH(CH_3)-\overset{+}{O}(H)-\overset{-}{B}F_3$ ⇌ $CH_3\overset{+}{C}H(CH_3)$ + $HO-\overset{-}{B}F_3$

Lewis base **Lewis acid** **Complex** **Isopropyl cation (electrophile)**

Benzene + $\overset{+}{C}H(CH_3)_2$ ⟶ [arenium ion with H and $CH(CH_3)_2$] $\xrightarrow{HO-\overset{-}{B}F_3}$ Cumene ($C_6H_5CH(CH_3)_2$) + H_2O + BF_3

Benzene **Electrophile** **An arenium ion** **Cumene**

Problems in Friedel-Crafts alkylations

Two major **problems** that may arise in Friedel-Crafts alkylations are **polyalkylation** and **rearrangement**. The following examples illustrate these problems.

The Friedel-Crafts alkylation of benzene with methyl chloride (CH_3Cl) in the presence of anhydrous $AlCl_3$ gives toluene. But the alkyl group (CH_3) in toluene activates the ring toward further alkylation. As a result, toluene undergoes further alkylation and finally a mixture of toluene and polymethylbenzenes are formed.

Benzene $\xrightarrow[AlCl_3]{CH_3Cl}$ Toluene $\xrightarrow[\textit{Polyalkylation}]{CH_3Cl/AlCl_3}$ p-xylene + 1,2,4-trimethylbenzene etc.

Benzene **Toluene** **Polymethylbenzenes**

In general, **polyalkylation** is a problem in the Friedel-Crafts alkylation which reduces the synthetic utility of the reaction. However, the use of an *excess* of benzene can reduce the chance of polyalkylation and monoalkylation can be favoured.

To illustrate the problem of **rearrangement**, consider the Friedel-Crafts alkylation of benzene with propyl bromide in the presence of anhydrous $AlCl_3$. Here the primary propyl group rearranges to a secondary isopropyl group to give isopropylbenzene as the *major product*. The minor product is propylbenzene.

Benzene + $CH_3CH_2CH_2-Br$ (1°) $\xrightarrow{AlCl_3}$ $C_6H_5CH(CH_3)CH_3$ (2°) + $C_6H_5CH_2CH_2CH_3$ (1°)

Benzene **Propyl bromide** **Isopropylbenzene (rearranged product)** **Propylbenzene**

Major product ***Minor product***

$$\left[\begin{array}{ccccc} CH_3CH_2CH_2Br & \xrightarrow{AlCl_3} & CH_3\underset{H}{C}H-\overset{+}{C}H_2 & \xrightarrow{\textit{Rearrangement}} & CH_3\overset{+}{C}H-CH_3 \\ & & 1^\circ & & 2^\circ \\ \textbf{Propyl bromide} & & \textbf{Propyl cation} & & \textbf{Isopropyl cation} \\ & & \textit{Less stable} & & \textit{More stable} \end{array}\right]$$

The rearrangement occurs because the primary carbocation (propyl cation) generated from propyl bromide can form a *more stable* secondary carbocation (isopropyl cation) by a shift of a hydrogen atom to the adjacent carbocation centre (for details, see later Section 16.2.1). The more stable carbocation acts as the preferred electrophile and leads to the formation of the rearranged product as the major product.

13.6.2 The Friedel-Crafts Acylation

The Friedel-Crafts acylation introduces an **acyl group** (–COR) into a benzene ring. The acylating reagent is usually an **acyl halide** or an **acid anhydride**.

The Friedel-Craft acylation of **benzene** with an acyl halide in the presence of a Lewis acid catalyst such as anhydrous $AlCl_3$ produces an **aromatic ketone** (C_6H_5COR). This is illustrated with two examples.

$$C_6H_5\text{–}H + R\text{–}\overset{O}{\overset{\|}{C}}\text{–}Cl \xrightarrow{\text{Ahnydrous } AlCl_3} C_6H_5\text{–}\overset{O}{\overset{\|}{C}}\text{–}R + HCl$$

Benzene **Acyl chloride** **Aromatic ketone**

Examples

$$C_6H_6 + CH_3\overset{O}{\overset{\|}{C}}\text{–}Cl \xrightarrow{AlCl_3} C_6H_5\overset{O}{\overset{\|}{C}}CH_3 + HCl$$

Benzene **Acetyl chloride** **Acetophenone**

$$C_6H_6 + CH_3\overset{O}{\overset{\|}{C}}\text{–}O\overset{O}{\overset{\|}{C}}CH_3 \xrightarrow{AlCl_3} C_6H_5\overset{O}{\overset{\|}{C}}CH_3 + CH_3CO_2H$$

Benzene **Acetic anhydride** **Acetophenone**

Mechanism

A mechanism for the Friedel-Crafts acylation is shown in Figure 13.8.

The **electrophile** is an **acylium ion** which is generated by dissociation of the complex formed from acyl chloride (Lewis base) and $AlCl_3$ (Lewis acid). **The acylium ion is stabilized by resonance.**

The Friedel Crafts acylation is an electrophilic aromatic substitution. The acylium ion as an electrophile attacks the benzene ring to form an arenium ion which then loses a proton to produce an aromatic ketone.

However, unlike the Friedel-Crafts alkylation when a catalytic or small quantity of $AlCl_3$ is needed, the Friedel-Crafts acylation requires an *excess* of $AlCl_3$ (>1 mol) because the product ketone (a Lewis base) complexes with $AlCl_3$ (a Lewis acid) in a 1:1 molar ratio. As a result of the complex formation, **the Friedel-Crafts acylation becomes irreversible.**

The product ketone is liberated by an aqueous work-up (i.e. by treating the complex with water).

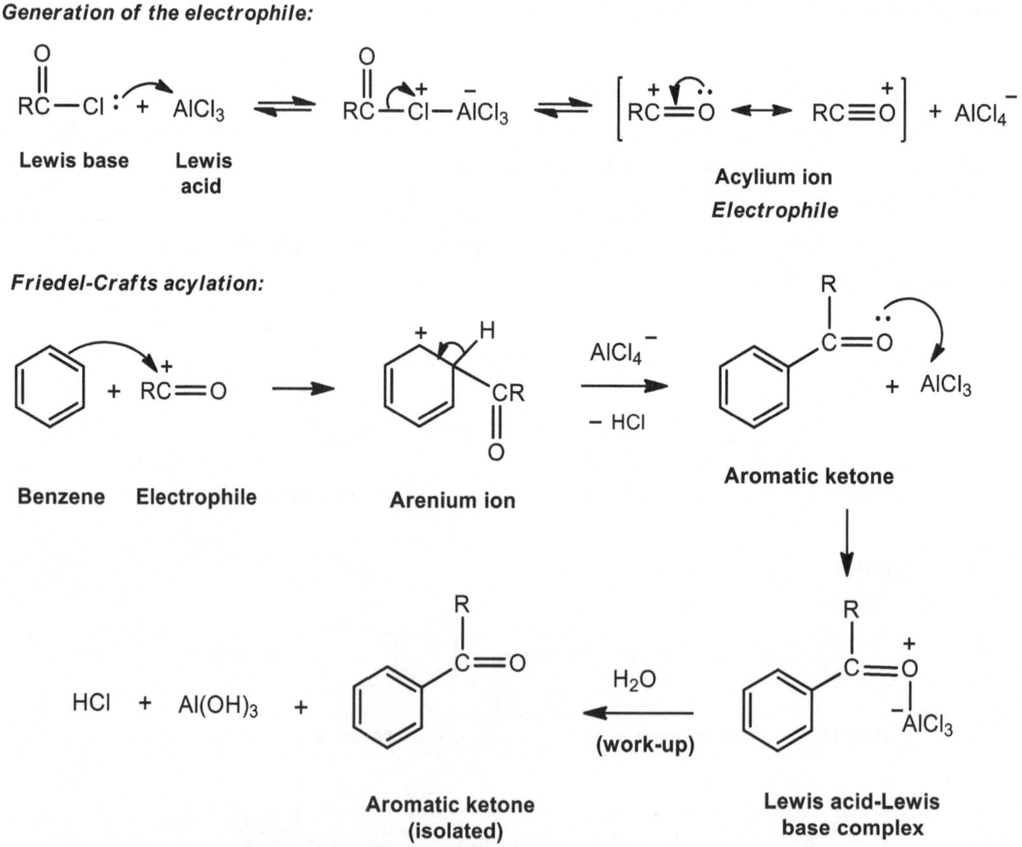

Generation of the electrophile:

Lewis base, Lewis acid; Acylium ion, *Electrophile*

Friedel-Crafts acylation:

Benzene, Electrophile, Arenium ion, $AlCl_4^-$, − HCl, Aromatic ketone, + $AlCl_3$, H_2O (work-up), HCl + $Al(OH)_3$ +, Aromatic ketone (isolated), Lewis acid-Lewis base complex

Figure 13.8: A mechanism for the Friedel-Crafts acylation of benzene.

Advantages

Unlike Friedel-Crafts alkylations, the **Friedel-Crafts acylation does not involve polyacylation**. This is because the introduction of an electron-withdrawing acyl group (COR) and its conversion to a complex with $AlCl_3$ *strongly deactivates* the ring toward further electrophilic substitution (acylation).

Also, there is **no rearrangement** in the acylation reaction. Thus Friedel-Crafts acylations are synthetically much more useful than Friedel-Crafts alkylations.

In fact, the **synthesis of alkylbenzenes** can be accomplished in an efficient manner through the Friedel-Crafts acylation followed by the Clemmensen or the Wolff-Kishner reduction of the carbonyl group (C=O) to a methylene group (CH_2) (see Section 9.7).

Using this sequence of reactions, the synthesis of ethylbenzene and propylbenzene from benzene is shown here.

$$\text{Benzene} + CH_3C(=O)Cl \xrightarrow{AlCl_3} \text{Acetophenone } (C_6H_5C(=O)CH_3) \xrightarrow[\Delta]{Zn(Hg)/HCl} \text{Ethylbenzene } (C_6H_5CH_2CH_3)$$

Benzene **Acetyl chloride** **Acetophenone** ***Clemmensen reduction*** **Ethylbenzene**

$$\text{Benzene} + CH_3CH_2C(=O)Cl \xrightarrow{AlCl_3} \text{Ethyl phenyl ketone } (C_6H_5C(=O)CH_2CH_3) \xrightarrow[\Delta]{H_2NNH_2/KOH} \text{Propylbenzene } (C_6H_5CH_2CH_2CH_3)$$

Benzene **Propionyl chloride** **Ethyl phenyl ketone** ***Wolff-Kishner reduction*** **Propylbenzene**

Note: The synthesis of ethylbenzene by a Friedel-Crafts alkylation encounters a problem of *polyalkylation*, whereas the synthesis of propylbenzene by a Friedel-Crafts alkylation encounters a problem of *rearrangement* (see Section 13.6.1).

PROBLEM 13.3: Write the structures for compounds **A–D** in the following scheme:

$$\text{Benzene} + \text{succinic anhydride} \xrightarrow{AlCl_3} \textbf{A} \xrightarrow[\Delta]{Zn(Hg)/HCl} \textbf{B} \xrightarrow{SOCl_2} \textbf{C} \xrightarrow{AlCl_3} \textbf{D}$$

Answer

The Friedel-Crafts acylation of benzene with a cyclic anhydride (succinic anhydride) gives (**A**). Clemmensen reduction of (**A**) gives a carboxylic acid (**B**), which on treatment with $SOCl_2$ forms an acid chloride (**C**). An intramolecular Friedel-Crafts acylation of (**C**) produces a cyclic ketone, α-tetralone (**D**).

$$\text{Benzene} + \text{Succinic anhydride} \xrightarrow{AlCl_3} C_6H_5C(=O)CH_2CH_2COOH \xrightarrow[\Delta]{Zn(Hg)/HCl} C_6H_5CH_2CH_2CH_2COOH$$

Benzene **Succinic anhydride** **3-Benzoylpropanoic acid (A)** **4-Phenylbutanoic acid (B)**

SOCl$_2$

AlCl$_3$

Cl

O

α-Tetralone

(D)

4-Phenylbutanoyl chloride

(C)

13.7 NUCLEOPHILIC AROMATIC SUBSTITUTION

We have previously learned that alkyl halides can undergo nucleophilic substitution (S_N2 or S_N1) readily.

But **aryl halides** such as chlorobenzene or bromobenzene are relatively **unreactive** toward nucleophilic substitution. This is because the benzene ring prevents backside attack of the nucleophile and the S_N2 reaction does not occur. An S_N1 reaction is also unfavourable because the phenyl cation is very unstable.

Nu:$^-$ Cl → No S_N2 reaction

Chlorobenzene (an aryl halide)

S_N1

Cl + X$^-$ No S_N1 reaction

Chlorobenzene

Phenyl cation *Highly unstable*

There are also other reasons for the relative unreactivity of the aryl halide toward nucleophilic substitution. The more electronegative **sp^2** carbon in the C–X bond in an aryl halide draws the electrons closer to the carbon nucleus, and therefore the halogen atom is displaced less easily (cf. the C–X bond in an alkyl halide).

Further, the +*M* effect of the halogen atom introduces a double bond character to the C–X bond in an aryl halide. As a result, the halogen atom is more tightly held to the carbon atom and cannot be displaced easily.

sp^2

X

Aryl halide

+*M*

X ⟷ =X$^+$

Aryl halide

Nucleophilic aromatic substitution under forcing conditions

Since simple aryl halides are relatively unreactive, they can undergo nucleophilic substitution only under highly vigorous conditions.

For example, chlorobenzene reacts with aqueous NaOH at 350°C under high pressure. The reaction produces sodium phenoxide. Acidification of the mixture yields phenol.

This method is known as the **Dow process**, which has been used for the **industrial preparation of phenol**.

Another commercial method for the synthesis of phenol involves the nucleophilic substitution of an anionic group (SO_3^-), which is lost as SO_3^{2-} (sulphite ion).

Sodium benzenesulphonate (prepared from benzenesulphonic acid) is fused with NaOH at 350°C to produce sodium phenoxide which, on acidification, yields phenol.

13.7.1 Activated Nucleophilic Aromatic Substitution: The Addition–Elimination Mechanism (S_NAr)

A strong electron-withdrawing group on the benzene ring makes the aromatic ring highly electron-deficient and thereby promotes the nucleophilic attack on the ring carbon of the C–X bond in an aryl halide. Thus **electron-withdrawing groups act as activating substituents in nucleophilic aromatic substitution**.

When **electron-withdrawing substituents** (e.g. NO_2) are placed at the ***ortho*** or ***para*** **position** with respect to the halogen atom (X), the nucleophilic substitution of aryl halides (ArX) occurs easily.

Further, with increasing number of *ortho* and *para* nitro groups on the benzene ring, the nucleophilic aromatic substitution becomes more facile.

To illustrate, consider the following examples:

p-Chloronitrobenzene + OH^- —($NaOH/H_2O$, 160°C)→ *p*-Nitrophenol

2,4-Dinitrochlorobenzene + OH^- —($NaHCO_3/H_2O$, 100°C)→ 2,4-Dinitrophenol

2,4,6-Trinitrochlorobenzene + H_2O —(Warm)→ 2,4,6-Trinitrophenol (picric acid)

p-Chloronitrobenzene reacts with aqueous NaOH at 160°C, whereas 2,4-dinitrochlorobenzene reacts under milder conditions with aqueous $NaHCO_3$ at 100°C. In comparison, 2,4,6-trinitrochlorobenzene reacts under extremely mild conditions by simply warming with water.

In each case, the product is the corresponding phenolic compound. 2,4,6-Trinitrophenol is known as **picric acid**.

Note: The role of an electron-withdrawing group to activate nucleophilic aromatic substitution is *opposite* to the case of electrophilic aromatic substitution when an electron-withdrawing group deactivates the ring.

The addition–elimination mechanism (S_NAr)

The **activated nucleophilic aromatic substitution** (when a strong electron-withdrawing substituent is present at the *ortho* or *para* to the leaving group) involves two steps. The first step is the *addition* of the nucleophile and the second step is the *elimination* of the leaving group. The mechanism is therefore called an **addition–elimination mechanism**.

The mechanism is also called the **S_NAr** mechanism, which distinguishes this mechanism from the S_N2 or S_N1 mechanism for nucleophilic substitution of alkyl halides.

An addition–elimination mechanism (S_NAr) for the nucleophilic substitution of ***p*-chloronitrobenzene** with hydroxide ion is shown in Figure 13.9.

Step 1 involves an attack of the nucleophile (OH^-) to the ring carbon bearing the chlorine atom to form a **carbanion** intermediate, which is *stabilized by resonance*. In **step 2**, the elimination of chloride ion (Cl^-) regenerates the aromaticity of the benzene ring and produces the substitution product (*p*-nitrophenol).

Cl, NO_2, ^-OH, Step 1, Addition, Cl, OH, NO_2, $- Cl^-$, Step 2, Elimination, OH, NO_2

p-Chloronitrobenzene

Carbanion intermediate (Meisenheimer complex)

p-Nitrophenol

Resonance stabilization of the carbanion intermediate:

Cl, OH, NO_2 ⟷ Cl, OH, NO_2 ⟷ Cl, OH, -M, N^+, O, O^- ⟷ Cl, OH, N^+, ^-O, O^-

Four resonance structures (most important)

Figure 13.9: An addition–elimination mechanism (S_NAr) for nucleophilic aromatic substitution.

The carbanion intermediate is known as the **Meisenheimer complex** after the German chemist, Jacob Meisenheimer. The resonance structures for the carbanion intermediate show that the *negative charge* on the ring carbon is at the *ortho* or *para* position with respect to the leaving group. This is why an electron-withdrawing NO_2 group can stabilize the intermediate by its $-M$ effect when it is placed at either an *ortho* or a *para* position with respect to the leaving group (Cl), and thereby facilitates the reaction.

PROBLEM 13.4: Predict the product in each of the following reactions:

(a) 2,4-Dinitrochlorobenzene $\xrightarrow{NH_2NH_2}$

(b) 2,4-Dinitrofluorobenzene $\xrightarrow{CH_3NH_2}$

Answer

Each reaction involves an activated nucleophilic aromatic substitution.

(a) The product is 2,4-dinitrophenylhydrazine.

$$O_2N{-}C_6H_3(NO_2){-}Cl \xrightarrow{NH_2NH_2} O_2N{-}C_6H_3(NO_2){-}NHNH_2 + HCl$$

2,4-Dinitrochlorobenzene → **2,4-Dinitrophenylhydrazine**

(b) The product is *N*-methyl-2,4-dinitroaniline.

$$O_2N{-}C_6H_3(NO_2){-}F \xrightarrow{CH_3NH_2} O_2N{-}C_6H_3(NO_2){-}NHCH_3 + HF$$

2,4-Dinitrofluorobenzene (DNFB) → ***N*-Methyl-2,4-dinitroaniline**

Note: 2,4-Dinitrofluorobenzene (DNFB) is *more reactive* than 2,4-dinitrochlorobenzene toward nucleophilic substitution by the S_NAr mechanism. This is because F is more electronegative than Cl, and the greater electron-withdrawing inductive effect (–*I*) of F stabilizes the carbanion intermediate more effectively than Cl DNFB is a useful reagent for the determination of structure of peptides and proteins.

13.7.2 Unactivated Nucleophilic Aromatic Substitution: The Elimination–Addition or Benzyne Mechanism

Aryl halides that *do not* possess strong electron-withdrawing substituents are **unactivated** toward nucleophilic attack, and they do not undergo nucleophilic substitution under ordinary conditions.

However, **unactivated aryl halides** such as chlorobenzene and bromobenzene can undergo nucleophilic aromatic substitution with a *very strong base* such as **sodium amide ($NaNH_2$) in liquid ammonia**. In this reaction, the *base* (NH_2^-) also acts as a *nucleophile*.

For example, bromobenzene reacts with $NaNH_2$ in liquid NH_3 to produce aniline.

$$C_6H_5Br + NaNH_2 \xrightarrow[-33^oC]{NH_3\,(l)} C_6H_5NH_2 + NaBr$$

Bromobenzene → **Aniline**

The elimination–addition or benzyne mechanism

The nucleophilic aromatic substitution of an *unactivated* aryl halide (e.g. bromobenzene) with NH_2^- takes place through an alternative mechanism called the **elimination–addition mechanism**, as shown in Figure 13.10.

The reaction involves an initial **elimination** of HBr by the strong base (NH_2^-) to form an unstable and highly reactive intermediate called **benzyne**. The elimination can occur in two steps: the amide ion as a base abstracts a relatively more acidic *ortho* hydrogen atom to form a carbanion which then loses a bromide ion (Br^-) to form benzyne.

The highly reactive benzyne then reacts with NH_2^- by a two-step **addition** to produce **aniline**. In this addition, the attack of NH_2^- as a nucleophile can take place at either end of the triple bond of benzyne to form a carbanion, which then abstracts a proton from the solvent (NH_3) to form aniline.

Bromobenzene —($^-NH_2$, − NH_3)→ Carbanion —(− Br^-)→ Benzyne —($^-NH_2$)→ Carbanion —(NH_3, − NH_2^-)→ Aniline

Elimination (− HBr) | Addition (+ NH_3)

Figure 13.10: An elimination–addition mechanism (or benzyne mechanism) for nucleophilic aromatic substitution.

Structure of benzyne

Benzyne contains a **triple bond** within the benzene ring but this triple bond differs from the normal triple bond of an alkyne. The extra bond in benzyne results from the overlap of two **sp²** orbitals on the adjacent carbon atoms of the benzene ring.

This extra bond in benzyne is much weaker and hence is *much more reactive* than the normal π bond in an alkyne formed by the overlap of two p orbitals.

sp² sp²

This bond results from overlap of two sp² orbitals

Benzyne

Note: In the Dow process (see above), chlorobenzene (an unactivated aryl halide) reacts with aqueous NaOH under *forcing conditions* to yield phenol. This reaction is likely to occur through an elimination–addition or benzyne mechanism.

PROBLEM 13.5: Predict the product(s) of the following reaction:

4-Chlorotoluene (Cl, CH_3) —($NaNH_2$ / NH_3 (*l*))→

Answer

This is a nucleophilic aromatic substitution of an unactivated aryl halide (*p*-chlorotoluene) with a strong base ($NaNH_2$ in liquid NH_3), which proceeds through an elimination–addition or benzyne mechanism. The same benzyne intermediate is formed by the abstraction of either *ortho* hydrogen atom by the amide ion in the elimination of HCl. But in the two-step addition, the amide ion (NH_2^-) can attack the benzyne intermediate in two possible ways. Attack at either end of the triple bond followed by the abstraction of proton from the solvent gives *p*-methylaniline (*p*-toluidine) and *m*-methylaniline (*m*-toluidine) as products.

p-Chlorotoluene $\xrightarrow{-\,HCl}$ 4-Methylbenzyne $\xrightarrow{NH_2^-}$ $\xrightarrow{NH_3}$ *p*-Methylaniline (*p*-Toluidine) + *m*-Methylaniline (*m*-Toluidine)

13.8 RADICAL SUBSTITUTION OF THE AROMATIC SIDE CHAIN

Aromatic hydrocarbons usually contain a benzene ring and alkyl groups attached to the ring. The **alkyl part** of an aromatic hydrocarbon is commonly called the **side chain.**

For example, methyl group (CH_3) is the side chain of toluene and ethyl group (CH_2CH_3) is the side chain of ethylbenzene.

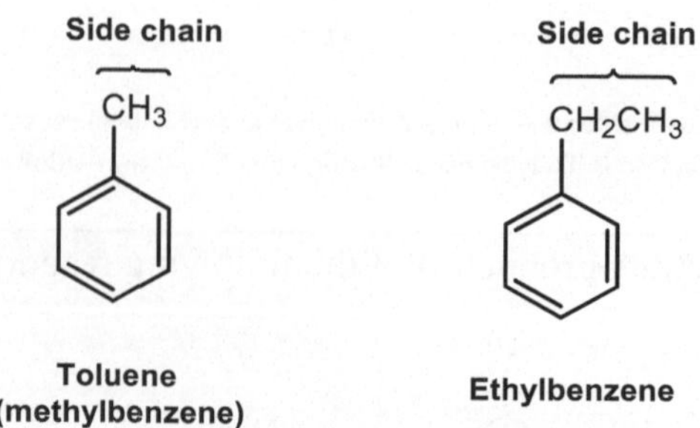

The carbon atom of an alkyl side chain which is *directly* attached to the benzene ring is called the **benzylic carbon**, and the hydrogen atoms attached to the benzylic carbon are called **benzylic hydrogen atoms** (cf. allylic hydrogen atoms, Section 11.1.4).

Benzylic carbon

A benzylic hydrogen

The substitution of a hydrogen atom of the side chain of an aromatic hydrocarbon by a halogen atom (Cl or Br) is called the **side chain halogenation.**

The side chain halogenation takes place when the halogenation reaction is carried out under conditions that promote the formation of **radicals** (see Section 11.1).

13.8.1 Side Chain Chlorination

Toluene can undergo side chain chlorination when it reacts with chlorine in the gas phase at 400–600°C or in the presence of UV light. The product is **benzyl chloride.**

CH_3 + Cl_2 —Heat or light ($h\nu$)→ CH_2Cl + HCl

Toluene Benzyl chloride

Mechanism

The **side chain chlorination** of toluene proceeds by a **radical chain mechanism** which is similar to the radical chain mechanism for allylic chlorination of propene (see Section 11.1.4).

The mechanism involves as usual chain initiation (**step 1**) to form chlorine radicals and then chain propagation (**steps 2 and 3**). In **step 2**, the carbon radical generated by the abstraction of a hydrogen atom from the methyl group is a **benzyl radical,** which is *stabilized by resonance.* In **step 3**, the benzyl radical abstracts a chlorine atom from a chlorine molecule to produce **benzyl chloride.**

Chain Initiation

Cl–Cl —Δ or $h\nu$, *Step 1*→ Cl· + Cl·

Chlorine radicals

Chain Propagation

Cl· + H–CH_2 (Toluene) —*Step 2*→ HCl + [·CH_2 ↔ CH_2 ↔ CH_2 ↔ CH_2]

Benzyl radical

Stabilized by resonance

·CH_2 (Benzyl radical) + Cl–Cl —*Step 3*→ CH_2Cl (Benzyl chloride) + Cl·

Multiple chlorinations

With *excess* of chlorine, multiple chlorinations of side chain can take place. Thus **benzyl chloride** (from chlorination of toluene) can undergo further chlorination to form **benzal chloride**, which on further chlorination gives **benzotrichloride**.

CH_3 —(Cl_2, $h\nu$)→ CH_2Cl —(Cl_2, $h\nu$)→ $CHCl_2$ —(Cl_2, $h\nu$)→ CCl_3

Toluene Benzyl chloride Benzal chloride Benzotrichloride

Monochlorination of ethylbenzene

Monochlorination of ethylbenzene in the presence of UV light produces 1-chloro-1-phenylethane as the **major product** and 1-chloro-2-phenylethane as the **minor product**.

(benzylic carbon)

CH_2CH_3 —(Cl_2, $h\nu$)→ $CHCH_3$ (Cl) + CH_2CH_2Cl

Ethylbenzene

1-Chloro-1-phenylethane
Major product

1-Chloro-2-phenylethane
Minor product

The **chlorination** of the side chain occurs predominantly at the **benzylic carbon** because the corresponding **radical** is a secondary as well as benzylic radical that is *stabilized by resonance*. Thus the major product is 1-chloro-1-phenyethane.

(Note that the *benzylic* radical is a general term; the parent benzylic radical from toluene is called *benzyl* radical).

The chlorination can also take place at the *terminal methyl carbon* but this gives the minor product (1-chloro-2-phenyethane) because the corresponding radical is a *less stable* primary radical.

2°, benzylic

CH_2CH_3 —($\cdot Cl$, $-HCl$)→ $\dot{C}HCH_3$ —($Cl—Cl$)→ $CHCH_3$ (Cl) + $\cdot Cl$

Benzylic radical
More stable

Major product

CH_2CH_3 —($\cdot Cl$, $-HCl$)→ $CH_2\dot{C}H_2$ (1°) —($Cl—Cl$)→ CH_2CH_2Cl + $\cdot Cl$

Primary radical
Less stable

Minor product

13.8.2 Side Chain Bromination

Bromination of the aromatic side chain can be accomplished with ***N*-bromosuccinimide** (NBS) (cf. allylic bromination, Section 11.1.4).

For example, when **toluene** is treated with **NBS** in CCl_4 in the presence of light, the product is **benzyl bromide**.

CH_3 + N—Br (O, O) —$h\nu$ / CCl_4→ CH_2Br

Toluene **N-Bromosuccinimide (NBS)** **Benzyl bromide**

PROBLEM 13.6: Write the structures for compounds **A** and **B** in the following scheme:

CH_2CH_3 —NBS (2 moles), $h\nu$, CCl_4→ A —1. $NaNH_2$, mineral oil, heat; 2. H^+→ B

Answer

Reaction of ethylbenzene with 2 moles of NBS leads to double bromination at the benzylic carbon to produce a *gem*-dibromide (**A**) which then undergoes double dehydrohalogenation on heating with $NaNH_2$ in mineral oil to yield phenylacetylene (**B**) after acid work-up (see Section 10.8).

CH_2CH_3 —NBS (2 moles), *Double bromination*→ C(Br)(Br)—CH_3 —1. $NaNH_2$; 2. H^+, *Double dehydrohalogenation*→ C≡CH

Ethylbenzene **A *gem*-Dibromide (A)** **Phenylacetylene (B)**

13.9 OXIDATION OF AROMATIC COMPOUNDS

A benzene ring is generally *inert* to oxidation. However, the **side chain** of a benzene ring can be **oxidized** by strong oxidizing agents such as potassium permanganate ($KMnO_4$).

For example, **toluene** is oxidized to **benzoic acid** by the action of hot alkaline $KMnO_4$ followed by acidification (work-up).

CH_3 —$KMnO_4$, NaOH, Δ→ C(=O)—ONa —H_3O^+ (work-up)→ C(=O)—OH

Toluene **Sodium benzoate** **Benzoic acid**

The **side chain oxidation** takes place at the **benzylic carbon**. Thus, an alkylbenzene with a *longer* side chain is ultimately oxidized to **benzoic acid**.

For example, ethylbenzene is oxidized to benzoic acid by hot alkaline $KMnO_4$.

(benzylic carbon)

CH_2CH_3 — 1. $KMnO_4/OH^-$ heat; 2. H_3O^+ → $C(=O)$—OH

Ethylbenzene **Benzoic acid**

Halogen and nitro compounds

Substituents such as **halogen and nitro groups** are usually stable to such oxidation and therefore remain **unchanged** after oxidation.

Besides alkaline $KMnO_4$, other oxidizing agents such as sodium or potassium dichromate ($Na_2Cr_2O_7$ or $K_2Cr_2O_7$) under acidic conditions have also been used for the side chain oxidation.

The following are two examples:

CH_3, Cl — 1. $KMnO_4/OH^-$ heat; 2. H_3O^+ → CO_2H, Cl

o-Chlorotoluene **o-Chlorobenzoic acid**

CH_3, NO_2 — $Na_2Cr_2O_7$, H_2SO_4, Δ → CO_2H, NO_2

***p*-Nitrotoluene** ***p*-Nitrobenzoic acid**

Phenolic and amino compounds

Oxidation of **phenol** or **aniline** affects the aromatic ring and forms a **nonaromatic compound (*p*-benzoquinone)** with sodium dichromate and sulphuric acid.

Mild oxidizing agents such as $FeCl_3$ can oxidize **hydroquinone** (*p*-dihydroxybenzene) to ***p*-benzoquinone**.

OH NH$_2$

or

$Na_2Cr_2O_7$ / H_2SO_4

Phenol Aniline

p-Benzoquinone

OH

Fe^{3+} / $(FeCl_3)$

OH

Hydroquinone *p*-Benzoquinone

Ozonolysis

Ozonolysis of benzene cleaves the benzene ring. The reaction of benzene with ozone (3 moles) followed by reduction with zinc and water gives **glyoxal** (3 moles).

1. 3 O_3
2. Zn/H_2O

Ozonolysis

3 CHO–CHO

Benzene Glyoxal

PROBLEM 13.7: How would you carry out the following transformation?

CH_3 → CO_2H, NO_2

Answer

CH_3 is an *ortho*/*para* director, whereas CO_2H is a *meta* director. Therefore, the transformation can be accomplished by performing oxidation first, and then nitration.

CH_3 — 1. $KMnO_4/OH^-$ heat; 2. H_3O^+ → CO_2H — HNO_3/H_2SO_4, heat → CO_2H, NO_2

Toluene Benzoic acid *m*-Nitrobenzoic acid

Note: If the sequence is reversed, i.e. nitration first and then oxidation, the product will be a mixture of *o*-nitrobenzoic acid and *p*-nitrobenzoic acid.

13.10 REDUCTION OF AROMATIC COMPOUNDS

A very useful reduction reaction in aromatic chemistry is the reduction of **aromatic nitro compounds**. A **nitro group** (NO_2) can be easily converted into an **amino group** (NH_2) by reduction.

The most common **reducing reagent** is tin or iron in dilute HCl. **Hydrogenation** with a palladium catalyst supported on charcoal (Pd/C) also brings about this reduction.

NO_2 — Sn or Fe / dilute HCl → NH_2

Nitrobenzene **Aniline**

NO_2 — H_2 / Pd/C → NH_2

Nitrobenzene **Aniline**

Importantly, this reduction turns a *meta* director (NO_2) into an *ortho/para* director (NH_2).

Synthetic applications

The following **schemes** illustrate the use of reduction of a nitro group to an amino group in synthesis:

Scheme 1:

Benzene — HNO_3 / H_2SO_4 → **Nitrobenzene** (NO_2) — Br_2 / Fe → ***m*-Bromonitrobenzene** (NO_2, Br) — Sn / HCl → ***m*-Bromoaniline** (NH_2, Br)

Scheme 2:

Toluene (CH_3) — HNO_3 / H_2SO_4 →

→ ***p*-Nitrotoluene** ***(separate)*** (CH_3, NO_2) — Sn / HCl → ***p*-Toluidine** (CH_3, NH_2)

→ ***o*-Nitrotoluene** (CH_3, NO_2) — Sn / HCl → ***o*-Toluidine** (CH_3, NH_2)

Selective reduction

When there are **two nitro groups** on the benzene ring, one of the nitro groups can be **selectively** reduced.

For example, *m*-dinitrobenzene is reduced by sodium sulphide (Na_2S) or ammonium sulphide [$(NH_4)_2S$] to *m*-nitroaniline.

NO_2 / NO_2 —(Na_2S or $(NH_4)_2S$)→ NH_2 / NO_2

***m*-Dinitrobenzene** → ***m*-Nitroaniline**

The reduction sequence

In general, the reduction of an aromatic nitro compound occurs through the following stages:

($C_6H_5NO_2$) **Nitrobenzene** → (C_6H_5NO) **Nitrosobenzene** → (C_6H_5NHOH) **Phenylhydroxylamine** → ($C_6H_5NH_2$) **Aniline**

In *strongly acidic solution,* the reduction proceeds all the way to an **amine**.

Reduction to hydroxylamine

It is possible to stop the reduction of a nitro group at the **hydroxylamine** stage in *nearly neutral solution.*

For example, the reduction of nitrobenzene with zinc dust and aqueous NH_4Cl produces phenylhydroxylamine.

NO_2 —(Zn, aq. NH_4Cl)→ NHOH

Nitrobenzene → **Phenylhydroxylamine**

Tollens' test for the detection of nitro group

The reduction of a nitro compound with Zn/NH_4Cl followed by the reaction with **Tollens' reagent** (see Section 9.10) is used as a useful test for the detection of nitro group in qualitative analysis.

Reduction with Zn/NH_4Cl produces a *hydroxylamine* which reduces Tollens' reagent [$Ag(NH_3)_2^+$] to *metallic silver* and is itself oxidized to a *nitroso* compound. The metallic silver is deposited on the walls of the test tube as a silver mirror (under the proper conditions) or as grey to black precipitate.

NO_2 — Zn, aq. NH_4Cl → NHOH — $Ag(NH_3)_2^+$, *Tollens' reagent* → NO + Ag↓

Nitrobenzene **Phenylhydroxylamine** **Nitrosobenzene** ***Silver mirror or grey precipitate***

This test is also known as the **Mulliken-Barker test.** In this test, the appearance of a silver mirror or grey precipitate indicates the presence of a nitro group.

13.10.1 Reduction of the Benzene Ring: The Birch Reduction

The benzene ring is usually resistant to most reduction methods. However, **catalytic hydrogenation** of **benzene** under pressure and at an elevated temperature leads to a *complete* reduction to form **cyclohexane.**

+ 3 H_2 — Ni, 10 atm, 150°C →

Benzene **Cyclohexane**

The Birch reduction

A *partial reduction* of the benzene ring is also possible. The most important such reduction is known as the **Birch reduction** after the Australian chemist A. J. Birch.

In this method, **benzene** is treated with an alkali metal (Na, Li or K) dissolved in liquid ammonia in the presence of a proton source such as alcohol. The product is **1,4-cyclohexadiene** (a nonconjugated diene).

The Birch reduction represents a case of **dissolving metal reduction.**

Na, liq. NH_3, C_2H_5OH →

5 1 4 2 3

Benzene **1,4-Cyclohexadiene**

A nonconjugated diene

Mechanism of the Birch reduction

The **mechanism** of the Birch reduction involves a sequence of **electron transfers** from the metal and **proton transfers** from the alcohol.

An electron transfer to **benzene** produces a **radical anion** (a species having an unpaired electron on a carbon atom and a negative charge on another carbon atom). Protonation of the radical anion gives a **cyclohexadienyl radical** which then accepts another electron from the metal to form the corresponding anion, **cyclohexadienyl anion**. Finally, protonation of the cyclohexadienyl anion produces **1,4-cyclohexadiene**.

Na C_2H_5OH ·Na C_2H_5OH

Benzene Benzene radical anion Cyclohexadienyl radical Cyclohexadienyl anion 1,4-Cyclohexadiene

Effects of substituents on the Birch reduction

Electron-donating substituents or **electron-withdrawing substituents** on the benzene ring can influence the course of the Birch reduction.

For example, the Birch reduction of **anisole** having an electron-donating methoxy group (OCH_3) produces **1-methoxy-1,4-cyclohexadiene** in which the substituent (OCH_3) is attached to a double bond.

Hydrolysis of the product in aqueous acid gives **2-cyclohexenone.**

Thus the birch reduction of anisole followed by hydrolysis provides a useful method for the synthesis of 2-cyclohexenone.

OCH_3 Na NH_3/C_2H_5OH OCH_3

Anisole (methoxybenzene) 1-Methoxy-1,4-cyclohexadiene

Hydrolysis H_2O/H^+

O + CH_3OH

2-Cyclohexenone

On the other hand, the Birch reduction of **benzoic acid** having an electron-withdrawing CO_2H group gives **1,4-dihydrobenzoic acid** (2,5-cyclohexadiene-1-carboxylic acid) in which the substituent (CO_2H) is *not* attached to any double bond.

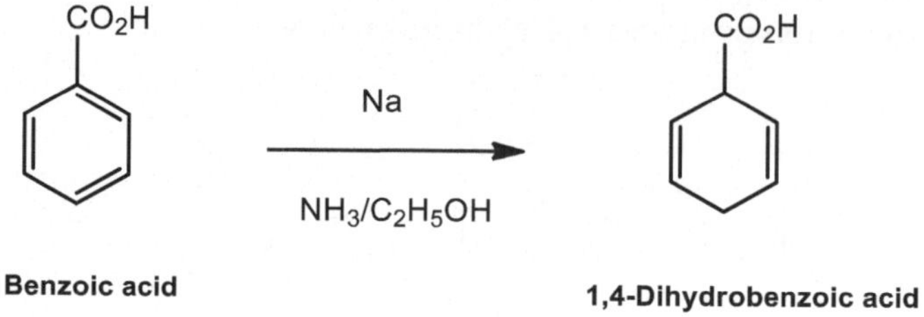

CO_2H CO_2H

Na

NH_3/C_2H_5OH

Benzoic acid **1,4-Dihydrobenzoic acid**

SUPPLEMENTARY PROBLEMS

13.8 Predict the major product in each of the following nitration reactions:

(a) CF_3 — HNO_3 / H_2SO_4 →

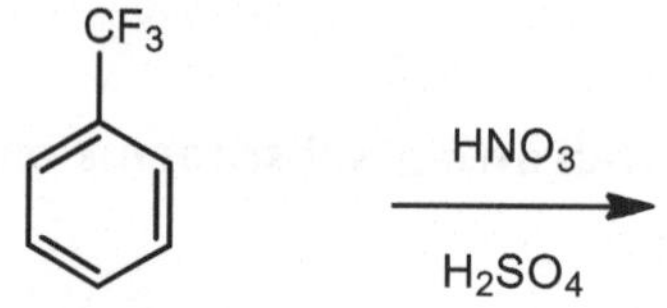

(b) CH_3, CH_3 — HNO_3 / H_2SO_4 →

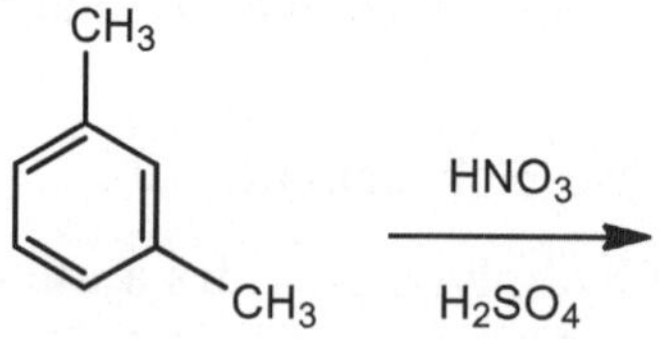

[*Ans:* (a) *m*-nitro(trifluoromethyl)benzene; (b) 2,4-dimethylnitrobenzene]

13.9 How would you carry out the following conversion? Draw structures in bond-line formulas.

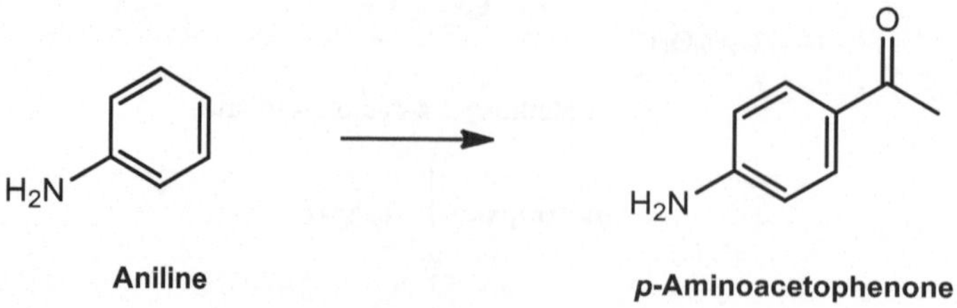

H_2N → O, H_2N

Aniline ***p*-Aminoacetophenone**

13.10 Predict the product of the following reaction:

p-Toluenesulphonic acid —NaOH / 300°C→

[*Ans: p*-Cresol (after acid work-up)]

13.11 2-Bromo-1,3-dimethylbenzene does not undergo nucleophilic substitution with sodium amide in liquid ammonia. Explain.

13.12 How would you carry out the following transformations?

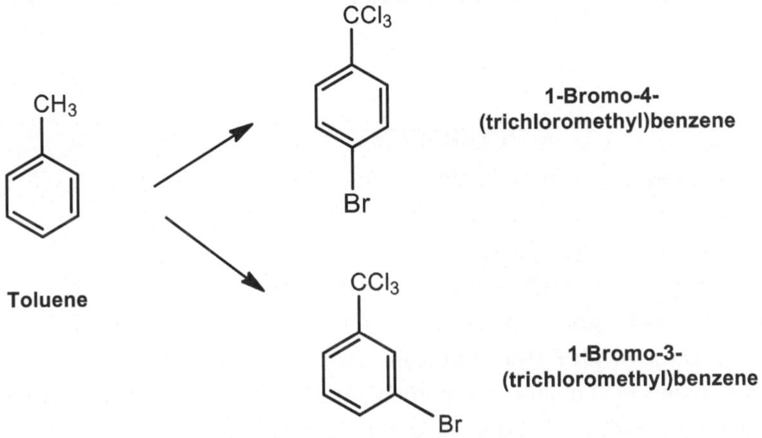

CCl_3

CH_3

1-Bromo-4-(trichloromethyl)benzene

Br

Toluene

CCl_3

1-Bromo-3-(trichloromethyl)benzene

Br

[*Hints:* use ring bromination and side chain chlorination in proper order]

13.13 Identify compounds **A–C** in the following scheme:

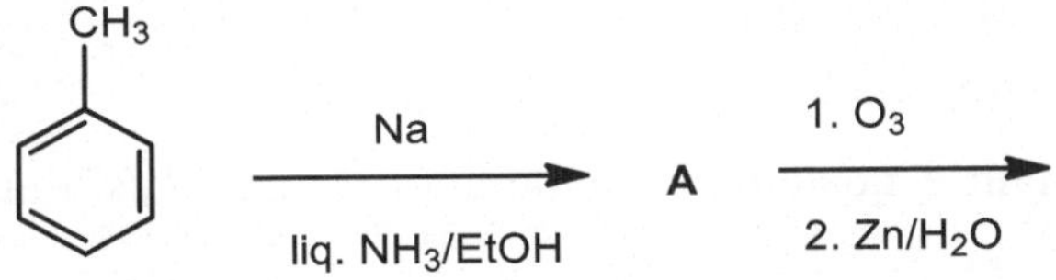

CH_3

Na, liq. NH_3/EtOH → A; 1. O_3, 2. Zn/H_2O → B + C

[*Ans:* **A** = 1-methyl-1,4-cyclohexadiene, **B** = 3-oxobutanal, **C** = 1,3-propanedial]

CHAPTER 14

Amines, Diazonium Salts and Phenols: Substitution and Elimination Reactions

IN THIS CHAPTER

- Preparation of amines by nucleophilic substitution: The Gabriel synthesis
- Reactions of amines with sulphonyl chlorides: The sulpha drugs
- Reactions of amines with nitrous acid
- Aromatic diazonium salts: Substitution of diazonium group
- Aromatic diazonium salts: Coupling reactions
- Diazomethane and the Arndt-Eistert synthesis
- The Hofmann elimination of quaternary ammonium salts and the Cope elimination
- The Reimer Tiemann reaction and the Kolbe reaction of phenols

Amines are derivatives of ammonia obtained by replacement of one, two or three hydrogen atoms by alkyl or aryl groups. They are classified as **primary (1°)**, **secondary (2°)** and **tertiary (3°)** amines based on the number of alkyl or aryl groups attached to the nitrogen atom (see Section 2.3.8).

Quaternary ammonium salts contain a quaternary ammonium cation (R_4N^+) and an anion (X^-) as a counter ion.

Amines are weak organic bases. The relative basicities of aliphatic and aromatic amines have been described previously (see Sections 5.6.1 and 5.6.2).

As bases, amines react readily with strong inorganic acids (such as HCl, HBr or HI) to form salts. For example, **ethylamine** forms an **amine salt** (ethylaminium chloride) with aqueous HCl.

$$CH_3CH_2\ddot{N}H_2 + HCl \xrightarrow{H_2O} CH_3CH_2\overset{+}{N}H_3\ Cl^-$$

Ethylamine → **Ethylaminium chloride**, ***An amine salt***

Amines can be prepared by a variety of methods. We have described earlier the preparation of amines by reduction methods. These include reduction of amides (Section 12.8.3), reduction of nitriles (Section 12.10.2), and reduction of nitro compounds (Section 13.10). Amines can also be prepared by rearrangement reactions (see later Sections 16.3.1 and 16.3.2).

In this chapter, we will describe briefly the preparation of amines through nucleophilic substitution reactions of alkyl halides.

14.1 NUCLEOPHILIC SUBSTITUTION OF ALKYL HALIDES: PREPARATION OF AMINES

Nucleophilic substitution of an **alkyl halide** (RX) with ammonia produces a **primary amine** (RNH_2). Initially, the substitution reaction (S_N2) gives an amine salt, which then releases the primary amine on reaction with a base (ammonia itself act as the base).

$$\underset{\text{Nucleophile}}{H_3\ddot{N}} + \underset{\text{Alkyl halide}}{R{-}X} \xrightarrow{S_N2} \underset{\text{An amine salt}}{R{-}\overset{+}{N}H_3\ X^-} \xrightarrow[\text{(base)}]{NH_3} \underset{1^\circ\ \text{Amine}}{R{-}NH_2} + NH_4^+\ X^-$$

Example

$$\underset{\text{Ethyl bromide}}{CH_3CH_2Br} \xrightarrow{NH_3} CH_3CH_2\overset{+}{N}H_3\ Br^- \xrightarrow{NH_3} \underset{\text{Ethylamine}}{CH_3CH_2NH_2} + NH_4Br$$

This method represents **alkylation of ammonia** (replacement of one of the hydrogen atoms of ammonia by an alkyl group).

One **problem** of this method is that the nucleophilic substitution can proceed further and lead to **multiple alkylations**. The product then is a mixture of primary amines with some secondary and tertiary amines and even some quaternary ammonium salts. Therefore, this method is not quite useful for the general synthesis of primary amines.

$$\underset{1^\circ\ \text{Amine}}{RNH_2} \xrightarrow{RX} R_2\overset{+}{N}H_2\ X^- \xrightarrow{NH_3} \underset{2^\circ\ \text{Amine}}{R_2NH} + NH_4X$$

$$R_2NH \xrightarrow{RX} R_3\overset{+}{N}H\ X^- \xrightarrow{NH_3} \underset{3^\circ\ \text{Amine}}{R_3N} + NH_4X$$

$$R_3N \xrightarrow{RX} \underset{\text{Quaternary ammonium salt}}{R_4\overset{+}{N}\ X^-}$$

However, in some cases, it is possible to *minimize* multiple alkylations by using a large *excess of ammonia*. An example is the **synthesis of alanine** (an α-amino acid) from α-bromopropionic acid:

$$\underset{\alpha\text{-Bromopropionic acid}}{CH_3\underset{\alpha}{C}H(Br)CO_2H} + NH_3\ \text{(excess)} \longrightarrow \underset{\text{Alanine (}\alpha\text{-Aminopropionic acid) (70\%)}}{CH_3\underset{\alpha}{C}H(NH_2)CO_2H}$$

14.1.1 The Gabriel Synthesis of Primary Amines

A specific route to the synthesis of primary amines through nucleophilic substitution is known as the **Gabriel phthalimide synthesis**. This method avoids the problem of multiple alkylations that occur when alkyl halides react with ammonia.

Phthalimide is a cyclic imide which is acidic in nature. It reacts with KOH to form a potassium salt of phthalimide.

In the Gabriel synthesis, the **phthalimide anion** of the salt is the nucleophile which reacts with an **alkyl halide** (RX) to form an ***N*-alkylphthalimide**.

Hydrolysis of *N*-alkylphthalimide with aqueous acid or aqueous base can form the primary amine but sometimes hydrolysis may be difficult.

It is better to treat the ***N*-alkylphthalimide** with **hydrazine** (NH_2NH_2) in ethanol to produce the **primary amine** (RNH_2).

KOH
R—X
(alkyl halide)
S_N2
N—R + KX

Phthalimide

Phthalimide anion
Nucleophile

N-Alkylphthalimide

NH_2NH_2,
C_2H_5OH
Δ

+ R—NH_2

1° Amine

Phthalazine-1,4-dione

Note: The Gabriel synthesis is not useful with tertiary alkyl halides because the reaction then proceeds with elimination rather than substitution.

PROBLEM 14.1: Outline a synthesis of the following amine using the Gabriel method:

CH_2NH_2

Benzylamine

Answer

The synthesis involves the reaction of potassium phthalimide (obtained by the treatment of phthalimide with KOH) with benzyl bromide and subsequent treatment of the resulting *N*-benzylphthalimide with hydrazine.

KOH; $C_6H_5CH_2Br$ (benzyl bromide); Δ, NH_2NH_2

Phthalimide → Potassium phthalimide → *N*-Benzylphthalimide → Benzylamine ($C_6H_5CH_2NH_2$)

14.2 REACTIONS OF AMINES WITH BENZENESULPHONYL CHLORIDE

Primary amines react with **benzenesulphonyl chloride** ($C_6H_5SO_2Cl$) in aqueous potassium hydroxide (KOH) to form ***N*-alkylbenzenesulphonamides.**

The *hydrogen* attached to the electronegative nitrogen in the sulphonamide is quite *acidic* because the benzenesulphonyl group is strongly *electron withdrawing.* Thus an *N*-alkylsulphonamide is **soluble in aqueous base** (KOH) by forming its potassium salt.

Acidification of this solution gives a precipitate of the *water insoluble N*-alkylsulphonamide.

(acidic hydrogen)

OH^-; $-$ HCl; KOH; H_3O^+ *Acidification*

($C_6H_5SO_2Cl$)

1° Amine + Benzenesulphonyl chloride → *N*-Alkylbenzenesulphonamide → Potassium salt of the sulphonamide *Soluble* → Water insoluble sulphonamide *Precipitate*

Secondary amines react with benzenesulphonyl chloride in aqueous KOH to form ***N,N*-dialkylbenzenesulphonamides.**

In *N,N*-dialkylsulphonamides the nitrogen atom bears *no* acidic hydrogen. Thus an *N,N*-dialkylsulphonamide is **insoluble in aqueous KOH** and therefore forms a precipitate.

(N bears no acidic hydrogen)

2° Amine + Benzenesulphonyl chloride $\xrightarrow[- HCl]{OH^-}$ ***N,N*-Dialkylbenzenesulphonamide**

Insoluble in KOH; forms a precipitate

Tertiary amines *do not react* with benzenesulphonyl chloride in aqueous KOH. Acidification of the mixture causes the tertiary amine to dissolve and form a water soluble amine salt.

3° Amine + Benzenesulphonyl chloride $\xrightarrow{OH^-}$ *No reaction* → Unreacted 3° Amine (*Insoluble*) $\xrightarrow{HCl}$ *Acidification* → Soluble amine salt

Hinsberg test

The Hinsberg test is based on sulphonamide formation and is used to distinguish between primary, secondary and tertiary amines.

In this test, the amine is treated with benzenesulphonyl chloride in aqueous KOH solution. The following observations indicate whether the amine is primary, secondary or tertiary.

- If there is *no* apparent change, the amine is **tertiary**. On acidification of the mixture, the tertiary amine dissolves because it forms a water soluble salt.
- If a sulphonamide is formed and it is *soluble* in aqueous KOH, the amine is **primary**. Visibly, there is a clear solution containing water soluble potassium salt of the sulphonamide. Acidification of the solution with aqueous HCl gives a precipitate of the sulphonamide.
- If a sulphonamide is formed but it is *insoluble* in aqueous KOH, the amine is **secondary**. Visibly, there is a precipitate of the sulphonamide which remains unchanged even after acidification of the mixture.

Regeneration of 1° and 2° amines from the corresponding sulphonamides

Primary and secondary amines can be regenerated from their sulphonamide derivatives by **hydrolysis** with aqueous acid.

The amine formed is protonated in the acidic medium; therefore, a *base work-up* (treatment with OH^-) is needed to produce the free amine.

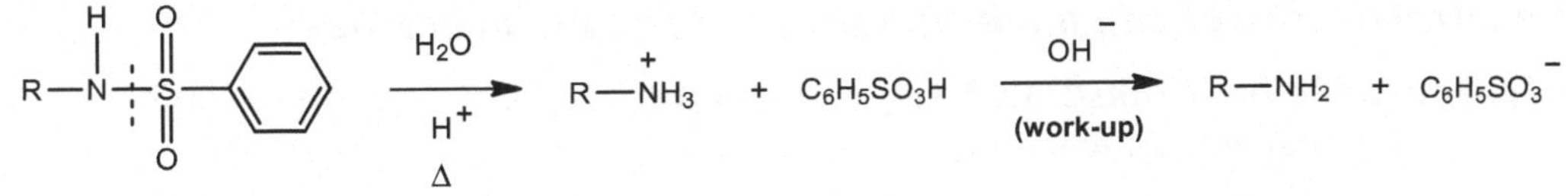

N-Alkylbenzenesulphonamide → (H_2O, H^+, Δ) **Protonated 1° amine** ($R{-}NH_3^+$) + $C_6H_5SO_3H$ → (OH^- (work-up)) **1° Amine** ($R{-}NH_2$) + $C_6H_5SO_3^-$

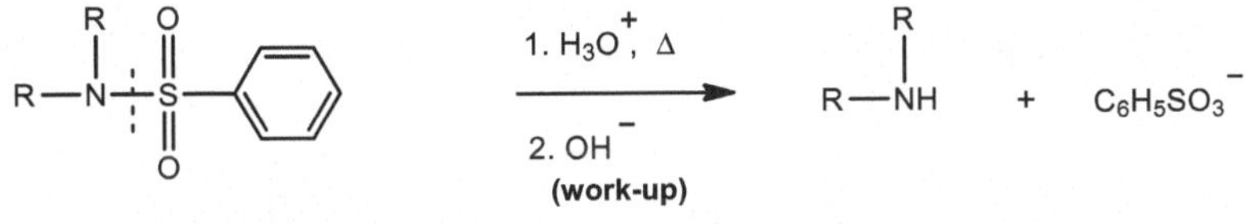

N,N-Dialkylbenzenesulphonamide → (1. H_3O^+, Δ; 2. OH^- (work-up)) **2° Amine** (R_2NH) + $C_6H_5SO_3^-$

14.2.1 The Sulpha Drugs: Sulphanilamides

Sulphanilic acid is *p*-aminobenzenesulphonic acid, which is a derivative of benzenesulphonic acid. The amide of sulphanilic acid (*p*-aminobenzenesulphonamide) is called **sulphanilamide**.

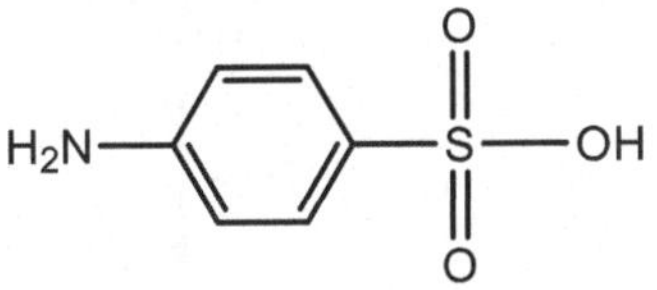

Sulphanilic acid (*p*-aminobenzenesulphonic acid)

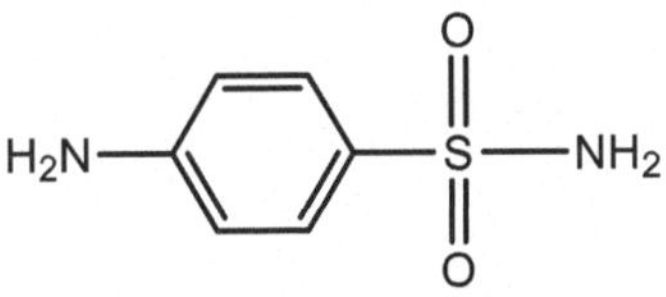

Sulphanilamide (*p*-aminobenzenesulphonamide)

Sulphanilamide is one of the medicinal drugs that have been effective against a broad range of bacterial infections. It was considered to be the first 'wonder drug'.

Some substituted sulphanilamides such as **sulphadiazine**, **sulphathiazole** and **sulphamethoxazole** have been found to be more effective. Sulphanilamide and substituted sulphanilamides with antibacterial properties are collectively known as **sulpha drugs**.

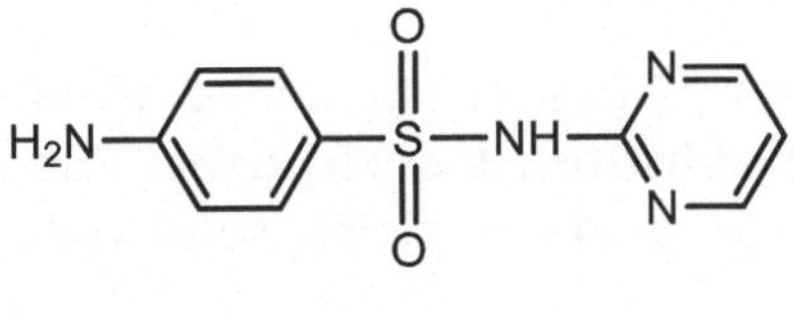

Sulphadiazine

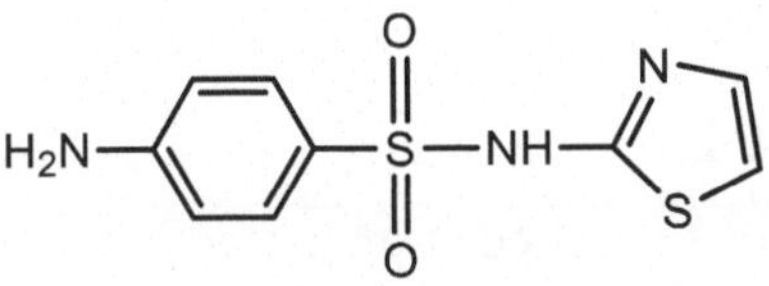

Sulphathiazole

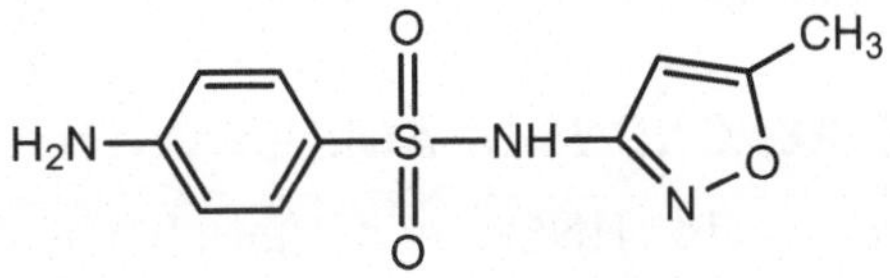

Sulphamethoxazole

Preparation of sulphanilamide and substituted suphanilamides

Sulphanilamide or ***N*-substituted sulphanilamides** can be prepared from **aniline** using the following sequence of reactions:

NH_2
Aniline
$(CH_3CO)_2O$
Acetylation
$NHCOCH_3$
Acetanilide
$ClSO_3H$
Chlorosulphonation
$NHCOCH_3$
SO_2Cl
p-Acetamidobenzene-sulphonyl chloride

RNH_2
$NH-COCH_3$
SO_2-NHR
N-substituted *p*-acetamidobenzenesulphonamide
Hydrolysis
1. H_3O^+
2. OH^-
NH_2
SO_2NHR
N-Substituted sulphanilamide

NH_3
$NH-COCH_3$
SO_2-NH_2
p-Acetamidobenzene-sulphonamide
Hydrolysis
1. H_3O^+
2. OH^-
NH_2
SO_2NH_2
Sulphanilamide

Aniline is first acetylated with acetic anhydride to form **acetanilide** which is then reacted with chlorosulphonic acid ($ClSO_3H$) to produce ***p*-acetamidobenzenesulphonyl chloride** (see Section 13.5.3).

Thereafter, addition of ammonia (NH_3) or an amine (RNH_2) gives ***p*-acetamido-benzenesulphonamide** or its ***N*-substituted derivatives**. Subsequent hydrolysis with aqueous HCl followed by a base work-up yields **sulphanilamide** or ***N*-substituted sulphanilamides.**

Note: The *selective* removal of the acetyl group ($COCH_3$) by hydrolysis is possible because an amide group (CO–NH) is more easily hydrolysed than a sulphonamide group (SO_2–NH).

14.3 REACTIONS OF AMINES WITH NITROUS ACID

Nitrous acid (HNO_2) is an unstable acid and cannot be stored. For a reaction with an amine, it is usually generated in the presence of the amine from an **aqueous solution of sodium nitrite** ($NaNO_2$) and a **strong acid** (e.g. HCl or H_2SO_4).

$HNO_2 \equiv HO{-}NO \equiv HO{-}N{=}O$

Nitrous acid
Unstable

$$NaNO_2 + HCl \xrightarrow{H_2O} HNO_2 + NaCl$$

$$2\,NaNO_2 + H_2SO_4 \xrightarrow{H_2O} 2\,HNO_2 + Na_2SO_4$$

Primary, secondary and tertiary amines react with nitrous acid to give different kinds of products.

14.3.1 Reactions of Primary Amines with Nitrous Acid

The reactions of **primary aliphatic amines** and **primary aromatic amines** with nitrous acid lead to different products.

Primary aliphatic amines

Primary aliphatic amines (RNH_2; R = alkyl) react with nitrous acid to form **aliphatic diazonium salts** which are **highly unstable**. Even at low temperatures, they break down spontaneously with evolution of nitrogen gas and yield a mixture of products (alkyl halides, alcohols and alkenes).

For example, isopropylamine reacts with an aqueous solution of $NaNO_2$ and HCl to give a mixture of isopropyl chloride, isopropanol and propene with the evolution of nitrogen.

The **instability** of the aliphatic diazonium salt is mainly attributed to the **departure of nitrogen (N_2) as an excellent leaving group**. The loss of nitrogen forms a **carbocation** that can react with a halide ion or water to give an **alkyl halide** or **alcohol**. The carbocation can also eliminate a proton to give an **alkene**.

$$R{-}NH_2 \xrightarrow[HCl]{NaNO_2} R{-}\overset{+}{N}{\equiv}N\ Cl^- \longrightarrow N_2 + \text{Alkyl halides, alcohols, alkenes}$$

(R = alkyl) — **1° Aliphatic amine**

$(RN_2^+Cl^-)$ — **Aliphatic diazonium chloride** ***Highly unstable***

Example

$$(CH_3)_2CHNH_2 \xrightarrow[HCl]{NaNO_2} (CH_3)_2CH{-}\overset{+}{N}{\equiv}N\ Cl^- \longrightarrow CH_3\overset{+}{C}HCH_3 + N_2$$

Isopropylamine → **Isopropyldiazonium chloride** → **Carbocation (isopropyl cation)** + N_2 ***Excellent leaving group***

Carbocation → H_2O, $-H^+$ → $CH_3CH(OH)CH_3$ **Isopropanol**

Carbocation → Cl^- → $CH_3CH(Cl)CH_3$ **Isopropyl chloride**

Carbocation → $-H^+$ → $CH_3CH{=}CH_2$ **Propene**

Since the reaction produces a complex mixture of products, it is of little synthetic importance. However, the **evolution of nitrogen gas is quantitative**, and this feature was used in the analysis of amino acids.

Primary aromatic amines

Primary aromatic amines ($ArNH_2$; Ar = aryl e.g. phenyl) react with nitrous acid to produce **aromatic diazonium salts**, which are **much more stable** than aliphatic diazonium salts.

Aromatic diazonium salts do not decompose when the **temperature** of the reaction is kept **low** (0–5°C).

For example, **aniline** reacts with an aqueous solution of $NaNO_2$ and HCl at 0–5°C to form **benzenediazonium chloride** ($C_6H_5N_2^+\ Cl^-$).

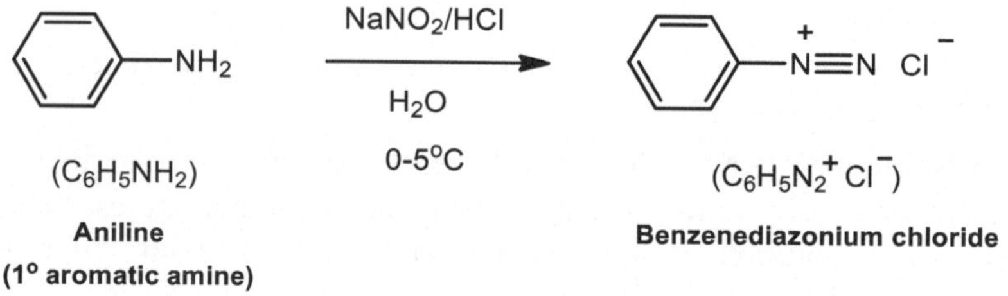

$NaNO_2$/HCl, H_2O, 0-5°C

($C_6H_5NH_2$) **Aniline** **(1° aromatic amine)** → ($C_6H_5N_2^+\ Cl^-$) **Benzenediazonium chloride**

In an **aromatic diazonium cation** (ArN_2^+), the aryl group (e.g. C_6H_5) can enter into **resonance** with the diazonium group (N_2^+) and thereby **stabilize** it. In contrast, there is no such resonance stabilization in the case of an aliphatic diazonium cation (RN_2^+). Thus an **aromatic diazonium cation is more stable than an aliphatic diazonium cation**.

$R-\overset{+}{N}\equiv N$

(no resonance involving alkyl group)

Aliphatic diazonium cation

Less stable

(stabilized by resonance involving phenyl group)

Aromatic diazonium cation

More stable

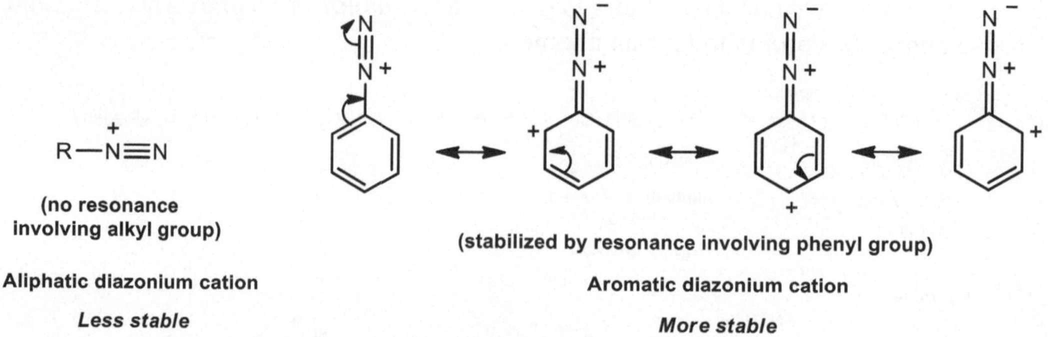

It may be mentioned that many diazonium salts are unstable above 5–10°C, and some may explode when dry.

Diazotization

The reaction of a primary aliphatic or aromatic amine with nitrous acid to form a diazonium salt is called **diazotization**.

A **mechanism** for the diazotization of a primary amine is shown in Figure 14.1. The mechanism involves a series of steps.

Generation of nitrosonium ion:

$HO{-}N{=}O \xrightarrow{H^+} H_2\overset{+}{O}{-}N{=}O \xrightarrow{-H_2O} \overset{+}{N}{=}O$

(HNO_2) (^+NO)

Nitrous acid **Nitrosonium ion**

Formation of diazonium cation:

$R{-}NH_2 + \overset{+}{N}{=}O \longrightarrow R{-}\overset{+}{N}H_2{-}N{=}O \xrightarrow{-H^+} R{-}NH{-}N{=}O$

(R = alkyl or aryl)

1° Amine ***N*-Nitrosoamine**

Tautomerization

$R{-}\overset{+}{N}{\equiv}N \xleftarrow{-H_2O} R{-}N{=}N{-}\overset{+}{O}H_2 \xleftarrow{H^+} R{-}N{=}N{-}OH$

Aliphatic or aromatic diazonium cation **Diazohydroxide**

Figure 14.1: A mechanism for the diazotization of a primary aliphatic or aromatic amine.

In the presence of a strong acid, nitrous acid produces a **nitrosonium ion** (^+NO) as an electrophile. The nucleophile is the primary amine which combines with the nitrosonium ion followed by the loss of a proton to form an ***N*-nitrosoamine**. Tautomerization of the *N*-nitrosoamine to a **diazohydroxide** and subsequent loss of water in the presence of acid produces the **diazonium cation.**

14.3.2 Reactions of Secondary Amines with Nitrous Acid

Both secondary **aliphatic** and **aromatic** amines react with nitrous acid to form ***N*-nitrosoamines**, which usually separate from the reaction mixture as yellow oily liquids.

The reaction proceeds through the combination of nitrosonium ion (^+NO) with the secondary amine followed by the loss of a proton to give the *N*-nitrosoamine product (cf. Figure 14.1 and draw steps).

This reaction, called a nitrosation, occurs on the amine nitrogen, and is termed an ***N*-nitrosation**. The following are two examples:

$(CH_3)_2NH \xrightarrow[H_2O]{NaNO_2/HCl} (CH_3)_2N{-}N{=}O$

N-Nitrosation

Dimethylamine ***N*-Nitrosodimethylamine**

$NHCH_3$ → ($NaNO_2/HCl$, H_2O; ***N-Nitrosation***) → $H_3C—N—N=O$

***N*-Methylaniline** → ***N*-Nitroso-*N*-methylaniline**

Note: The *N*-nitrosoamine has *no* hydrogen atom attached to the nitrogen. So it cannot tautomerize and a diazonium cation cannot be formed. Remember that *N*-nitrosoamines are carcinogenic and they should be handled with caution.

14.3.3 Reactions of Tertiary Amines with Nitrous Acid

Tertiary aliphatic amines react with nitrosonium ion (^{+}NO) generated from nitrous acid to form *N*-nitrosoammonium salts. But these salts are **unstable** and decompose into a variety of compounds including aldehydes or ketones. So these reactions are not useful in synthesis.

$R_3N:$ → ($NaNO_2$, HCl) → $R_3N^{+}—N=O\ Cl^{-}$

3° Aliphatic amine → ***N*-Nitrosoammonium chloride** ***Unstable***

Tertiary aromatic amines can undergo **electrophilic aromatic substitution** with nitrosonium ion (^{+}NO) as the *electrophile* generated from nitrous acid. Here the nitrosation (introduction of a nitroso group) occurs at the ring carbon. So the reaction is termed a ***C*-nitrosation**.

For example, ***N,N*-dimethylaniline** undergoes a *C*-nitrosation on treatment with an aqueous solution of $NaNO_2$ and HCl and produces mainly ***p*-nitroso-*N,N*-dimethylaniline**.

$H_3C—\ddot{N}—CH_3$ → ($NaNO_2/HCl$, 8°C; ***C-Nitrosation***) → $\ddot{N}(CH_3)_2$ … NO

***N,N*-Dimethylaniline** → ***p*-Nitroso-*N,N*-dimethylaniline**

Mechanism:

+*M*

$(H_3C)_2\ddot{N}$–C₆H₅ + $\overset{+}{N}$=O —(*para* attack)→ $(H_3C)_2\overset{+}{N}$=C₆H₅(H)(N=O)

N,N-Dimethylaniline **Electrophile** **An arenium ion**

(most stable resonance structure involving +*M* effect of $N(CH_3)_2$ group)

↓ $-H^+$

$(H_3C)_2\ddot{N}$–C₆H₄–NO

***p*-Nitroso-*N*,*N*-dimethylaniline**

Note: The dimethylamino group [$N(CH_3)_2$] is an *ortho/para* director; however, the electrophilic aromatic substitution occurs mainly at the *para* position because of the steric interference of the dimethylamino group for the *ortho* attack.

14.4 AROMATIC DIAZONIUM SALTS: SUBSTITUTION OF DIAZONIUM GROUP

The **diazonium group** (N_2^+) of aromatic diazonium salts can be substituted by a number of other atoms or groups including halogens (Cl, Br, I, F), hydrogen (H), hydroxyl (OH), nitrile (CN) and nitro (NO_2) groups.

The **substitution reactions** occur with **evolution of nitrogen gas (N_2)**. Most substitution reactions are carried out without isolating diazonium salts prepared by diazotization of primary aromatic amines at low temperatures (0–5°C).

14.4.1 Substitution by Cl, Br or CN: Sandmeyer Reactions

The substitution of the diazonium group (N_2^+) by Cl, Br and CN takes place when aromatic diazonium salts are treated with cuprous chloride (CuCl), cuprous bromide (CuBr) and cuprous cyanide (CuCN) respectively. These reactions are known as **Sandmeyer reactions**.

The following examples illustrate Sandmeyer reactions:

o-Toluidine (o-methylaniline) —($NaNO_2/HCl$, H_2O, (0-5°C))→ $\overset{+}{N_2}\ Cl^-$ (o-CH₃) —(CuCl, Δ)→ **o-Chlorotoluene** + N_2

***m*-Toluidine (*m*-methylaniline)** —($NaNO_2/HCl$, H_2O, (0-5°C))→ $\overset{+}{N_2}\ Cl^-$ (m-CH₃) —(CuBr, Δ)→ ***m*-Bromotoluene** + N_2

NH2 → NaNO2/HCl, H2O (0-5°C) → N2+ Cl− → CuCN, Δ → CN + N2

CH3 CH3 CH3

p-Toluidine (p-methylaniline) **p-Tolunitrile (p-methylbenzonitrile)**

14.4.2 Substitution by I

The substitution of the diazonium group (N_2^+) by I does *not* require the use of cuprous iodide. The diazonium salt is simply mixed with potassium iodide (KI) and allowed to react. The product is an aryl iodide.

To illustrate, consider a preparation of *m*-iodonitrobenzene from *m*-nitroaniline:

NH2, NO2 → NaNO2/H2SO4, H2O (0-5°C) → N2+ HSO4−, NO2 → KI, − N2 → I, NO2

***m*-Nitroaniline** ***m*-Nitrobenzenediazonium hydrogen sulphate** ***m*-Iodonitrobenzene**

14.4.3 Substitution by F or NO_2

For the substitution of the diazonium group (N_2^+) by F, the **diazonium salt is isolated as a stable diazonium fluoroborate**. On heating, the diazonium fluoroborate decomposes to yield an aryl fluoride.

The diazonium fluoroborate can be prepared by adding fluoroboric acid (HBF_4) to the solution of a diazonium salt obtained in the usual way (i.e. by treating an aromatic amine with $NaNO_2$ and HCl). The diazonium fluoroborate is precipitated which is then isolated and dried.

The method is illustrated with the conversion of aniline to fluorobenzene.

NH2 → NaNO2/HCl, H2O → N2+ Cl− → HBF4 → N2+ BF4− → Δ → F + N2 + BF3

Aniline **Benzenediazonium chloride** **Benzenediazonium fluoroborate** **Fluorobenzene**

Isolated as a crystalline salt

When a **diazonium fluoroborate** is decomposed with aqueous **sodium nitrite** ($NaNO_2$) containing **copper powder**, the diazonium group (N_2^+) is replaced by a nitro group (NO_2). The product is an aromatic nitro compound.

This is illustrated with the preparation of *p*-dinitrobenzene from *p*-nitroaniline:

$$p\text{-}O_2N\text{-}C_6H_4\text{-}NH_2 \xrightarrow[\text{2. } HBF_4]{\text{1. } NaNO_2/HCl} p\text{-}O_2N\text{-}C_6H_4\text{-}\overset{+}{N_2}\,\overset{-}{BF_4} \xrightarrow[Cu]{NaNO_2} p\text{-}O_2N\text{-}C_6H_4\text{-}NO_2 + N_2 + NaBF_4$$

***p*-Nitroaniline** → ***p*-Nitrobenzenediazonium fluoroborate** → ***p*-Dinitrobenzene**

14.4.4 Substitution by OH

The substitution of the diazonium group (N_2^+) by OH can be accomplished by simply heating a diazonium salt in water. This represents the hydrolysis of a diazonium salt to the corresponding phenol.

To illustrate, consider the following conversion of *o*-toluidine to *o*-cresol.

$$o\text{-}CH_3\text{-}C_6H_4\text{-}NH_2 \xrightarrow[H_2O\ (0\text{-}5^oC)]{NaNO_2/H_2SO_4} o\text{-}CH_3\text{-}C_6H_4\text{-}\overset{+}{N_2}\,\overset{-}{HSO_4} \xrightarrow[\Delta]{H_2O} o\text{-}CH_3\text{-}C_6H_4\text{-}OH + N_2$$

***o*-Toluidine** → ***o*-Toluenediazonium hydrogen sulphate** → ***o*-Cresol (*o*-methylphenol)**

Another method of substituting the diazonium group by a hydroxyl group is to add cuprous oxide (Cu_2O) and a large excess of cupric nitrate [$Cu(NO_3)_2$] to a solution of the diazonium salt.

This seems to be a better method and is illustrated with the following conversion of *p*-toluidine to *p*-cresol.

$$p\text{-}CH_3\text{-}C_6H_4\text{-}NH_2 \xrightarrow[H_2O\ (0\text{-}5^oC)]{NaNO_2/H_2SO_4} p\text{-}CH_3\text{-}C_6H_4\text{-}\overset{+}{N_2}\,\overset{-}{HSO_4} \xrightarrow[Cu(NO_3)_2,\ H_2O]{Cu_2O} p\text{-}CH_3\text{-}C_6H_4\text{-}OH + N_2$$

***p*-Toluidine** → ***p*-Toluenediazonium hydrogen sulphate** → ***p*-Cresol (*p*-methylphenol)**

14.4.5 Substitution by H: Deamination of Aromatic Amines

The substitution of the diazonium group (N_2^+) by a hydrogen atom (H) takes place when a diazonium salt is **reduced** with **hypophosphorous acid** (H_3PO_2). Hypophosphorous acid itself is then oxidized to phosphorous acid (H_3PO_3).

For example, consider the following conversion of aniline to benzene:

Deamination protocol

Aniline $\xrightarrow[H_2O\ (0\text{-}5^\circ C)]{NaNO_2/H_2SO_4}$ Benzenediazonium hydrogen sulphate ($N_2^+\ HSO_4^-$) $\xrightarrow{H_3PO_2}$ Benzene + N_2 + H_3PO_3

This method for removing an amino group (NH_2) from the benzene ring via the formation of a diazonium salt is called the **deamination**. The deamination is a useful protocol in synthesis.

Use of deamination protocol in synthesis

In planning a synthesis, we can introduce an **amino group** (NH_2) into a benzene ring to make use of its *ortho/para* orientation for a **subsequent reaction**. Thereafter, we can remove the amino group by the **deamination protocol** to get the desired product.

To illustrate, consider the following synthesis of 1,3,5-tribromobenzene from benzene:

Benzene $\xrightarrow[c.\ H_2SO_4]{c.\ HNO_3}$ Nitrobenzene $\xrightarrow[HCl]{Sn}$ Aniline $\xrightarrow{Br_2/H_2O}$ 2,4,6-Tribromoaniline $\xrightarrow[H_2O\ (0\text{-}5^\circ C)]{NaNO_2/HCl}$ A diazonium salt ($N_2^+\ Cl^-$) $\xrightarrow[-\ N_2]{H_3PO_2}$ 1,3,5-Tribromobenzene

Introduction of an NH_2 group into the benzene ring (via nitration) strongly activates the ring for **tribromination** at two *ortho* positions and one *para* position, thereby providing the desired relationships of three Br atoms on the benzene ring for the target product. Then, a simple **deamination protocol** gives the desired product (1,3,5-tribromobenzene).

Note: Direct bromination of benzene gives bromobenzene which on further bromination cannot lead to the desired product because Br is a deactivating *ortho/para* director.

PROBLEM 14.2: How would you carry out the following transformations?

(a) NO_2-benzene ⟶ 3-bromophenol (OH, Br)

(b) 4-methylaniline (NH_2, CH_3) ⟶ 3-bromotoluene (Br, CH_3)

Answer

(a) NO_2 is a *meta*-directing group. Therefore, bromination of nitrobenzene gives *m*-bromonitrobenzene. It is then possible to convert the NO_2 group to OH through a sequence of reduction, diazotization and substitution by OH. The product is *m*-bromophenol.

Nitrobenzene $\xrightarrow[Fe]{Br_2}$ ***m*-Bromonitrobenzene** $\xrightarrow[HCl]{Sn}$ ***m*-Bromoaniline** $\xrightarrow[0\text{-}5^{o}C]{NaNO_2/HCl}$ $\overset{+}{N_2}\,Cl^-$ (*m*-bromobenzenediazonium chloride) $\xrightarrow[\Delta,\ -N_2]{H_2O/H^+}$ ***m*-Bromophenol**

(b) Direct bromination of *p*-toluidine would lead to bromination at both *ortho* positions (*p*-position being blocked) with respect to the NH_2 group, which is a strongly activating *ortho*/*para* director. Therefore, the activating effect of the NH_2 group needs to be reduced by acetylation with acetic anhydride. Bromination of the resulting *N*-acetyl derivative followed by hydrolysis (which removes the acetyl group) gives 2-bromo-4-methylaniline. Then deamination via a diazonium salt produces *m*-bromotoluene.

NH_2, CH_3 — ($CH_3CO)_2O$ → $NHCOCH_3$, CH_3 — Br_2 → $NHCOCH_3$, Br, CH_3

p-Toluidine

↓ OH^-/H_2O, Δ

NH_2, Br, CH_3 — $NaNO_2$, H_2SO_4, 0-5°C → $\overset{+}{N_2}\ HSO_4^-$, Br, CH_3 — H_3PO_2, $-\ N_2$ → Br, CH_3

m-Bromotoluene

2-Bromo-4-methylaniline

14.5 AROMATIC DIAZONIUM SALTS: COUPLING REACTIONS

An aromatic diazonium cation (ArN_2^+), also called an **arenediazonium cation**, can act as an **electrophile** in electrophilic aromatic substitution.

In the diazonium cation, the **terminal nitrogen atom** with sextet of electrons is the **electrophilic site**.

Octet (not electrophilic site) **Sextet (electrophilic site)**

$$Ar-\overset{+}{N}\equiv N: \longleftrightarrow Ar-\ddot{N}=\overset{+}{N}:$$

Arenediazonium cation

A weak electrophile

The arenediazonium cation is a *weak* electrophile and requires highly activated aromatic compounds such as **phenols** and **aromatic amines** for electrophilic aromatic substitutions. These electrophilic aromatic substitutions are known as **diazo coupling reactions**.

Coupling reactions with phenols

Coupling reactions between phenols and diazonium salts take place more readily in a *slightly alkaline solution*. The coupling takes place on the carbon atom of the aromatic ring, called the ***C*-coupling**. The *C*-coupling produces an **azo compound** in which two aromatic rings are held together by an **azo linkage** (–N=N–).

For example, the coupling reaction between phenol and benzene diazonium chloride produces ***p*-hydroxyazobenzene**. Note that azobenzene is C_6H_5–N=N–C_6H_5.

In the presence aqueous NaOH, phenol (C_6H_5OH) is converted to **phenoxide ion** ($C_6H_5O^-$) which provides a **more strongly activated ring** than phenol for electrophilic substitution. This is because O^- (which has $+I$ and a relatively stronger $+M$ effect) is more electron-donating than OH (having $-I$ and a relatively weaker $+M$ effect).

The **major** azo product results from the electrophilic attack of the *bulky* diazonium cation at the *para* position, since the *ortho* attack encounters a steric interference.

Note: Azo compounds are generally intensely coloured and they are used extensively as **dyes**. The product *p*-hydroxyazobenzene has an orange colour.

Coupling reactions with aromatic amines

Coupling reactions between aromatic amines and diazonium salts are carried out in a *slightly acidic solution*, when the concentration of the diazonium cation (electrophile) is relatively high.

With **tertiary aromatic amines**, the ***C*-coupling** takes place (cf. Section 14.3.3).

For example, the coupling reaction between *N,N*-dimethylaniline and benzene diazonium chloride produces ***p-N,N*-dimethylaminoazobenzene**, which has a yellow colour.

Benzenediazonium chloride + *N,N*-Dimethylaniline —(0°C, *C-Coupling*)→ *p-N,N*-Dimethylaminoazobenzene (*Yellow colour*)

Mechanism:

N,N-Dimethylaniline (+M) + Benzenediazonium cation —(*para* attack)→ An arenium ion —($-H^+$)→ *p-N,N*-Dimethylaminoazobenzene

With **primary and secondary aromatic amines**, the diazonium cation reacts preferentially at the nitrogen atom, called the ***N*-coupling** (cf. reactions with nitrous acid, Section 14.3.1 and 14.3.2).

For example, the coupling reaction of aniline (a primary amine) with benzene diazonium chloride produces **diazoaminobenzene**, which is yellow in colour.

Benzenediazonium chloride + Aniline —(CH_3CO_2Na, *N*-Coupling)→ Diazoaminobenzene (*Yellow colour*)

Mechanism:

Aniline + Benzenediazonium cation → (C_6H_5–$\overset{+}{N}H_2$–N=N–C_6H_5) —($-H^+$)→ Diazoaminobenzene

The *N*-coupled product (diazoaminobenzene) can be **rearranged** to the *C*-coupled product (*p*-aminoazobenzene) by treatment with acid (HCl).

$$C_6H_5{-}N{=}N{-}NH{-}C_6H_5 \xrightarrow{H^+} C_6H_5{-}N{=}N{-}C_6H_4{-}NH_2$$

Diazoaminobenzene *Yellow* → ***p*-Aminoazobenzene** *Orange*

Azo dye test for primary aromatic amine

The coupling reaction of an **aromatic diazonium salt** with an *alkaline* solution of **β-naphthol** (a phenolic compound) is used as a laboratory test for detecting a primary aromatic amine. This test is known as the **azo dye test**.

β-Naphthol is 2-hydroxynaphthalene or 2-naphthol (1 and 2 positions of the naphthalene ring are denoted as α and β respectively).

1- or α-position; 2- or β-position

Naphthalene **β-Naphthol (2-naphthol)**

In this test, the suspected amine is treated with aqueous HCl and $NaNO_2$ under cold conditions. The resulting solution is then poured into a cold solution of β-naphthol in sodium hydroxide. The appearance of a **red or orange dye** indicates a positive test for primary aromatic amine.

In the coupling reaction, the ***C*-coupling** occurs at the **1- or α-position** of the naphthalene ring.

The following scheme shows the formation of an azo dye taking aniline as a primary aromatic amine:

$$C_6H_5{-}NH_2 \xrightarrow[H_2O\ (0\text{-}5^oC)]{NaNO_2/HCl} C_6H_5{-}\overset{+}{N}{\equiv}N\ \overset{-}{Cl} \xrightarrow[\text{NaOH, } H_2O]{\beta\text{-naphthol}} \text{Azo dye}$$

Aniline → **Benzenediazonium chloride** → (*C-Coupling at α position*) **Azo dye** *Red or orange*

PROBLEM 14.3: Methyl orange is an acid-base indicator (red in acid and yellow in base). Outline a synthesis of methyl orange.

$Na^+ \ ^-O_3S-C_6H_4-N{=}N-C_6H_4-N(CH_3)_2$

Methyl orange

Answer

Methyl orange can be prepared from sulphanilic acid (*p*-aminobenzenesulphonic acid) and *N,N*-dimethylaniline.

Sulphanilic acid (a primary aromatic amine) is diazotized with $NaNO_2$ and HCl, and the resulting solution is coupled with *N,N*-dimethylanilne. Subsequent treatment with NaOH gives methyl orange as the sodium salt of the azo compound.

$HO_3S-C_6H_4-NH_2$ (**Sulphanilic acid**) $\xrightarrow[H_2O\ (0\text{-}5^oC)]{NaNO_2/HCl}$ $HO_3S-C_6H_4-\overset{+}{N}{\equiv}N\ Cl^-$

$\downarrow$ $C_6H_5-N(CH_3)_2$ **(*N,N*-dimethylaniline)**

$HO_3S-C_6H_4-N{=}N-C_6H_4-N(CH_3)_2 \xrightarrow{NaOH} Na^+ \ ^-O_3S-C_6H_4-N{=}N-C_6H_4-N(CH_3)_2$

Methyl orange

14.6 ALIPHATIC DIAZO COMPOUNDS: DIAZOMETHANE

Methanediazonium cation ($CH_3N_2^+$) is an **aliphatic diazonium cation,** which is highly unstable. Removal of a proton (H^+) from the methyl group (CH_3) gives a neutral compound called **diazomethane** (CH_2N_2).

Like benzenediazonium cation ($C_6H_5N_2^+$), diazomethane is also *stabilized by resonance* involving the carbanion carbon. Thus **diazomethane (CH_2N_2) is much more stable than methanediazonium ion ($CH_3N_2^+$).**

$H_3C-\overset{+}{N}{\equiv}N$ ($CH_3N_2^+$)

Methanediazonium cation

Highly unstable

$H_2\overset{-}{C}-\overset{+}{N}{\equiv}N \longleftrightarrow H_2C{=}\overset{+}{N}{=}\overset{-}{N}$ (CH_2N_2)

Diazomethane

Stabilized by resonance

Diazomethane is a toxic gas and should be handled very carefully. It is normally **used in ether solution at a low temperature**. Diazomethane with a carbanion centre can act as a *nucleophile* as well as a *base*. It is highly useful in synthesis.

Reactions of diazomethane: Methylation of carboxylic acids and phenols

Diazomethane as a **base** can readily abstract an acidic hydrogen atom from a carboxylic acid or a phenol, and is converted to a very reactive methanediazonium ion ($CH_3N_2^+$) that acts as a methylating agent.

For example, diazomethane converts a **carboxylic acid** to its **methyl ester**.

$$R{-}C(=O){-}O{-}H + H_2\overset{-}{C}{-}\overset{+}{N}{\equiv}N \longrightarrow R{-}C(=O){-}O^- + CH_3{-}\overset{+}{N}{\equiv}N$$

(R = alkyl or aryl) (CH_2N_2) ($CH_3N_2^+$)

Carboxylic acid **Diazomethane** **Carboxylate ion** **Methanediazonium cation**

$$\downarrow$$

$$R{-}C(=O){-}OCH_3 + N_2$$

Methyl ester

Example

$$C_6H_5{-}CO_2H + CH_2N_2 \xrightarrow[0^oC]{Et_2O} C_6H_5{-}CO_2CH_3 + N_2$$

Benzoic acid **Diazomethane** **Methyl benzoate**

For methylation of phenols, see Problem 14.4a.

14.6.1 The Arndt-Eistert Synthesis

The **Arndt-Eistert synthesis** is a method for converting a **carboxylic acid** (RCO_2H) to its **next higher homologue** (RCH_2CO_2H). Diazomethane (CH_2N_2) is employed in this synthesis.

An **acid chloride** (RCOCl), derived from a carboxylic acid (RCO_2H), reacts with **diazomethane (2 moles)** to give an **α-diazoketone** ($RCOCHN_2$).

In this reaction, **one molecule of diazomethane** acts as a **nucleophile** for nucleophilic substitution at the carbonyl group of the acid chloride to form an intermediate and the **second molecule of diazomethane** acts as a **base** to remove an acidic hydrogen atom from the intermediate to form the **α-diazoketone** ($RCOCHN_2$) and **methanediazonium ion** ($CH_3N_2^+$).

Note that the methylene hydrogen atom in the intermediate is *acidic* because it is flanked by two electron-withdrawing groups (C=O and N_2^+). Further, methanediazonium ion formed reacts immediately with the chloride ion to form CH_3Cl and N_2.

The **α-diazoketone**, on being heated in the presence of silver oxide as catalyst, undergoes the **Wolff rearrangement** (see Section 16.2.3) to produce a **ketene** with loss of nitrogen (N_2).

When the rearrangement is carried out in an **aqueous medium**, the ketene is converted directly into a **carboxylic acid**.

$$R{-}\overset{O}{\overset{\|}{C}}{-}OH \xrightarrow{SOCl_2} R{-}\overset{O}{\overset{\|}{C}}{-}Cl \xrightarrow[\textit{(nucleophile)}]{\overset{-}{C}H_2{-}\overset{+}{N}{\equiv}N} R{-}\overset{O}{\overset{\|}{C}}{-}CH(H){-}\overset{+}{N}{\equiv}N + Cl^-$$

(RCO₂H) Carboxylic acid; Acid chloride; (acidic) H

$$\xrightarrow[\textit{(base)}]{\overset{-}{C}H_2{-}\overset{+}{N}{\equiv}N} R{-}\overset{O}{\overset{\|}{C}}{-}\overset{-}{C}H_{\alpha}{-}\overset{+}{N}{\equiv}N + [CH_3{-}N_2^+ \xrightarrow{Cl^-} CH_3Cl + N_2]$$

(RCOCHN₂) An α-Diazoketone

$$\xrightarrow[\Delta,\ -N_2]{Ag_2O} R{-}CH{=}C{=}O \xrightarrow{H_2O} R{-}CH_2{-}\overset{O}{\overset{\|}{C}}{-}OH$$

A Ketene; (RCH₂CO₂H) Carboxylic acid, *Next higher homologue*

To illustrate, consider the following conversion of **benzoic acid** to its next higher homologue, **phenylacetic acid**:

$$C_6H_5{-}CO_2H \xrightarrow{SOCl_2} C_6H_5{-}COCl \xrightarrow[\text{(2 moles)}]{CH_2N_2} C_6H_5{-}COCHN_2 + CH_3Cl + N_2$$

Benzoic acid; Benzoyl chloride; An α-diazoketone

$$\xrightarrow[-N_2]{Ag_2O/H_2O} C_6H_5{-}CH_2CO_2H$$

Phenylacetic acid

PROBLEM 14.4: Identify compounds **A–C** and write their structures in bond-line formulas.

(a) $C_6H_5OH \xrightarrow{CH_2N_2}$ **A**

(b) $H_3C{-}C(CH_3)_2{-}Cl \xrightarrow[\text{2. } CO_2 \text{ then } H_3O^+]{\text{1. Mg/Et}_2\text{O}}$ **B** $\xrightarrow[\text{2. } CH_2N_2 \text{ (excess)} \quad \text{3. } Ag_2O,\ H_2O]{\text{1. } PCl_5}$ **C**

Answer

(a) Since phenol has an acidic hydroxyl group (OH), diazomethane methylates phenol to produce anisole (**A**).

$$C_6H_5OH \xrightarrow[\text{(diazomethane)}]{\overset{-}{C}H_2{-}\overset{+}{N}{\equiv}N} C_6H_5OCH_3 + N_2$$

(C_6H_5OH) **Phenol**; ($C_6H_5OCH_3$) **Anisole** **(A)**

(b) The scheme first involves a Grignard reaction with carbon dioxide to produce pivalic acid or 2,2-dimethylpropanoic acid (**B**) (see Section 9.8.2). Thereafter, an Arndt-Eistert synthesis yields the next higher homologue, 3,3-dimethylbutanoic acid (**C**).

t-Butyl chloride —(Mg / Et_2O)→ t-Butylmagnesium chloride —(CO_2, then H_3O^+)→ 2,2-Dimethylpropanoic acid (B) —(PCl_5)→ 2,2-Dimethylpropanoyl chloride —($CH_2=\overset{+}{N}=N^-$ (2 moles))→ An α-diazoketone —(Ag_2O/H_2O, $-N_2$)→ 3,3-Dimethylbutanoic acid (C)

14.7 REACTIONS OF QUATERNARY AMMONIUM SALTS: THE HOFMANN ELIMINATION

The most important reaction of a quaternary ammonium salt is **E2 elimination** (see Section 10.3.1). Usually, a **quaternary ammonium hydroxide** is used for the elimination reaction.

The quaternary ammonium hydroxide provides both the **substrate** (a positively charged quaternary ammonium ion) and the **base** (hydroxide ion). Therefore, *no* external base is required and the elimination is simply accomplished by heating the quaternary ammonium hydroxide.

The **elimination** involves the **loss** of a **tertiary amine** (leaving group) and a **β-hydrogen atom** from the quaternary ammonium ion to produce an **alkene**.

(quaternary ammonium ion; acts as a substrate) (hydroxide ion; acts as a base)

$H-\overset{|}{\underset{|}{C}}-\overset{|}{\underset{|}{C}}-\overset{+}{N}R_3\ \ \overset{-}{O}H$

Quaternary ammonium hydroxide

Elimination:

$HO^- + H-C_\beta-C_\alpha-\overset{+}{N}R_3 \xrightarrow[E2]{\Delta} >C=C< + NR_3 + H_2O$

Quaternary ammonium hydroxide — Alkene — 3° Amine

Preparation of quaternary ammonium hydroxides

Quaternary ammonium hydroxides can be prepared from quaternary ammonium halides (see Section 14.1) and aqueous silver oxide. The precipitated silver halide is removed by filtration.

$$RCH_2CH_2-\overset{+}{N}(CH_3)_3\ X^- \xrightarrow[H_2O]{Ag_2O} RCH_2CH_2-\overset{+}{N}(CH_3)_3\ OH^- + AgX\downarrow$$

(X = Cl, Br, I)

Quaternary ammonium halide → **Quaternary ammonium hydroxide** (β, α carbons marked) + AgX **(silver halide)**

Hofmann elimination

The **major product** from the E2 elimination of a quaternary ammonium hydroxide is the **less substituted alkene** which conforms to the **Hofmann rule** (see Section 10.4). In fact, August W. von Hofmann proposed his rule in 1851 based on his studies of elimination from quaternary ammonium salts.

The following are two examples of the Hofmann elimination of quaternary ammonium hydroxides:

Example 1:

$$HO^- + H^{\beta}-CH_2-CH_2-\overset{+}{N}(CH_2CH_3)_3 \xrightarrow{\Delta} CH_2{=}CH_2 + N(CH_2CH_3)_3 + H_2O$$

Tetraethylammonium hydroxide → **Ethylene** + **Triethylamine**

Example 2:

$$CH_3\underset{(b)\,H^{\beta}}{CH}-\underset{^{+}N(CH_3)_3}{CH}-\underset{H^{\beta}\,(a)}{CH_2}\ OH^- \xrightarrow[-\,N(CH_3)_3,\ -\,H_2O]{\Delta}$$

s-Butyltrimethylammonium hydroxide

Loss of H(*a*) → $CH_3CH_2CH{=}CH_2$ **1-Butene** ***Major product (95%)***

Loss of H(*b*) → $CH_3CH{=}CHCH_3$ **2-Butene** ***Minor product (5%)***

In the first example, **tetraethylammonium hydroxide** has four identical ethyl groups attached to the quaternary nitrogen atom. Thus, heating tetraethylammonium hydroxide produces *only* **ethylene** as the alkene product.

In the second example, ***s*-butyltrimethylammonium hydroxide** possesses two alternative β-hydrogens (marked *a* and *b*) that can be eliminated. The **major** product from the Hofmann elimination is **1-butene** (less substituted alkene).

Hofmann exhaustive methylation or Hofmann degradation

Before modern spectroscopic techniques became available, the elimination of quaternary ammonium hydroxides was widely used in structural elucidation of nitrogen-containing natural products. This method is known as the **Hofmann exhaustive methylation** or the **Hofmann degradation**.

In this method, the amino nitrogen of the compound is exhaustively or completely methylated to form a quaternary ammonium salt. Treatment of the salt with aqueous silver oxide gives the corresponding quaternary ammonium hydroxide which, on heating, accomplishes the Hofmann elimination.

Figure 14.2 illustrates the method involving complete degradation of a cyclic secondary amine (piperidine).

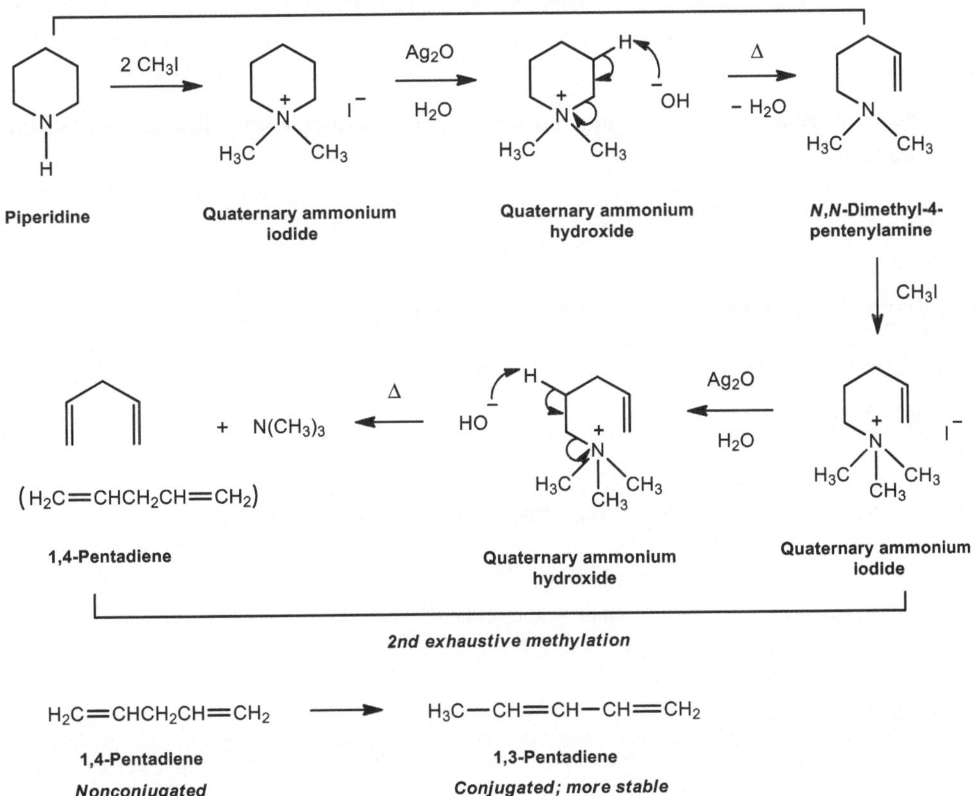

Figure 14.2: Hofmann exhaustive methylation method for complete degradation of piperidene.

The scheme involves *two* exhaustive methylations. The **first exhaustive methylation** cleaves the piperidine ring but an amino group still remains a part of the elimination product, which is ***N,N*-dimethyl-4-pentenylamine.**

Therefore, a **second exhaustive methylation** is required for complete degradation of piperidine. The final *nitrogen-free* elimination product is **1,4-pentadiene.** However, under the conditions, the nonconjugated 1,4-pentadiene can rearrange to a more stable conjugated diene (**1,3-pentadiene**).

14.8 REACTIONS OF TERTIARY AMINE OXIDES: THE COPE ELIMINATION

Oxidation of tertiary amines with hydrogen peroxide (H_2O_2) or a peroxy acid (RCO_3H) gives **tertiary amine oxides.**

$$R_3N: \xrightarrow[H_2O_2 \text{ or } RCO_3H]{[O]} R_3\overset{+}{N}{-}O^-$$

Tertiary amine → **Tertiary amine oxide**

Tertiary amine oxides with a *β-hydrogen atom* with respect to the *N*-oxide group can undergo **pyrolytic elimination** (see Section 10.7). This elimination is called the **Cope elimination.**

The Cope elimination of a tertiary amine oxide produces an **alkene** and a dialkyl-hydroxylamine.

$$R{-}\underset{\beta}{CH}(H){-}\underset{\alpha}{CH_2}{-}\overset{+}{N}(O^-)(CH_3){-}CH_3 \xrightarrow[150^\circ C]{\Delta} RCH{=}CH_2 + (CH_3)_2N{-}OH$$

(leaving group: O^-)

Tertiary amine oxide → **Alkene** + **Dimethylhydroxylamine**

Example

$$CH_3CH_2CH_2N(CH_3)_2 \xrightarrow[CH_3OH]{H_2O_2} CH_3CH(H)CH_2\overset{+}{N}(CH_3)_2(O^-) \xrightarrow{150^\circ C} CH_3CH{=}CH_2 + (CH_3)_2NOH$$

Dimethylpropylamine → **Dimethylpropylamine oxide** → **Propene**

Mechanism

The Cope elimination proceeds by a **pericyclic mechanism** involving cyclic movement of electrons in a single step. The elimination is **syn** because the β-hydrogen atom and amine oxide group depart from the same side of the molecule.

R(H)CH–CH(H)–N⁺(CH₃)₂–O⁻ **Pericyclic mechanism** → **syn Elimination** RCH=CH₂ + $(CH_3)_2NOH$

Tertiary amine oxide **Alkene**

PROBLEM 14.5: How would you accomplish the following conversion?

Cyclohexyl–$CH_2N(CH_3)_2$ → Cyclohexylidene=CH_2

Answer

The substrate (*N,N*-dimethylcyclohexylmethylamine) is a tertiary amine. Oxidation of the tertiary amine with H_2O_2 gives a tertiary amine oxide (*N,N*-dimethylcyclohexylmethylamine oxide) which undergoes the Cope elimination to produce methylenecyclohexane.

Cyclohexyl–$CH_2N(CH_3)_2$ $\xrightarrow{H_2O_2}$ Cyclohexyl(H)–CH_2–$\overset{+}{N}(CH_3)_2$–O^- $\xrightarrow[-\ (CH_3)_2NOH]{\Delta}$ Cyclohexylidene=CH_2

***N,N*-Dimethylcyclohexylmethylamine** ***N,N*-Dimethylcyclohexylmethylamine oxide** **Methylenecyclohexane**

14.9 PHENOLIC ALDEHYDES AND PHENOLIC ACIDS

Phenols are more acidic than alcohols but less acidic than carboxylic acids. The relative acidities of phenols have been described earlier in Chapter 5 (see Section 5.5.2).

A simple **laboratory test** for the **phenolic –OH group** is based on the reaction with aqueous ferric chloride ($FeCl_3$), and is known as **ferric chloride ($FeCl_3$) test**. When *neutral* $FeCl_3$ solution is added to a phenolic compound, the solution turns violet, blue or green. The appearance of the characteristic colour is due to the formation of a complex of the phenolic compound with ferric ion (Fe^{3+}).

$$6\ ArOH + Fe^{3+} \longrightarrow [Fe(OAr)_6]^{3-} + 6\ H^+$$

(Ar = phenyl or substituted phenyl)

Phenolic compound **Complex**

Violet, blue or green colour

Electrophilic aromatic substitution reactions of phenols such as bromination, nitration and sulphonation have been described previously in Chapter 13.

In this section, we will describe the Reimer-Tiemann reaction which can produce phenolic aldehydes, and the Kolbe reaction that produces phenolic acids.

14.9.1 The Reimer-Tiemann Reaction

Phenols react with chloroform ($CHCl_3$) in the presence of aqueous NaOH to produce **phenolic aldehydes**. This reaction is known as the **Reimer-Tiemann reaction**. An *ortho* isomer usually predominates. After the reaction, acidification with aqueous acid (work-up) yields the product.

For example, **phenol** itself undergoes the Reimer-Tiemann reaction to produce **salicylaldehyde** (*o*-hydroxybenzaldehyde).

OH
Phenol
+ $CHCl_3$
NaOH
H_2O
$O^- Na^+$
CHO
H_3O^+
(work-up)
OH
CHO
Salicylaldehyde
(o-hydroxybenzaldehyde)

The Reimer-Tiemann reaction introduces a **formyl group** (CHO) into the benzene ring. This is called **formylation**.

Mechanism

A mechanism of the Reimer-Tiemann reaction is shown in Figure 14.3. The reaction is an electrophilic aromatic substitution.

The **electrophile** is an electron-deficient **dichlorocarbene** ($:CCl_2$) which is a Lewis acid electrophile. Dichlorocarbene is generated from the base-induced elimination of HCl from chloroform. This type of elimination in which two groups (H and Cl) depart from the same carbon atom is called the **1,1- or α-elimination**.

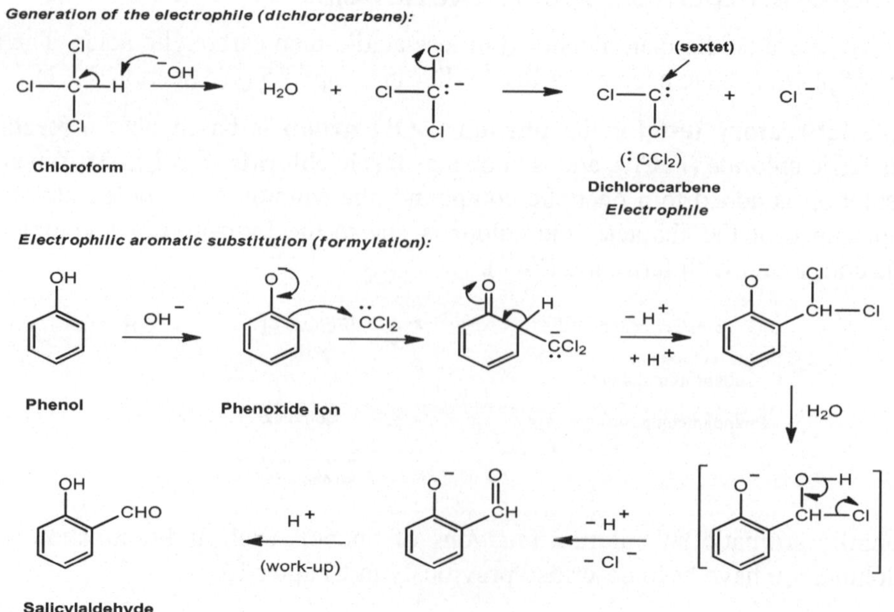

Figure 14.3: A mechanism for the Reimer-Tiemann reaction.

In the presence of a base, phenol is converted to the more nucleophilic or electron-rich **phenoxide ion** which attacks the dichlorocarbene mainly through its *ortho*-position. Proton transfers regenerate the aromaticity of the benzene ring. The $CHCl_2$ group on the ring is finally hydrolysed to CHO under the basic condition.

14.9.2 The Kolbe Reaction

The **Kolbe reaction** provides a method to prepare **phenolic acids** from phenols. The reaction introduces a **carboxyl group** (CO_2H) into the benzene ring, which is called **carboxylation**.

In the Kolbe reaction, the phenoxide ion (which is more reactive than phenol) is used *directly*. For example, **sodium phenoxide** is heated with carbon dioxide under pressure (4-7 atm) to produce predominantly the *ortho* isomer (sodium salicylate). Acidification of the mixture yields **salicylic acid** (*o*-hydroxybenzoic acid).

O^-Na^+ + CO_2 —(125°C, 4-7 atm)→ OH, $CO_2^-Na^+$ —(H_3O^+, (work-up))→ OH, CO_2H

Sodium phenoxide — Sodium salicylate — Salicylic acid (o-hydroxybenzoic acid)

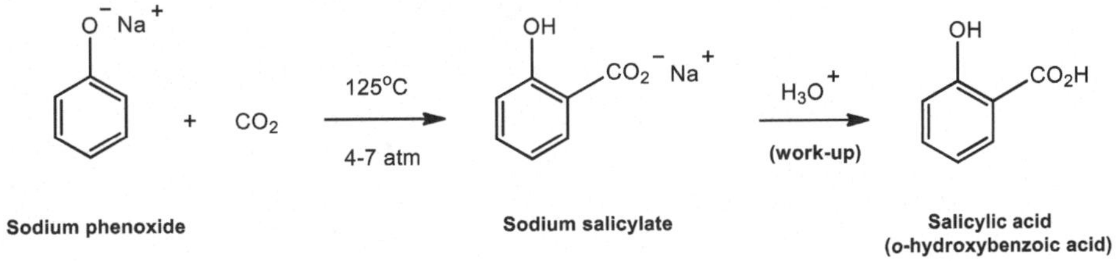

Mechanism

A mechanism for the Kolbe reaction is shown in Figure 14.4.The reaction is an electrophilic aromatic substitution.

The **electrophile** is **carbon dioxide** which is a π bond electrophile. The highly nucleophilic **phenoxide ion** attacks carbon dioxide through its *ortho*-position. Proton transfers regenerate the aromaticity of the benzene ring producing salicylate ion. Acid work-up yields salicylic acid.

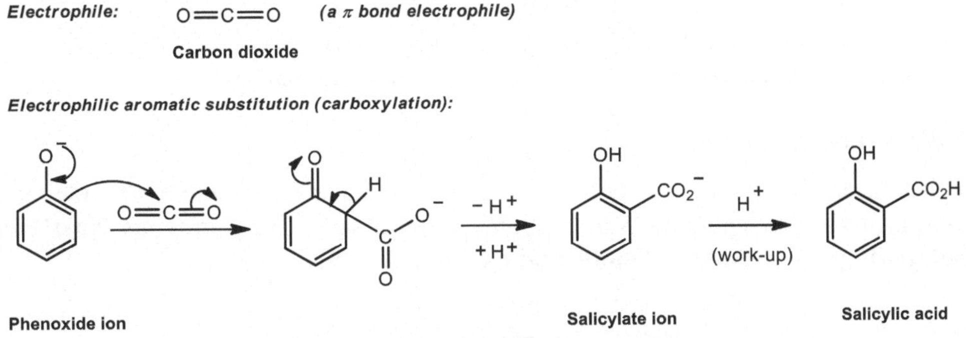

Electrophile: O=C=O *(a π bond electrophile)*

Carbon dioxide

Electrophilic aromatic substitution (carboxylation):

Phenoxide ion + O=C=O → (H, C(=O)O⁻ intermediate) —($-H^+$, $+H^+$)→ OH, CO_2^- —(H^+, (work-up))→ OH, CO_2H

Phenoxide ion — Salicylate ion — Salicylic acid

Figure 14.4: A mechanism for the Kolbe reaction.

14.10 DISTINGUISHING AMINES, PHENOLS AND CARBOXYLIC ACIDS

Amines are *basic* while phenols and carboxylic acids are *acidic*. They can be distinguished by their **solubility behaviour** toward aqueous acids and bases.

Solubility in aqueous acids

Aromatic amines are generally insoluble in water but they dissolve in aqueous hydrochloric acid (HCl) forming a water-soluble salt.

$$\underset{\textbf{Aromatic amines}}{Ar{-}NH_2} + HCl \xrightarrow{H_2O} \underset{\substack{\textbf{Amine salts}\\ \textit{Soluble}}}{Ar{-}\overset{+}{N}H_3\ Cl^-}$$

$Ar{-}OH$	$Ar{-}CO_2H$
Phenols	**Aromatic carboxylic acids**
Not soluble in aqueous HCl	***Not soluble in aqueous HCl***

In contrast, phenols and aromatic carboxylic acids are *not* soluble in aqueous HCl. Thus, the solubility behaviour in aqueous HCl can distinguish aromatic amines from phenols and carboxylic acids.

It is possible to **regenerate the amines** from their acidic solutions by basification with aqueous NaOH. The regenerated amine is insoluble and separates out from the solution. If the aromatic amine is a solid, then it will be precipitated.

$$\underset{\substack{\textbf{Protonated amine}\\ \textit{In solution}}}{Ar{-}\overset{+}{N}H_3} \xrightarrow[\text{(NaOH)}]{OH^-} \underset{\substack{\textbf{Regenerated amine}\\ \textit{Separates out}}}{Ar{-}NH_2} + H_2O$$

Solubility in aqueous bases

Both phenols and aromatic carboxylic acids dissolve in aqueous sodium hydroxide (NaOH) by forming their corresponding water-soluble sodium salts. Thus **aqueous NaOH *alone* cannot distinguish between phenols and carboxylic acids.**

$$\underset{\textbf{Phenols}}{Ar{-}OH} + NaOH \xrightarrow{H_2O} \underset{\substack{\textbf{Sodium phenoxides}\\ \textit{Soluble}}}{Ar{-}O^-\ Na^+} + H_2O$$

$$\mathrm{Ar{-}CO_2H} \quad + \quad \mathrm{NaOH} \xrightarrow{H_2O} \mathrm{Ar{-}CO_2^-\,Na^+} \quad + \quad \mathrm{H_2O}$$

Carboxylic acids — **Sodium carboxylates** *Soluble*

However, carboxylic acids are more acidic than phenols. As such, **carboxylic acids dissolve in aqueous sodium bicarbonate** ($NaHCO_3$) with evolution of carbon dioxide (CO_2). In contrast, **phenols are *not* soluble in aqueous $NaHCO_3$** which is a weaker base than NaOH.

Thus the solubility behaviour in aqueous $NaHCO_3$ can distinguish phenols from carboxylic acids.

$$\mathrm{Ar{-}CO_2H} \quad + \quad \mathrm{NaHCO_3} \xrightarrow{H_2O} \mathrm{Ar{-}CO_2^-\,Na^+} \quad + \quad \mathrm{CO_2}\uparrow \quad + \quad \mathrm{H_2O}$$

Carboxylic acids — **Sodium carboxylates** *Soluble* — ***Used as a test for CO_2H group***

$$\mathrm{Ar{-}OH} \quad + \quad \mathrm{NaHCO_3} \xrightarrow{H_2O}\!\!\!\!\!\!\!\times \quad \text{Phenols are not soluble}$$

Phenols

The evolution of CO_2 gas in the reaction of carboxylic acids with $NaHCO_3$ is used as a **qualitative test for a carboxyl group** (CO_2H).

Notably, **CO_2 is evolved from $NaHCO_3$**, *not* from carboxylic acids (RCO_2H). Besides $-CO_2H$, other functional groups with comparable acidity (e.g. –OH in nitrophenols) or stronger acidity (e.g. $-SO_3H$) will also respond to this test.

It is possible to **regenerate phenols** from their alkaline solutions in NaOH by acidifying with aqueous acid (HCl or H_2SO_4). On the other hand, **carboxylic acids can be regenerated** from their basic solutions in NaOH or $NaHCO_3$ by acidification with aqueous acid.

$$\mathrm{Ar{-}O^-\,Na^+} \xrightarrow{HCl} \mathrm{Ar{-}OH} \quad + \quad \mathrm{NaCl}$$

Sodium phenoxides *In NaOH solution* — **Phenols** *Separates out*

$$\mathrm{Ar{-}CO_2^-\,Na^+} \xrightarrow{HCl} \mathrm{Ar{-}CO_2H} \quad + \quad \mathrm{NaCl}$$

Sodium carboxylates ***In NaOH or $NaHCO_3$ solution*** — **Carboxylic acids** *Separates out*

Note: Some aliphatic amines (e.g. ethylamine) and some aliphatic carboxylic acids (e.g. acetic acid) are *soluble* in water and they will automatically dissolve in aqueous acid or base. So they cannot be distinguished by their solubility behaviour.

SUPPLEMENTARY PROBLEMS

14.6 Write the structures of compounds **A–D** in the following reactions:

(a) C_6H_5–CH_2Br $\xrightarrow{NaN_3}$ **A** $\xrightarrow{LiAlH_4}$ **B**

(b) C_6H_5–C(H)=O + NH_3 $\xrightarrow{H_2/Ni}$ **C**

(c) C_6H_{11}–C(H)=O + $(CH_3)_2NH$ $\xrightarrow{NaBH_3CN}$ **D**

[*Hints:* (a) nucleophilic substitution followed by reduction; (b) reductive amination; (c) reductive amination (the reducing agent is sodium cyanoborohydride, a milder reducing agent than sodium borohydride)]

14.7 How would you carry out the following transformations?

(a) C_6H_5–CH_3 ⟶ HO_2C–C_6H_4–CH_3

Toluene **p-Toluic acid**

(b) O_2N–C_6H_4–NH_2 ⟶ 1,2,3-$C_6H_3Br_3$

p-Nitroaniline **1,2,3-Tribromobenzene**

14.8 Outline a synthesis of the following azo dye.

N=N–C_6H_4–SO_3^- Na^+ (on 2-hydroxynaphthalene, OH)

Orange II

14.9 Write steps for the complete Hofmann degradation of the following compound (2-methylpiperidine).

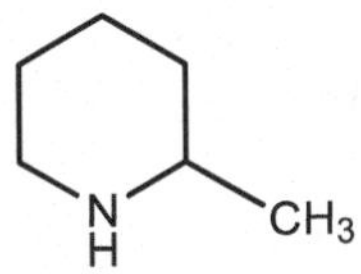

N
H
CH_3

[*Hint:* the nitrogen-free elimination product is 1,4-hexadiene]

14.10 Vanillin is found in vanilla beans and is the main component of the flavour of vanilla. Write the structure of vanillin in the following Reimer-Tiemann reaction:

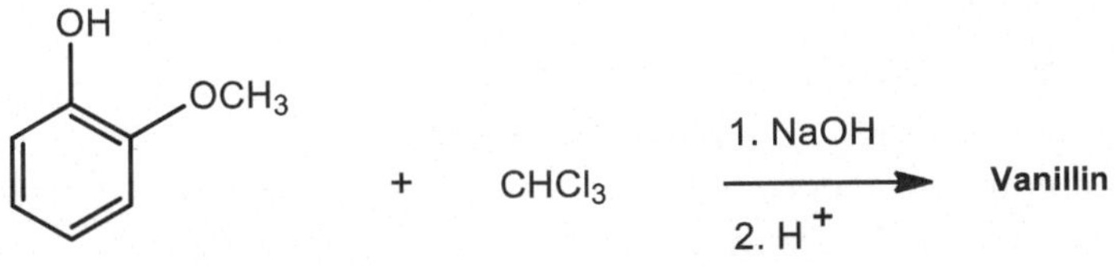

OH
OCH_3

+ $CHCl_3$ $\xrightarrow[2.\ H^+]{1.\ NaOH}$ **Vanillin**

Guaiacol
(*o*-Methoxyphenol)

[*Hint:* formylation occurs at the *para* position]

14.11 A laboratory test for a phenolic compound is known as the **back dye test**. In this test, aniline is diazotized with HCl and $NaNO_2$ under cold conditions and the resulting diazotized solution is then added to a cold alkaline solution of the sample. The appearance of a red azo dye indicates the presence of a phenolic –OH group.

Write the reactions for the back dye test using *p*-cresol (*p*-methylphenol) as the sample.

[*Hint:* coupling occurs at an *ortho* position of the phenolic compound because the *para* position is blocked by a methyl substituent]

CHAPTER 15

Reactions of Enolates: The Aldol and Claisen Reactions

IN THIS CHAPTER

- Keto–enol tautomerism; enolate ions
- α-Halogenation of ketones, haloform reaction and the Hell-Volhard-Zelinsky reaction
- The aldol, crossed aldol and directed aldol reactions
- The aldol condensation and the aldol cyclization
- The Cannizzaro reaction
- The Claisen condensation, crossed Claisen condensations and the Dieckmann cyclization
- Active methylene compounds: Acetoacetic ester and malonic ester
- Acetoacetic ester synthesis of ketones
- Malonic ester synthesis of carboxylic acids
- The Michael reaction and the Robinson annulation
- Barbiturates

We have described previously the reactions of carbonyl compounds that include nucleophilic addition to aldehydes and ketones (Chapter 9) and nucleophilic substitution of carboxylic acids and their derivatives (Chapter 12).

In this chapter, we will describe another important chemistry of **carbonyl compounds** which is based on the **acidity of α hydrogens**. α Hydrogens are the hydrogen atoms that are placed on carbon atoms *adjacent* to the carbonyl group (see Section 5.5.4).

The acidity of α hydrogens gives the **keto–enol tautomerism** and leads to the formation of an **enolate anion** in the presence of a base.

15.1 KETO–ENOL TAUTOMERISM

The reaction whereby an α-hydrogen atom of a carbonyl compound *moves* to the carbonyl oxygen atom is called **enolization** or **keto–enol tautomerism**.

Keto-enol tautomerism:

$$\overset{\alpha}{\text{H}}-\overset{\alpha}{\text{C}}-\text{C}=\text{O} \;\rightleftharpoons\; -\text{C}=\text{C}-\text{O}-\text{H}$$

Keto form **Enol form**

In the term 'keto-enol', **keto** refers to ketones as well as aldehydes, carboxylic acids, acyl halides and esters, and **enol** comes from ene (C=C) + ol (OH). Keto and enol forms are called **tautomers**, and their interconversion is called **tautomerization**.

Note: The keto and enol forms represent *constitutional isomers* and are clearly different from resonance structures that represent a single compound. The tautomerization is a reaction and is catalysed by both acid and base. It rapidly reaches a state of equilibrium.

Aldehydes and ketones

For simple aldehydes and ketones such as acetaldehyde, acetone and cyclohexanone, the **keto form is highly favoured** and the proportion of enol form at equilibrium is very small.

This is mainly because the C=O bond in the keto form is much stronger than the C=C bond in the enol form.

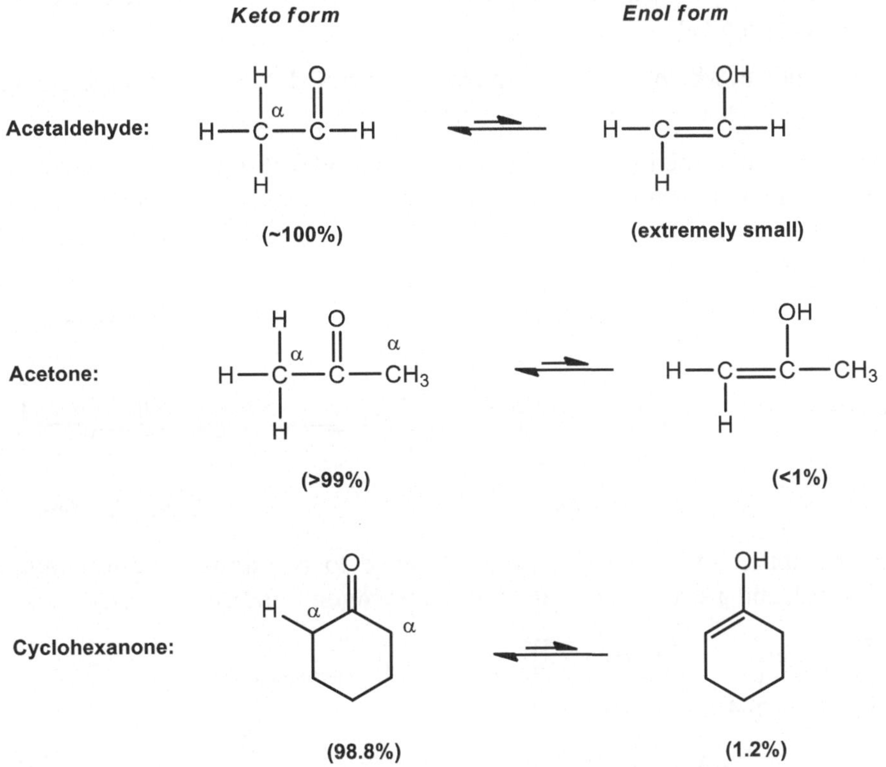

	Keto form		***Enol form***
Acetaldehyde:	$H{-}C(H)_2{-}C({=}O){-}H$ (α carbon)	⇌	$H{-}CH{=}C(OH){-}H$
	(~100%)		**(extremely small)**
Acetone:	$H{-}C(H)_2{-}C({=}O){-}CH_3$ (α carbons)	⇌	$H{-}CH{=}C(OH){-}CH_3$
	(>99%)		**(<1%)**
Cyclohexanone:	cyclohexanone (α carbons marked, α-H shown)	⇌	cyclohex-1-en-1-ol (OH)
	(98.8%)		**(1.2%)**

Note: Acetaldehyde has a single α carbon atom and the α-CH_3 group leads to only one enol form. But there are two α carbon atoms for acetone or cyclohexanone. However, two α-CH_3 groups in acetone are equivalent, so are the two α-CH_2 groups in cyclohexanone. Therefore, a single enol form is possible for acetone or cyclohexanone.

Phenols

When the enol form has the *high stability of an aromatic ring*, the proportion of the keto form is extremely small.

For example, **2,4-cyclohexadienone** exists almost exclusively in the **enol form** which has a benzene ring. The enol form is **phenol**.

Keto form ***Enol form***

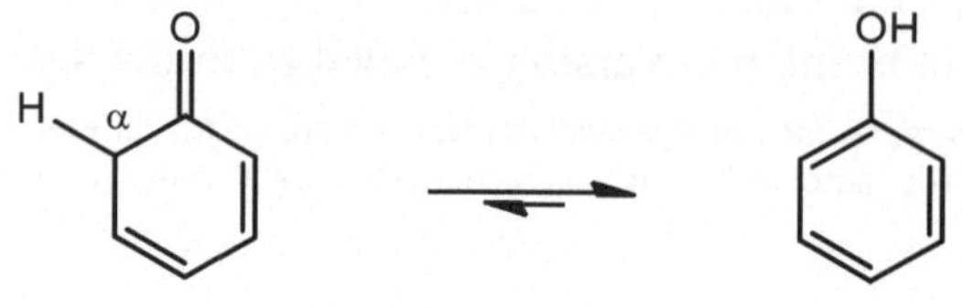

(negligible) (~100%)

2,4-Cyclohexadienone **Phenol**

Active methylene compounds

The acidity of an α-hydrogen atom is greatly enhanced when it is flanked by two carbonyl groups (see Section 5.5.4).

These compounds in which two C=O groups are separated by a methylene group (CH_2) are called **active methylene compounds** (see later Section 15.8).

An example of active methylene compounds is **acetylacetone** (2,4-pentanedione). Here an **enol form** can arise from the movement of an *active methylene hydrogen atom* to the carbonyl oxygen atom. The **enol content at equilibrium is much higher** (76%) than that for a simple ketone (e.g. acetone).

Keto form ***Enol form***

Acetylacetone (2,4-pentanedione)

$CH_3-CO-CH_2-CO-CH_3 \rightleftharpoons CH_3-CO-CH=C(OH)-CH_3$

(active methylene)

(24%) **(76%)**

The greater stability of the enol form is attributed to **resonance** due to conjugation of C=C with C=O, and a strong **intramolecular hydrogen bonding**.

$[CH_3-CO-CH=C(OH)-CH_3 \leftrightarrow CH_3-C(O^-)=CH-C(=OH^+)-CH_3]$

(hydrogen bond)

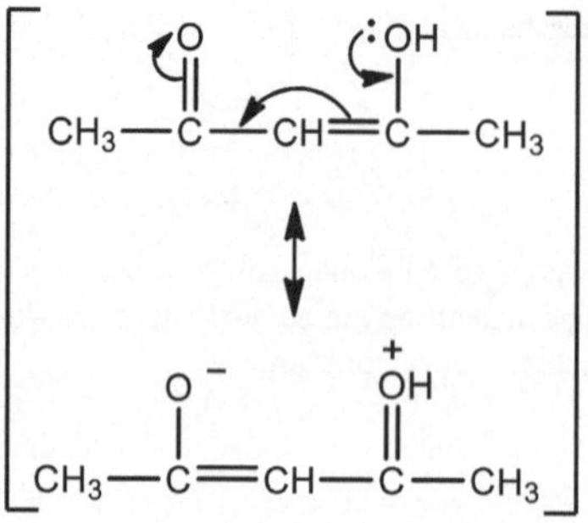

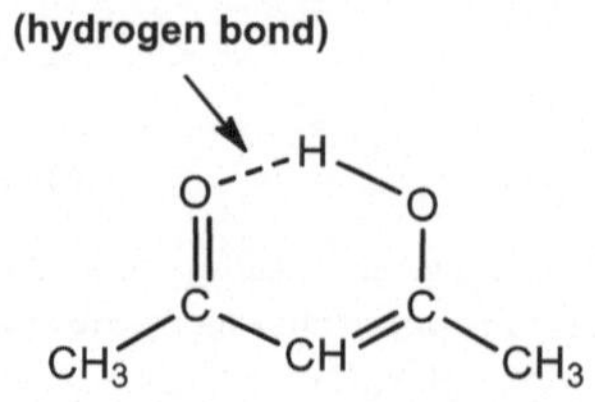

Enol form is stabilized by resonance

Enol form is also stabilized by intramolecular hydrogen bonding

Note: Another enol form of acetylacetone is possible by the movement of a *methyl hydrogen atom* to the carbonyl oxygen. But this enol form resembles the enol form of acetone and hence is present in negligible amount at equilibrium. Since two terminal methyl groups of acetylacetone are equivalent, both lead to this same enol.

Acetylacetone: $CH_2(H)-CO-CH_2-CO-CH_3$ ⇌ $H_2C{=}C(OH)-CH_2-CO-CH_3$

Keto form ***Another enol form*** **(negligible)**

15.1.1 Enolate Ions

When a base abstracts an α proton of carbonyl compounds (aldehydes, ketones or esters), an **α-carbanion** is formed, which is **stabilized by resonance** involving the carbonyl group.

B^- (base) + H–Cα–C=O → (− BH) [$^-$Cα–C=O (α-carbanion) ⟷ –C=C–O$^-$ (anion of enol form)]

Carbonyl compound (aldehyde, ketone or ester)

Enolate ion ***Stabilized by resonance***

In one resonance structure, the negative charge is on **α carbon**, whereas in the other resonance structure, the negative charge is on **oxygen** which represents an **anion of the enol form** of the carbonyl compound.

This type of **α-carbanion** stabilized by resonance with a carbonyl group is called an **enolate anion (or enolate ion)**.

Thus the enolate ion distinguishes itself from other ordinary carbanions that are not stabilized by resonance. Enolate ions can act as *ambident nucleophiles* with two nucleophilic sites, i.e. carbon and oxygen (see Section 7.5).

In this chapter, we will describe some important **reactions of enolate ions** in which they react **through the α carbon atom**.

15.2 HALOGENATION OF KETONES

Ketones with α-hydrogens react with halogens (Cl_2, Br_2 or I_2) in the presence of a *base* or an *acid* to form **α-halo ketones**. The reaction is called **α-halogenation of ketones**.

$$-\overset{O}{\overset{\|}{C}}-\overset{H}{\overset{|}{C}}_{\alpha}- \;+\; X_2 \xrightarrow{OH^-\text{ or }H^+} -\overset{O}{\overset{\|}{C}}-\overset{X}{\overset{|}{C}}_{\alpha}- \;+\; HX$$

Ketone (X = Cl, Br, I) **α-Halo ketone**

Example

$$CH_3-\overset{O}{\overset{\|}{C}}-CH_2H \;+\; Br_2 \xrightarrow[\text{or }H^+]{OH^-} CH_3-\overset{O}{\overset{\|}{C}}-CH_2Br$$

Acetone **Bromoacetone**

Mechanism

In the presence of a base, α-halogenation occurs through the formation of an enolate ion. The enolate ion then reacts as a carbon nucleophile with halogen to give the product (α-halo ketone).

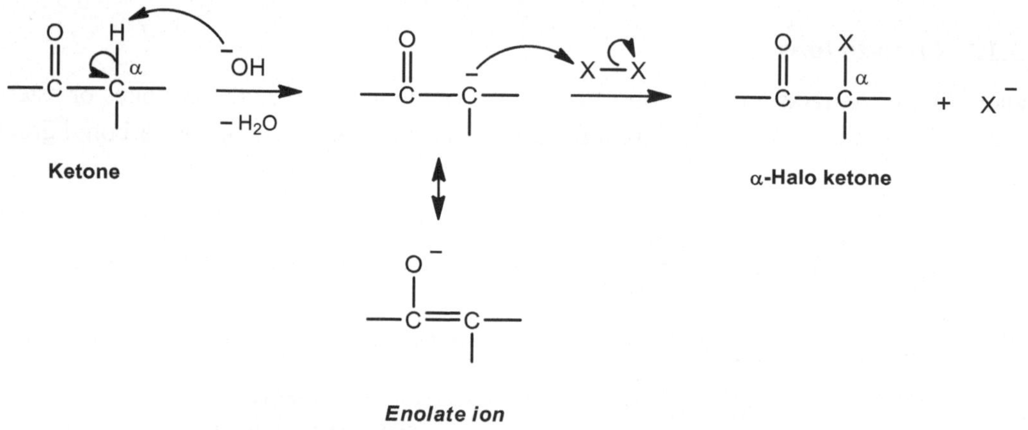

Ketone

Enolate ion

α-Halo ketone

But **in the presence of an acid, α-halogenation proceeds through the formation of an enol.** The enol then reacts with halogen through the α carbon followed by the loss of a proton to give the product.

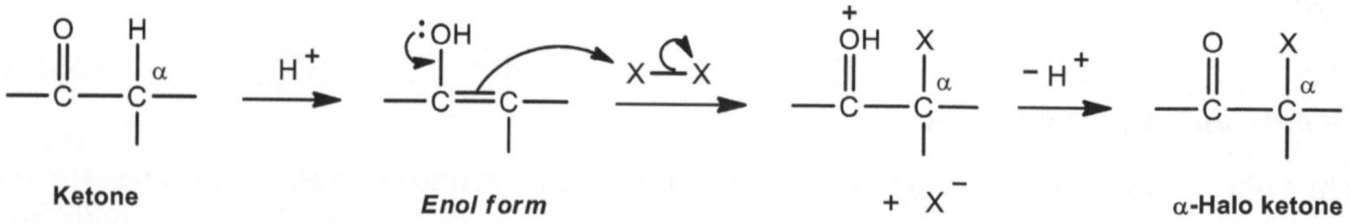

Ketone

Enol form

α-Halo ketone

15.2.1 The Haloform Reaction

In the presence of a *base*, **multiple halogenations** can occur at the **same α-carbon** because the introduction of the halogen atom (an electron-withdrawing substituent) makes the remaining hydrogen atoms at that α-carbon more acidic.

Thus **methyl ketones** (R–$COCH_3$) react with a halogen (e.g. Cl_2) in the presence of an aqueous base to form a **trichloro derivative** (R–$COCCl_3$).

But under basic conditions, the trichloro derivative (RCO–CCl_3) can react further with the base (OH^-) by **nucleophilic substitution at the carbonyl group** (see Section 12.1). This is because *trichloromethyl ion* (CCl_3^-) acts as a *good leaving group* since it is stabilized by dispersal of its negative charge by the three electronegative chlorine atoms. Then a proton transfer produces **chloroform** ($CHCl_3$) and a **carboxylate ion** (RCO_2^-).

(electron-withdrawing)

$R{-}C(=O){-}CH_2{-}H$ (R = H, alkyl or aryl) Methyl ketones $\xrightarrow[\alpha\text{-Halogenation}]{Cl_2/OH^-}$ $R{-}C(=O){-}CHCl{-}H$ (acidity is increased) $\xrightarrow[\alpha\text{-Halogenation}]{Cl_2/OH^-}$ $R{-}C(=O){-}CHCl_2$ $\xrightarrow[\alpha\text{-Halogenation}]{Cl_2/OH^-}$ $R{-}C(=O){-}CCl_3$ Trichloro derivative

$R{-}C(=O){-}CCl_3$ $\xrightarrow[\text{Nucleophilic substitution}]{OH^-}$ $R{-}C(=O){-}OH$ + $^-CCl_3$ (A good leaving group) $\xrightarrow[\text{Proton transfer}]{-H^+, +H^+}$ $R{-}C(=O){-}O^-$ (Carboxylate ion) + $CHCl_3$ (Chloroform)

In general, the whole process is known as the **haloform reaction**. Besides chloroform, **bromoform** ($CHBr_3$) and **iodoform** (CHI_3) can also be produced by treating methyl ketones with bromine and iodine respectively in the presence of aqueous sodium hydroxide.

$$R{-}C(=O){-}CH_3 + 3\,Br_2 + 4\,NaOH \longrightarrow R{-}C(=O){-}ONa + CHBr_3 + 3\,NaBr + 3\,H_2O$$

Methyl ketones — Sodium carboxylate — Bromoform

$$R{-}C(=O){-}CH_3 + 3\,I_2 + 4\,NaOH \longrightarrow R{-}C(=O){-}ONa + CHI_3 + 3\,NaI + 3\,H_2O$$

Methyl ketones — Sodium carboxylate — Iodoform

Synthetic application

An important synthetic application of the haloform reaction is to **convert a methyl ketone to a carboxylic acid**.

For example, when **acetophenone** ($C_6H_5COCH_3$) is reacted with I_2 and aqueous NaOH, **benzoic acid** ($C_6H_5CO_2H$) is produced after acid work-up.

$$C_6H_5{-}C(=O){-}CH_3 \xrightarrow[2.\ H_3O^+\ \text{(work-up)}]{1.\ I_2/NaOH} C_6H_5{-}C(=O){-}OH + CHI_3$$

Acetophenone — Benzoic acid

15.2.2 Iodoform Test

The **iodoform reaction** with iodine and aqueous sodium hydroxide is used as a **test** for methyl ketones (R–$COCH_3$).

The appearance of a bright yellow precipitate of iodoform (CHI_3) indicates a **positive iodoform test** for the presence of $COCH_3$ (acetyl) group.

$$-\overset{O}{\overset{\|}{C}}-CH_3 \xrightarrow[NaOH]{I_2} -\overset{O}{\overset{\|}{C}}-ONa + CHI_3\downarrow$$

Yellow precipitate

Besides methyl ketones, **alcohols containing $CHOHCH_3$ group** such as ethanol (CH_3CH_2OH) and isopropanol ($CH_3CHOHCH_3$) will give a **positive iodoform test.**

This is because I_2 and aqueous NaOH form an oxidizing solution which can **oxidize** $CHOHCH_3$ group to $COCH_3$ group. (Remember that halogens are oxidizing agents.)

$$-\overset{OH}{\overset{|}{C}H}-CH_3 \xrightarrow{I_2,\ 2\ OH^-} \left[-\overset{O}{\overset{\|}{C}}-CH_3\right] + 2\,I^- + 2\,H_2O \xrightarrow[OH^-]{I_2} -\overset{O}{\overset{\|}{C}}-O^- + CHI_3\downarrow$$

Iodoform

Examples

$$H-\overset{OH}{\overset{|}{C}H}CH_3 \xrightarrow[NaOH]{I_2} \left[H-\overset{O}{\overset{\|}{C}}CH_3\right] \longrightarrow CHI_3 + HCO_2Na$$

(CH_3CH_2OH) Ethanol; (acetadehyde); Iodoform

$$H_3C-\overset{OH}{\overset{|}{C}H}CH_3 \xrightarrow[NaOH]{I_2} \left[H_3C-\overset{O}{\overset{\|}{C}}CH_3\right] \longrightarrow CHI_3 + CH_3CO_2Na$$

($CH_3CHOHCH_3$) Isopropanol; (acetone); Iodoform

It is of note that a halogen (X_2) dissolved in aqueous NaOH produces a solution containing sodium hypohalite (NaOX), and this solution is also called a *hypohalite solution.*

PROBLEM 15.1: Which of the following compounds would give a positive iodoform test?
1-Butanol 2-Butanol Butanone 3-Pentanone 1-Phenylethanol

Answer

Only compounds containing $COCH_3$ group or $CHOHCH_3$ group will give a positive iodoform test. Thus butanone which has $COCH_3$ group and 2-butanol and 1-phenylethanol (each of which possesses $CHOHCH_3$ group) will give a positive iodoform test.

But 1-butanol or 3-pentanone does not have either $COCH_3$ or $CHOHCH_3$ group and hence they will give a negative iodoform test.

Compounds that give positive iodoform test:

$CH_3CH_2C(=O)CH_3$ — **Butanone**

$CH_3CH_2CH(OH)CH_3$ — **2-Butanol**

$C_6H_5CH(OH)CH_3$ — **1-Phenylethanol**

Compounds that give negative iodoform test:

$CH_3CH_2CH_2CH_2OH$ — **1-Butanol**

$CH_3CH_2C(=O)CH_2CH_3$ — **3-Pentanone**

15.3 HALOGENATION OF CARBOXYLIC ACIDS: THE HELL-VOLHARD-ZELINSKY REACTION

Carboxylic acids with α-hydrogens react with bromine or chlorine in the presence of phosphorus to form **α-halo acids.**

This reaction for α-halogenation of carboxylic acids is known as the Hell-Volhard-Zelinsky (HVZ) reaction.

$$\underset{\textbf{Carboxylic acid}}{-\overset{H}{\underset{|}{\overset{|}{C}}}{}_{\alpha}-\overset{O}{\overset{\|}{C}}-OH} + \underset{(X = Br, Cl)}{X_2} \xrightarrow{P} \left[-\overset{X}{\underset{|}{\overset{|}{C}}}{}_{\alpha}-\overset{O}{\overset{\|}{C}}-X\right] \xrightarrow{H_2O} \underset{\alpha\textbf{-Halo acid}}{-\overset{X}{\underset{|}{\overset{|}{C}}}{}_{\alpha}-\overset{O}{\overset{\|}{C}}-OH}$$

Example

$$\underset{\textbf{Acetic acid}}{\overset{\alpha}{CH_3}CO_2H} \xrightarrow[2.\ H_2O]{1.Br_2/P} \underset{\textbf{Bromoacetic acid}}{\overset{Br}{\overset{|}{C}}H_2CO_2H}$$

Mechanism

A mechanism for the HVZ reaction is shown in Figure 15.1.

The reaction of phosphorus with bromine forms phosphorus tribromide (PBr_3) which converts a carboxylic acid to the corresponding acyl bromide. Importantly, an **acyl bromide can form an enol** but a carboxylic acid itself does not form an enol easily.

The **enol form** of the acyl bromide then reacts with bromine to give an **α-bromo acyl bromide**. Subsequent hydrolysis with water produces an **α-bromo acid**.

$$RCH_2{-}C(=O){-}OH \xrightarrow[(P + Br_2)]{PBr_3} RCH(H){-}C(=O){-}Br \rightleftharpoons RCH{=}C(OH){-}Br$$

Carboxylic acid → **Acyl bromide** ⇌ ***Enol form***

Bromination ($Br{-}Br$)

$$RCH(Br){-}C(=\overset{+}{O}H){-}Br + Br^- \xrightarrow{-H^+} RCH(Br){-}C(=O){-}Br \xrightarrow[\textit{Hydrolysis}]{H_2O} RCH(Br){-}C(=O){-}OH$$

α-Bromo acid ← **α-Bromo acyl bromide**

Figure 15.1: A mechanism for the Hell-Volhard-Zelinsky (HVZ) reaction.

Synthetic applications

α-Bromo acids or α-chloro acids are useful intermediates in the synthesis of **α-hydroxy acids** and **α-amino acids.**

The following examples illustrate the synthesis of **lactic acid** (an α-hydroxy acid) from propionic acid and the synthesis of **glycine** (an α-amino acid) from acetic acid.

$$CH_3CH_2CO_2H \xrightarrow[2.\ H_2O]{1.\ Br_2/P} CH_3CH(Br)CO_2H \xrightarrow[2.\ H^+\ (\text{work-up})]{1.\ OH^-} CH_3CH(OH)CO_2H$$

Propionic acid → **α-Bromopropionic acid** → **Lactic acid**

$$CH_3CO_2H \xrightarrow[2.\ H_2O]{1.\ Cl_2/P} Cl{-}CH_2CO_2H \xrightarrow[(\text{excess})]{NH_3} H_2N{-}CH_2CO_2H$$

Acetic acid → **Chloroacetic acid** → **Glycine**

15.4 THE ALDOL REACTION: ADDITION OF ENOLATE IONS TO ALDEHYDES AND KETONES

In general, an enolate ion as a nucleophile can add to the electrophilic carbonyl group (C=O) of an aldehyde or a ketone to form an addition product.

The **nucleophilic addition of an enolate ion** to aldehydes or ketones is called the **aldol reaction**. Here the **enolate ion** is derived from an aldehyde or a ketone that possesses **α-hydrogen atoms**.

The product of the aldol reaction is a **β-hydroxy aldehyde or ketone**. As the product is both an **ald**ehyde/ketone and an alcoh**ol**, it is called an **aldol**.

General reaction:

Aldehyde or ketone + H—En (En = enolate) —Addition→ Product

Aldol reaction:

Aldehyde or ketone + Aldehyde or ketone with α-hydrogen (H—En) → β-Hydroxy aldehyde or ketone *Aldol*

Example 1

When **acetaldehyde** (CH_3CHO) is treated with *dilute* sodium hydroxide (NaOH) at room temperature, the reaction produces **3-hydroxybutanal** ($CH_3CHOHCH_2CHO$) as an **aldol product**. The aldol is a **β-hydroxy aldehyde**.

The reaction involves *two* molecules of the same aldehyde (acetaldehyde) that has α hydrogens. *One molecule* of the aldehyde acts as the **source of enolate ion** by the loss of an α hydrogen, while *another molecule* acts as the **acceptor** of the enolate ion to the carbonyl group (C=O). The addition of two molecules gives the aldol.

$$2\,CH_3CHO \xrightarrow[H_2O]{NaOH} CH_3CH(OH)CH_2CHO$$

Acetaldehyde → 3-Hydroxybutanal, *A β-hydroxy aldehyde or aldol*

⇓

$$CH_3CHO + CH_3CHO \xrightarrow[H_2O]{OH^-} CH_3-CH(OH)-CH_2-C(=O)-H$$

Acceptor (CH_3CHO) + *Enolate ion source* (CH_3CHO) → *Aldol*

Example 2

The aldol reaction is *reversible.* For ketones which are less reactive than aldehydes, the equilibrium for the aldol reaction is usually unfavourable. This problem can be overcome by carrying out the reaction in a special apparatus (Soxhlet apparatus) that allows the removal of the aldol product from contact with the base as it is formed.

By this method, **acetone** is converted to **4-hydroxy-4-methyl-2-pentanone** as an **aldol product** in the presence of barium hydroxide [$Ba(OH)_2$] in good yield (80%). Here the aldol is a **β-hydroxy ketone**.

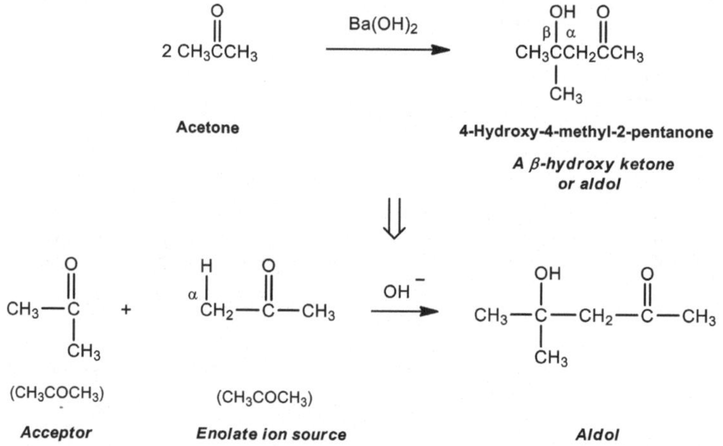

$$2\ CH_3COCH_3 \xrightarrow{Ba(OH)_2} CH_3C(OH)(CH_3)CH_2COCH_3$$

Acetone → 4-Hydroxy-4-methyl-2-pentanone (*A β-hydroxy ketone or aldol*)

⇓

$$CH_3COCH_3 + CH_3COCH_3 \xrightarrow{OH^-} CH_3{-}C(OH)(CH_3){-}CH_2{-}CO{-}CH_3$$

(CH_3COCH_3) *Acceptor* + (CH_3COCH_3) *Enolate ion source* → *Aldol*

Mechanism

A mechanism for the **base-catalysed aldol reaction** using acetaldehyde is shown in Figure 15.2.

The reaction proceeds through the formation of an **enolate ion** from **one molecule of acetaldehyde** and base (hydroxide ion). Then the enolate ion as a nucleophile attacks the carbonyl group (C=O) of **another molecule of acetaldehyde**. The resulting alkoxide ion abstracts a proton from water to give the aldol.

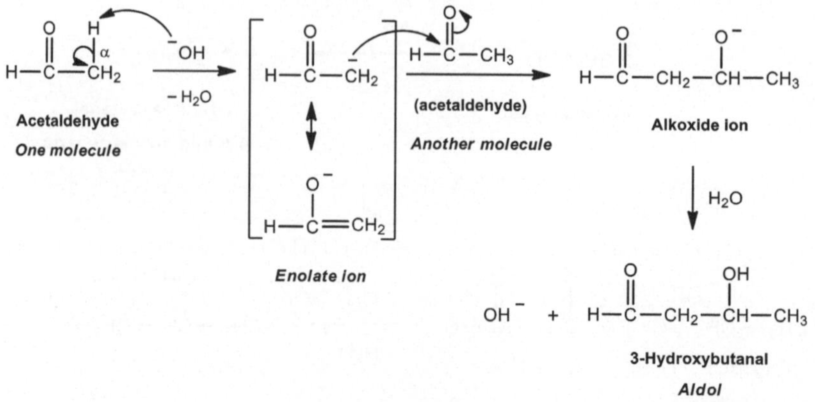

$$H{-}CO{-}CH_3 \xrightarrow[-H_2O]{^-OH} \left[H{-}CO{-}\bar{C}H_2 \leftrightarrow H{-}C(O^-){=}CH_2\right] \xrightarrow{H{-}CO{-}CH_3} H{-}CO{-}CH_2{-}CH(O^-){-}CH_3 \xrightarrow{H_2O} OH^- + H{-}CO{-}CH_2{-}CH(OH){-}CH_3$$

Acetaldehyde (*One molecule*); *Enolate ion*; (acetaldehyde) *Another molecule*; Alkoxide ion; 3-Hydroxybutanal (*Aldol*)

Figure 15.2: A mechanism for the base-catalysed aldol reaction.

15.4.1 Crossed Aldol Reactions

An aldol reaction with *two different* aldehydes or ketones is termed a **crossed aldol reaction**.

For example, consider an aldol reaction using both **acetaldehyde** (CH_3CHO) and **propionaldehyde** (CH_3CH_2CHO) in aqueous NaOH solution. The aldol reaction may involve two *identical* aldehyde molecules, called a **self-aldol** reaction or two *different* aldehyde molecules, called a **crossed aldol** reaction.

Between two identical (or different) aldehyde molecules, one molecule acts as a **source of enolate ion** by the loss of an α proton, while the other molecule acts as an **acceptor** of the enolate ion to its carbonyl group (C=O).

As a result, **two** self-aldol reactions and **two** crossed aldol reactions are possible, resulting in the formation of a complex mixture of **four aldol products**. This type of aldol reactions is of little synthetic value.

$$\underset{\text{Acetaldehyde}}{CH_3CHO} + \underset{\text{Propionaldehyde}}{CH_3CH_2CHO} \xrightarrow[H_2O]{NaOH} \text{A mixture of four aldols}$$

Acceptor ***Enolate ion source***

$$\underset{\text{Acetaldehyde}}{CH_3\overset{O}{\overset{\|}{C}}H + \overset{H}{\overset{|\alpha}{C}}H_2CHO} \xrightarrow{\textit{Self-aldol}} CH_3\overset{OH}{\overset{|}{C}}HCH_2CHO \quad \text{(3-hydroxybutanal)}$$

$$\underset{\text{Propionaldehyde}}{CH_3CH_2\overset{O}{\overset{\|}{C}}H + CH_3\overset{H}{\overset{|\alpha}{C}}HCHO} \xrightarrow{\textit{Self-aldol}} CH_3CH_2\overset{OH}{\overset{|}{C}}H\underset{CH_3}{\underset{|}{C}}HCHO \quad \text{(3-hydroxy-2-methylpentanal)}$$

$$\underset{\text{Acetaldehyde}}{CH_3\overset{O}{\overset{\|}{C}}H} + \underset{\text{Propionaldehyde}}{CH_3\overset{H}{\overset{|\alpha}{C}}HCHO} \xrightarrow{\textit{Crossed aldol}} CH_3\overset{OH}{\overset{|}{C}}H\underset{CH_3}{\underset{|}{C}}HCHO \quad \text{(3-hydroxy-2-methylbutanal)}$$

$$\underset{\text{Propionaldehyde}}{CH_3CH_2\overset{O}{\overset{\|}{C}}H} + \underset{\text{Acetaldehyde}}{\overset{H}{\overset{|\alpha}{C}}H_2CHO} \xrightarrow{\textit{Crossed aldol}} CH_3CH_2\overset{OH}{\overset{|}{C}}HCH_2CHO \quad \text{(3-hydroxypentanal)}$$

Aldehydes with no α hydrogens as one component

When *one* of the aldehydes has **no α hydrogen atom**, it **cannot act as a source of enolate ion** and therefore it can act only as an **acceptor**.

For example, consider a crossed aldol reaction using **formaldehyde** (with no α H atom) and **isobutyraldehyde** (with an α H atom).

Here only **isobutyraldehyde** can act as a **source of enolate ion**. Although both isobutyraldehyde and formaldehyde can act as an acceptor, formaldehyde is a **much better acceptor** because the carbonyl group (C=O) in formaldehyde is much more electrophilic.

Thus the crossed aldol reaction between formaldehyde (acceptor) and isobutyraldehyde (enolate ion source) produces the **major** aldol product (3-hydroxy-2,2-dimethylpropanal).

$$HCH(=O) + CH_3C(H)(CH_3)CHO \xrightarrow[\text{Crossed aldol}]{NaOH/H_2O} CH_2(OH)C(CH_3)_2CHO$$

Formaldehyde *Acceptor* + **Isobutyraldehyde** (α H) *Enolate ion source* → **3-Hydroxy-2,2-dimethylpropanal** *Major product*

PROBLEM 15.2: Give the structure of the product in the following crossed aldol reaction:

$$\underset{\text{(excess)}}{HCHO} + CH_3CHO \xrightarrow[H_2O]{Na_2CO_3}$$

Answer

A crossed aldol reaction between formaldehyde (acceptor) and acetaldehyde (enolate ion source) in aqueous Na_2CO_3 gives 3-hydroxypropanal ($HOCH_2CH_2CHO$).

But formaldehyde is highly reactive and tends to react further with 3-hydroxypropanal when formaldehyde is present in excess. 3-Hydroxypropanal possesses two α H atoms and can act as a source of enolate ion. The second crossed aldol reaction between 3-hydroxypropanal (enolate ion source) and formaldehyde (acceptor) gives a product that still possesses one α H atom. Thus a third crossed aldol reaction with formaldehyde is possible which gives the final aldol product called **pentaerythrose**.

Since pentaerythrose has no α-hydrogen atom, it cannot undergo further aldol reaction with formaldehyde.

$$\underset{\text{Formaldehyde}}{\text{HCH}(=\text{O})} + \underset{\text{Acetaldehyde}}{\text{CH}_2(\text{H}^{\alpha})\text{CHO}} \xrightarrow[\textit{First aldol}]{\text{Base}} \text{CH}_2(\text{OH})\text{CH}_2\text{CHO} \equiv \underset{\text{3-Hydroxypropanal}}{\text{HOCH}_2\text{CH}(\text{H}^{\alpha})\text{CHO}}$$

$$\text{3-Hydroxypropanal} \xrightarrow[\textit{Second aldol}]{\text{CH}_2\text{=O, base}} \text{HOCH}_2\text{C}(\text{CH}_2\text{OH})(\text{H}^{\alpha})\text{CHO} \xrightarrow[\textit{Third aldol}]{\text{CH}_2\text{=O, base}} \underset{\text{Pentaerythrose}}{\text{HOCH}_2\text{C}(\text{CH}_2\text{OH})_2\text{CHO}}$$

Note: Aqueous Na_2CO_3 is a weak base; under strongly basic condition in aqueous NaOH, pentaerythrose can undergo a crossed Cannizzaro reaction with formaldehyde (see Problem 15.5).

15.4.2 Directed Aldol Reactions

It is possible to direct a crossed aldol reaction to proceed in *one direction* only to yield a **single aldol product**. This is called a **directed aldol reaction.**

The formation of enolate is a reversible reaction and the amount of enolate formed depends on the **strength of the base** employed. For example, when acetone is treated with aqueous NaOH, the equilibrium lies to the left and the *amount of enolate formed is small.*

$$\underset{\text{Acetone}}{\text{CH}_3\text{—C}(=\text{O})\text{—CH}_3} + \text{Na}^+\text{OH}^- \rightleftharpoons \underset{\text{Sodium enolate}}{\text{CH}_3\text{—C}(\text{O}^-\text{Na}^+)\text{=CH}_2} + \text{H}_2\text{O}$$

But if acetone is treated with a *very strong base*, the **equilibrium can lie far to the right.**

Lithium diisopropylamide (LDA) as a very strong base

Lithium diisopropylamide (LDA) is a very strong base. It can be prepared by treating diisopropyl amine with methyllithium in an ether solvent such as tetrahydrofuran (THF).

$$\underset{\substack{(i\text{-Pr}_2\text{NH}) \\ \text{Diisopropylamine}}}{(\text{CH}_3)_2\text{CH—NH—CH(CH}_3)_2} + \underset{\substack{(\text{MeLi}) \\ \text{Methyllithium}}}{\text{CH}_3\text{—Li}} \xrightarrow{\text{THF}} \underset{\substack{(i\text{-Pr}_2\text{N}^-\text{Li}^+) \\ \text{Lithium diisopropylamide} \\ \text{(LDA)}}}{(\text{CH}_3)_2\text{CH—N}^-(\text{CH(CH}_3)_2)\ \text{Li}^+} + \text{CH}_4$$

When **acetone** is treated with **LDA** in THF, the reaction is essentially irreversible and the formation of **lithium enolate** is almost quantitative.

$$CH_3COCH_3 + [(CH_3)_2CH]_2N^-Li^+ \xrightarrow{THF} CH_3C(O^-Li^+){=}CH_2 + [(CH_3)_2CH]_2NH$$

Acetone **Lithium diisopropylamide (LDA)** **Lithium enolate**

Use of preformed lithium enolates in directed aldol reactions

To bring about a **directed aldol reaction**, a **lithium enolate** is first produced from a *specific* aldehyde (or ketone) with LDA at a low temperature in tetrahydrofuran (THF). Then a **second aldehyde (or ketone)** is added which reacts with the **preformed enolate** to undergo an aldol reaction. The product is initially a **lithium alkoxide**, which gives the **aldol** by treatment with water.

To illustrate, consider the following synthesis of a target aldol, **4-hydroxy-2-pentanone** using acetone and acetaldehyde.

$$CH_3COCH_3 \xrightarrow[\text{THF, -78°C}]{i\text{-}Pr_2N^-Li^+ \text{ (LDA)}} CH_3C(O^-Li^+){=}CH_2 \xrightarrow{H{-}CO{-}CH_3} CH_3COCH_2{-}CH(O^-Li^+)CH_3 \xrightarrow[\text{(work-up)}]{H_2O} CH_3COCH_2CH(OH)CH_3$$

Acetone

(first form the enolate of acetone)

Acetaldehyde

(then add acetaldehyde)

Lithium alkoxide

4-Hydroxy-2-pentanone

Target aldol

Note: The target aldol (4-hydroxy-2-pentanone) cannot be prepared by the usual crossed aldol reaction with acetone and acetaldehyde in aqueous NaOH. This is because a self-aldol reaction of the more reactive aldehyde (acetaldehyde) would take place primarily, which would give 3-hydroxybutanal as the predominant aldol in a mixture of products.

A directed aldol reaction provides one of the most effective ways to design a crossed aldol reaction in synthesis.

PROBLEM 15.3: Predict the product of the following directed aldol reaction:

$$CH_3CH_2COCH_3 \xrightarrow[\text{THF, -78°C}]{LDA} \xrightarrow{CH_3CHO} \xrightarrow{H_2O}$$

Answer

Butanone is an unsymmetrical ketone. It can produce two possible enolates via abstraction of an α-hydrogen atom from the methyl group (CH_3) or from the methylene group (CH_2).

Abstraction of α-H from CH_3 group → $CH_3CH_2{-}C(O^-){=}CH_2$ *Kinetic enolate*

Butanone: $CH_3CH_2{-}C({=}O){-}CH_3$ (α carbons on both sides) —Base→

Abstraction of α-H from CH_2 group → $CH_3CH{=}C(O^-){-}CH_3$ *Thermodynamic enolate*

The enolate with the more substituted double bond is thermodynamically more stable, and is called the **thermodynamic enolate**. On the other hand, the enolate with the less substituted double bond is formed faster because its formation involves the abstraction of the less hindered hydrogen atom from the less substituted methyl α carbon. This enolate is called the **kinetic enolate**.

LDA is a strong but sterically hindered base with bulky isopropyl substituents on the basic nitrogen atom. Thus LDA removes the *less hindered* methyl hydrogen atom faster because the removal of the more hindered methylene hydrogen atom encounters a severe steric hindrance. As a result, **the formation of the kinetic enolate is favoured using LDA as the base.**

The kinetic enolate from butanone then reacts with acetaldehyde to form a lithium alkoxide, which yields the aldol (5-hydroxy-3-hexanone) on aqueous work-up.

$CH_3CH_2C({=}O)CH_3$ (Butanone) —LDA, THF, -78°C→ $CH_3CH_2{-}C(O^-Li^+){=}CH_2$ (*Kinetic enolate*) —$CH_3CH{=}O$ (Acetaldehyde)→ $CH_3CH_2C({=}O)CH_2CH(O^-Li^+)CH_3$ (A lithium alkoxide) —H_2O→ $CH_3CH_2C({=}O)CH_2CH(OH)CH_3$ (5-Hydroxy-3-hexanone)

15.5 THE ALDOL CONDENSATION: ADDITION-DEHYDRATION

Aldols contain an alcohol OH group. Therefore, **dehydration** of aldols can occur. When heated in base, aldols undergo dehydration to produce **α,β-unsaturated aldehydes or ketones.**

The loss of water is favoured by the formation of a *more stable conjugated product* in which C=C is conjugated with C=O.

$$2\ CH_3CHO \xrightarrow[10^\circ C]{NaOH/H_2O} CH_3CH(OH)-CH(H)CHO \xrightarrow[\Delta,\ -H_2O]{NaOH/H_2O} CH_3CH\overset{\beta}{=}\overset{\alpha}{CH}-C(=O)-H$$

Aldol addition — *Dehydration*

Acetaldehyde — **3-Hydroxybutanal** — **Crotonaldehyde (2-butenal)** *α,β-Unsaturated aldehyde*

If the α,β-unsaturated product is particularly stable (for example, when a benzene ring is a part of the conjugated system) dehydration occurs so readily that the unsaturated compound is normally the product isolated. In such a case, the reaction is called an **aldol condensation** since a condensation reaction involves combination of two molecules with elimination of a small molecule such as water.

Example

Consider an aldol condensation between benzaldehyde and acetaldehyde in aqueous sodium hydroxide. The product isolated is **cinnamaldehyde** ($C_6H_5CH{=}CHCHO$).

Here the α,β-unsaturated product is more stabilized by additional conjugation of C=C with the phenyl ring.

$$C_6H_5-C(=O)-H + CH_3CHO \xrightarrow[25^\circ C]{NaOH/H_2O} [C_6H_5-CH(OH)-CH_2CHO] \xrightarrow{-H_2O} C_6H_5-CH{=}CH-C(=O)-H$$

Benzaldehyde — **Acetaldehyde** — *(not isolated)*

Cinnamaldehyde
Aldol condensation product

Claisen-Schmidt reaction

The **Claisen-Schmidt reaction** is a crossed aldol condensation in which a **ketone having α-hydrogen atoms is used as one of the partners**. The reaction is named after the German chemists Ludwig Claisen and J. G. Schmidt.

The reaction is useful in synthesis because a ketone does not self-condense appreciably under the reaction conditions.

For example, the Claisen-Schmidt reaction between benzaldehyde and acetone in aqueous sodium hydroxide yields **benzalacetone** (4-phenyl-3-buten-2-one).

$$C_6H_5\text{-}CH{=}O + CH_3\text{-}\overset{O}{\overset{\|}{C}}\text{-}CH_3 \xrightarrow[-\,H_2O]{NaOH/H_2O} C_6H_5\text{-}CH{=}CH\text{-}\overset{O}{\overset{\|}{C}}\text{-}CH_3$$

Benzaldehyde **Acetone** **Benzalacetone (4-phenyl-3-buten-2-one)**

Reduction of aldols and α,β-unsaturated carbonyl compounds

Besides a hydroxyl (OH) group, an **aldol** also contains an aldehyde (CHO) or a keto (CO) group which can be easily **reduced**. The product is a **1,3-diol**.

For example, 3-hydroxybutanal (an aldol) can be reduced with $NaBH_4$ to produce 1,3-butanediol.

$$CH_3\overset{OH}{\overset{|}{C}}HCH_2\overset{O}{\overset{\|}{C}}H \xrightarrow{NaBH_4} CH_3\overset{OH}{\overset{|}{C}}HCH_2\overset{OH}{\overset{|}{C}}H_2$$

3-Hydroxybutanal *Aldol* **1,3-Butanediol** *1,3-Diol*

α,β-Unsaturated carbonyl compounds can also be easily converted to other compounds by **reduction**.

This is illustrated with the reduction of 2-butenal (an α,β-unsaturated aldehyde) with a variety of reducing agents to form different products:

$CH_3CH{=}CH\text{-}CH{=}O$ **2-Butenal** *α,β-Unsaturated aldehyde*

$\xrightarrow{LiAlH_4}$ $CH_3CH{=}CH\text{-}CH_2OH$ **2-Buten-1-ol** *Allylic alcohol*

$\xrightarrow{H_2/Pd\text{-}C}$ $CH_3CH_2\text{-}CH_2\text{-}CH{=}O$ **Butanal** *Saturated aldehyde*

$\xrightarrow[\text{high pressure}]{H_2/Ni}$ $CH_3CH_2\text{-}CH_2\text{-}CH_2OH$ **1-Butanol** *Saturated alcohol*

PROBLEM 15.4: Write structures for compounds **A–C** in the following scheme for the synthesis of lily aldehyde (**C**), a compound used in perfumes.

$$p\text{-}(CH_3)_3C\text{-}C_6H_4\text{-}CH_2OH \xrightarrow[CH_2Cl_2]{PCC} \textbf{A} \xrightarrow[OH^-]{CH_3CH_2CHO} \textbf{B} \xrightarrow{H_2/Pd\text{-}C} \textbf{C}$$

Lily aldehyde

Answer

Oxidation of the primary alcohol (–CH_2OH) with pyridinium chlorochromate (PCC) in dichloromethane gives an aromatic aldehyde (**A**), which undergoes an aldol condensation with propionaldehyde to produce an α,β-unsaturated aldehyde (**B**). Finally, hydrogenation of (**B**) in the presence of Pd-C catalyst reduces C=C to yield the lily aldehyde (**C**).

CH_2OH / $C(CH_3)_3$ —PCC, CH_2Cl_2→ CH=O / $C(CH_3)_3$ —CH_3CH_2CHO, OH^-, *Aldol condensation*→ $CH{=}C(CH_3)CHO$ / $C(CH_3)_3$ —H_2/Pd-C→ $CH_2CH(CH_3)CHO$ / $C(CH_3)_3$

p-t-Butylbenzyl alcohol **(A)** **(B)** **(C)** **Lily aldehyde**

15.5.1 Intramolecular Aldol Condensation: The Aldol Cyclization

If the substrate for the aldol condensation contains two carbonyl groups, an intramolecular aldol condensation can take place to form cyclic compounds.

This is called the **aldol cyclization** which has been broadly used for the synthesis of five- and six-membered rings that are relatively unstrained.

For example, **5-oxohexanal** (a keto aldehyde) cyclizes on treatment with a base and produces **2-cyclohexenone**.

$CH_3COCH_2CH_2CH_2CHO$ ≡ (Enolate ion source / Acceptor) —OH^-, Intramolecular aldol addition→ [aldol] —Dehydration, $-H_2O$→ 2-Cyclohexenone

5-Oxohexanal

A keto aldehyde

Enolate ion source ***Acceptor***

2-Cyclohexenone

In this reaction, the aldol addition involves the ketone side of the molecule as a source of enolate ion and the more electrophilic aldehyde group as an acceptor. The intramolecular aldol addition forms a **six-membered ring**. Then dehydration occurs to yield the cyclization product (2-cyclohexenone).

Note: On the ketone side, the α-CH_3 group (not the α-CH_2) is involved in the aldol addition because it forms an *unstrained* six-membered ring.

15.6 THE CANNIZZARO REACTION

Aldehydes that have **no α-hydrogen atom** cannot form an enolate ion with a base. But when they are treated with a *concentrated* base, they undergo an intermolecular oxidation-reduction, known as the **Cannizzaro reaction.**

The Cannizzaro reaction is usually carried out by heating aldehydes (having no α-hydrogen) with concentrated sodium or potassium hydroxide. Two aldehyde molecules react when *one molecule* of the aldehyde is **oxidized to a carboxylic acid** and the *other molecule* is **reduced to an alcohol**. In the basic medium, a carboxylate ion is formed which on acidification gives the carboxylic acid.

Examples

The following are two examples of the Cannizzaro reaction with formaldehyde (having no α hydrogen, not even α carbon) and benzaldehyde (having no α hydrogen).

$$2\,H{-}\overset{O}{\overset{\|}{C}}{-}H \xrightarrow[\Delta]{\text{Concentrated NaOH}} H{-}\overset{O}{\overset{\|}{C}}{-}O^- + H{-}\overset{OH}{\underset{H}{C}}{-}H \xrightarrow[\text{(work-up)}]{H_3O^+} HCO_2H + CH_3OH$$

(no α-hydrogen, not even α-carbon)

Formaldehyde **Formic acid** **Methanol**

$$2\,C_6H_5{-}\overset{O}{\overset{\|}{C}}{-}H \xrightarrow[\text{2. } H_3O^+ \text{ (work-up)}]{\text{1. Conc. KOH, }\Delta} C_6H_5{-}\overset{O}{\overset{\|}{C}}{-}OH + C_6H_5{-}CH_2OH$$

(no α-hydrogen)

Benzaldehyde **Benzoic acid** **Benzyl alcohol**

Note: The Cannizzaro reaction represents a disproportionation reaction. Disproportionation implies that one half of the aldehyde is oxidized by the other half which itself is reduced.

Mechanism

A mechanism for the Cannizzaro reaction using benzaldehyde is shown in Figure 15.3.

Hydroxide ion (OH^-) is the initial nucleophile which attacks the aldehyde carbonyl group to form a **tetrahedral intermediate**. Then a **second hydroxide ion** deprotonates this intermediate and forms a **dianion.**

The **dianion** is highly unstable and readily **donates a hydride ion to another molecule of aldehyde** to form a carboxylate ion and an alkoxide ion. Acid work-up yields the **products** (benzoic acid and benzyl alcohol).

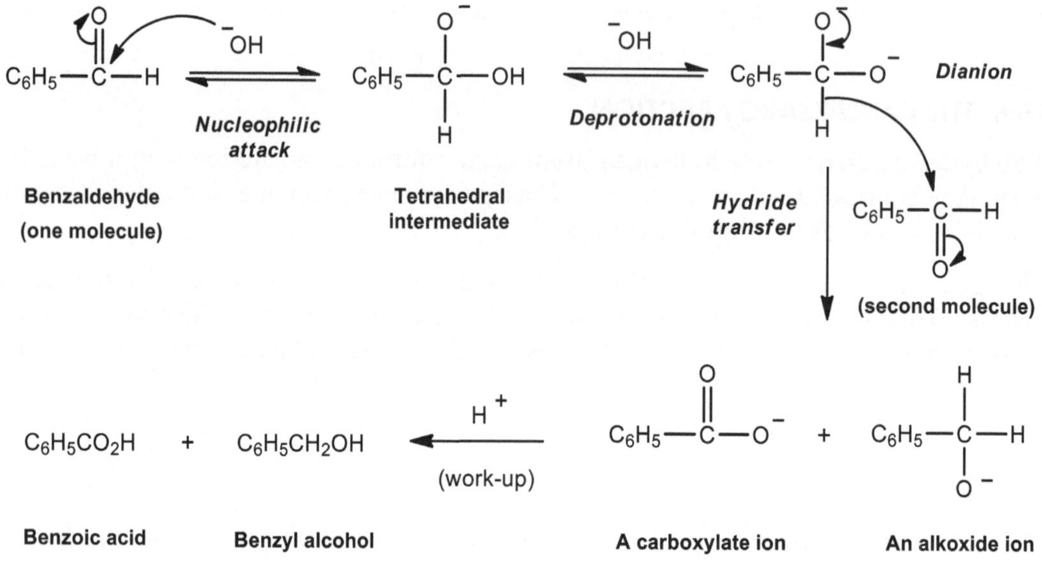

$C_6H_5-C(=O)-H$ (Benzaldehyde (one molecule)) $+\ {}^{-}OH \rightleftharpoons$ (Nucleophilic attack) $C_6H_5-C(O^-)(H)-OH$ (Tetrahedral intermediate) $+\ {}^{-}OH \rightleftharpoons$ (Deprotonation) $C_6H_5-C(O^-)(H)-O^-$ (Dianion)

Hydride transfer to $C_6H_5-C(=O)-H$ (second molecule) gives:

$C_6H_5-C(=O)-O^-$ (A carboxylate ion) $+$ $C_6H_5-CH_2-O^-$ (An alkoxide ion) $\xrightarrow[\text{(work-up)}]{H^+}$ $C_6H_5CO_2H$ (Benzoic acid) $+$ $C_6H_5CH_2OH$ (Benzyl alcohol)

Figure 15.3: A mechanism for the Cannizzaro reaction.

PROBLEM 15.5: Write the structure of the product (pentaerythritol) in the following reaction:

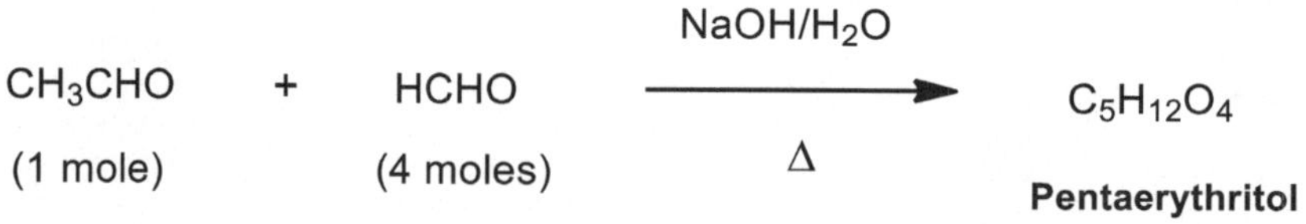

$$\underset{\text{(1 mole)}}{CH_3CHO} + \underset{\text{(4 moles)}}{HCHO} \xrightarrow[\Delta]{NaOH/H_2O} \underset{\textbf{Pentaerythritol}}{C_5H_{12}O_4}$$

Answer

One mole of acetaldehyde reacts with *three moles* of formaldehyde via a sequence of crossed aldol reactions to form pentaerythrose which has no α-hydrogen atom (see Problem 15.2).

Under strongly basic conditions in the presence of NaOH, a *fourth molecule* of formaldehyde can undergo a crossed Cannizzaro reaction with pentaerythrose to produce pentaerythritol and formate ion.

Formaldehyde carbonyl is much more electrophilic than pentaerythrose carbonyl; thus formaldehyde readily reacts with hydroxide ion leading to the formation of the corresponding dianion which reduces pentaerythrose to pentaerythritol by hydride ion transfer and formaldehyde itself is oxidized to formate ion.

$$CH_3CHO + HCHO \xrightarrow[\text{Crossed aldol}]{OH^-} HOH_2C{-}C(CH_2OH)_2{-}CHO \xrightarrow[\text{Crossed Cannizzaro}]{HCHO,\ OH^-} HOH_2C{-}C(CH_2OH)_2{-}CH_2OH + HCO_2^-$$

(1 mole) (3 moles)

Acetaldehyde Formaldehyde → Pentaerythrose (reduced) + HCHO (oxidized) → Pentaerythritol + Formate ion

Note: In general, a *more electrophilic aldehyde* acts as the *reducing agent* and itself is oxidized to a carboxylic acid in a crossed Cannizzaro reaction.

15.7 THE CLAISEN CONDENSATION: REACTION OF ENOLATE IONS WITH ESTERS

The Claisen condensation is a nucleophilic substitution of esters by an enolate ion. The enolate ion is derived from an ester possessing α-hydrogen atoms. The product is a β-keto ester.

General reaction:

$$-\overset{O}{\overset{\|}{C}}-OR + H-En \xrightarrow{\text{Substitution}} -\overset{O}{\overset{\|}{C}}-En + ROH$$

Ester (En = enolate) Product

Claisen condensation:

$$-\overset{O}{\overset{\|}{C}}-OR + H-\overset{|}{\underset{|}{C}}_{\alpha}-\overset{O}{\overset{\|}{C}}-OR \longrightarrow -\overset{O}{\overset{\|}{C}}{}_{\beta}-\overset{|}{\underset{|}{C}}_{\alpha}-\overset{O}{\overset{\|}{C}}-OR + ROH$$

Ester; Ester with α-hydrogen (H—En); *β-Keto ester*; (alcohol)

Since the Claisen reaction involves the combination of *two* molecules of ester with *elimination of a small molecule* (alcohol), it is called the **Claisen condensation**.

Example

When **ethyl acetate** (an ester having α-hydrogens) is treated with sodium ethoxide, it undergoes a Claisen condensation to produce **ethyl acetoacetate** (commonly known as **acetoacetic ester**). Ethyl acetoacetate is a β-keto ester.

Note that ethoxide ion ($C_2H_5O^-$) is used as a base rather than hydroxide ion (HO^-) as hydroxide can hydrolyse the ester.

The product is isolated after acidification of the reaction mixture. Removing ethanol (formed in the reaction) through distillation improves the yield of the product.

$$2\ CH_3\overset{O}{\overset{\|}{C}}OC_2H_5 \xrightarrow[\text{2. } H_3O^+ \text{ (acidification)}]{\text{1. } NaOC_2H_5} CH_3\overset{O}{\overset{\|}{C}}CH_2\overset{O}{\overset{\|}{C}}OC_2H_5 + C_2H_5OH$$

Ethyl acetate → $(CH_3COCH_2CO_2C_2H_5)$ **Ethyl acetoacetate (acetoacetic ester) (75%)** + C_2H_5OH **(remove as formed)**

⇓

$$CH_3-\overset{O}{\overset{\|}{C}}-OC_2H_5 + H-\overset{\alpha}{C}H_2-\overset{O}{\overset{\|}{C}}-OC_2H_5 \xrightarrow{C_2H_5O^-} CH_3-\overset{O}{\overset{\|}{C}}{}^{\beta}-\overset{\alpha}{C}H_2-\overset{O}{\overset{\|}{C}}-OC_2H_5 + C_2H_5OH$$

Ethyl acetate (one molecule) + **Ethyl acetate (another molecule)** ***Enolate ion source*** → **Ethyl acetoacetate** ***A β-keto ester***

Mechanism

A mechanism for the Claisen condensation using ethyl acetate is shown in Figure 15.4.

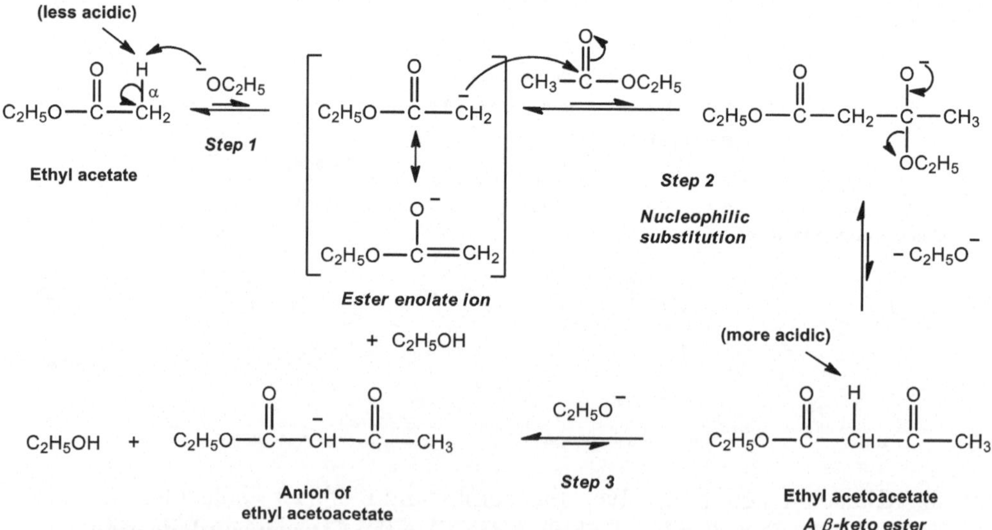

Figure 15.4: A mechanism for the Claisen condensation in the presence of ethoxide ion.

The reaction proceeds as follows. In **step 1**, *one molecule* of ethyl acetate forms an **ester enolate ion** by abstraction of an α hydrogen atom by the base (ethoxide ion).

Step 2 involves a **nucleophilic substitution** at the carbonyl group of a *second molecule* of ethyl acetate by the enolate ion to form **ethyl acetoacetate**. This step indicates the nucleophilic attack of the enolate ion followed by the loss of ethoxide ion.

However, the **equilibrium reaction** for the formation of ester enolate ion in **step 1** is quite **unfavourable** because the α-hydrogens of an ester are very less acidic (pK_a of ethyl acetate = 25.1). As a result, only a very small amount of the ester enolate is formed. Since

step 2 (nucleophilic substitution) leading to ethyl acetoacetate is also **reversible**, very little product could be formed under equilibrium.

The overall equilibrium for the Claisen condensation is driven to the product side through step 3. Here **ethyl acetoacetate** (β-keto ester) is converted to **its anion** by reaction with the base (ethoxide ion), and this equilibrium is highly favourable because the methylene hydrogen atoms of ethyl acetoacetate are much more acidic ($pK_a = 10.7$) (see Section 5.5.4).

Isolation of product

The product, **ethyl acetoacetate** is recovered by acidification of the sodium salt of ethyl acetoacetate present in the reaction mixture.

$$C_2H_5O-\overset{O}{\overset{\|}{C}}-\overset{Na^+}{\overset{-}{C}H}-\overset{O}{\overset{\|}{C}}-CH_3 \xrightarrow{HCl} C_2H_5O-\overset{O}{\overset{\|}{C}}-CH_2-\overset{O}{\overset{\|}{C}}-CH_3 + NaCl$$

Sodium salt of ethyl acetoacetate **Ethyl acetoacetate**

PROBLEM 15.6: The following ester possesses an α-hydrogen atom but it does not undergo the Claisen condensation in the presence of ethoxide ion. Explain.

$$(CH_3)_2CH\overset{O}{\overset{\|}{C}}OC_2H_5 \xrightarrow[\times]{C_2H_5O^-} \text{No Claisen condensation}$$

Ethyl isobutyrate

Answer

The driving force for the Claisen condensation in the presence of ethoxide ion as a base is the formation of anion of the β-keto ester product.

The ester (ethyl isobutyrate) has only one α-hydrogen atom. If it undergoes the Claisen condensation, it would produce a β-keto ester with *no* acidic hydrogen and therefore the formation of anion of the β-keto ester is not possible. Therefore, the Claisen condensation becomes unfavourable and does not take place.

$$H_3C-\overset{CH_3}{\overset{|}{C}H}-\overset{O}{\overset{\|}{C}}-OC_2H_5 + H-\underset{CH_3}{\overset{CH_3}{\overset{|}{\underset{|}{C}}}}{}_{\alpha}-\overset{O}{\overset{\|}{C}}-OC_2H_5 \underset{}{\overset{C_2H_5O^-}{\rightleftharpoons}} H_3C-\overset{CH_3}{\overset{|}{C}H}-\overset{O}{\overset{\|}{C}}{}_{\beta}-\underset{CH_3}{\overset{CH_3}{\overset{|}{\underset{|}{C}}}}{}_{\alpha}-\overset{O}{\overset{\|}{C}}-OC_2H_5$$

(no acidic H atom)

$$\xrightarrow[\times]{C_2H_5O^-}$$

β-Keto ester anion cannot form

Note: The Claisen condensation of ethyl isobutyrate can be carried out in the presence of a very strong base such as lithium diisopropylamide (LDA). LDA converts the ester to its enolate ion almost completely. Since *this step is essentially* *irreversible in the forward direction*, the Claisen condensation takes place to produce the β-keto ester. The β-keto ester cannot form its anion but this step is *not* needed to drive the reaction to the product side in this case.

$$H_3C-\underset{}{\overset{CH_3}{CH}}-CO_2C_2H_5 \xrightarrow[\text{THF, -78°C}]{\text{LDA}} H_3C-\overset{CH_3}{\underset{Li^+}{\bar{C}}}-CO_2C_2H_5 \quad \textbf{Lithium enolate}$$

Claisen condensation with $(CH_3)_2CHC(=O)-OC_2H_5$:

$$(CH_3)_2CH\overset{O}{\overset{\|}{C}}-\overset{CH_3}{\underset{CH_3}{C}}-CO_2C_2H_5 + C_2H_5O^- Li^+$$

A β-keto ester

15.7.1 Crossed Claisen Condensations

Like crossed aldol reactions (Section 15.4.1), **crossed Claisen condensations** become practical when **one ester component has no α hydrogen**. This ester component having no α hydrogen cannot form an enolate ion but can undergo **nucleophilic substitution** by an **enolate ion** generated from the other ester component having α hydrogens.

For example, **ethyl acetate** (having α-hydrogens) condenses with **ethyl benzoate** (having no α-hydrogen) to produce the β-keto ester **ethyl benzoylacetate**.

$$C_6H_5-\overset{O}{\overset{\|}{C}}-OC_2H_5 + H-\overset{\alpha}{C}H_2-\overset{O}{\overset{\|}{C}}-OC_2H_5 \xrightarrow[2.\ H^+]{1.\ NaOC_2H_5} C_6H_5-\overset{O}{\overset{\|}{C}}{}^{\beta}-\overset{\alpha}{C}H_2-\overset{O}{\overset{\|}{C}}-OC_2H_5$$

($C_6H_5CO_2C_2H_5$) **Ethyl benzoate**; ($CH_3CO_2C_2H_5$) **Ethyl acetate**; ($C_6H_5COCH_2CO_2C_2H_5$) **Ethyl benzoylacetate** ***A β-keto ester***

A crossed Claisen condensation between an ester and a ketone is also possible. Here the *enolate ion* is generated from the *ketone* which is more acidic than ester.

For example, **ethyl acetate** condenses with **acetone** in the presence of sodium hydride (a strong base) to yield **acetylacetone**. Here the product (acetylacetone) is a **β-diketone** (1,3-diketone), *not* a β-keto ester.

$$CH_3-\overset{O}{\overset{\|}{C}}-OC_2H_5 + H-\overset{\alpha}{C}H_2-\overset{O}{\overset{\|}{C}}-CH_3 \xrightarrow[2.\ H^+]{1.\ NaH} CH_3-\underset{3}{\overset{O}{\overset{\|}{C}}}{}^{\beta}-\underset{2}{CH_2}-\underset{1}{\overset{O}{\overset{\|}{C}}}-CH_3$$

Ethyl acetate; **Acetone** ***Enolate ion source***; **Acetylacetone** ***A β- or 1,3-diketone***

15.7.2 Intramolecular Claisen Condensation: The Dieckmann Cyclization

Like intramolecular aldol condensations (Section 15.5.1), **intramolecular Claisen condensations** can take place with **diesters** (esters containing two ester groups). The intramolecular Claisen condensation is known as the **Dieckmann condensation.**

The Dieckmann condensation produces a **cyclic β-keto ester**. It is also called the **Dieckmann cyclization** which is useful for the preparation of relatively unstrained five- and six-membered rings.

For example, **diethyl adipate** (a diester of adipic acid) undergoes the Dieckmann cyclization in the presence of sodium ethoxide. The product is **2-carbethoxy-cyclopentanone** which is a cyclic β-keto ester with a five-membered ring.

$C_2H_5O-\overset{O}{\overset{\|}{C}}-CH_2CH_2CH_2CH_2-\overset{O}{\overset{\|}{C}}-OC_2H_5$

Diethyl adipate
(diethyl hexanedioate)

Dieckmann cyclization — 1. $NaOC_2H_5$ 2. H^+

+ C_2H_5OH

$CO_2C_2H_5$

2-Carbethoxycyclopentanone
A cyclic β-keto ester

15.8 ACTIVE METHYLENE COMPOUNDS: ACETOACETIC ESTER AND MALONIC ESTER

Compounds in which a **methylene** (CH_2) group is connected to **two electron-withdrawing groups** (e.g. two C=O groups) are called **active methylene compounds**. The reactivity of the methylene group in these compounds arises from the fact that their *methylene hydrogen atoms are appreciably acidic.*

Acetoacetic ester (ethyl acetoacetate) and **malonic ester (diethyl malonate)** are two active methylene compounds that are widely used in synthesis. Since the methylene hydrogens of acetoacetic ester or malonic ester are appreciably acidic (see Section 5.5.4), the formation of their enolate anions is relatively simple.

One or both methylene hydrogens of acetoacetic ester and malonic ester can be easily substituted by alkyl groups to produce their **monoalkyl and dialkyl derivatives.**

$Z'—CH_2—Z$ (Z and Z' are electron-withdrawing groups, e.g. C=O groups)

Active methylene compounds

Examples

$CH_3—CO—CH_2—CO—OC_2H_5$

Acetoacetic ester (ethyl acetoacetate)

$C_2H_5O—CO—CH_2—CO—OC_2H_5$

Malonic ester (diethyl malonate)

Monoalkyl derivatives:

$CH_3—CO—CH(R)—CO—OC_2H_5$

Monoalkylacetoacetic ester

$C_2H_5O—CO—CH(R)—CO—OC_2H_5$

Monoalkymalonic ester

Dialkyl derivatives:

$CH_3—CO—C(R)(R')—CO—OC_2H_5$

Dialkylacetoacetic ester

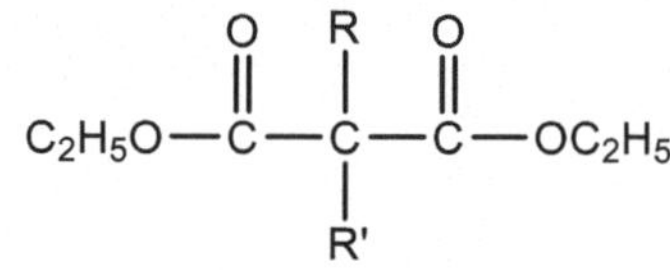

Dialkymalonic ester

15.8.1 Hydrolysis and Decarboxylation

In the synthetic applications of acetoacetic ester or malonic ester, a *sequence* of hydrolysis and decarboxylation is generally employed.

Acetoacetic ester and its alkyl derivatives

Acetoacetic ester and its alkyl derivatives are **β-keto esters** which can be **hydrolysed** with dilute sodium or potassium hydroxide. Subsequent acidification of the mixture gives a **β-keto acid**.

$$CH_3—CO—CH(R)—CO—OC_2H_5 \xrightarrow[\Delta]{\text{dil. NaOH}} CH_3—CO—CH(R)—CO—O^- Na^+$$

Alkylacetoacetic ester

A β-keto ester

(hydrolysis of ester group)

(acidification) H_3O^+

$$CH_3—\overset{\beta}{CO}—\overset{\alpha}{CH}(R)—CO—OH$$

A β-keto acid

When heated, **β-keto acids** undergo **decarboxylation** (loss of carbon dioxide). The product is a **ketone**. Notably, some β-keto acids even decarboxylate at room temperature.

$$CH_3-\overset{O}{\overset{\|}{C}}-\overset{R}{\overset{|}{CH}}-\overset{O}{\overset{\|}{C}}-OH \xrightarrow[\text{Decarboxylation}]{\Delta} CH_3-\overset{O}{\overset{\|}{C}}-CH_2-R + CO_2$$

β-Keto acid — Ketone

Decarboxylation of a β-keto acid occurs by a **pericyclic mechanism** producing an enol, which then tautomerizes to a ketone.

$$\text{β-Keto acid} \xrightarrow[\text{Pericyclic mechanism}]{-CO_2} \text{Enol } (CH_3-C(OH)=CH-R) \xrightarrow{\text{Tautomerization}} CH_3-\overset{O}{\overset{\|}{C}}-CH_2-R \text{ (Ketone)}$$

β-Keto acid — Enol — Ketone

Malonic ester and its alkyl derivatives

Malonic ester and its alkyl derivatives are **β-diesters (1,3-diesters)** which can also be **hydrolysed** with aqueous sodium or potassium hydroxide. Subsequent acidification of the mixture gives a ***gem*-diacid (1,1-diacid)**. In the *gem*-diacid, two CO_2H groups are attached to the same carbon atom.

$$C_2H_5O-\overset{O}{\overset{\|}{C}}-\overset{R}{\overset{|}{CH}}-\overset{O}{\overset{\|}{C}}-OC_2H_5 \xrightarrow[2.\ H_3O^+]{1.\ NaOH,\ H_2O,\ \Delta} HO-\overset{O}{\overset{\|}{C}}-\overset{R}{\overset{|}{CH}}-\overset{O}{\overset{\|}{C}}-OH$$

(hydrolysis of ester groups)

Alkymalonic ester — *A β-diester*

Alkymalonic acid — *A gem- or 1,1-diacid*

Malonic acid and alkylmalonic acids (*gem*-diacids) undergo **decarboxylatipon** when they are heated. The product is a **monocarboxylic acid**.

$$HO-\overset{O}{\overset{\|}{C}}-\overset{R}{\overset{|}{CH}}-\overset{O}{\overset{\|}{C}}-OH \xrightarrow[-CO_2]{\Delta} R-CH_2-\overset{O}{\overset{\|}{C}}-OH$$

Alkymalonic acid — *Decarboxylation* — Monocarboxylic acid

PROBLEM 15.7: The aldol condensation of an active methylene compound with an aldehyde or a ketone is known as the **Knoevenagel condensation**. The reaction is catalysed by weak bases such as diethylamine [$(C_2H_5)_2NH$].

The following scheme outlines a synthesis of cinnamic acid using the Knoevenagel condensation. Write the structures for compounds **A, B** and **cinnamic acid.**

C_6H_5–CHO + $CH_2(CO_2C_2H_5)_2$ $\xrightarrow{(C_2H_5)_2NH}$ **A**

A $\xrightarrow{1.\ OH^-,\ H_2O;\ 2.\ H_3O^+}$ **B**

B $\xrightarrow{\Delta}$ **Cinnamic acid**

Answer

The Knoevenagel condensation of benzaldehyde with malonic ester gives the compound (**A**). Hydrolysis of **A** gives (**B**). When heated, **B** undergoes decarboxylation to yield **cinnamic acid**.

C_6H_5–CH(=O) + H–CH(α)($CO_2C_2H_5$)–$CO_2C_2H_5$ $\xrightarrow{(C_2H_5)_2NH}$ *Knoevenagel condensation* [C_6H_5–CH(OH)–CH($CO_2C_2H_5$)–$CO_2C_2H_5$]

Benzaldehyde **Malonic ester**

$\xrightarrow{-\ H_2O}$ C_6H_5–CH=C($CO_2C_2H_5$)–$CO_2C_2H_5$

(A)

Knoevenagel condensation product

Hydrolysis → C_6H_5–CH=C(CO_2H)–CO_2H

(B)

$\xrightarrow{\Delta,\ -\ CO_2}$ *Decarboxylation* → C_6H_5–CH=CH–CO_2H

(C)

Cinnamic acid

15.9 ALKYLATION OF ENOLATE IONS: USE OF ACETOACETIC ESTER AND MALONIC ESTER IN SYNTHESIS

An enolate ion is an **ambident nucleophile** with two nucleophilic sites or atoms (carbon and oxygen). It is possible to introduce an alkyl group at either site, called the C-alkylation or the O-alkylation.

Here we will describe briefly the **synthetic uses** of acetoacetic ester and malonic ester involving their **C-alkylation**.

15.9.1 Acetoacetic Ester Synthesis of Ketones

Acetoacetic ester has been used extensively in synthesis. One important use of acetoacetic ester is the **synthesis of ketones that are monosubstituted or disubstituted derivatives of acetone**.

The synthesis involves the **C-alkylation at the methylene carbon** of acetoacetic ester.

In monosubstituted and disubstituted acetone, the acetone fragment comes from acetoacetic ester and the alkyl substitution comes from alkyl halides.

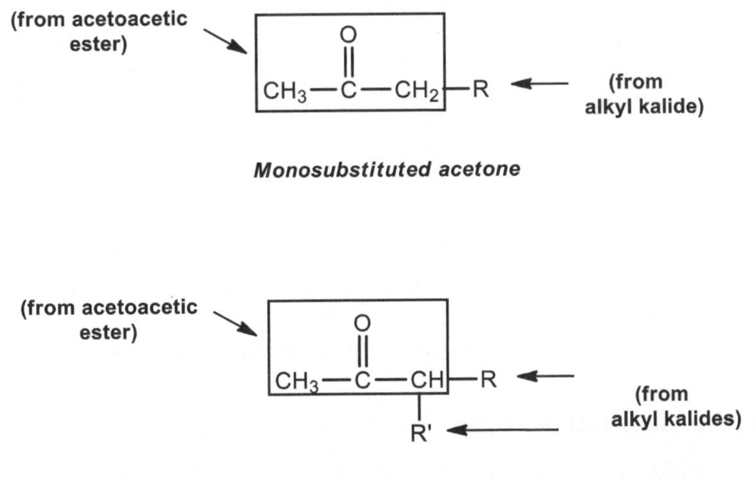

Disubstituted acetone

For the synthesis, **acetoacetic ester** is first converted to its **enolate ion** using sodium ethoxide as a base. The enolate ion as a carbanion nucleophile then reacts with an **alkyl halide** in an S_N2 reaction to give the **C-alkylated product**. Finally, a sequence of **hydrolysis and decarboxylation** affords the desired **ketone**.

Note: The use of primary alkyl halides gives best yields. Secondary alkyl halides give lower yields and tertiary alkyl halides lead to elimination instead of substitution.

Synthesis of monosubstituted acetone

Consider, for example, the synthesis of **2-pentanone**, which is a monosubstituted acetone (CH_3COCH_2**R**, where **R** = CH_2CH_3).

$$CH_3-\overset{\overset{\large O}{\|}}{C}-CH_2-\overbrace{CH_2CH_3}^{\mathbf{R}}$$

2-Pentanone

The **alkyl halide** needed for alkylation is **ethyl bromide** (CH_3CH_2Br).

The following scheme outlines the synthesis:

Acetoacetic ester $CH_3COCH_2COOC_2H_5$ —(C_2H_5ONa, C_2H_5OH)→ [Enolate ion (sodioacetoacetic ester) $CH_3COCH^-COOC_2H_5 \leftrightarrow CH_3C(O^-)=CHCOOC_2H_5$] Na^+ —(CH_3CH_2Br, S_N2, *C-Alkylation*, $-NaBr$)→ Ethyl derivative of acetoacetic ester $CH_3COCH(CH_2CH_3)COOC_2H_5$ —(1. dil. NaOH, 2. H_3O^+, *Hydrolysis*)→ $CH_3COCH(CH_2CH_3)COOH$ —(Δ, $-CO_2$, *Decarboxylation*)→ 2-Pentanone $CH_3COCH_2CH_2CH_3$

Treatment of acetoacetic ester with sodium ethoxide gives an enolate ion as the sodium salt of acetoacetic ester (sodioacetoacetic ester) which then undergoes an S_N2 reaction with ethyl bromide (a primary halide). The ethyl derivative of acetoacetic ester is then subjected to a sequence of hydrolysis and decarboxylation to produce 2-pentanone.

Synthesis of disubstituted acetone

To illustrate, consider the synthesis of **3-benzyl-2-butanone**, which is a disubstituted acetone (CH_3COCH**RR′**, where. **R** = CH_3 and **R′** = $CH_2C_6H_5$).

$CH_3COCH(CH_3)(CH_2C_6H_5)$ (R = CH_3, R' = $CH_2C_6H_5$)

3-Benzyl-2-butanone

The synthesis requires **two successive alkylation protocols**, as shown in the following scheme:

$CH_3COCH_2COOC_2H_5$ (**Acetoacetic ester**) $\xrightarrow[C_2H_5OH]{C_2H_5ONa}$ $CH_3CO\overset{-}{C}HCOOC_2H_5\ Na^+$ $\xrightarrow[\text{First alkylation}]{C_6H_5CH_2Br}$ $CH_3COCH(CH_2C_6H_5)COOC_2H_5$

$CH_3COCH(CH_2C_6H_5)COOC_2H_5$ $\xrightarrow[C_2H_5OH]{C_2H_5ONa}$ $CH_3CO\overset{-}{C}(CH_2C_6H_5)COOC_2H_5\ Na^+$ $\xrightarrow[\text{Second alkylation}]{CH_3I}$ $CH_3COC(CH_3)(CH_2C_6H_5)COOC_2H_5$

Hydrolysis: 1. dil. NaOH 2. H_3O^+

$CH_3COC(CH_3)(CH_2C_6H_5)COOH$ $\xrightarrow[-\ CO_2]{\Delta}$ (**Decarboxylation**) $CH_3COCH(CH_3)CH_2C_6H_5$

3-Benzyl-2-butanone

The *larger benzyl group* (**R′** = $CH_2C_6H_5$) is introduced *first* on the less hindered methylene carbon. The alkyl halide used is benzyl bromide ($C_6H_5CH_2Br$), a primary halide that undergoes a facile S_N2 reaction. The benzyl derivative of acetoacetic ester is then treated with sodium ethoxide for the second alkylation with methyl iodide. Subsequent hydrolysis and decarboxylation produces 3-benzyl-2-butanone.

15.9.2 Malonic Ester Synthesis of Carboxylic Acids

Malonic ester has been useful for the synthesis of carboxylic acids that are monosubstituted and disubstituted derivatives of acetic acid.

The synthesis involves the C-alkylation of malonic ester.

In monosubstituted and disubstituted acetic acid, the acetic acid fragment comes from malonic ester and the alkyl substitution from alkyl halides.

(from alkyl kalide) → R–[CH_2–C(=O)–OH] ← (from malonic ester)

Monosubstituted acetic acid

(from alkyl kalides) → R–[CH(R′)–C(=O)–OH] ← (from malonic ester)

Disubstituted acetic acid

The malonic ester synthesis is analogous to acetoacetic ester synthesis. Malonic ester is first converted to its enolate anion using a base such as sodium ethoxide and then alkylated by treating with an alkyl halide in an S_N2 reaction. Subsequent hydrolysis and decarboxylation produces a substituted acetic acid.

Synthesis of monosubstituted acetic acid

Consider the synthesis of **butanoic acid** ($CH_3CH_2CH_2CO_2H$). This is a monosubstituted acetic acid (**R**CH_2CO_2H, where **R** = CH_3CH_2).

Here the **alkyl halide** required for alkylation is **ethyl bromide** (CH_3CH_2Br).

The following scheme outlines the synthesis of butanoic acid:

$C_2H_5O{-}CO{-}CH_2{-}CO{-}OC_2H_5$ (**Malonic ester**) $\xrightarrow[C_2H_5OH]{C_2H_5ONa}$ $[(C_2H_5OCO)_2CH^- \leftrightarrow$ enolate$]\ Na^+$ (***Enolate ion***)

$\xrightarrow[-\ NaBr]{CH_3CH_2{-}Br}$ (***S_N2 C-Alkylation***) $(C_2H_5OCO)_2CH{-}CH_2CH_3$

$\xrightarrow[2.\ H_3O^+]{1.\ OH^-/H_2O}$ (***Hydrolysis***) $(HOCO)_2CH{-}CH_2CH_3$ $\xrightarrow[-\ CO_2]{\Delta}$ (***Decarboxylation***) $HO{-}CO{-}CH_2{-}CH_2CH_3$ (**Butanoic acid**)

Synthesis of disubstituted acetic acid

Consider the synthesis of **2-methylpropanoic acid**. This is a disubstituted acetic acid in which the two alkyl substituents are the same (**RR′**$CHCO_2H$, where **R** = **R′** = CH_3).

$\underset{R}{H_3C}{-}CH(\underset{R'}{CH_3}){-}CO{-}OH$ **(R = R' = CH_3)**

2-Methylpropanoic acid

The following scheme outlines the synthesis of 2-methylpropanoic acid.

$C_2H_5O-C(=O)-CH_2-C(=O)-OC_2H_5$ (Malonic ester) $\xrightarrow[\text{2. } CH_3I]{\text{1. } C_2H_5ONa/C_2H_5OH}$ (First alkylation) $C_2H_5O-C(=O)-CH(CH_3)-C(=O)-OC_2H_5$

Second alkylation: 1. C_2H_5ONa/C_2H_5OH 2. CH_3I → $C_2H_5O-C(=O)-C(CH_3)_2-C(=O)-OC_2H_5$

Hydrolysis: 1. OH^-/H_2O 2. H_3O^+ → $HO-C(=O)-C(CH_3)_2-C(=O)-OH$

Decarboxylation: Δ, $-CO_2$ → $HO-C(=O)-CH(CH_3)-CH_3$

2-Methylpropanoic acid

15.10 THE MICHAEL REACTION

We have described earlier the conjugate nucleophilic addition to α,β-unsaturated aldehydes and ketones (see Section 9.11).

When the nucleophile is an **enolate ion**, the conjugate addition to α,β-unsaturated carbonyl compounds is known as the **Michael reaction**, after the American chemist Arthur Michael who discovered the reaction in 1887.

A general Michael reaction between an enolate ion and an α,β-unsaturated carbonyl compound is shown in Figure 15.5.

The reaction involves the **attack of the enolate ion** as a carbanion nucleophile to the **β-carbon atom of the α,β-unsaturated carbonyl compound**. The conjugate addition gives an **enol** which rapidly tautomerizes to a **keto form**. The product is a **1,5-dicarbonyl compound.**

Note: The net result of the Michael reaction is an addition to the carbon-carbon double bond (C=C) of the α,β-unsaturated carbonyl compounds.

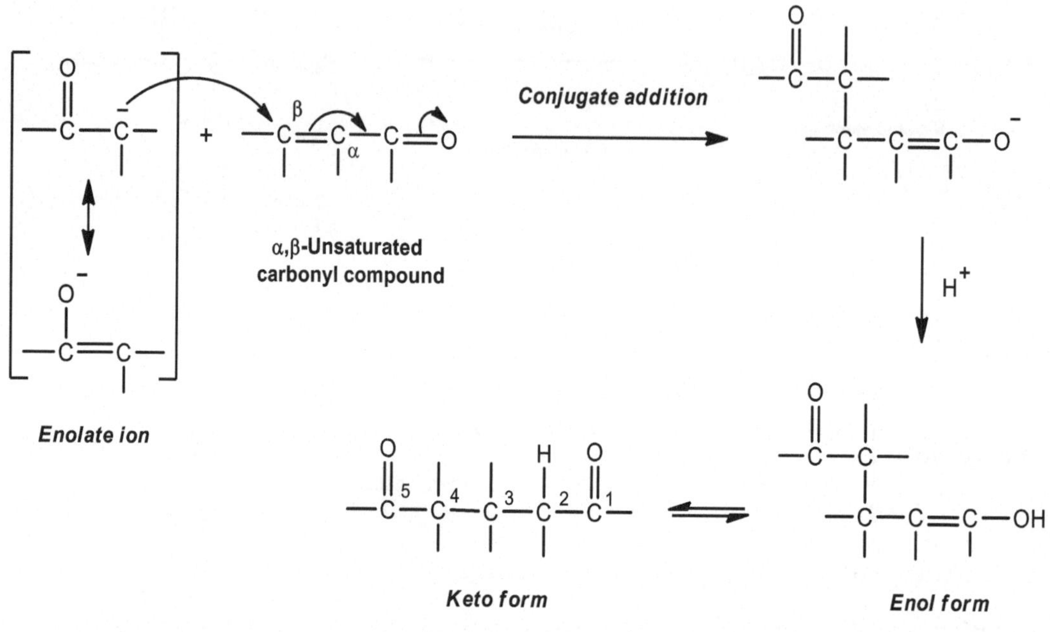

Figure 15.5: A general Michael reaction between an enolate ion and an α,β-unsaturated carbonyl compound.

Example

In the Michael reaction, the α,β-unsaturated carbonyl compound may be a ketone or an ester and the enolate ion is usually derived from an active methylene compound such as acetoacetic ester or malonic ester.

The following is an example of a Michael reaction between **ethyl 3-methyl-2-butenoate** (an α,β-unsaturated ester) and **malonic ester**. The product (a Michael adduct) results from the addition of malonic ester to C=C of the α,β-unsaturated ester.

The mechanism of the reaction involves a conjugate addition of the enolate ion derived from malonic ester to ethyl 3-methyl-2-butenoate followed by a rapid tautomerization to give the product.

$$\underset{\beta\ \ \ \ \alpha}{H_3C-\overset{CH_3}{\overset{|}{C}}=CH-\overset{O}{\overset{\|}{C}}-OC_2H_5} + H-\underset{CO_2C_2H_5}{\underset{|}{CH}}-CO_2C_2H_5 \xrightarrow[C_2H_5OH]{C_2H_5ONa} H_3C-\overset{CH_3}{\overset{|}{C}}-CH_2-\overset{O}{\overset{\|}{C}}-OC_2H_5 \text{ with } -CH(CO_2C_2H_5)-CO_2C_2H_5 \text{ on the quaternary C}$$

Ethyl 3-methyl-2-butenoate
An α,β-Unsaturated ester

Malonic ester

Michael adduct

Mechanism:

$$H-\underset{CO_2Et}{\underset{|}{CH}}-CO_2Et \xrightarrow{EtO^-} \bar{C}H(CO_2Et)-CO_2Et + EtOH$$

(Et = C_2H_5)

Malonic ester

$$\bar{C}H(CO_2Et)_2 + (CH_3)_2C=CH-\overset{O}{\overset{\|}{C}}-OEt \xrightarrow[\textbf{\textit{Conjugate addition}}]{} CH_3\overset{CH_3}{\overset{|}{C}}(CH(CO_2Et)_2)-CH=\overset{O^-}{\overset{|}{C}}-OEt$$

$$\xrightarrow{EtOH} CH_3\overset{CH_3}{\overset{|}{C}}(CH(CO_2Et)_2)-CH=\overset{OH}{\overset{|}{C}}-OEt + EtO^-$$

$$\xrightarrow{\textbf{Tautomerization}} CH_3\overset{CH_3}{\overset{|}{C}}(CH(CO_2Et)_2)-CH_2-\overset{O}{\overset{\|}{C}}-OEt$$

Michael adduct

Synthetic applications

The Michael reaction is useful for the synthesis of **1,5-dicarbonyl compounds**.

For example, the **Michael adduct** from ethyl 3-methyl-2-butenoate and malonic ester (see above) undergoes **hydrolysis and decarboxylation** to produce **3,3-dimethylglutaric acid**, a 1,5-dicarboxylic acid.

Note that 1,5 positions of the dicarboxylic acid indicate the relative positions of the two carbonyl groups.

$$H_3C-C(CH_3)(CH(CO_2C_2H_5)CO_2C_2H_5)-CH_2-C(=O)-OC_2H_5 \xrightarrow[\text{2. } H_3O^+]{\text{1. } OH^-/H_2O} H_3C-C(CH_3)(CH(CO_2H)CO_2H)-CH_2-C(=O)-OH \xrightarrow[-CO_2]{\Delta} H_3C-C(CH_3)(CH_2-C(=O)-OH)-CH_2-C(=O)-OH$$

Michael adduct

Ester hydrolysis

Decarboxylation

3,3-Dimethylglutaric acid

1,5-Dicarboxylic acid

15.10.1 The Robinson Annulation

A sequence of Michael addition and aldol condensation has been successfully used for the construction of cyclic structures in the synthesis of many natural products. This procedure is known as the **Robinson annulation**, after the English chemist Robert Robinson, who was awarded the Nobel Prize in 1947 for his research on natural products. The term *annulation* implies ring formation.

The following is an example of the Robinson annulation:

(active methylene hydrogen)

2-Methyl-1,3-cyclohexanedione

Active methylene compound

+ $CH_2=CH-C(=O)-CH_3$

Methyl vinyl ketone

α,β-Unsaturated ketone

$\xrightarrow[CH_3OH]{KOH}$ *Michael reaction*

$-CH_2-CH_2-C(=O)-CH_3$

Michael adduct

Pyrrolidine (base)

Aldol condensation

$-H_2O$

A bicyclic product

The first step is a **Michael reaction** between 2-methyl-1,3-cyclohexanedione (an active methylene compound) and methyl vinyl ketone (an α,β-unsaturated ketone). The resulting Michael adduct then undergoes an **aldol condensation** in the presence of pyrrolidine as a base to produce a bicyclic product.

The overall reaction is a **Robinson annulation** to form a *new* six-membered ring.

15.11 MALONIC ESTER CONDENSATION WITH UREA: BARBITURATES

Malonic ester condenses with **urea** in the presence of sodium ethoxide with elimination of two molecules of ethanol. The product is **barbituric acid**.

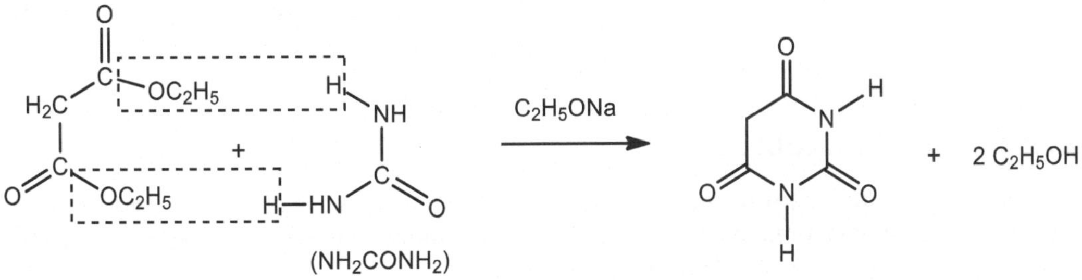

C_2H_5ONa ; (NH_2CONH_2) ; $+ \; 2\,C_2H_5OH$

Malonic ester (diethyl malonate) **Urea** **Barbituric acid**

As the name implies, barbituric acid is an acid, which is even *stronger* than acetic acid. The greater acidity of C–H bonds is due to the stabilization of the conjugate base (anion) of barbituric acid by resonance with adjacent carbonyl groups.

Barbituric acid has a six-membered heterocyclic ring with two nitrogen atoms and it can tautomerize to an aromatic heterocyclic ring called a **pyrimidine ring**.

Tautomerization

Barbituric acid or 2,4,6-trihydroxypyrimidine

Derivatives of barbituric acid are called **barbiturates** which are used in medicine as **sleep producing drugs**. Two such drugs are **veronal** (5,5-diethylbarbituric acid) and **phenobarbital** (5-ethyl-5-phenylbarbituric acid).

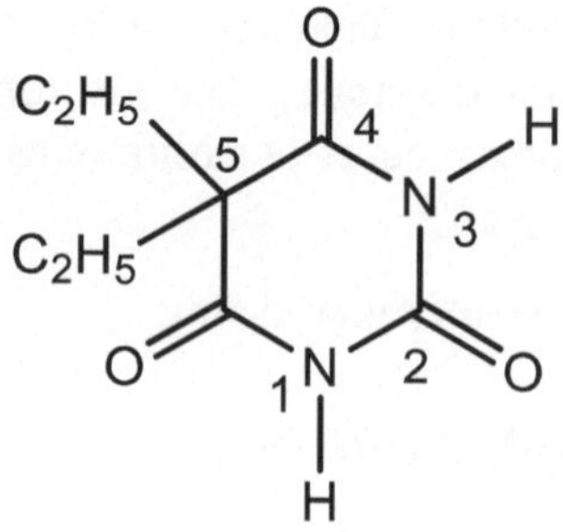

Veronal
(5,5-diethylbarbituric acid)

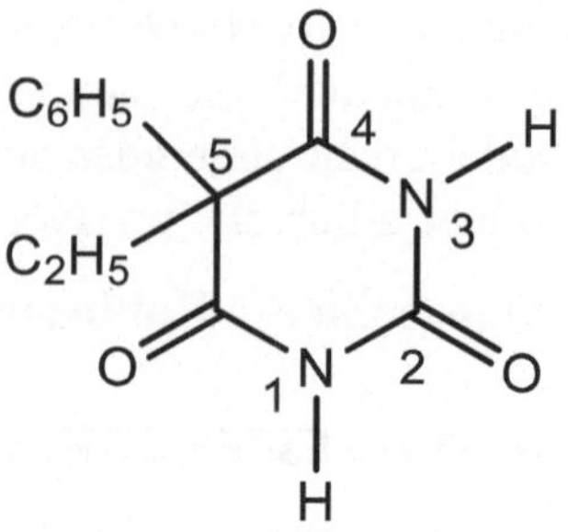

Phenobarbital
(5-ethyl-5-phenylbarbituric acid)

Note: Barbiturates are *addictive* and an overdose could lead to fatal results.

SUPPLEMENTARY PROBLEMS

15.8 Draw keto–enol equilibrium for acetoacetic ester (ethyl acetoacetate) and malonic ester (diethyl malonate) respectively. In which case, more than one enol form is possible? Indicate the predominant enol form in such a case.

Why acetoacetic ester has higher enol content than malonic ester?

15.9 Predict the product in each of the following reactions:

(a) 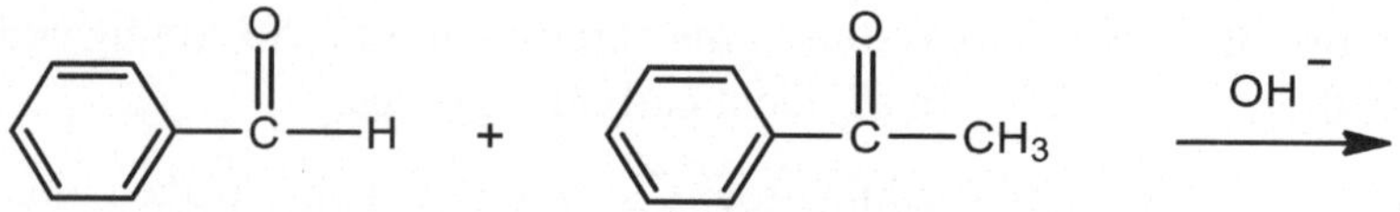 C_6H_5–C(=O)–H + C_6H_5–C(=O)–CH_3 $\xrightarrow{OH^-}$

Benzaldehyde **Acetophenone**

(b) 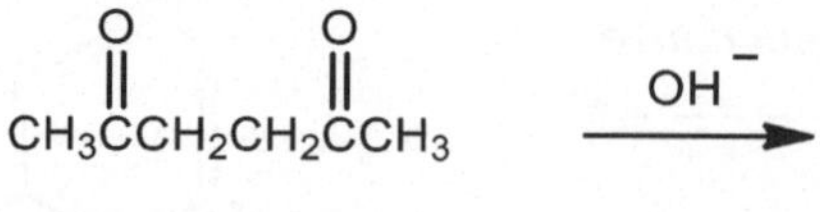 $CH_3\overset{O}{\overset{\|}{C}}CH_2CH_2\overset{O}{\overset{\|}{C}}CH_3$ $\xrightarrow{OH^-}$

2,5-Hexanedione

[*Ans:* (a) benzalacetophenone (1,3-diphenyl-2-propen-1-one); (b) 3-methyl-2-cyclopentenone]

15.10 Outline a directed aldol synthesis of the following compound:

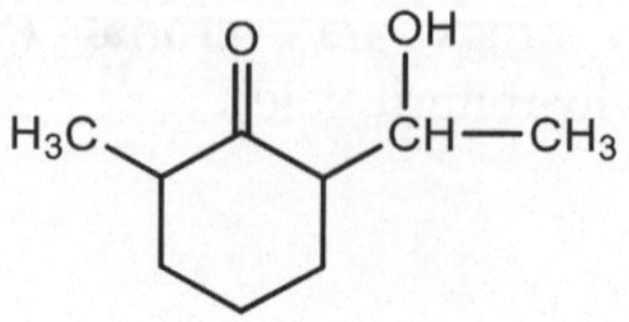

(2-(1-hydroxyethyl)-6-methylcyclohexanone: ring C=O with H_3C on one α-carbon and CH(OH)–CH_3 on the other)

[*Hint:* start with 2-methylcyclohexanone and prepare a kinetic enolate with LDA]

15.11 Write the products formed in the following Cannizzaro reactions:

(a) C_6H_5CHO + $HCHO$ $\xrightarrow[\text{2. } H_3O^+]{\text{1. Conc. NaOH, } \Delta}$

Benzaldehyde Formaldehyde

(b) $HC(=O)-C(=O)H$ $\xrightarrow[\text{2. } H_3O^+]{\text{1. Conc. NaOH, } \Delta}$

Glyoxal

[*Ans:* (a) benzyl alcohol and formic acid; (b) glycolic acid (hydroxyacetic acid)]

15.12 Give the structures of the products that are formed in the following crossed Claisen condensations:

(a) $CH_3CO_2C_2H_5$ + $HCO_2C_2H_5$ $\xrightarrow[\text{2. } H^+]{\text{1. } C_2H_5ONa}$

Ethyl acetate Ethyl formate

(b) $CH_3CH_2CO_2C_2H_5$ + $CO_2C_2H_5$–$CO_2C_2H_5$ $\xrightarrow[\text{2. } H^+]{\text{1. } C_2H_5ONa}$

Ethyl propanoate Diethyl oxalate

15.13 How would you carry out the following conversion?

$C_2H_5O_2C-CH_2CH_2CH_2CH_2CH_2-CO_2C_2H_5 \longrightarrow$ (cyclohexanone, $C_6H_{10}=O$)

Diethyl pimelate (diethyl heptanedioate) Cyclohexanone

[*Hint:* use the Dieckmann cyclization as the first step]

15.14 Predict the product in each of the following malonic ester syntheses:

(a) $CH_2(CO_2C_2H_5)_2$ $\xrightarrow[\text{2. } BrCH_2CH_2CH_2Br]{\text{1. } NaOC_2H_5}$ $\xrightarrow{NaOC_2H_5}$ $\xrightarrow[\text{2. } \Delta]{\text{1. } HCl/H_2O}$

Malonic ester

(b) $CH_2(CO_2C_2H_5)_2$ $\xrightarrow[\text{2. } CH_2I_2]{\text{1. } NaOC_2H_5 \text{ (2 moles)}}$ $\xrightarrow[\text{2. } \Delta]{\text{1. } HCl/H_2O}$

Malonic ester

[*Ans:* (a) cyclobutanecarboxylic acid; (b) glutaric acid ($HO_2CCH_2CH_2CH_2CO_2H$)]

15.15 Write structures for compounds **A–C** in the following scheme:

Acetoacetic ester $\xrightarrow[\text{2. } BrCH_2CO_2Et \text{ (ethyl bromoacetate)}]{\text{1. NaOEt}}$ **A** $\xrightarrow[\text{2. } H_3O^+]{\text{1. dil. NaOH}}$ **B** $\xrightarrow{\Delta}$ **C**

Ans:

A = $CH_3COCH(CO_2Et)CH_2CO_2Et$

B = $CH_3COCH(CO_2H)CH_2CO_2H$

C = $CH_3COCH_2CH_2CO_2H$ (γ, β, α labelled)

4-Oxopentanoic acid
(a γ-keto acid)

CHAPTER 16

Rearrangement Reactions: 1,2-Shifts and Aromatic Rearrangements

IN THIS CHAPTER

- 1,2-Rearrangements or 1,2-shifts
- Wagner-Meerwein rearrangement
- Pinacol rearrangement
- Wolff rearrangement
- Benzilic acid rearrangement
- Hofmann and Curtius rearrangements
- Beckmann rearrangement
- Baeyer-Villiger and Hock rearrangements
- Aromatic rearrangements with migration from nitrogen: Bamberger and benzidine rearrangements
- Aromatic rearrangements with migration from oxygen: Fries and Claisen rearrangements

In a variety of organic reactions described in the previous chapters, the molecular skeleton remains intact and changes occur at the functional groups.

However, there are other reactions in which atoms or groups migrate within molecules and carbon skeletons are modified. These reactions are called molecular rearrangements, which constitute a separate class of reactions.

Most commonly, a rearrangement reaction involves the **migration of a group from one atom to the next atom** within the molecule, which is called the **1,2-rearrangement** or **1,2-shift**.

A rearrangement can also occur within an **aromatic ring**, which is known as the **aromatic rearrangement**.

16.1 1,2-REARRANGEMENTS OR 1,2-SHIFTS

The most common type of 1,2-rearrangements is the rearrangement in which a group migrates with its bonding electron pair to an adjacent atom that is electron-deficient.

This is schematically represented as follows:

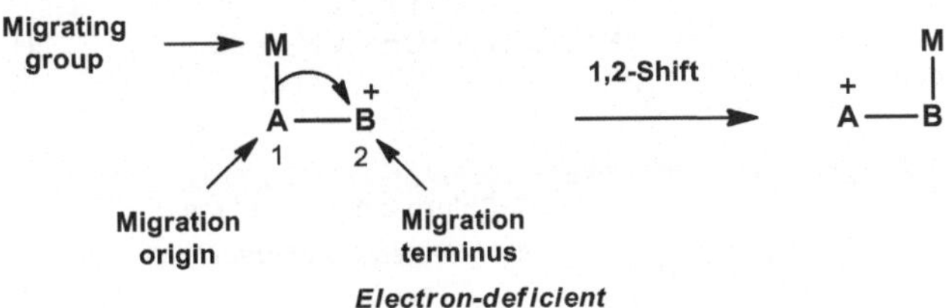

Here **M** is the **migrating group** which may be hydrogen, alkyl or aryl group. The atom **A** at the 1-position from which migration begins is called the **migration origin**, and the adjacent atom **B** at the 2-position to which the migrating group moves is called the **migration terminus**. Note that the numbers 1, 2 denote relative positions.

In this chapter, we will describe some important 1,2-rearrangements in which the migration origin is carbon atom (C) but the electron-deficient migration terminus may be carbon atom (C), nitrogen atom (N) or oxygen atom (O).

16.2 MIGRATION TO ELECTRON-DEFICIENT CARBON ATOM

In this case of 1,2-shift, a migrating group moves from a carbon atom (migration origin) to an adjacent carbon atom (migration terminus) which can be **electron-deficient** as a positively charged **carbocation carbon**, a neutral **carbene carbon** with sextet of electrons or an electron-deficient **carbonyl carbon**.

16.2.1 The Wagner-Meerwein Rearrangement

Carbocations have an electron-deficient, positively charged carbon atom with sextet of electrons. The 1,2-migrations to an electron-deficient carbon atom of a carbocation are broadly known as **Wagner-Meerwein rearrangements**.

A carbocation can be generated from an alcohol, alkene, alkyl halide or amine. To illustrate the Wagner-Meerwein rearrangements, we consider several examples as described below.

A. *Reaction of alcohols with hydrogen halide*

Some primary and secondary alcohols can undergo a Wagner-Meerwein rearrangement on reaction with hydrogen halide.

For example, when **neopentyl alcohol** (a primary alcohol) is treated with HCl, a Wagner-Meerwein rearrangement takes place. The expected product (neopentyl chloride) is *not* formed. Instead, a **rearranged alkyl chloride** (2-chloro-2-methylbutane) is produced along with an **alkene** (2-methyl-2-butene).

$(CH_3)_3C{-}CH_2{-}OH \xrightarrow{HCl}$ ~~$(CH_3)_3C{-}CH_2{-}Cl$~~ *Not formed*

$(CH_3)_3C{-}CH_2{-}OH \xrightarrow{HCl} CH_3{-}C(Cl)(CH_3){-}CH_2{-}CH_3 + CH_3{-}C(CH_3){=}CH{-}CH_3$

Neopentyl alcohol (2,2-dimethyl-1-propanol)

2-Chloro-2-methylbutane *Rearranged alkyl chloride*

2-Methyl-2-butene

Mechanism

Protonation of neopentyl alcohol followed by the loss of water gives a **primary carbocation** (neopentyl cation) which rearranges to a **more stable tertiary carbocation** by a 1,2-shift of a methyl group. **The conversion of a less stable primary carbocation to a more stable tertiary carbocation provides the driving force for the rearrangement reaction.**

The rearranged carbocation then adds to a chloride ion (nucleophile) to form the **rearranged chloride** (2-chloro-2-methylbutane) or loses a proton to give a **more substituted or more stable alkene** (2-methyl-2-butene) rather than a less substituted alkene (2-methyl-1-butene).

$$(CH_3)_3C{-}CH_2{-}OH \xrightarrow{H^+} (CH_3)_3C{-}CH_2{-}\overset{+}{O}H_2 \xrightarrow{-\,H_2O} (CH_3)_3C{-}\overset{+}{C}H_2$$

Neopentyl alcohol → **1° Carbocation** *Less stable*

1,2-Shift

$$CH_3{-}\overset{+}{C}(CH_3){-}CH_2{-}H \;(\text{with } CH_3)$$

3° Carbocation *More stable*

$- H^+$: $CH_3{-}C(CH_3){=}CH{-}CH_3$

2-Methyl-2-butene *More substituted alkene*

Cl^-: $CH_3{-}C(Cl)(CH_3){-}CH_2{-}CH_3$

2-Chloro-2-methylbutane

B. *Dehydration of Alcohols*

A Wagner-Meerwein rearrangement can take place during **dehydration** of some **primary** and **secondary alcohols**.

For example, dehydration of **3,3-dimethyl-2-butanol** (a secondary alcohol) in the presence of acid catalyst (H_3PO_4) involves a Wagner-Meerwein rearrangement to produce **2,3-dimethyl-2-butene** as the **major** product. The minor product is 2,3-dimethyl-1-butene.

$$CH_3{-}C(CH_3)_2{-}CH(OH){-}CH_3 \xrightarrow[\Delta]{H_3PO_4} CH_3{-}C(CH_3){=}C(CH_3){-}CH_3 + H_2C{=}C(CH_3){-}CH(CH_3){-}CH_3$$

3,3-Dimethyl-2-butanol → **2,3-Dimethyl-2-butene** *Major* + **2,3-Dimethyl-1-butene** *Minor*

Mechanism

The dehydration proceeds through the formation of the protonated alcohol which loses water to form a **secondary carbocation**. The less stable secondary carbocation then rearranges to a **more stable tertiary carbocation** by a 1,2-shift of a methyl group.

Finally, the loss of a proton from the rearranged carbocation produces the **more substituted or more stable alkene** (2,3-dimethyl-2-butene) as the **major** product. The less substituted or less stable alkene (2,3-dimethyl-1-butene) is formed as the minor product.

$$CH_3C(CH_3)_2CH(OH)CH_3 \xrightarrow{H^+} CH_3C(CH_3)_2CH(\overset{+}{O}H_2)CH_3 \xrightarrow{-H_2O} CH_3C(CH_3)_2\overset{+}{C}HCH_3$$

3,3-Dimethyl-2-butanol — **Protonated alcohol** — **2° Carbocation** ***Less stable***

1,2-Shift

$$\overset{(b)}{H}CH_2\overset{+}{C}(CH_3)C(CH_3)\overset{(a)}{H}CH_3$$

3° Carbocation ***More stable***

(b) $-H^+$ → $H_2C{=}C(CH_3)CH(CH_3)CH_3$ **2,3-Dimethyl-1-butene** ***Less substituted: minor***

(a) $-H^+$ → $CH_3C(CH_3){=}C(CH_3)CH_3$ **2,3-Dimethyl-2-butene** ***More substituted: major***

C. *Reaction of Alkenes with Hydrogen Halide*

Some alkenes can undergo a Wagner-Meerwein rearrangement when treated with hydrogen halide. In this case, a *carbocation is generated from the alkene by protonation*.

For example, when 3,3-dimethyl-1-butene is treated with HCl, the products are 2-chloro-2,3-dimethylbutane and 2,3-dimethyl-2-butene.

$$CH_3C(CH_3)_2CH{=}CH_2 \xrightarrow{HCl} CH_3C(Cl)(CH_3)CH(CH_3)CH_3 + CH_3C(CH_3){=}C(CH_3)CH_3$$

3,3-Dimethyl-1-butene — **2-Chloro-2,3-dimethylbutane** — **2,3-Dimethyl-2-butene**

Mechanism

Protonation of the alkene gives a **secondary carbocation**. The less stable secondary carbocation then rearranges to a **more stable tertiary carbocation** by a 1,2-methyl shift.

The rearranged carbocation then combines with chloride ion to form 2-chloro-2,3-dimethylbutane or loses a proton to give the more substituted alkene, 2,3-dimethyl-2-butene.

$CH_3C(CH_3)_2CH{=}CH_2$ (3,3-Dimethyl-1-butene) $\xrightarrow{H^+}$ $CH_3C(CH_3)_2\overset{+}{C}HCH_3$ (2° Carbocation, *Less stable*) $\xrightarrow{\text{1,2-Shift}}$ $CH_3\overset{+}{C}(CH_3)CH(CH_3)CH_3$ (3° Carbocation, *More stable*)

3° Carbocation $\xrightarrow{Cl^-}$ $CH_3CCl(CH_3)CH(CH_3)CH_3$ (2-Chloro-2,3-dimethylbutane)

3° Carbocation $\xrightarrow{-H^+}$ $CH_3C(CH_3){=}C(CH_3)CH_3$ (2,3-Dimethyl-2-butene, *More substituted alkene*)

Note: A Wagner-Meerwein rearrangement can also occur during *hydration of some alkenes* (see Problem 16.8b).

PROBLEM 16.1: Explain the formation of product in the following reaction:

$CH_3CH(OH)CH(CH_3)CH_3$ (3-Methyl-2-butanol) $\xrightarrow{HBr}$ $CH_3CBr(CH_3)CH_2CH_3$ (2-Bromo-2-methylbutane)

Answer

Protonation of the alcohol (3-methyl-2-butanol) and subsequent loss of water form a less stable secondary carbocation, which can rearrange to a more stable tertiary carbocation by a 1,2-shift of hydrogen with its bonding electron pair (i.e. as a hydride). So this is a case of **1,2-hydride shift**. The rearranged carbocation then combines with bromide ion to yield the product (2-bromo-2-methylbutane).

$CH_3CH(CH_3)CH(OH)CH_3$ (3-Methyl-2-butanol) $\xrightarrow{H^+}$ $CH_3CH(CH_3)CH(\overset{+}{O}H_2)CH_3$ $\xrightarrow{-H_2O}$ $CH_3CH(CH_3)\overset{+}{C}HCH_3$ (2° Carbocation, *Less stable*)

$\xrightarrow{\text{1,2-Hydride shift}}$ $CH_3\overset{+}{C}(CH_3)CH_2CH_3$ (3° Carbocation, *More stable*) $\xrightarrow{Br^-}$ $CH_3CBr(CH_3)CH_2CH_3$ (2-Bromo-2-methylbutane)

16.2.2 The Pinacol Rearrangement

The dihydric alcohols with two OH groups on adjacent carbons are called **1,2-diols,** which are known as **pinacols.**

When a **pinacol** (1,2-diol) is treated with acid (e.g. H_2SO_4), a 1,2-rearrangement occurs with the *loss of water* and formation of a **ketone or aldehyde** as the product. This rearrangement of a pinacol in the presence of acid is called the **pinacol rearrangement.**

Preparation of pinacols

Pinacols are readily prepared from **ketones** by reductive dimerization with magnesium amalgam [Mg(Hg)] followed by hydrolysis.

For example, acetone produces the pinacol (2,3-dimethyl-2,3-butanediol) as follows:

$$CH_3-C(CH_3)=O \;+\; O=C(CH_3)-CH_3 \xrightarrow{Mg(Hg)} CH_3-C(CH_3)(O)-C(CH_3)(O)-CH_3\;[O{-}Mg{-}O] \xrightarrow{H_2O} CH_3-C(CH_3)(OH)-C(CH_3)(OH)-CH_3$$

Acetone **(2 molecules)** → **Pinacol (2,3-Dimethyl-2,3-butanediol)**

An example of the pinacol rearrangement

When the **pinacol (2,3-dimethyl-2,3-butanediol)** derived from acetone is treated with sulphuric acid, a pinacol rearrangement takes place producing a ketone called **pinacolone (3,3-dimethyl-2-butanone).**

$$CH_3-C(CH_3)(OH)-C(CH_3)(OH)-CH_3 \xrightarrow{H_2SO_4} CH_3-C(=O)-C(CH_3)_2-CH_3$$

Pinacol (2,3-Dimethyl-2,3-butanediol) → **Pinacolone (3,3-dimethyl-2-butanone)**

Mechanism of the pinacol rearrangement

Protonation of *one* of the two hydroxyl groups of the pinacol and subsequent loss of water gives a **tertiary carbocation**. But this carbocation, although tertiary, can rearrange by the 1,2-shift of a methyl group to give a **new carbocation** that is **more stabilized by resonance** involving a lone pair on oxygen atom.

Finally, the loss of a proton gives **pinacolone** (a stable ketone) as the product.

$$(CH_3)_2C(OH)-C(OH)(CH_3)_2 \xrightarrow{H^+} (CH_3)_2C(OH)-C(CH_3)_2(\overset{+}{O}H_2) \xrightarrow{-H_2O} (CH_3)_2C(OH)-\overset{+}{C}(CH_3)_2$$

Pinacol

3° Carbocation
Less stable

1,2-Shift

$$CH_3-\overset{+}{C}(OH)-C(CH_3)_3 \longleftrightarrow CH_3-C(=\overset{+}{O}H)-C(CH_3)_3 \xrightarrow{-H^+} CH_3-C(=O)-C(CH_3)_3$$

(stabilized by resonance)

New Carbocation
More stable

Pinacolone

Application: Synthesis of pivalic acid from acetone

The following scheme delineates a synthesis of **pivalic acid** (trimethylacetic acid) starting from acetone. In the final step, pinacolone undergoes the iodoform reaction to produce pivalic acid and iodoform.

$$2\ CH_3-C(=O)-CH_3 \xrightarrow[\text{2. } H_2O]{\text{1. Mg(Hg)}} (CH_3)_2C(OH)-C(OH)(CH_3)_2 \xrightarrow[\text{Pinacol rearrangement}]{H_2SO_4} (CH_3)_3C-C(=O)-CH_3$$

Acetone — Pinacol — Pinacolone

Iodoform reaction: 1. $I_2/NaOH$; 2. H_3O^+ (work-up)

$$CHI_3 + (CH_3)_3C-C(=O)-OH$$

Pivalic acid
(trimethylacetic acid)

Rearrangement with unsymmetrical pinacols

The pinacol (2,3-dimethyl-2,3-butanediol) shown above is a *symmetrical pinacol*. Thus *any* of the two OH groups can be lost after protonation during the rearrangement.

But if a pinacol is unsymmetrical, the question arises as to which OH group is lost in the pinacol rearrangement?

In general, for the unsymmetrical pinacol, the **OH group is lost from the carbon atom that will give the more stable *initial* carbocation**, which then rearranges and leads to the product.

To illustrate, consider the pinacol rearrangement of the unsymmerical pinacol, **2-methyl-1,1-diphenyl-1,2-propanediol**:

2-Methyl-1,1-diphenyl-1,2-propanediol

H^+

$-\ H_2O$

Less stable (not formed)

$-\ H_2O$

More stable

1,2-Shift

$-\ H^+$

3,3-Diphenyl-2-butanone

Here the rearrangement proceeds through the **loss of the OH group** from the **carbon atom attached to two phenyl substituents** because the resulting *initial carbocation* is **more stable** by conjugation with phenyl groups. (Note that an alternative loss of the OH group from the carbon bearing two methyl groups gives a less stable carbocation and this becomes unfavourable.)

Then a 1,2-shift of a methyl group leads to **3,3-diphenyl-2-butanone** as the product.

Note: Some pinacol rearrangements may involve two possible migrating groups. In such cases, **migratory aptitudes** (migratory tendencies) of the two groups will determine the course of the rearrangement reaction. The group that has a greater migratory aptitude will migrate. For example, phenyl (C_6H_5) has a greater migratory aptitude than methyl (CH_3) (see Problem 16.9c).

PROBLEM 16.2: Predict the product in the following reaction:

$$(C_6H_5)_2C(OH)-CH(NH_2)-CH_3 \xrightarrow{HNO_2}$$

2-Amino-1,1-diphenyl-1-propanol

Answer

Diazotization of the α-amino alcohol (2-amino-1,1-diphenyl-1-propanol) with nitrous acid produces an aliphatic diazonium cation which is unstable and readily loses nitrogen to form a *carbocation*, similar to the initial carbocation formed from a 1,2-diol (pinacol). Then 1,2-shift of a phenyl group leads to the product (1,2-diphenyl-1-propanone).

This reaction resembles the pinacol rearrangement and is often called the **semipinacol rearrangement**.

$$C_6H_5-C(OH)(C_6H_5)-CH(NH_2)-CH_3 \xrightarrow{HNO_2} C_6H_5-C(OH)(C_6H_5)-CH(N_2^+)-CH_3 \xrightarrow{-N_2} C_6H_5-C(OH)(C_6H_5)-\overset{+}{C}H-CH_3$$

2-Amino-1,1-diphenyl-1-propanol
An α-amino alcohol

Aliphatic diazonium cation
Unstable

Carbocation

1,2-Shift

$$\left[C_6H_5-\overset{+}{C}(OH)-CH(C_6H_5)-CH_3 \longleftrightarrow C_6H_5-C(=\overset{+}{O}H)-CH(C_6H_5)-CH_3 \right] \xrightarrow{-H^+} C_6H_5-C(=O)-CH(C_6H_5)-CH_3$$

Stabilized by resonance

1,2-Diphenyl-1-propanone

16.2.3 The Wolff Rearrangement

We have previously introduced the Wolff rearrangement in the **Arndt-Eistert synthesis** for converting a carboxylic acid to its next higher homologue (see Section 14.6.1).

The **Wolff rearrangement** is a 1,2-rearrangement in which the migrating group moves to the **electron-deficient carbon atom** of an intermediate **carbene**. The carbene carbon is a neutral (uncharged) carbon atom with sextet of electrons.

The following is a Wolff rearrangement in a general case:

$$R-\overset{O}{\overset{\|}{C}}-\ddot{C}H^{-}-\overset{+}{N}\equiv N \xrightarrow[-N_2]{Ag_2O} O=\underset{}{\overset{R}{C}}-\ddot{C}H \xrightarrow{1,2\text{-Shift}} \left[O=\overset{+}{C}-\ddot{C}H^{-}-R \longleftrightarrow O=C=CH-R \right]$$

(R = alkyl or aryl)

α-Diazoketone — **Keto-carbene** (**sextet**) — **A ketene**

The intermediate carbene is generated by heating an **α-diazoketone** ($RCOCHN_2$) in the presence of silver oxide as catalyst. The *loss of nitrogen* from the α-diazoketone gives a **keto-carbene** which undergoes a **1,2-shift** of an alkyl or aryl group. The product is a **ketene** (RCH=C=O).

Note: It is also possible that the loss of nitrogen (N_2) and migration of the group (R) in the Wolff rearrangement is concerted or simultaneous.

$$O=\overset{R}{C}-\ddot{C}H^{-}-\overset{+}{N}\equiv N \xrightarrow[\text{Concerted}]{Ag_2O} O=C=CH-R + N_2$$

α-Diazoketone — **A ketene**

16.2.4 The Benzilic acid Rearrangement

When **benzil** ($C_6H_5COCOC_6H_5$) is heated with ethanolic potassium hydroxide, it undergoes a rearrangement to produce **benzilic acid** [$(C_6H_5)_2C(OH)CO_2H$].

The product (benzilic acid) is isolated after acidification of the reaction mixture.

$$C_6H_5-\overset{O}{\overset{\|}{C}}-\overset{O}{\overset{\|}{C}}-C_6H_5 \xrightarrow[C_2H_5OH,\ \Delta]{KOH} (C_6H_5)_2C(OH)-\overset{O}{\overset{\|}{C}}-O^{-}K^{+} \xrightarrow[\text{(work-up)}]{H_3O^+} (C_6H_5)_2C(OH)-\overset{O}{\overset{\|}{C}}-OH$$

Benzil
An α-diketone or 1,2-diketone

Benzilic acid
An α-hydroxy acid

Benzil is an α-diketone (1,2-diketone) while benzilic acid is an α-hydroxy acid. In general, the **conversion of α-diketones to α-hydroxy acids** is known as the **benzilic acid rearrangement**.

Mechanism

The benzilic acid rearrangement involves a **1,2-shift of an aryl group** (e.g. C_6H_5) to an electron-deficient carbon of a **carbonyl group** (C=O).

Attack of the hydroxide ion (OH^-) to *one* of the carbonyl groups of **benzil** is followed by a 1,2-shift of a phenyl group (C_6H_5) to the adjacent carbonyl carbon. Then an exchange of proton (H^+) between OH and O^- gives the anion of benzilic acid, which on acidification produces **benzilic acid**.

C_6H_5—CO—CO—C_6H_5 (**Benzil**) + ^-OH ⇌ C_6H_5—CO—C(O^-)(OH)—C_6H_5 —**1,2-Shift**→ C_6H_5—C(O^-)(C_6H_5)—CO—OH

⇅ **Proton exchange**

C_6H_5—C(OH)(C_6H_5)—CO—O^- (**Benzilic acid anion**) —H_3O^+ (work-up)→ C_6H_5—C(OH)(C_6H_5)—CO—OH (**Benzilic acid**)

Note: The benzilic acid rearrangement generally takes place with *aromatic* α-diketones (such as benzil).

PROBLEM 16.3: Write structures for compounds **A–D** in the following scheme:

C_6H_5—CHO (**Benzaldehyde**) —KCN→ **A** —HNO_3→ **B** —1. KOH; 2. H^+→ **C** —HI, P→ **D**

Answer

The reaction of two molecules of benzaldehyde in the presence of cyanide ion produces benzoin (**A**), an α-hydroxyketone. This reaction of aromatic aldehydes with cyanide ion is called the **benzoin condensation**. Cyanide ion is required for the reaction to proceed but it is regenerated at the end of the reaction.

Benzoin is oxidized by nitric acid to benzil **(B)**. Benzil on treatment with KOH undergoes the **benzilic acid rearrangement** to form benzilic acid **(C)** after acid work-up. Benzilic acid is reduced with hydroiodic acid and red phosphorus to diphenylacetic acid **(D)**.

$$C_6H_5{-}\overset{\overset{O}{\|}}{C}{-}H + H{-}\overset{\overset{O}{\|}}{C}{-}C_6H_5 \xrightarrow[\text{Benzoin condensation}]{CN^-} C_6H_5{-}\overset{\overset{O}{\|}}{C}{-}\overset{\overset{OH}{|}}{CH}{-}C_6H_5$$

Benzaldehyde (2 molecules) → Benzoin (A)

Benzoin (A) —Oxidation, HNO_3→ Benzil (B)

$$C_6H_5{-}\overset{\overset{O}{\|}}{C}{-}\overset{\overset{O}{\|}}{C}{-}C_6H_5 \xrightarrow[\text{Benzilic acid rearrangement}]{1.\ KOH\quad 2.\ H^+} C_6H_5{-}\overset{\overset{OH}{|}}{\underset{\underset{C_6H_5}{|}}{C}}{-}\overset{\overset{O}{\|}}{C}{-}OH \xrightarrow[\text{Reduction}]{HI/P} C_6H_5{-}\overset{\overset{H}{|}}{\underset{\underset{C_6H_5}{|}}{C}}{-}\overset{\overset{O}{\|}}{C}{-}OH$$

Benzil (B) → Benzilic acid (C) → Diphenylacetic acid (D)

16.3 MIGRATION TO ELECTRON-DEFICIENT NITROGEN ATOM

An alkyl or aryl group can migrate from carbon (migration origin) to electron-deficient nitrogen atom (migration terminus). A family of closely related **1,2-rearrangements** in which a group migrates from an **acyl carbon** to **nitrogen** is well-known.

Two such rearrangements are the **Hofmann rearrangement** and the **Curtius rearrangement**, as shown below.

The **distinction** between these rearrangements lies in the nature of the **leaving group** (L). In the Hofmann rearrangement, L = Br while L = N_2^+ for the Curtius rearrangement.

The product is an **isocyanate**.

(acyl carbon) (leaving group)

$$R{-}\overset{\overset{O}{\|}}{C}{-}\ddot{N}{-}L \xrightarrow{L\,=\,Br} R{-}\ddot{N}{=}C{=}O$$

(R = alkyl or aryl) Isocyanate — ***Hofmann rearrangement***

$$R{-}\overset{\overset{O}{\|}}{C}{-}\ddot{N}{-}L \xrightarrow{L\,=\,N_2^+} R{-}\ddot{N}{=}C{=}O$$

(R = alkyl or aryl) Isocyanate — ***Curtius rearrangement***

If the leaving group departs before migration, a species with an electron deficient nitrogen atom having a sextet of electrons would result. This species is called a **nitrene** which resembles the carbene in the Wolff rearrangement (see Section 16.2.3). A 1,2-shift of an alkyl or aryl group from the carbonyl carbon to nitrogen can then lead to the product.

$$R-\overset{O}{\overset{\|}{C}}-\ddot{\underset{..}{N}}-L \xrightarrow[\text{(loss of leaving group)}]{-L^-} R-\overset{O}{\overset{\|}{C}}-\ddot{N} \xrightarrow{\text{1,2-Shift}} O{=}C{=}\underset{..}{N}-R$$

A nitrene — Isocyanate

However, a nitrene as an intermediate has not been detected in these rearrangements.

Thus, a more **likely mechanism** is that the migrating group (R) moves with its bonding electron pair to the nitrogen atom at the same time the leaving group (L) departs, i.e. **the movement of the migrating group and the departure of the leaving group are simultaneous or concerted**.

$$R-\overset{O}{\overset{\|}{C}}-\ddot{N}^{-}-L \xrightarrow[\text{(concerted loss of L and migration of R)}]{\text{1,2-Shift}} R-\ddot{N}{=}C{=}O + L^-$$

Isocyanate

16.3.1 The Hofmann Rearrangement

Amides ($RCONH_2$) react with bromine in aqueous sodium hydroxide (or aqueous sodium hypobromite) to yield **amines** (RNH_2).

In this reaction, the carbonyl group (C=O) of the amide is lost and an amine possessing one less carbon atom is produced.

The reaction is called the **Hofmann degradation** or the **Hofmann rearrangement** (since the key step of the reaction involves a rearrangement).

$$R-\overset{O}{\overset{\|}{C}}-NH_2 \xrightarrow[\text{NaOH}]{Br_2} R-NH_2 + CO_2 + NaBr$$

(R = alkyl or aryl)

Amide — Amine

Balanced equation:

$$R-\overset{O}{\overset{\|}{C}}-NH_2 + Br_2 + 4\,NaOH \longrightarrow R-NH_2 + Na_2CO_3 + 2\,NaBr + 2\,H_2O$$

The following are two examples of the Hofmann rearrangement:

$$CH_3CH_2\overset{O}{\overset{\|}{C}}-NH_2 \xrightarrow[\text{NaOH}]{Br_2} CH_3CH_2-NH_2 + CO_2$$

Propanamide — Ethylamine

$$\text{Nicotinamide} \xrightarrow[H_2O]{NaOBr} \text{3-Aminopyridine} + CO_2$$

Nicotinamide

3-Aminopyridine

Note: The Hofmann rearrangement provides a suitable method for the preparation of pure **primary amines,** *not* contaminated with 2° or 3° amines.

Mechanism

The Hofmann rearrangement proceeds through the following mechanism:

$$R{-}CO{-}NH_2 \underset{}{\overset{OH^-}{\rightleftharpoons}} H_2O + R{-}CO{-}\ddot{N}H^- \xrightarrow[-Br^-]{Br{-}Br} R{-}CO{-}NH{-}Br$$

Amide

N-Bromoamide

$$R{-}NH_2 \xleftarrow[-CO_2]{H_2O} O{=}C(O^-){-}NH{-}R \xleftarrow[H_2O]{OH^-} O{=}C{=}N{-}R \xleftarrow[-Br^-]{\text{1,2-Shift}} R{-}CO{-}\ddot{N}^-{-}Br$$

Amine

Carbamate ion

Isocyanate

N-Bromoamide anion

Treatment of the **amide** ($RCONH_2$) with Br_2 in NaOH causes a base-promoted bromination to form an ***N*-bromoamide**. The *N*-bromoamide then reacts with hydroxide ion to form its anion, ***N*-bromoamide anion**. The *N*-bromoamide anion then undergoes a 1,2-shift with the loss of a bromide ion to give an **isocyanate**.

However, the *isocyanate is not the isolated product* in this reaction. In the presence of the aqueous base, the isocyanate is converted to a **carbamate ion** which spontaneously loses carbon dioxide to form the **amine** as the product.

N-Substituted amides

The *N*-substituted amides of the type RCONHR′ or RCONR′R″ do not undergo the Hofmann rearrangement.

$$R{-}CO{-}NH{-}R' \text{ or } R{-}CO{-}NR'{-}R'' \xrightarrow[NaOH]{Br_2} \text{No Hofmann rearrangement}$$

N-Substituted amide

N,N-Disubstituted amide

This is because **two hydrogen atoms** on the **nitrogen atom** of the amide need to be lost to form the ***N*-bromoamide anion** which actually undergoes the rearrangement (see the mechanism above). Thus the *N*-substituted amides (RCONHR′) having **one** hydrogen atom or *N,N*-disubstituted amides (RCONR′R″) having **no** hydrogen atom on the N atom fails to undergo the rearrangement.

PROBLEM 16.4: Predict the product of the following reaction:

1. Br_2/OH^-

2. H_3O^+

Phthalimide

Answer

Phthalimide is hydrolysed in aqueous base to the half-amide of phthalic acid. (A half-amide of a dicarboxylic acid such as phthalic acid possesses one amide group and one carboxyl group, whereas a full amide has two amide groups).

The half-amide then undergoes the Hofmann rearrangement with bromine and hydroxide ion. Acidification of the reaction mixture produces anthranilic acid (*o*-aminobenzoic acid).

OH^-, H_2O ***Hydrolysis***

Half-amide

Br_2, OH^- ***Hofmann rearrangement***

H_3O^+ **(work-up)**

Anthranilic acid

16.3.2 The Curtius Rearrangement

The Curtius rearrangement occurs with **acyl azides** ($RCON_3$). Acyl azides can be easily prepared by the reaction of acyl chlorides (RCOCl) with sodium azide (NaN_3).

$$R{-}\overset{O}{\overset{\|}{C}}{-}Cl \xrightarrow[(NaN_3)]{N_3^-} R{-}\overset{O}{\overset{\|}{C}}{-}N_3 \equiv R{-}\overset{O}{\overset{\|}{C}}{-}\ddot{\overset{-}{N}}{-}\overset{+}{N}{\equiv}N$$

Acyl chloride **Acyl azide**

Acyl azides on heating undergo the **Curtius rearrangement** to give **isocyanates**. The **isocyanates can be isolated** by carrying out the reaction in an aprotic solvent such as chloroform.

$$R{-}\overset{O}{\overset{\|}{C}}{-}\ddot{\overset{-}{N}}{-}\overset{+}{N}{\equiv}N \xrightarrow[CHCl_3]{\Delta} R{-}\ddot{N}{=}C{=}O + N_2$$

Acyl azide — ***Curtius rearramgement*** — **Isocyanate** ***Can be isolated***

Mechanism

Like the Hofmann rearrangement, the **Curtius rearrangement** involves the migration of a group from the acyl carbon to the nitrogen atom as the leaving group departs.

The **departing group is N_2** which is a highly stable molecule. Thus the reaction occurs with the evolution of nitrogen gas.

$$R{-}\overset{O}{\overset{\|}{C}}{-}\ddot{\overset{-}{N}}{-}\overset{+}{N}{\equiv}N \xrightarrow[\text{1,2-Shift}]{\Delta} O{=}C{=}\ddot{N}{-}R + N_2$$

Acyl azide **Isocyanate**

Transformation of isocyanates

Isocyanates can be hydrolysed to **amines**. They can also be reacted with alcohols to form esters of carbamic acids called **urethanes**.

$$R{-}N{=}C{=}O \xrightarrow{H_2O} \left[R{-}\underset{H}{\underset{|}{N}}{-}\overset{OH}{\overset{|}{C}}{=}O\right] \xrightarrow{-\,CO_2} R{-}NH_2$$

Isocyanate **Carbamic acid (unstable)** **Primary amine**

$$R{-}N{=}C{=}O \xrightarrow{C_2H_5OH} R{-}\underset{H}{\underset{|}{N}}{-}\overset{OC_2H_5}{\overset{|}{C}}{=}O$$

Isocyanate **Ethyl carbamate** ***Urethane***

The following **scheme** illustrates how an acyl chloride (**benzoyl chloride**) can be converted to an aromatic amine (**aniline**) via a Curtius rearrangement.

C_6H_5COCl (**Benzoyl chloride**) —NaN_3→ $C_6H_5CON_3$ (**Benzoyl azide**) —Δ, $CHCl_3$→ $C_6H_5N{=}C{=}O$ (**Phenyl isocyanate**) + N_2

Phenyl isocyanate —H_2O→ $C_6H_5NH_2$ (**Aniline**) + CO_2

16.3.3 The Beckmann Rearrangement

When the **oxime** of a ketone is treated with concentrated sulphuric acid (c. H_2SO_4), it undergoes a **1,2-rearrangement** to produce an ***N*-substituted amide**. This is known as the **Beckmann rearrangement**.

In the Beckmann rearrangement, **the group *anti* (or *trans*) to the hydroxyl group of the oxime migrates**. As a result, the *migrating group* (i.e. the group *anti* to OH) is attached to the *nitrogen atom* of the *amide* product.

The Beckmann rearrangement is thus stereospecific.

Migrating group (*anti* to OH) → R(R')C=N–OH —c. H_2SO_4→ R'–C(=O)–N(H)–R **(migrating group becomes attached to N)**

Oxime (R is *anti* to OH) **N-Substituted amide**

Remember that the oxime can be readily prepared from the corresponding ketone by its reaction with hydroxylamine (NH_2OH) (see Section 9.5.2).

Example

Acetophenone on treatment with hydroxylamine forms the *more stable* stereoisomer of **acephenone oxime** in which the large phenyl group (C_6H_5) is *anti* or *trans* to OH.

The oxime then undergoes the Beckmann rearrangement in the presence of concentrated H_2SO_4 to produce ***N*-phenylacetamide (acetanilide)**.

Acetophenone $\xrightarrow{NH_2OH}$ Acetophenone oxime (C_6H_5 is *anti* to OH) $\xrightarrow[\text{Beckmann rearrangement}]{H_2SO_4}$ *N*-Phenylacetamide (acetanilide)

Mechanism

Protonation of the oxime OH converts it to a good leaving group. Then the **1,2-migration** of the ***anti*-group (phenyl)** to **nitrogen** occurs at the same time the leaving group departs as water.

The resulting cation is **stabilized by resonance** involving the lone pair on nitrogen. The reaction of the cation with water followed by the loss of a proton and tautomerization gives *N*-phenylacetamide.

Acetophenone oxime $\xrightarrow{H^+}$ $C_6H_5(CH_3)C{=}N{-}OH_2^+$ $\xrightarrow[-H_2O]{\text{1,2-Shift}}$ [$CH_3{-}\overset{+}{C}{=}N{-}C_6H_5$ ↔ $CH_3{-}C{\equiv}\overset{+}{N}{-}C_6H_5$] Stabilized by resonance $\xrightarrow{H_2O}$ $CH_3{-}C(\overset{+}{O}H_2){=}N{-}C_6H_5$ $\xrightarrow{-H^+}$ $CH_3{-}C(OH){=}N{-}C_6H_5$ $\xrightarrow{\text{Tautomerization}}$ $CH_3{-}C(=O){-}NH{-}C_6H_5$

N-Phenylacetamide

Note: The Beckmann rearrangement can also be carried out with reagents other than proton acids. One such reagent is phosphorus pentachloride (PCl_5). PCl_5 converts OH to $OPCl_4$ which is a good leaving group that departs as $OPCl_4^-$.

$$-OH + PCl_5 \rightarrow -OPCl_4 + HCl$$

PROBLEM 16.5: Write the structures for compounds **A–C** in the following scheme:

Cyclohexanone $\xrightarrow{NH_2OH}$ **A** $\xrightarrow{H_2SO_4}$ **B** $\xrightarrow{\Delta}$ **C**

Answer

Cyclohexanone reacts with hydroxylamine to form cyclohexanone oxime (**A**) which undergoes the Beckmann rearrangement in the presence of sulphuric acid. The rearrangement involves the migration of a **ring CH_2 group** to **nitrogen** leading to an expansion of the six-membered ring to a seven-membered ring.

The resulting cyclic amide is called caprolactam (**B**). When heated, caprolactam produces a polymer of the nylon group, called nylon 6 (**C**).

NH_2OH → (A) Cyclohexanone oxime → H_2SO_4, Beckmann rearrangement → (B) Caprolactam → Δ → (C) Nylon 6: $-[NH-C(=O)-(CH_2)_5]_n-$

Mechanism of the Beckmann rearrangement:

Oxime + H^+ → 1,2-Shift → H_2O, $-H^+$ → Tauto. → caprolactam

16.4 MIGRATION TO ELECTRON-DEFICIENT OXYGEN ATOM

The most general rearrangement involving migration of a group to an electron-deficient oxygen atom is the **Baeyer-Villiger rearrangement.**

The acid-catalysed rearrangement of a tertiary hydroperoxide is similar to the Baeyer-Villiger rearrangement, and is known as the **Hock rearrangement**.

16.4.1 The Baeyer-Villiger Rearrangement

In the Baeyer-Villiger rearrangement, a **ketone** (RCOR) is converted to an **ester** (RCO_2R) by treatment with a peroxy acid ($R'CO_3H$).

$$\underset{\text{Ketone}}{R-\overset{O}{\overset{\|}{C}}-R} \xrightarrow[\text{(peroxy acid)}]{R'CO_3H} \underset{\text{Ester}}{R-\overset{O}{\overset{\|}{C}}-O-R} + R'CO_2H$$

The rearrangement is an **oxidation** in which an oxygen atom is inserted next to the carbonyl group of the ketone, and is also known as the **Baeyer –Villiger oxidation.**

The Baeyer-Villiger oxidation has already been introduced in Chapter 9 (see Section 9.10.1).

Here we will consider the **mechanism** of the Baeyer-Villiger rearrangement.

Mechanism

The reaction is initiated by the formation of an **adduct** between the ketone and the peroxy acid, which then undergoes the rearrangement.

Protonation of the **ketone** (RCOR) followed by the nucleophilic attack of the **peroxy acid** ($R'CO_3H$) to the protonated carbonyl group leads to the **adduct**. A carboxylate ion ($R'CO_2^-$) then departs as a *good leaving group* and at the same time **1,2-migration** takes place from the carbon to the electron-deficient oxygen that is being created.

The resulting **carbocation** is **stabilized by resonance** involving a lone pair on oxygen. Finally, the loss of a proton produces the **ester**.

Ketone

H^+

(peroxy acid)

$-H^+$

1,2-Shift

$-R'CO_2^-$

Stabilized by resonance

Adduct

$-H^+$

Ester

16.4.2 The Hock Rearrangement

The **acid-catalysed rearrangement** of a **hydroperoxide** to an alcohol and a carbonyl compound is known as the **Hock rearrangement.**

H^+

A hydroperoxide

Alcohol

Ketone

The Hock rearrangement is a key step in the **cumene process**, which is used for the **industrial production of phenol**.

Cumene process

Cumene is **isopropylbenzene** which is cheaply available from the Friedel-Crafts alkylation of benzene with propene. It readily forms **cumene hydroperoxide** by oxidation with oxygen in air. When cumene hydroperoxide is treated with aqueous acid (10% H_2SO_4), it undergoes the **Hock rearrangement** to produce **phenol** and **acetone**.

$CH_3CH{=}CH_2$ + Benzene → (H_3PO_4, 250°C, pressure; *Friedel-Crafts alkylation*) → Cumene (isopropylbenzene)

Benzene

Propene

Cumene (isopropylbenzene) → (95-135°C, O_2) → Cumene hydroperoxide

Cumene hydroperoxide → (H_3O^+; *Hock rearrangement*) → Acetone ($CH_3—C(=O)—CH_3$) + Phenol

Acetone

Phenol

Cumene hydroperoxide

Mechanism of the Hock rearrangement

Protonation of cumene hydroperoxide converts OH to a good leaving group that will depart as water. The **protonated hydroperoxide** resembles the **adduct** in the Baeyer-Villiger rearrangement.

The loss of water from the protonated hydroperoxide and the **1,2-migration** of the adjacent **phenyl** group to the electron-deficient oxygen atom that is being created are simultaneous or concerted. Note that a *phenyl group* (C_6H_5) has a *much greater tendency to migrate* than a methyl group (CH_3).

The **carbocation** resulting from phenyl migration then reacts with water to give a **hemiacetal** which is hydrolysed under the reaction conditions to yield phenol and acetone.

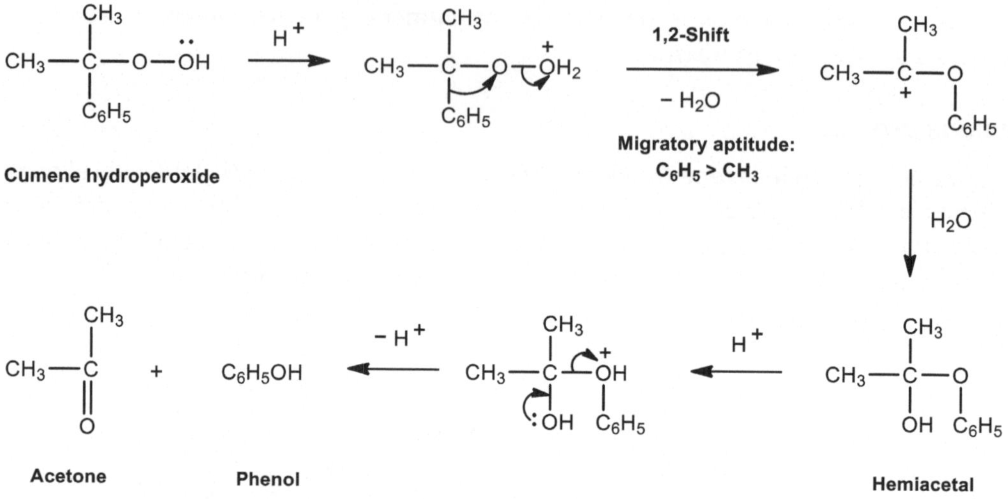

Cumene hydroperoxide

1,2-Shift

$- H_2O$

Migratory aptitude: $C_6H_5 > CH_3$

Acetone Phenol Hemiacetal

16.5 Aromatic Rearrangements

Many rearrangements can occur involving the **aromatic ring**. In general, these aromatic rearrangements are of the following two types:

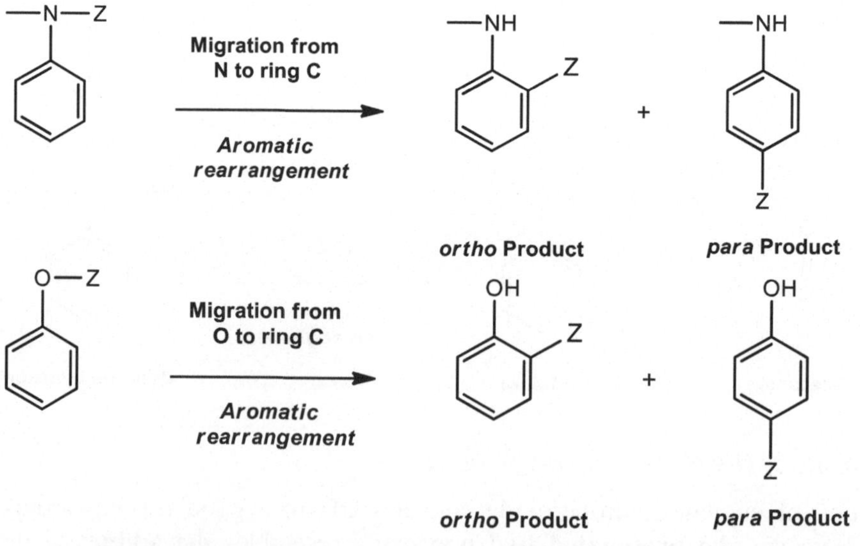

Migration from N to ring C

Aromatic rearrangement

ortho **Product** ***para*** **Product**

Migration from O to ring C

Aromatic rearrangement

ortho **Product** ***para*** **Product**

In one type, the **migrating group** (Z) migrates from **nitrogen** (migration origin) to the ***ortho*** **or** ***para*** **carbon** of the aromatic ring (migration terminus), while in the other type, the **migrating group** (Z) migrates from **oxygen** (migration origin) to the ***ortho*** **or** ***para*** **carbon** of the aromatic ring (migration terminus).

Intermolecular and intramolecular rearrangements

In some rearrangements, **migrations can occur between molecules**, which are called **intermolecular rearrangements**. In these rearrangements, the **migrating group** is completely detached from a molecule and is **free** for migration to another molecule.

In some other rearrangements, **migrations can occur within the molecule**, which are termed **intramolecular reaarangements.** In intramolecular rearrangements, the **migrating group** is **never free** and remains attached to the molecule by partial bonds during migration.

16.5.1 Intermolecular Migration from Nitrogen to Carbon: The Bamberger Rearrangement

When **phenylhydroxylamine** is treated with aqueous sulphuric acid, it rearranges to ***p*-aminophenol**. This rearrangement is known as the **Bamberger rearrangement.**

Phenylhydroxylamine is easily prepared by reduction of nitrobenzene with zinc dust and aqueous NH_4Cl (see Section 13.10). In general, arylhydroxylamines can be prepared by reduction of the corresponding aromatic nitro compounds.

NO_2 | Zn / NH_4Cl | NH—OH | H_2O/H^+ *Bamberger rearrangement* | NH_2 … OH

Nitrobenzene **Phenylhydroxylamine** ***p*-Aminophenol**

Mechanism

Protonation of the OH group of phenylhydroxylamine and subsequent loss of water can form a **carbocation centre** at the ***para*** **position** of the benzene ring. Then a nucleophilic attack by water and the loss of proton from the *para* position regenerates the aromatic ring and produces *p*-aminophenol.

Since the migrating group (OH) is *lost* from the phenylhydroxylamine molecule, the rearrangement is **intermolecular**.

NH—OH | H^+ | $NH-\overset{+}{O}H_2$ | $- H_2O$ | [$\overset{+}{N}H$ ↔ NH +] | $H_2\ddot{O}$

Phenylhydroxylamine

NH | $- H^+$ | $H_2\overset{+}{O}$ H | NH | HO H | $- H^+, + H^+$ (aromatization) | NH_2 … OH

***p*-Aminophenol**

16.5.2 Intramolecular Migration from Nitrogen to Carbon: The Benzidine Rearrangement

Hydrazobenzene on treatment with acid rearranges to **benzidine**, a biphenyl derivative (4,4′-diaminobiphenyl). This rearrangement is known as the **benzidine rearrangement.**

Hydrazobenzene can be easily prepared by reducing nitrobenzene in basic solution with zinc dust and aqueous sodium hydroxide.

NO_2 Zn NaOH NH—NH H⁺ (HCl) **Benzidine rearrangement** NH_2 4 1 1' 4' NH_2

Nitrobenzene **Hydrazobenzene** **Benzidine (4,4'-diaminobiphenyl)**

Mechanism

The benzidine rearrangement proceeds through a **diprotonated species** (both nitrogen atoms of hydrazobenzene are protonated). Then the breaking of the N–N single bond and formation of the C–C single bond between two *para* positions occurs in a single step through a **pericyclic mechanism**. Finally, the loss of a proton from each ring regenerates the aromaticity and produces benzidine.

In this rearrangement, the migrating group (NHC_6H_5) cannot be free and the rearrangement is **intramolecular**.

NH——NH $2H^+$ $\overset{+}{N}H_2$—$\overset{+}{N}H_2$ **Pericyclic mechanism** $\overset{+}{N}H_2$ $\overset{+}{N}H_2$ H H

Hydrazobenzene **Diprotonated hydrazobenzene**

(aromatization) $-2H^+$

H_2N—⟨⟩—⟨⟩—NH_2

Benzidine

16.5.3 Intermolecular Migration from Oxygen to Carbon: The Fries Rearrangement

When **phenyl acetate** (an aryl ester) is treated with anhydrous aluminium chloride (a Lewis acid), it produces a mixture of ***o*-hydroxyacetophenone and *p*-hydroxyacetophenone.** The products are isolated after aqueous work-up.

The reaction involves the **migration of acetyl group** ($COCH_3$) from **oxygen atom** to ***ortho* and *para* positions** of the benzene ring. This rearrangement is known as the **Fries rearrangement**.

Phenyl acetate can be easily prepared by acetylation of phenol with acetic anhydride.

OH
$(CH_3CO)_2O$
O—C(=O)—CH_3
1. $AlCl_3$
2. H_2O (work-up)
Fries rearrangement
Phenol
Phenyl acetate
OH, C=O, CH_3 + OH, O=C, CH_3
o-Hydroxyacetophenone *p*-Hydroxyacetophenone

Mechanism

The mechanism of the Fries rearrangement is similar to that of the Friedel-Crafts acylation reaction.

Phenyl acetate (an ester) forms a **complex** with the Lewis acid ($AlCl_3$) which then eliminates **acetyl cation** (an acylium ion) as an **electrophile**. The acetyl cation is *stabilized by resonance* involving the lone pair on oxygen atom.

The acetyl cation then attacks the *ortho* and *para* positions of the aromatic ring in an **electrophilic aromatic substitution** (the $OAlCl_2$ substituent being an *ortho*/*para* director). Subsequent aqueous work-up gives *o*-hydroxyacetophenone and *p*-hydroxyacetophenone.

Since the acylium cation is eliminated from the molecule and is thus free during the course of the reaction, the rearrangement is **intermolecular**.

Phenyl acetate

$\xrightarrow{AlCl_3}$

Complex

$\xrightarrow{-Cl^-}$

($OAlCl_2$: *o-/p*-director)

Acetyl cation (electrophile)

$-H^+$ *Electrophilic aromatic substitution*

(stabilized by chelation)

(formed faster)

$\xleftarrow{H_2O}$

o-Hydroxyacetophenone p-Hydroxyacetophenone

Product composition

In general, **low temperatures** (~60°C) favour the formation of the ***para*** **product**. This is because the *para* isomer is formed faster (kinetic control) since the attack at the *para* position is not hindered by the $OAlCl_2$ substituent.

However, **high temperatures** (~160°C) favour the ***ortho*** **product** because the *ortho* isomer is more stable (thermodynamic control) by *chelation* involving the coordination of oxygen lone pair to aluminium atom (see above).

PROBLEM 16.6: Predict the products of the following reaction:

Phenyl acetate ($OCOCH_3$) + Toluene (CH_3) $\xrightarrow{AlCl_3}$

Answer

The Fries rearrangement of **phenyl acetate** is carried out in the presence of **toluene** which is capable of undergoing the Friedel-Crafts acylation.

Since the Fries rearrangement is *intermolecular*, the acetyl cation generated in the rearrangement reaction can also attack toluene to give mainly the *para* isomer, *p*-acetyltoluene (*p*-methylacetophenone). Thus the reaction gives **three products**: *o*-hydroxyacetophenone and *p*-hydroxyacetophenone (from the Fries rearrangement of phenyl acetate) and *p*-methylacetophenone (from the Friedel-Crafts acylation of toluene).

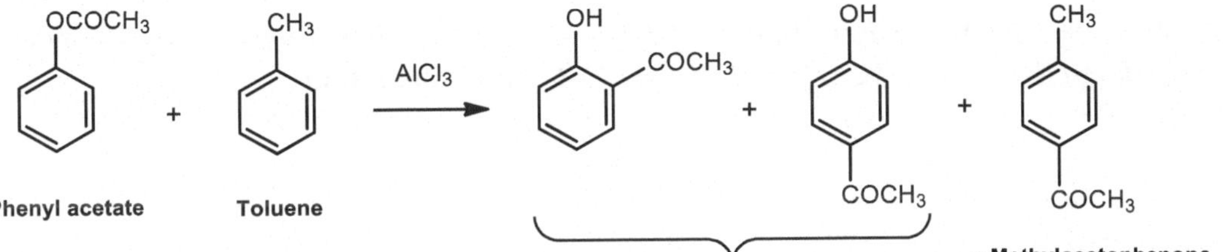

$OCOCH_3$ CH_3 $AlCl_3$ OH $COCH_3$ OH CH_3

Phenyl acetate + **Toluene** → + $COCH_3$ + $COCH_3$

From Fries rearrangement ***p*-Methylacetophenone** ***Crossover product***

Note: The experiment of the above type is called a **crossover experiment** since it determines whether any part (the migrating group) of one molecule crosses over to a molecule of another compound.

In the present case, the migrating group ($COCH_3$) of phenyl acetate crosses over to toluene (a different compound) to give *p*-methylacetophenone, which is called a crossover product.

16.5.4 Intramolecular Migration from Oxygen to Carbon: The Claisen Rearrangement

When **allyl phenyl ether** is heated to 200°C, it undergoes a rearrangement to give ***o*-allylphenol**. This is called the **Claisen rearrangement.**

Normally, the allyl group migrates from the oxygen atom to the *ortho* position of the benzene ring.

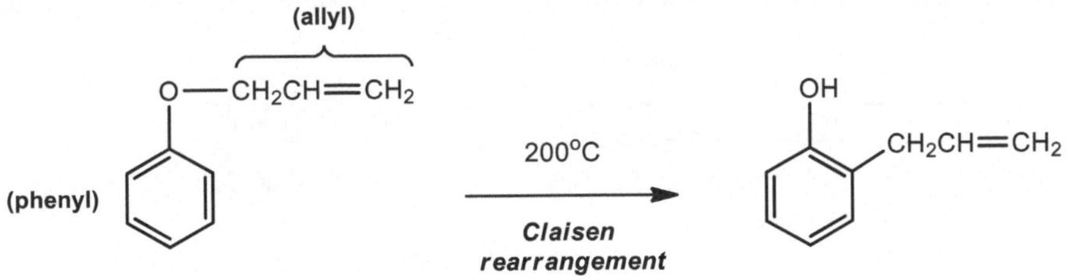

(allyl) $O—CH_2CH{=}CH_2$ **(phenyl)** → 200°C ***Claisen rearrangement*** OH $CH_2CH{=}CH_2$

Allyl phenyl ether **o-Allylphenol**

Allyl phenyl ether can be prepared through a Williamson synthesis from phenol and allyl bromide.

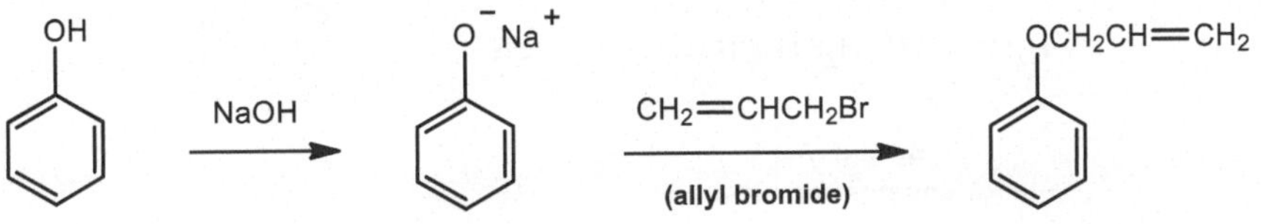

OH → NaOH → $O^-\ Na^+$ → $CH_2{=}CHCH_2Br$ **(allyl bromide)** → $OCH_2CH{=}CH_2$

Phenol **Sodium phenoxide** **Allyl phenyl ether**

Mechanism

The Claisen rearrangement proceeds through a **pericyclic mechanism** involving a C–O bond breaking and a C–C bond formation in a concerted fashion. The initial product of the rearrangement is an *unstable* intermediate which rapidly tautomerizes (or aromatizes) to *o*-allylphenol.

Notably, the **alkene carbon atom (C3)** *farthest from the ether oxygen atom* is bonded to the ***ortho* position** of the benzene ring in the **C–C bond formation**.

Pericyclic mechanism

Tautomerization

$CH_2CH{=}CH_2$

Allyl phenyl ether

Unstable intermediate

***o*-Allylphenol**

Since the migrating allyl group is not dissociated from the allyl phenyl ether molecule during the reaction, the Claisen rearrangement is **intramolecular**.

Aliphatic Claisen rearrangment

A Claisen rearrangement can also occur when **allyl vinyl ethers** are heated. In this ether, a vinyl group (an aliphatic group) replaces a phenyl group (an aryl group).

Unlike the initial unstable intermediate in the **aromatic Claisen rearrangement** of allyl phenyl ether, the **aliphatic Claisen rearrangement** of allyl vinyl ether produces straightway a **stable** compound (4-pentenal), which is a **γ,δ-unsaturated aldehyde**.

(vinyl) **(allyl)**

$CH_2{=}CH{-}O{-}CH_2CH{=}CH_2$ ≡

Allyl vinyl ether

Δ ***Pericyclic mechanism***

$H_2C{=}CHCH_2CH_2CH{=}O$ ≡

4-Pentenal
(a γ,δ-unsaturated aldehyde)

Note: An aliphatic Claisen rearrangement yields a γ,δ-unsaturated carbonyl compound, whereas an aromatic Claisen rearrangement produces an allylphenol.

PROBLEM 16.7: Give the product of the following Claisen rearrangement of the ^{14}C-labeled allyl phenyl ether:

$OCH_2CH{=}^{14}CH_2$ (attached to benzene) $\xrightarrow{\Delta}$

Answer

The observed product is ^{14}C-labeled *o*-allylphenol in which ^{14}C (C3) of the allyl group is directly attached to the benzene ring.

No other product with ^{14}C label at the terminal carbon of the allyl group is formed. This suggests that the allyl group is not free during the reaction.

Thus, the labelling experiment demonstrates that the rearrangement is **intramolecular** in consistent with the pericyclic mechanism.

^{14}C-labeled allyl phenyl ether $\xrightarrow{\Delta}$ [intermediate: $H_2C{=}CH$, O, $^{14}CH_2$, H] $\xrightarrow{\text{Tautomerization}}$ OH, $^{14}CH_2CH{=}CH_2$

Labels: 1, CH_2, 2, CH, $^{14}CH_2$, 3

Observed product

$O{-}CH_2{-}CH{=}^{14}CH_2$ (attached to benzene) $\xrightarrow{\Delta}$ ✗ OH, $CH_2CH{=}^{14}CH_2$

^{14}C-labeled allyl phenyl ether ***Not formed***

SUPPLEMENTARY PROBLEMS

16.8 Each of the following reactions involves a Wagner-Meerwein rearrangement. Draw the structure of the expected product(s) in each case.

(a) Neopentyl bromide (Br) $\xrightarrow[\text{EtOH}]{Ag^+}$

Neopentyl bromide

(b) [3,3-Dimethyl-1-butene] $\xrightarrow[H_2O]{H^+}$

3,3-Dimethyl-1-butene

[*Ans:*

(a) (2-ethoxy-2-methylbutane) and (2-methyl-2-butene) (b) (2,3-dimethyl-2-butanol)]

16.9 Write structures for compounds **A–D** in the following reactions:

(a) Cyclopentanone $\xrightarrow[\text{2. } H_2O]{\text{1. Mg(Hg)}}$ **A** $\xrightarrow{H_2SO_4}$ **B**

(b) $CH_3CH_2C(OH)(CH_3)CH_2OH$ $\xrightarrow{H_2SO_4}$ **C**

(c) $C_6H_5C(CH_3)(OH)C(CH_3)(OH)C_6H_5$ $\xrightarrow{H_2SO_4}$ **D**

[*Hints:* **A** is a pinacol; **B** results from ring carbon migration, **C** from hydride migration, and **D** from phenyl migration in pinacol rearrangements]

16.10 *N*-Methylpropanamide does not undergo the Hofmann rearrangement when treated with bromine in aqueous sodium hydroxide. Explain.

16.11 How would you carry out the following transformation?

Cyclohexanone ⟶ A lactone

16.12 Predict the product in each of the following Claisen rearrangements:

(a) $\xrightarrow{\Delta}$

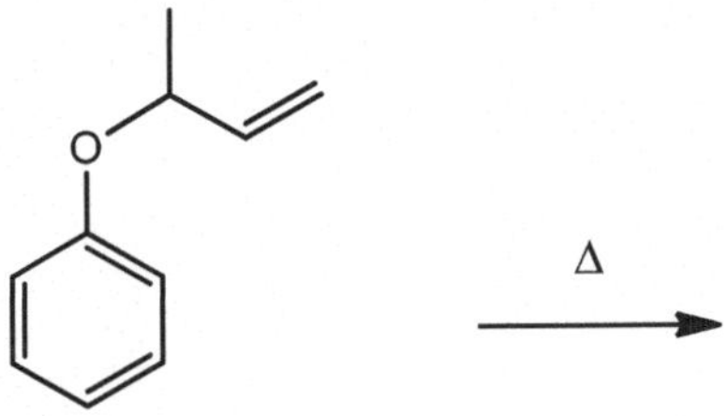

(b) $\xrightarrow{\Delta}$

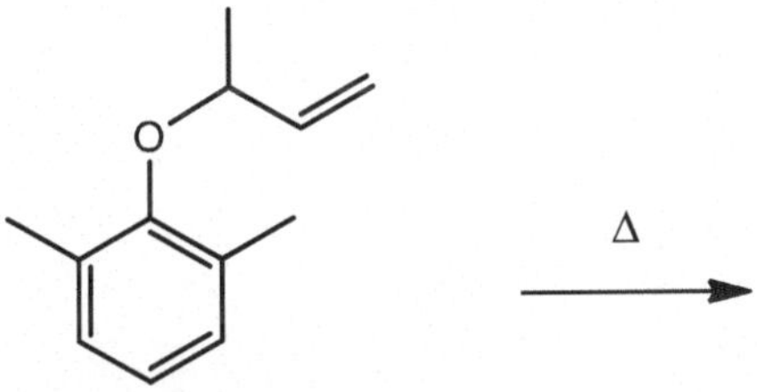

[*Hints:* (a) migration to an *ortho* position with C3 carbon (farthest from the ether oxygen) bonded to the benzene ring; (b) migration to the *para* position with C1 carbon (nearest to the ether oxygen) bonded to the benzene ring]

PART-III

Synthesis of Organic Compounds

CHAPTER 17

Synthetic Strategy: Retrosynthetic Analysis

IN THIS CHAPTER

- Retrosynthetic strategy
- Functional group interconversion (FGI)
- Disconnection, synthon, and synthetic equivalent (or reagent)
- Latent polarity, natural synthons, and unnatural (or umpolung) synthons
- C–O disconnections for alcohols, ethers and esters
- C–N disconnections for amines and amides
- C–C disconnections for alkynes, alcohols and carbonyl compounds
- C–C disconnections for β-hydroxy carbonyl compounds and α,β-unsaturated carbonyl compounds
- C–C disconnections for 1,3-dicarbonyl compounds and 1,5-dicarbonyl compounds

Organic chemistry is above all a creative science. The **synthesis** of a particular organic compound from *easily available starting materials* is fundamental to all organic chemistry.

Most of the organic chemistry we have described so far has concentrated on **reactions** which can convert starting materials to desirable products. But, faced with the problem of preparing a *new* compound, how could we find out which materials or reagents to start with?

In fact, there is a general strategy to design logically the synthesis of organic compounds, which is known as the **retrosynthetic analysis**.

17.1 RETROSYNTHETIC ANALYSIS: WORKING BACKWARD

The molecule to be synthesized is called the **target molecule (TM)**. In planning a synthesis, we need to think about **starting materials**. For this, we look at **reactions in reverse**. This means that we **work backward** from the product (TM) towards the starting materials.

This 'working backward' process is called a **retrosynthetic analysis**, and we represent the process as

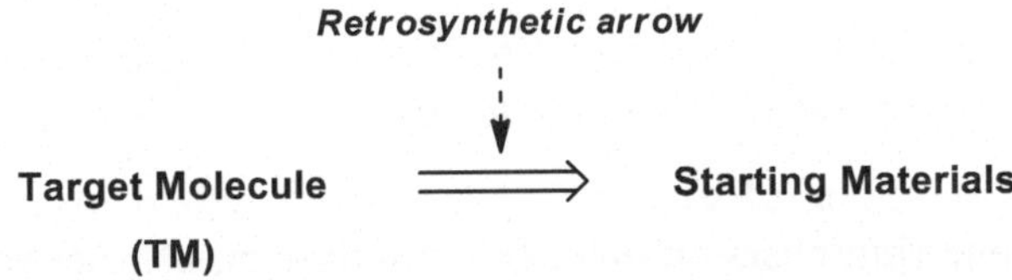

The **open arrow** (⇒) is a special symbol for a retro or reverse reaction. This arrow which denotes the 'implies' symbol from logic is called a **retrosynthetic arrow**. E.J. Corey

introduced the term retrosynthetic analysis formally, although organic chemists have used 'working backward' approach intuitively for many years.

Illustrations

1. Consider **ethyl acetate** (an ester) as the **target molecule (TM)**. We know that an ester can be made by a reaction between a carboxylic acid and an alcohol. Imagining the reaction in reverse, we can represent the **retrosynthetic analysis** of ethyl acetate as follows:

$$CH_3-\overset{O}{\overset{\|}{C}}-O-CH_2CH_3 \Longrightarrow CH_3-\overset{O}{\overset{\|}{C}}-OH + HO-CH_2CH_3$$

Ethyl acetate (TM) **Acetic acid** **Ethanol**

Readily available starting materials

   The **starting materials** are acetic acid and ethanol, which are readily available.

2. Consider a case when **more than one reaction** is required to bring about a **synthesis**. In this case, we **repeat** the analytical process until we reach the level of the compounds that we have as readily available starting materials.

   For example, consider ***p*-aminobenzoic acid** as the **target molecule (TM)**. We know that an amino group can be made by **reduction** of a nitro group and an aromatic carboxylic acid can be prepared by **oxidation** of a side chain methyl group. A nitro group can be easily introduced into an aromatic ring by **nitration** as an electrophilic aromatic substitution reaction.

   Imagining these reactions in reverse in a *proper sequence*, we represent the **retrosynthetic analysis** of *p*-aminobenzoic acid (TM) as follows:

$$p\text{-}H_2N\text{-}C_6H_4\text{-}CO_2H \xRightarrow{\text{Reduction}} p\text{-}O_2N\text{-}C_6H_4\text{-}CO_2H \xRightarrow{\text{Oxidation}} p\text{-}O_2N\text{-}C_6H_4\text{-}CH_3 \xRightarrow{\text{Nitration}} C_6H_5\text{-}CH_3$$

p-Aminobenzoic acid (TM) **p-Nitrobenzoic acid** **p-Nitrotoluene** **Toluene** ***Readily available starting material***

   In the retrosynthetic analysis, the *type of reaction* (such as reduction, oxidation or nitration) can be written above or below a retrosynthetic arrow.

   Note carefully the **sequence or order of events**. Oxidation of methyl group is a *proper* choice when the benzene ring bears an electron-withdrawing NO_2 group. (In the presence of electron-donating NH_2, oxidation will be a bad choice.)

   The **starting material** derived is toluene, which is readily available.

Two important terms in retrosynthetic analysis

We know that a reaction involves the *conversion of one functional group into another functional group* or the *formation of new C–O, C–N or C–C bonds*. In the retrosynthetic analysis, we imagine these processes in reverse.

The conversion of one functional group into another in the retrosynthetic analysis is known as a **functional group interconversion (FGI)**, while mentally breaking a bond into component fragments is called a **disconnection**.

Using these terms, the retrosynthetic analysis is defined thus:

The retrosynthetic analysis is the process of breaking down a target molecule (TM) into available starting materials by functional group interconversion (FGI) and disconnection.

17.2 FUNCTIONAL GROUP INTERCONVERSION (FGI)

We know that a functional group can be converted into another functional group by using an appropriate reagent in a reaction. The reverse of this operation used in a retrosynthetic analysis is called the **functional group interconversion (FGI)**.

These reactions mostly fall into the categories of *substitution*, *reduction* and *oxidation*.

17.2.1 FGI by Substitution

Substitution is a useful reaction for converting a functional group to another. The following are two common cases of substitution and their use as FGI in retrosynthetic analyses.

Substitution at the carbonyl group

A carboxylic acid can be transformed into an acyl chloride using phosphorus pentachloride (PCl_5) or thionyl chloride ($SOCl_2$) by a substitution reaction at the carbonyl group (see Section 12.5).

$$\underset{\textbf{Carboxylic acid}}{R-\overset{\overset{\displaystyle O}{\|}}{C}-OH} \xrightarrow[\text{or } SOCl_2]{PCl_5} \underset{\textbf{Acyl chloride}}{R-\overset{\overset{\displaystyle O}{\|}}{C}-Cl}$$

The **reverse** of this transformation in a retrosynthetic analysis denotes the transformation of an **acyl chloride** to a **carboxylic acid**, which is a **functional group interconversion (FGI)**.

The following is an example:

$$\underset{\textbf{Benzoyl chloride}}{C_6H_5-\overset{\overset{\displaystyle O}{\|}}{C}-Cl} \xRightarrow[\textbf{substitution}]{\textbf{FGI}} \underset{\textbf{Benzoic acid}}{C_6H_5-\overset{\overset{\displaystyle O}{\|}}{C}-OH}$$

The term FGI is usually written above or below the retrosynthetic arrow.

Substitution at the saturated carbon

An alkyl halide can be converted to the corresponding alcohol with hydroxide ion by a substitution reaction (see Section 11.2).

$$\underset{\textbf{Alkyl chloride}}{RCH_2{-}Cl} \xrightarrow{OH^-} \underset{\textbf{Alcohol}}{RCH_2{-}OH}$$

In a retrosynthetic analysis, the transformation of an **alcohol** to an **alkyl chloride** is a **functional group interconversion (FGI)**.

An example is the following:

$$\underset{\textbf{Benzyl alcohol}}{C_6H_5{-}CH_2{-}OH} \underset{\textbf{substitution}}{\overset{\textbf{FGI}}{\Longrightarrow}} \underset{\textbf{Benzyl chloride}}{C_6H_5{-}CH_2{-}Cl}$$

17.2.2 FGI by Reduction

Reduction is a highly useful method for converting a functional group to another. The following are some important cases of reduction and their use as FGI in retrosynthetic analyses.

Reduction of carbonyl compounds

Carbonyl compounds such as ketones and esters can be reduced to alcohols using lithium aluminium hydride ($LiAlH_4$) (see Sections 9.6 and 12.8.1). Amides or substituted amides are reduced to amines (see Section 12.8.3).

$$\underset{\textbf{Ketone}}{R{-}\overset{O}{\overset{\|}{C}}{-}R} \xrightarrow{LiAlH_4} \underset{\textbf{Alcohol}}{R{-}\overset{OH}{\overset{|}{C}H}{-}R}$$

$$\underset{\textbf{Ester}}{R{-}\overset{O}{\overset{\|}{C}}{-}OR'} \xrightarrow{LiAlH_4} \underset{\textbf{Alcohol}}{R{-}CH_2{-}OH} + R'OH$$

$$R-\overset{\overset{O}{||}}{C}-NHR' \xrightarrow{LiAlH_4} R-CH_2-NHR'$$

***N*-Substituted amide** **2° Amine**

Thus, in a retrosynthetic analysis, the transformation of an **alcohol** to a **ketone** (or an **ester**) is a **functional group interconversion (FGI)** involving reduction.

This is illustrated below with the retrosynthetic analysis of **1,3-butanediol** (a dihydric alcohol).

$$CH_3-\overset{\overset{OH}{|}}{CH}-CH_2-CH_2-OH \xRightarrow[\text{reduction}]{\text{FGI}} CH_3-\overset{\overset{O}{||}}{C}-CH_2-\overset{\overset{O}{||}}{C}-OC_2H_5$$

(keto carbonyl) **(ester carbonyl)**

1,3-Butanediol — ***A 1,3-diol***

Ethyl acetoacetate — ***A β-keto ester***

Similarly, a retrosynthetic transformation of ***N*-methylethylamine** (a secondary amine) to ***N*-methylacetamide** (a substituted amide) is an **FGI** involving reduction.

$$CH_3CH_2-NHCH_3 \xRightarrow[\text{reduction}]{\text{FGI}} CH_3\overset{\overset{O}{||}}{C}-NHCH_3$$

N*-Methylethylamine** — ***A 2° amine

N*-Methylacetamide** — ***A substituted amide

Reduction of nitriles

Nitriles can be reduced to primary amines using lithium aluminium hydride or by catalytic hydrogenation (see Section 12.10.2).

$$R-C\equiv N \xrightarrow{LiAlH_4} R-CH_2-NH_2$$

Nitrile **1° Amine**

Thus, a retrosynthetic transformation of a **primary amine** to a **nitrile** is an **FGI** involving reduction.

$$CH_3CH_2CH_2-NH_2 \xRightarrow[\text{reduction}]{\text{FGI}} CH_3CH_2C\equiv N$$

Propylamine — ***A 1° amine***

Propionitrile — ***A nitrile***

Reduction of aromatic nitro compounds

Reduction of an aromatic nitro group to an amino group is usually accomplished with metal and acid (see Section 13.10).

NO_2 Sn HCl NH_2

Aromatic nitro compound **Aromatic amine**

Thus, in a retrosynthetic analysis, the transformation of a **primary aromatic amine** (e.g. *p*-aminophenol) to a **nitro compound** (*p*-nitrophenol) is an **FGI** involving reduction.

NH_2 OH **FGI** **reduction** NO_2 OH

***p*-Aminophenol** ***p*-Nitrophenol**

A 1° aromatic amine ***An aromatic nitro compound***

Reduction of alkynes to alkenes

Catalytic hydrogenation of an alkyne can reduce it to an alkene or alkane depending on the metal catalyst used (see Section 8.15).

$$R{-}C{\equiv}C{-}R \xrightarrow[\text{Lindlar's catalyst}]{H_2} R{-}CH{=}CH{-}R$$

Alkyne **Alkene**

Thus, the retrosynthetic transformation of an **alkene** (1-butene) to an **alkyne** (1-butyne) is an **FGI** involving reduction.

$$CH_3CH_2CH{=}CH_2 \xRightarrow[\text{reduction}]{\text{FGI}} CH_3CH_2C{\equiv}CH$$

1-Butene **1-Butyne**

An alkene ***An alkyne***

17.2.3 FGI by Oxidation

Oxidation is another useful reaction for converting a functional group to another. Consider the following cases of oxidation and their use as FGI in retrosynthetic analyses.

Oxidation of alcohols

A simple approach to the synthesis of aldehyde and ketones is the oxidation of the corresponding alcohols (see Section 10.9). Alcohols can also be oxidized to carboxylic acids.

$$R-CH_2-OH \xrightarrow{PCC} R-\overset{O}{\overset{\|}{C}}-H$$

1° Alcohol → **Aldehyde**

$$R-\overset{OH}{\overset{|}{C}H}-R \xrightarrow[H_2SO_4]{CrO_3} R-\overset{O}{\overset{\|}{C}}-R$$

2° Alcohol → **Ketone**

$$R-CH_2-OH \xrightarrow[H_2SO_4]{CrO_3} R-\overset{O}{\overset{\|}{C}}-OH$$

1° Alcohol → **Carboxylic acid**

Thus, we can write the retrosynthetic analysis of an **aldehyde**, a **ketone** or a **carboxylic acid** using **FGI involving oxidation**. To illustrate, consider the following examples:

$$CH_3\overset{CH_3}{\overset{|}{C}H}-\overset{O}{\overset{\|}{C}}-H \xrightarrow[\text{oxidation}]{\text{FGI}} CH_3\overset{CH_3}{\overset{|}{C}H}-CH_2OH$$

2-Methylpropanal — ***An aldehyde*** ⟹ **2-Methyl-1-propanol** — ***A 1° alcohol***

$$CH_3CH_2-\overset{O}{\overset{\|}{C}}-CH_3 \xrightarrow[\text{oxidation}]{\text{FGI}} CH_3CH_2-\overset{OH}{\overset{|}{C}H}-CH_3$$

Butanone — ***A ketone*** ⟹ **2-Butanol** — ***A 2° alcohol***

$$CH_3\overset{CH_3}{\overset{|}{C}H}CH_2CO_2H \xrightarrow[\text{oxidation}]{\text{FGI}} CH_3\overset{CH_3}{\overset{|}{C}H}CH_2CH_2OH$$

3-Methylbutanoic acid — ***A carboxylic acid*** ⟹ **3-Methyl-1-butanol** — ***A 1° alcohol***

Oxidation of aromatic side chain

The side chain alkyl group (e.g. methyl) of a benzene ring can be oxidized to a carboxyl group, which gives a convenient access to aromatic carboxylic acids (see Section 13.9).

CH_3 (benzene ring) $\xrightarrow[OH^-]{KMnO_4}$ CO_2H (benzene ring)

Aromatic compound with methyl side chain → **Aromatic carboxylic acid**

Thus, in a retrosynthetic analysis, the transformation of an **aromatic carboxyl group** to a **methyl group** is an **FGI** involving oxidation.

CO_2H / NO_2 (p-substituted benzene) $\xRightarrow[\text{oxidation}]{\text{FGI}}$ CH_3 / NO_2 (p-substituted benzene)

***p*-Nitrobenzoic acid** ⇒ ***p*-Nitrotoluene**

17.3 DISCONNECTION

Besides FGI, another important way to perform retrosynthetic analysis is to look for the **disconnection** in the target molecule (TM). **Disconnection is the imagined cleavage of a bond to break the molecule into its component parts.**

For choosing a suitable disconnection, consider the following:

- A disconnection must correspond to a known and reliable reaction.
- If the compound consists of two parts joined by a heteroatom (such as N or O), make a disconnection next to the heteroatom.

In simple cases, the disconnection may straightway lead to the *possible starting materials.*

To illustrate, consider the following examples:

Example 1

Benzocaine is a local anaesthetic, which is an **ester**. We know that esters can be made from carboxylic acids and alcohols.

So the best disconnection in the retrosynthetic analysis of benzocaine is the **C–O disconnection** (a disconnection next to oxygen), which gives ***p*-aminobenzoic acid** (a carboxylic acid) and **ethanol** (an alcohol) as starting materials.

Benzocaine (ethyl *p*-aminobenzoate) — *Ester* ⟹ (C—O ester) *p*-Aminobenzoic acid — *Carboxylic acid* + Ethanol ($HO—C_2H_5$) — *Alcohol*

Note that the *site* of disconnection is usually indicated by a *wavy line*, and often the type of disconnection (**C–O ester**) is indicated above or below the retrosynthetic arrow.

Example 2

Paracetamol is a common antipyretic drug, which is an **amide**. We know that amides can be made from amines and acyl chlorides or acid anhydrides.

So a reasonable disconnection in the retrosynthetic analysis of paracetamol is the **C–N disconnection** (a disconnection next to nitrogen), which gives ***p*-aminophenol** (an amine) and **acetyl chloride** (an acyl chloride) or **acetic anhydride** (an acid anhydride) as starting materials.

Paracetamol (*p*-hydroxyacetanilide) — *Amide* ⟹ (C—N amide) *p*-Aminophenol — *Amine* + Acetyl chloride ($Cl—COCH_3$) — *Acyl chloride* *or* Acetic anhydride ($CH_3CO—O—COCH_3$) — *Acid anhydride*

Here the type of disconnection is indicated as **C–N amide**.

17.4 SYNTHONS AND SYNTHETIC EQUIVALENTS

In a retrosynthetic step, we may **disconnect** (or break) the bond **electronically** as for an ionic reaction. The **fragments** of this disconnection are usually a **cation** and an **anion**. These fragments are called **synthons**.

To illustrate, we consider two examples:

Example 1

Acetophenone ($C_6H_5COCH_3$) is a **ketone**. Consider the following retrosynthetic analysis of acetophenone using a **C–C disconnection** that corresponds to a **Friedel-Crafts acylation**.

C—C

Friedel-Crafts

Acetophenone

Synthons

Benzene

Acetyl chloride

Synthetic equivalents (or reagents)

The disconnection gives **two synthons** with associated polarity (a cationic synthon and an anionic synthon). **But synthons cannot be found as laboratory chemicals.**

Thus what we need are the **synthetic equivalents** of the synthons, which are the actual **reagents** to use.

For the **anionic synthon**, the **synthetic equivalent or reagent** is the corresponding hydrocarbon, **benzene**. This is reasonable because benzene is electron-rich and can act as a nucleophile.

For the **cationic synthon**, the **synthetic equivalent or reagent** is the corresponding halide, **acetyl chloride** in which the carbonyl carbon is electrophilic.

Note: A synthon is an idealised fragment resulting from a disconnection, which helps us to work out which synthetic equivalent or reagent to use.

Example 2

1-Phenylethanol ($C_6H_5CHOHCH_3$) is an **alcohol**. It should be appreciated that often more than one correct retrosynthetic analysis is possible for a target molecule.

Here we delineate two reasonable retrosynthetic analyses of 1-phenylethanol.

1-Phenylethanol: Retrosyntheic analysis 1

In this retrosynthetic analysis, we choose a **C–C disconnection** of the **carbon–methyl bond** that leads to suitable synthetic equivalents or reagents.

OH | CH–CH_3 (1-Phenylethanol) $\xrightarrow{\text{C—C}}$ OH | $\overset{+}{C}H$ (phenyl) + $\bar{C}H_3$

Synthons

1-Phenylethanol

≡ Benzaldehyde (Ph–C(=O)–H) ≡ CH_3—MgI Methylmagnesium iodide

Synthetic equivalents (or reagents)

The disconnection gives two synthons: a **carbocation** directly attached to OH and a simple **carbanion.**

We know that addition of a proton to the carbonyl group of aldehyde or ketone forms a protonated carbonyl compound (see Section 9.1, Figure 9.2). *A protonated aldehyde resembles the cationic synthon.*

Thus the **synthetic equivalent or reagent** for the **cationic synthon** is an aldehyde, **benzaldehyde.**

The anionic synthon (methyl anion) is a carbanion. Organometallic compounds such as Grignard reagents act as a carbon nucleophile or carbanion donor in their reaction with aldehydes or ketones (see Section 9.8.2).

Therefore, as a **synthetic equivalent or reagent** for the **anionic synthon** we can use a Grignard reagent, **methylmagnesium iodide.**

1-Phenylethanol: Retrosyntheic analysis 2

Another equally possible retrosynthetic analysis of 1-phenylethanol involves a **C–C disconnection** of the **carbon–phenyl bond.**

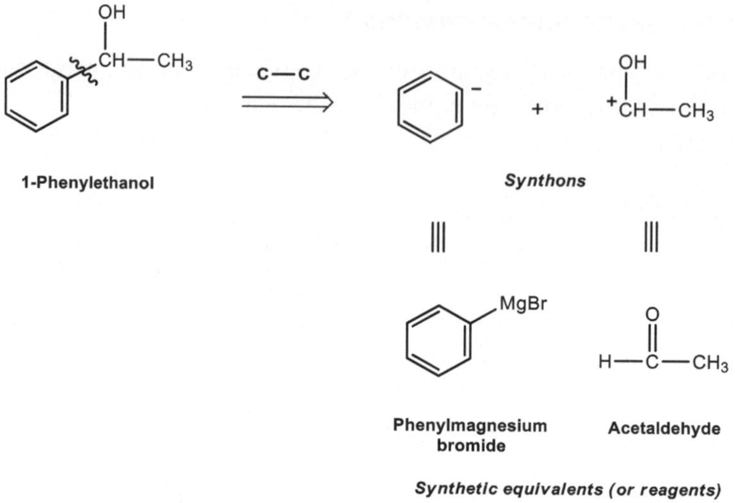

1-Phenylethanol

$\overset{C-C}{\Longrightarrow}$

Synthons: $C_6H_5^-$ + $^+CH(OH)-CH_3$

Phenylmagnesium bromide (C_6H_5MgBr) — Acetaldehyde ($H-C(=O)-CH_3$)

Synthetic equivalents (or reagents)

Here the synthetic equivalent or reagent for the cationic synthon is acetaldehyde. As a synthetic equivalent or reagent for the anionic synthon, we can use phenylmagnesium bromide as a Grignard reagent.

Note: Although both retrosynthetic analyses are reasonable, the preferred analysis for the synthesis of the target molecule depends on several factors including the cost and availability of the reagents.

17.4.1 Examples of Synthons and Their Synthetic Equivalents (or reagents)

In general, retrosynthetic analyses of organic compounds may involve a variety of synthons and their synthetic equivalents.

A cationic synthon is an **electrophilic synthon**, also called an **acceptor synthon** since it accepts electrons. On the other hand, an anionic synthon is a **nucleophilic synthon**, also termed a **donor synthon** because it donates electrons.

Some common synthons and their synthetic equivalents (or reagents) are shown in Figure 17.1.

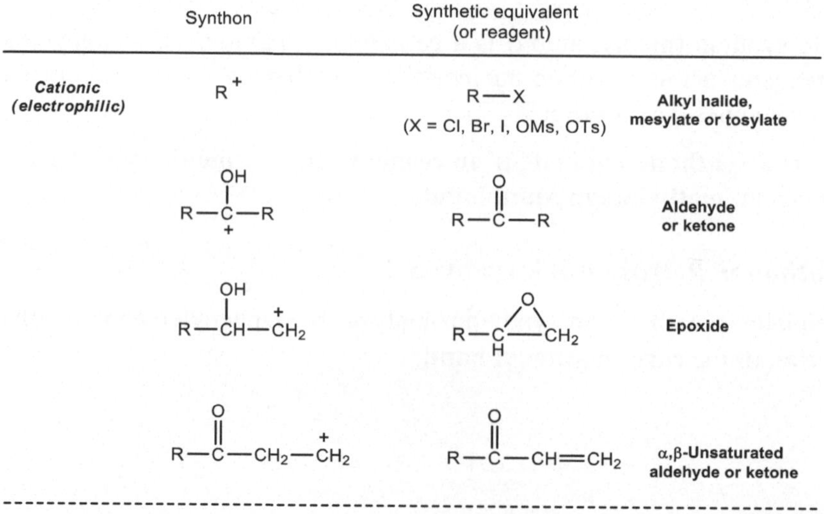

	Synthon	Synthetic equivalent (or reagent)	
Cationic (electrophilic)	R^+	$R-X$ (X = Cl, Br, I, OMs, OTs)	**Alkyl halide, mesylate or tosylate**
	$R-\overset{+}{C}(OH)-R$	$R-C(=O)-R$	**Aldehyde or ketone**
	$R-CH(OH)-\overset{+}{C}H_2$	$R-CH(-O-)CH_2$	**Epoxide**
	$R-C(=O)-CH_2-\overset{+}{C}H_2$	$R-C(=O)-CH{=}CH_2$	**α,β-Unsaturated aldehyde or ketone**

	Synthon	Synthetic equivalent	
Anionic (nucleophilic)	R^-	$RMgX$, RLi, R_2CuLi	**Grignard reagent, organolithium or organocopper reagent**
	$R-C(=O)-\bar{C}H_2$	$R-C(=O)-CH_3$	**Ketone**
		or	
		$R-C(=O)-CH_2-C(=O)-OC_2H_5$	**β-Keto ester**
	$C_2H_5O-C(=O)-\bar{C}H_2$	$C_2H_5O-C(=O)-CH_3$	**Ester**
		or	
		$C_2H_5O-C(=O)-CH_2-C(=O)-OC_2H_5$	**Malonic ester**

Figure 17.1: Some common synthons and their synthetic equivalents (or reagents).

17.5 C–X DISCONNECTIONS

Disconnection of a bond between carbon (C) and a heteroatom (X) is called the **C–X disconnection**. The heteroatom is commonly **oxygen (O)** and **nitrogen (N)**.

This disconnection can be easily chosen since we need to recognize **only one heteroatom-containing functional group**.

We have already introduced a C–O disconnection for ester and a C–N disconnection for amide in Section 17.3.

Here we will describe C–O and C–N disconnections involving synthons. As a result of C–O or C–N disconnection, the heteroatom (O or N) will get a negative charge, forming an anionic synthon.

17.5.1 C–O Disconnections

The retrosynthetic analysis of alcohols, ethers or esters can involve a **C–O disconnection**.

Alcohols

We know that alcohols are easily prepared by nucleophilic substitution of alkyl halides with sodium or potassium hydroxide. The reaction forms a C–O bond.

Therefore, we can write a retrosynthetic analysis of an alcohol, for example, **1-propanol** with a **C–O disconnection** as follows:

$$\underset{\textbf{1-Propanol}}{CH_3CH_2CH_2 \wr OH} \xRightarrow[\text{alcohol}]{C-O} CH_3CH_2\overset{+}{C}H_2 + \bar{O}H$$

Synthons		Reagents
$CH_3CH_2\overset{+}{C}H_2$	≡	$CH_3CH_2CH_2Br$ **Propyl bromide**
+		
$\bar{O}H$	≡	$NaOH$

Here the disconnection is indicated as **C–O alcohol**. The anionic synthon will get a negative charge on oxygen, for which the reagent is NaOH. For the cationic synthon, the reagent is propyl bromide.

Ethers

Ethers are usually prepared by the Williamson synthesis involving a nucleophilic substitution of an alkyl halide with an alkoxide ion, which forms a C–O bond. Therefore, a reasonable retrosynthetic analysis of ethers involves a **disconnection of the C–O bond.**

Consider, for example, the retrosynthetic analysis of **ethoxybenzene** ($C_6H_5OC_2H_5$). As the ether is *unsymmetrical*, **two C–O disconnections** (marked *a*, *b*) are possible:

a
O–CH_2CH_3
Ethoxybenzene
C—O
ether
Good choice
O^- + $\overset{+}{C}H_2CH_3$
Synthons
≡
OH + base
Phenol
$BrCH_2CH_3$
Ethyl bromide
Reagents
A reliable S_N2 reaction

b
O—CH_2CH_3
Ethoxybenzene
C—O
ether
Bad choice
+ + $^-O—CH_2CH_3$
Synthons
≡
Cl
Chlorobenzene
NaO—CH_2CH_3
Sodium ethoxide
Reagents
No reaction

The **disconnection *a*** corresponds to a reaction between phenoxide ion (from phenol and base) and ethyl bromide, which is a **reliable S_N2 reaction** and is therefore a **good choice**.

In contrast, the **disconnection *b* is a bad choice** because we know that chlorobenzene does not undergo an aromatic nucleophilic substitution with ethoxide ion because chlorobenzene bears no activating electron-withdrawing substituents such as nitro group.

Therefore, a reasonable retrosynthetic analysis of ethoxybenzene involves the C–O ether disconnection *a*.

Note: Disconnections must correspond to known and reliable reactions, which would help us decide *where* to make the disconnection.

Esters

In the retrosynthetic analysis of esters, we almost always choose a **C–O disconnection** (see also Section 17.3).

For example, consider a retrosynthetic analysis of **benzyl benzoate** ($C_6H_5CO_2CH_2C_6H_5$). The **C–O ester disconnection** gives suitable synthons, for which their synthetic equivalents or reagents (benzoyl chloride and benzyl alcohol) are readily available.

$C_6H_5C(=O)\text{-}\!\!\!\!\!\!\sim OCH_2C_6H_5$ (Benzyl benzoate) $\xRightarrow[\text{ester}]{\text{C—O}}$ $C_6H_5C^+(=O)$ + $^-OCH_2C_6H_5$ (Synthons) $\equiv$ $C_6H_5C(=O)Cl$ (Benzoyl chloride) + $C_6H_5CH_2OH$ (Benzyl alcohol) (Reagents)

Benzyl benzoate

Synthons

Reagents

17.5.2 C–N Disconnections

The retroynthetic analyses of amines and amides can involve a **C–N disconnection**. The C–N disconnection for amides is a useful strategy (see Section 17.3).

Amines

Nucleophilic substitution of alkyl halides with ammonia or an amine can produces a new amine. The reaction forms a C–N bond.

However, the C–N amine disconnection in a retrosynthetic analysis is often not satisfactory.

The problem is that the product amine (say, a secondary amine) is usually more reactive than the starting material (a primary amine) and **multiple alkylations** are possible when the reagents (a primary amine and an alkyl halide) are employed in the synthesis (see also Section 14.1).

$R\text{—}NH\text{-}\!\!\sim CH_3$ (2° Amine) $\xRightarrow[\text{amine}]{\text{C—N}}$ $R\text{—}\bar{N}H$ + CH_3^+ (Synthons) $\equiv$ $R\text{—}NH_2$ (1° Amine) + $CH_3\text{—}I$ (Methyl iodide) (Reagents) ***Not useful***

Problem of multiple alkylations in synthesis:

$$R{-}NH_2 \xrightarrow{CH_3I} R{-}NHCH_3 \xrightarrow{CH_3I} R{-}N(CH_3)_2 \xrightarrow{CH_3I} R{-}\overset{+}{N}(CH_3)_3\ I^-$$

1° Amine **2° Amine** **3° Amine** **Quaternary ammonium salt**

In some cases where the product amine (e.g. glycine) is **less reactive** than the starting material (ammonia), the C–N disconnection for amines is useful.

$$H_2N{-}CH_2{-}CO_2H \xRightarrow[\text{amine}]{C-N} NH_3 + Cl{-}CH_2CO_2H \quad \textit{Useful}$$

Glycine **Chloroacetic acid**

A 1° Amine ***Reagents***

Synthesis:

$$Cl{-}CH_2CO_2H \xrightarrow{NH_3} H_2N{-}CH_2{-}CO_2H \xrightarrow[\times]{ClCH_2CO_2H} \text{No multiple alkylations}$$

Note that glycine is less reactive as a nucleophile because it possesses an electron-withdrawing carboxyl group.

FGI followed by C–N amide (or C=N imine) disconnection

One solution to the problem of amine synthesis is to perform a **functional group interconversion (FGI)** to convert the **target amine** to the *less reactive* **amide** and then apply the **C–N amide disconnection.**

This is illustrated with the following retrosynthetic analysis of **diethylamine**:

$$CH_3CH_2{-}NH{-}CH_2CH_3 \xRightarrow[\text{reduction}]{FGI} CH_3CH_2{-}NH{-}\overset{O}{\overset{\|}{C}}CH_3 \xRightarrow[\text{amide}]{C-N} CH_3CH_2\overset{-}{N}H + {}^{+}\overset{O}{\overset{\|}{C}}CH_3$$

Diethylamine ***N*-Ethylacetamide** ***Synthons***

$$\equiv \qquad \equiv$$

$$CH_3CH_2NH_2 \qquad Cl{-}\overset{O}{\overset{\|}{C}}CH_3$$

Ethylamine **Acetyl chloride**

Reagents

Another method is to perform an **FGI** to convert the **target amine** to an **imine**, and then apply a **C=N imine disconnection**, which gives an aldehyde or ketone and an amine as starting materials.

This approach, called the **reductive amination**, is illustrated with the following retrosynthetic analysis of a secondary (2°) amine.

$RCH_2—NH—CH_2R'$ (A 2° Amine) $\xrightarrow[\text{reduction}]{\text{FGI}}$ $RCH_2—N{=}CHR'$ (An imine) $\xrightarrow[\text{imine}]{\text{C=N}}$ RCH_2NH_2 (1° Amine) + $R'—CH{=}O$ (Aldehyde)

Reagents

Amino Alcohols: 1,2-diX Disconnection

Amino alcohols are amines containing an alcoholic hydroxyl group. Here we consider **1,2-amino alcohols (or β-amino alcohols)** that contain alcohol and amine functional groups in a relative 1,2-position. They are referred to as **1,2-diX (X = N, O) compounds.**

The C–N disconnection for 1,2-amino alcohols is decided by **both functional groups,** and this **two-group C–X disconnection** is denoted by **1,2-diX disconnection.**

The 1,2-diX disconnection involves the disconnection of C–N bond between the β carbon (C2) and nitrogen.

This is illustrated with the following retrosynthetic analysis of ***N*-methylethanolamine**:

$CH_3—NH—\overset{2}{\underset{\beta}{C}}H_2—\overset{1}{\underset{\alpha}{C}}H_2—OH$ $\xrightarrow{\text{1,2-diX}}$ $CH_3—NH^-$ + $\overset{+}{C}H_2—CH_2OH$

N-Methylethanolamine

A 1,2- or β-amino alcohol

Synthons: $CH_3—NH^-$ ≡ CH_3NH_2 (Methylamine); $\overset{+}{C}H_2—CH_2(OH)$ ≡ $CH_2—CH_2$ bridged by O (Ethylene oxide (an epoxide))

Synthons *Reagents*

Note that the synthetic equivalent or reagent for the cationic synthon is an epoxide (see Figure 17.1).

17.6 C–C DISCONNECTIONS: TARGET MOLECULES WITH ONE FUNCTIONAL GROUP

The **C–C disconnection** is by far the most important disconnection in the retrosynthetic analysis because the carbon framework of organic molecules is built by forming C–C bonds. Since target molecules contain *many* C–C bonds, it is often difficult to decide *where* to disconnect.

If a **target molecule** contains **only one functional group,** the situation becomes relatively simple. We can use the chemistry of that functional group to determine which C–C bond to disconnect. This type of C–C disconnection with respect to *one functional group* is called **one-group C–C disconnection.**

A C–C disconnection gives an **electrophilic carbon** and a **nucleophilic carbon** as **synthons**. The synthetic equivalents or reagents for these synthons are generally available.

Here we will describe **one group C–C disconnection** for alkynes, alcohols and carbonyl compounds (ketones and carboxylic acids).

17.6.1 Alkynes

The **functional group** of an alkyne is a **triple bond**. We know that alkylation of alkynes (i.e. substitution of a relatively acidic hydrogen atom attached to the triple bond by an alkyl group) is a reliable reaction (see Section 11.2). Therefore, a reasonable **C–C disconnection** of alkynes is **next to the triple bond**.

As a result of this one group C–C disconnection, the alkyne carbon (sp) will get a negative charge forming the **anionic synthon**.

To illustrate, consider the following retrosynthetic analysis of **1-butyne**.

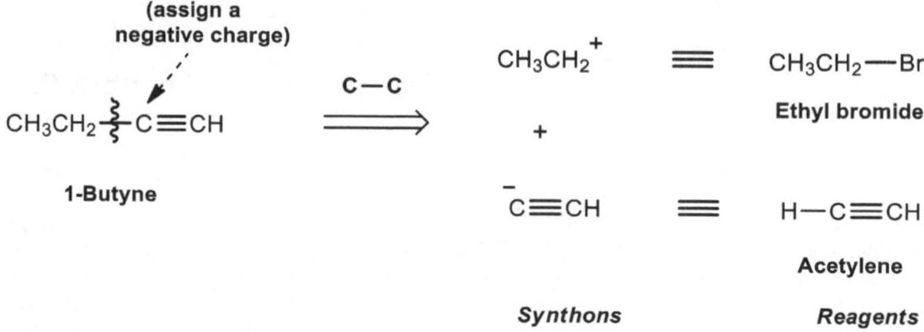

For the *anionic synthon*, the **synthetic equivalent or reagent** is **acetylene** whereas for the *cationic carbocation synthon*, the **reagent** is **ethyl bromide**. These regents are readily available.

17.6.2 Alcohols

The **functional group** of an alcohol is a **hydroxyl group (OH)**. A reasonable C–C disconnection for alcohols is **next to the OH group**. This type of C–C disconnection is termed a **1,1 C–C disconnection**.

An alternative C–C disconnection for alcohols is also possible, which is indicated as a **1,2 C–C disconnection**.

1,1 C–C Disconnection

A C–C disconnection next to the OH group is labelled as 1,1 C–C because the C–C bond being disconnected is attached to **C1**, which is the *same* carbon atom (**C1**) that bears the OH functional group. Note that the carbon atom bearing OH is assigned number 1.

As a result of 1,1 C–C disconnection, the carbon (C1) bearing OH will get a positive charge, forming the cationic synthon.

The 1,1 C–C disconnection for alcohols is illustrated with the following examples.

1. Consider ***t*-butanol** as the target molecule. The 1,1-C–C disconnection gives **two synthons** for which the reagents are **acetone** and a Grignard reagent, **methylmagnesium bromide** (see Figure 17.1).

$$(CH_3)_2C(OH)\text{–}CH_3 \;\xrightarrow{1,1\ C-C}\; CH_3\text{–}\overset{+}{C}(OH)\text{–}CH_3 + \overset{-}{C}H_3$$

t-Butanol

Synthons: $CH_3\text{–}\overset{+}{C}(OH)\text{–}CH_3 \equiv CH_3\text{–}C(=O)\text{–}CH_3$ (Acetone); $\overset{-}{C}H_3 \equiv CH_3\text{–}MgBr$ (Methylmagnesium bromide)

Synthons *Reagents*

2. Consider **3-methyl-2-pentanol** as the target molecule. This is a tertiary alcohol with two *identical* ethyl groups.

   We know that tertiary alcohols with *two identical alkyl groups* next to the hydroxyl group can be prepared by the reaction of an ester with two moles of a Grignard reagent (see Section 12.9.1).

   Thus we can write the retrosynthetic analysis of 3-methyl-2-pentanol as follows:

$$CH_3\text{–}C(OH)(CH_2CH_3)\text{–}CH_2CH_3 \;\xrightarrow{1,1\ C-C}\; CH_3\text{–}C(=O)\text{–}CH_2CH_3 + CH_3CH_2MgBr$$

3-Methyl-2-pentanol → Butanone + Ethylmagnesium bromide

$$CH_3\text{–}C(=O)\text{–}CH_2CH_3 \;\xrightarrow{1,1\ C-C}\; CH_3\text{–}\overset{+}{C}(=O) + CH_3CH_2MgBr$$

(Ethylmagnesium bromide)

$$CH_3\text{–}\overset{+}{C}(=O) \equiv CH_3\text{–}C(=O)\text{–}OC_2H_5$$

Ethyl acetate

The analysis involves two consecutive 1,1 C–C disconnections. Note that the C–C bond being disconnected is between C1 and an *ethyl* group. The second 1,1 C–C for butanone (a ketone) gives ethyl acetate and ethylmagnesium bromide as the reagents. Overall, the analysis leads to an ester (**ethyl acetate**) and *two molecules* of a Grignard reagent (**ethylmagnesium bromide**) as the starting materials.

Alternatively, we can also use an *equivalent short cut*, a **double disconnection (2 × 1,1 C–C)** that leads *directly* to **ethyl acetate** (an ester) and 2 molecules of **ethylmagnesium bromide** (a Grignard reagent), as shown below.

$$CH_3-\overset{OH}{\underset{CH_2CH_3}{C}}-CH_2CH_3 \xRightarrow[1,1\ C-C]{2\ x} CH_3-\overset{O}{\overset{\|}{C}}-OC_2H_5 + 2\ CH_3CH_2MgBr$$

3-Methyl-2-pentanol; Ethyl acetate; Ethylmagnesium bromide

1,2 C–C disconnection

A C–C disconnection is labelled as 1,2 C–C when the C–C bond being disconnected is attached to **C2**, which is adjacent to **C1** *bearing the OH group.*

As a result of 1,2 C–C disconnection, C2 gets a positive charge, forming the cationic synthon.

To illustrate, consider a retrosynthetic analysis of **1-phenyl-2-propanol** involving a **1,2 C–C disconnection.**

$$C_6H_5-CH_2-\overset{OH}{CH}-CH_3 \xRightarrow{1,2\ C-C} \overset{+}{C}H_2-\overset{OH}{CH}-CH_3 \equiv CH_2-CH-CH_3\ (\text{An epoxide})$$

$$+\quad C_6H_5^{-} \equiv C_6H_5-MgBr\ (\text{Phenylmagnesium bromide})$$

1-Phenyl-2-propanol

Synthons *Reagents*

The 1,2 C–C disconnection gives two synthons. For the *cationic synthon*, the reagent is an **epoxide** (see Figure 17.1), while for the *anionic synthon*, the reagent is **phenylmagnesium bromide** (a Grignard reagent).

The epoxide (**2-methyloxirane**) can be easily prepared by oxidation of propene with a peroxy acid.

Note: The numbering (1, 2) relating to disconnection must not be confused with the IUPAC numbering for the target molecule.

17.6.3 Carbonyl Compounds

The retrosynthetic analysis of carbonyl compounds is described here with ketones and carboxylic acids.

Ketones

The retrosynthetic analysis of ketones can involve a 1,1 C–C disconnection or 1,2 C–C disconnection.

1,1 C–C disconnection

The functional group of a ketone is a **carbonyl group (C=O)**. The disconnection of the C–C bond attached to the carbonyl group is a **1,1 C–C disconnection**.

In this disconnection, the carbonyl carbon (C1) will get a positive charge, forming a cationic synthon.

To illustrate, consider a retrosynthetic analysis of **acetophenone** as the target molecule.

The **1,1 C–C disconnection** gives an acyl cation and a carbanion as the synthons, for which the usual reagents are an ester and a Grignard reagent respectively. But we know that esters react with Grignard reagents to give tertiary alcohols, *not* ketones. Therefore, *esters and Grignard reagents are not useful reagents in this case.*

$C_6H_5-C(=O)\text{-}\!\!\wr\!\!-CH_3$ (Acetophenone) $\xRightarrow{1,1\ C-C}$ $C_6H_5-\overset{+}{C}(=O)$ + $\overset{-}{C}H_3$ (Synthons)

$C_6H_5-C(=O)^{+} \equiv C_6H_5-C(=O)-OC_2H_5$ (Ester)

$^{-}CH_3 \equiv CH_3-MgBr$ (Grignard reagent)

Reagents — **Not useful**

However, a less reactive organocadmium reagent reacts with an acid chloride to produce a ketone (see Section 12.9.3).

Therefore, for the acyl cation (benzoyl cation), we can choose **benzoyl chloride** and for the methyl carbanion, we can choose **dimethylcadmium** as suitable reagents.

$C_6H_5-C(=O)\text{-}\!\!\wr\!\!-CH_3$ (Acetophenone) $\xRightarrow{1,1\ C-C}$ $C_6H_5-\overset{+}{C}(=O)$ + $\overset{-}{C}H_3$ (Synthons)

$C_6H_5-C(=O)^{+} \equiv C_6H_5-C(=O)-Cl$ (Benzoyl chloride)

$^{-}CH_3 \equiv CH_3-Cd-CH_3$ (Dimethylcadmium)

Reagents — **Useful as a reliable reaction**

1,2 C–C disconnection

The retrosynthetic analysis of a ketone can also involve a **1,2 C–C disconnection**. In this disconnection, the **carbon atom C2** (adjacent to the carbonyl carbon C1) **will get a negative charge**, forming an **anionic synthon**. Note that C2 is the α carbon with respect to the carbonyl group.

This is illustrated with the following retrosynthetic analysis of **4-phenyl-2-butanone** involving a 1,2 C–C disconnection.

The **1,2 C–C disconnection between α carbon (C2) and β carbon** gives a negative charge on the α carbon forming an **α-carbanion** as the anionic synthon. The cationic synthon is a **carbocation**.

(assign a negative charge)

$C_6H_5{-}CH_2 \wr CH_2{-}C({=}O){-}CH_3$ (β, α; 2, 1)

4-Phenyl-2-butanone

1,2 C—C ⟹ (disconnect between α and β)

$C_6H_5{-}\overset{+}{C}H_2 \equiv C_6H_5{-}CH_2Br$ Benzyl bromide

+

$\overset{-}{C}H_2{-}C({=}O){-}CH_3 \equiv C_2H_5O{-}C({=}O){-}CH_2{-}C({=}O){-}CH_3$

(an α-Carbanion) Acetoacetic ester

Synthons *Reagents*

For the cationic synthon (**benzyl cation**), we can use **benzyl bromide** as the reagent. For the **α-carbanion synthon**, a suitable reagent is **acetoacetic ester** (see Figure 17.1) because acetoacetic ester synthesis of ketones is a reliable method (see Section 15.9.1).

Carboxylic acids

A **carboxylic acid** is also a carbonyl compound. Consider **butanoic acid** as the target molecule.

The retrosynthetic analysis of butanoic acid can be performed using a **1,2 C–C disconnection** (cf. the analysis of 4-phenyl-2-butanone above).

(assign a negative charge)

$CH_3{-}CH_2 \wr CH_2{-}C({=}O){-}OH$ (β, α; 2, 1)

Butanoic acid

1,2 C—C ⟹ (disconnect between α and β)

$CH_3{-}\overset{+}{C}H_2 \equiv H_3C{-}CH_2Br$ Ethyl bromide

+

$\overset{-}{C}H_2{-}C({=}O){-}OH \equiv C_2H_5O{-}C({=}O){-}CH_2{-}C({=}O){-}OC_2H_5$

(an α-carbanion) Malonic ester

Synthons *Reagents*

The reagent for the cationic synthon is **ethyl bromide**. For the anionic synthon (an α-carbanion), the reagent of choice is **malonic ester** (see Figure 17.1) because malonic ester synthesis of carboxylic acids is a reliable method (see Section 15.9.2).

17.7 C–C DISCONNECTIONS: TARGET MOLECULES WITH TWO FUNCTIONAL GROUPS

If a target molecule contains *two* functional groups, **both functional groups together** would determine **which C–C bond to disconnect**. This type of C–C disconnection with respect to two functional groups is called a **two-group C–C disconnection**.

We will describe **two group C–C disconnections** for the following difunctionalized compounds:

(a) β-hydroxy carbonyl compounds or aldols
(b) α,β-unsaturated carbonyl compounds
(c) 1,3-dicarbonyl compounds
(d) 1,5-dicarbonyl compounds

17.7.1 Latent Polarity: Natural Synthons and Unnatural (or umpolung) Synthons

We have learned that disconnections must correspond to known reliable reactions. Thus a C–C disconnection on a target molecule must provide viable synthons for which corresponding reagents are available.

The concept of **latent polarity** can assist us in the identification of a **suitable C–C disconnection** on a target molecule.

Latent polarity

Latent polarity is the imaginary pattern of alternating positive and negative charges on a target molecule *as we move away from a functional group* (OH or C=O).

The pattern starts with a **positive charge on the carbon atom bearing the oxygen atom**, as shown for alcohols and ketones here.

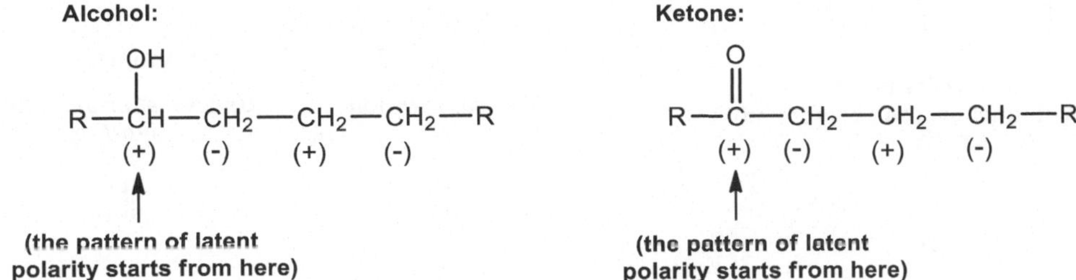

Note that the latent polarity is indicated by a bracketed charge, (+) or (–) to distinguish it from the full charge. It signifies the **natural polarity** of the molecule.

Natural synthons

Consider a **β-hydroxy ketone** in which hydroxyl (OH) and carbonyl (C=O) functional groups are in a **1,3-relationship**. Thus a β-hydroxy ketone is a **1,3-difunctionalized compound**.

A C–C disconnection, *as shown below*, gives two synthons which correspond to the latent or natural polarity of the molecule. These synthons are called natural or logical synthons.

The reagents for these natural synthons (an aldehyde and a ketone) are readily available.

OH O
3 | β 2 || 1
R—CH—CH$_2$—C—R
(+) (-) (+)

C—C ⟹

OH
|
R—CH
+
Natural synthon

+

O
– ||
CH$_2$—C—R
Natural synthon

β-Hydroxy ketone
(1,3-difunctionalized compound)

|||

O
||
R—CH
Aldehyde

|||

O
||
CH$_3$—C—R
Ketone

Reagents

Unnatural or umpolung synthons

Consider an **α-hydroxy ketone** in which hydroxyl (OH) and carbonyl (C=O) functional groups are in a **1,2-relationship**. Thus an α-hydroxy ketone is a **1,2-difunctionalized compound**.

A C–C disconnection, *as shown below*, gives a cationic synthon and an anionic synthon.

OH O
2 | α || 1
R—CH—C—R

C—C ⟹

OH
|
R—CH
+
Natural synthon

+

O
– ||
C—R
(acyl anion)
Unnatural or umpolung synthon

α-Hydroxy ketone
(1,2-difunctionalized compound)

|||

O
||
R—CH
Aldehyde

|||

?

Reagents

The **cationic synthon** which correspond to the latent or natural polarity of the molecule is a **natural synthon.**

But the anionic synthon (acyl anion) has a polarity *opposite* to the natural polarity (+) of the carbonyl group. The acyl anion is therefore an unnatural or illogical synthon, which is also called an umpolung synthon (from the German *umpolung*, meaning inverse polarity).

Note: In general, the C–C disconnection for compounds with 1,2- or 1,4-related functional groups gives an umpolung synthon. *Equivalent reagents for umpolung synthons are indeed available*, and they are the key to the synthesis of 1,2-dicarbonyl compounds (RCOCOR) and 1,4-dicarbonyl compounds ($RCOCH_2CH_2COR$). However, these syntheses are beyond the scope of our discussion.

17.7.2 β-Hydroxy Carbonyl Compounds or Aldols

An **aldol** is a **β-hydroxy aldehyde or ketone** (see Section 15.4) in which two functional groups (OH and C=O) are in a **1,3-relationship**.

Consider the following **aldol or β-hydroxy ketone** (a 1,3-difunctionalized compound) as the target molecule and examine its pattern of latent polarities.

(latent polarities from hydroxyl group)

$C_6H_5-CH(OH)-CH_2-C(=O)-CH_3$ with polarities (+) on CH, (−) on CH_2, (+) on C from the hydroxyl group, and (+) on CH, (−) on CH_2, (+) on C from the carbonyl group

(latent polarities from carbonyl group)

An overlapping pattern of latent polarities for a β-hydroxy ketone or aldol

The latent polarities from the hydroxyl group and the latent polarities from the carbonyl group **overlap**. When such overlapping pattern in a 1,3-difunctionalized compound is present, a reasonable **C–C disconnection** gives two **natural synthons**, and a simple synthesis of the target molecule is possible.

Illustrative example

Consider a retrosynthetic analysis of **4-hydroxy-4-phenyl-2-butanone** (a β-hydroxy ketone or aldol).

A reasonable **C–C disconnection** involves a disconnection of the bond **between the β carbon and the α carbon** to give an **α-carbanion as an anionic synthon**.

(assign a negative charge)

$C_6H_5-CH(OH)-CH_2-C(=O)-CH_3$ (β at CH, α at CH_2)

4-Hydroxy-4-phenyl-2-butanone

An aldol or β-hydroxy ketone

1,3-diO ⟹ (disconnect between α and β)

$C_6H_5-\overset{+}{C}H(OH)$ ≡ $C_6H_5-CH=O$ Benzaldehyde

\+

$^{-}CH_2-C(=O)-CH_3$ (an α-carbanion) ≡ $CH_3-C(=O)-CH_3$ Acetone

Natural synthons *Reagents*

This C–C disconnection is labelled as a **1,3-diO disconnection**, which indicates a **C–C disconnection** for 1,3-difunctionalized compounds with **two oxygen-containing functional groups** (OH and C=O).

The **1,3-diO** disconnection gives two **natural synthons** for which the synthetic equivalents or reagents (benzaldehyde and acetone) are readily available.

17.7.3 α,β-Unsaturated Carbonyl Compounds

β-Hydroxy carbonyl compounds often *dehydrate* easily to form α,β-unsaturated carbonyl compounds (see Section 15.5).

Thus the retrosynthetic analysis of an **α,β-unsaturated ketone** (e.g. mesityl oxide) involves an **FGI dehydration** followed by a **1,3-diO disconnection**, as shown below.

(add OH) (add H)

$CH_3-C(CH_3)=CH-C(=O)-CH_3$ $\xRightarrow[\text{dehydration}]{\text{FGI}}$ $CH_3-C(OH)(CH_3)-CH_2-C(=O)-CH_3$

Mesityl oxide

α,β-Unsaturated ketone

A β-Hydroxy ketone

⇓ **1,3-diO**

$CH_3-C(=O)-CH_3 + CH_3-C(=O)-CH_3$

Acetone (2 molecules)

The analysis leads to *two molecules* of **acetone** as the **starting material.**

Note that **FGI** denotes the retrosynthetic transformation of an alkene to an alcohol, which is the reverse of the dehydration of alcohol.

A shorthand disconnection (α,β)

In an aldol condensation (see Sections 15.5), dehydration of the aldol occurs so readily that the product normally isolated is an α,β-unsaturated carbonyl compound.

Thus, in the retrosynthetic analysis of **α,β-unsaturated carbonyl compounds**, we can also write a **shorthand disconnection** which leads *directly* to the two carbonyl components.

The shorthand expression is denoted by **α,β**.

This is illustrated with **cinnamaldehyde** (an α,β-unsaturated aldehyde) as the target molecule.

Add O (as C=O) Add 2 H's

$$C_6H_5{-}\overset{\beta}{CH}{=}\overset{\alpha}{CH}{-}\overset{O}{\overset{\|}{C}}{-}H \xrightarrow{\alpha,\beta} C_6H_5{-}CH{=}O + CH_3{-}\overset{O}{\overset{\|}{C}}{-}H$$

Cinnamaldehyde — ***An α,β-unsaturated aldehyde*** | **Benzaldehyde** | **Acetaldehyde**

In the **α,β** disconnection, we disconnect the C=C double bond completely and add components of water. So we write a **C=O at the β carbon** and add **two H atoms at the α carbon**.

17.7.4 1,3-Dicarbonyl Compounds

We know that the Claisen condensation gives a β-keto ester or a β-diketone (see Section 15.7), which are **1,3-dicarbonyl compounds** in which two C=O groups are in a 1,3-relationship.

The **C–C disconnection** for 1,3-dicarbonyl compounds (β-keto esters or β-diketones) thus corresponds to a Claisen condensation.

Illustrative examples

1. Consider **acetoacetic ester** (a β-keto ester) as the target molecule.

   The retrosynthetic analysis involves a **C–C disconnection between α carbon and β carbon** so as to produce an **ester α-carbanion** as an anionic synthon.

   This C–C disconnection for the β-keto ester with **two oxygen-containing C=O groups** in a 1,3-relationship is also a **1,3-diO** disconnection.

   More specifically, it is called a **1,3-diCO disconnection**.

(assign a negative charge)

$$CH_3{-}\overset{O}{\overset{\|}{\underset{\beta}{C}}}{-}\underset{\alpha}{CH_2}{-}\overset{O}{\overset{\|}{C}}{-}OC_2H_5 \xrightarrow[\text{(disconnect between }\alpha\text{ and }\beta)]{\text{1,3-diCO}} CH_3{-}\overset{O}{\overset{\|}{C}}{}^{+} + \overset{-}{C}H_2{-}\overset{O}{\overset{\|}{C}}{-}OC_2H_5$$

Acetoacetic ester — ***A β-keto ester***

$CH_3{-}\overset{O}{\overset{\|}{C}}{}^{+} \equiv CH_3{-}\overset{O}{\overset{\|}{C}}{-}OC_2H_5$ **Ethyl acetate**

$\overset{-}{C}H_2{-}\overset{O}{\overset{\|}{C}}{-}OC_2H_5$ **(ester α-carbanion)** $\equiv CH_3{-}\overset{O}{\overset{\|}{C}}{-}OC_2H_5$ **Ethyl acetate**

Natural synthons | ***Reagents***

The **1,3-diCO** disconnection gives two **natural synthons** which give *two molecules* of the same ester (ethyl acetate) as the starting material.

2. Consider **acetylacetone** (a β-diketone or 1,3-diketone) as the target molecule.

   The retrosynthetic analysis involves a **1,3-diCO disconnection** which leads to ethyl acetate and acetone as starting materials.

(assign a negative charge)

$CH_3-C(=O)-CH_2-C(=O)-CH_3$ (β, α, β)

Acetylacetone

A β-diketone

1,3-diCO ⟹ **(disconnect between α and β)**

$CH_3-C^+(=O)$ ≡ $CH_3-C(=O)-OC_2H_5$ **Ethyl acetate**

\+

$^-CH_2-C(=O)-CH_3$ **(an α-carbanion)** ≡ $CH_3-C(=O)-CH_3$ **Acetone**

Natural synthons ***Reagents***

17.7.5 1,5-Dicarbonyl Compounds

1,5-Dicarbonyl compounds are 1,5-difunctionalized compounds in which two C=O groups are in a 1,5-relationship.

We know that the Michael reaction gives 1,5-dicarbonyl compounds from the conjugate addition of an α-carbanion derived from an active methylene compound to α,β-unsaturated carbonyl compounds (see Section 15.10).

Thus the C–C disconnection for 1,5-dicarbonyl compounds can correspond to a Michael reaction.

Latent polarities

Similar to 1,3-difunctionalized compounds (such as aldols, β-keto esters and β-diketones), the **latent polarities from two C=O groups** in 1,5-dicarbonyl compounds are also **superimposed**.

$R-\overset{5}{C}(=O)-\overset{4}{C}H_2-\overset{3}{C}H_2-\overset{2}{C}H_2-\overset{1}{C}(=O)-R$

(+) (-) (+) (-) (+)

Latent polarities of two carbonyl groups are superimposed

Therefore, a reasonable **C–C disconnection** for 1,5-dicarbonyl compounds will give two **natural synthons**, for which reagents are easily available.

Illustrative examples

1. Consider **2,6-heptanedione** (a 1,5-diketone) as the target molecule.

   The retrosynthetic analysis involves a **C–C disconnection** *nearer the centre of the target molecule* to give an **α-carbanion** as an anionic synthon.

   This C–C disconnection for the 1,5-dicarbonyl compound is labelled as a **1,5-diCO disconnection**.

(assign a negative charge)

$CH_3-CO-CH_2-CH_2 \wr CH_2-CO-CH_3$ (β, α)

2,6-Heptanedione

A 1,5-diketone

$\xrightarrow{\text{1,5-diCO}}$ (disconnect between α and β)

$CH_3-CO-CH_2-\overset{+}{C}H_2 \equiv CH_3-CO-CH=CH_2$ **Methyl vinyl ketone**

\+

$\overset{-}{C}H_2-CO-CH_3$ (an α-carbanion) $\equiv$ $CH_2(CO_2C_2H_5)-CO-CH_3$ **Acetoacetic ester**

Natural synthons ***Reagents***

The **1,5-diCO** disconnection gives two **natural synthons**. The reagent for the cationic synthon is **methyl vinyl ketone** (an α,β-unsaturated ketone), while the reagent for the α-carbanion synthon is **acetoacetic ester** (see Figure 17.1) as an active methylene compound.

2. Consider **3-carboxymethylcyclohexanone** as the target molecule. This is a 1,5-dicarbonyl compound containing a cyclic keto group and a carboxyl group.

   The disconnection of the C–C bond between the ring and the chain provides the *most reasonable* **1,5-diCO disconnection** because this disconnection does not affect the ring.

(cyclohexanone, positions 1, 2, 3; α CH_2 (4) $-$ C(=O) (5) $-$OH)

3-Carboxymethylcyclohexanone

A 1,5-dicarbonyl compound cotaining a cyclic keto and carboxyl groups

$\xrightarrow{\text{1,5-diCO}}$

(cyclohexanone cation at C-3) $\equiv$ **2-Cyclohexenone**

\+

$\overset{-}{C}H_2-CO-OH$ (an α-carbanion) $\equiv$ $CH_2(CO_2C_2H_5)-CO-OC_2H_5$ **Malonic ester**

Natural synthons ***Reagents***

This 1,5-diCO disconnection gives two **natural synthons** for which the reagents are readily available. For the cationic synthon, the reagent is **2-cyclohexenone** (an α,β-unsaturated ketone) and for the α-carbanion synthon, the reagent is **malonic ester** (see Figure 17.1) as an active methylene compound.

CHAPTER 18

Synthesis of Organic Compounds: The Retrosynthetic Approach

IN THIS CHAPTER

- Synthesis of different classes of organic compounds with illustrative examples: Alcohols, aldehydes, ketones, carboxylic acids, esters, amides, amines and ethers
- Synthesis of hydrocarbons: Alkanes, alkenes and alkynes
- Synthesis of aromatic compounds
- Synthesis of aldols and α,β-unsaturated carbonyl compounds
- Synthesis of β-keto esters and β-diketones
- Synthesis of 1,5-dicarbonyl compounds and cyclic ketones
- Synthesis of cyclic compounds by the Diels-Alder reaction
- Synthesis of some simple drugs: Benzocaine, aspirin, paracetamol, propranolol, salbutamol, ibuprofen, oblivon and terodilin

In the **retrosynthetic analysis** (Chapter 17), first we recognize the functional groups in the target molecule (TM) and then work *backward* through functional group interconversion (FGI) and disconnection of bonds (C–O, C–N or C–C) corresponding to known and reliable reactions. These exercises offer both intellectual challenge and reward in planning an organic synthesis. Finally, we reach the available starting materials.

For a **synthesis** of the target molecule, we draw a scheme in the *forward* direction according to the retrosynthetic analysis, adding reagents and conditions. This is how the synthesis is carried out in the laboratory.

In this chapter, we will give a brief description of the synthesis of organic molecules using the retrosynthetic approach.

It is pertinent to mention here that after the reaction is over, the *work-up* of the reaction mixture with aqueous acid or aqueous base is a *routine* method to isolate the product. For instance, in Grignard reactions or lithium aluminium hydride reductions, aqueous or acid work-up releases the product *but this last step is not always explicitly written and often considered to be implied in the synthetic scheme.*

18.1 SYNTHESIS OF ALCOHOLS

We have learned in the previous chapter (Section 17.6.2) that **1,1 C–C disconnection** of an alcohol (i.e. disconnection of the C–C bond next to hydroxyl group) gives an aldehyde or ketone and a Grignard reagent as starting materials.

A **1,2 C–C disconnection** of an alcohol is also possible, which gives an epoxide and a Grignard reagent.

Note: A Grignard reagent is not a reagent to be found in the shelf of a laboratory. When required, it is usually prepared by the reaction of an alkyl or aryl halide and magnesium metal in an anhydrous ether solvent (see Section 9.8.1).

The **synthesis** of alcohols is illustrated here with *three* examples using the retrosynthetic approach.

A. *1-Phenylethanol*

1-Phenylethanol is a **secondary alcohol**. *Two* **1,1 C–C** disconnections (marked *a* and *b*) are possible for the target molecule (TM):

1-Phenylethanol (TM)

Retrosynthetic analysis 1 (involving disconnection *a*) gives benzaldehyde (an aldehyde) and methylmagnesium iodide (a Grignard reagent) as starting materials.

Thus, according to this analysis, 1-phenylethanol can be synthesized by a Grignard reaction (see Section 9.8.2) using benzaldehyde and methylmagnesium iodide (**synthesis 1**).

Methylmagnesium iodide (CH_3MgI) can be prepared from methyl iodide (CH_3I) and magnesium metal (Mg) in anhydrous diethyl ether (Et_2O).

1-Phenylethanol: Retrosynthetic analysis 1

1-Phenylethanol (TM) $\xrightarrow{\text{1,1 C—C}}$ $C_6H_5\overset{+}{C}H(OH)$ + $\overset{-}{C}H_3$

Use: **Benzaldehyde** (C_6H_5CHO); Use: CH_3MgI **Methylmagnesium iodide**

1-Phenylethanol: Synthesis 1

$$CH_3I \xrightarrow[Et_2O]{Mg} CH_3MgI \xrightarrow[\text{2. } H_3O^+ \text{ (work-up)}]{\text{1. } C_6H_5CHO} C_6H_5CH(OH)CH_3$$

Methyl iodide → **Methylmagnesium iodide** → **1-Phenylethanol**

On the other hand, **retrosynthetic analysis 2** (involving disconnection *b*) reveals that 1-phenylethanol can also be synthesized by a Grignard reaction using acetaldehyde and phenylmagnesium bromide as starting materials (**synthesis 2**).

Phenylmagnesium bromide is prepared from phenyl bromide (bromobenzene) and magnesium metal in anhydrous ether.

1-Phenylethanol: Retrosynthetic analysis 2

1-Phenylethanol (TM) — 1,1 C—C ⟹ $C_6H_5^-$ + $^+CH(OH)CH_3$

Use: Phenylmagnesium bromide; Use: Acetaldehyde ($H—C(=O)—CH_3$)

1-Phenylethanol: Synthesis 2

Bromobenzene $\xrightarrow[Et_2O]{Mg}$ Phenylmagnesium bromide (—MgBr) $\xrightarrow[\text{2. } H_3O^+ \text{ (work-up)}]{\text{1. } CH_3CHO}$ 1-Phenylethanol ($C_6H_5—CH(OH)—CH_3$)

Note: In reality, both syntheses are likely to proceed without difficulty and give the desired alcohol in good yield.

B. *2-Methyl-2-butanol*

2-Methyl-2-butanol is a **tertiary alcohol**. Using retrosynthetic analyses, the target molecule (TM) can be synthesized in different ways.

The **retrosynthetic analysis 1** reveals that 2-methyl-2-butanol can be synthesized from acetone as a ketone and ethylmagnesium bromide as a Grignard reagent (**synthesis 1**).

Ethylmagnesium bromide (CH_3CH_2MgBr) can be prepared from ethyl bromide (CH_3CH_2Br) and magnesium metal (Mg) in anhydrous ether (not shown in the synthetic scheme).

2-Methyl-2-butanol: Retrosynthetic analysis 1

2-Methyl-2-butanol (TM) ($CH_3CH_2—C(CH_3)(OH)—CH_3$) — 1,1 C—C ⟹ $CH_3CH_2^-$ + $^+C(CH_3)(OH)—CH_3$

$CH_3CH_2^-$ ≡ (Use) CH_3CH_2MgBr

$^+C(CH_3)(OH)—CH_3$ ≡ (Use) $CH_3—C(=O)—CH_3$

2-Methyl-2-butanol: Synthesis 1

$$CH_3COCH_3 \text{ (Acetone)} + CH_3CH_2MgBr \text{ (Ethylmagnesium bromide)} \xrightarrow[\text{2. } H_3O^+ \text{ (work-up)}]{\text{1. } Et_2O} CH_3CH_2C(CH_3)(OH)CH_3 \text{ (2-Methyl-2-butanol)}$$

Using the **retrosynthetic analysis 2**, the alcohol can be prepared from butanone and methylmagnesium iodide (**synthesis 2**).

2-Methyl-2-butanol: Retrosynthetic analysis 2

$$CH_3CH_2{-}C(CH_3)(OH){\wr}CH_3 \text{ (2-Methyl-2-butanol (TM))} \xrightarrow{1,1\ C-C} CH_3CH_2{-}C^+(CH_3)(OH) \overset{\text{Use}}{\equiv} CH_3CH_2{-}C(CH_3){=}O \;+\; {}^-CH_3 \overset{\text{Use}}{\equiv} CH_3MgI$$

2-Methyl-2-butanol: Synthesis 2

$$CH_3CH_2COCH_3 \text{ (Butanone)} + CH_3MgI \text{ (Methylmagnesium iodide)} \xrightarrow[\text{2. } H_3O^+ \text{ (work-up)}]{\text{1. } Et_2O} CH_3CH_2C(CH_3)(OH)CH_3 \text{ (2-Methyl-2-butanol)}$$

The **retrosynthetic analysis 3** uses a double disconnection (**2 × 1,1 C–C**) (see Section 17.6.2) because the tertiary alcohol contains two identical methyl groups. Thus, 2-methyl-2-butanol can be synthesized by a Grignard reaction using ethyl acetate as an ester and two molar equivalents of methylmagnesium iodide as a Grignard reagent (**synthesis 3**).

2-Methyl-2-butanol: Retrosynthetic analysis 3

$$CH_3CH_2{-}C(CH_3)(OH){\wr}CH_3 \text{ (2-Methyl-2-butanol (TM))} \xrightarrow{2\times\ 1,1\ C-C} CH_3CH_2{-}C(=O){-}OC_2H_5 + 2\ CH_3MgI$$

2-Methyl-2-butanol: Synthesis 3

$$CH_3CH_2COOC_2H_5 \text{ (Ethyl propanoate)} + CH_3MgI \text{ (Methylmagnesium iodide)} \xrightarrow{Et_2O} [CH_3CH_2COCH_3] \text{ (Butanone)} \xrightarrow[H_3O^+]{CH_3MgI;} CH_3CH_2C(CH_3)(OH)CH_3 \text{ (2-Methyl-2-butanol)}$$

Note: All three routes for the synthesis of 2-methyl-2-butanol appear to be satisfactory. However, the final choice might be determined by the availability of starting materials.

C. 2-Phenylethanol

2-Phenylethanol is a **primary alcohol**. A **retrosynthetic analysis** of the alcohol (TM) might involve a **1,2 C–C disconnection** (see Section 17.6.2).

The analysis reveals that 2-phenylethanol can be **synthesized** from ethylene oxide as an epoxide and phenylmagnesium bromide as a Grignard reagent (see Section 11.9.1).

2-Phenylethanol: Retrosynthetic analysis

$$C_6H_5 \overset{2}{\wr} \overset{1}{CH_2{-}CH_2OH} \xrightarrow[]{1,2\ C{-}C} C_6H_5^- \ \overset{Use}{\equiv}\ C_6H_5MgBr \quad + \quad \overset{+}{C}H_2{-}CH_2OH \ \overset{Use}{\equiv}\ CH_2{-}CH_2 \text{ (epoxide, O bridging)}$$

2-Phenylethanol (TM)

2-Phenylethanol: Synthesis

$$\underset{\text{Ethylene oxide (an epoxide)}}{CH_2{-}CH_2\ (O)} + \underset{\text{Phenylmagnesium bromide}}{C_6H_5MgBr} \longrightarrow C_6H_5CH_2CH_2OMgBr \xrightarrow[\text{(work-up)}]{H_3O^+} \underset{\text{2-Phenylethanol}}{C_6H_5CH_2CH_2OH}$$

PROBLEM 18.1: The following alcohol has a characteristic smell. Using a retrosynthetic analysis, outline the synthesis of the alcohol starting from an ester and other reagents as necessary.

$$C_6H_5CH_2{-}C(CH_3)(OH){-}CH_3$$

2-Methyl-1-phenyl-2-propanol

Answer

The target molecule is a tertiary alcohol containing two identical methyl groups. Therefore, a double disconnection (2 × 1,1 C–C) can lead to an ester (ethyl phenylacetate) as one of the starting materials. The other reagent is methylmagnesium iodide, two molar equivalents of which are needed.

2-Methyl-1-phenyl-2-propanol: Retrosynthetic analysis

$$\underset{\text{(TM)}}{C_6H_5CH_2{-}C(\wr CH_3)(OH)\wr CH_3} \xrightarrow{2\text{x } 1,1\ C{-}C} C_6H_5CH_2{-}\overset{O}{\overset{\|}{C}}{-}OC_2H_5 + 2\ CH_3MgI$$

2-Methyl-1-phenyl-2-propanol: Synthesis

$$C_6H_5CH_2C(=O)OC_2H_5 + 2\ CH_3MgI \xrightarrow[2.\ H_3O^+]{1.\ Et_2O} C_6H_5CH_2C(CH_3)(OH)CH_3$$

Ethyl phenylacetate **Methylmagnesium iodide** **2-Methyl-1-phenyl-2-propanol**

18.2 SYNTHESIS OF ALDEHYDES, KETONES AND CARBOXYLIC ACIDS

18.2.1 Aldehydes and Ketones

Aldehydes and ketones can be prepared by *oxidation* of the corresponding alcohols. The **retrosynthetic analysis** of an aldehyde or a ketone can thus involve an **FGI oxidation** followed by a **1,1 C–C** or **1,2 C–C disconnection.**

The **synthesis** of an aldehyde and a ketone is described here using the retrosynthetic approach.

Aldehyde: 2-Methylpropanal

A **retrosynthetic analysis** of 2-methylpropanal (TM) using an **FGI oxidation** gives 2-methylpropanol (a primary alcohol). Then a **1,1 C–C** disconnection leads to formaldehyde and isopropylmagnesium bromide (a Grignard reagent) as the starting materials.

In the forward **synthesis**, a Grignard reaction between isopropylmagnesium bromide and formaldehyde is followed by the oxidation of the resultant primary alcohol (2-methylpropanol). The oxidation is usually accomplished with pyridinium chlorochromate (PCC) in dichloromethane (see Section 10.9).

The Grignard reagent can be prepared from isopropyl bromide and magnesium metal in anhydrous ether (not shown in the synthetic scheme).

2-Methylpropanal: Retrosynthetic analysis

$$CH_3-CH(CH_3)-C(=O)-H \xRightarrow[\text{oxidation}]{\text{FGI}} CH_3-CH(CH_3)-CH_2OH \xRightarrow{1,1\ C-C} CH_3-\overset{-}{C}H(CH_3) + H_2\overset{+}{C}-OH$$

2-Methylpropanal (TM)

Use: $CH_3-CH(CH_3)MgBr$; Use: $H_2C{=}O$

2-Methylpropanal: Synthesis

$$CH_3CH(CH_3)MgBr + H-C(H){=}O \xrightarrow[2.\ H_3O^+]{1.\ Et_2O} CH_3CH(CH_3)CH_2OH \xrightarrow[CH_2Cl_2]{PCC} CH_3CH(CH_3)CHO$$

Isopropylmagnesium bromide **Formaldehyde** **2-Methylpropanol (a 1° alcohol)** **2-Methylpropanal**

Ketone: 1-Phenyl-2-butanone

A **retrosynthetic analysis** of 1-phenyl-2-butanone (TM) using an **FGI oxidation** and a **1,2 C–C** disconnection leads to phenylmagnesium bromide as a Grignard reagent and 2-methyloxirane as an epoxide. A further analysis of the epoxide involving epoxidation gives propene as a simple alkene.

In the forward **synthesis**, the reaction of propene with a peroxy acid (*m*-CPBA) gives 2-methyloxirane (see Section 8.7), which then undergoes a Grignard reaction with phenylmagnesium bromide to produce 1-phenyl-2-propanol (a secondary alcohol). The oxidation of the secondary alcohol to the target ketone is carried out with chromic acid (CrO_3/H_2SO_4).

1-Phenyl-2-propanone: Retrosynthetic analysis

$C_6H_5—CH_2—C(=O)—CH_3$ ⟹ (FGI oxidation) C_6H_5 ≀ $CH_2—CH(OH)—CH_3$ ⟹ (1,2 C—C) $H_2\overset{+}{C}—CH(OH)—CH_3$ + $C_6H_5^-$

1-Phenyl-2-propanone (TM)

Use: $CH_2—CH—CH_3$ (epoxide, O bridging) ; Use: C_6H_5MgBr

$H_2C{=}CHCH_3$ ⟸ (Epoxidation) $CH_2—CH—CH_3$ (epoxide)

1-Phenyl-2-propanone: Synthesis

$H_2C{=}CHCH_3$ —(*m*-CPBA)→ $CH_2—CHCH_3$ (epoxide) —(1. C_6H_5MgBr; 2. H_3O^+)→ $C_6H_5CH_2CH(OH)CH_3$ —(CrO_3 / H_2SO_4)→ $C_6H_5CH_2C(=O)CH_3$

Propene — **2-Methyloxirane** — **1-Phenyl-2-propanol (a 2° alcohol)** — **1-Phenyl-2-propanone**

18.2.2 Carboxylic Acids

Carboxylic acids can be prepared by the hydrolysis of cyanides or using the Grignard reaction with carbon dioxide (see Section 9.8.2). The **retrosynthetic analysis** of a carboxylic acid can then involve an **FGI hydrolysis** and/or **C–C disconnection.**

Using the retrosynthetic approach, the **synthesis** of carboxylic acids is illustrated with *two* examples.

A. *Phenylacetic Acid*

A **retrosynthetic analysis** of phenylacetic acid (TM) using an **FGI hydrolysis** gives a cyanide (benzyl cyanide). Then a **C–C disconnection** gives the corresponding chloride (benzyl chloride) and sodium cyanide.

The **synthesis** of the target molecule can be easily accomplished starting from benzyl chloride. Treating benzyl chloride with NaCN followed by hydrolysis with a base produces phenylacetic acid.

Phenylacetic acid: Retrosynthetic analysis

$$C_6H_5-CH_2-\underset{\underset{O}{\|}}{C}-OH \xrightarrow[\text{hydrolysis}]{\text{FGI}} C_6H_5-CH_2 \wr C\equiv N \xrightarrow{C-C} C_6H_5-\overset{+}{C}H_2 + \overset{-}{C}\equiv N$$

Use: $C_6H_5-CH_2-Cl$; Use: NaCN

Phenylacetic acid
(TM)

Phenylacetic acid: Synthesis

$$C_6H_5CH_2Cl \xrightarrow[H_2O]{NaCN} C_6H_5CH_2CN \xrightarrow[\text{2. } H^+ \text{ (work-up)}]{\text{1. } OH^-/H_2O} C_6H_5CH_2CO_2H$$

Benzyl chloride — **Benzyl cyanide** — **Phenylacetic acid**

Note: The anionic synthon (cyanide ion) is an **unnatural or umpolung** synthon because the carbon attached to nitrogen has a (–) polarity, which is opposite to its latent or natural polarity (+) (see Section 17.7.1).

B. *Pivalic Acid (trimethylacetic acid)*

Pivalic acid is a ***tertiary*-alkyl acid**. Therefore, the cyanide method is *not* suitable in this case because the elimination reaction of a *tertiary*-alkyl halide with basic cyanide ion would compete. But the Grignard method could work very well.

A **retrosynthetic analysis** of pivalic acid (TM) involving a **C–C disconnection** gives *t*-butylmagnesium chloride (a Grignard reagent) and CO_2 as the reagents.

In the **synthesis**, the Grignard reagent (prepared from *t*-butyl chloride and metallic magnesium in anhydrous ether) reacts with CO_2 to produce pivalic acid (trimethylacetic acid).

Pivalic acid: Retrosynthetic analysis

$$(CH_3)_3C \wr C(=O)-OH \xrightarrow{C-C} \overset{+}{C}(=O)-OH \;+\; (CH_3)_3C^-$$

$\overset{+}{C}(=O)-OH$ ≡ (Use) $O=C=O$

$(CH_3)_3C^-$ ≡ (Use) $(CH_3)_3C-MgCl$

Pivalic acid
(TM)

Pivalic acid: Synthesis

$$(CH_3)_3C-Cl \xrightarrow[Et_2O]{Mg} (CH_3)_3C-MgCl \xrightarrow[\text{2. } H_3O^+ \text{ (work-up)}]{\text{1. } CO_2} (CH_3)_3C-CO_2H$$

***t*-Butyl chloride** — ***t*-Butylmagnesium chloride** — **Pivalic acid (trimethylacetic acid)**

PROBLEM 18.2: Give a retrosynthetic analysis and then outline a synthesis of the following compound:

H_3C–C_6H_4–CO_2H

p-Methylbenzoic acid

Answer

p-Methylbenzoic acid (TM) can be synthesized by the Grignard method. The C–C disconnection gives a Grignard reagent (*p*-tolylmagnesium bromide) and CO_2 as the reagents. The Grignard reagent can be prepared from *p*-bromotoluene.

p-Methylbenzoic acid: Retrosynthetic analysis

$$H_3C\text{–}C_6H_4 \,\wr\, CO_2H \xRightarrow{\text{C–C}} H_3C\text{–}C_6H_4\text{–}MgBr + O{=}C{=}O$$

p-Methylbenzoic acid
(TM)

p-Methylbenzoic acid: Synthesis

$$H_3C\text{–}C_6H_4\text{–}Br \xrightarrow[Et_2O]{Mg} H_3C\text{–}C_6H_4\text{–}MgBr \xrightarrow[2.\ H_3O^+]{1.\ CO_2} H_3C\text{–}C_6H_4\text{–}CO_2H$$

p-Bromotoluene **p-Tolylmagnesium bromide** **p-Methylbenzoic acid**

18.2.3 Acetoacetic Ester Synthesis of Ketones and Malonic Ester Synthesis of Carboxylic Acids

Acetoacetic ester is a reagent for an α-carbanion synthon from a *ketone* (see Section 17.6.3). Actually it stands for an **activated α-carbanion synthon** in which the α-carbanion is attached to an electron-withdrawing ester group (CO_2Et).

$$\bar{C}H_2\text{–}C(=O)\text{–}CH_3 \xrightarrow{\textit{Activate}} \bar{C}H(CO_2Et)\text{–}C(=O)\text{–}CH_3 \equiv CH_2(CO_2Et)\text{–}C(=O)\text{–}CH_3$$

An α-carbanion from a ketone — ***Synthon***

Activated α-carbanion — ***Activated synthon***

Acetoacetic ester — ***Reagent***

On the other hand, **malonic ester** is a reagent for an α-carbanion synthon from a *carboxylic acid* (see Section 17.6.3). In fact, it stands for a **protected and activated α-carbanion synthon** in which CO_2H is protected as an ester group (CO_2Et) and the α-carbanion is activated by an electron-withdrawing CO_2Et group.

$\overset{-}{C}H_2-\overset{O}{\overset{\|}{C}}-OH$ —*Protect CO_2H*→ $\overset{-}{C}H_2-\overset{O}{\overset{\|}{C}}-OEt$ —*Activate*→ $\overset{-}{C}H(CO_2Et)-\overset{O}{\overset{\|}{C}}-OEt$ ≡ $CH_2(CO_2Et)-\overset{O}{\overset{\|}{C}}-OEt$

An α-carbanion from a carboxylic acid — *Synthon*; **Protected α-carbanion** — *Protected synthon*; **Protected and activated α-carbanion** — *Protected and activated synthon*; **Malonic ester** — *Reagent*

The **acetoacetic ester synthesis of a ketone** is performed by alkylation of acetoacetic ester followed by hydrolysis and decarboxylation (see Section 15.9.1). Note that *hydrolysis and decarboxylation are necessary to remove the ester group (CO_2Et) added for activation.*

The **malonic ester synthesis of a carboxylic acid** is accomplished by alkylation of malonic ester followed by hydrolysis and decarboxylation (see Section 15.9.2). Notably, *hydrolysis and decarboxylation remove protection as CO_2Et as well as CO_2Et added for activation.*

Ketone: 4-Phenyl-2-butanone

A **retrosynthetic analysis** of 4-phenyl-2-butanone (TM) involving a **1,2 C–C** disconnection gives acetoacetic ester and benzyl bromide as the reagents.

In the forward **synthesis**, alkylation of acetoacetic ester with benzyl bromide in the presence of a base (sodium ethoxide in ethanol) followed by hydrolysis and decarboxylation produces the target ketone.

4-Phenyl-2-butanone: Retrosynthetic analysis

$C_6H_5-CH_2 \wr CH_2-\overset{O}{\overset{\|}{C}}-CH_3$ (**4-Phenyl-2-butanone (TM)**) $\xRightarrow{\text{1,2 C–C}}$ $C_6H_5-\overset{+}{C}H_2$ + $\overset{-}{C}H_2-\overset{O}{\overset{\|}{C}}-CH_3$ (↑ **(activate)**)

$C_6H_5-\overset{+}{C}H_2$ ≡ (**Use**) $C_6H_5-CH_2Br$ **Benzyl bromide**

$\overset{-}{C}H_2-\overset{O}{\overset{\|}{C}}-CH_3$ ≡ (**Use**) $CH_2(CO_2Et)-\overset{O}{\overset{\|}{C}}-CH_3$ + Base **Acetoacetic ester**

4-Phenyl-2-butanone: Synthesis

$CH_3-\overset{O}{\overset{\|}{C}}-CH_2-\overset{O}{\overset{\|}{C}}-OEt$ (**Acetoacetic ester**) → [1. C_2H_5ONa/C_2H_5OH; 2. $C_6H_5CH_2Br$; ***Alkylation***] → $CH_3-\overset{O}{\overset{\|}{C}}-CH(CH_2C_6H_5)-\overset{O}{\overset{\|}{C}}-OEt$

→ ***Hydrolysis*** [1. dil. NaOH; 2. H_3O^+] → $CH_3-\overset{O}{\overset{\|}{C}}-CH(CH_2C_6H_5)-\overset{O}{\overset{\|}{C}}-OH$

→ [Δ, $- CO_2$; ***Decarboxylation***] → $CH_3-\overset{O}{\overset{\|}{C}}-CH_2-CH_2C_6H_5$ **4-Phenyl-2-butanone**

Carboxylic Acid: 3-Phenylpropanoic Acid

A **retrosynthetic analysis** of 3-phenylpropanoic acid (TM) involving a **1,2 C–C** disconnection leads to malonic ester and benzyl bromide as the reagents.

In the forward **synthesis**, after alkylation of malonic ester with benzyl bromide in the presence sodium ethoxide as a base, hydrolysis and decarboxylation are needed to produce the target carboxylic acid.

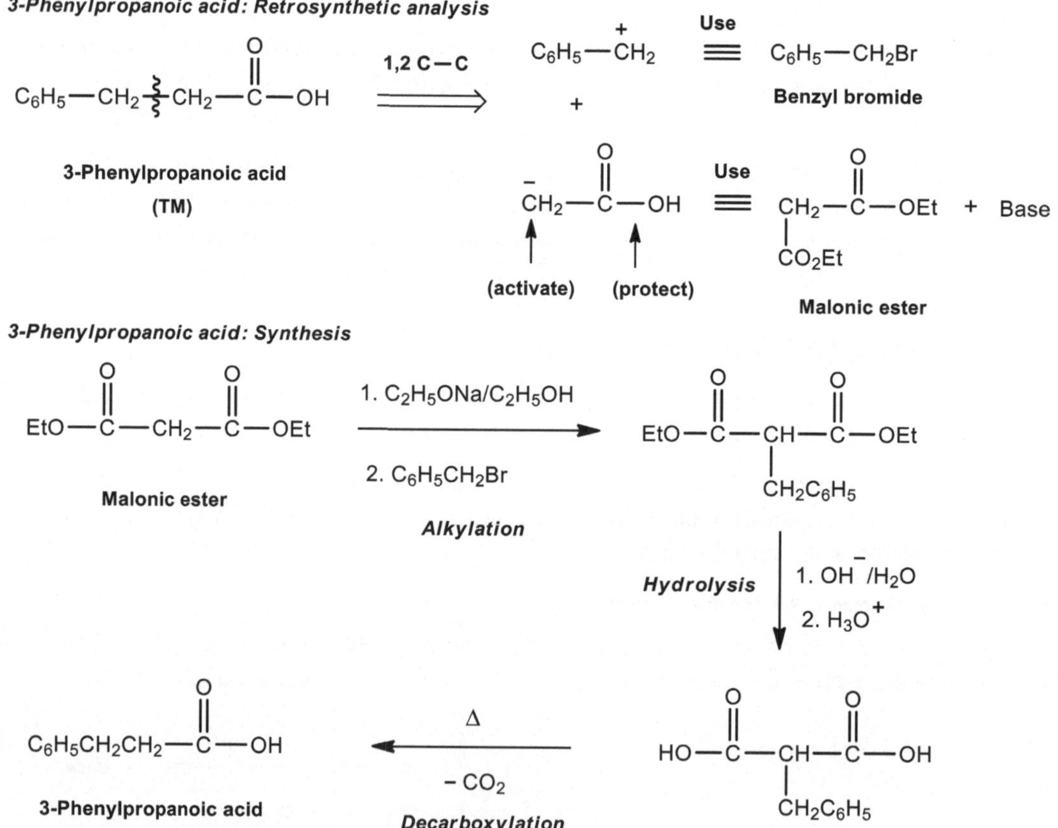

18.3 SYNTHESIS OF ESTERS AND AMIDES

The retrosynthetic approach to the synthesis of esters involves a C–O disconnection, while the retrosynthetic analysis of amides uses a C–N disconnection (see Section 17.5). The **C–O ester** or **C–N amide** disconnection leads to readily available starting materials, and a suitable synthesis of an ester or an amide can be accomplished.

Some esters and amides are useful common drugs. The synthesis of some of these drugs is included in Section 18.12.

In this section, the **synthesis** of an ester and an amide is illustrated using the retrosynthetic approach.

Ester: Benzyl Benzoate

Benzyl benzoate is an ester which is used as an *insect repellent.*

The **retrosynthetic analysis** of benzyl benzoate (TM) involving C–O ester disconnection gives benzoyl chloride and benzyl alcohol as readily available starting materials.

In the forward **synthesis**, treatment of benzyl alcohol with benzoyl chloride in the presence of a base (pyridine) produces the target ester.

Benzyl benzoate: Retrosynthetic analysis

Benzyl benzoate (TM) $\xrightarrow[\text{ester}]{\text{C—O}}$ $C_6H_5CO^+$ + $^-OCH_2C_6H_5$; Use ≡ C_6H_5COCl; Use ≡ $C_6H_5CH_2OH$

Benzyl benzoate: Synthesis

$$C_6H_5\text{—}\overset{O}{\overset{\|}{C}}\text{—}Cl + C_6H_5CH_2OH \xrightarrow{\text{Pyridine}} C_6H_5\text{—}\overset{O}{\overset{\|}{C}}\text{—}OCH_2C_6H_5$$

Benzoyl chloride **Benzyl alcohol** **Benzyl benzoate** ***An insect repellent***

Amide: Propanil

Propanil [*N*-(3,4-dichlorophenyl)propanamide] is an amide, which is a *weedkiller* used in rice fields.

The **C–N amide disconnection** of propanil gives propanoyl chloride and 3,4-dichloroaniline. A further analysis of 3,4-dichloroaniline (an aromatic amine) using an FGI reduction and a C–N disconnection (nitration) gives *o*-dichlorobenzene as the starting material.

In the forward **synthesis**, nitration of *o*-dichlorobenzene followed by reduction (catalytic hydrogenation) and treatment with propanoyl chloride produces propanil.

Propanil: Retrosynthetic analysis

Propanil [*N*-(3,4-dichlorophenyl) propanamide] (TM) $\xrightarrow[\text{amide}]{\text{C—N}}$ $Cl\text{—}\overset{O}{\overset{\|}{C}}CH_2CH_3$ + 3,4-dichloroaniline (NH_2, Cl, Cl) $\xrightarrow[\text{reduction}]{\text{FGI}}$ 3,4-dichloronitrobenzene (NO_2, Cl, Cl) $\xrightarrow[\text{nitration}]{\text{C—N}}$ *o*-dichlorobenzene (Cl, Cl) + $\overset{+}{N}O_2$; Use ≡ HNO_3/H_2SO_4

Propanil: Synthesis

o-Dichlorobenzene → (c. HNO_3, c. H_2SO_4) → 3,4-Dichloronitrobenzene → (H_2, Pd/C) → 3,4-Dichloroaniline → (CH_3CH_2COCl (propanoyl chloride)) → Propanil ($NHCOCH_2CH_3$)

Propanil
A weedkiller

PROBLEM 18.3: Saccharin is an artificial sweetener. Propose a retrosynthetic analysis of saccharin and then outline its synthesis.

Saccharin

Answer

Saccharin is an imide. The C–N amide disconnection gives a sulphonamide intermediate. Analysis of the intermediate by an FGI oxidation, a S–N sulphonamide disconnection and a C–S disconnection gives toluene as the starting material. Note that the C–S disconnection corresponds to a chlorosulphonation reaction (see Section 13.5.3).

In the forward synthesis, chlorosulphonation of toluene with chlorosulphonic acid ($ClSO_3H$) is followed by the treatment of ammonia to produce *o*-toluenesulphonamide. Subsequent oxidation of the side chain methyl group of the sulphonamide with $KMnO_4$ yields saccharin.

Saccharin: Retrosynthetic analysis

Saccharin (TM) ⟹ (C—N amide) ⟹ (SO_2NH_2, COOH) ⟹ (FGI oxidation) ⟹ (SO_2NH_2, CH_3) ⟹ (S—N sulphonamide) ⟹ (SO_2Cl, CH_3) + NH_3 ⟹ (C—S chlorosulphonation) ⟹ CH_3 (toluene)

Saccharin: Synthesis

Toluene →($HO-SO_2Cl$ ($ClSO_3H$)) o-Toluenesulphonyl chloride (*separate from para isomer*) →(NH_3) o-Toluenesulphonamide →($KMnO_4$) [2-sulphamoylbenzoic acid (CO_2H, SO_2NH_2)] → Saccharin

Saccharin
An artificial sweetener

18.4 SYNTHESIS OF AMINES

Amines can be synthesized using the retrosynthetic analysis involving a C–N disconnection (see Section 17.5.2). However, a simple **C–N amine** disconnection often gives unsatisfactory results because of the problem of multiple alkylations. A better method is to involve an FGI before a C–N disconnection.

The **synthesis** of amines is described here with *three* examples using the retrosynthetic approach involving a suitable FGI and consequent C–N disconnection.

A. *Diethylamine*

Diethylamine is a **secondary amine**. We can consider its synthesis by the **amide reduction approach** (reduction of the corresponding amide).

A **retrosynthetic analysis** of diethylamine (TM) using an **FGI reduction** followed by a **C–N amide** disconnection gives ethylamine and acetyl chloride as starting materials.

In the forward **synthesis**, treatment of ethylamine with acetyl chloride gives *N*-ethylacetamide, which is then reduced with $LiAlH_4$ to produce diethylamine.

Diethylamine: Retrosynthetic analysis

$$CH_3CH_2-NH-CH_2CH_3 \xRightarrow[\text{reduction}]{\text{FGI}} CH_3CH_2-NH \wr C(=O)CH_3 \xRightarrow{\text{C-N amide}} CH_3CH_2NH_2 + Cl-C(=O)CH_3$$

Diethylamine (TM)

Diethylamine: Synthesis

$$CH_3CH_2NH_2 \xrightarrow[\text{(acetyl chloride)}]{CH_3COCl} CH_3CH_2NH-C(=O)-CH_3 \xrightarrow{LiAlH_4} CH_3CH_2NH-CH_2CH_3$$

Ethylamine — ***N*-Ethylacetamide** — **Diethylamine**

B. *Benzylamine*

Benzylamine is a **primary amine**. The **reductive amination approach** (see Section 17.5.2) can be employed for a suitable synthesis of the amine.

A **retrosynthetic analysis** of benzylamine (TM) involving an FGI reduction followed by a C=N imine disconnection gives benzaldehyde and ammonia as readily available starting materials.

In the forward **synthesis**, the reduction is carried out *without* isolating the intermediate imine formed from benzaldehyde and ammonia. Sodium borohydride ($NaBH_4$) or sodium cyanoborohydride ($NaCNBH_3$) or H_2/catalyst is commonly used as a reducing agent.

Benzylamine: Retrosynthetic analysis

C_6H_5–CH_2—NH_2 ⟹ (FGI, reduction) C_6H_5–CH=NH ⟹ (C=N imine) C_6H_5–CH=O + NH_3

Benzylamine (TM)

Benzylamine: Synthesis (reductive amination)

C_6H_5–CH=O + NH_3 → (H_2, Ni) C_6H_5–CH_2NH_2

Benzaldehyde **Benzylamine**

C. *N-Ethylpiperidine*

N-Ethylpiperidine is a **cyclic tertiary amine**. It can be easily synthesized by the **amide reduction approach** (cf. the synthesis of diethylamine above) because piperidine is readily available as a starting material.

N-Ethylpiperidine: Retrosynthetic analysis

(piperidinyl)N—CH_2CH_3 ⟹ (FGI, reduction) (piperidinyl)N–C(=O)CH_3 ⟹ (C—N amide) (piperidine)NH + $CH_3C(=O)$—Cl

***N*-Ethylpiperidine (TM)**

N-Ethylpiperidine: Synthesis

(piperidine)NH → (CH_3COCl) (piperidinyl)N—C(=O)CH_3 → ($LiAlH_4$) (piperidinyl)NCH_2CH_3

Piperidine ***N*-Acetylpiperidine** ***N*-Ethylpiperidine**

18.5 SYNTHESIS OF ETHERS

Ethers are conveniently prepared by the S_N2 reaction of alkoxide ion with alkyl halides, known as the Williamson synthesis (see Section 11.8.1).

The retrosynthetic approach involves a C–O disconnection of ethers which gives the starting materials. However, it should be kept in mind that *disconnection on either side of the oxygen atom may not always correspond to a reliable S_N2 reaction.*

The **synthesis** of ethers is described here with *two* examples using the retrosynthetic approach.

A. *t-Butyl methyl ether*

The **retrosynthetic analysis** of *t*-butyl methyl ether (TM) can involve two possible **C–O disconnections** (marked *a*, *b*). The disconnection (marked *a*) gives methyl bromide and *t*-butoxide ion, whereas the disconnection (marked *b*) gives *t*-butyl bromide and methoxide ion.

t-Butyl methyl ether: Retrosynthetic analysis

$$(CH_3)_3C\text{–O}\overset{a}{\wr}CH_3 \xRightarrow[a]{\text{C—O ether}} (CH_3)_3C\text{–}O^- + CH_3Br$$

t-Butyl methyl ether (TM) → **t-Butoxide ion** + **Methyl bromide** — **Good choice (a reliable S_N2)**

$$(CH_3)_3C\overset{b}{\wr}\text{O–}CH_3 \xRightarrow[b]{\text{C—O ether}} (CH_3)_3C\text{–Br} + CH_3O^-$$

t-Butyl methyl ether (TM) → **t-Butyl bromide** + **Methoxide ion** — **Bad choice (not a reliable S_N2)**

t-Butyl methyl ether: Synthesis (according to disconnection a)

$$(CH_3)_3CO^-\ K^+ + CH_3\text{–Br} \xrightarrow{S_N2} (CH_3)_3COCH_3$$

Potassium t-butoxide + **Methyl bromide** → **t-Butyl methyl ether**

We know that an S_N2 reaction on a *tertiary substrate* is unfavourable. So the reaction of methoxide ion with *t*-butyl bromide would lead to E2 elimination instead of S_N2 substitution. Therefore, the **disconnection *b* is a bad choice** because it cannot lead to a successful synthesis of the target molecule by an S_N2 reaction between *t*-butyl bromide and methoxide ion.

In contrast, the **disconnection *a* is a good choice** since an S_N2 reaction is highly reliable on a methyl halide, although *t*-butoxide ion is a bulky nucleophile in this case.

Hence the forward **synthesis** is carried out with potassium *t*-butoxide and methyl bromide as the reagents to produce *t*-butyl methyl ether.

B. *p-Methylanisole*

p-Methylanisole is a wallflower perfume compound.

The C–O ether disconnection (labelled *a*) on the methyl side of *p*-methylanisole (TM) is a **good choice** because it gives *p*-methylphenoxide ion and methyl bromide (or dimethyl sulphate) that can undergo a reliable S_N2 reaction on a methyl carbon.

The alternative C–O disconnection (labelled *b*) on the aryl side is a **bad choice** because *p*-chlorotoluene does not undergo nucleophilic aromatic substitution with methoxide ion as the benzene ring is not activated by strong electron-withdrawing substituents.

Therefore, the **synthesis** is performed according to the disconnection *a* with *p*-methylphenoxide ion generated from *p*-methylphenol in NaOH solution. Dimethyl sulphate is generally used for methylation of phenols in alkaline solution.

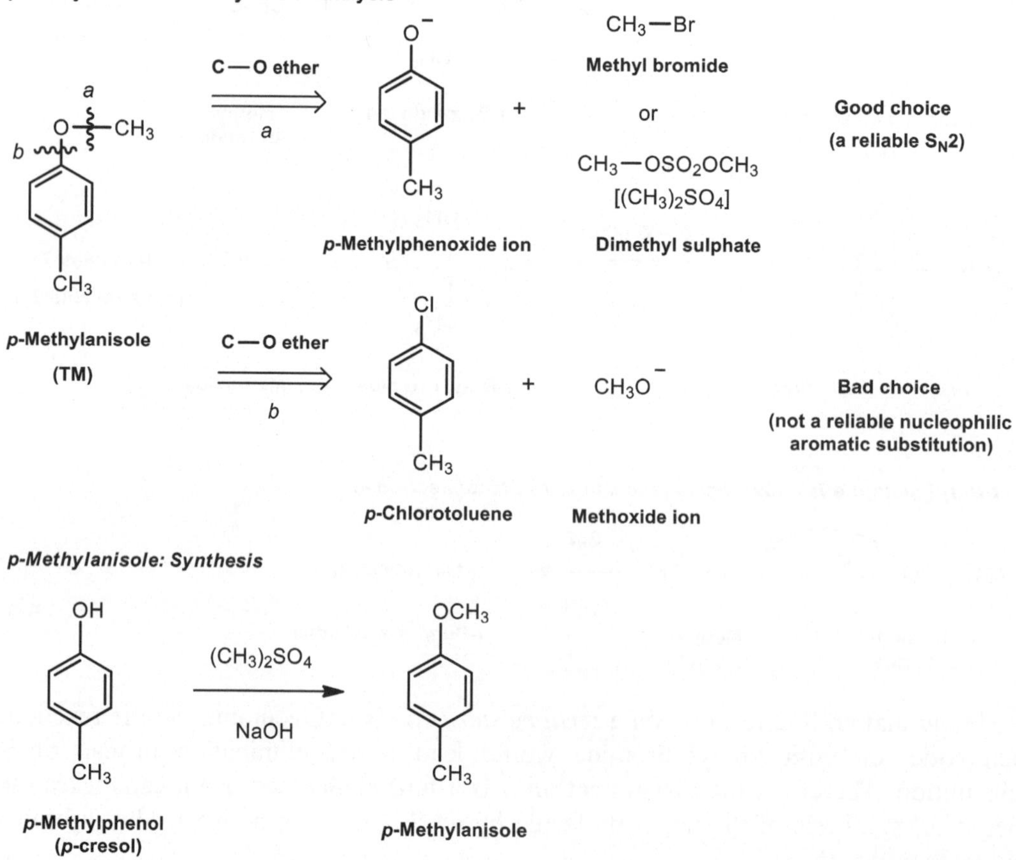

PROBLEM 18.4: The compound shown below is an herbicide (2,4-D). Give a retrosynthetic analysis and then outline the synthesis of the compound.

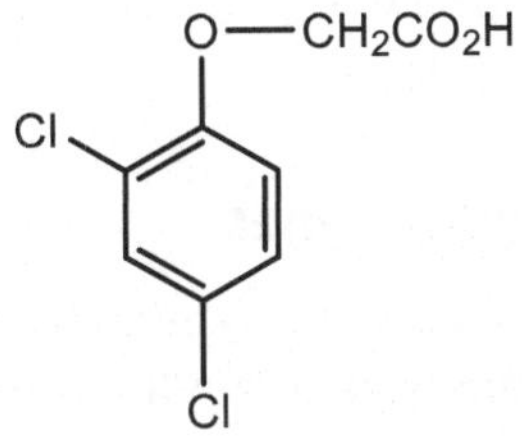

O—CH_2CO_2H
Cl
Cl

2,4-D
(2,4-dichlorophenoxyacetic acid)

Answer

The herbicide (2,4-D or 2.4-dichlorophenoxyacetic acid) is an ether in which the oxygen atom is bonded to an alkyl carbon and an aryl carbon. The C–O disconnection on the alkyl side of the oxygen atom is a reasonable choice because it corresponds to a reliable S_N2 reaction.

The synthesis is performed by the reaction of 2,4-dichlorophenol with chloroacetic acid in alkaline solution.

2,4-D: Retrosynthetic analysis

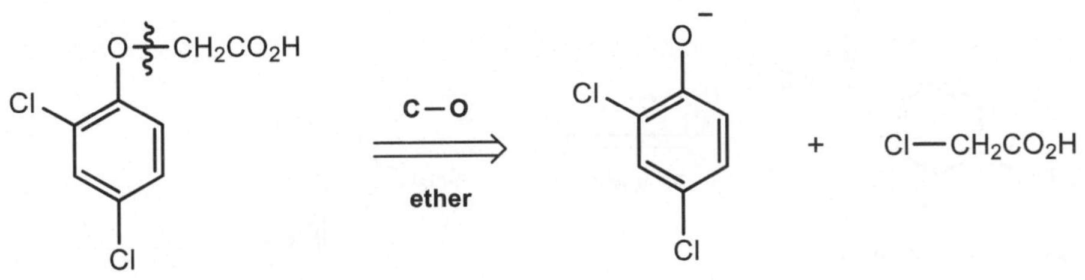

O—CH_2CO_2H, Cl, Cl
2,4-D
(TM)
⟹ **C–O** **ether**
O^-, Cl, Cl + Cl—CH_2CO_2H

2,4-D: Synthesis

OH, Cl, Cl + $ClCH_2CO_2H$ —NaOH→ OCH_2CO_2H, Cl, Cl

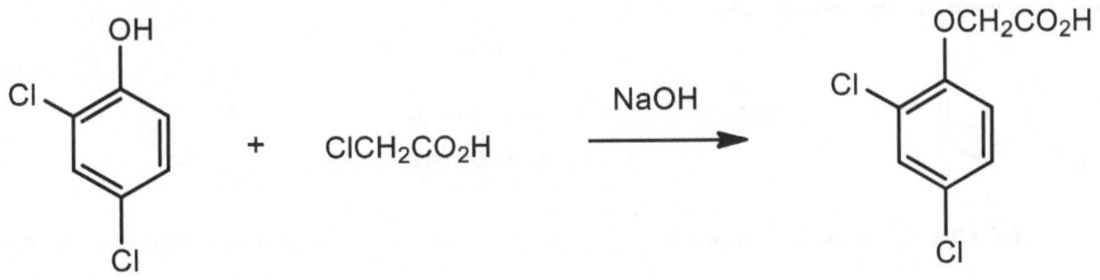

2,4-Dichlorophenol **Chloroacetic acid** **2,4-D**
An herbicide

18.6 SYNTHESIS OF AROMATIC COMPOUNDS

The retrosynthetic approach to the synthesis of aromatic compounds generally involves the disconnection of a C–C bond or a C–X bond (X = N, O or halogen atom) between the *aromatic ring* and the *rest* of the molecule.

18.6.1 Acylbenzenes and Alkylbenzenes

Acylbenzenes and alkylbenzenes can be synthesized by Friedel-Crafts reactions (see Section 13.6). The retrosynthetic analysis of acylbenzenes or alkylbenzenes thus involves a C–C (or Ar–C) Friedel-Crafts disconnection.

The **synthesis** of acylbenzenes and alkylbenzenes is described here with *four* examples using the retrosynthetic approach.

A. *p-Methoxyacetophenone*

p-Methoxyacetophenone is an **acylbenzene** and is a *perfume compound.*

The **retrosynthetic analysis** of *p*-methoxyacetophenone (TM) involving a C–C Friedel-Crafts disconnection gives anisole and acetyl chloride as the reagents.

Since anisole is available as a starting material, the **synthesis** is just a one-step Friedel-Crafts acylation reaction.

p-Methoxyacetophenone: Retrosynthetic analysis

CH_3O–C_6H_4–C(=O)–CH_3 (**p-Methoxyacetophenone (TM)**) $\xRightarrow[\text{Friedel-Crafts}]{\text{C—C}}$ CH_3O–$C_6H_4^-$ + $^+$C(=O)–CH_3

Use: CH_3O–C_6H_5 (**Anisole**) and Cl–C(=O)–CH_3 (**Acetyl chloride**)

p-Methoxyacetophenone: Synthesis

CH_3O–C_6H_5 (**Anisole**) + CH_3COCl (**Acetyl chloride**) $\xrightarrow[\textit{Friedel-Crafts acylation}]{\text{Anhydrous } AlCl_3}$ CH_3O–C_6H_4–$COCH_3$

p-Methoxyacetophenone
A perfume compound

B. *BHT (butylated hydroxytoluene)*

BHT (butylated hydroxytoluene) is an **alkylbenzene** and is an *antioxidant used in foods.*

We know that a problem in the Friedel-Crafts alkylation is polyalkylation or rearrangement (see Section 13.6.1), but with a *tertiary* alkyl halide the Friedel-Crafts alkylation is a useful reaction.

The **retrosynthetic analysis** of BHT (TM) involving a Friedel-Crafts C–C disconnection gives *p*-hydroxytoluene and a *t*-butyl cation. The reagent for the *t*-butyl cation can be *t*-butyl chloride and anhydrous $AlCl_3$ or isobutene (a readily available alkene) and a protic acid.

As shown, the **synthesis** is a one-step Friedel-Crafts alkylation of *p*-hydroxytoluene with isobutene in the presence of HF.

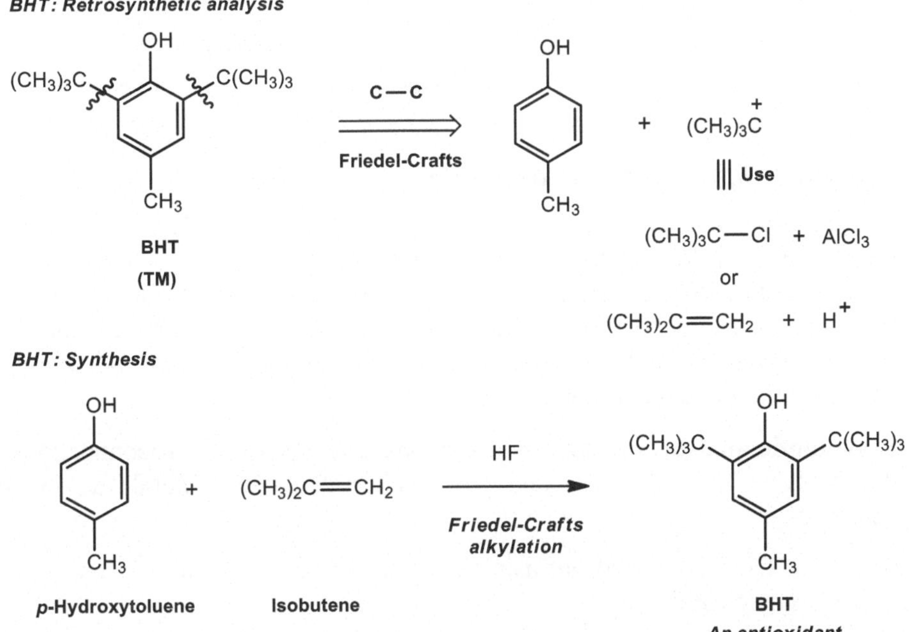

C. *Isobutylbenzene*

Isobutylbenzene is an **alkylbenzene**. However, the Friedel-Crafts alkylation using isobutyl chloride (a primary alkyl halide) is not useful in this case because of the associated problem of rearrangement or polyalkylation.

But the **Friedel-Crafts acylation** does not suffer from these disadvantages. Therefore, an alternative and efficient route to the synthesis of alkylbenzenes uses the Friedel-Crafts acylation followed by reduction of the C=O group to CH_2 group.

The **retrosynthetic analysis** of isobutylbenzene (TM) thus involves an FGI reduction followed by a Friedel-Crafts C–C disconnection, which gives benzene and isobutyryl chloride as the starting materials.

The forward **synthesis** is a Friedel-Crafts acylation of benzene with isobutyryl chloride followed by the Clemmensen reduction of the resulting phenyl isopropyl ketone to produce isobutylbenzene.

Isobutylbenzene: Retrosynthetic analysis

Isobutylbenzene (TM) $\xRightarrow[\text{reduction}]{\text{FGI}}$ phenyl ketone $\xRightarrow[\text{Friedel-Crafts}]{\text{C—C}}$ benzene + $Cl—C(=O)—CH(CH_3)CH_3$

Isobutylbenzene: Synthesis

Benzene + Isobutyryl chloride $(CH_3)_2CHCOCl$ $\xrightarrow[\text{Friedel-Crafts acylation}]{AlCl_3}$ Phenyl isopropyl ketone ($O{=}CCH(CH_3)_2$) $\xrightarrow[\text{HCl, Clemmensen reduction}]{Zn(Hg)}$ Isobutylbenzene ($CH_2CH(CH_3)_2$)

D. *DDT (dichlorodiphenyltrichloroethane)*

DDT (dichlorodiphenyltrichloroethane) is a powerful *insecticide* which was used extensively throughout the world from the 1940s to the early 1970s, when it was banned in many countries due to its toxic effects. DDT is insoluble in water but soluble in fats; it tends to accumulate and persist in the fatty tissues of most animals. As a result, the food chain that runs from plankton to small animals to birds to larger animals including humans tends to build up the toxic quantities of this compound.

The **retrosynthetic analysis** of DDT involves two C–C (or Ar–C) disconnections, which give chlorobenzene (2 molecules) and trichloroacetaldehyde (chloral) as inexpensive starting materials.

The **synthesis** of DDT is carried out using chlorobenzene and chloral in the presence of sulphuric acid as acid catalyst.

DDT: *Retrosynthetic analysis*

DDT (dichlorodiphenyltrichloroethane) (TM) $\xRightarrow{\text{C—C}}$ Chlorobenzene + $HC(OH)(CCl_3)—C_6H_4—Cl$ $\xRightarrow{\text{C—C}}$ Chloral ($HC(=O)—CCl_3$) Use $\equiv$ $HC^+(OH)CCl_3$ + Chlorobenzene

DDT: *Synthesis*

2 Chlorobenzene + Chloral ($HC(=O)—CCl_3$) $\xrightarrow{H_2SO_4}$ DDT ($Cl—C_6H_4—CH(CCl_3)—C_6H_4—Cl$)

18.6.2 Aromatic Compounds via Diazonium Salts

Diazonium salts are highly useful intermediates in the synthesis of a variety of aromatic compounds since the diazonium group (N_2^+) can be replaced by a number of other atoms or groups (see Section 14.4).

The **synthesis** is illustrated with *two* examples using the retrosynthetic approach.

A. *p*-Cresol

The **retrosynthetic analysis** of *p*-cresol or *p*-methylphenol (TM) involves a sequence of FGI substitution via diazonium salt, FGI reduction and C–N (nitration) disconnection to give toluene as a readily available starting material.

In the forward **synthesis**, the reaction sequence is therefore nitration of toluene, reduction of nitro group, diazotization of amino group and substitution of diazonium group by hydroxyl group.

***p*-Cresol: Retrosynthetic analysis**

OH, CH_3 — FGI (substitution via diazonium salt) ⟹ NH_2, CH_3 — FGI (reduction) ⟹ NO_2, CH_3 — C—N (nitration) ⟹ NO_2^+ + CH_3

***p*-Cresol** **(TM)**

***p*-Cresol: Synthesis**

CH_3 — HNO_3 / H_2SO_4 → CH_3, NO_2 — Sn / HCl → CH_3, NH_2 — $NaNO_2$ / H_2SO_4 (0-5°C) → CH_3, N_2^+ HSO_4^- — H_2O / Δ → CH_3, OH

Toluene — ***p*-Nitrotoluene (separate from *ortho* isomer)** — ***p*-Toluidine** — ***p*-Toluenediazonium hydrogen sulphate** — ***p*-Cresol**

B. *o*-Tolunitrile

The **retrosynthetic analysis** of *o*-tolunitrile or *o*-methylbenzonitrile (TM) involves a similar sequence as above (i.e. FGI substitution via diazonium salt, FGI reduction and C–N nitration), and gives toluene as a starting material.

Thus, in the forward **synthesis**, the reaction sequence is: nitration, reduction, diazotization and substitution of diazonium group by cyano group.

o-Tolunitrile: Retrosynthetic analysis

o-Tolunitrile (TM) $\xRightarrow[\text{substitution via diazonium salt}]{\text{FGI}}$ o-toluidine $\xRightarrow[\text{reduction}]{\text{FGI}}$ o-nitrotoluene $\xRightarrow[\text{nitration}]{\text{C—N}}$ NO_2^+ + toluene

o-Tolunitrile: Synthesis

Toluene $\xrightarrow[H_2SO_4]{HNO_3}$ **o-Nitrotoluene** (separate from *para* isomer) $\xrightarrow[Pd/C]{H_2}$ **o-Toluidine** $\xrightarrow[\text{HCl (0-5°C)}]{NaNO_2}$ **o-Toluenediazonium chloride** ($N_2^+ Cl^-$) $\xrightarrow[\Delta]{CuCN}$ **o-Tolunitrile**

18.7 SYNTHESIS OF ALKANES

The retrosynthetic approach to the synthesis of alkanes involves a C–C disconnection. The reagent for the cationic synthon is usually an alkyl halide, while that for the anionic synthon is an organometallic reagent.

R—R (**Alkane**) $\xRightarrow{\text{C—C}}$ R^+ + R^-

Synthons		***Reagents***	
R^+	**Use** ≡	R—X	**Alkyl halide**
R^-	**Use** ≡	R—M (M = metal)	**Organometallic reagent**

The Wurtz Reaction

Coupling of alkyl halides in the presence of sodium metal is known as the **Wurtz reaction**. This is one of the oldest organic reactions for the synthesis of alkanes.

The Wurtz reaction is usually carried out by adding sodium metal to an alkyl halide in an inert solvent such as diethyl ether. The **alkyl halide** reacts with sodium to form an **organosodium compound** which then reacts with unreacted alkyl halide to produce an **alkane**.

$$\underset{\substack{(X = Cl, Br, I) \\ \textbf{Alkyl halide}}}{R{-}X} \xrightarrow[Et_2O]{Na} \underset{\textbf{Organosodium compound}}{R^- Na^+} \xrightarrow{R-X} \underset{\textbf{Alkane}}{R{-}R}$$

Overall equation:

$$2\,R{-}X + 2\,Na \longrightarrow R{-}R + 2\,NaX$$

Example

$$\underset{\textbf{Ethyl bromide}}{CH_3CH_2Br} \xrightarrow[Et_2O]{Na} \underset{\textbf{Ethylsodium}}{CH_3CH_2Na} \xrightarrow{CH_3CH_2Br} \underset{\textbf{Butane}}{CH_3CH_2CH_2CH_3}$$

The Wurtz reaction has a **limited synthetic utility**. For example, the reaction between two *different* alkyl halides is synthetically useless because it produces a mixture of alkanes that are difficult to separate.

$$\underset{\textbf{Ethyl bromide}}{CH_3CH_2Br} + \underset{\textbf{Propyl bromide}}{CH_3CH_2CH_2Br} \xrightarrow[Et_2O]{Na} \underset{\textbf{Butane}}{CH_3CH_2CH_2CH_3} + \underset{\textbf{Pentane}}{CH_3CH_2CH_2CH_2CH_3}$$

$$+ \underset{\textbf{Hexane}}{CH_3CH_2CH_2CH_2CH_2CH_3}$$

A mixture of alkanes; difficult to separate

18.7.1 The Corey-House Synthesis

A highly versatile method for the synthesis of alkanes is known as the **Corey-House synthesis,** named after the American chemists E.J. Corey and Herbert O. House who developed this method during the late 1960s.

In the Corey-House synthesis, the organometallic reagent for the anionic synthon from a **C–C disconnection** (see above) is **lithium dialkylcuprate (R_2CuLi)**.

Preparation of lithium dialkylcuprate (R_2CuLi)

Lithium dialkylcuprate (R_2CuLi) is prepared from an alkyl halide (RX) in two steps. First, the alkyl halide is treated with lithium metal in an ether solvent to form an alkyllithium (RLi). Then the alkyllithium is treated with cuprous iodide (CuI) to produce a lithium dialkylcuprate (R_2CuLi).

$$\underset{\textbf{Alkyl halide}}{R{-}X} + 2\,Li \xrightarrow{Et_2O} \underset{\textbf{Alkyllithium}}{RLi} + LiX$$

$$\underset{\textbf{Alkyllithium}}{2\,RLi} + CuI \longrightarrow \underset{\textbf{Lithium dialkylcuprate}}{R_2CuLi} + LiI$$

Coupling of lithium dialkylcuprate with alkyl halide

The Corey-House synthesis involves coupling of lithium dialkylcuprate (R_2CuLi) with another alkyl halide (R′X) to produce an alkane (RR′). Coupling takes place between two alkyl groups, one from R_2CuLi and another from R′X.

$$\underset{\substack{\text{(R = any alkyl or cycloalkyl)} \\ \textbf{Lithium dialkylcuprate}}}{R_2CuLi} + \underset{\substack{\text{(R' = CH}_3\text{, 1}^\circ\text{ alkyl or 2}^\circ\text{ cycloalkyl)} \\ \textbf{Alkyl halide}}}{R'{-}X} \xrightarrow{\textbf{Corey-House synthesis}} \underset{\textbf{Alkane}}{R{-}R'} + RCu + LiX$$

Note: In the Corey-House synthesis, the *alkyl group* in lithium dialkylcuprate could be any alkyl or cycloalkyl. However, the *alkyl group* in the alkyl halide is usually methyl, 1° alkyl or 2° cycloalkyl (e.g. cyclohexyl).

Synthesis of alkanes

The **retrosynthetic analysis** for the synthesis of alkanes involves a C–C disconnection that corresponds to a Corey-House synthesis.

Using the retrosynthetic analysis, the **synthesis** of alkanes is illustrated with *two* examples.

A. *Isobutane (2-methylpropane)*

The **retrosynthetic analysis** involving a C–C disconnection for isobutane or 2-methylpropane (TM) discloses two possibilities, as shown below.

The disconnection that gives methyl iodide (a *methyl* halide) is a **good choice** but the other disconnection is a **bad choice** because it gives isopropyl bromide, which is a *secondary* alkyl halide. (Remember that the alkyl group in lithium dialkylcuprate could be any alkyl group; thus both lithium diisopropylcuprate and lithium dimethylcuprate are reasonable reagents).

Therefore, the **synthesis** is performed using methyl iodide and lithium diisopropylcuprate, which gives a good yield of the product. Lithium diisopropylcuprate can be prepared from isopropyl bromide (see above for the general preparation of lithium dialkylcuprate).

Isobutane: Retrosynthetic analysis

$$CH_3-\overset{CH_3}{\overset{|}{CH}}\text{-}\xi\text{-}CH_3 \quad \xrightarrow[]{C-C} \quad \left(CH_3-\overset{CH_3}{\overset{|}{CH}}\right)_2CuLi + CH_3-I$$

Isobutane (TM) → **Lithium diisopropylcuprate** + **Methyl iodide** — **Good choice (a methyl halide)**

$$\xrightarrow[]{C-C} \quad CH_3-\overset{CH_3}{\overset{|}{CH}}-Br \; (2^o) + (CH_3)_2CuLi$$

Isopropyl bromide + **Lithium dimethylcuprate** — **Bad choice (a 2° alkyl halide)**

Isobutane: Synthesis

$$[(CH_3)_2CH]_2CuLi + CH_3-I \xrightarrow{Et_2O} CH_3\overset{CH_3}{\overset{|}{C}}HCH_3$$

Lithium diisopropylcuprate + **Methyl iodide** → **Isobutane**

[***Preparation of lithium diisopropylcuprate:***

$$(CH_3)_2CHBr \xrightarrow[Et_2O]{Li} (CH_3)_2CHLi \xrightarrow{CuI} [(CH_3)_2CH]_2CuLi$$
]

B. *Butane*

Two different C–C disconnections are possible for butane (TM). The **retrosynthetic analysis 1** involving C2–C3 disconnection gives ethyl bromide (a 1° alkyl halide) and lithium diethylcuprate as starting materials.

Thus the **synthesis 1** using ethyl bromide and lithium diethylcuprate is likely to succeed and will give a good yield of the product.

Butane: Retrosynthetic analysis 1

$$CH_3-\overset{3}{CH_2}\text{-}\xi\text{-}\overset{2}{CH_2}-CH_3 \quad \xrightarrow[]{C-C} \quad CH_3CH_2-Br + (CH_3CH_2)_2CuLi$$

Butane (TM)

Butane: Synthesis 1

$$CH_3CH_2-Br + (CH_3CH_2)_2CuLi \xrightarrow{Et_2O} CH_3CH_2CH_2CH_3$$

Ethyl bromide + **Lithium diethylcuprate** → **Butane**

The **retrosynthetic analysis 2** involving C1–C2 disconnection reveals two possibilities. In one case, the alkyl halide is propyl bromide (a 1° alkyl halide), while in the other case, the alkyl halide is methyl iodide (a methyl halide). Thus both possibilities offer a reasonable choice.

So the **synthesis 2** can be performed using propyl bromide and lithium dimethylcuprate or using methyl iodide and lithium dipropylcuprate.

Butane: Retrosynthetic analysis 2

$$CH_3-\overset{2}{CH_2}-CH_2 \wr \overset{1}{CH_3} \Rightarrow CH_3CH_2CH_2-Br + (CH_3)_2CuLi$$

$$\Rightarrow (CH_3CH_2CH_2)_2CuLi + CH_3-I$$

Butane

(TM)

Butane: Synthesis 2

$$CH_3CH_2CH_2-Br + (CH_3)_2CuLi \xrightarrow{Et_2O} CH_3CH_2CH_2CH_3$$

Propyl bromide **Lithium dimethylcuprate** **Butane**

OR

$$(CH_3CH_2CH_2)_2CuLi + CH_3-I \xrightarrow{Et_2O} CH_3CH_2CH_2CH_3$$

Lithium dipropylcuprate **Methyl iodide** **Butane**

PROBLEM 18.5: Perform a retrosynthetic analysis and then outline a Corey-House synthesis of isopropylcyclohexane.

$$C_6H_{11}-CH(CH_3)-CH_3$$

Isopropylcyclohexane

Answer

The C–C disconnection between the ring carbon and the alkyl group discloses two possibilities. The disconnection that gives cyclohexyl bromide (a 2° cycloalkyl halide) is a good choice because the alkyl halide could be a **2° cycloalkyl halide**, but *not* a 2° alkyl halide. The other possibility that gives isopropyl bromide (a 2° alkyl halide) is a bad choice.

Therefore, the Corey-House synthesis using cyclohexyl bromide and lithium diisopropylcuprate is likely to succeed and gives a good yield of isopropylcyclohexane.

Isopropylcyclohexane: Retrosynthetic analysis

Isopropylcyclohexane (TM) $[C_6H_{11}CH(CH_3)_2]$ ⟹ Cyclohexyl bromide (2° cycloalkyl) + $[(CH_3)_2CH]_2CuLi$ — **Good choice**

Isopropylcyclohexane (TM) ⟹ $(C_6H_{11})_2CuLi$ + $(CH_3)_2CHBr$ (2° alkyl) — **Bad choice**

Isopropylcyclohexane: Synthesis

$$\text{Cyclohexyl bromide} + [(CH_3)_2CH]_2CuLi \xrightarrow{Et_2O} C_6H_{11}CH(CH_3)_2$$

Cyclohexyl bromide + **Lithium diisopropylcuprate** → **Isopropylcyclohexane**

18.8 SYNTHESIS OF ALKENES

Alkenes can be prepared by elimination reactions such as dehydration of alcohols (see Section 10.6) or dehydrohalogenation of alkyl halides (see Section 10.2.1).

The retrosynthetic approach to the synthesis of alkenes thus involves an FGI dehydration (or FGI dehydrohalogenation) followed by a C–C disconnection.

Using the retrosynthetic approach, the **synthesis** of alkenes is illustrated with *two* examples.

A. *2-Methyl-2-butene*

A **retrosynthetic analysis** of 2-methyl-2-butene (TM) using an FGI dehydration gives 2-methyl-2-butanol (a tertiary alcohol). Then a 1,1 C–C disconnection leads to acetone and ethylmagnesium bromide as the starting materials.

In the forward **synthesis**, after the Grignard reaction between acetone and ethylmagnesium bromide, dehydration of the resulting alcohol is carried out with sulphuric acid.

2-Methyl-2-butene: Retrosynthetic analysis

$$CH_3CH{=}C(CH_3){-}CH_3 \xrightarrow[\text{dehydration}]{\text{FGI}} CH_3CH_2{-}C(CH_3)(OH){-}CH_3 \xrightarrow{1,1\ C-C} CH_3CH_2MgBr + CH_3{-}C(=O){-}CH_3$$

2-Methyl-2-butene (TM) — **(a 3° alcohol)**

2-Methyl-2-butene: Synthesis

$$CH_3COCH_3 \text{ (Acetone)} + CH_3CH_2MgBr \text{ (Ethylmagnesium bromide)} \xrightarrow[2.\ H_3O^+]{1.\ Et_2O} CH_3CH_2C(CH_3)(OH)CH_3 \text{ (2-Methyl-2-butanol)} \xrightarrow[\Delta,\ -H_2O]{H_2SO_4} CH_3CH{=}C(CH_3){-}CH_3 \text{ (2-Methyl-2-butene)}$$

Note: In FGI, the hydroxyl group (OH) can be put at either end of the double bond (C=C). But putting OH at the branch carbon is better because it gives a tertiary alcohol which can be disconnected to relatively simpler starting materials.

B. *Styrene*

Besides dehydration, *dehydrohalogenation* of alkyl halides provides another method for alkene synthesis. Alkyl halides are easily prepared from alcohols.

The following **retrosynthetic analysis** of styrene (TM) involves a sequence of two FGIs to give 2-phenylethanol (a primary alcohol). Then a 1.2 C–C disconnection gives an epoxide (ethylene oxide) and phenylmagnesium bromide as the starting materials (see Section 18.1).

In the forward **synthesis**, a Grignard reaction between ethylene oxide and phenylmagnesium bromide gives 2-phenylethanol, which on treatment with phosphorus tribromide forms 1-bromo-2-phenylethane as a primary alkyl halide. The dehydrohalogenation of the primary alkyl halide is best carried out with potassium *t*-butoxide (a sterically hindered base) to produce styrene.

Styrene: Retrosynthetic analysis

$$C_6H_5{-}CH{=}CH_2 \text{ (Styrene (TM))} \xRightarrow[\text{dehydrohalogenation}]{\text{FGI}} C_6H_5{-}CH_2{-}CH_2Br \xRightarrow[\text{substitution}]{\text{FGI}} C_6H_5 \wr CH_2{-}CH_2OH \xRightarrow{1,2\ C{-}C} C_6H_5MgBr + CH_2{-}CH_2 \text{ (epoxide, O)}$$

Styrene: Synthesis

$$CH_2{-}CH_2 \text{ (O; Ethylene oxide)} + C_6H_5MgBr \text{ (Phenylmagnesium bromide)} \xrightarrow[2.\ H_3O^+]{1.\ Et_2O} C_6H_5CH_2CH_2OH \text{ (2-Phenylethanol)} \xrightarrow{PBr_3} C_6H_5CH_2CH_2Br \text{ (1-Bromo-2-phenylethane)} \xrightarrow{(CH_3)_3CO^-\ K^+} C_6H_5CH{=}CH_2 \text{ (Styrene)}$$

18.8.1 The Wittig Synthesis

A highly versatile method for alkene synthesis employs the Wittig reaction between an aldehyde or ketone and a phosphorus ylide (see Section 9.9). A great advantage of the

Wittig synthesis is that there is no ambiguity as to the location of the double bond; thus the Wittig synthesis produces a *single alkene*.

The **retrosynthetic analysis** of an alkene using a Wittig reaction involves a disconnection at the C=C bond which gives an aldehyde or ketone and a phosphorus ylide as the reagents. The ylide is easily prepared by the reaction of an alkyl halide with triphenylphosphine (Ph_3P) followed by deprotonation with a base such as butyllithium (BuLi).

Using the retrosynthetic approach involving a C=C Wittig disconnection, the **synthesis** of alkenes is described here with *two* examples.

A. *2-Methyl-1-phenyl-1-propene*

2-Methyl-1-phenyl-1-propene is an *unsymmetrical* alkene. In general, for an unsymmetrical alkene, the Wittig disconnection of C=C reveals two possibilities because either end of the double bond can come from the carbonyl compound or the phosphorus ylide.

Thus the **retrosynthetic analysis** involving a C=C Wittig disconnection for 2-methyl-1-phenyl-1-propene (TM) discloses two possible routes, as shown below.

In one route, the reagents are benzaldehyde (C_6H_5CHO) and a phosphorus ylide. Further analysis of the ylide gives isopropyl bromide [$(CH_3)_2CHBr$]. Both benzaldehyde and isopropyl bromide are available as starting materials.

In the other route, the reagents are acetone (CH_3COCH_3) and another phosphorus ylide which on further analysis gives benzyl bromide ($C_6H_5CH_2Br$).

Both routes lead to the successful **synthesis** of the target alkene. In one synthesis, the phosphorus ylide is prepared by the reaction of isopropyl bromide with triphenylphosphine (Ph_3P) followed by treatment with butyllithium (BuLi) as a base.

In the other synthesis, the ylide is prepared by the reaction of benzyl bromide with Ph_3P followed by treatment with BuLi.

2-Methyl-1-phenyl-1-propene: Retrosynthetic analysis

$$C_6H_5CH{=}C(CH_3)-CH_3 \xRightarrow[\text{Wittig}]{C=C} C_6H_5CH{=}O + Ph_3\overset{+}{P}-\overset{-}{C}(CH_3)-CH_3 \Longrightarrow Br-CH(CH_3)CH_3$$

2-Methyl-1-phenyl-1-propene (TM) → **Benzaldehyde** + ***Phosphorus ylide*** ⇒ **Isopropyl bromide**

$$C_6H_5CH{=}C(CH_3)-CH_3 \xRightarrow[\text{Wittig}]{C=C} C_6H_5\overset{-}{C}H-\overset{+}{P}Ph_3 + O{=}C(CH_3)-CH_3$$

2-Methyl-1-phenyl-1-propene (TM) → ***Phosphorus ylide*** + **Acetone**

$$C_6H_5\overset{-}{C}H-\overset{+}{P}Ph_3 \Longrightarrow C_6H_5CH_2-Br$$

Phosphorus ylide ⇒ **Benzyl bromide**

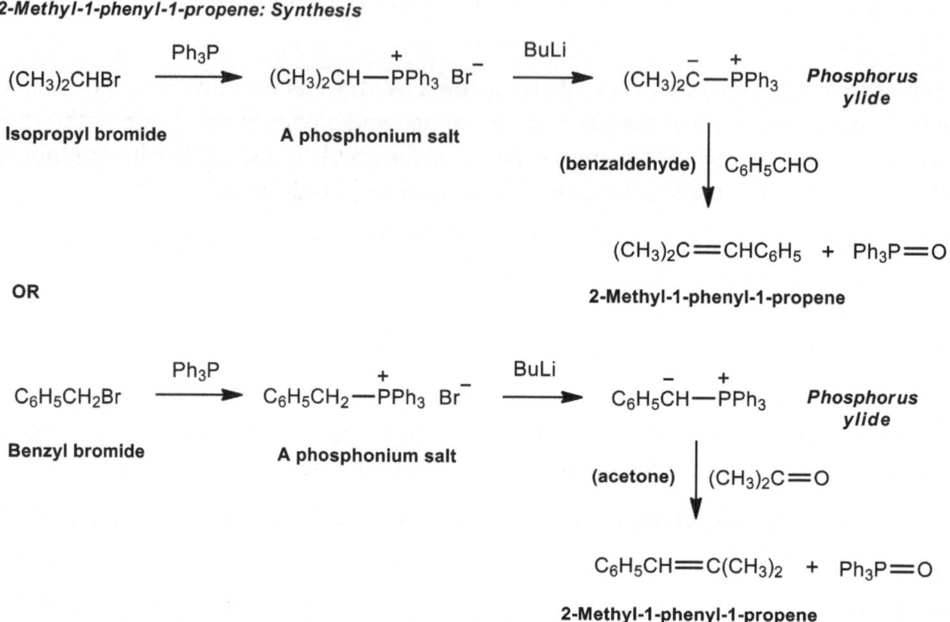

B. *Methylenecyclohexane*

The Wittig C=C disconnection for methylenecyclohexane (TM) offers two possible routes (marked *a*, *b*) as shown below.

In the route (*a*), the reagents are cyclohexanone and a phosphorus ylide which on further analysis gives methyl iodide, while in the route (*b*) the reagents are formaldehyde and a phosphorus ylide which on further analysis gives cyclohexyl bromide.

Either route can lead to a successful synthesis of the target alkene. However, the route (*a*) is preferred because cyclohexanone is easier to handle than formaldehyde.

The forward **synthesis** is shown according to the route (*a*). In this synthesis, the phosphorus ylide is prepared by the reaction of methyl iodide with Ph_3P followed by treatment with BuLi.

Methylenecyclohexane: Retrosynthetic analysis

C=C Wittig *a* ⟹ Cyclohexanone + $Ph_3\overset{+}{P}—\overset{-}{C}H_2$ (*Phosphorus ylide*) ⟹ I—CH_3 (Methyl iodide)

Methylenecyclohexane (TM) [$=CH_2$]

C=C Wittig *b* ⟹ $H_2C=O$ (Formaldehyde) + cyclohexyl^{-}—$\overset{+}{P}Ph_3$ (*Phosphorus ylide*) ⟹ cyclohexyl—Br (Cyclohexyl bromide)

Methylenecyclohexane: Preferred synthesis (route a)

$$CH_3I \xrightarrow{Ph_3P} CH_3\overset{+}{-}PPh_3\ I^- \xrightarrow{BuLi} \overset{-}{C}H_2-\overset{+}{P}Ph_3 \xrightarrow[\text{(cyclohexanone)}]{} \text{Methylenecyclohexane} (=CH_2)$$

Methyl iodide — ***Phosphorus ylide*** — **Methylenecyclohexane**

PROBLEM 18.6: Give a retrosynthetic analysis and then outline a Wittig synthesis of the following diene:

$$C_6H_5-CH=CH-CH=CH_2$$

1-Phenyl-1,3-butadiene

Answer

Since the target molecule (TM) has two double bonds, either double bond can be disconnected. However, the Wittig disconnection of the internal double bond (*a*) leads to readily available starting materials (benzaldehyde and allyl bromide).

This route thus provides an easy **synthesis** of 1-phenyl-1,3-butadiene. In the synthesis, the phosphorus ylide is prepared by the reaction of allyl bromide with Ph_3P followed by treatment with BuLi.

1-Phenyl-1,3-butadiene: Retrosynthetic analysis

$$C_6H_5CH \overset{a}{=} CH-CH=CH_2 \xRightarrow[\text{Wittig}]{C=C} C_6H_5CH=O + Ph_3\overset{+}{P}-\overset{-}{C}H-CH=CH_2$$

1-Phenyl-1,3-butadiene (TM) — **Benzaldehyde** — ***Phosphorus ylide***

$$\Downarrow$$

$$BrCH_2-CH=CH_2$$

Allyl bromide

1-Phenyl-1,3-butadiene: Synthesis

$$CH_2=CHCH_2Br \xrightarrow{Ph_3P} CH_2=CHCH_2-\overset{+}{P}Ph_3\ Br^- \xrightarrow{BuLi} CH_2=CH\overset{-}{C}H-\overset{+}{P}Ph_3$$

Allyl bromide — ***Phosphorus ylide***

$$\downarrow C_6H_5CHO \text{ (benzaldehyde)}$$

$$CH_2=CHCH=CHC_6H_5$$

1-Phenyl-1,3-butadiene

18.9 SYNTHESIS OF ALKYNES

Double dehydrohalogenation of 1,2-dihaloalkanes provides suitable routes to the synthesis of alkynes (see Section 10.8). 1,2-dihaloalkanes can be prepared from the corresponding alkene.

Thus the retrosynthetic approach to the synthesis of alkynes can involve a sequence of FGIs. This is illustrated with the synthesis of the following alkyne.

1-Butyne

A **retrosynthetic analysis** of 1-butyne (TM) involving two consecutive FGI eliminations (dehydrohalogenation) gives 1,2-dibromobutane. Then an FGI addition (bromination) gives 1-butene as the starting material.

In the forward **synthesis**, after bromination of 1-butene, both dehydrohalogenations are accomplished in a single step using $NaNH_2$ in liquid NH_3.

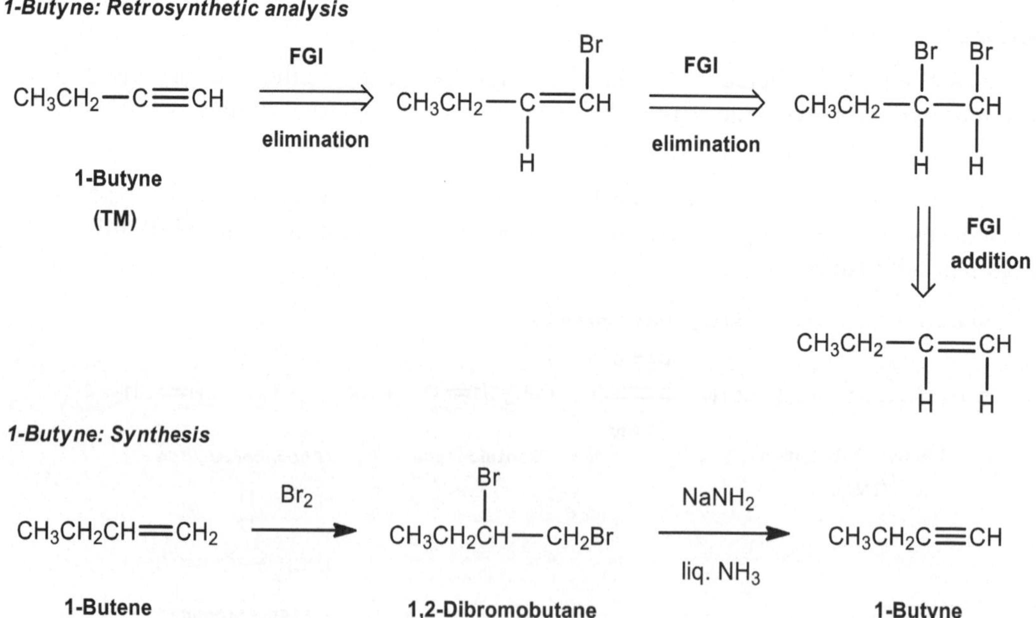

18.9.1 The Use of Acetylene

Acetylene is readily available. It is relatively acidic and has the ability to form a carbanion called an acetylide ion ($HC{\equiv}C^-$) in the presence of a strong base, usually $NaNH_2$ in liquid NH_3. The acetylide ion reacts with alkyl halides (RX) by forming a new C–C bond to give a monosubstituted acetylene ($HC{\equiv}CR$) (see Section 11.2), which can be alkylated further to produce a disubstituted acetylene ($RC{\equiv}CR$).

The retrosynthetic approach to the synthesis of alkynes thus involves one or two C–C disconnections. This is illustrated with the synthesis of the following alkyne.

2-Butyne

2-Butyne is a disubstituted acetylene. A **retrosynthetic analysis** of 2-butyne (TM) involves two successive C–C disconnections. The first disconnection of the C–C bond attached to the triple bond gives propyne and methyl iodide. Then a second C–C disconnection gives acetylene and the same alkyl halide (methyl iodide). Thus both disconnections lead to acetylene and methyl iodide (2 molecules) as starting materials.

In the forward **synthesis**, $NaNH_2$ in liquid NH_3 is used to form an alkynide ion. Acetylene reacts with $NaNH_2$ in liquid NH_3 to form sodium acetylide which then reacts with methyl iodide to form propyne. The reaction of propyne with $NaNH_2$ in liquid NH_3 followed by treatment with a second mole of methyl iodide produces 2-butyne.

2-Butyne: Retrosynthetic analysis

$$CH_3{-}C{\equiv}C{-}\!\!\wr\!\!{-}CH_3 \xRightarrow{C-C} CH_3{-}\!\!\wr\!\!{-}C{\equiv}C{-}H + CH_3{-}I$$

2-Butyne
(TM)

$$\Downarrow \; C-C$$

$$HC{\equiv}CH + CH_3{-}I$$

2-Butyne: Synthesis

$$\underset{\textbf{Acetylene}}{H{-}C{\equiv}C{-}H} \xrightarrow[\text{liq. } NH_3]{NaNH_2} \underset{\textbf{Sodium acetylide}}{H{-}C{\equiv}\overset{-}{C}\;Na^+} \xrightarrow[\text{(methyl iodide)}]{CH_3-I} \underset{\textbf{Propyne}}{H{-}C{\equiv}C{-}CH_3}$$

$$\downarrow \; NaNH_2/\text{liq. } NH_3$$

$$\underset{\textbf{2-Butyne}}{CH_3{-}C{\equiv}C{-}CH_3} \xleftarrow{CH_3-I} \underset{\textbf{Sodium propynide}}{Na^+\;\overset{-}{C}{\equiv}C{-}CH_3}$$

Synthesis of ketones via alkynes

Alkynes can serve as intermediates in the synthesis of ketones (see Section 8.13). This is illustrated with the synthesis of the following ketone using a retrosynthetic approach.

2-Pentanone

2-Pentanone is a *methyl ketone* which can be prepared by hydration of a terminal alkyne (monosubstituted acetylene). Thus the **retrosynthetic analysis** of 2-pentanone involves an FGI hydration and a C–C disconnection, which gives acetylene and propyl bromide as starting materials.

In the forward **synthesis,** hydration is carried out with aqueous sulphuric acid in the presence of mercuric ion.

2-Pentanone: Retrosynthetic analysis

$CH_3CH_2CH_2-C(=O)-CH_3 \xrightarrow[\text{hydration}]{\text{FGI}} CH_3CH_2CH_2 \wr C\equiv CH \xrightarrow{C-C} CH_3CH_2CH_2-Br + H-C\equiv CH$

2-Pentanone *(a methyl ketone)* (TM); Propyl bromide; Acetylene

2-Pentanone: Synthesis

$HC\equiv CH \xrightarrow[\text{2. } CH_3CH_2CH_2Br]{\text{1. } NaNH_2/\text{liq. } NH_3} CH_3CH_2CH_2-C\equiv CH \xrightarrow[HgSO_4/H_2SO_4]{H_2O} CH_3CH_2CH_2-C(=O)CH_3$

Acetylene; 1-Pentyne; 2-Pentanone

PROBLEM 18.7: Suggest a retrosynthetic analysis and then outline a synthesis of the following compound:

CH=CH$_2$

1-Vinylcyclohexene

Answer

The acetylide ion (from acetylene) can add to a ketone to form an **acetylenic alcohol** that contains a triple bond and an alcoholic OH group.

1-Vinylcyclohexene is a diene. Acetylene can be used to make the double bond *outside* the ring by reduction of the triple bond. The double bond *inside* the ring can be formed by dehydration of alcohol.

The retrosynthetic analysis of 1-vinylcyclohexene (TM) thus involves two FGIs (partial reduction and dehydration) followed by a C–C disconnection, which gives acetylene and cyclohexanone as starting materials.

In the forward synthesis, partial reduction of the triple bond to a double bond is accomplished by hydrogenation in the presence of Lindlar's catalyst and dehydration of the alcohol is best carried out in the presence of potassium hydrogen sulphate ($KHSO_4$) as an efficient solid acid catalyst.

1-Vinylcyclohexene: Retrosynthetic analysis

1-Vinylcyclohexene (TM) $\xrightarrow[\text{dehydration}]{\text{FGI}}$ 1-vinylcyclohexanol (HO, CH=CH$_2$) $\xrightarrow[\text{reduction}]{\text{FGI}}$ 1-ethynylcyclohexanol (HO, C≡CH) $\xrightarrow{1,1\ C-C}$ cyclohexanone + $HC\equiv CH$

1-Vinylcyclohexene: Synthesis

HC≡CH (Acetylene) —$NaNH_2$, liq. NH_3→ HC≡C⁻ Na⁺ —(cyclohexanone)→ 1-ethynylcyclohexanol (*An acetylenic alcohol*)

—Lindlar's catalyst (H_2, $Pd/CaCO_3$, lead acetate, quinoline)→ 1-vinylcyclohexanol (HO, CH=CH₂)

—H^+ ($KHSO_4$)→ **1-Vinylcyclohexene**

18.10 SYNTHESIS OF ALDOLS

Aldols are β-hydroxy aldehydes or β-hydroxy ketones in which two functional groups (OH and C=O) are in a 1,3-relationship. Aldols are prepared by the aldol reaction of aldehydes or ketones (see Section 15.4).

The retrosynthetic approach to aldol synthesis involves a simple **1,3-diO disconnection** which gives two carbonyl compounds (aldehyde/ketone) as starting materials (see Section 17.7.2). The two carbonyl compounds may be the same or different.

Using the retrosynthetic analysis, the **synthesis** of aldols is described here with *two* examples.

A. *3-Hydroxy-2-methylpentanal*

The target aldol (3-hydroxy-2-methylpentanal) is a β-hydroxy aldehyde.

The **retrosynthetic analysis** of 3-hydroxy-2-methylpentanal (TM) involves a 1,3-diO disconnection which gives *two molecules* of propanal (an aldehyde) as the starting material.

The **synthesis** is an aldol reaction with propanal in dilute sodium hydroxide. Since the aldol reaction involves a single aldehyde, no special precautions need to be taken to prevent side reactions as in the case of a crossed aldol reaction.

3-Hydroxy-3-methylpentanal: Retrosynthetic analysis

$CH_3CH_2CH(OH)$–⁄⁄–$CH(CH_3)$–C(=O)–H (**3-Hydroxy-2-methylpentanal (TM)**) ⟹ (1,3-diO) CH_3CH_2CH=O + H–$CH(CH_3)$–C(=O)–H (**Two molecules of propanal**)

3-Hydroxy-3-methylpentanal: Synthesis

$$2\ CH_3CH_2\overset{O}{\overset{\|}{C}}H \xrightarrow[H_2O]{NaOH} CH_3CH_2\overset{OH}{\overset{|}{C}H}\underset{CH_3}{\underset{|}{C}H}CHO$$

Aldol reaction

Propanal → **3-Hydroxy-2-methylpentanal**

B. *4-Hydroxy-4-methyl-2-pentanone*

The target aldol (4-hydroxy-4-methyl-2-pentanone) is a β-hydroxy ketone.

The **retrosynthetic analysis** of the target molecule (TM) involving a 1,3-diO disconnection reveals that *two molecules* of the same ketone (acetone) are required as the starting material.

In the forward **synthesis**, the aldol reaction with acetone is conveniently carried out with barium hydroxide as an insoluble basic catalyst.

4-Hydroxy-4-methyl-2-pentanone: Retrosynthetic analysis

$$CH_3-\underset{CH_3}{\underset{|}{\overset{OH}{\overset{|}{C}}}}\wr CH_2-\overset{O}{\overset{\|}{C}}-CH_3 \overset{\textbf{1,3-diO}}{\Longrightarrow} CH_3-\underset{CH_3}{\underset{|}{\overset{O}{\overset{\|}{C}}}} + H-CH_2-\overset{O}{\overset{\|}{C}}-CH_3$$

4-Hydroxy-4-methyl-2-pentanone (TM) **Two molecules of acetone**

4-Hydroxy-4-methyl-2-pentanone: Synthesis

$$2\ CH_3\overset{O}{\overset{\|}{C}}CH_3 \xrightarrow{Ba(OH)_2} CH_3\underset{CH_3}{\underset{|}{\overset{OH}{\overset{|}{C}}}}CH_2\overset{O}{\overset{\|}{C}}CH_3$$

Aldol reaction

Acetone → **4-Hydroxy-4-methyl-2-pentanone**

PROBLEM 18.8: 2-Ethyl-1,3-hexanediol is an insect repellent. Give a retrosynthetic analysis and then outline the synthesis of the diol. Draw structures in bond-line formulas.

2-Ethyl-1,3-hexanediol
An insect repellent

[$CH_3CH_2CH_2CHOHCH(CH_2CH_3)CH_2OH$]

Answer

The target molecule (TM) is a 1,3-diol which can be made from an aldol by reduction.

Thus the retrosynthetic analysis of 2-ethyl-1,3-hexanediol involves an FGI reduction followed by a 1,3-diO disconnection. The starting material is butanal.

In the forward synthesis, an aldol reaction of butanal with dilute sodium hydroxide gives 2-ethyl-3-hydroxyhexanal. Subsequent reduction of the aldol is carried out with sodium borohydride to produce the target diol.

2-Ethyl-1,3-hexanediol: Retrosynthetic analysis

OH OH FGI reduction 1,3-diO O + O

2-Ethyl-1,3-hexanediol (TM)

Two molecules of butanal

2-Ethyl-1,3-hexanediol: Synthesis

2 O NaOH H_2O *Aldol reaction* OH O $NaBH_4$ *Reduction* OH OH

Butanal **2-Ethyl-3-hydroxyhexanal** **2-Ethyl-1,3-hexanediol**

18.10.1 α,β-Unsaturated Carbonyl Compounds

α,β-Unsaturated carbonyl compounds can be prepared by the dehydration of aldols.

In aldol condensations, dehydration of the aldol occurs spontaneously to give the α,β-unsaturated carbonyl product that is stabilized by additional conjugation with, for example, a phenyl group (see Section 15.5).

The retrosynthetic approach to the synthesis of α,β-unsaturated aldehydes or ketones involves either an **FGI dehydration** followed by a **1,3-diO disconnection** or a shorthand **α,β disconnection** (see Section 17.7.3).

Using the retrosynthetic approach, the **synthesis** of α,β-unsaturated carbonyl compounds is illustrated with *two* examples.

A. *Mesityl oxide*

Mesityl oxide (4-methyl-3-penten-2-one) is an α,β-unsaturated ketone.

A **retrosynthetic analysis** of mesityl oxide (TM) involving FGI dehydration gives an aldol (4-hydroxy-4-methyl-2-pentanone, see above). Then a 1,3-diO disconnection gives acetone (two molecules) as the starting material.

In the forward **synthesis**, the dehydration of the aldol to mesityl oxide can be carried out by heating in aqueous base.

However, if the aldol reaction of acetone is brought about with acid catalysis, the reaction *directly* leads to mesityl oxide via the dehydration of the initially formed aldol.

Mesityl oxide: Retrosynthetic analysis

$$CH_3-\underset{\underset{CH_3}{|}}{C}=CH-\overset{\overset{O}{\|}}{C}-CH_3 \xRightarrow[\text{dehydration}]{\text{FGI}} CH_3-\underset{\underset{CH_3}{|}}{\overset{\overset{OH}{|}}{C}}\wr CH_2-\overset{\overset{O}{\|}}{C}-CH_3 \quad \textit{An aldol}$$

Mesityl oxide
(4-methyl-3-penten-2-one)
(TM)

$$\Downarrow \text{1,3-diO}$$

$$CH_3-\underset{\underset{CH_3}{|}}{\overset{\overset{O}{\|}}{C}} + CH_3-\overset{\overset{O}{\|}}{C}-CH_3$$

Two molecules of acetone

Mesityl oxide: Synthesis

$$2\,CH_3\overset{\overset{O}{\|}}{C}CH_3 \xrightarrow[\textit{Base-catalysed aldol reaction}]{Ba(OH)_2} CH_3\underset{\underset{CH_3}{|}}{\overset{\overset{OH}{|}}{C}}CH_2\overset{\overset{O}{\|}}{C}CH_3 \xrightarrow[\Delta,\ \textit{Dehydration}]{OH^-/H_2O} CH_3\underset{\underset{CH_3}{|}}{C}=CH\overset{\overset{O}{\|}}{C}CH_3 + H_2O$$

Acetone **Mesityl oxide**

OR

$$2\,CH_3\overset{\overset{O}{\|}}{C}CH_3 \xrightarrow[-\,H_2O,\ \textit{Acid-catalysed aldol reaction}]{HCl} CH_3\underset{\underset{CH_3}{|}}{C}=CH\overset{\overset{O}{\|}}{C}CH_3$$

Acetone **Mesityl oxide**

B. *Benzalacetophenone*

Benzalacetophenone (1,3-diphenyl-2-propen-1-one) is a stable α,β-unsaturated ketone that can be obtained directly by the aldol condensation.

The **retrosynthetic analysis** of benzalacetophenone (TM) involving a shorthand α,β disconnection leads directly to benzaldehyde and acetophenone as the starting materials. In this case, the two carbonyl components are different.

The **synthesis** is an aldol condensation between benzaldehyde and acetophenone with dilute sodium hydroxide. Here the dehydration of the initially formed aldol takes place spontaneously.

Benzalacetophenone: Retrosynthetic analysis

$$C_6H_5\text{—}CH\text{=}CH\text{—}\overset{O}{\overset{\|}{C}}\text{—}C_6H_5 \overset{\alpha,\beta}{\Longrightarrow} C_6H_5\text{—}CH\text{=}O + CH_3\text{—}\overset{O}{\overset{\|}{C}}\text{—}C_6H_5$$

Benzalacetophenone (1,3-diphenyl-2-propen-1-one) (TM) | **Benzaldehyde** | **Acetophenonne**

Two different carbonyl compounds

Benzalacetophenone: Synthesis

$$C_6H_5\overset{O}{\overset{\|}{C}}H + CH_3\overset{O}{\overset{\|}{C}}C_6H_5 \xrightarrow{NaOH/H_2O} \left[C_6H_5\overset{OH}{\overset{|}{C}}H\text{—}CH_2\overset{O}{\overset{\|}{C}}C_6H_5\right] \xrightarrow{-H_2O} C_6H_5CH\text{=}CH\overset{O}{\overset{\|}{C}}C_6H_5$$

Benzaldehyde **Acetophenonne** — ***Aldol condensation*** — ***Not isolated*** — **Benzalacetophenone**

PROBLEM 18.9: 2-Ethyl-2-hexenal is an intermediate in the synthesis of oxanamide (a tranquilizer). Write down a retrosynthetic analysis and outline its synthesis. Draw structures in bond-line formulas.

Oxanamide intermediate

2-Ethyl-2-hexenal

Answer

2-Ethyl-2-hexenal is an α,β-unsaturated aldehyde. A retrosynthetic analysis of 2-ethyl-2-hexenal (TM) involving a shorthand α,β disconnection gives two molecules of butanal as the starting material (cf. Problem 18.8).

In the forward synthesis, the dehydration of the aldol (2-ethyl-3-hydroxyhexanal) to the α,β-unsaturated aldehyde (2-ethyl-2-hexenal) does not occur spontaneously since it is not stabilized by additional conjugation. Thus the dehydration is carried out by heating in the NaOH solution.

2-Ethyl-2-hexenal: Retrosynthetic analysis

2-Ethyl-2-hexenal (TM) $\overset{\alpha,\beta}{\Longrightarrow}$ **Two molecules of butanal**

2-Ethyl-2-hexenal: Synthesis

2 **Butanal** $\xrightarrow{NaOH/H_2O}$ (***Aldol reaction***) **2-Ethyl-3-hydroxyhexanal** $\xrightarrow[-H_2O]{NaOH/H_2O,\ \Delta}$ (***Dehydration***) **2-Ethyl-2-hexenal**

18.11 SYNTHESIS OF β-KETO ESTERS AND β-DIKETONES

β-Keto esters and β-diketones are **1,3-dicarbonyl compounds** in which two C=O groups are in a 1,3-relationship. The Claisen condensation provides a suitable route to the synthesis of these compounds (see Section 15.7).

The retrosynthetic approach to the synthesis of β-keto esters or β-diketones involves a C–C disconnection called a **1,3-diCO** disconnection (see Section 17.7.4).

Using the retrosynthetic approach, the **synthesis** of β-keto esters and β-diketones is described with *three* examples.

A. *Ethyl acetoacetate*

Ethyl acetoacetate, commonly called acetoacetic ester, is a β-keto ester.

The **retrosynthetic analysis** of ethyl acetoacetate (TM) involving a 1,3-diCO disconnection gives *two molecules* of the same ester (ethyl acetate) as the starting material.

The **synthesis** is a Claisen condensation of ethyl acetate with sodium ethoxide as a base. An acid work-up releases the product.

Ethyl acetoacetate: Retrosynthetic analysis

$$CH_3-\overset{O}{\overset{\|}{C}}\wr CH_2-\overset{O}{\overset{\|}{C}}-OC_2H_5 \xRightarrow{\text{1,3-diCO}} CH_3-\overset{O}{\overset{\|}{C}}-OC_2H_5 + H-CH_2-\overset{O}{\overset{\|}{C}}-OC_2H_5$$

Ethyl acetoacetate
(a β-keto ester)
(TM)

Two molecules of ethyl acetate

Ethyl acetoacetate: Synthesis

$$2\ CH_3\overset{O}{\overset{\|}{C}}-OC_2H_5 \xrightarrow[\text{2. } H_3O^+ \text{ (work-up)}]{\text{1. } C_2H_5ONa} CH_3\overset{O}{\overset{\|}{C}}CH_2\overset{O}{\overset{\|}{C}}OC_2H_5 + C_2H_5OH$$

Claisen condensation

Ethyl acetate **Ethyl acetoacetate**

B. *Ethyl benzoylacetate*

Ethyl benzoylacetate is another β-keto ester.

The **retrosynthetic analysis** of ethyl benzoylacetate (TM) involves a 1,3-diCO disconnection which gives two different esters (ethyl benzoate and ethyl acetate) as the starting materials.

The **synthesis** is a crossed Claisen condensation with sodium ethoxide as a base. An acid work-up releases the product.

Ethyl benzoylacetate: Retrosynthetic analysis

$$C_6H_5-\overset{O}{\overset{\|}{C}}\wr CH_2-\overset{O}{\overset{\|}{C}}-OC_2H_5 \xRightarrow{\text{1,3-diCO}} C_6H_5-\overset{O}{\overset{\|}{C}}-OC_2H_5 + H-CH_2-\overset{O}{\overset{\|}{C}}-OC_2H_5$$

Ethyl benzoylacetate (another β-keto ester) (TM) — **Ethyl benzoate** + **Ethyl acetate** (***Two different esters***)

Ethyl benzoylacetate: Synthesis

$$C_6H_5-\overset{O}{\overset{\|}{C}}-OC_2H_5 + CH_3-\overset{O}{\overset{\|}{C}}-OC_2H_5 \xrightarrow[\text{Crossed Claisen condensation}]{\text{1. } C_2H_5ONa \;\; \text{2. } H_3O^+ \text{ (work-up)}} C_6H_5\overset{O}{\overset{\|}{C}}-CH_2\overset{O}{\overset{\|}{C}}OC_2H_5 + C_2H_5OH$$

Ethyl benzoate + **Ethyl acetate** → **Ethyl benzoylacetate**

C. Benzoylacetone

Benzoylacetone (1-phenyl-1,3-butanedione) is a β-diketone (or 1,3-diketone). This is a perfumery compound with fragrant balsamic odour.

The **retrosynthetic analysis** of benzoylacetone (TM) uses a 1,3-diCO disconnection which gives ethyl benzoate (an ester) and acetone (a ketone) as the starting materials.

The **synthesis** is a crossed Claisen condensation between ethyl benzoate and acetone with sodium ethoxide. An acid work-up releases the product.

Benzoylacetone: Retrosynthetic analysis

$$C_6H_5-\overset{O}{\overset{\|}{C}}\wr CH_2-\overset{O}{\overset{\|}{C}}-CH_3 \xRightarrow{\text{1,3-diCO}} C_6H_5-\overset{O}{\overset{\|}{C}}-OC_2H_5 + H-CH_2-\overset{O}{\overset{\|}{C}}-CH_3$$

Benzoylacetone (a β-diketone) (TM) — **Ethyl benzoate** ***An ester*** + **Acetone** ***A ketone***

Benzoylacetone: Synthesis

$$C_6H_5-\overset{O}{\overset{\|}{C}}-OC_2H_5 + CH_3-\overset{O}{\overset{\|}{C}}-CH_3 \xrightarrow[\text{Crossed Claisen condensation}]{\text{1. } C_2H_5ONa \;\; \text{2. } H_3O^+ \text{ (work-up)}} C_6H_5\overset{O}{\overset{\|}{C}}-CH_2\overset{O}{\overset{\|}{C}}CH_3 + C_2H_5OH$$

Ethyl benzoate + **Acetone** → **Benzoylacetone**

PROBLEM 18.10: Give a retrosynthetic analysis of the following cyclic β-keto ester and outline its synthesis.

$$\text{cyclopentanone}-\overset{O}{\overset{\|}{C}}-OEt$$

2-Carbethoxycyclopentanone

Answer

A retrosynthetic analysis of 2-carbethoxycyclopentanone (TM) using a 1,3-diCO disconnection gives diethyl adipate (a symmetrical diester) as the readily available starting material.

The synthesis is an intramolecular Claisen condensation (known as the Dieckmann condensation, see Section 15.7.2) of diethyl adipate with sodium ethoxide as a base. The intramolecular synthesis is fast and clean.

2-Carbethoxycyclopentanone: Retrosynthetic analysis

2-Carbethoxycyclopentanone (TM) ⟹ (1,3-diCO) ≡ (Use) **Diethyl adipate** + Base

2-Carbethoxycyclopentanone: Synthesis

$(EtO_2CCH_2CH_2CH_2CH_2CO_2Et)$ **Diethyl adipate** → (1. C_2H_5ONa; 2. H^+ (work-up); *Dieckmann condensation*) **2-Carbethoxycyclopentanone** + EtOH

18.12 SYNTHESIS OF 1,5-DICARBONYL COMPOUNDS

The Michael addition of an α-carbanion to an α,β-unsaturated carbonyl compound provides a useful route to the synthesis of 1,5-dicarbonyl compounds in which two C=O groups are in a 1,5-relationship (see Section 15.10). The α-carbanion is usually derived from an active methylene compound such as acetoacetic ester or malonic ester.

The retrosynthetic approach to the synthesis of 1,5-dicarbonyl compounds involves a C–C disconnection, termed a **1,5-diCO** disconnection (see Section 17.7.5).

Using the retrosynthetic approach, the **synthesis** of 1,5-dicarbonyl compounds is described here with *two* examples..

A. 2,6-HEPTANEDIONE

2,6-Heptanedione is a 1,5-diketone. (Remember that the numbers 1, 5 denote the relative position of two C=O groups). The diketone is symmetrical and therefore, *only one* 1,5-diCO disconnection is possible.

The **retrosynthetic analysis** of 2,6-heptanedione (TM) involving the 1,5-diCO disconnection gives methyl vinyl ketone and acetoacetic ester as the starting materials.

The **synthesis** involves a Michael reaction between acetoacetic ester (an active methylene compound) and methyl vinyl ketone (an α,β-unsaturated ketone) in the presence of sodium ethoxide as a base. Hydrolysis and decarboxylation of the Michael adduct gives 2,6-heptanedione.

2,6-Heptanedione: Retrosynthetic analysis

$CH_3{-}\overset{O}{\overset{\|}{C}}{-}CH_2{-}CH_2 \wr CH_2{-}\overset{O}{\overset{\|}{C}}{-}CH_3$ (carbons numbered 5, 4, 3, 2, 1)

2,6-Heptanedione
(TM)

$\xRightarrow{\text{1,5-diCO}}$ $CH_3{-}\overset{O}{\overset{\|}{C}}{-}CH_2{-}\overset{+}{C}H_2$ + $\overset{-}{C}H_2{-}\overset{O}{\overset{\|}{C}}{-}CH_3$

Use ⇓ $CH_3{-}\overset{O}{\overset{\|}{C}}{-}CH{=}CH_2$ **Methyl vinyl ketone**

(activate) **Use** ⇓ $\underset{CO_2C_2H_5}{CH_2}{-}\overset{O}{\overset{\|}{C}}{-}CH_3$ **Acetoacetic ester** + Base

2,6-Heptanedione: Synthesis

$CH_3{-}\overset{O}{\overset{\|}{C}}{-}CH{=}CH_2$ (**Methyl vinyl ketone**) + $\underset{CO_2C_2H_5}{CH_2}{-}\overset{O}{\overset{\|}{C}}{-}CH_3$ (**Acetoacetic ester**) $\xrightarrow[C_2H_5OH]{C_2H_5ONa}$ (***Michael reaction***) $CH_3{-}\overset{O}{\overset{\|}{C}}{-}CH_2{-}CH_2{-}\underset{CO_2C_2H_5}{CH}{-}\overset{O}{\overset{\|}{C}}{-}CH_3$ (***Michael adduct***)

Hydrolysis ↓ 1. dil. NaOH 2. H_3O^+

$CH_3{-}\overset{O}{\overset{\|}{C}}{-}CH_2{-}CH_2{-}\underset{CO_2H}{CH}{-}\overset{O}{\overset{\|}{C}}{-}CH_3$ $\xrightarrow[-CO_2]{\Delta}$ (***Decarboxylation***) $CH_3{-}\overset{O}{\overset{\|}{C}}{-}CH_2{-}CH_2{-}CH_2{-}\overset{O}{\overset{\|}{C}}{-}CH_3$

2,6-Heptanedione

B. 3-Carboxymethylcyclohexanone

3-Carboxymethylcyclohexanone (TM) is 1,5-keto acid in which the keto carbonyl (C=O) and a carboxyl carbonyl (C=O) are in a 1,5-relationship. In this case, two 1,5-diCO disconnections, marked 2–3 (*a*) and 3–4 (*b*), are possible.

However, the best disconnection is at the C–C bond (*b*) between the ring and the chain (see also Section 17.7.5).

The **retrosynthetic analysis** using the 1,5-diCO disconnection (*b*) leads to 2-cyclohexenone (in which the ring remains intact) and malonic ester (a suitable and readily available active methylene compound).

Thus the **synthesis** corresponds to the disconnection (*b*), and involves a Michael reaction between malonic ester and 2-cyclohexenone in the presence of sodium ethoxide.

Hydrolysis and decarboxylation of the Michael adduct produces 3-carboxymethylcyclohexanone.

3-Carboxymethylcyclohexanone: Retrosynthetic analysis

3-Carboxymethylcyclohexanone (TM)

a: 1,5-diCO — Not good

b: 1,5-diCO — (protect), (activate), Use, Use

2-Cyclohexenone; $CH_2(CO_2C_2H_5)_2$ + Base — Malonic ester

3-Carboxymethylcyclohexanone: Synthesis

2-Cyclohexenone + Malonic ester $\xrightarrow[C_2H_5OH]{C_2H_5ONa}$ (*Michael reaction*) *Michael adduct*

Hydrolysis: 1. KOH 2. H^+

Decarboxylation: Δ, $-CO_2$ → TM

18.12.1 The Robinson Annulation

An important method for the construction of cyclic structures uses the Robinson annulation which involves a Michael addition followed by an intramolecular aldol condensation (see Section 15.10.1).

Using a retrosynthetic approach involving the Robinson annulation, the **synthesis** of the following cyclic ketone is described.

6-Carbethoxy-3,5-diphenyl-2-cyclohexenone

6-Carbethoxy-3,5-diphenyl-2-cyclohexenone (TM) is a six-membered cyclic ketone.

The **retrosynthetic analysis** of the target molecule first involves an α,β disconnection corresponding to an aldol condensation, which gives a 1,5-diketone. Then a **1,5-diCO** disconnection corresponding to a Michael reaction leads to acetoacetic ester and benzalacetophenone as the starting materials.

(If benzalacetophenone is not available, it can be easily prepared from benzaldehyde and acetophenone, see Section 18.10.1).

The **synthesis** is carried out by allowing benzalacetophenone to react with acetoacetic ester in the presence of sodium ethoxide as a base. The initial Michael adduct spontaneously cyclizes by an intramolecular aldol condensation to produce the target cyclic ketone.

6-Carbethoxy-3,5-diphenyl-2-cyclohexenone: Retrosynthetic analysis

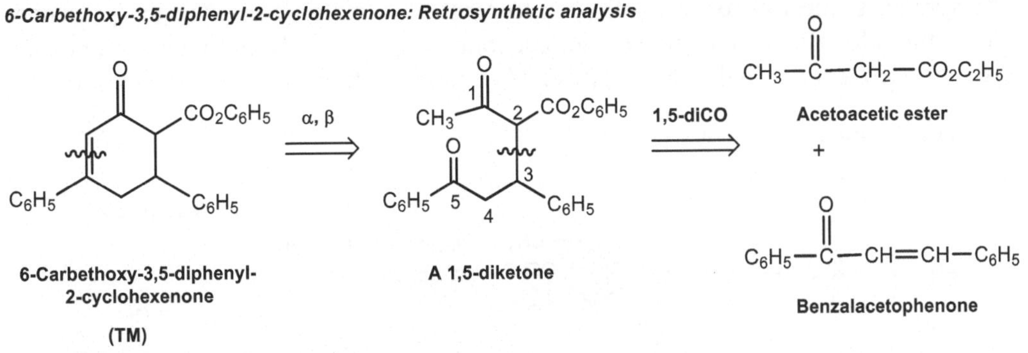

$CO_2C_6H_5$ C_6H_5 C_6H_5 O

6-Carbethoxy-3,5-diphenyl-2-cyclohexenone (TM)

α, β ⟹

O 1 2 $CO_2C_6H_5$ CH_3 O 3 C_6H_5 5 4 C_6H_5

A 1,5-diketone

1,5-diCO ⟹

$CH_3-\overset{O}{\overset{\|}{C}}-CH_2-CO_2C_2H_5$

Acetoacetic ester

\+

$C_6H_5-\overset{O}{\overset{\|}{C}}-CH{=}CH-C_6H_5$

Benzalacetophenone

6-Carbethoxy-3,5-diphenyl-2-cyclohexenone: Synthesis

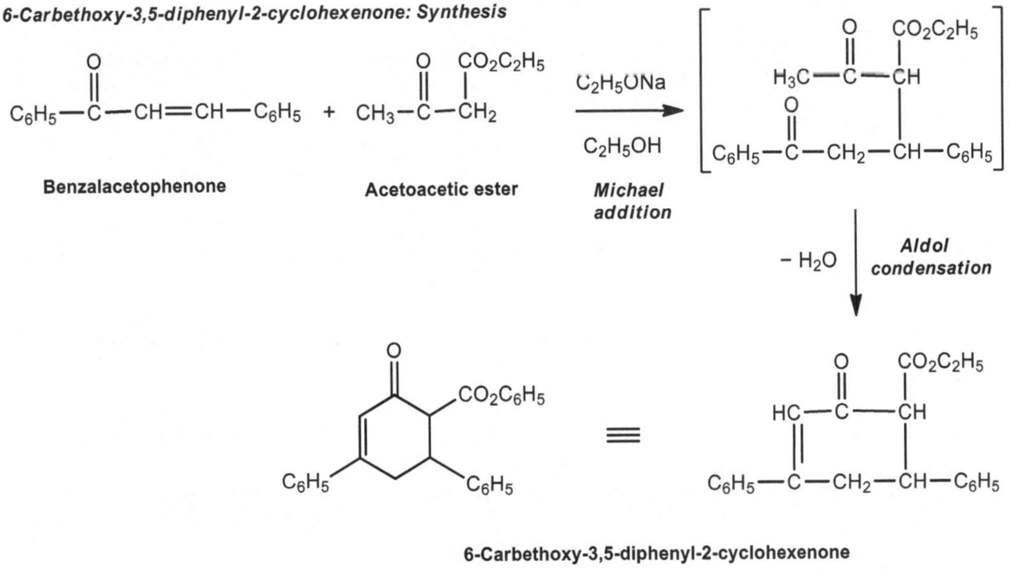

$C_6H_5-\overset{O}{\overset{\|}{C}}-CH{=}CH-C_6H_5$ + $CH_3-\overset{O}{\overset{\|}{C}}-CH_2-CO_2C_2H_5$ $\xrightarrow[C_2H_5OH]{C_2H_5ONa}$ [Michael adduct: $H_3C-CO-CH(CO_2C_2H_5)-CH(C_6H_5)-CH_2-CO-C_6H_5$]

Benzalacetophenone **Acetoacetic ester** ***Michael addition***

$-H_2O$ ***Aldol condensation*** ↓

$CO_2C_6H_5$ C_6H_5 C_6H_5 O ≡ HC—C(=O)—CH($CO_2C_2H_5$), C_6H_5—C(=HC)—CH_2—CH—C_6H_5

6-Carbethoxy-3,5-diphenyl-2-cyclohexenone

PROBLEM 18.11: Suggest a retrosynthetic analysis and then outline a synthesis of dimedone:

Dimedone
(5,5-dimethyl-1,3-cyclohexanedione)

Answer

Dimedone (5,5-dimethyl-1,3-cyclohexanedione) is a cyclic 1,3-diketone.

Thus the retrosynthetic analysis of dimedone (TM) first involves a 1,3-diCO disconnection, which gives a keto ester as a 1,5-dicarbonyl compound. Then a 1,5-diCO disconnection leads to mesityl oxide and malonic ester as the starting materials.

In the synthesis, mesityl oxide is allowed to react with malonic ester in the presence of sodium ethoxide. The initial Michael adduct spontaneously cyclizes by an intramolecular Claisen condensation. Finally, hydrolysis and decarboxylation produces dimedone.

Dimedone: Retrosynthetic analysis

Dimedone (TM) ⟹ (1,3-diCO) A keto ester ⟹ (1,5-diCO) Mesityl oxide + (activate) Use ≡ $CH_2(CO_2Et)$–C(=O)–OEt Malonic ester + Base

Dimedone: Synthesis

$(CH_3)_2C{=}CH{-}C(=O){-}CH_3$ (Mesityl oxide) + $CH_2(CO_2Et){-}C(=O){-}OEt$ (Malonic ester) → (C_2H_5ONa, C_2H_5OH; *Michael addition*) [$(CH_3)_2C(CH_2{-}C(=O)CH_3){-}CH(CO_2Et){-}C(=O){-}OEt$]

→ (− EtOH; *Intramolecular Claisen condensation*) A β-keto ester ($(CH_3)_2C{-}CH_2{-}C{=}O$, CH_2, $EtO_2C{-}CH_\alpha{-}C_\beta{=}O$) ≡ cyclic β-keto ester with CO_2Et

→ (1. dil. NaOH; 2. H^+, heat; *Hydrolysis and decarboxylation*) Dimedone

18.13 THE DIELS-ALDER SYNTHESIS OF CYCLIC MOLECULES

The Diels-Alder reaction is one of most elegant reactions employed in organic synthesis. The reaction makes two C–C bonds in one step to form a cyclohexene product by a pericyclic mechanism (see Section 8.17).

The retrosynthetic approach involves a **Diels-Alder (D–A) disconnection** by drawing the *reaction mechanism in reverse*. Thus the disconnection is performed by **drawing three arrows round the cyclohexene ring starting from the double bond**. The arrows can be drawn in either direction, clockwise or anticlockwise. The D–A disconnection gives a **diene** and a **dienophile** (alkene or alkyne) as the starting materials.

Using the retrosynthetic approach, the **synthesis** of six-membered cyclohexene rings can be accomplished in a straightforward manner by simply mixing and heating the diene and dienophile components together.

The synthesis is illustrated here with *three* target molecules (TM 1, TM 2 and TM 3).

TM 1: 3-Cyclohexene-1-carboxaldehyde

The **retrosynthetic analysis** of TM 1 (3-cyclohexene-1-carboxaldehyde) involves a **Diels-Alder (D-A) disconnection**, which gives 1,3-butadiene (as a diene) and acrolein (as a dienophile).

Note that the *alkene* dienophile (acrolein) bears an electron-withdrawing group (CHO) to make the reaction favourable.

The **synthesis** is a Diels-Alder reaction between 1,3-butadiene and acrolein (propenal) to produce TM 1.

TM 1: Retrosynthetic analysis

Arrow starts here

O H

D-A

+

O H

TM 1

1,3-Butadiene *Diene*

Acrolein *Dienophile*

TM 1: Synthesis

CHO

+

Δ

Diels-Alder reaction

CHO

1,3-Butadiene

Acrolein (propenal)

TM 1

TM 2: 1,2-Dicarbomethoxy-1,4-cyclohexadiene

TM 2 (1,2-dicarbomethoxy-1,4-cyclohexadiene) contains two double bonds in the ring, *one* of which is attached to electron-withdrawing CO_2CH_3 groups.

In the **retrosynthetic analysis** of TM 2, the three arrows are best drawn starting from the double bond which is *not* connected to electron-withdrawing groups. The Diels-Alder (D-A) disconnection then gives 1,3-butadiene (as a diene) and dimethyl acetylenedicarboxylate (as a dienophile).

Note that the *alkyne* dienophile (dimethyl acetylenedicarboxylate) bears two electron withdrawing CO_2CH_3 groups to make the reaction favourable.

The **synthesis** is a Diels-Alder reaction of 1,3-butadiene with dimethyl acetylenedicarboxylate (DMAD) to produce TM 2. In this reaction, one π bond of the triple bond of the dienopile is involved and the other π bond is retained in the product.

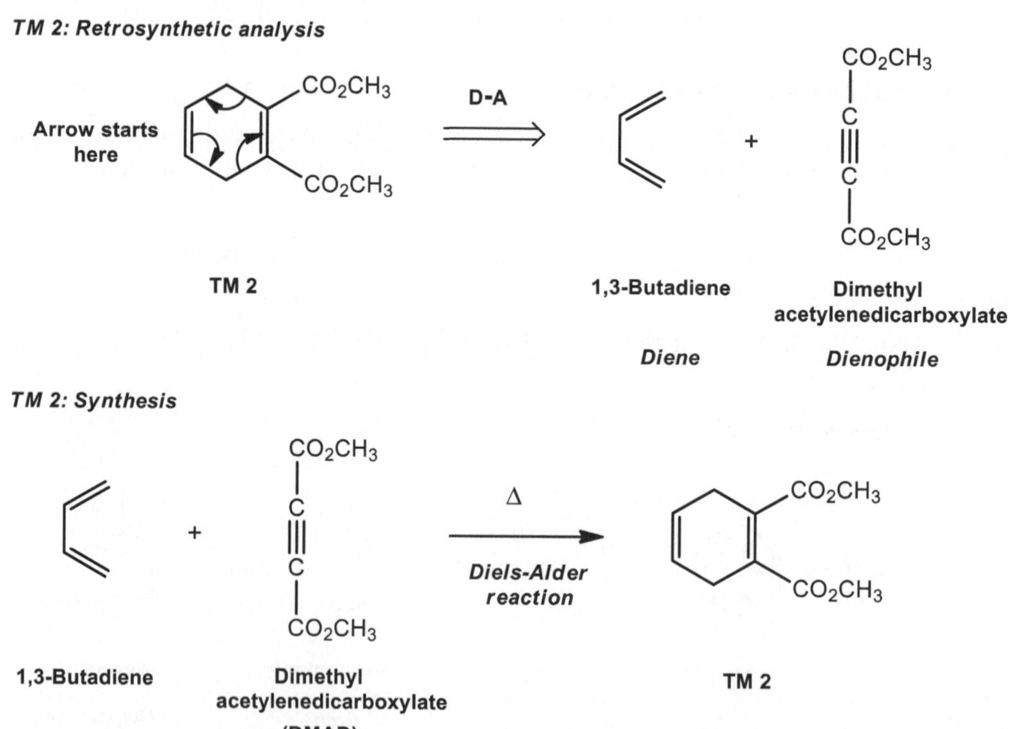

TM 3: trans-1,2-Dicarbomethoxy-4,5-dimethyl-4-cyclohexene

The target molecule (TM 3) has two CO_2Me groups that are *trans* to each other.

The **retrosynthetic analysis** of TM 3 involving a Diels-Alder (D-A) disconnection gives 2,3-dimethyl-1,3-butadiene (as a diene) and dimethyl fumarate (as a dienophile). In dimethyl fumarate, the two CO_2Me groups are *trans*.

The Diels-Alder **synthesis** of TM 3 with the diene and the dienophile is stereospecific, i.e. *trans* stereochemistry of the alkene dienophile is retained in the product.

TM 3: Retrosynthetic analysis

Me, Me, H, CO_2Me, CO_2Me, H

D-A

Me, Me

+

MeO_2C, H, H, CO_2Me

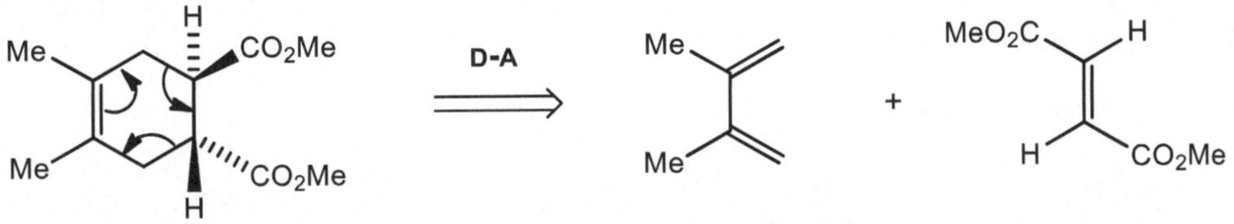

(two CO_2Me groups are *trans*)

TM 3

2,3-Dimethyl-1,3-butadiene

Diene

Dimethyl fumarate (*trans*)

Dienophile

TM 3: Synthesis

Me, Me

+

MeO_2C, H, H, CO_2Me

Δ

Diels-Alder reaction

Me, Me, H, CO_2Me, CO_2Me, H

2,3-Dimethyl-1,3-butadiene

Dimethyl fumarate

(*trans*-product)

TM 3

PROBLEM 18.12: Limonene is a cyclic compound that has the natural smell of citrus fruits. Suggest a retrosynthetic analysis of limonene and then outline its synthesis.

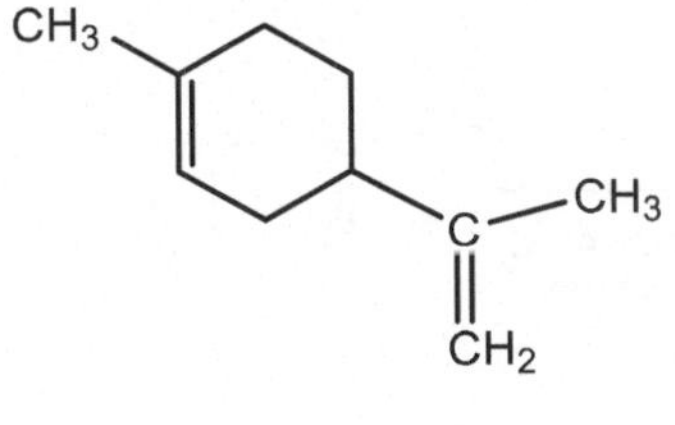

CH_3, C, CH_3, CH_2

Limonene

Found in citrus fruits

Answer

Limonene has a C=C double bond outside the ring. Thus a reasonable retrosynthetic analysis of limonene first involves a Wittig C=C disconnection which gives a ketone. Then a Diels-Alder (D-A) disconnection leads to 2-methyl-1,3-butadiene (as a diene) and methyl vinyl ketone (as a dienophile).

Note that the dienophile bears a favourable electron-withdrawing C=O group, which results from the Wittig disconnection.

The synthesis of limonene is accomplished by a Diels-Alder reaction between the diene and the dienophile followed by a Wittig reaction using a phosphorus ylide that can be prepared from methyl iodide.

Importantly, the 2-substituted diene produces the desired '*para*' adduct in the Diels-Alder reaction (see Section 8.17).

Limonene: Retrosynthetic analysis

Limonene (TM) ⇒ (C=C, Wittig) A ketone + $Ph_3\overset{+}{P}—\overset{-}{C}H_2$ (A phosphorus ylide)

A ketone ⇒ (D-A) 2-Methyl-1,3-butadiene (*Diene*) + Methyl vinyl ketone (*Dienophile*)

A phosphorus ylide ⇒ I—CH_3 (Methyl iodide)

Limonene: Synthesis

$CH_3I \xrightarrow{Ph_3P} Ph_3\overset{+}{P}—CH_3\ I^- \xrightarrow{BuLi} Ph_3\overset{+}{P}—\overset{-}{C}H_2$ (Phosphorus ylide)

2-Methyl-1,3-butadiene + Methyl vinyl ketone $\xrightarrow{\Delta}$ (*Diels-Alder reaction*) '*para*' *Adduct* $\xrightarrow{Ph_3\overset{+}{P}—\overset{-}{C}H_2}$ (*Wittig reaction*) Limonene

18.14 SYNTHESIS OF SOME SIMPLE DRUGS

Finally, in this section, we will introduce some simple and common drugs and describe their synthesis using the retrosynthetic approach.

A. *Benzocaine*

Benzocaine is used as a local anaesthetic. It is an **ester** (ethyl *p*-aminobenzoate).

The retrosynthetic analysis of benzocaine (TM) involves a C–O ester disconnection, which suggests that benzocaine can be synthesized from *p*-aminobenzoic acid and ethanol.

A further analysis of *p*-aminobenzoic acid using an FGI reduction and FGI oxidation gives *p*-nitroaniline. Then a C–N (nitration) disconnection gives toluene as a readily available starting material.

Thus the synthesis of benzocaine is accomplished using a proper sequence of nitration, oxidation, reduction and esterification, starting from toluene.

Benzocaine: Retrosynthetic analysis

Benzocaine (TM) $\xrightarrow[\text{ester}]{\text{C–O}}$ *p*-Aminobenzoic acid + Ethanol (C_2H_5OH)

p-Aminobenzoic acid $\Downarrow$ FGI reduction

p-Nitrobenzoic acid $\xrightarrow[\text{oxidation}]{\text{FGI}}$ *p*-Nitrotoluene $\xrightarrow[\text{nitration}]{\text{C–N}}$ Toluene + NO_2^+ (Use $\equiv$ HNO_3/H_2SO_4)

Benzocaine: Synthesis

Toluene $\xrightarrow[\text{c. } H_2SO_4]{\text{c. } HNO_3}$ (*Nitration*) *p*-Nitrotoluene (separate from *o*-nitrotoluene) $\xrightarrow[2.\ H_3O^+]{1.\ KMnO_4/OH^-}$ (*Oxidation*) *p*-Nitrobenzoic acid $\xrightarrow[HCl]{Sn}$ (*Reduction*) *p*-Aminobenzoic acid $\xrightarrow[H^+]{C_2H_5OH}$ (*Esterification*) Benzocaine (ethyl *p*-aminobenzoate)

B. *Aspirin*

Aspirin is an analgesic and anti-inflammatory drug. It acts by inhibiting the synthesis of prostaglandins involved in allergic and inflammation responses.

Aspirin is an **ester**. The retrosynthetic analysis of aspirin (TM) therefore uses a C–O ester disconnection which gives salicylic acid and acetic anhydride as the reagents. A further analysis of salicylic acid involving a C–C (Ar–C) disconnection leads to phenol and carbon dioxide as the starting materials.

Thus the synthesis of aspirin (acetylsalicylic acid) is performed using a Kolbe reaction of sodium phenoxide with CO_2 followed by acetylation with acetic anhydride.

Aspirin: Retrosynthetic analysis

Aspirin (TM) $\xrightarrow[\text{ester}]{C-O}$ $^{+}C(=O)-CH_3$ (Use ≡ $CH_3C(=O)-O-C(=O)CH_3$, Acetic anhydride) + Salicylic acid

Salicylic acid $\xrightarrow{C-C}$ Phenol + CO_2

Aspirin: Synthesis

Phenol $\xrightarrow{NaOH}$ Sodium phenoxide (O^-Na^+) $\xrightarrow[\text{2. } H_3O^+]{\text{1. } CO_2}$ (*Kolbe reaction*) Salicylic acid $\xrightarrow{(CH_3CO)_2O}$ (*Acetylation*) Aspirin (acetylsalicylic acid) ($OCOCH_3$, CO_2H)

C. *Paracetamol*

Paracetamol (also known as acetaminophen) is the commonly used analgesic and antipyretic drug. It is an **amide** (*p*-hydroxyacetanilide).

The retrosynthetic analysis of paracetamol (TM) involves a C–N amide disconnection which reveals that the drug can be prepared by acetylation of *p*-aminophenol with acetic anhydride. A further analysis of *p*-aminophenol involving an FGI reduction followed by a C–N (nitration) disconnection gives phenol as a readily available starting material.

The synthetic sequence for paracetamol is therefore nitration, reduction and acetylation, starting from phenol.

Note that the nitration of phenol is carried out with dilute nitric acid, instead of a mixture of concentrated nitric acid and concentrated sulphuric acid (see Section 13.4.1).

Paracetamol: Retrosynthetic analysis

Paracetamol (TM) $\overset{C-N}{\underset{amide}{\Longrightarrow}}$ Acetic anhydride ($CH_3COOCCH_3$, $CH_3C(=O)-O-C(=O)CH_3$) + NH_2–C₆H₄–OH (*p*-Aminophenol) $\overset{FGI}{\underset{reduction}{\Longrightarrow}}$ NO_2–C₆H₄–OH $\overset{C-N}{\underset{nitration}{\Longrightarrow}}$ $^+NO_2$ + Phenol

Paracetamol: Synthesis

Phenol $\xrightarrow[\textit{Nitration}]{\text{dil. } HNO_3}$ *p*-Nitrophenol (separate from ortho isomer) $\xrightarrow[\text{Pd/C, } \textit{Reduction}]{H_2}$ *p*-Aminophenol $\xrightarrow[\textit{Acetylation}]{(CH_3CO)_2O}$ Paracetamol (*p*-hydroxyacetanilide, $NHCOCH_3$ / OH)

D. *Propranolol*

Propranolol is a beta-blocker drug that reduces blood pressure. It is an **amino alcohol** in which amine and alcohol functional groups are in a relative 1,2 position.

Therefore, a reasonable retrosynthetic analysis of propranolol (TM) involves a 1,2-diX disconnection (see Section 17.5.2). The 1,2-diX disconnection gives isopropylamine (a secondary amine) and an epoxide. The epoxide contains a simple C–O ether bond; thus a further analysis of the epoxide using a C–O ether disconnection gives 1-naphthol and epichlorohydrin as common and readily available starting materials.

In the forward synthesis, 1-naphthol is converted to sodium 1-naphthoxide which is then allowed to react with epichlorohydrin in an S_N2 reaction. The resulting epoxide then reacts with isopropylamine in a ring-opening S_N2 reaction to produce propranolol.

Propranolol: Retrosynthetic analysis

OH

OCH_2CHCH_2–$NHCH(CH_3)_2$

1 2

Propranolol

(TM)

1,2-diX

O–CH_2–CH–CH_2 (O)

\+ $(CH_3)_2CHNH_2$

An epoxide

Isopropylamine

C—O ether

OH

\+ Cl—CH_2—CH—CH_2 (O)

1-Naphthol

Epichlorohydrin

Propranolol: Synthesis

OH

NaOH

O^- Na^+

H_2C—CH—CH_2Cl (O)

S_N2

OCH_2—CH—CH_2 (O)

1-Naphthol

Sodium 1-naphthoxide

S_N2 $(CH_3)_2CHNH_2$

OH

$OCH_2CHCH_2NHCH(CH_3)_2$

Propranolol

E. *Salbutamol*

Salbutamol is an anti-asthma drug.

Like propranolol, it is also an **amino alcohol** in which an amine and alcohol functional groups are in a relative 1,2 position. Thus a 1,2-diX disconnection of salbutamol can lead to a successful synthesis but it involves methods that are beyond our scope.

Alternatively, a retrosynthetic analysis of salbutamol (TM) involving an FGI reduction followed by a 1,2-diX disconnection gives an α-bromo ketone and *t*-butylamine. Note that the FGI involves both CH_2OH and CHOH groups.

A further analysis of the α-bromo ketone using a C–Br disconnection and then a C–C (Ar–C) disconnection corresponding to the Fries rearrangement (see Section 16.5.3) gives aspirin as the starting material.

In the forward synthesis, aspirin is treated with anhydrous aluminium chloride in a Fries rearrangement. Then bromination of the resulting ketone is followed by a nucleophilic substitution (S_N2) with *t*-butylamine. Finally, *both* ketone and carboxyl groups are reduced with $LiAlH_4$ to produce salbutamol.

Salbutamol: Retrosynthetic analysis

Salbutamol (TM) ⟹ (FGI, reduction) ⟹ (1,2-diX) An α-bromo ketone + *t*-Butylamine ⟹ (C–Br) ⟹ (C–C, Fries rearrangement) Aspirin

Salbutamol: Synthesis

Aspirin → (1. $AlCl_3$, 2. H_2O; *Fries rearrangement*) → (Br_2, $CHCl_3$; *Bromination*) → (S_N2, $(CH_3)_3CNH_2$) → ($LiAlH_4$; *Reduction of both C=O and CO_2H*) → Salbutamol

Note: The present scheme represents an elegant synthesis of the drug salbutamol from another drug aspirin. Aspirin is readily available as a starting material; it can also be prepared easily (see above).

F. *Ibuprofen*

Ibuprofen is a common anti-inflammatory drug. It is a **carboxylic acid**.

A retrosynthetic analysis of ibuprofen (TM) involving a Grignard C–C disconnection gives a Grignard reagent and CO_2 as the reagents. An analysis of the Grignard reagent using a series of FGIs leads to an aromatic ketone. Then a Friedel-Crafts C–C disconnection gives isobutylbenzene and acetyl chloride as the starting materials.

In the forward synthesis, isobutylbenzene is treated with acetyl chloride in a Friedel-Crafts acylation reaction to produce a ketone, which is then converted to the corresponding Grignard reagent through reduction, substitution and treatment with magnesium. Finally, the Grignard reaction with carbon dioxide yields ibuprofen.

Ibuprofen: Retrosynthetic analysis

Ibuprofen (TM) $\xRightarrow[\text{Grignard}]{C-C}$ A Grignard reagent + CO_2

A Grignard reagent ⇓

Aryl–CH(CH$_3$)–Cl $\xRightarrow[\text{substitution}]{\text{FGI}}$ Aryl–CH(CH$_3$)–OH

⇓ FGI reduction

An aromatic ketone $\xRightarrow[\text{Friedel-Crafts}]{C-C}$ Isobutylbenzene + CH_3COCl (Acetyl chloride)

Ibuprofen: Synthesis

Isobutylbenzene, $(CH_3)_2CHCH_2$–C$_6$H$_4$ $\xrightarrow[AlCl_3]{CH_3COCl}$ (***Friedel-Crafts acylation***) $(CH_3)_2CHCH_2$–C$_6$H$_4$–C(=O)–CH_3

Reduction $NaBH_4$ → $(CH_3)_2CHCH_2$–C$_6$H$_4$–CH(OH)CH$_3$

$\xrightarrow{HCl}$ (***Substitution***) $(CH_3)_2CHCH_2$–C$_6$H$_4$–CH(Cl)CH$_3$

Mg, Et_2O → $(CH_3)_2CHCH_2$–C$_6$H$_4$–CH(MgCl)CH$_3$

$\xrightarrow{1.\ CO_2;\ 2.\ H_3O^+}$ (***Grignard reaction***) $(CH_3)_2CHCH_2$–C$_6$H$_4$–CH(CH$_3$)CO$_2$H

Ibuprofen

Note: If isobutylbenzene is not readily available, it can be easily prepared from benzene (see Section 18.6.1).

G. *Oblivon*

Oblivon is a tranquilizer. It is an **acetylenic alcohol** (3-methyl-1-pentyn-3-ol).

A retrosynthetic analysis of oblivion (TM) involving a 1,1 C–C disconnection gives butanone and acetylene as the starting materials.

The synthesis is straightforward. Acetylene is treated with $NaNH_2$ in liquid NH_3 to form acetylide ion (sodium acetylide) which is then added to butanone to produce oblivion.

Oblivon: Retrosynthetic analysis

$$CH_3CH_2-\overset{OH}{\underset{CH_3}{C}}-C\equiv CH \xRightarrow{1,1\ C-C} CH_3CH_2-\overset{O}{\overset{\|}{\underset{CH_3}{C}}} + H-C\equiv CH$$

Oblivon **(TM)** | **Butanone** | **Acetylene**

Oblivon: Synthesis

$$HC\equiv CH \xrightarrow[\text{liq. } NH_3]{NaNH_2} HC\equiv C^- Na^+ \xrightarrow[\text{2. } H_3O^+ \text{ (work-up)}]{\text{1. } CH_3CH_2\overset{O}{\overset{\|}{C}}CH_3} CH_3CH_2\overset{OH}{\underset{CH_3}{C}}-C\equiv CH$$

Acetylene | **Oblivon (3-Methyl-1-pentyn-3-ol)**

SUPPLEMENTARY PROBLEMS

18.13 Give a retrosynthetic analysis and then outline the synthesis of each of the following alcohols:

(a) $CH_3CH_2-\overset{C_6H_5}{\underset{OH}{C}}-CH_2CH_3$

3-Phenyl-3-pentanol

(b) OH

1-Phenylcyclohexanol

[*Hints*: (a) use 2 × 1,1 C–C disconnection; (b) use 1,1 C–C disconnection]

Problem 18.14: Using the retrosynthetic approach, outline a synthesis of the following compound from benzene:

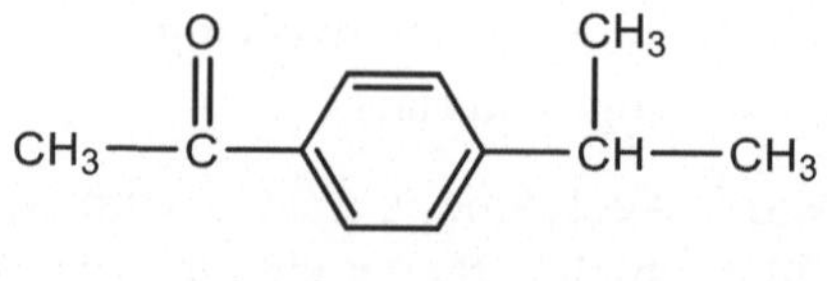

p-Isopropylacetophenone

[*Hint*: use Friedel-Crafts C–C disconnections: acylation first, then alkylation. Check that disconnections in the reverse order (alkylation first, then acylation) would give a wrong orientation in the synthesis because a ketone group is *meta*-orienting]

18.15 (a) Suggest a retrosynthetic analysis and then outline an acetoacetic ester synthesis of the following compound:

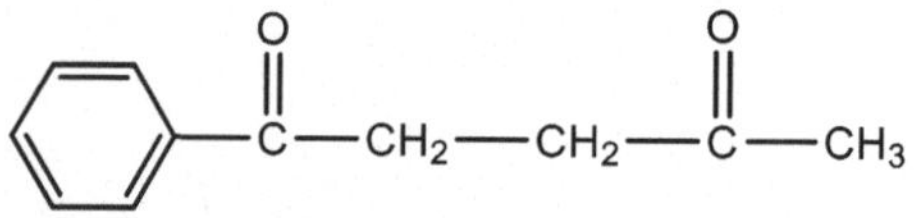

1-Phenyl-1,4-pentanedione

(b) Give a retrosynthetic analysis and then outline a synthesis of the following compound from malonic ester and any other reagents as necessary:

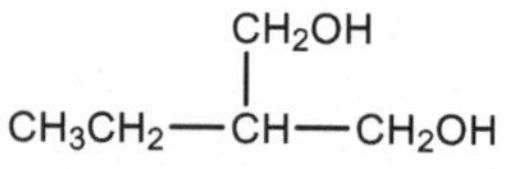

2-Ethyl-1,3-propanediol

[*Hints*: (a) use 1,2 C–C disconnection; (b) use an FGI reduction, then 1,2 C–C disconnection]

18.16 Write down a retrosynthetic analysis and outline the synthesis of the following ester:

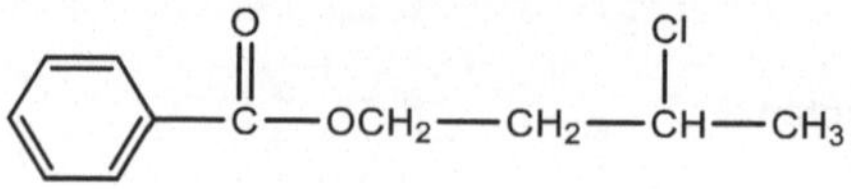

3-Chlorobutyl benzoate

[*Hint*: use C–O ester disconnection; then apply on the alcohol component a sequence of FGI reduction and a C–Cl disconnection to obtain an α,β-unsaturated acid (crotonic acid) as the starting material]

18.17 Terodilin is an antispasmodic drug used in urology. Give a retrosynthetic analysis of terodilin involving reductive amination and then outline its synthesis.

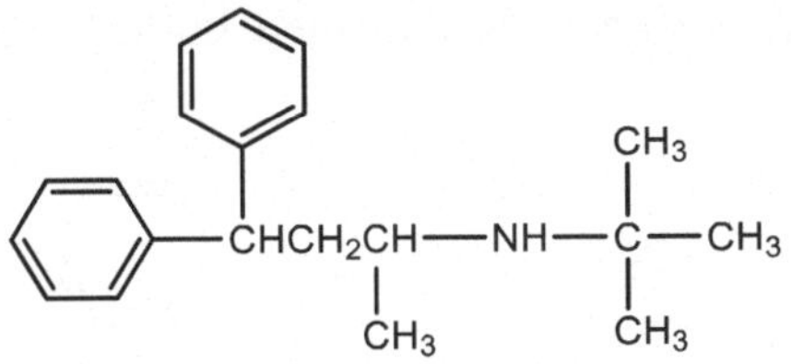

$$\mathrm{(C_6H_5)_2CHCH_2CH(CH_3)\text{—}NH\text{—}C(CH_3)_3}$$

Terodilin

[*Hint*: use an FGI reduction and a C=N imine disconnection]

18.18 Write down a retrosynthetic analysis and then outline the synthesis of each of the following compounds as indicated:

(a) 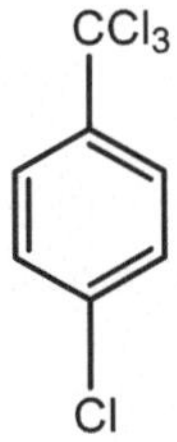 CCl_3–C_6H_4–Cl (para) **from toluene**

***p*-Chlorobenzotrichloride**

(b) 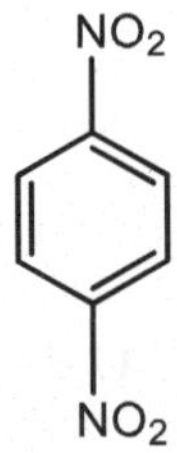 NO_2–C_6H_4–NO_2 (para) **from aniline**

***p*-Dinitrobenzene**

[*Hints*: (a) use an FGI side chain chlorination and C–Cl disconnection (ring chlorination); (b) use an FGI substitution via diazonium salt and a C–N disconnection (nitration); include amine protection and deprotection protocol in the synthetic sequence]

18.19 Using a suitable retrosynthetic analysis outline a Corey-House synthesis of the following alkane:

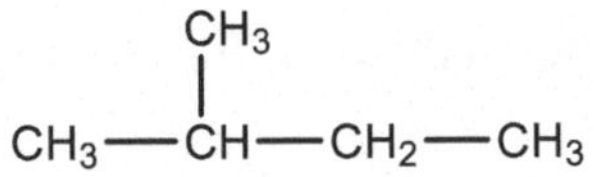

$$CH_3\text{—}\underset{}{\overset{CH_3}{\overset{|}{C}H}}\text{—}CH_2\text{—}CH_3$$

2-Methylbutane

[*Hint*: use any C–C disconnection; however, disconnection near the middle (C2–C3) gives simpler starting materials. Remember that 2° or 3° alkyl halide gives poor yield]

18.20 Propose a retrosynthetic analysis and then outline the synthesis of each of the following alkenes:

(a) C_6H_5—CH=CH$_2$ **Styrene**

(b) C_6H_5—CH=CH—C_6H_5 **Stilbene**

[*Hints*: use a C=C Wittig disconnection in each case]

18.21 The following alkynyl bromide is an important intermediate in the synthesis of *cis*-jasmone, a fragrant constituent found in the oil of jasmine flowers. Suggest a retrosynthetic analysis and outline the synthesis of the alkynyl bromide.

$CH_3CH_2—C\equiv C—CH_2Br$

1-Bromo-2-pentyne

[*Hint*: use an FGI substitution and a C–C disconnection; in the synthesis, protection of the acetylenic alcohol is not necessary: with excess base (such as butyllithium) a dianion is formed, alkylation then occurs mainly on the carbon]

18.22 Give a retrosynthetic analysis and then outline the synthesis of the following:

$CH_3CH_2—CH=C(CH_3)—CH_2OH$

2-Methyl-2-penten-1-ol

[*Hint*: use a sequence of FGI reduction, FGI dehydration and 1,3-diO disconnection]

18.23 Flosal is a French perfume ingredient. Suggest a retrosynthetic analysis and then outline its synthesis.

$C_6H_5—CH=C(CHO)—CH_2CH_2CH_2CH_2CH_3$

Flosal

[*Hint*: use an α,β disconnection]

18.24 Write down a retrosynthetic analysis of the following diketone and then outline its synthesis.

(cyclopentanone bearing a $C(=O)CH_3$ group at C-2)

2-Acetylcyclopentanone

[*Hint*: use a 1,3-diCO disconnection (cf. Dieckmann condensation)]

18.25 Using a retrosynthetic analysis, outline a synthesis of the following dicarboxylic acid starting from malonic ester and other reagents as necessary.

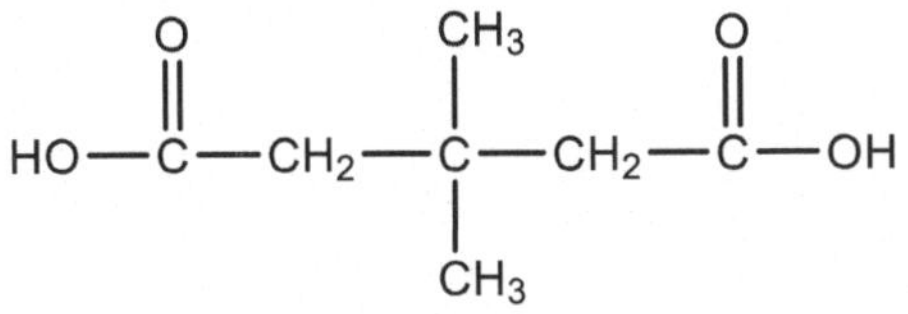

$$HO-\overset{O}{\overset{\|}{C}}-CH_2-\underset{CH_3}{\overset{CH_3}{C}}-CH_2-\overset{O}{\overset{\|}{C}}-OH$$

3,3-Dimethylglutaric acid

[*Hint*: use a 1,5-diCO disconnection to get an α,β-unsaturated ester (ethyl 3-methyl-2-butenoate) and malonic ester as the starting materials; include hydrolysis and decarboxylation of the Michael adduct in the synthetic scheme]

18.26 Give a retrosynthetic analysis of the following bicyclic compound and then outline its synthesis involving a Robinson annulation.

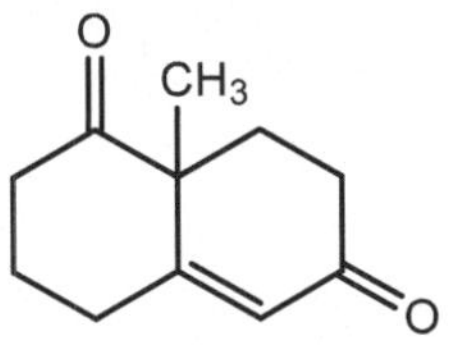

O, CH_3, O

[*Hint*: use an α,β disconnection and then a 1,5-diCO disconnection (without breaking the ring) to get 2-methyl-1,3-cyclohexanedione and methyl vinyl ketone as the starting materials]

18.27 Using a Diels-Alder disconnection, suggest which diene and dienophile would you employ to synthesize each of the following compounds?

(a) H_3C, H_3C, H, H, CO_2CH_3, CO_2CH_3

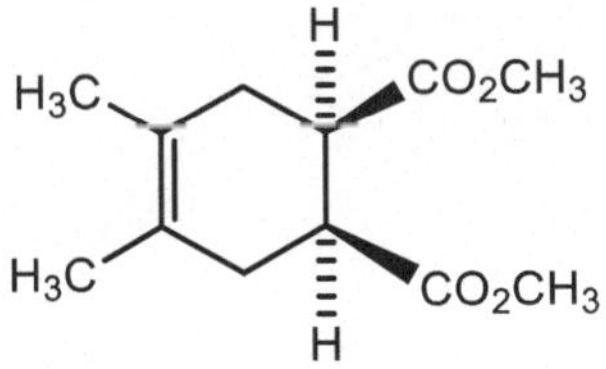

(b) O, O, O

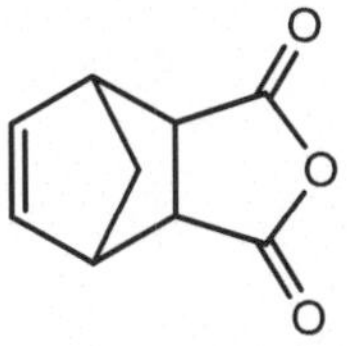

[*Ans*: (a) 2,3-dimethyl-1,3-butadiene and dimethyl maleate; (b) cyclopentadiene and maleic anhydride]

PART-IV

Special Topics

CHAPTER 19

Polymers and Polymerization Reactions

IN THIS CHAPTER

- Homopolymers and copolymers
- Polymerization reactions: Radical, ionic and Zeigler-Natta polymerizations
- Copolymerization reactions
- Addition or chain-growth polymers
- Stereochemistry of addition polymers: Atactic, isotactic and syndiotactic
- Condensation or step-growth polymers
- Natural fibres, natural rubber and synthetic rubber
- Biodegradable polymers

So far, we have dealt with small and simple molecules. But there are also very large and complex molecules with high molecular mass, called **macromolecules**. Macromolecules that play an important role in the chemistry of life processes are called **biomolecules**. These biomolecules will be described in the next and final chapter (Chapter 20).

Macromolecules of *synthetic origin* also have a tremendous impact on our day-to-day living. The materials known as plastics seem indispensable to the modern world. They are used as structural materials, fibres and protective films.

A **polymer** is a macromolecule which is made up of a *repeating* sequence of smaller units called **monomers**. The term 'polymer' comes from two Greek words (*poly* means many and *mer* means unit or part). The process by which polymers are created from respective monomers is called **polymerization**.

Polymers can be divided into natural, synthetic and semisynthetic polymers. **Natural polymers** are derived from plants, and they include cotton, resins and rubber. **Synthetic polymers** of a wide variety are made on a large scale in industry, which include plastics (e.g. polythene), synthetic fibres (e.g. nylon 6,6) and synthetic rubbers (e.g. Buna–S). **Semisynthetic polymers** are also available, such as cellulose acetate (rayon).

Polymers are also classified based on their behaviour on heating. Polymers that melt or soften on heating and then solidify and regain their properties on cooling are called **thermoplastic polymers**. On the other hand, polymers that soften or melt and then become infusible solids on heating are called **thermosetting** polymers.

19.1 HOMOPOLYMERS AND COPOLYMERS

Polymers are formed by chemical combination of many monomers. When all monomer units in a polymer are identical, the polymer is called a **homopolymer**.

On the other hand, a polymer consisting of more than one kind of monomer unit is called a **copolymer**.

$$n\,\mathrm{A} \longrightarrow \mathrm{A{-}A{-}A{-}A{-}A{-}A{-}A} \equiv \mathrm{A}\!\left(\mathrm{A}\right)_{n-2}\!\mathrm{A}$$

Monomer **Homopolymer**

$$n\,\mathrm{A} + n\,\mathrm{B} \longrightarrow \mathrm{A{-}B{-}A{-}B{-}A{-}B{-}A{-}B} \equiv \mathrm{A}\!\left(\mathrm{B{-}A}\right)_{n-1}\!\mathrm{B}$$

Two different monomers **Copolymer**

The number *n* indicates the number of monomer molecules that combine to form a particular polymer, that is, ***n* is the degree of polymerization**.

Examples

An example of a **homopolymer** is **polythene**. It is formed by polymerization of one kind of monomeric species (ethylene).

$$n\,CH_2{=}CH_2 \xrightarrow{\text{Polymerization}} {-}CH_2{-}CH_2\!\left(CH_2{-}CH_2\right)_{n-2}\!CH_2{-}CH_2{-} \equiv \left(CH_2{-}CH_2\right)_n$$

Ethylene

Monomeric units

Polythene
A *homopolymer*

Buna–S is an example of a **copolymer** which is formed by polymerization of two kinds of monomeric species (1,3-butadiene and styrene).

$$n\,CH_2{=}CH{-}CH{=}CH_2 + n\,C_6H_5{-}CH{=}CH_2$$

1,3-Butadiene **Styrene**

Polymerization ↓

Butadiene unit ***Styrene unit***

$$\left(CH_2{-}CH{=}CH{-}CH_2{-}\underset{\displaystyle C_6H_5}{\underset{|}{CH}}{-}CH_2\right)_n$$

Butadiene-styrene copolymer
(Buna-S)

Branched polymers and cross-liking

A polymer chain can be linear or branched. **Branched polymers** contain covalent bonds between polymer chains, known as **cross-links**.

In some cases, the cross-links are formed by the *functional groups* of the monomers, while in others additional *small molecules* (X) are added to participate in the cross-linking process (see later Section 19.7.2).

A **schematic representation** of cross-linking between polymer chains is shown below. In the first case, the two identical groups (A, A) are cross-linked directly.

In the second case, the two identical groups (A, A) are cross-linked via an additional small molecule (X).

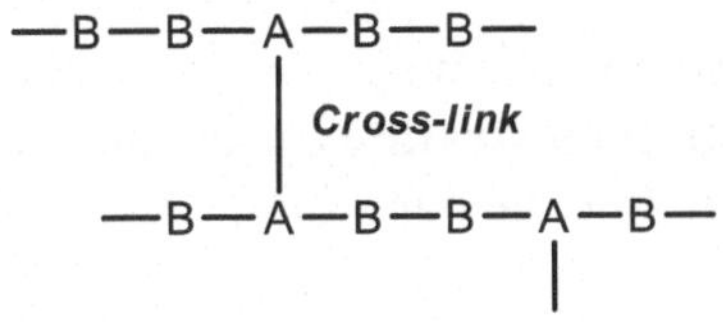

—B—B—A—B—B—
Cross-link
—B—A—B—B—A—B—

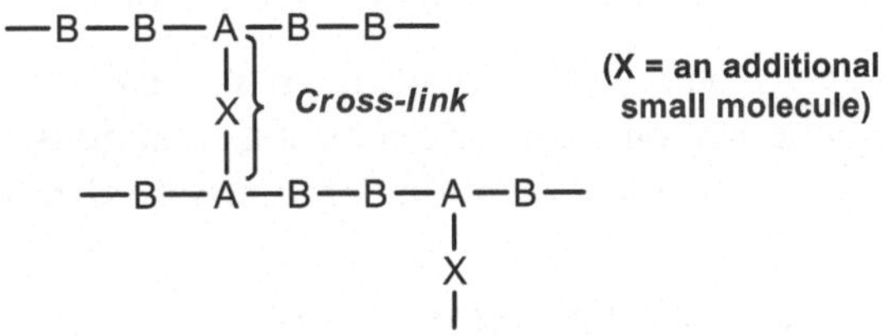

—B—B—A—B—B—
X } Cross-link
—B—A—B—B—A—B—
X

(X = an additional small molecule)

Problem 19.1: Indicate which of the following polymers is a homopolymer or a copolymer. Identify the monomer or monomers in each case.

(a) $\left[CH_2-CH(C_6H_5) \right]_n$

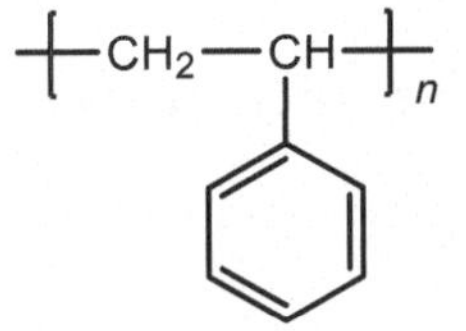

(b) $\left[CH_2-CH=CH-CH_2-CH_2-CH(CN) \right]_n$

Answer

(a) It is a homopolymer. The **monomer** from which it is created is **styrene**. The polymer is known as **polystyrene**. It is used for making insulators, wrapping materials, toys, and radio and television cabinets.

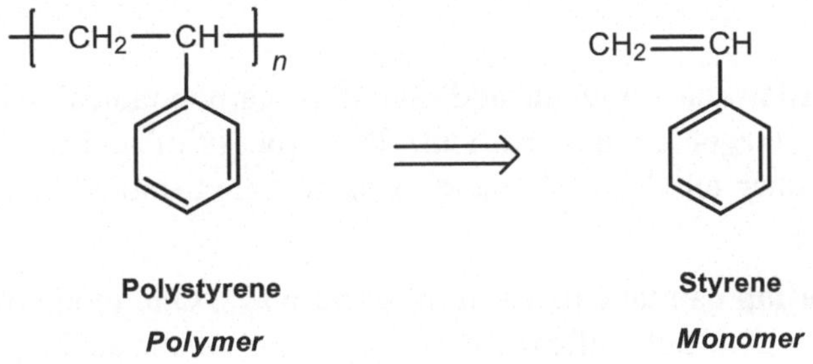

$\left[CH_2-CH(C_6H_5) \right]_n \Longrightarrow CH_2=CH(C_6H_5)$

Polystyrene — ***Polymer***

Styrene — ***Monomer***

(b) It is a copolymer. It consists of two different monomeric species, which are **1,3-butadiene** and **acrylonitrile**. The polymer is a synthetic rubber, known as **Buna–N** (see Section 19.7.3).

$$\left[\!-CH_2-CH{=}CH-CH_2-CH_2-\underset{\displaystyle CN}{\underset{|}{CH}}-\!\right]_n \Longrightarrow CH_2{=}CH-CH{=}CH_2 + CH_2{=}\underset{\displaystyle CN}{\underset{|}{CH}}$$

Buna-N *Copolymer* — **1,3-Butadiene** *Monomer* — **Acrylonitrile** *Monomer*

19.2 POLYMERIZATION REACTIONS

In general, polymerization reactions can take place by *addition* or *condensation*. Here we will consider the addition polymerization.

Alkenes are useful starting materials for the preparation of polymers and their polymerizations involve addition reactions. **Addition polymerizations** can proceed through a **radical or ionic chain mechanism** depending on how they are initiated.

19.2.1 Radical Polymerization

The radical polymerization of alkenes is most common. The polymerization is often *initiated* by organic peroxides (such as benzoyl peroxide, acetyl peroxide or *t*-butyl peroxide) that possesses a weak O–O bond.

The **radical polymerization** of alkenes is described here using **ethylene** and **styrene** respectively.

Polymerization of ethylene

Ethylene is a *symmetrical* alkene. When ethylene is heated in the presence of a small amount of benzoyl peroxide as a radical initiator, it polymerizes to produce a polymer called **polythene**.

A **radical chain mechanism** for the polymerization of ethylene is shown in Figure 19.1. The chain mechanism involves *chain initiation*, *chain propagation* and *chain termination* steps.

Benzoyl peroxide dissociates to form **phenyl radicals**, which then add to the ethylene double bond to generate a **new carbon radical**. This carbon radical **initiates a chain reaction**.

The **chain propagation** starts with the addition of the carbon radical to another molecule of ethylene to form a **larger carbon radical**. The process of addition of newer and still larger radicals to other ethylene molecules gets repeated until the chain propagation is stopped.

The **chain termination** can take place in different ways. One mode of termination is the **combination of two product radicals**, which gives polythene as the polymer.

Chain Initiation

$$C_6H_5-\overset{O}{\overset{\|}{C}}-O-O-\overset{O}{\overset{\|}{C}}-C_6H_5 \xrightarrow{\text{Heat}} 2\ C_6H_5-\overset{O}{\overset{\|}{C}}-\dot{O} \xrightarrow{-2\,CO_2} 2\ C_6H_5^{\bullet}$$

Benzoyl peroxide **Phenyl radical**

$$C_6H_5^{\bullet} + H_2C{=}CH_2 \longrightarrow C_6H_5-CH_2-\dot{C}H_2$$

Ethylene **New carbon radical (chain initiator)**

Chain Propagation

$$C_6H_5-CH_2-\dot{C}H_2 + H_2C{=}CH_2 \longrightarrow C_6H_5-CH_2-CH_2-CH_2-\dot{C}H_2$$

Carbon radical **Ethylene** **Larger carbon radical**

$$\downarrow H_2C{=}CH_2 \text{ (many molecules)}$$

$$C_6H_5-(CH_2-CH_2)_m-CH_2-\dot{C}H_2$$

(*m* is a large number)

Chain Termination

$$C_6H_5-(CH_2-CH_2)_m-CH_2-\dot{C}H_2 + C_6H_5-(CH_2-CH_2)_m-CH_2-\dot{C}H_2 \xrightarrow{\text{Combination}} \begin{matrix} C_6H_5-(CH_2-CH_2)_m-CH_2-CH_2 \\ | \\ C_6H_5-(CH_2-CH_2)_m-CH_2-CH_2 \end{matrix}$$

$$\equiv C_6H_5-(CH_2-CH_2)_n-C_6H_5$$

[*n* = 2 (*m* + 1)]

Polythene

Figure 19.1: Radical polymerization of ethylene.

Polymerization of styrene

Styrene is an *unsymmetrical* alkene. When styrene is heated in the presence of a peroxide initiator, the radical formed by the dissociation of the peroxide adds to the **less hindered or less substituted end** of styrene to generate a **more stable radical**, which is secondary

(2°) and benzylic in nature. This new radical **initiates** chains. (Check that addition to the more substituted end of styrene gives a less stable primary radical).

Chains then **propagate** by adding successive styrene molecules, with **radical addition** occurring to the **less substituted end** of styrene. Besides electronic factors, the radical addition to the less substituted alkene carbon is also favoured by *steric factors*.

The **more substituted end** of an alkene is sometimes designated as the **'Head'** and the **less substituted end** as the **'Tail'**. Thus the radical polymerization of styrene takes place in a **head-to-tail** fashion, *not* in a head-to head fashion.

Finally, **chain termination** by the combination of two product radicals produces polystyrene as the polymer.

Chain initiation

R—C(=O)—O—O—C(=O)—R —Heat→ 2 R—C(=O)—O· —(− 2 CO_2)→ 2 R·

Peroxide **Radical**

R· + CH_2=CH(Ph) → R—CH_2—ĊH(Ph)

(less substituted or less hindered) **(2° and benzylic)**

Styrene **New radical (more stable)**

Chain propagation

Head **Tail**

R—CH_2—ĊH(Ph) + CH_2=CH(Ph) —**Head-to-tail**→ R—CH_2—CH(Ph)—CH_2—ĊH(Ph) —**etc.**→

More stable radical **Styrene** **Larger radical**

Chain termination

R—[CH_2—CH(Ph)]$_n$—R

Polystyrene

OR

Head **Head**

R—CH_2—ĊH(Ph) + CH(Ph)=CH_2 —**Head-to-head**—✕→ R—CH_2—CH(Ph)—CH(Ph)—ĊH_2

Not formed

19.2.2 Ionic Polymerization

The **polymerization** of alkenes can be initiated by **ions** (cations or anions) instead of radicals, which also proceeds by a **chain mechanism**. However, the ionic polymerization is not as common as the radical polymerization.

Cationic polymerization

The cationic polymerization is initiated by acids (typically a Lewis acid such as BF_3 or $AlCl_3$) and a *trace of water.*

To illustrate, consider the polymerization of **isobutene** in the presence of BF_3 and a small amount of water. Figure 19.2 shows a **cationic chain mechanism** for the polymerization of isobutene.

BF_3 reacts with water to release a proton that attacks isobutene to form a stable ***t*-butyl carbocation**. The polymerization is **initiated** by the *t*-butyl cation. The growing chains in the **propagation** steps are **large carbocations**. The reaction is **terminated** by the loss of a proton to produce **polyisobutene** as the polymer.

Initiation

$$BF_3 + H_2O \rightleftharpoons H^+ + BF_3(OH)^-$$

$$CH_3-C(CH_3)=CH_2 + H^+ \longrightarrow CH_3-\overset{+}{C}(CH_3)-CH_3$$

Isobutene → *t*-Butyl carbocation (*Initiator*)

Propagation

$$(CH_3)_3\overset{+}{C} + H_2C=C(CH_3)_2 \longrightarrow (CH_3)_3C-CH_2-\overset{+}{C}(CH_3)_2 \xrightarrow{\text{etc.}} (CH_3)_3C{+}\left[CH_2-C(CH_3)_2\right]_n{-}CH_2-\overset{+}{C}(CH_3)_2$$

Isobutene; Large carbocation; (*n* is a large number)

Termination

$$(CH_3)_3C\left[CH_2-C(CH_3)_2\right]_n-CH_2-\overset{+}{C}(CH_3)-CH_2-H \xrightarrow{-H^+} (CH_3)_3C\left[CH_2-C(CH_3)_2\right]_n-CH_2-C(CH_3)=CH_2$$

Polyisobutene

Figure 19.2 Cationic polymerization of isobutene.

Anionic polymerization

The anionic polymerization is initiated by strong bases (such as $NaNH_2$ in liquid ammonia). The polymerization generally occurs with *alkenes containing electron-withdrawing groups*.

To illustrate, consider **acrylonitrile** having an electron-withdrawing CN group. The polymerization occurs when it is treated with $NaNH_2$ in liquid ammonia. A **chain mechanism** for the anionic polymerization of acrylonitrile is shown in Figure 19.3.

A nucleophilic addition of amide ion to acrylonitrile produces a **carbanion** that is *stabilized by resonance* with the electron-withdrawing CN group. The nucleophilic addition can also be regarded as a conjugate addition. The carbanion acts as the **initiator**.

The growing chains in the **propagation** steps are **large carbanions** formed by successive additions of acrylonitrile molecules. The polymerization is **terminated** by the abstraction of a proton from ammonia to produce **polyacrylonitrile** as the polymer.

Initiation

NH_2^- + $H_2C{=}CH{-}C{\equiv}N$ ⟶ [$H_2N{-}CH_2{-}\overset{-}{C}H{-}C{\equiv}N$ ⟷ $H_2N{-}CH_2{-}CH{=}C{=}N^-$]

($NaNH_2$) Acrylontrile Carbanion *Initiator*

Propagation

$H_2N{-}CH_2{-}\overset{-}{C}H(CN)$ + $H_2C{=}CH{-}CN$ ⟶ $H_2N{-}CH_2{-}CH(CN){-}CH_2{-}\overset{-}{C}H(CN)$

Acrylontrile Large carbanion

↓ etc.

$H_2N{+}CH_2{-}CH(CN){+}_n CH_2{-}\overset{-}{C}H(CN)$

(*n* is a large number)

Termination

$H_2N{+}CH_2{-}CH(CN){+}_n CH_2{-}\overset{-}{C}H(CN)$ $\xrightarrow{NH_3}$ $H_2N{+}CH_2{-}CH(CN){+}_n CH_2{-}CH_2(CN)$ + NH_2^-

Polyacrylontrile

Figure 19.3: Anionic polymerization of acrylonitrile.

Note: The polymerization of acrylonitrile can also be carried out by the radical polymerization method (see Section 19.4). For commercial production of polyacrylonitrile, the anionic polymerization is less important than the radical polymerization.

Problem 19.2: 'Super glue' is a solution containing highly pure methyl α-cyanoacrylate. Explain how the super glue acts as a remarkable adhesive.

Answer

Methyl α-cyanoacrylate possesses two electron-withdrawing groups (CN and CO_2CH_3). Thus the anionic polymerization of methyl α-cyanoacrylate can occur with weaker bases such as hydroxide ion, or even with traces of water found on the surfaces of the two objects that are being glued together. Thus the adhesive property of super glue is a result of the anionic polymerization of methyl α-cyanoacrylate.

$$n\ CH_2{=}C(CN)(CO_2CH_3) \xrightarrow[\text{Anionic polymerization}]{H_2O} \left[CH_2-C(CN)(CO_2CH_3) \right]_n$$

Methyl α-cyanoacrylate **Poly(methyl α-cyanoacrylate)**

19.2.3 Zeigler–Natta Polymerization

The Zeigler-Natta polymerization does not involve radical or ionic initiators. It is accomplished using catalysts called **Zeigler–Natta catalysts,** named after the German chemist, Karl Zeigler and the Italian chemist, Giulio Natta. They shared the Nobel Prize in chemistry in 1963 for their discovery.

Zeigler–Natta catalysts are organometallic complexes of transition metals. The most commonly used **Zeigler-Natta catalyst** is prepared from **titanium tetrachloride** ($TiCl_4$) and **triethylaluminium** [$(C_2H_5)_3Al$]. It is an insoluble material and is classified as a *heterogeneous catalyst*. The catalyst is generally used as a suspended solid in a hydrocarbon solvent.

Using the Zeigler-Natta catalyst, **ethylene** is polymerized to **polythene** as

$$CH_2{=}CH_2 \xrightarrow[\text{Zeigler-Natta catalyst}]{TiCl_4,\ (C_2H_5)_3Al} \sim CH_2-CH_2-CH_2-CH_2 \sim$$

Ethylene **Polythene**

Mechanism

The details of the mechanism are beyond our scope. A simplified version is presented here.

The Zeigler-Natta catalyst formed from $TiCl_4$ and $(C_2H_5)_3Al$ provides an **octahedral complex** of Ti with a **vacant coordination site**. In forming the complex, $(C_2H_5)_3Al$ donates a C_2H_5 group to the Ti atom and the Al atom is coordinated to Cl atoms.

The **polymerization** is initiated when the alkene monomer initially forms a **π complex** by coordination to the vacant site of the Ti atom in the octahedral complex, which is followed by the **insertion** of the monomer between the Ti atom and the ethyl group.

After insertion, **one coordination site of titanium is vacated**. The process is then repeated and the polymerization propagates through *insertion* of ethylene molecules between the Ti atom and the growing polymer chain.

Finally, the polymerization can be **terminated** in several ways, such as by adding hydrogen (i.e. hydrogenation), thereby producing *linear* polythene.

$(C_2H_5)_3Al$ + $TiCl_4$ → (vacant coordination site)

An octahedral complex of titanium with a vacant coordination site

Initiation $H_2C{=}CH_2$ **(monomer)**

(π complex)

Insertion of monomer

$CH_2—CH_2—CH_2CH_3$ **(vacant coordination site)**

Propagation $H_2C{=}CH_2$ **(many molecules)**

$CH_2—CH_2—CH_2—CH_2$〰 **(H_2)**

H_2 ***Termination***

$L_nTi—H$ + $CH_3—CH_2—CH_2—CH_2$〰

(L_n = other groups or ligands)

Polythene

19.3 COPOLYMERIZATION

Copolymerization is a polymerization of two or more different monomers so that each kind of monomer enters the polymer chain. Copolymers have properties that are quite different from those of individual homopolymers.

The following is an example of radical copolymerization of 1,3-butadiene with styrene in the presence of a peroxide.

$$CH_2{=}CH{-}CH{=}CH_2 + C_6H_5{-}CH{=}CH_2 \xrightarrow{\text{Peroxide}} \left[CH_2{-}CH{=}CH{-}CH_2{-}CH_2{-}\underset{C_6H_5}{CH} \right]_n$$

1,3-Butadiene **Styrene**

Butadiene-styrene copolymer

(Buna-S)

The **butadiene-styrene copolymer** (known as **Buna–S**) has a very high strength and is a good substitute for natural rubber (see Section 19.7.3).

For the preparation of copolymers by condensation reactions, see later Section 19.6.

19.4 ADDITION POLYMERS OR CHAIN-GROWTH POLYMERS

We have previously seen that addition polymerization of alkenes proceeds by a chain mechanism. As the *growing* polymer chain develops from a *chain* reaction, **addition polymers** are also called **chain-growth polymers**.

In this section, we will focus on some typical addition polymers.

A. *Polythene*

Polythene is a very familiar plastic. It has been produced commercially since 1943.

Polythene can be prepared by **radical polymerization** of ethylene in the presence of a peroxide catalyst or oxygen under very high pressure (about 2000 atm) and at a temperature near 200°C.

$$n\, CH_2{=}CH_2 \xrightarrow[\text{High pressure}]{\text{Peroxide}} \left[CH_2{-}CH_2 \right]_n$$

Radical polymerization

Ethylene **Low density polythene**

Polythene thus produced has a **branched structure**. It has a **low density** (0.92 g/mL) since the molecules are not quite closely packed due to branching. It is capable of softening on heating and has a softening point at about 110°C. It is a thermoplastic. **Low density polythene** is widely used in manufacturing flexible bottles, films, sheets, and insulation for electric wires.

On the other hand, polythene can also be prepared at low pressure (6-7 atm) using the **Ziegler–Natta catalyst**.

$$CH_2{=}CH_2 \xrightarrow[\text{Low pressure}]{TiCl_4/(C_2H_5)_3Al} \sim\!\sim CH_2{-}CH_2{-}CH_2{-}CH_2\sim\!\sim$$

Ethylene — ***Zeigler-Natta polymerization*** — **High density polythene**

Polythene thus produced has a **linear structure**. It has a **high density** (0.96 g/mL) due to close packing. It has also a higher melting point and greater strength, and is chemically inert. **High density polythene** is used for making bottles, buckets, dustbins, pipes, etc.

B. *Poly(vinyl chloride) (PVC)*

Poly(vinyl chloride) is one of the major plastics produced by the chemical industry.

Radical polymerization of vinyl chloride (chloroethene) in the presence of a peroxide produces **poly(vinyl chloride)**, known as **PVC**.

$$n\ CH_2{=}CH(Cl) \xrightarrow[\text{Radical polymerization}]{\text{Peroxide}} \left[CH_2{-}CH(Cl) \right]_n$$

Vinyl chloride — **Poly(vinyl chloride) (PVC)**

PVC is a hard, brittle and rigid material, and is often used for making pipes, rods and compact discs.

PVC can be softened by blending it with additives, called **plasticizers**. A common plasticizer is diisooctyl phthalate (an ester). The softer polymer is used for making plastic raincoats, shower curtains and garden hoses.

C. *Polyacrylonitrile (PAN)*

Radical polymerization of acrylonitrile (vinyl cyanide) in the presence of a mixture of ferrous sulphate and hydrogen peroxide produces **polyacrylonitrile (PAN)**.

$FeSO_4$ react with H_2O_2 to form *hydroxyl radicals* (·OH), which act as chain initiators.

$$n\ CH_2{=}CH(CN) \xrightarrow[H_2O_2]{FeSO_4} \left[CH_2{-}CH(CN) \right]_n$$

Radical polymerization

Acrylonitrile — **Polyacrylonitrile**

$$\left(\text{Initiation: } Fe^{2+} + HO{-}OH \longrightarrow Fe^{3+} + {\cdot}OH + OH^-\right)$$

Polyacrylonitrile is used for making commercial acrylic fibres; the trade name for which is **Orlon**. It is used as a substitute for wool, and the fibres are also used in making carpets and clothing.

D. *Polytetrafluoroethene (Teflon)*

Teflon is made by radical polymerization of tetrafluoroethylene in aqueous suspension containing ferrous sulphate and hydrogen peroxide. The reaction is exothermic, and water helps dissipate the heat produced.

$$n\ CF_2{=}CF_2 \xrightarrow[H_2O_2,\ H_2O]{FeSO_4} \left[CF_2{-}CF_2\right]_n$$

Tetrafluoroethylene — ***Radical polymerization*** — Teflon

Teflon is chemically inert and resistant to attack by corrosive reagents. It is used in greaseless bearings, and in non-stick coatings for pans and other utensils.

E. *Poly(methyl methacrylate)*

Radical polymerization of methyl methacrylate (methyl 2-methylpropenoate) in the presence of a peroxide produces **poly(methyl methacrylate)**. It is a hard and highly transparent polymer.

$$n\ CH_2{=}C(CH_3)(CO_2CH_3) \xrightarrow[\text{Radical polymerization}]{\text{Peroxide}} \left[CH_2{-}C(CH_3)(CO_2CH_3)\right]_n$$

Methyl methacrylate — **Poly(methyl methacrylate)**

The monomer can also be partially polymerized to a viscous liquid which is then poured into sheets. Further heating completes the polymerization to give a clear plastic sheet marketed under the names Lucite, Plexiglas and Perspex.

PROBLEM 19.3: Identify each of the following polymers and give the monomer from which it is made.

(a) $\left[CH_2{-}CH(CH_3){-}CH_2{-}CH(CH_3)\right]_n$

(b) $\left[CH_2{-}CH(O{-}C(=O){-}CH_3){-}CH_2{-}CH(O{-}C(=O){-}CH_3)\right]_n$

Answer

(a) The polymer is polypropylene. The monomer from which the polymer is made is propylene.

$$\left[CH_2-\underset{CH_3}{\underset{|}{CH}}-CH_2-\underset{CH_3}{\underset{|}{CH}} \right]_n \Longrightarrow n\ CH_2{=}\underset{CH_3}{\underset{|}{CH}}$$

Polypropylene — ***Polymer*** ⟹ **Propylene** — ***Monomer***

(b) The polymer is poly(vinyl acetate) and the monomer is vinyl acetate. Poly(vinyl acetate) is used as an emulsion in water-based paints.

$$\left[CH_2-\underset{OCOCH_3}{\underset{|}{CH}}-CH_2-\underset{OCOCH_3}{\underset{|}{CH}} \right]_n \Longrightarrow n\ H_2C{=}\underset{OCOCH_3}{\underset{|}{CH}}$$

(In the polymer, each side group is drawn as O–C(=O)–CH_3.)

Poly(vinyl acetate) — ***Polymer*** ⟹ **Vinyl acetate** — ***Monomer***

19.5 STEREOCHEMISTRY OF ADDITION POLYMERS

Addition polymers can exist in **stereoisomeric forms**. To illustrate, consider the stereoisomers possible for **polypropylene**.

Polymerization of propylene ($CH_3CH{=}CH_2$) in a **head-to-tail** fashion (cf. polymerization of styrene, Section 19.2.1) produces a polymer in which **C2** of *each* monomer unit becomes a **chiral centre** (*). (Remember that the more substituted alkene carbon is the 'head' and the less substituted alkene carbon is the 'tail'.)

Tail (→ C1) **Head** (→ C2)

$$\underset{1}{CH_2}{=}\underset{\underset{3}{CH_3}}{\underset{|2}{CH}} \xrightarrow{\text{Head-to-tail polymerization}} -CH_2-\overset{*}{\underset{CH_3}{\underset{2|}{CH}}}-CH_2-\overset{*}{\underset{CH_3}{\underset{2|}{CH}}}-CH_2-\overset{*}{\underset{CH_3}{\underset{2|}{CH}}}-$$

(* denotes a chiral centre)

Propylene → **Polypropylene**

Three spatial arrangements of the chiral centres are possible for the polymer, namely, **atactic**, **isotactic** and **syndiotactic**.

Atactic polymer

The stereoisomer of the polymer in which the **configurations at chiral centres are random** along the polymer chain is called an **atactic polymer**. The word atactic comes from *a* (without) and the Greek *taktikos*, meaning order.

In **atactic polypropylene**, the methyl groups are randomly disposed up or down along the polymer chain.

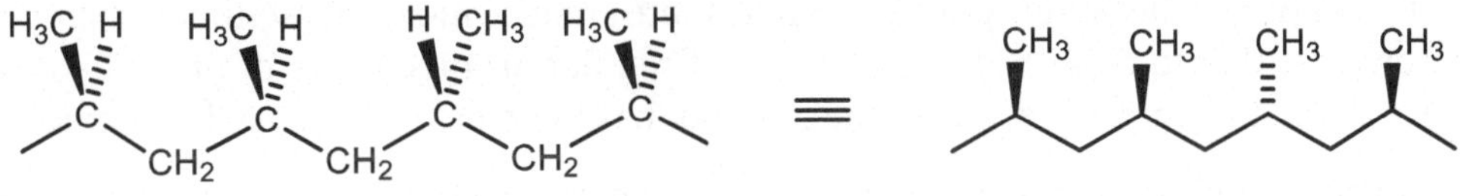

(H's are omitted for clarity)

Atactic polypropylene

Isotactic polymer

In the **isotactic polymer**, the **configurations at all chiral centres are identical**. The word isotactic comes from *iso* (same) and the Greek *taktikos,* meaning order.

In **isotactic polypropylene**, the methyl groups are on the same side (either all up or all down) along the polymer chain.

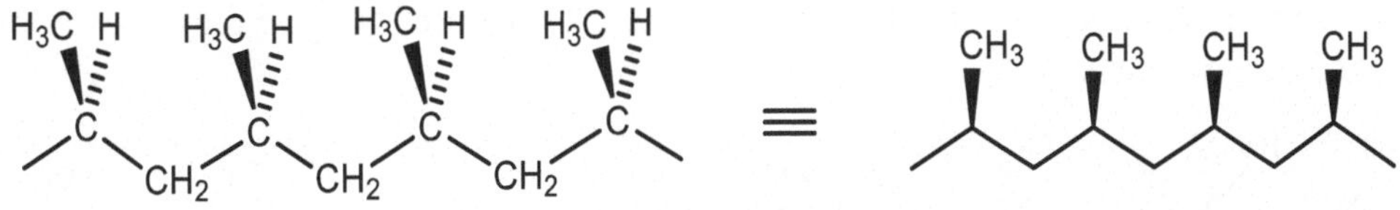

(H's are omitted for clarity)

Isotactic polypropylene

Syndiotactic polymer

The stereoisomer in which the **configurations at chiral centres alternate** along the polymer chain is called a **syndiotactic polymer**. The word *syndiotactic* comes from *syndio* (two together) and the Greek *taktikos,* meaning order.

In **syndioatactic polypropylene**, the methyl groups are alternately up and down along the polymer chain.

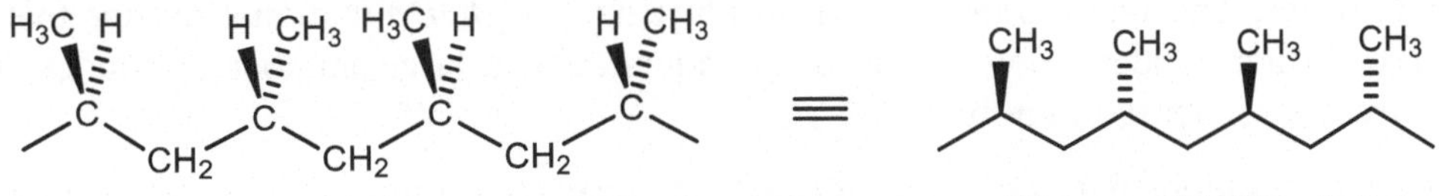

(H's are omitted for clarity)

Syndiotactic polypropylene

Physical properties

The stereochemistry of polymer molecules has a critical effect on the physical properties of the polymers.

Atactic polypropylene is a *soft, noncrystalline and elastic material*. This is because of the random configurations at the chiral centres along the atactic polymer chain. Polypropylene obtained by radical polymerization at high pressure is atactic.

Isotactic and syndiotactic polypropylenes are much more crystalline than atactic polypropylene due to close packing because of regular arrangements of groups along the polymer chain. In general, a syndiotactic polymer has the most rigid, crystalline structure.

Both isotactic and syndiotactic polypropylenes have been made using Zeigler–Natta catalysts at much lower pressures. Only the isotactic form of polypropylene is marketed in large quantities.

Problem 19.4: In contrast to polypropylene, polyisobutene does not exist in atactic, isotactic or syndiotactic form. Explain.

Answer

Head-to-tail polymerization of isobutene (2-methylpropene) gives polyisobutene in which C2 of the monomeric unit is *not* a chiral centre because four groups attached to C2 are not all different (two are identical methyl groups). Therefore, the polymer does not exist in stereoisomeric forms.

$$\underset{\text{Isobutene (2-methylpropene)}}{\overset{1}{CH_2}{=}\overset{2}{C}(\overset{3}{CH_3}){-}CH_3} \xrightarrow{\text{Head-to-tail polymerization}} \underset{\text{Polyisobutene}}{-CH_2-C(CH_3)_2-CH_2-C(CH_3)_2-CH_2-C(CH_3)_2-}$$

Tail (C1), Head (C2)

(C2 is not a chiral centre)

19.6 CONDENSATION POLYMERS OR STEP-GROWTH POLYMERS

Condensation polymers are **copolymers** that are formed from two different monomers, each containing *more than one* functional group.

The **polymerization** involves a **condensation reaction** that eliminates a simple molecule such as water. The initial condensation product has *residual* functional groups that can undergo further condensation reactions. A sequence of a large number of condensation steps gives a condensation polymer.

Since condensation polymers are formed in **discrete condensation steps** that are independent of each other, and are not formed by a chain reaction, they are also called **step-growth polymers.**

Here we will consider some important condensation polymers.

A. *Polyesters*

Polyesters constitute a main class of **synthetic fibres**. Polyesters are produced by heating dicarboxylic acids (or diesters) with diols.

Of the polyester fibres, **Dacron or Terylene** is the most prevalent. It is manufactured by heating a mixture of **ethylene glycol** (1.2-ethanediol) and **terephthalic acid** (1,4-benzenedicarboxylic acid) in the presence of zinc acetate-antimony trioxide catalyst.

$$n\ HO{-}CH_2{-}CH_2{-}OH \;+\; n\ HO{-}\overset{O}{\overset{\|}{C}}{-}C_6H_4{-}\overset{O}{\overset{\|}{C}}{-}OH \xrightarrow[Sb_2O_3]{Zn(OAc)_2}$$

Ethylene glycol **Terephthalic acid**

$$\left[O{-}CH_2{-}CH_2{-}\underbrace{O{-}\overset{O}{\overset{\|}{C}}}_{\text{(ester linkage)}}{-}C_6H_4{-}\overset{O}{\overset{\|}{C}} \right]_n + n\ H_2O$$

Terylene or Dacron

Note that dimethyl terephthalate (i.e. dimethyl ester of terephthalic acid) can also be used instead of terephthalic acid.

The terylene fibre is crease resistant. It is used in blending with cotton, which provides a fabric with high durability and excellent ease-of-care properties. It is also used in blending with wool fibres, and as glass reinforcing materials in safety helmets.

B. *Polyamides*

Nylons are **polyamides** which constitute another main class of **synthetic fibres**. Of the nylons, nylon 6,6 and nylon 6 are the principal commercial products.

Nylon 6,6 is prepared by condensation polymerization of **adipic acid** (a **6**-carbon dicarboxylic acid) and **hexamethylenediamine** (a **6**-carbon diamine) under high pressure and at a temperature of about 200°C.

$$n\ H_2N{-}(CH_2)_6{-}NH_2 \;+\; n\ HO{-}\overset{O}{\overset{\|}{C}}{-}(CH_2)_4{-}\overset{O}{\overset{\|}{C}}{-}OH \xrightarrow[\text{High pressure}]{\sim 200^{o}C}$$

Hexamethylenediamine **Adipic acid**

$$\left[NH{-}(CH_2)_6{-}\underbrace{NH{-}\overset{O}{\overset{\|}{C}}}_{\text{(amide linkage)}}{-}(CH_2)_4{-}\overset{O}{\overset{\|}{C}} \right]_n + n\ H_2O$$

Nylon 6,6

Nylon 6,6 has a high tensile strength which is attributed to strong *intermolecular hydrogen bonding* between polyamide chains. Nylon 6,6 is used as textile fibres, and in making sheets and bristles for brushes.

Nylon 6 is prepared commercially by heating **caprolactam** (a **6**-carbon cyclic amide) with water and a small amount of acetic acid at a high temperature. Caprolactam undergoes ring-opening polymerization to produce the polyamide.

$$\text{Caprolactam} \xrightarrow[\text{CH}_3\text{CO}_2\text{H},\ \sim 250^\circ\text{C}]{\text{H}_2\text{O}} \left[\text{NH(CH}_2)_5-\underbrace{\overset{\text{O}}{\overset{\|}{\text{C}}}-\text{NH}}_{\text{(amide linkage)}}(\text{CH}_2)_5-\overset{\text{O}}{\overset{\|}{\text{C}}} \right]_n$$

Caprolactam

Nylon 6

Nylon 6 is used in making fabrics, tyre cords and ropes.

C. *Polyurethanes*

Urethanes ($RNHCO_2R'$) are esters of carbamic acids ($RNHCO_2H$), which are formed when an isocyanate reacts with an alcohol (see Section 16.3.2).

Polyurethanes can be prepared by step polymerization of a **diisocyanate** with a **diol**. Although no small molecules such as water is eliminated as in a condensation reaction, the urethane links are formed in discrete steps. Thus polyurethanes belong to step-growth polymers.

$$\text{O}=\text{C}=\text{N}-\text{C}_6\text{H}_4-\text{CH}_2-\text{C}_6\text{H}_4-\text{N}=\text{C}=\text{O} + \text{HO}-\text{CH}_2-\text{CH}_2-\text{OH}$$

A diisocyanate

Ethylene glycol

$$\downarrow$$

$$\left[\overset{\text{O}}{\overset{\|}{\text{C}}}-\text{HN}-\text{C}_6\text{H}_4-\text{CH}_2-\text{C}_6\text{H}_4-\underbrace{\text{NH}-\overset{\text{O}}{\overset{\|}{\text{C}}}-\text{O}}_{\text{(urethane linkage)}}-\text{CH}_2-\text{CH}_2-\text{O} \right]_n$$

Polyurethane

The reaction of a diisocyanate with a diol usually leads to a *linear* copolymer, but triisocyanates and triols produce *branched* copolymers.

The most familiar applications of polyurethanes are polyurethane foams, used as paddings and cushions. Addition of a small amount of water during polymerization hydrolyses some of the isocyanate groups to release CO_2 which expands the mixture to a sponge-like material.

Polyurethanes are also used in making rigid foams, varnishes and coatings, adhesives and synthetic fibres such as spandex.

D. *Phenol-formaldehyde polymers: Bakelite*

Bakelite is a phenol-formaldehyde copolymer and is one of the oldest commercial plastics.

Bakelite is made by condensation polymerization of **phenol** with **formaldehyde** under acidic or basic conditions.

The **initial reaction** between phenol and formaldehyde introduces a hydroxymethyl group (CH_2OH) at the *ortho* or *para* position of phenol.

Phenol **Formaldehyde** ***o*-Hydroxymethylphenol** ***p*-Hydroxymethylphenol**

The **polymerization** then proceeds by forming a **methylene (CH_2) bridge** between two phenol rings, and leads to a branched and highly cross-linked polymer as an infusible solid mass, called **Bakelite**.

Bakelite

Bakelite is a thermosetting plastic. It is used for making combs, electrical switches, handles of utensils and computer discs.

Problem 19.5: Draw a mechanism for the following reaction sequence:

Phenol →(CH_2O, OH^-) o-Hydroxymethylphenol →(phenol, OH^-) Methylenediphenol

Answer

The first step is an electrophilic aromatic substitution of phenol in basic medium to form *o*-hydroxymethylphenol.

The second step proceeds by the following mechanism: the loss of hydroxide ion leads to an α,β-unsaturated ketone, then a nucleophilic attack by phenoxide ion and subsequent aromatization gives the product.

Mechanism of the first step:

OH, OH^-, $CH_2{=}O$, H, $CH_2{-}O^-$, O^-, $CH_2{-}OH$

o-Hydroxymethylphenol anion

Mechanism of the second step:

O^-, $CH_2{-}OH$, $- OH^-$, O, CH_2, O, O^-, CH_2, O, H, (aromatization), OH, CH_2, O^-

o-Hydroxymethylphenol anion

α.β-Unsaturated ketone

Methylenediphenol anion

19.7 NATURAL POLYMERS

Natural polymers are derived from plants and animals, which include **natural fibres** and **natural rubber**.

Rubber can also be prepared artificially, which is known as **synthetic rubber**. In this section, we consider both natural and synthetic rubbers.

19.7.1 Natural Fibres

The polymer molecules that make up a fibre must be relatively long. Individual polymer chains must also be sufficiently close to each other so that there is strong intermolecular force acting between them, which imparts a high tensile strength of the fibre.

Cotton is the most widely used fibre. The cotton fibres are essentially pure cellulose which is a linear polymer of D-glucose (see Section 20.1.11). Cotton fabrics are made from fibres from the cotton plant. They have excellent wearability and are pleasing in appearance.

Silk is a fibrous protein produced by insects. The most common type of silk is made by silkworms. The cocoons of various silkworms, after chemical treatment, provide a protein called fibroin which is used for most silk fabric. Silk is known for its shine and softness as a material.

Wool is a fibrous animal protein called keratin. It is obtained from the hairs of sheep, goats and other animals. The protein has α-helical structure which leads to fabric with excellent handling qualities. It is generally used to make outerwear and cold weather clothes like sweaters and coats.

Rayon is a *regenerated* natural fibre. It is made by treating cellulose (from cotton or wood pulp) with carbon disulphide (CS_2) in basic solution. Cellulose has many OH groups which react with CS_2, thereby converting cellulose to a soluble cellulose derivative called cellulose xanthate.

$$\text{Cellulose—OH} + \underset{(CS_2)}{S{=}C{=}S} \xrightarrow{\text{NaOH}} \underset{\textbf{Cellulose xanthate}}{\text{Cellulose—O—}\overset{\overset{S}{\|}}{C}\text{—}S^-\,Na^+} + H_2O$$

The solution of cellulose xanthate is then passed through a small orifice or slit into an acidic solution which regenerates cellulose as insoluble fibres. These fibres are **rayon** which is used as a fabric for clothing and the cord of tyres.

$$\underset{\textbf{Cellulose xanthate}}{\text{Cellulose—O—}\overset{\overset{S}{\|}}{C}\text{—}S^-\,Na^+} \xrightarrow[H_2O]{H^+} \underset{\textbf{Rayon}}{\text{Cellulose—OH}} + CS_2 + Na^+$$

The regenerated cellulose can also be produced as sheets. These sheets are **cellophane** which is used as the packaging material.

19.7.2 Natural Rubber

Rubber is a polymer with elastic properties. Natural rubber is obtained from rubber latex which is a colloidal suspension of rubber obtained from the rubber tree found in tropical regions such as India, Sri Lanka, Indonesia, Malaysia and South America.

Natural rubber is an **addition polymer of isoprene** (2-methyl-1,3-butadiene). The isoprene units of natural rubber are linked in a **head-to-tail** fashion and all the double bonds are ***cis***.

Thus natural rubber is a linear polymer, which is ***cis*-1,4-polyisoprene**.

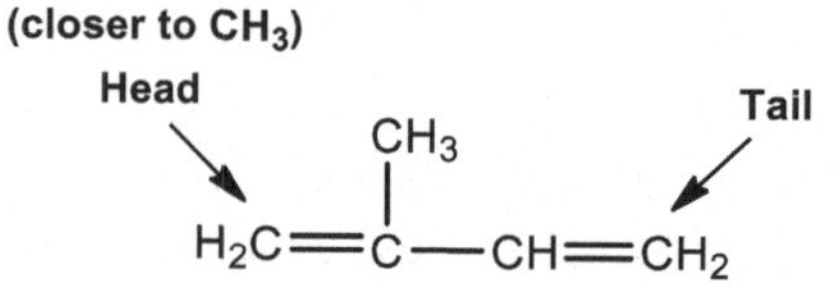

(closer to CH_3) Head **Tail**

$$H_2C{=}C(CH_3){-}CH{=}CH_2$$

Isoprene

$$\sim CH_2\,\text{(head)}{-}C(CH_3){=}CH{-}CH_2\,\text{(tail)}{-}CH_2\,\text{(head)}{-}C(CH_3){=}CH{-}CH_2\,\text{(tail)}{-}CH_2\,\text{(head)}{-}C(CH_3){=}CH{-}CH_2\,\text{(tail)}\sim \quad (\text{all } cis)$$

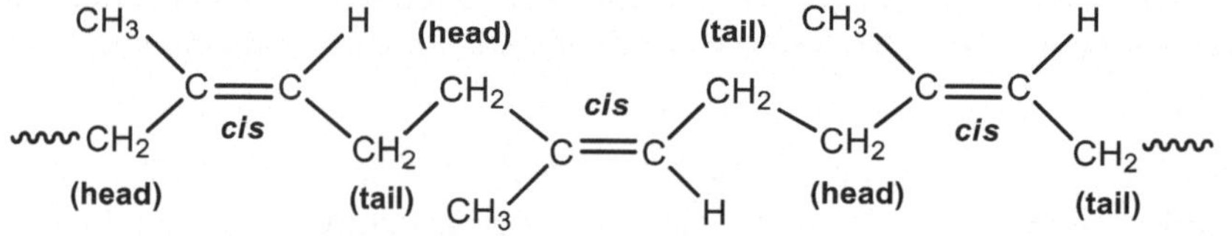

Natural rubber
(*cis*-1,4-polyisoprene)

Polyisoprene chains are randomly coiled and held together by weak intermolecular *van der Waals forces*. Thus rubber can be stretched like a spring by an external deforming force.

But pure natural rubber is soft and tacky. In order to make natural rubber more useful, a process known as **vulcanization of rubber** is carried out.

Vulcanization of rubber

In vulcanization, **natural rubber is heated with sulphur**. As a result, **sulphide or disulphide cross-links** are formed between the polyisoprene chains and makes the rubber harder. The vulcanization process is usually accelerated by the addition of a radical initiator.

In forming cross-links, sulphur reacts at the double bonds (C=C) as well as at the allylic hydrogen atoms.

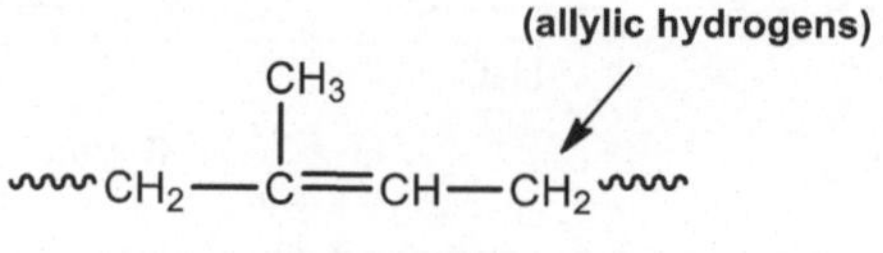

(allylic hydrogens)

$$\sim CH_2{-}C(CH_3){=}CH{-}CH_2\sim$$

Polymer chain

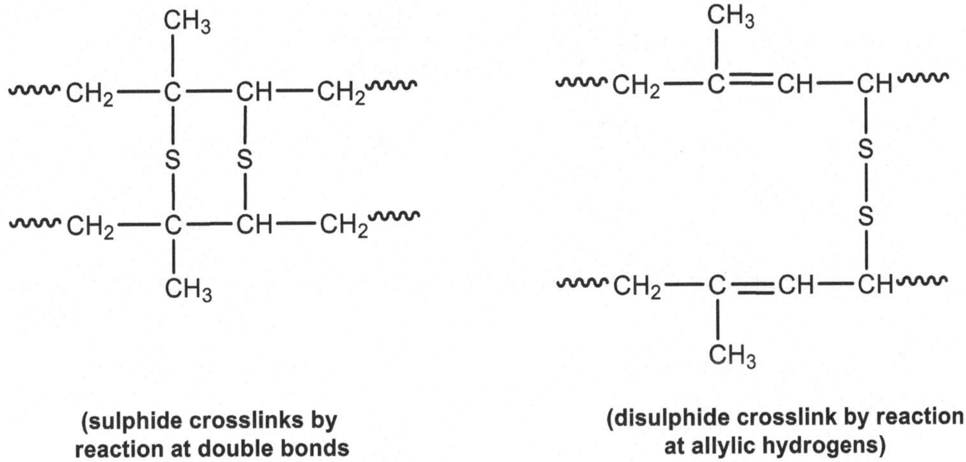

(sulphide crosslinks by reaction at double bonds)

(disulphide crosslink by reaction at allylic hydrogens)

Vulcanized rubber

Soft rubber contains 1–2% sulphur. For making **tyre rubber**, 5% sulphur is used as a cross-linking agent.

19.7.3 Synthetic Rubber

Synthetic rubber is an artificial elastomer. They are superior to natural rubber in respect of thermal stability and resistance to oils and oxidising agents.

There are different types of synthetic rubber that include **neoprene** and **Buna rubbers**.

Neoprene

Neoprene (or polychloroprene) is a homopolymer. It is made by radical polymerization of chloroprene (2-chloro-1,3-butadiene) in the presence of a peroxide.

$$n\,H_2C{=}C(Cl){-}CH{=}CH_2 \xrightarrow[\text{Radical polymerization}]{\text{Peroxide}} {-}\!\left[CH_2{-}C(Cl){=}CH{-}CH_2\right]\!_n{-}$$

Chloroprene (2-chloro-1,3-butadiene) **Neoprene**

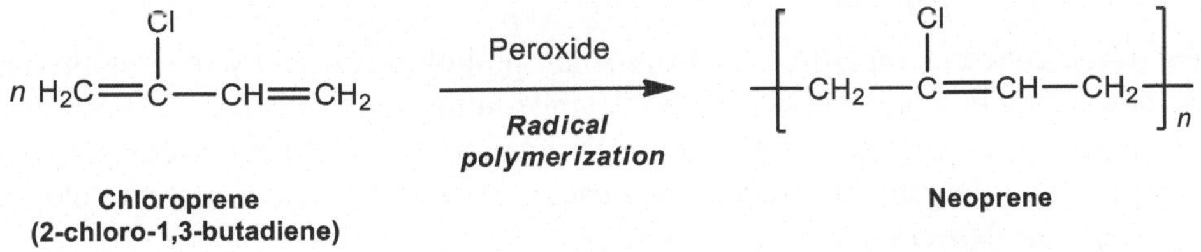

Neoprene has a high resistance to vegetable and mineral oils. It is used in making conveyor belts, gaskets and hoses.

Buna rubbers

Buna rubbers (Buna–S and Buna–N) are copolymers. Radical copolymerization of 1,3-butadiene with styrene in the presence of a peroxide produces **Buna–S** (see also Section 19.3).

$$n\,CH_2{=}CH{-}CH{=}CH_2 + n\,H_2C{=}CH(C_6H_5) \xrightarrow[\text{Copolymerization}]{\text{Peroxide}} {-}\![CH_2{-}CH{=}CH{-}CH_2{-}CH_2{-}CH(C_6H_5)]_n\!{-}$$

1,3-Butadiene **Styrene** **Buna-S**

Buna–S is used for making car tyres, floor tiles, cable insulation, footwear components, etc.

Buna–N is made by copolymerization of 1,3-butadiene with acrylonitrile in the presence of a peroxide.

$$n\,CH_2{=}CH{-}CH{=}CH_2 + n\,H_2C{=}CH(CN) \xrightarrow[\text{Copolymerization}]{\text{Peroxide}} {-}\![CH_2{-}CH{=}CH{-}CH_2{-}CH_2{-}CH(CN)]_n\!{-}$$

1,3-Butadiene **Acrylonitrile** **Buna-N**

Buna–N is resistant to petrol, lubricating oils and organic solvents. It is used in making fuel and oil handling hoses, oil seals, protective gloves, moulded goods, footwear and floor mats.

19.8 BIODEGRADABLE POLYMERS

Synthetic polymers have generally an undesirable influence on the environment because they are quite resistant to the environmental degradation processes. They tend to accumulate as solid waste and remain undegraded for a long time. This brings a lot of problems with waste disposal.

In view of the concerns regarding environmental pollution caused by the polymeric solid wastes, new synthetic polymers which can degrade in the environment have been designed and developed. These polymers which undergo deterioration and degrade completely when exposed to microorganisms (such as bacteria) in the environment are called **biodegradable polymers**.

Aliphatic polyesters constitute an important class of biodegradable polymers. Besides, **polyamides** have also been designed as polymers that are biodegradable.

Aliphatic polyesters

An example of biodegradable aliphatic polyesters is poly(3-**h**ydroxy**b**utyrate-*co*-3-hydroxy**v**alerate), commonly known as **PHBV**.

PHBV is made by condensation polymerization of 3-hydroxybutyric acid (3-hydroxybutanoic acid) and 3-hydroxyvaleric acid (3-hydroxypentanoic acid).

$$HO-CH(CH_3)-CH_2-C(=O)-OH + HO-CH(CH_2CH_3)-CH_2-C(=O)-OH \xrightarrow{\text{Condensation polymerization}} -[O-CH(CH_3)-CH_2-C(=O)-O-CH(CH_2CH_3)-CH_2-C(=O)]_n-$$

3-Hydroxybutyric acid; 3-Hydroxyvaleric acid; PHBV

PHBV is used in orthopaedic devices, in controlled release of drugs, and in speciality packaging. As a biodegradable polymer, it undergoes bacterial degradation in the environment.

Another example of biodegradable aliphatic polyesters is **polycaprolactone (PCL)**. It is prepared by ring-opening polymerization of caprolactone using a catalyst such as stannous octanoate (cf. nylon 6 synthesis, Section 19.6).

$$\text{Caprolactone} \xrightarrow[\text{Ring-opening polymerization}]{\text{Catalyst}} -[O-(CH_2)_5-C(=O)]_n-$$

Caprolactone; Polycaprolactone (PCL)

PCL is used for controlled release and targeted drug delivery, tissue regeneration and cell transplantation, and in food packaging.

Polyamides

An example of biodegradable polyamides is **nylon 2–nylon 6**. It is prepared by condensation polymerization of glycine (a **2**-carbon amino acid) with aminocaproic acid (a **6**-carbon amino acid).

$$H_2N-CH_2-C(=O)-OH + H_2N-(CH_2)_5-C(=O)-OH \xrightarrow{\text{Condensation polymerization}} -[NH-CH_2-C(=O)-NH-(CH_2)_5-C(=O)]_n-$$

Glycine (2-aminoethanoic acid); Aminocaproic acid (6-aminohexanoic acid); Nylon 2-nylon 6

Nylon 2–nylon 6 is used as a thread in bristles for toothbrushes, for making strings of musical instruments, and in the synthesis of artificial fibres.

SUPPLEMENTARY PROBLEMS

19.6 Identify the following polymer and outline its synthesis:

$$\left[CH_2-\underset{OCH_3}{\underset{|}{CH}}-CH_2-\underset{OCH_3}{\underset{|}{CH}} \right]_n$$

[*Hints:* poly(methyl vinyl ether), use radical polymerization]

19.7 The polymer *Saran Wrap* is used in food packaging. Is it a homopolymer or copolymer? Outline its synthesis.

$$\left[CH_2-\overset{Cl}{\overset{|}{\underset{Cl}{\underset{|}{C}}}}-CH_2-\underset{Cl}{\underset{|}{CH}} \right]_n$$

Saran Wrap

[*Hints:* use vinyl chloride and vinylidene chloride (1,1-dichloroethene) for synthesis]

19.8 The polymerization of phenol with formaldehyde can be carried out to produce an initial linear polymer (Novolac), used in paints. Write a structure for Novolac.

19.9 *Kodel* is a synthetic fibre that is widely used commercially. Write the structures of monomers for this polymer. Is *Kodel* an addition polymer or a condensation polymer?

$$\left[O-CH_2-C_6H_{10}-CH_2-O-\overset{O}{\overset{\|}{C}}-C_6H_4-\overset{O}{\overset{\|}{C}} \right]_n$$

Kodel

19.10 Identify the following polymers. Write the monomers from which each polymer is made. Mention their uses.

(a) $$\left[NH-\overset{O}{\overset{\|}{C}}-NH-CH_2 \right]_n$$

(b) $$\left[HN-C_3N_3(NH-)-NH-CH_2 \right]_n$$

[*Hints*: (a) Urea-formaldehyde resin; (b) melamine polymer].

CHAPTER 20

Biomolecules: Carbohydrates, Lipids, Proteins and Nucleic Acids

IN THIS CHAPTER

- Carbohydrates and their classification; reducing and nonreducing sugars
- Monosaccharides: Aldose and ketose
- D-Glucose: Open-chain and cyclic structures, mutarotation and glycoside formation
- D-Fructose: Open-chain and cyclic structures
- Reactions of monosaccharides: Oxidation, oxidative cleavage, reduction and osazone formation
- Glycosylamines and aminosugars
- Disaccharides and polysaccharides
- Lipids: Oils, fats and waxes; phospholipids and cell membranes
- Steroids: Cholesterol, sex hormones and vitamin D
- α-Amino acids: Classification, structures and isoelectric points
- Peptides: Structure and amino acid sequence
- Proteins: Primary, secondary, tertiary and quaternary structures; denaturation and renaturation; enzymes and their mechanism of action
- Nucleic acids: Nucleosides, nucleotides, RNA and DNA; nucleotide sequence and double-helix structure of DNA
- Genetic codes, transcription and translation
- Polymerase chain reaction (PCR); DNA finger printing

Biomolecules are organic molecules that are present in all living organisms, from the simplest single-cell bacteria to humans. Biomolecules include **carbohydrates**, **lipids**, **proteins** and **nucleic acids**. These are the molecules that run the chemistry of life.

Nucleic acids contain the genetic information of every organism and direct the synthesis of proteins in the cell. Proteins include *enzymes* that are all important to a cell's function because they catalyse the biological reactions. Carbohydrates and lipids are involved in storing energy and building membranes, and they also have a vital role in biological recognition and transport.

In this chapter, we will introduce the structures and properties of some carbohydrates, lipids, proteins and nucleic acids.

20.1 CARBOHYDRATES

Carbohydrates have the general formula $\mathbf{C_x(H_2O)_y}$. Thus they appear to be 'hydrates of carbon', and so the name carbohydrate is coined.

Simple carbohydrates are also called **sugars** or **saccharides** (from the Latin *saccharum*, meaning sugar). The names of most sugars end with **-ose**, such as **glucose** for blood sugar and **sucrose** for ordinary table sugar.

Carbohydrates are primarily derived from plants. We encounter carbohydrates almost always in our daily lives. The cotton of our clothes is largely cellulose, so is the paper used for printing this book. Starch is a major constituent of many foodstuffs such as rice, potatoes, corn and peas. They are an important source of energy for living organisms.

Chemically, carbohydrates are defined as polyhydroxy aldehydes and ketones or compounds that produce polyhydroxy aldehydes and ketones on hydrolysis.

20.1.1 Classification of Carbohydrates

Carbohydrates are generally classified based on their behaviour toward **hydrolysis**. There are three broad classes:

(a) **Monosaccharides**: Carbohydrates that cannot be hydrolysed further into simpler carbohydrates are called monosaccharides. A number of monosaccharides are known. Glucose, fructose and ribose are some common examples.

(b) **Oligosaccharides**: Carbohydrates that undergo hydrolysis to produce 2–10 molecules of monosaccharides are called oligosaccharides. They are further divided into **disaccharides, trisaccharides**, and so on.

Carbohydrates that hydrolyse to give **two molecules** of either the *same* monosaccharide or *different* monosaccharides are called **disaccharides**.

For example, **maltose** is a disaccharide because 1 mole of maltose on hydrolysis produces 2 moles of the same monosaccharide, glucose.

Sucrose is another disaccharide since 1 mole of sucrose hydrolyses to yield 1 mole of glucose and 1 mole of a different monosaccharide, fructose.

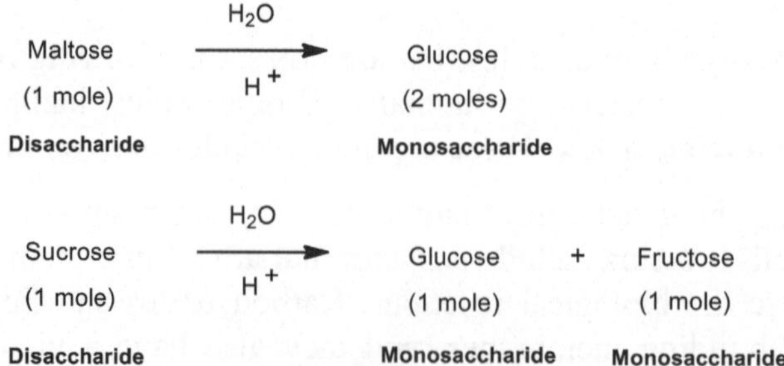

(c) **Polysaccharides**: Carbohydrates that undergo hydrolysis to produce a *large number* of monosaccharide molecules are called polysaccharides. Common examples are **cellulose** and **starch**.

Hydrolysis of either cellulose or starch yields a large number of glucose molecules.

$$\underset{\textbf{Polysaccharide}}{\text{Cellulose or Starch}} \xrightarrow[H^+]{H_2O} \underset{\textbf{Monosaccharide}}{n \text{ Glucose}} \quad \textbf{(}\boldsymbol{n}\textbf{ is a large number)}$$

20.1.2 Reducing and Nonreducing Sugars

Carbohydrates are also divided as reducing sugars and nonreducing sugars.

Reducing sugars

Carbohydrates that reduce Fehling's solution or Tollens' reagent are called reducing sugars.

Fehling's solution is an alkaline solution containing cupric ions (Cu^{2+}) complexed with tartrate ions. Complexing the cupric ion prevents the precipitation of cupric hydroxide in alkaline medium. Fehling's solution is **blue** in colour. It is an oxidizing agent. On reduction, the blue colour is discharged and a **brick-red cuprous oxide** precipitates.

The discharge of blue colour and the appearance of a red precipitate is a vivid indication of a **positive Fehling's test** for a **reducing sugar**.

All monosaccharides are reducing sugars. Thus glucose or fructose gives a positive Fehling's test.

Positive Fehling's test:

Glucose or Fructose	+	Cu^{2+} (tartrate complex)	⟶	Cu_2O ↓	+	Oxidation products
Reducing sugar		**Fehling's solution (blue)**		**Brick-red precipitate**		

Tollens' reagent contains the complex diamminesilver(I) ion, $Ag(NH_3)_2^+$ (see Section 9.10). Tollens' reagent is **colourless.** It is also an oxidizing agent. On reduction, Ag^+ is reduced to metallic silver which is deposited on the walls of the test tube as a **silver mirror**.

Glucose or fructose as a **reducing sugar** gives a **positive Tollens' test** (sometimes called a silver mirror test).

Positive Tollens' test:

Glucose or Fructose	+	$Ag(NH_3)_2^+$	⟶	Ag ↓	+	Oxidation products
Reducing sugar		**Tollens' reagent (colourless)**		**Silver mirror**		

Nonreducing sugars

Carbohydrates that do not reduce Fehling's solution or Tollens' reagent are called nonreducing sugars.

Some disaccharides are reducing sugars, while some others are nonreducing sugars. For example, **maltose is a reducing sugar** but **sucrose is a nonreducing sugar**.

Thus sucrose as a nonreducing sugar gives a negative Fehling's or Tollens' test.

Negative Fehling's or Tollens' test:

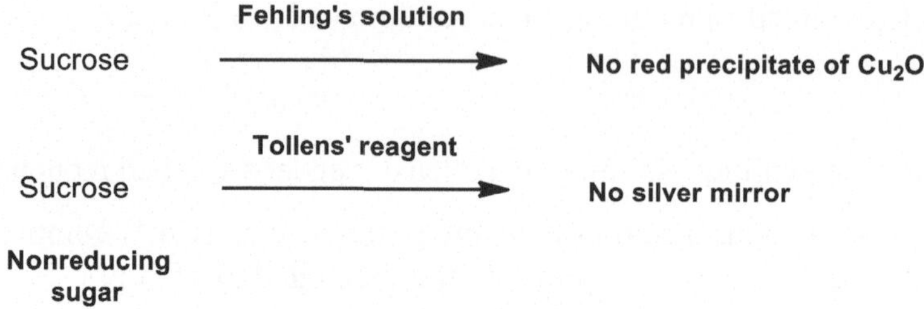

20.1.3 Monosaccharides: Aldose and Ketose

Monosaccharides are polyhydroxy aldehydes or polyhydroxy ketones. When a monosaccharide contains an **aldehyde** group, it is called an **aldose**. If a monosaccharide contains a **keto** group, it is called a **ketose**.

CHO
|
$(CHOH)_n$
|
CH_2OH

(polyhydroxy aldehyde)

Aldose

CH_2OH
|
C=O
|
$(CHOH)_m$
|
CH_2OH

(polyhydroxy ketone)

Ketose

Based on the **number of carbon atoms** present in the monosaccharide molecule, a monosaccharide is called a **triose** (three carbon atoms), **tetrose** (four carbon atoms), **pentose** (five carbon atoms), **hexose** (six carbon atoms), and so on.

Examples

Since a monosaccharide must contain more than one OH group, the *simplest* aldose is **aldotriose** which is glyceraldehyde, and the *simplest* ketose is **ketotriose** which is dihydroxyacetone.

CHO
|
CHOH
|
CH_2OH

Aldotriose
(glyceraldehyde)

Simplest aldose

CH_2OH
|
C=O
|
CH_2OH

Ketotriose
(dihydroxyacetone)

Simplest ketose

An aldose containing four carbon atoms is called an **aldotetrose**, i.e. it is a C_4 aldose. Thus, an **aldopentose** is a C_5 aldose, and an **aldohexose** is a C_6 aldose. Similarly, if a ketose contains six carbon atoms, it is called a **ketohexose** (a C_6 ketose).

The following are the structures of an aldopentose and a ketohexose:

CHO
|
CHOH
|
CHOH
|
CHOH
|
CH_2OH

Aldopentose
(C_5 aldose)

CH_2OH
|
C=O
|
CHOH
|
CHOH
|
CHOH
|
CH_2OH

Ketohexose
(C_6 ketose)

20.1.4 (+)-Glucose: Open-chain Structure and Configuration

Glucose (from the Greek *glykys*, meaning sweet) occurs in ripe grapes, honey and most sweet fruits, and is the major free sugar circulating in the blood of humans.

Naturally occurring **glucose** is **optically active** and is **dextrorotatory** (+), which is also known as dextrose. It is by far the most important and most abundant monosaccharide.

(+)-Glucose is an **aldohexose** and has the following structure:

CHO
|
CHOH
|
CHOH
|
CHOH
|
CHOH
|
CH_2OH

(+)-Glucose
An aldohexose

This structure of (+)-glucose is its **open-chain structure**, which is consistent with the following observations:

1. The molecular formula of glucose is $C_6H_{12}O_6$.
2. On prolonged heating with HI, glucose forms **hexane**, a straight-chain C_6 alkane. This indicates that *six carbon atoms of glucose are in a straight chain.*

$$\underset{\textbf{(+)-Glucose}}{\begin{array}{c} CHO \\ | \\ (CHOH)_4 \\ | \\ CH_2OH \end{array}} \xrightarrow[\Delta]{HI} \underset{\textbf{Hexane}}{CH_3CH_2CH_2CH_2CH_2CH_3}$$

3. Acetylation of glucose with acetic anhydride gives **glucose pentaacetate**. This indicates that *glucose contains five OH groups*.

   Since glucose is a stable compound, no carbon atom is attached to more than one OH group (because a *gem*-diol with two OH groups on the same carbon is unstable). Therefore, the five OH groups of glucose are attached to five different carbon atoms (see later Section 20.1.8).

$$\underset{\textbf{(+)-Glucose}}{\begin{array}{c} CHO \\ | \\ (CHOH)_4 \\ | \\ CH_2OH \end{array}} \xrightarrow{(CH_3CO)_2O} \textbf{Glucose pentaacetate}$$

$$\left[-OH \xrightarrow{(CH_3CO)_2O} \underset{\textbf{An acetate}}{-O-\overset{\overset{O}{\|}}{C}-CH_3} \right]$$

4. Glucose reacts with hydroxylamine to form an **oxime**. This indicates the *presence of a carbonyl group (C=O) in glucose.*

$$\underset{\textbf{(+)-Glucose}}{\begin{array}{c} H-C=O \\ | \\ (CHOH)_4 \\ | \\ CH_2OH \end{array}} \xrightarrow{NH_2OH} \underset{\textbf{Glucose oxime}}{\begin{array}{c} HC=NOH \\ | \\ (CHOH)_4 \\ | \\ CH_2OH \end{array}}$$

5. When glucose is oxidized with a mild oxidizing agent like bromine water, it forms **gluconic acid** (in which five OH groups remain intact). This indicates that the *carbonyl group in glucose is present as an aldehyde group (–CHO).*

$$\underset{\textbf{(+)-Glucose}}{\begin{array}{c} CHO \\ | \\ (CHOH)_4 \\ | \\ CH_2OH \end{array}} \xrightarrow[Br_2/H_2O]{[O]} \underset{\textbf{Gluconic acid}}{\begin{array}{c} CO_2H \\ | \\ (CHOH)_4 \\ | \\ CH_2OH \end{array}}$$

6. Both glucose and gluconic acid on oxidation with nitric acid gives **glucaric acid** (or **saccharic acid**). Glucaric acid is a dicarboxylic acid with terminal carboxyl groups. Therefore, *glucose has a terminal primary alcoholic group (CH_2OH).*

$$\underset{\text{(+)-Glucose}}{\begin{array}{c}CHO\\|\\(CHOH)_4\\|\\CH_2OH\end{array}} \xrightarrow[HNO_3]{[O]} \underset{\text{Glucaric acid (saccharic acid)}}{\begin{array}{c}CO_2H\\|\\(CHOH)_4\\|\\CO_2H\end{array}} \xleftarrow[HNO_3]{[O]} \underset{\text{Gluconic acid}}{\begin{array}{c}CO_2H\\|\\(CHOH)_4\\|\\CH_2OH\end{array}}$$

PROBLEM 20.1: Give structures for compounds **A–C** in the following scheme:

$$\text{(+)-Glucose} \xrightarrow{HCN} \textbf{A} \xrightarrow[H^+]{H_2O} \textbf{B} \xrightarrow[\Delta]{HI} \textbf{C}$$

Answer

Glucose has an aldehyde group and therefore reacts with HCN to form a cyanohydrin (**A**), which on hydrolysis gives a hexahydroxy acid (**B**). Reduction of **B** with HI produces heptanoic acid (**C**).

$$\underset{\text{(+)-Glucose}}{\begin{array}{c}CHO\\|\\(CHOH)_4\\|\\CH_2OH\end{array}} \xrightarrow{HCN} \underset{\textbf{(A)}}{\begin{array}{c}CN\\|\\CHOH\\|\\(CHOH)_4\\|\\CH_2OH\end{array}} \xrightarrow[H^+]{H_2O} \underset{\textbf{(B)}}{\begin{array}{c}CO_2H\\|\\CHOH\\|\\(CHOH)_4\\|\\CH_2OH\end{array}} \xrightarrow[\Delta]{HI} \underset{\textbf{(C)}\ \text{Heptanoic acid}}{\begin{array}{c}CO_2H\\|\\(CH_2)_5\\|\\CH_3\end{array}}$$

Configuration (+)-glucose

(+)-Glucose is an aldohexose. The aldohexose structure possesses four *unlike* chiral centres. Therefore, the **total number of stereoisomers** possible for the aldohexose structure = $\mathbf{2^4}$ = **16** (see Section 4.6.2).

$$\begin{array}{c}CHO\\|\\H-\overset{*}{C}-OH\\|\\H-\overset{*}{C}-OH\\|\\H-\overset{*}{C}-OH\\|\\H-\overset{*}{C}-OH\\|\\CH_2OH\end{array}$$

Four unlike chiral centres

No. of stereoisomers = 2^4 = 16

Aldohexose structure

The 16 stereoisomers arise from 8 chiral diastereomers, each of which has 2 enantiomers.

Glucose is *one* of the eight diastereomers, and it exists as **(+)-glucose** and **(–)-glucose**. Note that (–)-glucose is the enantiomer of the naturally occurring (+)-glucose.

(It may be mentioned here that the set of eight diastereomers consists of allose, altrose, glucose, mannose, gulose, idose, galactose and talose, each of which can exist in (+) and (–) forms.)

In general, the configuration of a sugar is designated by D or L based on the OH group at the highest numbered chiral centre in the Fischer projection formula (see Section 4.8).

The configuration of (+)-glucose has been assigned as D, i.e. the OH at C5 is on the right. Thus, the naturally occurring glucose is written as **D-(+)-glucose**, where D stands for configuration and (+) stands for optical rotation.

1 CHO
2 CHOH
3 CHOH
4 CHOH
H—5—OH ← **(on the right)**
CH_2OH

D Configuration of (+)-glucose

In 1888, Emil Fischer set out to work on the stereochemistry of (+)-glucose and finally assigned the **configuration of the chiral centres C2, C3 and C4** relative to the configuration at C5. He was awarded the Nobel Prize in 1902 for his work.

The complete configuration of D-(+)-glucose is

CHO
H—2—OH
HO—3—H
H—4—OH
H—5—OH
CH_2OH

D-(+)-Glucose

Complete Configuration

20.1.5 Epimers of D-glucose: D-Mannose, D-allose and D-galactose

By definition, epimers are diastereomers that differ in the configuration at *only one* chiral centre.

However, **D-aldohexoses** have the **same configuration at C5**. So the possible **epimers** of D-glucose can arise from the differing configuration only at C2 or C3 or C4.

The **C2 epimer** of D-glucose that differs only in the configuration at C2 is called **D-mannose.**

Similarly, **D-allose** is the **C3 epimer** that differs only in the configuration at C3, and **D-galactose** is the **C4 epimer** that differs only in the configuration at C4.

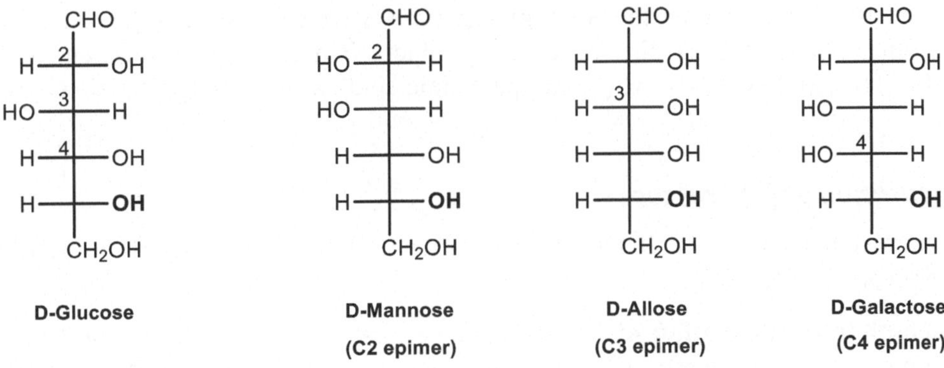

Three epimers of D-glucose

20.1.6 (–)-Fructose: Open-Chain Structure and Configuration

Fructose is a monosaccharide that occurs as a component of fruit juices and honey. It is formed along with glucose by hydrolysis of sucrose or cane sugar. Naturally occurring **fructose** is **optically active** and is **levorotatory (–)**.

(–)-Fructose has the same molecular formula ($C_6H_{12}O_6$) as for (+)-glucose. But unlike glucose which is an aldohexose, fructose is a ketohexose. The position of keto group in fructose is C2; therefore, (–)-fructose is a **2-ketohexose**.

Although fructose is a ketone (not an aldehyde), it *reduces Fehling's solution or Tollens' reagent* similar to (+)-glucose. This is because fructose is not an ordinary ketone but possesses OH group adjacent (or α) to the keto group, i.e. **fructose** is an **α-hydroxy ketone**. Like aldehydes, α-hydroxy ketones can also reduce Fehling's solution or Tollens' reagent. Therefore, fructose is a reducing sugar.

The **open-chain structure** of (–)-fructose as a 2-ketohexose is consistent with the following scheme (cf. Problem 20.1):

$^1CH_2OH$		CH_2OH		CH_2OH		CH_3
$^2C{=}O$		$HO{-}C{-}CN$		$HO{-}C{-}CO_2H$		$\overset{2}{H}C{-}\overset{1}{C}O_2H$
^3CHOH	$\xrightarrow{HCN}$	$CHOH$	$\xrightarrow[H^+]{H_2O}$	$CHOH$	$\xrightarrow[\Delta]{HI}$	CH_2
^4CHOH		$CHOH$		$CHOH$		CH_2
^5CHOH		$CHOH$		$CHOH$		CH_2
$^6CH_2OH$		CH_2OH		CH_2OH		CH_3
(-)-Fructose ***2-Ketohexose***		**A cyanohydrin**		**A polyhydroxy acid**		**2-Methylhexanoic acid**

Note that the placement of the **keto group (C=O) at C2** (not at C3) could lead to 2-methylhexanoic acid as the observed product. [Check that a different product, $CH_3CH_2CH_2CH(CH_2CH_3)CO_2H$ (2-ethylpentanoic acid) would result if the C=O group is at C3.]

Configuration of (–)-Fructose

(–)-Fructose belongs to the **D family** of sugars, i.e. the OH at C5 is on the right in the Fischer projection formula.

The **complete configuration of D-(–)-Fructose** is shown below. It has the same configurations at the chiral centres C3, C4 and C5 as for D-(+)-glucose (see Problem 20.3).

D-(-)-Fructose		D-(+)-Glucose
1 CH_2OH		CHO
2 $C{=}O$		$H{-}C{-}OH$
3 $HO{-}C^*{-}H$	cf.	3 $HO{-}C{-}H$
4 $H{-}C^*{-}OH$		4 $H{-}C{-}OH$
5 $H{-}C^*{-}OH$		5 $H{-}C{-}OH$
6 CH_2OH		CH_2OH

D- (-)-Fructose **D-(+)-Glucose**

20.1.7 Reactions of Monosaccharides: Oxidation, Oxidative Cleavage, Reduction and Osazone Formation

Oxidation

Dilute HNO_3 oxidizes *both* CHO and the terminal CH_2OH group of an aldose to CO_2H groups. These dicarboxylic acids are called **aldaric acids.**

For example, **D-glucose** on oxidation with HNO_3 gives **D-glucaric acid** (also called D-saccharic acid). D-Glucaric acid has no plane of symmetry and is **optically active**.

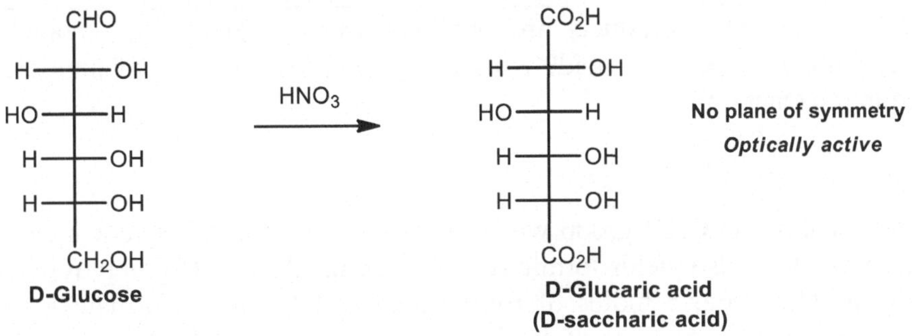

CHO / H—OH / HO—H / H—OH / H—OH / CH_2OH
D-Glucose

$\xrightarrow{HNO_3}$

CO_2H / H—OH / HO—H / H—OH / H—OH / CO_2H
D-Glucaric acid
(D-saccharic acid)

No plane of symmetry
Optically active

Oxidative cleavage

Oxidation of an aldose or a ketose with aqueous periodic acid (HIO_4) leads to cleavage of C–C bonds with formation of aldehydes, ketones or acids. **Oxidative cleavage** takes place when an OH group is **adjacent** to another OH group (or a CHO or C=O group).

For example, consider glyceraldehyde, the simplest aldose. Oxidation of **1 mole of glyceraldehyde** with periodic acid yields **2 moles of formic acid** and **1 mole of formaldehyde**. In the process, **2 moles of periodic acid** are consumed.

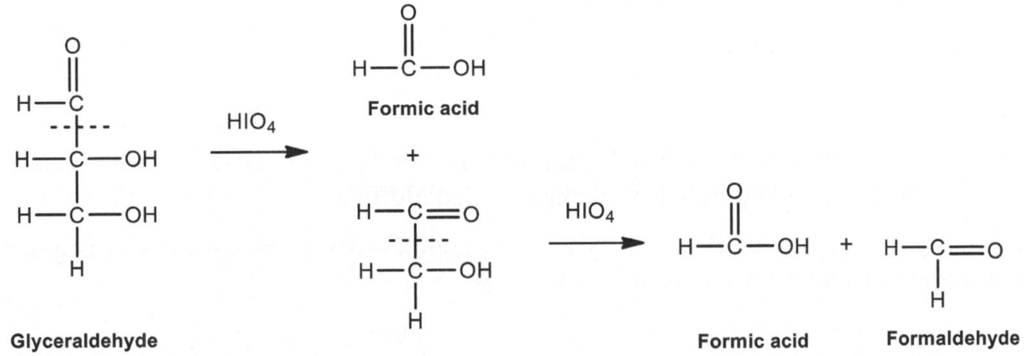

Glyceraldehyde (H—C(=O) - - - H—C—OH — H—C—OH — H) $\xrightarrow{HIO_4}$ H—C(=O)—OH (**Formic acid**) + H—C=O - - - H—C—OH — H $\xrightarrow{HIO_4}$ H—C(=O)—OH (**Formic acid**) + H—C(=O)—H (**Formaldehyde**)

In short, the above scheme is written as:

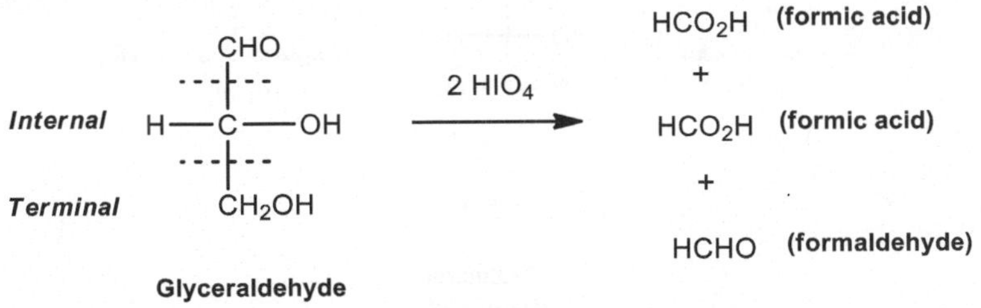

Internal / ***Terminal***: CHO - - - H—C—OH - - - CH_2OH (**Glyceraldehyde**) $\xrightarrow{2\ HIO_4}$ HCO_2H **(formic acid)** + HCO_2H **(formic acid)** + HCHO **(formaldehyde)**

In the oxidative cleavage, CHO is oxidized to formic acid and the terminal CH_2OH is oxidized to formaldehyde. Further, the internal CHOH is ultimately converted to formic acid.

This is how oxidative cleavage of glyceraldehyde can also be written (see above).

Note: The periodic acid oxidation of sugars is generally quantitative, and is used for structural determination of sugars by measuring the number of moles of periodic acid consumed and determining the yields of the products (formaldehyde and formic acid).

PROBLEM 20.2: What products are obtained when D-glucose is treated with an appropriate amount of periodic acid? How many molar equivalents of periodic acid are consumed in the reaction?

Answer

D-Glucose contains one CHO group which gives formic acid. It has four internal CHOH groups, each of which also yields formic acid. The terminal CH_2OH group of glucose gives formaldehyde. Therefore, 5 moles of formic acid and 1mole of formaldehyde will be produced. In the process, 5 moles of periodic acid are consumed for 1 mole of D-glucose.

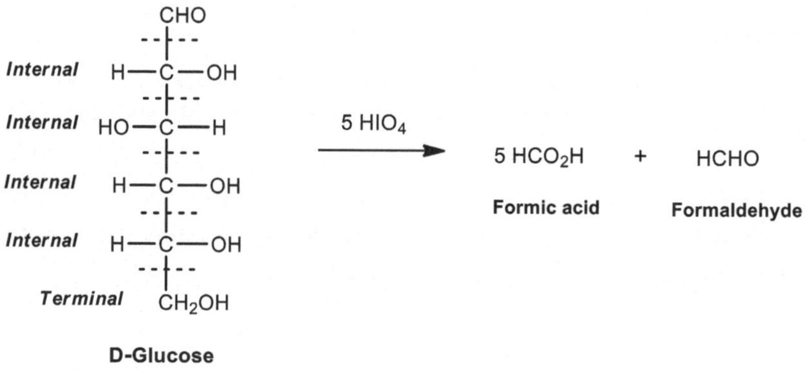

CHO
Internal H—C—OH
Internal HO—C—H
Internal H—C—OH
Internal H—C—OH
Terminal CH_2OH

D-Glucose

$\xrightarrow{5\ HIO_4}$ $5\ HCO_2H$ + HCHO

Formic acid Formaldehyde

Reduction

Aldoses or ketoses can be reduced with sodium borohydride ($NaBH_4$) or by catalytic hydrogenation (H_2/Pt) to polyhydric alcohols called **alditols**.

For example, reduction of **D-glucose** yields **D-glucitol** (also called D-sorbitol). D-glucitol has no plane of symmetry and is **optically active**.

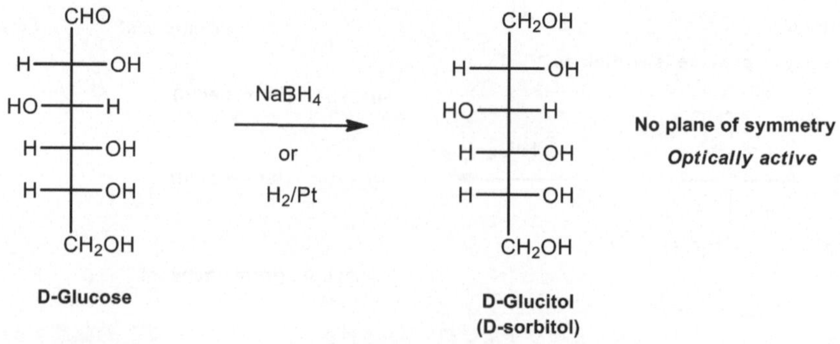

CHO / H—OH / HO—H / H—OH / H—OH / CH_2OH

D-Glucose

$\xrightarrow[\text{or } H_2/Pt]{NaBH_4}$

CH_2OH / H—OH / HO—H / H—OH / H—OH / CH_2OH

D-Glucitol
(D-sorbitol)

No plane of symmetry
Optically active

Osazone formation

Aldehydes or ketones react with phenylhydrazine to form phenylhydrazones. However, when an **aldose** containing an aldehyde group is treated with *excess* phenylhydrazine, a product called **osazone** is formed.

Osazone contains **two phenylhydrazone groups at C1 and C2**. Overall, the reaction consumes 3 moles of phenylhydrazine forming osazone, aniline and ammonia.

$$\underset{\text{Aldose}}{\overset{1}{C}H{=}O-\overset{2}{C}HOH-(CHOH)_n-CH_2OH} + \underset{\text{Phenylhydrazine}}{3\ C_6H_5NHNH_2} \longrightarrow \underset{\text{Osazone}}{\overset{1}{C}H{=}NNHC_6H_5-\overset{2}{C}{=}NNHC_6H_5-(CHOH)_n-CH_2OH} + C_6H_5NH_2 + NH_3 + 2\ H_2O$$

Osazones are crystalline derivatives of sugars which are useful for identifying them.

Since osazone formation involves C1 and C2 of an aldose, the reaction leads to a **loss of the chiral centre at C2** but does not affect other chiral centres. Thus two aldoses that differ in configuration only at C2 will give the same osazone.

For example, **D-glucose** and its C2 epimer **D-mannose** give the **same osazone** because they have the same configurations at C3, C4 and C5.

D-Glucose		Same osazone		D-Mannose
1 CHO		$CH{=}NNHC_6H_5$		1 CHO
H–2–OH	$3\ C_6H_5NHNH_2$ →	$C{=}NNHC_6H_5$	← $3\ C_6H_5NHNH_2$	HO–2–H
HO–3–H		HO–H		HO–3–H
H–4–OH		H–OH		H–4–OH
H–5–OH		H–OH		H–5–OH
CH_2OH		CH_2OH		CH_2OH

PROBLEM 20.3: D-Fructose reacts with excess phenylhydrazine to form the same osazone as that obtained from D-glucose. Predict the complete configuration of D-fructose.

Answer

Osazone formation affects only the chiral centre C2 of D-glucose. D-fructose has only three chiral centres at C3, C4 and C5. Therefore, since D-fructose and D-glucose yield the same osazone, D-fructose has the same configurations at C3, C4 and C5 as for D-glucose.

D-Fructose		Same osazone		D-Glucose
CH_2OH		$CH{=}NNHPh$		CHO
$C{=}O$	$PhNHNH_2$ → ($Ph = C_6H_5$)	$C{=}NNHPh$	← $PhNHNH_2$	H–OH
HO–3–H		HO–H		HO–3–H
H–4–OH		H–OH		H–4–OH
H–5–OH		H–OH		H–5–OH
CH_2OH		CH_2OH		CH_2OH

20.1.8 D-(+)-Glucose: Cyclic Structure, Mutarotation and Glycoside Formation

The open-chain structure of D-(+)-glucose can explain *some* of its properties but *not all* of its properties. The following are some observations that cannot be explained by the open-chain structure.

(a) Careful crystallization of glucose from water under different conditions leads to *two forms* of (+) glucose (α and β) that have different values for optical rotation.

Crystallization from aqueous solution below 50°C gives **α-D-glucose** (specific rotation $[\alpha]_D = +112$), whereas crystallization by evaporation of saturated aqueous solution kept above 98°C produces **β-D-glucose** ($[\alpha]_D = +18.7$).

(b) The pentaacetate of glucose does not form an oxime with NH_2OH, which indicates the *absence of free aldehyde group (–CHO)* in the pentaacetate.

(c) The reaction of glucose with methanol under acidic conditions gives *two isomeric products*, which are acetals.

To account for the above observations, a **cyclic structure** for D-(+)-glucose has been proposed.

Cyclic structure

The **cyclic structure** of D-(+)-glucose can be formed by an intramolecular **hemiacetal reaction** between the OH group at C5 and the aldehyde group. The reaction creates a **new chiral centre at C1** and **two cyclic forms** are thus possible, differing only in the configuration at C1.

These cyclic forms are two diastereomers that differ only in the configuration at C1, i.e. they are **C1 epimers**. In carbohydrate chemistry, the **hemiacetal carbon** (C1) is called the **anomeric carbon**, and the diastereomers or epimers differing only in the configuration at the anomeric carbon (C1) are called **anomers**.

The two cyclic forms of D-(+)-glucose are called **α anomer** (i.e. **α-D-(+)-glucose**) and **β anomer** (i.e. **β-D-(+)-glucose**). Note that in the Fischer projection formula, OH at C1 is on the right in the α anomer while OH at C1 is on the left in the β anomer.

In aqueous solution, α and β anomers of glucose exist in equilibrium with the open-chain structure.

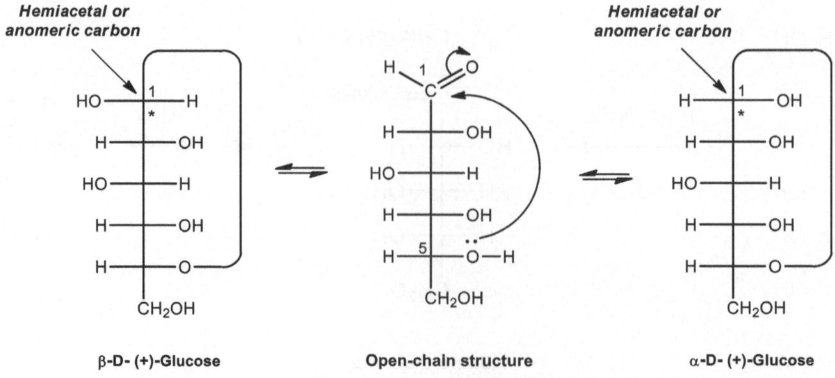

Haworth formulas

The cyclic structure of glucose is depicted above as a Fischer projection. The cyclic structure can also be represented using a hexagon, called a **Haworth formula**, named after the English chemist Walter N. Haworth who received the Nobel Prize in chemistry in 1937 for his work in carbohydrate chemistry.

In the Haworth formula, the **anomeric carbon** (C1) is located *clockwise* from the ring oxygen, and the ring is numbered in a clockwise direction from C1. The **substituents** are placed on *vertical bond lines* passing through each carbon atom.

The relationships between the Fischer projection and the Haworth formula of **α-D-(+)-glucose** are shown in Figure 20.1.

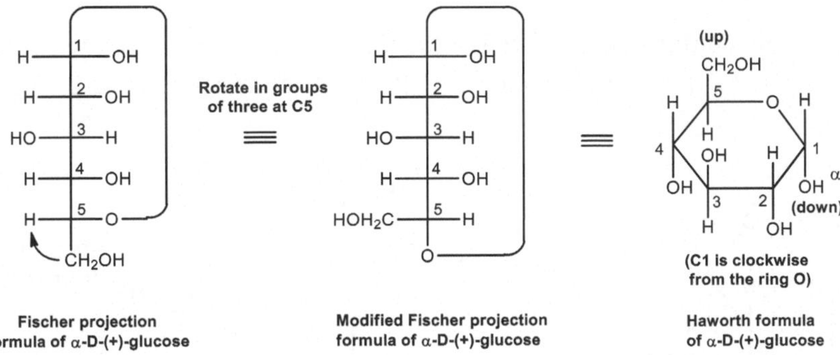

Figure 20.1: Relationships between the Fischer projection formula and the Haworth formula of α-D-(+)-glucose.

As shown, the atoms or groups on the same side (right side) of **modified Fischer projection formula** remain on the same side (lower face) of the ring in the **Haworth formula**.

Thus, in the **α anomer**, OH at C1 is *down* while CH_2OH at C5 is *up* in the Haworth formula, i.e. they are *trans*. (Consequently, in the β anomer, OH at C1 will be *up*, being *cis* to CH_2OH at C5).

To designate the **ring size** of a monosaccharide, a system of **nomenclature** has been introduced. If a monosaccharide ring is **six-membered**, it is called a **pyranose** in analogy with pyran (a six-membered heterocyclic compound with one oxygen atom in the ring).

The six-membered cyclic structure of glucose is therefore called a **pyranose structure**. Thus the full names of the anomers of glucose are **α-D-(+)-glucopyranose** and **β-D-(+)-glucopyranose**.

Pyran | α-D-(+)-Glucopyranose | β-D-(+)-Glucopyranose

PROBLEM 20.4: Draw the conformational formulas of α-D-(+)-glucose and β-D-(+)-glucose.

Answer

Like cyclohexane, the six-membered ring of glucose is best represented by a **chair conformation** (see Section 4.13.3). The relationships between the Haworth formulas and the conformational formulas are shown below.

In the chair conformation, the anomeric carbon (C1) is located *clockwise* from the ring oxygen, as in the Haworth formula. Then the atoms or groups which are **up** in the Haworth formula are also drawn **up** in the conformational formula, and similarly for the down atoms or groups.

Thus, in the chair conformation of **α-D-glucose**, the OH at C1 is down which is axial, whereas the OH at C1 is up and equatorial in the chair conformation of **β-D-glucose**.

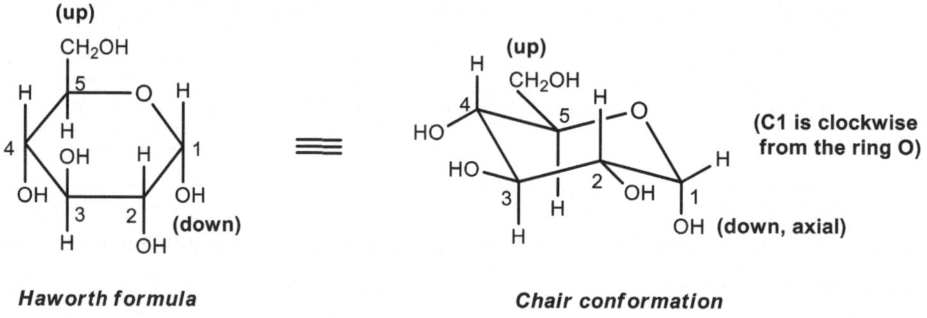

(up) CH_2OH … OH (down) ≡ (up) CH_2OH … (C1 is clockwise from the ring O) … OH (down, axial)

Haworth formula *Chair conformation*

α-D-(+)-Glucose

(up) CH_2OH … (up) OH ≡ (up) CH_2OH … (up, equatorial) OH

Haworth formula *Chair conformation*

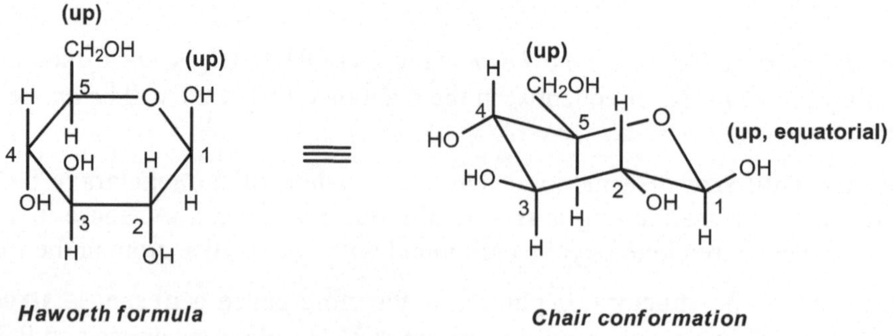

β-D-(+)-Glucose

Mutarotation

Both α anomer and β anomer of D-(+)-glucose have been isolated. Either anomer of glucose slowly **changes optical rotation** in aqueous solution to the same final value.

A solution of **α-D-(+)-glucose** has an initial specific rotation of +112. When the aqueous solution is allowed to stand, its specific rotation *decreases* and reaches a final value of

+52.7. On the other hand, an aqueous solution of **β-D-(+)-glucose** has an initial specific rotation of +18.7. On standing, its specific rotation slowly *increases* to +52.7. This change of rotation of either form of glucose toward a final value is called **mutarotation**.

Mutarotation occurs in aqueous solution because the α anomer is converted to the β anomer (or the β anomer is converted to the α anomer) *through the open-chain form* (see above). An equilibrium is ultimately established between the α form and the β form. It is the specific rotation of the *equilibrium mixture* of α and β forms that is +52.7.

The concentration of the open-chain form at equilibrium is extremely small. Calculation shows that about 36% α anomer and 64% β anomer are present at equilibrium.

Glycoside formation

D-Glucose reacts with methanol in the presence of HCl to form **acetals**. The carbohydrate acetals are called **glycosides**. A glycoside of glucose is also known as a **glucoside**.

The **glycoside formation** involves the cyclic structure of D-glucose, which itself is a hemiacetal. Therefore, only *one mole* of methanol is required to form glycosides (acetals).

The reaction between the anomeric OH (or hemiacetal OH) attached to the anomeric carbon (C1) of glucose and methanol (CH_3OH) in the presence of HCl produces a methyl glucoside (or methyl acetal).

Since a solution of D-glucose contains a mixture of α-D-glucopyranose and β-D-glucopyranose, the product is a mixture of **methyl α-D-glucopyranoside** and **methyl β-D-glucopyranoside**.

Nonreducing sugars

Note: Methyl glucosides do not give a positive test with Fehling's solution or Tollens' reagent; therefore, they are *nonreducing sugars*. This is because glycosides as acetals are stable in the basic solution of the test reagents; hence they cannot undergo ring opening to form the open-chain structure having the free CHO group responsible for the reducing property of sugars.

Formation of glucose pentaacetate

D-Glucose reacts with acetic anhydride to form **glucose pentaacetate**. The reaction takes place via the cyclic structure of glucose and involves the acetylation of all OH groups including the anomeric OH.

The product is a mixture of α and β anomers of the pentaacetate resulting from a mixture of α-D-glucose and β-D-glucose

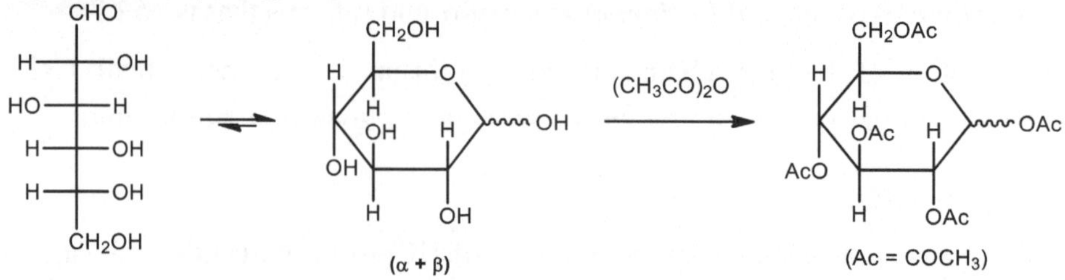

$(CH_3CO)_2O$

(α + β)

(Ac = $COCH_3$)

D-Glucose

Glucose pentaacetate
(α + β)

Note: Since the anomeric OH is protected by acetylation, the ring opening of glucose pentaacetate does not take place to form the open-chain structure with the free aldehyde group. This is why glucose pentaacetate does not react with NH_2OH to form an oxime.

20.1.9 D-(−)-Fructose: Cyclic Structure

D-(−)-Fructose can exist in two cyclic forms that result from intramolecular hemiacetal reaction of the OH group at C5 with the keto group (C=O) at C2. Thus **the hemiacetal or anomeric carbon of fructose is C2**.

The cyclic structure of D-fructose has a **five-membered ring**. The two cyclic forms are α and β anomers, i.e. **α-D-(−)-fructose** and **β-D-(−)-fructose**.

These are shown below in Fischer projection formulas.

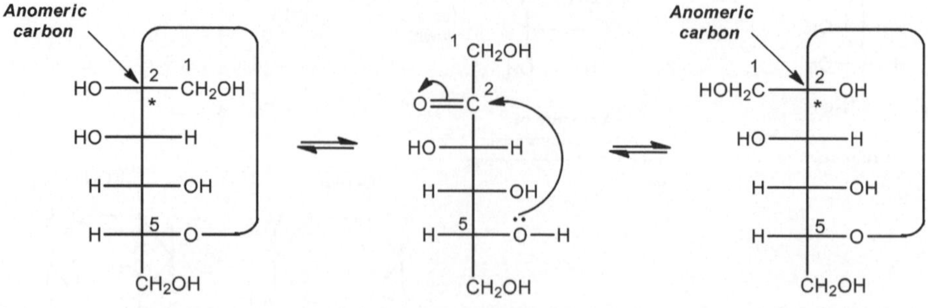

Anomeric carbon

β-D-(-)-Fructose **Open-chain structure** **α-D-(-)-Fructose**

Haworth formula

The five-membered cyclic structure of fructose can be represented by a **Haworth formula** using a pentagon, in which the anomeric carbon (C2) is located *clockwise* from the ring oxygen.

The relationships between the Fischer projection and the Haworth formula of **α-D-(−)-fructose** are shown below (cf. Figure 20.1).

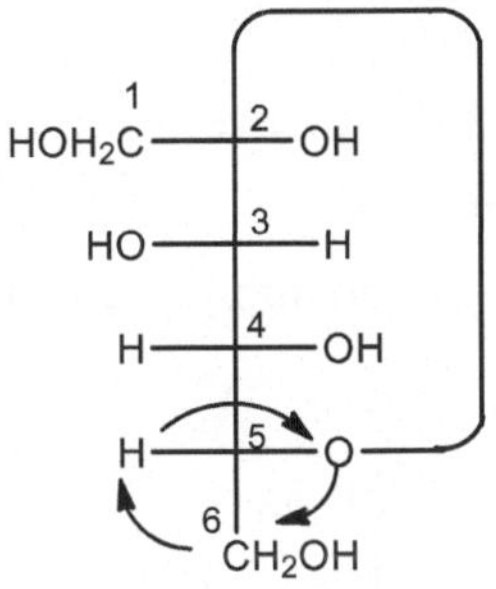

Fischer projection of α-D-(-)-Fructose

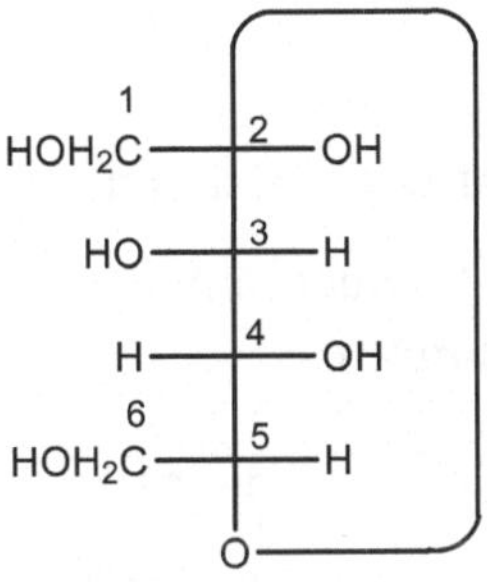

Modified Fischer projection of α-D-(-)-Fructose

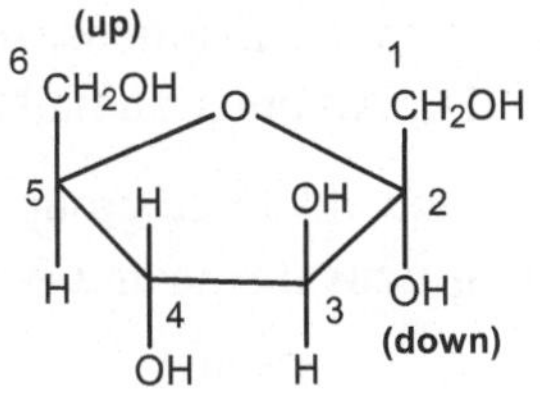

(C2 is clockwise from the ring O)

Haworth formula of α-D-(-)-Fructose

In **α-D-(−)-fructose**, OH at C2 (anomeric carbon) is *down* while CH_2OH at C5 is *up* in the Haworth formula, i.e. they are *trans*. (Consequently, in the β anomer, OH at C2 will be *up*, being *cis* to CH_2OH at C5).

The five-membered ring in fructose is called a **furanose** ring, in analogy with furan. Thus the five-membered cyclic structure of fructose is called a **furanose structure**.

The full names of the two anomers of fructose are **α-D-(−)-fructofuranose** and **β-D-(−)-fructofuranose**.

Furan

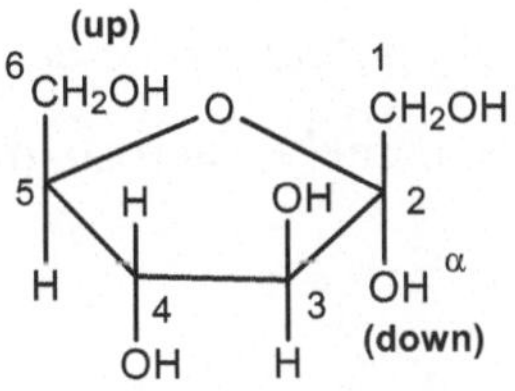

α-D-(-)-Fructofuranose

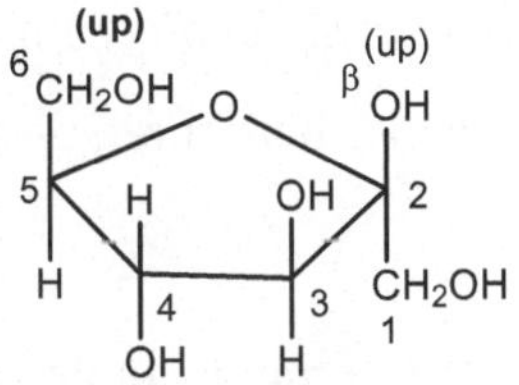

β-D-(-)-Fructofuranose

20.1.10 Glycosylamines and Amino Sugars

A glycosylamine is a sugar in which the anomeric OH is replaced by an amino group (NH_2).

For example, the replacement of the anomeric OH of α-D-glucose and β-D-glucose by NH_2 gives **α-D-glucosylamine** and **β-D-glucosylamine** respectively.

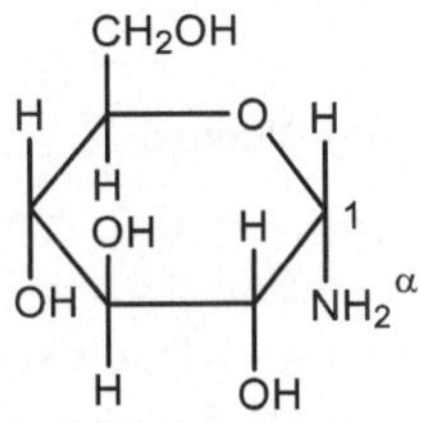

α-D-Glucosylamine

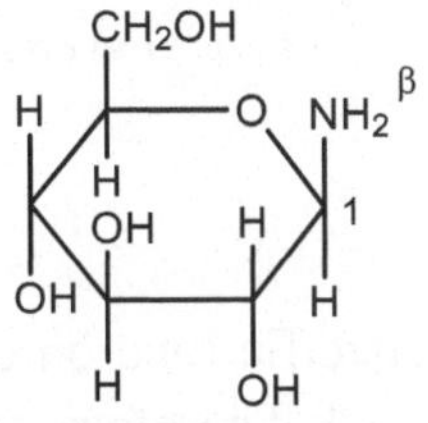

β-D-Glucosylamine

Amino sugars

An amino sugar is a sugar in which a nonanomeric OH is replaced by an amino group (NH_2).

For example, **D-glucosamine** is an amino sugar in which OH at **C2** of glucose is replaced by NH_2. An open-chain structure of D-glucosamine is shown below.

Since the aldehyde group is free in D-glucosamine, it can also exist in two anomeric forms, **α-D-glucosamine** and **β-D-glucosamine**.

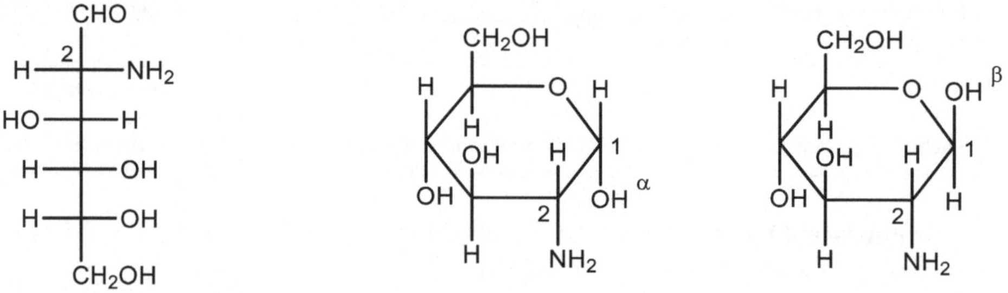

D-Glucosamine

Open-chain structure

α-D-Glucosamine

β-D-Glucosamine

20.1.11 Disaccharides: Sucrose, Maltose and Lactose

In **disaccharides**, *two* monosaccharide units are held together by a glycosidic linkage. A **glycosidic linkage** (–O–) is formed by a reaction of the **anomeric OH** of *one monosaccharide molecule* with **any OH** of *another molecule.*

We will consider here three disaccharides, namely, sucrose, maltose and lactose.

Sucrose

Sucrose is the common table sugar. It is the most widely occurring disaccharide and is commercially obtained from sugar cane or sugar beets.

Sucrose, also called **cane sugar,** has the molecular formula $C_{12}H_{22}O_{11}$ and is **dextrorotatory** (+). It has a specific rotation of +66.5.

Hydrolysis of (+)-sucrose with aqueous acid or by the action of the enzyme *invertase* (from yeast) gives equal amounts of **D-(+)-glucose** and **D-(–)-fructose.**

$$C_{12}H_{22}O_{11} \xrightarrow[\textit{Inversion of sucrose}]{H_3O^+ \text{ or } \textit{Invertase}} C_6H_{12}O_6 + C_6H_{12}O_6$$

(+)-Sucrose	**(+)-Glucose**	**(-)-Fructose**
$[\alpha]_D$ **= + 66.5**	$[\alpha]_D$ **= + 52.7**	$[\alpha]_D$ **= - 92.4**

D-(+)-glucose has a specific rotation of +52.7, while D-(–)-fructose has a large negative specific rotation of –92.4. Therefore, the mixture of D-(+)-glucose and D-(–)-fructose has a net negative rotation and is **levorotatory**. Since the hydrolysis causes a *change* in the sign of rotation from positive to negative, the reaction is often called the **inversion of sucrose**.

The levorotatory mixture is known as **invert sugar**. Honey is mostly invert sugar; the enzyme *invertase* is supplied by the bees.

In sucrose, glucose and fructose units are held together by a **glycosidic linkage** between **C1** (anomeric carbon) of **α-D-glucose** and **C2** (anomeric carbon) of **β-D-fructose**. This glycosidic linkage is designated as **α1,β2** linkage.

The following is the structure of sucrose:

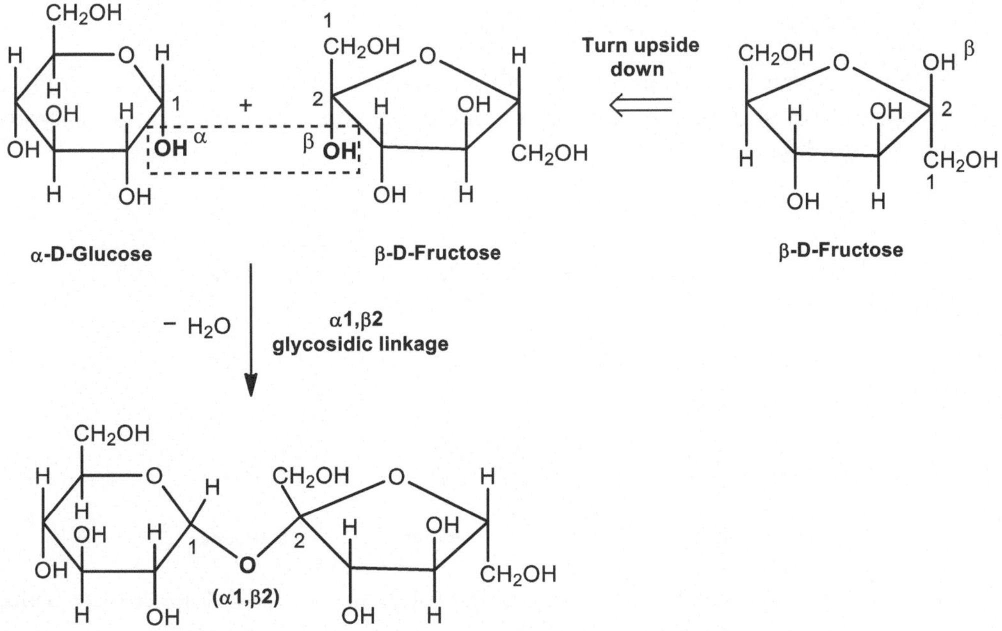

α-D-Glucose + **β-D-Fructose** ⟸ Turn upside down **β-D-Fructose**

$- H_2O$ ↓ **α1,β2 glycosidic linkage**

(α1,β2)

Sucrose

Since sucrose has *no* free anomeric OH for either glucose or fructose unit, it cannot exist in equilibrium with the open-chain structure of glucose or fructose unit having the free aldehyde or α-hydroxy keto group. Therefore, sucrose is a **nonreducing sugar**.

Maltose

Maltose is another disaccharide of molecular formula $C_{12}H_{22}O_{11}$, which is **dextrorotatory (+)**. It can be obtained by hydrolysis of starch by the action of the enzyme *diastase*.

Hydrolysis of (+)-maltose ($C_{12}H_{22}O_{11}$) with aqueous acid gives **two molecules of D-(+)-glucose** ($C_6H_{12}O_6$) (see Section 20.1.1).

Two glucose units in maltose are held together by an **α1,4 glycosidic linkage**, i.e. the linkage between **C1** of **α-D-glucose** and **C4** of a *second* **D-glucose (α or β anomer)**.

Therefore, maltose can exist as two anomers, **α-(+)-maltose** (in which the second glucose is α-D-glucose) and **β-(+)-maltose** (in which the second glucose is β-D-glucose).

The structure of the β anomer of maltose, i.e. **β-(+)-maltose** is shown below. (Try to draw the structure α-(+)-maltose).

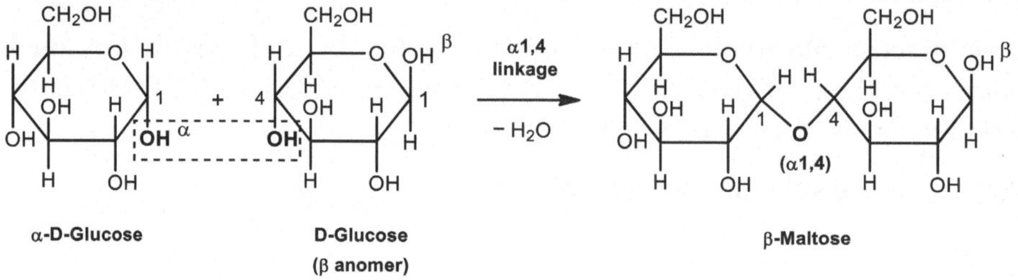

α-D-Glucose + D-Glucose (β anomer) —(α1,4 linkage, − H_2O)→ β-Maltose

Since maltose has one glucose unit with a free anomeric OH, it can exist in equilibrium with the open-chain structure of that unit having the free aldehyde group. Therefore, maltose is a **reducing sugar**.

Lactose

Lactose is a disaccharide present in milk, and is commonly known as milk sugar. It is **dextrorotatory (+)**.

Hydrolysis of (+)-lactose with aqueous acid produces **D-(+)-galactose** (see Section 20.1.5) and **D-(+)-glucose**.

$$C_{12}H_{22}O_{11} \xrightarrow[H^+]{H_2O} C_6H_{12}O_6 + C_6H_{12}O_6$$

(+)-Lactose → D-Galactose + D-Glucose

In lactose, the two monosaccharide units are held together by a **β1,4 glycosidic linkage** between **C1** of **β-D-galactose** and **C4** of **D-glucose (α or β anomer)**. Thus lactose, like maltose, also exists as α and β anomers, i.e. **α-(+)-lactose** and **β-(+)-lactose**.

The structure of **α-(+)-lactose** is shown below. (Draw the structure of β-(+)-lactose).

β-D-Galactose + D-Glucose (α anomer) —(β1,4 linkage, − H_2O)→ α-Lactose

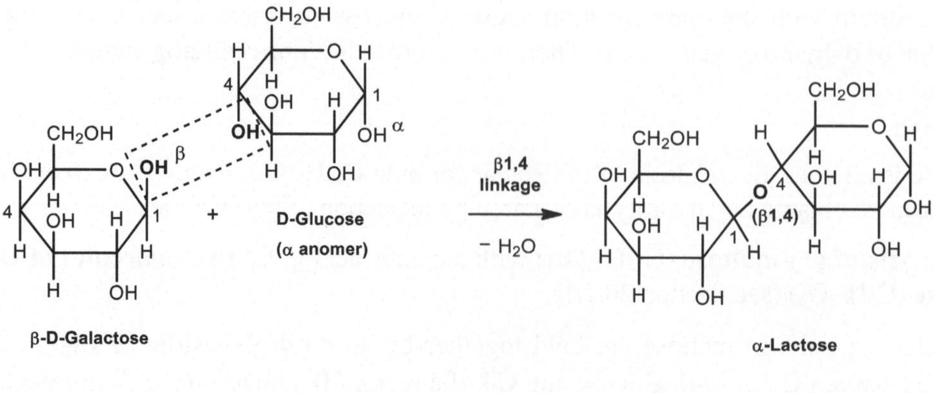

Since lactose contains a free hemiacetal OH on the glucose unit, it can exist in equilibrium with the open-chain structure of glucose unit having the free aldehyde group. Therefore, lactose is a **reducing sugar**.

20.1.12 Polysaccharides: Cellulose, Starch and Glycogen

Polysaccharides consist of a large number of monosaccharide units held together by glycosidic linkages. They are also known as **glycans**. The most important polysaccharides are cellulose, starch and glycogen.

Cellulose

Cellulose is the chief constituent of wood and plant fibres such as cotton. It occurs exclusively in plants and is a major component of cell wall of plant cells.

Cellulose has the molecular formula $(C_6H_{10}O_5)_n$. **Complete hydrolysis** of cellulose with aqueous acid yields **D-(+)-glucose** as the only monosaccharide.

The structure of cellulose is a *linear chain* of D-glucose units joined together by **β1,4 glycosidic linkages**, i.e. the linkage between C1 of one β-D-glucose unit and C4 of the next β-D-glucose unit.

β-D-Glucose + β-D-Glucose + etc. —β1,4 linkage→ Cellulose

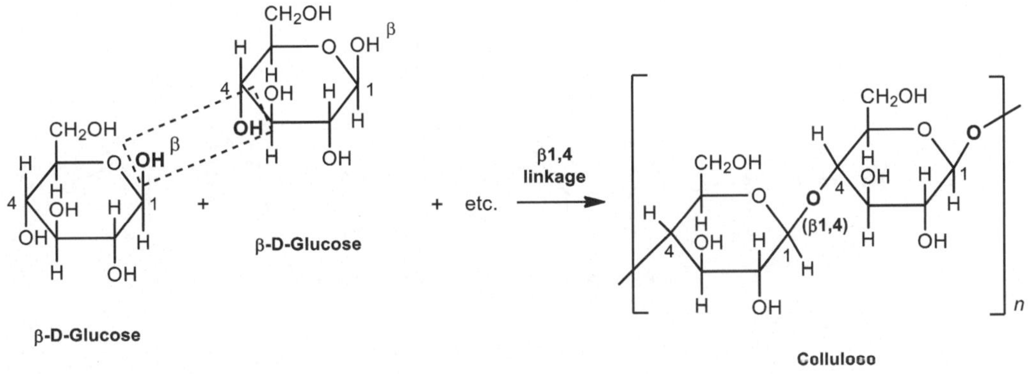

Starch

Starch is the principal food storage of plants. It occurs in the roots, tubers and seeds of plants. The important sources of starch are rice, wheat, potatoes and corn. It is the most important dietary source of humans.

Heating **starch** with water produces a colloidal suspension which consists of two main components, **amylose** and **amylopectin**. Amylose is *soluble* in water and constitutes 10–20% of most starches, whereas amylopectin is *insoluble* in water and constitutes 80–90% of starch.

Amylose is a polymer of **α-D-glucose** and its structure is a long *unbranched* chain of D-glucose units held together by **α1,4 glycosidic linkages**, i.e. the linkage between C1 of one α-D-glucose unit and C4 of the next α-D-glucose unit (cf. cellulose).

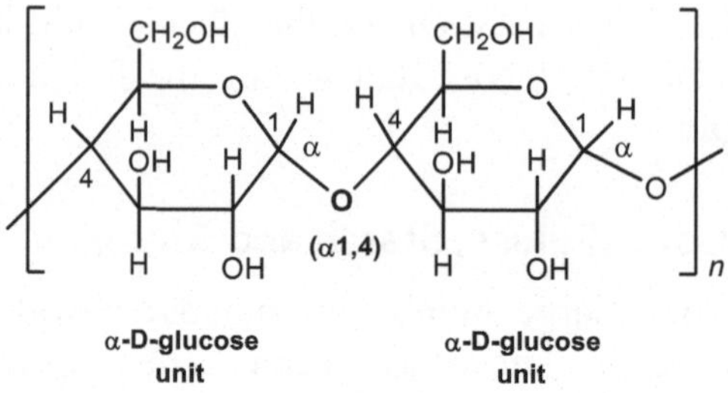

α-D-glucose unit α-D-glucose unit

Amylose

Amylopectin has **α1,4** linked chains of D-glucose units similar to amylose, but the chains are *branched.* **Branching** occurs by **α1,6 glycosidic linkages** between C1 of one glucose unit and C6 of another, at intervals of 20-25 glucose units.

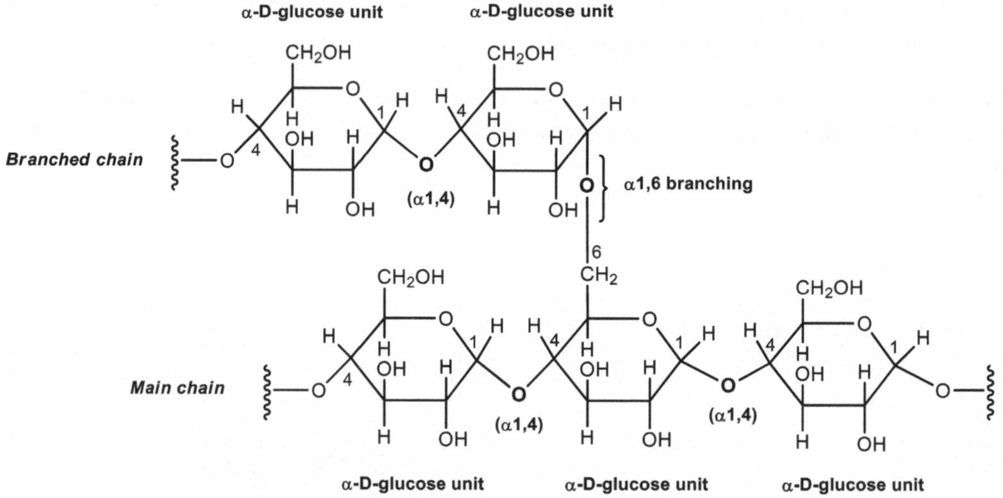

Branched chain

Main chain

α1,6 branching

Amylopectin

Glycogen

Glycogen is the animal counterpart of starch. It is found in muscles and liver, and in this form carbohydrates are stored in animals. Glycogen remains inside the cell where it is needed as an energy source.

Glycogen has a structure very similar to that of amylopectin but chains in glycogen are more *highly branched.* It appears that the highly branched structure of glycogen allows a more specific fit between the enzyme and glycogen (substrate), and leads to ready formation of glucose when needed by the body.

20.2 LIPIDS

Lipids are organic compounds found in cells and tissues that are **soluble in nonpolar solvents** such as chloroform or diethyl ether but **insoluble in water**. The name lipid comes from the Greek *lipos*, meaning fat.

Lipids exhibit a variety of structural types. Here we will consider two classes of lipids, namely, **glycerol lipids** and **steroids**.

20.2.1 Glycerol Lipids: Introduction

The most abundant lipids of plant and animal origin are derivatives of **glycerol** (a trihydric alcohol). **Oils and fats** are *triesters* of glycerol called **triacylglycerols** or **triglycerides.**

Another type of glycerol lipids is called **phospholipids**. Phospholipids are *mixed* esters of glycerol in which *one terminal hydroxyl group* is esterified with **phosphoric acid**.

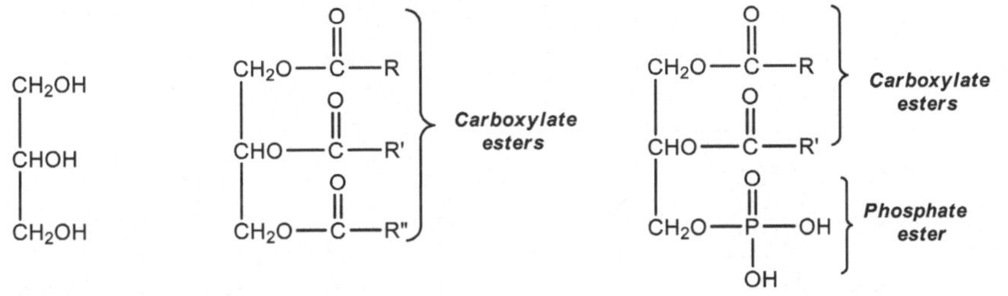

CH_2OH / CHOH / CH_2OH

Glycerol

$CH_2O-C(=O)-R$ / $CHO-C(=O)-R'$ / $CH_2O-C(=O)-R''$ — *Carboxylate esters*

Oil or fat

Triacylglycerol or triglyceride

$CH_2O-C(=O)-R$ / $CHO-C(=O)-R'$ — *Carboxylate esters*; $CH_2O-P(=O)(OH)-OH$ — *Phosphate ester*

Phospholipid

20.2.2 Oils and Fats

Triacylglycerols (or triglycerides) that are **liquids** at room temperature are known as **oils.** Some examples are sunflower oil, peanut oil, coconut oil, olive oil and corn oil.

On the other hand, **triacylglycerols** that are **solids** at room temperature are called **fats,** such as butter and tallow.

Triacylglycerols with three *identical* acyl groups are called **simple triglycerides**, while triacylglycerols with *different* acyl groups are called **mixed triglycerides.**

Alkaline hydrolysis

Alkaline hydrolysis of a fat or oil produces **glycerol** and **salts of three long-chain carboxylic acids**. These carboxylic acids are known as **fatty acids**, and their salts are **soaps**. The reaction is called **saponification**.

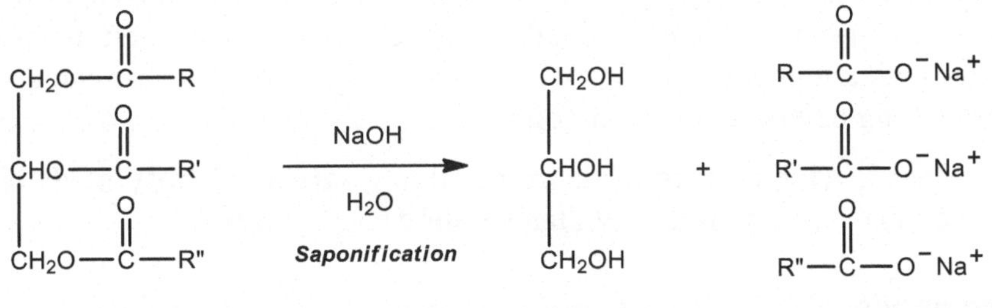

$$\begin{matrix} CH_2O-C(=O)-R \\ CHO-C(=O)-R' \\ CH_2O-C(=O)-R'' \end{matrix} \xrightarrow[H_2O]{NaOH} \begin{matrix} CH_2OH \\ CHOH \\ CH_2OH \end{matrix} + \begin{matrix} R-C(=O)-O^-Na^+ \\ R'-C(=O)-O^-Na^+ \\ R''-C(=O)-O^-Na^+ \end{matrix}$$

Saponification

Oil or fat — **Glycerol** — **Salts of fatty acids** ***Soap***

Fatty acids

Most natural fatty acids are unbranched and contain an **even number of carbon atoms.** They can be **saturated** or **unsaturated.**

Some unsaturated fatty acids contain **two or three double bonds** in which all double bonds are ***cis***.

The following are examples of some common saturated and unsaturated fatty acids:

Saturated fatty acids:

$CH_3(CH_2)_{10}CO_2H$	$CH_3(CH_2)_{12}CO_2H$	$CH_3(CH_2)_{14}CO_2H$	$CH_3(CH_2)_{16}CO_2H$
Lauric acid	**Myristic acid**	**Palmitic acid**	**Stearic acid**
(C_{12})	(C_{14})	(C_{16})	(C_{18})

Unsaturated fatty acids:

$CH_3(CH_2)_5$–CH=CH–$(CH_2)_7CO_2H$ (*cis*; C10=C9)

Palmitoleic acid

(C_{16}; double bond between C9 and C10)

$CH_3(CH_2)_7$–CH=CH–$(CH_2)_7CO_2H$ (*cis*; C10=C9)

Oleic acid

(C_{18}; double bond between C9 and C10)

Polyunsaturated fatty acids:

$CH_3(CH_2)_4$–CH=CH–CH_2–CH=CH–$(CH_2)_7CO_2H$ (*cis*; C13=C12, *cis*; C10=C9)

Linoleic acid

(C_{18}; double bonds: C9-C10, C12-C13)

CH_3CH_2–CH=CH–CH_2–CH=CH–CH_2–CH=CH–$(CH_2)_7CO_2H$ (*cis*; C16=C15, *cis*; C13=C12, *cis*; C10=C9)

Linolenic acid

(C_{18}; double bonds: C9-C10, C12-C13, C15-C16)

The ***cis*** **stereochemistry** of the double bond of an unsaturated fatty acid prevents the molecules from packing closely together, which *lowers* its melting point. Thus **triglycerides** with a higher proportion of **unsaturated** and **polyunsaturated fatty acids** have lower melting points, and they are **oils.**

On the other hand, **triglycerides** made up of largely **saturated fatty acids** have higher melting points. They are solids at room temperature and are **fats.**

Hardening of oils

Catalytic hydrogenation of carbon-carbon double bonds (C=C) in vegetable oils is carried out industrially to produce cooking fats. The process is known as the **hardening**

of oils. However, complete hydrogenation is avoided because it produces a very hard and brittle fat.

The vegetable oil is hydrogenated partially to obtain a **solid or semisolid fat** such as margarine having a consistency similar to that of butter.

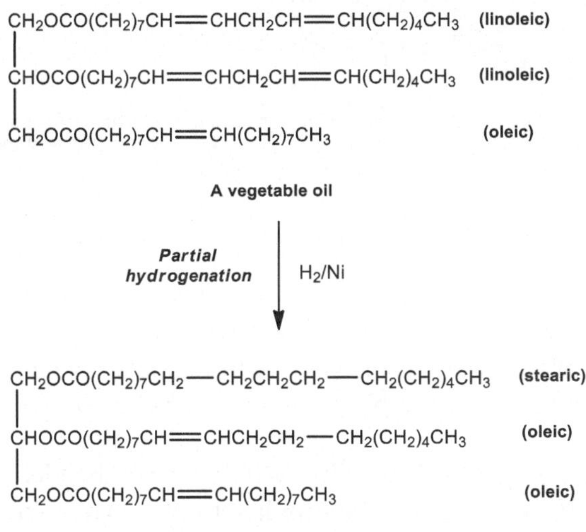

The process of **partial hydrogenation** offers several advantages. It gives a fat with a longer shelf life and increased stability toward oxidation which causes the polyunsaturated oil to become rancid.

However, in partial hydrogenation, the catalyst can *isomerize* some of the remaining double bonds from the ***cis*** to the ***trans*** configuration. The ***trans* fats** have serious implications for human health since they appear to be associated with an increased risk of heart disease.

20.2.3 Soaps and Detergents

Soaps are a mixture of sodium (or potassium) salts of long-chain fatty acids and are almost completely miscible with water. But unlike true solutions, soap is dispersed in water in **spherical clusters** of hundreds of fatty acid chains, called **micelles**. Sodium ions (Na^+) are scattered throughout the solution as solvated ions.

A soap micelle is shown in Figure 20.2.

A **fatty acid chain** has a **polar end**, $-CO_2^-$, and a **nonpolar hydrocarbon end**. The polar end is water soluble (hydrophilic) and the nonpolar end is water insoluble (hydrophobic).

In terms of the rule 'like dissolves like' (see Section 5.2.3), each nonpolar end seeks a nonpolar environment, and they are packed together in the **interior** of the **micelle**. On the other hand, the polar carboxylate ions project **outward** to the polar solvent (water) and stud the surface of the **micelle**. The individual micelles repel each other due to their similar charges and keep them dispersed.

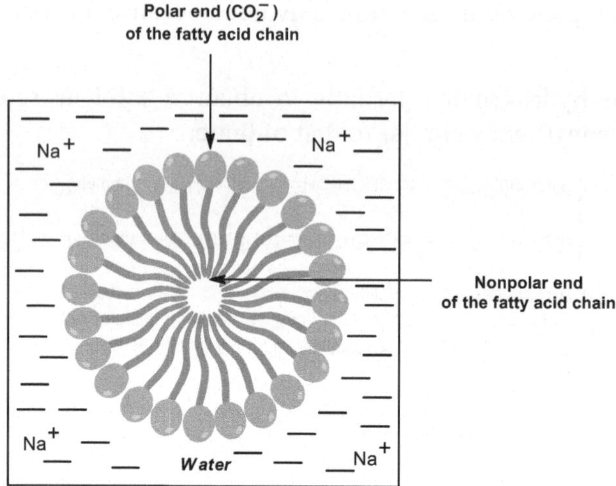

Figure 20.2: A soap micelle showing its interface with the polar medium (water). The nonpolar hydrocarbon chains are in the interior of the micelle and the polar CO_2^- groups are at the interface.

Soaps serve as cleansing agents. Most dirt particles are embedded in a layer of grease or fat. Water alone is unable to penetrate the oily layer but a soap solution can. The nonpolar ends of soap dissolve in the oily layer and each dirt particle develops an outer layer of carboxylate ions projecting into the surrounding water layer. The individual dirt globules then repel each other and become dispersed throughout the aqueous phase, and can be removed from the surface being cleaned.

Hard water is *less effective* for cleaning purposes because calcium and magnesium ions present in hard water react with soap to form *insoluble* calcium and magnesium salts. The following is the reaction between soap (sodium stearate) and calcium ion:

$$2\ CH_3(CH_2)_{16}CO_2Na \quad + \quad Ca^{2+} \longrightarrow [CH_3(CH_2)_{16}CO_2]_2Ca\downarrow \quad + \quad 2\ Na^+$$

Sodium stearate — ***Soluble*** ; **Calcium stearate** — ***Insoluble***

Detergents

Detergents act in the same way as soaps. They have long **nonpolar hydrocarbon chains** with polar groups at one end. The **polar group** of most synthetic detergents is a sulphate ion ($-OSO_3^-$) or a sulphonate ion ($-SO_3^-$) as the sodium salt.

The following are two examples of detergents:

$$CH_3(CH_2)_{10}CH_2-OSO_3^-\ Na^+$$

Sodium lauryl sulphate

$$CH_3CH_2(CH_2)_{10}CH(CH_3)-C_6H_4-SO_3^-\ Na^+$$

A sodium alkylbenzenesulphonate

Detergents have an advantage over soaps because they function well in hard water because the corresponding calcium or magnesium salts are generally soluble in water.

20.2.4 Waxes

Lipid waxes are *not* glycerol lipids. They are **esters of long-chain fatty acids and long-chain alcohols**. The carboxylic acid and alcohol residues of these esters are usually saturated and contain an *even number* of carbon atoms.

Lipid waxes are found as protective coatings on the skin of animals and on the leaves and fruits of plants. Several esters have been isolated from waxes.

For example, the esters isolated from **beeswax** are the following:

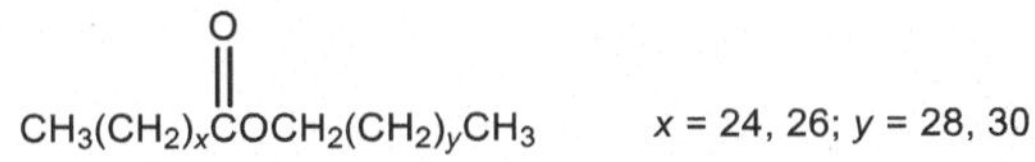

$$CH_3(CH_2)_xC(=O)OCH_2(CH_2)_yCH_3 \qquad x = 24, 26;\ y = 28, 30$$

Esters from beeswax

Note: Lipid waxes differ from paraffin waxes. Paraffin waxes are high molecular weight hydrocarbons derived from petroleum.

20.2.5 Phospholipids

Phospholipids constitute another class of lipids that possess a **phosphate ester linkage**.

Most phospholipids are **glycerol lipids** which are derived from the *parent lipid* called **phosphatidic acid**. In phosphatidic acid, one terminal hydroxyl group of glycerol is esterified with phosphoric acid and the remaining two hydroxyl groups are esterified with fatty acids.

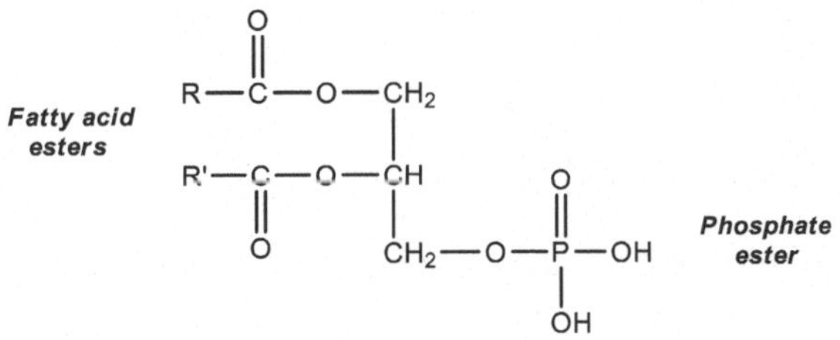

Fatty acid esters: $R-C(=O)-O-CH_2$, $R'-C(=O)-O-CH$; $CH_2-O-P(=O)(OH)-OH$ *Phosphate ester*

Phosphatidic acid

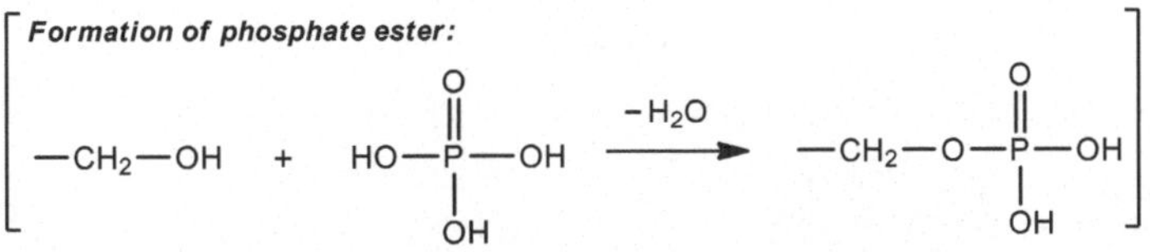

Formation of phosphate ester:

$$-CH_2-OH + HO-P(=O)(OH)-OH \xrightarrow{-H_2O} -CH_2-O-P(=O)(OH)-OH$$

Phosphatides or phosphoglycerides

Phosphatides (or phosphoglycerides) are derivatives of phosphatidic acid, in which the phosphate group is further esterified with a nitrogen containing alcohol such as **choline** and **ethanolamine**.

The structures of choline and ethanolamine are as follows:

$HOCH_2{-}CH_2\overset{+}{N}(CH_3)_3\ \ OH^-$

Choline

$HOCH_2{-}CH_2NH_2$

Ethanolamine

The most important phosphatides are **lecithins** (also called choline phosphoglycerides) and **cephalins** (also known as ethanolamine phosphoglycerides).

$R{-}\overset{O}{\overset{\|}{C}}{-}O{-}CH_2$
$R'{-}\underset{O}{\underset{\|}{C}}{-}O{-}CH$
$CH_2{-}O{-}\overset{O}{\overset{\|}{P}}(O^-){-}O{-}CH_2CH_2\overset{+}{N}(CH_3)_3$

Lecithins
(choline phosphoglycerides)

$R{-}\overset{O}{\overset{\|}{C}}{-}O{-}CH_2$
$R'{-}\underset{O}{\underset{\|}{C}}{-}O{-}CH$
$CH_2{-}O{-}\overset{O}{\overset{\|}{P}}(O^-){-}O{-}CH_2CH_2\overset{+}{N}H_3$

Cephalins
(ethanolamine phosphoglycerides)

Like soaps and detergents, lecithins and cephalins contain a **hydrophilic polar group** and a **hydrophobic component**.

The following is an example of a lecithin showing polar and nonpolar groups:

Nonpolar group | *Polar group*

(stearic) $CH_3(CH_2)_{15}CH_2{-}\overset{O}{\overset{\|}{C}}{-}O{-}CH_2$
(oleic) $CH_3(CH_2)_7CH{=}CH(CH_2)_6CH_2{-}\underset{O}{\underset{\|}{C}}{-}O{-}CH$
$CH_2{-}O{-}\overset{O}{\overset{\|}{P}}(O^-){-}O{-}CH_2CH_2\overset{+}{N}(CH_3)_3$

A lecithin

Lipid bilayers

Phosphatides such as lecithins and cephalins are constituents of cell walls and membranes, where they provide an interface between an organic and an aqueous environment by forming stacked bimolecular micelles, known as **lipid bilayers**.

A schematic diagram of a portion of a lipid bilayer is shown in Figure 20.3.

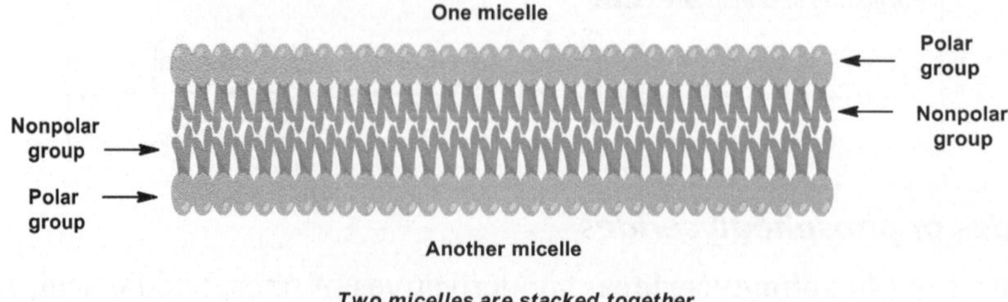

Figure 20.3: A portion of a lipid bilayer.

The cell membranes consisting of lipid bilayers not only enclose the cell but also mediate selective transport of various molecules and ions in and out of the cell.

20.2.6 Steroids: Introduction

Steroids are among the most important lipids. They are widely distributed in plants and animals. Steroids exert a variety of physiological functions.

Many **hormones** (chemical messengers secreted by glands in the body) are steroids. **D vitamins** are also derived from steroids.

The **structural feature** common to all steroids is a **tetracyclic ring system**. This ring system contains *three six-membered rings* designated by the letters A, B and C and a *five-membered ring* designated by the letter D.

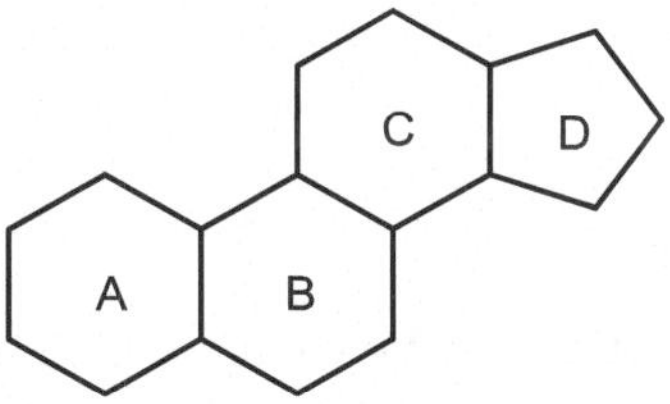

C D
A B

Cholesterol

Cholesterol is the most abundant steroid that is found in nearly all animal tissues. Human gallstones that may develop in the gall bladder are mainly cholesterol. Cholesterol is an intermediate in the biosynthesis of other steroids and is essential to life. Our body synthesizes cholesterol that we need.

However, cholesterol has a bad popular image because high levels of blood cholesterol can develop plaques that may block arteries of the heart and cause heart attacks.

Cholesterol has the following structure:

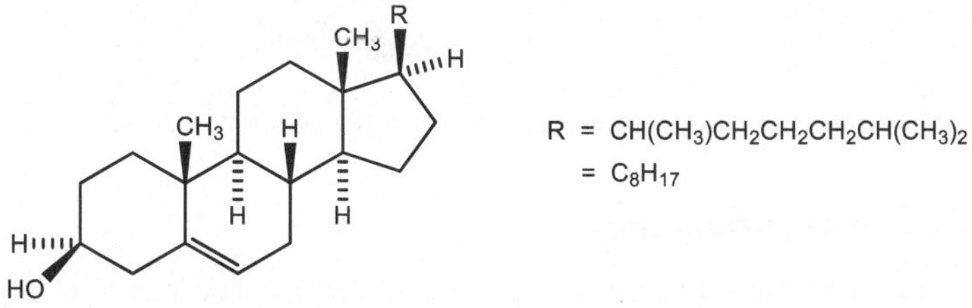

R CH_3 H CH_3 H H H H HO

R = $CH(CH_3)CH_2CH_2CH_2CH(CH_3)_2$
= C_8H_{17}

Cholesterol

20.2.7 Steroid Hormones

Most steroid hormones are highly specialized in their biological activities. An important class of steroid hormones constitutes **sex hormones** which are generated by gonads (testes in the male and ovaries in the female).

The sex hormones are of three major types: the **male hormones** (or androgens), the **female hormones** (or estrogens) and the **pregnancy hormones** (or progestins).

Male hormone or androgen

Testosterone is the major male hormone secreted by the testes. It promotes the development of secondary male characteristics such as the growth of facial and body hair, muscular development and the maturation of male sex organs.

The following is the structure of testosterone:

OH
CH_3
CH_3 H
H H
O

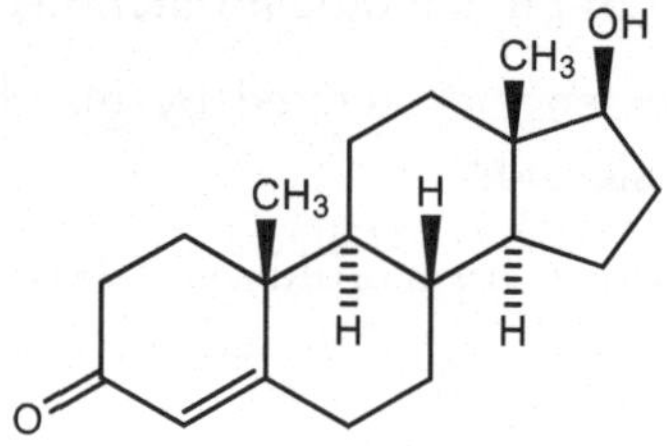

Testosterone

Female hormone or estrogen

Estradiol is the principal female sex hormone which promotes the development of secondary female characteristics at the onset of puberty. It also stimulates the development of the mammary glands during pregnancy.

Estradiol has the following structure:

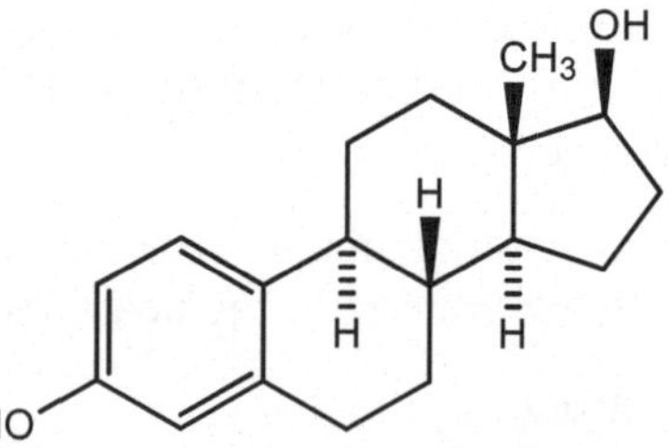

OH
CH_3
H
H H
HO

Estradiol

Note: Interestingly, the female sex hormone, estradiol is derived from the male hormone, testosterone.

Pregnancy hormone or progestin

Progesterone is an important pregnancy hormone whose secretion is continued till the completion of pregnancy. It also suppresses ovulation and inhibits further conception while pregnant. Recognition of this function has led to the development and synthesis of a number of oral contraceptives that have made a significant impact on world population control.

The structure of progesterone *resembles* the structure of testosterone, except that the hydroxyl group (OH) on the five-membered ring is replaced by an acetyl group ($COCH_3$).

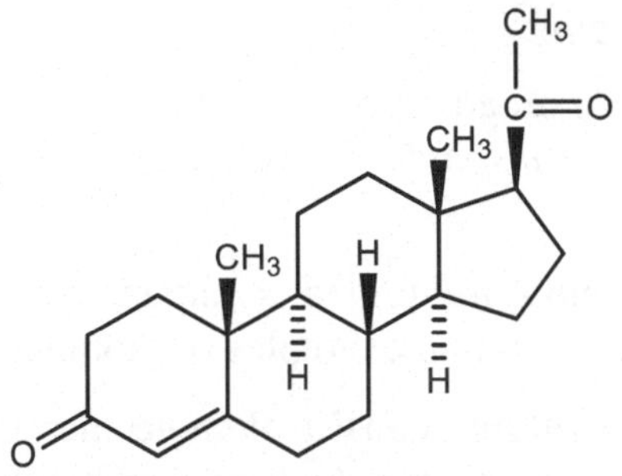

Progesterone

20.2.8 Vitamin D

Vitamin D is an essential vitamin that plays an important role in maintaining a proper bone structure. The deficiency of vitamin D in children causes rickets, a disease characterized by poor bone growth. Only a few foods like fatty fish naturally contain vitamin D, and it is difficult to get enough through food.

There are different forms of vitamin D including **vitamin D_2** and **vitamin D_3**.

The synthesis of vitamin D_3 occurs in the human body upon exposure to sunlight on the skin. The production of vitamin D_3 in the skin is the primary natural source of vitamin D, and it is sometimes called the sunshine vitamin. Upon exposure to sunlight, the steroid (7-dehydrocholesterol) present in human skin undergoes a photochemical reaction which opens the B ring of the steroid to give ultimately vitamin D_3.

Vitamin D_2 is similarly derived from a steroid (ergosterol) from yeast, and is found naturally in sun-exposed mushrooms.

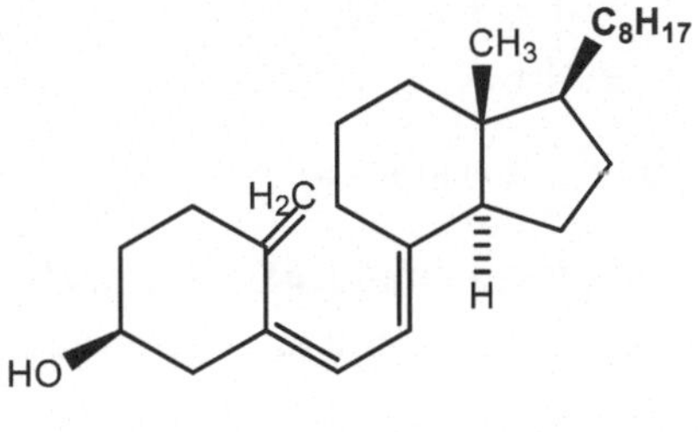

Vitamin D_3
Formed in human skin

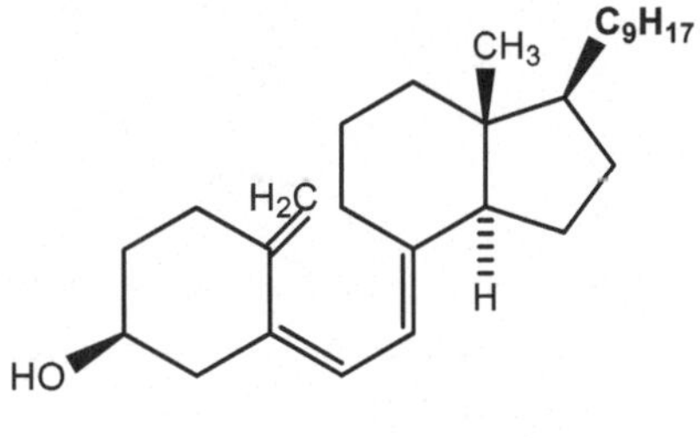

Vitamin D_2
Formed in mushrooms

20.3 PROTEINS

Proteins are the most abundant biomolecules and have the most diverse functions. The name protein originates from the Greek *proteios*, meaning first. Proteins are ranked first because they are the most important molecules of life. Proteins occur in all parts of living organisms and are associated with growth and maintenance of the body. They also catalyse and regulate all biological reactions.

Globular and fibrous proteins

Proteins are classified into two broad types: globular proteins, which are *soluble* in water, and fibrous proteins, which are *insoluble* in water. The difference in solubility is largely due to their shapes.

Molecules of **globular proteins** are folded to attain a spherical shape. Enzymes and albumins (present in eggs) are common examples of globular proteins.

On the other hand, **fibrous proteins** consist of molecules that are long and thread-like. They run side by side and are held together by hydrogen bonds and disulphide links to form fibres. Fibrous proteins are the main structural materials of animal tissues. Some examples are keratin (present in hair, nail, wool, etc.), collagen (present in tendons), and myosin (present in muscles).

Polyamides

Chemically, **proteins are polyamides**. Complete hydrolysis of proteins with 6M hydrochloric acid yields a mixture of different amino acids. These amino acids are **α-amino acids**, in which an amino group (NH_2) is connected to the α-carbon atom of a carboxylic acid. Thus, **α-amino acids are building blocks of proteins.**

20.3.1 α-Amino Acids

The *parent* α-amino acid is **glycine** (α-aminoacetic acid). Other **α-amino acids** are considered as derivatives of glycine by substitution of one of the α-hydrogen atoms by other groups (R).

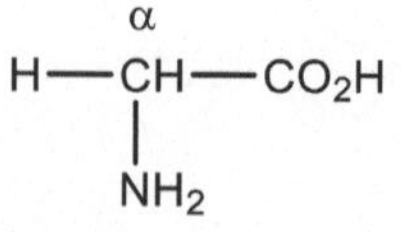

$$\underset{\text{Glycine}}{H-\overset{\alpha}{C}H(NH_2)-CO_2H} \qquad \underset{\text{Other }\alpha\text{-amino acids}}{R-\overset{\alpha}{C}H(NH_2)-CO_2H}$$

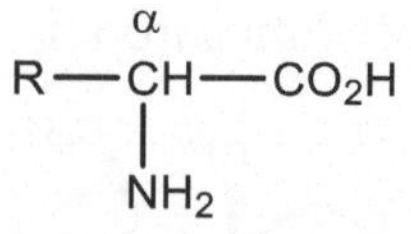

Glycine | **Other α-amino acids**

The α-carbon atom in glycine is *not* a chiral centre, and **glycine molecules are achiral.** In contrast, the **α-carbon atom** in **other amino acids** represents a **chiral centre**, and their molecules are **chiral.**

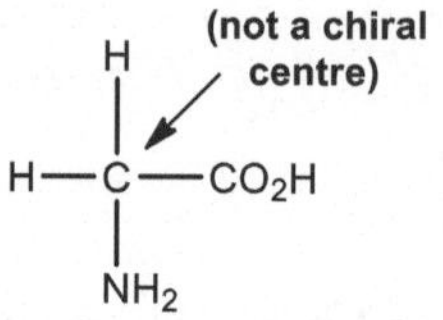

$$H-C(H)(NH_2)-CO_2H \qquad R-\overset{*}{C}(H)(NH_2)-CO_2H$$

(not a chiral centre) | (chiral centre)

Glycine *Achiral* | **Other α-amino acids** *Chiral*

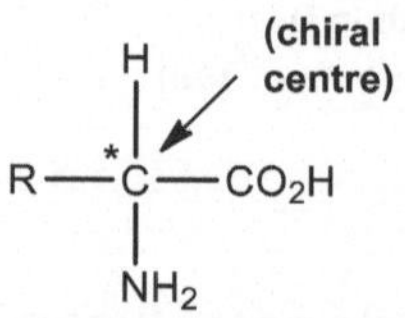

Almost all naturally occurring **chiral amino acids** have the **L configuration**, i.e. NH_2 at the α-carbon is on the left in the Fischer projection (see Section 4.8).

The following are a Fischer projection formula and an equivalent flying wedge formula for L-amino acids.

$$\begin{array}{c} CO_2H \\ H_2N-\overset{*}{C}-H \\ R \end{array} \equiv \text{(flying wedge: } H_2N, H \text{ on } \overset{*}{C}\text{; } R, CO_2H)$$

L-Amino acids

20.3.2 Amino Acids from Proteins

Twenty α-amino acids (or simply, amino acids) are normally found in **proteins**. They can be divided into three groups based on the structures of their **side chains** (R).

These are: (a) **neutral amino acids** (in which R contains no acidic carboxyl or basic nitrogen-containing group), (b) **acidic amino acids** (in which R contains an acidic carboxyl group), and (c) **basic amino acids** (in which R contains an amino or a basic nitrogen-containing group).

The **names and structures** of 20 amino acids are shown in Table 20.1.

The amino acids are generally denoted by **three-letter abbreviations** or **one-letter abbreviations**. For example, glycine is denoted by **Gly** or **G**.

In the table, the one-letter abbreviations are indicated in brackets.

Table 20.1: Natural amino acids

Name	Structure	Abbreviations
Neutral amino acids (15)		
Glycine	$H-CH(NH_2)CO_2H$	Gly (G)
Alanine	$CH_3-CH(NH_2)CO_2H$ (methyl)	Ala (A)
Valine	$(CH_3)CH_3CH-CH(NH_2)CO_2H$ (isopropyl)	Val (V)
Leucine	$(CH_3)CH_3CHCH_2-CH(NH_2)CO_2H$ (isobutyl)	Leu (L)

Name	Structure	Abbreviations
Isoleucine	$CH_3CH_2\overset{*}{C}H(CH_3)—CH(NH_2)CO_2H$ (*sec*-butyl)	Ile (I)
Serine	$HOCH_2—CH(NH_2)CO_2H$	Ser (S)
Threonine	$CH_3\overset{*}{C}H(OH)—CH(NH_2)CO_2H$	Thr (T)
Cysteine	$HSCH_2—CH(NH_2)CO_2H$	Cys (C)
Methionine	$CH_3SCH_2CH_2—CH(NH_2)CO_2H$	Met (M)
Phenylalanine	$C_6H_5—CH_2—CH(NH_2)CO_2H$	Phe (F)
Tyrosine	$HO—C_6H_4—CH_2—CH(NH_2)CO_2H$	Tyr (Y)
Tryptophan	(indol-3-yl)$—CH_2—CH(NH_2)CO_2H$ (ring N–H)	Trp (W)
Asparagine	$H_2NOC—CH_2—CH(NH_2)CO_2H$	Asn (N)
Glutamine	$H_2NOC—CH_2CH_2—CH(NH_2)CO_2H$	Gln (Q)
Proline	(pyrrolidine ring) $CHCO_2H$, NH	Pro (P)

Name	Structure	Abbreviations
Acidic amino acids (2)		
Aspartic acid	$HO_2CCH_2-CH(NH_2)CO_2H$	Asp (D)
Glutamic acid	$HO_2CCH_2CH_2-CH(NH_2)CO_2H$	Glu (E)
Basic amino acids (3)		
Lysine	$H_2NCH_2CH_2CH_2CH_2-CH(NH_2)CO_2H$	Lys (K)
Arginine	$H_2NC(=NH)NHCH_2CH_2CH_2-CH(NH_2)CO_2H$	Arg (R)
Histidine	(imidazol-4-yl)$-CH_2-CH(NH_2)CO_2H$	His (H)

Neutral amino acids

Of twenty amino acids, there are 15 neutral amino acids. **Glycine** is the parent amino acid. Among others, **alanine**, **valine**, **leucine** and **isoleucine** have *alkyl groups* in the side chain.

Serine and **threonine** contain alcoholic *hydroxyl group* in the side chain.

The *sulphur-containing* amino acids are **cysteine** and **methionine**.

Phenylalanine, **tyrosine** and **tryptophan** each contain an *aromatic group* in the side chain. Phenylalanine has a phenyl group, tyrosine has a phenolic group, and tryptophan has a heterocyclic indole ring (see below).

Asparagine and **Glutamine** contain a neutral *amide group* in the side chain.

Proline is a *cyclic* α-amino acid; the amino nitrogen is 2° and part of a five-membered pyrrolidine ring (see below).

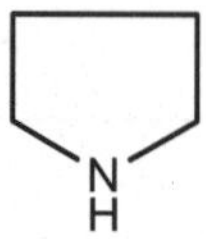

Pyrrolidine 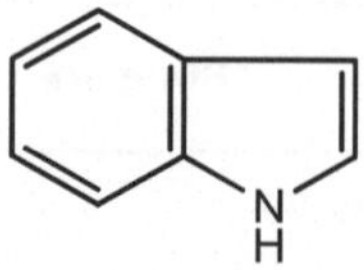 **Indole**

Note: There is an additional chiral centre (*) in the side chain of isoleucine and threonine.

Acidic amino acids

There are two acidic amino acids, namely, **aspartic acid** and **glutamic acid** which contain a *carboxyl group* in their side chains.

Basic amino acids

Of three basic amino acids, **lysine** has an *amino group*, **arginine** has a *guanidino group*, and **histidine** has a heterocyclic *imidazole ring* in their side chains.

The structures of guanidine and imidazole are given below.

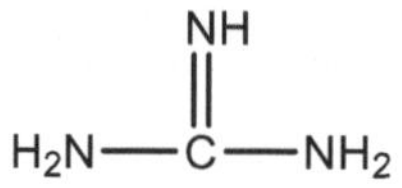

$H_2N-C(=NH)-NH_2$

Guanidine 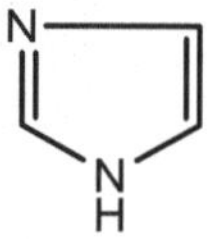 **Imidazole**

Two more amino acids: Hydroxyproline and cystine

There are two more amino acids, namely, **hydroxyproline** (obtained from proline) and **cystine** (obtained from cysteine) that are found in certain proteins.

Hydroxyproline has a *hydroxyl group* on the five-membered pyrrolidine ring of proline. Cystine has a *disulphide bond* which is formed by mild oxidation of the thiol group (SH) present in cysteine.

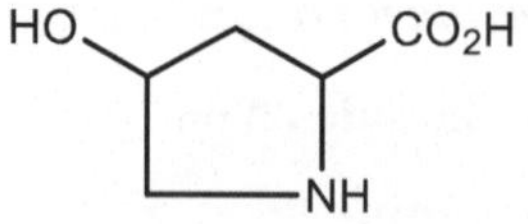

HO, CO_2H, NH

Hydroxyproline (Hyp)

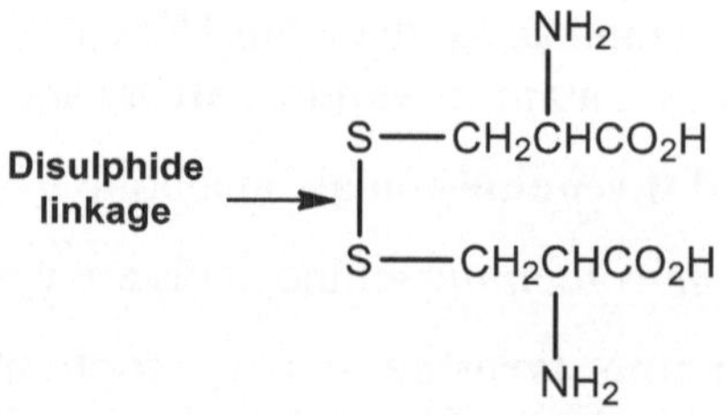

Disulphide linkage → $S-CH_2CH(NH_2)CO_2H$ / $S-CH_2CH(NH_2)CO_2H$

Cystine (Cys-Cys)

Essential amino acids

Of all twenty amino acids required to synthesize proteins for proper growth and function, certain amino acids cannot be synthesized by the body. Therefore, these amino acids need to be supplied through the diet. These are called **essential amino acids**.

There are nine essential amino acids: **valine, leucine, isoleucine, threonine, methionine, phenylalanine, tryptophan, lysine** and **histidine**. The best sources of essential amino acids are meat, eggs and the soy products.

Arginine is generally considered nonessential, but it becomes essential when the body cannot make as much as required in certain cases, such as during illness or pregnancy. Thus **arginine** is sometimes called a **conditionally essential amino acid**.

PROBLEM 20.5: Give the *R*/*S* configuration for L-serine and L-cysteine.

Answer

The L configuration for α-amino acids, except cysteine, corresponds to the *S* configuration at the α-carbon, as shown below for L-serine.

L-cysteine has the *R* configuration. This is because the order of priority for the four ligands in L-cysteine is different from the order of priority of the four ligands in L-serine. A C–S bond has a higher priority over a C–O bond because sulphur has higher atomic number than oxygen.

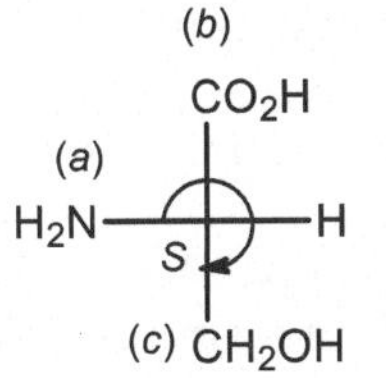

(*b*) CO_2H; (*a*) H_2N; H; *S*; (*c*) CH_2OH

Priority order: $NH_2 > CO_2H > CH_2OH > H$

L-Serine = (*S*)-Serine

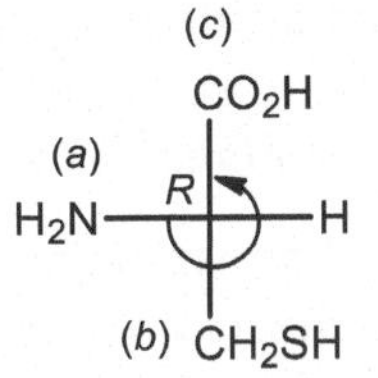

(*c*) CO_2H; (*a*) H_2N; H; *R*; (*b*) CH_2SH

Priority order: $NH_2 > CH_2SH > CO_2H > H$

L-Cysteine = (*R*)-Cysteine

20.3.3 Amino Acids as Dipolar Ions

Amino acids contain an acidic group (CO_2H) as well as a basic group (NH_2). Thus an intramolecular acid–base reaction between CO_2H (a proton donor) and NH_2 (a proton acceptor) can take place to form a **dipolar ion**, also called a **zwitterion**.

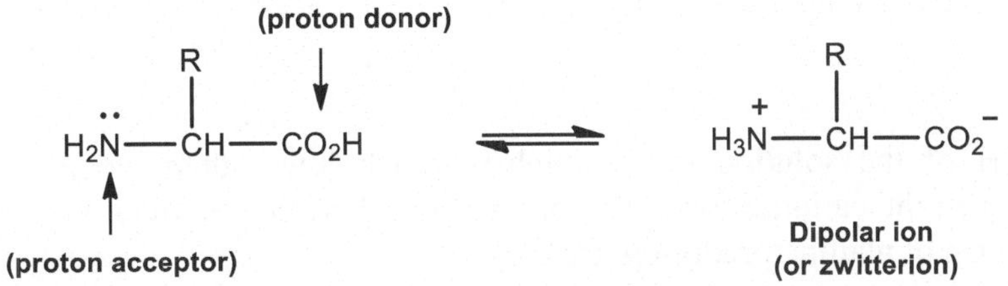

$$H_2\ddot{N}-\underset{}{\overset{R}{\overset{|}{C}H}}-CO_2H \rightleftharpoons H_3\overset{+}{N}-\overset{R}{\overset{|}{C}H}-CO_2^-$$

(proton donor) → CO_2H; (proton acceptor) → H_2N

Dipolar ion (or zwitterion)

In the zwitterion, the anionic part is a carboxylate ion (CO_2^-) and the cationic part is an aminium ion (NH_3^+).

In the *solid state*, amino acids exist as zwitterions. But in an *aqueous solution*, there is equilibrium between the zwitterion and the cationic form and the anionic form of an amino acid.

In **strongly acidic solutions** (low pH), amino acids are present primarily in the **cationic form**, whereas in **strongly basic solutions** (high pH), they exist mainly in the **anionic form**. At some **intermediate pH**, amino acids can exist predominantly as the **zwitterion.**

$$\underset{\textbf{Cationic form}}{H_3\overset{+}{N}-\underset{}{\overset{R}{\overset{|}{C}H}}-CO_2H} \underset{\textit{Low pH}}{\overset{H^+}{\leftrightharpoons}} \underset{\substack{\textbf{Zwitterion}\\ \textbf{\textit{Net charge is zero}}}}{H_3\overset{+}{N}-\overset{R}{\overset{|}{C}H}-CO_2^-} \underset{\textit{High pH}}{\overset{-H^+}{\rightleftharpoons}} \underset{\textbf{Anionic form}}{H_2N-\overset{R}{\overset{|}{C}H}-CO_2^-}$$

Isoelectric point

The **zwitterion** has a **zero net charge** from one positive charge and one negative charge, and is therefore electrically neutral.

The pH at which an amino acid carries no net charge is called its isoelectric point (p*I*). Each amino acid has a particular value of p*I*.

A neutral amino acid such as alanine has a pI = 6.0. The pI values of acidic amino acids are *lower* than those of neutral amino acids. For example, aspartic acid has a pI = 3.0. The basic amino acids have *higher* p*I* values; for example, lysine has a pI = 9.8.

If the pH of the solution of an amino acid is lower than its p*I*, i.e. pH < p*I*, the amino acid exists predominantly as the cation, and therefore in an electric field the amino acid migrates to the cathode (negative electrode).

At a pH > p*I*, the anion predominates, and the amino acid migrates to the anode (positive electrode).

But at a pH = p*I*, the concentration of the zwitterion is at its maximum and no migration occurs because the amino acid has no net charge.

PROBLEM 20.6: Write the predominant form of alanine at pH = 1 and pH =13 respectively. The p*I* of alanine is 6.0.

Answer

When the pH of the solution is 1, which is much lower than the pI, alanine exists predominantly in the cationic form. But at a pH of 13, which is much higher than the pI, the anionic form of alanine predominates.

$$\underset{\substack{\textbf{Zwitterion}\\ \textbf{at pH = p}\boldsymbol{I}\textbf{ = 6.0}}}{H_3\overset{+}{N}-\overset{CH_3}{\overset{|}{C}H}-CO_2^-} \xrightarrow[(pH\,=\,1)]{H^+} \underset{\substack{\textbf{Cationic form}\\ \textbf{\textit{Major}}}}{H_3\overset{+}{N}-\overset{CH_3}{\overset{|}{C}H}-CO_2H}$$

$$\overset{+}{H_3N}-\underset{}{CH}(CH_3)-CO_2^- \xrightarrow[\text{(pH = 13)}]{-H^+} H_2N-CH(CH_3)-CO_2^-$$

Zwitterion at pH = p*I* = 6.0 — **Anionic form *Major***

20.3.4 Peptides: Introduction

Peptides are linear polymers of α-amino acids. Intermolecular reaction between the carboxyl (CO_2H) group of *one* amino acid and the amino (NH_2) group of *another* amino acid can occur with the elimination of water to form an **amide bond** (–CO–NH–), called a **peptide bond** or **peptide linkage**.

The *amino acid units* in the peptide polymer are called **amino acid residues.**

$$H_2N-CH(R)-C(=O)-OH + H_2N-CH(R')-C(=O)-OH \xrightarrow{-H_2O} H_2N-CH(R)-C(=O)-NH-CH(R')-C(=O)-OH$$

One amino acid — **Another amino acid** — **(amino acid residue)** — **(amino acid residue)** — ***Peptide linkage***

A dipeptide

A polymer that contains *two* amino acid residues is called a **dipeptide**. Similarly, a polymer containing *three* amino acid residues is called a **tripeptide**, a polymer containing *four* amino acid residues a **tetrapeptide**, and so on.

A polymer that contains *many* amino acid residues is called a **polypeptide**. *The terms 'peptides' and 'polypeptides' have often been used interchangeably*. Typically, a **peptide** contains 2 to ~50 amino acid residues, and has molecular weight around 5000 or below.

Proteins are regarded as polypeptides of high molecular weight. By an arbitrary convention, a **protein** is a polypeptide that contains about 50 or more amino acid residues, and has molecular weight greater than 5000.

Examples: Peptide hormones

Some natural peptides act as **hormones. Oxytocin** is a peptide hormone which is involved in the control of uterine contraction during child birth, and in stimulating lactation afterwards. It is a peptide consisting of 9 amino acid residues, in which two cysteine residues are linked by a disulphide bond.

Insulin is a hormone secreted by pancreas. It is involved in the regulation of blood glucose levels. A deficiency of insulin production in humans can lead to diabetes. Insulin is therefore clinically used in the control of *diabetes mellitus*. It has 51 amino acid residues, and contains two polypeptide chains, called the A chain and the B chain. The chains are held together by two disulphide linkages.

20.3.5 Structure of Peptides: The Amino Acid Sequence

The structure of a peptide is the sequence of the component amino acids. One end of a peptide or polypeptide chain has an amino acid residue with the *free amino group*, called the **N-terminal amino acid residue**. The other end has an amino acid residue with the *free carboxyl group*, called the **C-terminal amino acid residue**.

(free NH_2) (free CO_2H)

$$H_2N-\underset{R}{CH}-\underset{O}{C}-NH-\underset{R'}{CH}-\underset{O}{C}-NH-\underset{R''}{CH}-\underset{O}{C}-OH$$

N-Terminal amino acid residue **C-Terminal amino acid residue**

By convention, the amino acid sequence of a peptide (or a protein) is drawn with the N-terminal amino acid residue on the left and the C-terminal amino acid residue on the right.

Dipeptides

Consider the following two dipeptides obtained from glycine and alanine.

One dipeptide has **glycine (Gly or G)** as the N-terminal amino acid residue (or simply, N-terminal residue), while the other dipeptide has **alanine (Ala or A)** as the N-terminal residue. They are different.

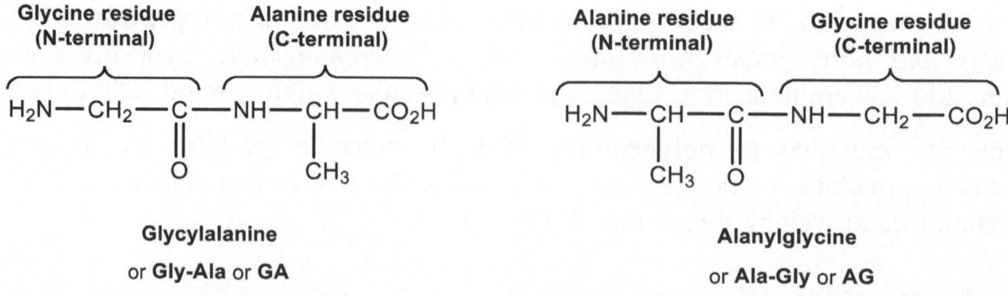

The dipeptides are named as follows:

In general, a peptide is named as a derivative of the C-terminal amino acid. Other amino acids are considered as *substituents* in a **sequence starting with the N-terminal residue**.

Therefore, the first dipeptide is named glycylalanine and the second dipeptide is named alanylglycine.

To simplify the nomenclature, abbreviations for amino acids are often used. Thus, Gly-Ala (or GA) stands for glycylalanine, while Ala-Gly (or AG) stands for alanylglycine.

Tripeptides

Consider the following tripeptide obtained from the amino acids valine, glycine and phenylalanine:

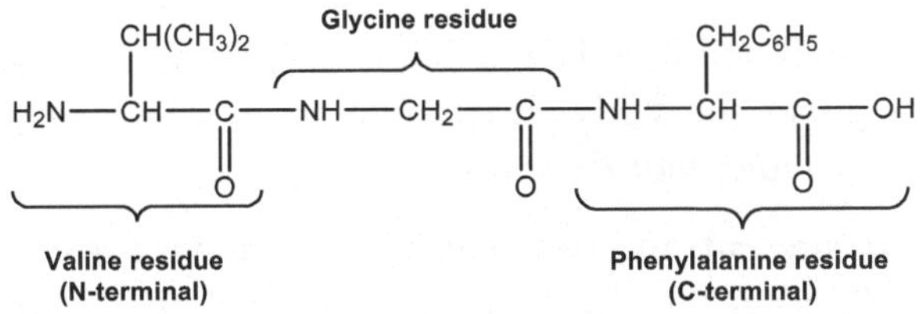

$$\underbrace{H_2N-\overset{CH(CH_3)_2}{\overset{|}{CH}}-\underset{O}{\underset{||}{C}}}_{\substack{\textbf{Valine residue}\\\textbf{(N-terminal)}}}-\overbrace{NH-CH_2-\underset{O}{\underset{||}{C}}}^{\textbf{Glycine residue}}-\underbrace{NH-\overset{CH_2C_6H_5}{\overset{|}{CH}}-\underset{O}{\underset{||}{C}}-OH}_{\substack{\textbf{Phenylalanine residue}\\\textbf{(C-terminal)}}}$$

Valylglycylphenylalanine

or Val-Gly-Phe or VGF

The tripeptide has valine (Val or V) as the **N-terminal residue** and phenylalanine (Phe or F) as the **C-terminal residue**.

The name of the tripeptide is **valylglycylphenylalanine** or **Val-Gly-Phe** or **VGF**.

PROBLEM 20.7: Aspartame is an artificial sweetener, which is nearly 200 times sweeter than sucrose. It is a dipeptide (Asp-Phe), in which the C-terminal residue is a methyl ester. Write the structure of aspartame.

Answer

In aspartame, the N-terminal amino acid residue is aspartic acid (Asp), and the C-terminal residue is the methyl ester of phenylalanine (Phe). Therefore, **aspartame** is Asp-PheOCH_3 and it has the following structure:

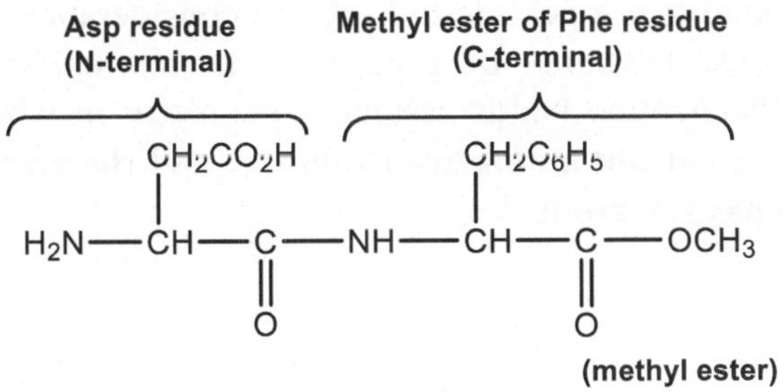

$$\overbrace{H_2N-\overset{CH_2CO_2H}{\overset{|}{CH}}-\underset{O}{\underset{||}{C}}}^{\substack{\textbf{Asp residue}\\\textbf{(N-terminal)}}}-\overbrace{NH-\overset{CH_2C_6H_5}{\overset{|}{CH}}-\underset{\substack{O\\\textbf{(methyl ester)}}}{\underset{||}{C}}-OCH_3}^{\substack{\textbf{Methyl ester of Phe residue}\\\textbf{(C-terminal)}}}$$

Asp-PheOCH_3

20.3.6 Structure of Proteins: Introduction and Primary Structure

The structure of a protein is very complex. Protein structures are generally divided into four levels: **primary**, **secondary**, **tertiary**, and **quaternary**. These levels are hierarchical, each level being more complex than the previous one.

The primary structure of a protein is the unique sequence of amino acids in its polypeptide chain (cf. Section 20.3.5). The final protein structure depends on this sequence.

Even a *single mutation or substitution* of amino acid residue in the sequence can *affect the structure and function* of proteins. An example is given below.

Example

Haemoglobin is a protein in red blood cells that makes possible for the blood to transport or carry oxygen throughout the body. Haemoglobin has four polypeptide chains, *two identical* α *chains* and *two identical* β *chains*.

Normal haemoglobin present in disc-shaped erythrocytes (red blood cells) has a *glutamic acid residue* (Glu) at position 6 in each of the β chains. In **sickle-cell anemia**, the haemoglobin has a *valine residue* (Val) instead of glutamic acid. Remarkably, a single mutation of Glu by Val in each of the β chains causes a change in the structure of the entire protein so that the mutated haemoglobin molecules start associating with one another, forming fibres of aggregated haemoglobins. These distort the red blood cells into *crescent or sickle shapes*, and causes anemia.

20.3.7 Secondary Structure of Proteins: α-Helix and β-sheet

The secondary structure of a protein is the local conformation of its polypeptide backbone. The local conformation implies the *structural unit or fold* in a section of the polypeptide chain. A regular repetition of particular structural units occurs with certain specific distances between them.

The peptide bond or linkage is the principal factor in determining the secondary conformation.

Peptide linkage and conformation

The carbon-nitrogen bond in the peptide linkage has considerable double bond character due to resonance. Thus, *rotations of groups about the carbon-nitrogen bond are hindered.* Consequently, peptide bonds assume a geometry in which the **amide group (CONH) and the two α carbon atoms are coplanar, and the amide hydrogen atom is *trans* to the carbonyl oxygen atom.**

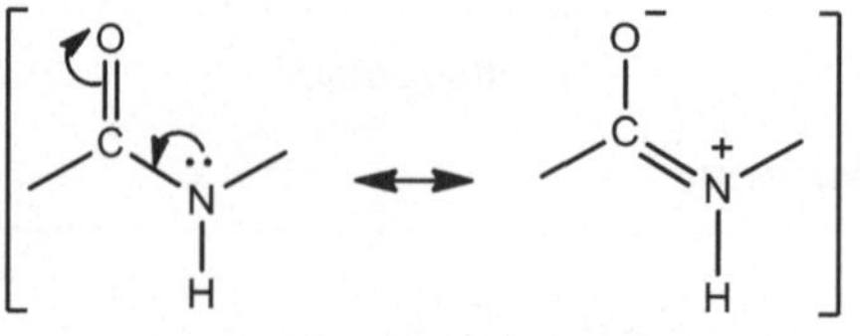

Resonance structures for peptide group

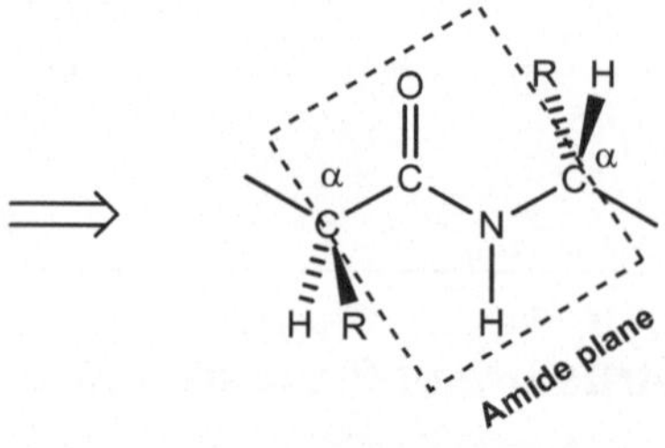

***trans*-Peptide group (O and H are *trans*)**

However, **rotations of α-carbon groups** attached to the *carbonyl carbon* and to the *amide nitrogen* are relatively free. These rotations allow peptide chains to form **different conformations as secondary structure motifs.**

These secondary structures are mainly **stabilized by hydrogen bonds** between the carbonyl oxygen (C=O) and the amide hydrogen (N–H) of two *different amino acid residues that are apart* in the peptide backbone.

The most common secondary structures of a protein are the **α-helix** and the **β-pleated sheet**.

α-Helix

In the α-helix conformation, the peptide backbone forms a **right-handed helix**.

'Right-handed' refers to the direction of turn (cf. a corkscrew or spiral staircase). When the thumb of the right hand is along the helix axis, the flexed fingers describe the twist of the peptide backbone, i.e. if the thumb is pointed upward, the turn is anticlockwise.

There are **3.6 amino acid residues per turn** in an α-helix, which means that there is one amino acid residue for every 100 degrees of rotation (360/3.6).

The **α-helix structure is stabilized by intramolecular hydrogen bonding** between the carbonyl (C=O) oxygen and the amide NH which is *four residues away*. The N, H and O lie approximately in a straight line, and the hydrogen bonds are nearly parallel to the axis of the helix.

The amino acid **side chains** (R) face *outward*, away from the helix axis. Thus, the R groups can stabilize or destabilize the helix from their interactions, but they are not essential for helix formation.

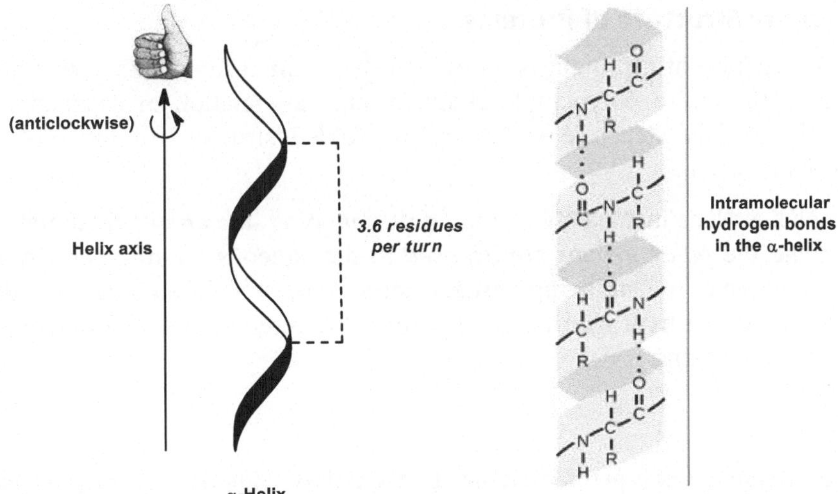

β-Pleated sheet

The second major secondary structure element in proteins is the **β-pleated sheet**. The β-pleated sheet is far more extended or stretched out than the α-helix.

A β-sheet consists of several **β-strands**, which are stretched segments of the polypeptide chain kept together by a network of hydrogen bonds between adjacent strands. The β-sheet is a **pleated sheet** instead of a flat-sheet structure so as to avoid the steric repulsion between the side chain R groups.

Each **β−strand is represented by an arrow**, which defines its direction starting from the N-terminus to the C-terminus. The interacting β-strands may be aligned either **antiparallel** (in opposite directions) or **parallel** (in the same direction).

Antiparallel β-sheets are slightly *more stable* than parallel β-sheets, and are more common. A characteristic feature of an **antiparallel β-sheet** is that **hydrogen bonds form 10-membered and 14-membered rings alternately**. Each ring consists of C, N, O and H atoms.

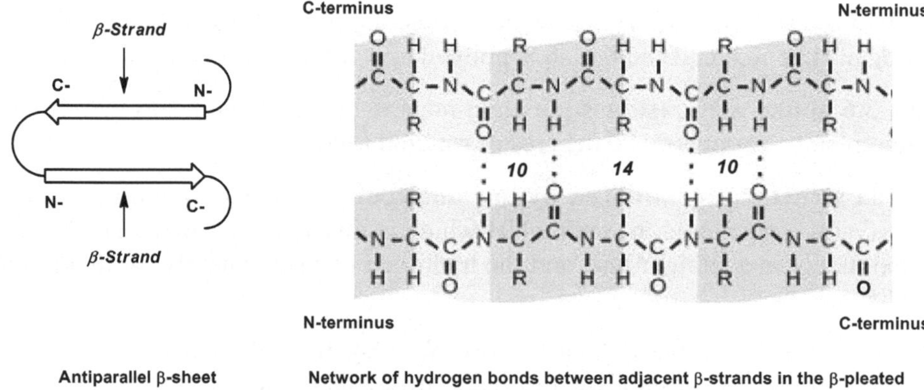

Antiparallel β-sheet

Network of hydrogen bonds between adjacent β-strands in the β-pleated sheet. Hydrogen bonds form 10- and 14-membered rings alternately

20.3.8 Tertiary Structure of Proteins

The tertiary structure of a protein is its overall three-dimensional shape that arises from the folding of the entire polypeptide chain through association of secondary structure elements. The tertiary structure is the *unique* folded structure of the protein, which determines its function.

The folding takes place in such a way that most *nonpolar side chains* lie in the *interior* of the protein and the *polar groups* are *exposed* to the aqueous environment. A variety of forces are involved in stabilizing tertiary structures, which include van der Waals interactions, ionic and hydrogen bonds, hydrophobic interactions, and disulphide bonds from cysteine side chains.

Examples

Myoglobin contains a polypeptide chain of 153 amino acids. It is a **conjugated protein** since it contains a nonprotein group, called a **prosthetic group**, as a part of its structure. The prosthetic group present in myoglobin is called **heme**. Heme is an iron-porphyrin, a complex of Fe^{2+} with porphyrin (a cyclic compound containing four pyrrole rings).

Myoglobin contains *only α-helices as secondary structure elements*, and its **tertiary structure** is shown in Figure 20.4. The heme prosthetic group is shown by a ball and stick model.

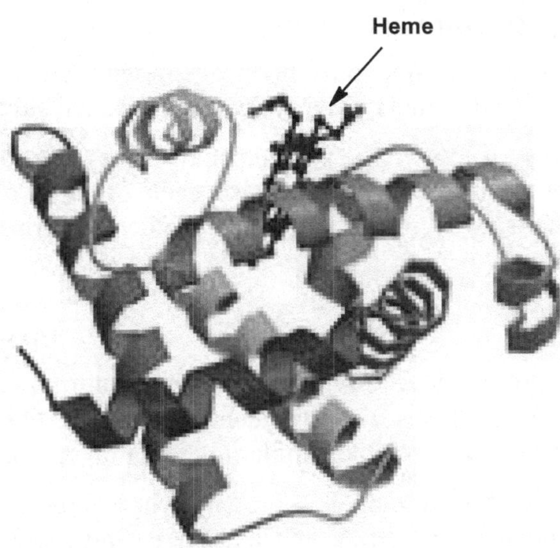

Figure 20.4: A schematic representation of the tertiary structure of myoglobin.

Carbonic anhydrase is a zinc-containing **metalloenzyme** that plays an important role in respiration by influencing transport of CO_2 in the blood. The protein contains *antiparallel β-sheets along with α-helices as secondary structure elements*.

Figure 20.5 shows the **tertiary structure** of carbonic anhydrase.

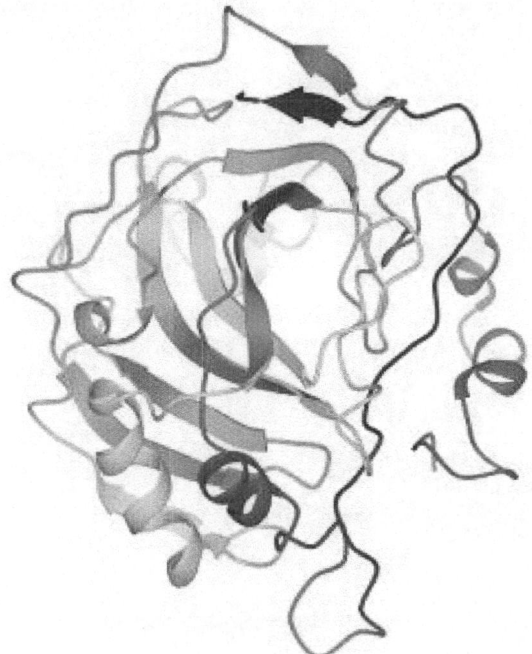

Figure 20.5: A schematic representation of the tertiary structure of carbonic anhydrase.

20.3.9 Quaternary structure of proteins

The **quaternary structure** is not inherent to all proteins but exists in those proteins which are composed of two or more polypeptide chains, called **subunits**.

A subunit has a *tertiary structure*. The subunits may or may not be identical. **The spatial arrangement of the subunits with respect to each other is called the quaternary structure**. The subunits of a protein are held together by hydrogen bonds and other noncovalent interactions.

Example

Haemoglobin is the prototypical example of a multisubunit protein. It consists of **four subunits**: two identical α chains and two identical β chains. Each subunit contains a **heme** group, and has a tertiary structure similar to that of myoglobin.

Myoglobin and haemoglobin were the first proteins whose structures were determined by John C. Kendrew and Max F. Perutz at Cambridge University in England by X-ray analysis. They received the Nobel Prize in chemistry in 1962 for their work.

The **quaternary structure** of haemoglobin is shown in Figure 20.6.

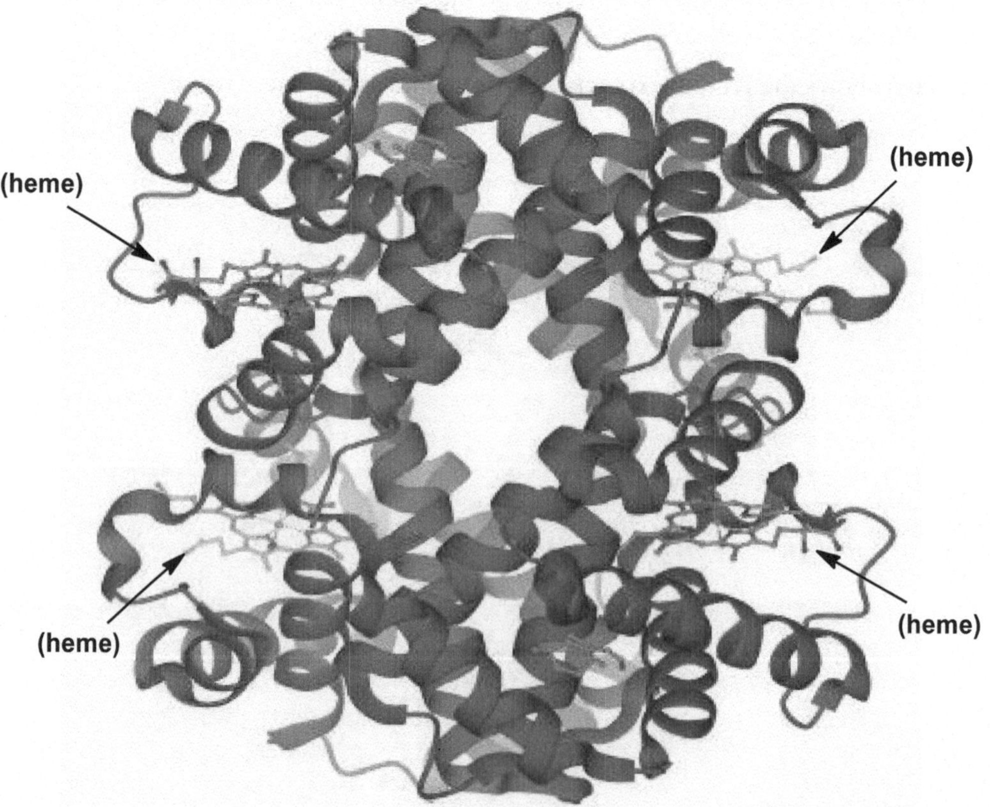

Figure 20.6: A schematic representation of the quaternary structure of haemoglobin.

20.3.10 Denaturation and renaturation of proteins

Proteins found in biological systems have unique three-dimensional structures, which determine their biological functions. Such biologically active proteins are called **native proteins**, and their structures are called native structures.

Denaturation

When native proteins are subjected to *heat*, *changes in pH* or the *addition of chemical compounds such as urea*, they lose their three-dimensional structures with concomitant losses of their biological activity. This process is known as **denaturation of proteins.**

Denaturation involves **unfolding** of proteins, when most of the secondary and tertiary structures are unfolded or lost but *primary structures remain intact.*

As a result of **denaturation**, proteins may be *coagulated* and precipitated from their aqueous solutions. The coagulation of egg white on boiling is a common example of denaturation.

Renaturation

Denaturation can be irreversible, but many proteins are known where the process of denaturation or unfolding can be *reversed.* This reverse process is called **renaturation or refolding of proteins.**

In a classic experiment, the American biochemist Christian B. Anfinsen demonstrated that an enzyme (ribonuclease A) refolds spontaneously from its denatured state to the native conformation after removal of urea used to denature the protein.

It was concluded that the information required to direct the folding of a protein to its native three-dimensional structure is entirely contained in its amino acid sequence or primary structure.

Anfinsen shared the Nobel Prize in chemistry in 1972 for his work on protein folding.

20.3.11 Enzymes and Their Mechanism of Action

Enzymes are catalytic proteins, which catalyse essentially all of the reactions that occur in living cells.

An **enzyme-catalysed reaction** is extremely fast and takes place quantitatively under very mild physiological conditions (approximately neutral pH and a temperature of about 37°C). In their catalytic activity, enzymes are extraordinarily **specific** for a particular substrate and for a particular type of reaction.

Nomenclature

Enzymes are commonly named by adding the suffix **–ase** to a root. The root may indicate the *substrate* on which the enzyme acts. For example, the enzyme that catalyses the hydrolysis of *maltose* is called **maltase**, whereas an enzyme that catalyses the hydrolysis of *proteins* is called a **proteinase** or **protease**.

The root may also indicate the *function* of the enzyme. For example, an enzyme that catalyses *oxidation-reduction* reaction is called an **oxidoreductase**.

Cofactors

Some enzymes require the presence of a non-protein organic molecule or metal ion, called a **cofactor**, for their function.

For an oxidoreductase enzyme, the cofactor is generally an organic molecule such as nicotinamide-adenine dinucleotide (NAD^+), called a **coenzyme**. NAD^+ is an oxidizing agent, which itself is reduced to NADH by accepting a hydride ion.

In some enzymes, the cofactor is a metal ion, such as zinc in carbonic anhydrase.

Mechanism of enzyme action

The action of an enzyme begins by forming a complex with the substrate. The **enzyme-substrate complex** forms at the **active site** of the enzyme. The active site is the specific region of the enzyme which combines with the substrate.

Binding of the substrate to the enzyme causes certain bonds of the substrate to become strained and thus easily broken during the reaction. The product of the reaction, which usually has a *different shape* from the substrate, is released from the enzyme surface to regenerate the enzyme for another reaction cycle.

$$\text{Enzyme} + \text{Substrate} \rightleftharpoons \text{Enzyme-substrate complex} \longrightarrow \text{Enzyme} + \text{Product}$$

The **enzyme-substrate interactions** are mediated by noncovalent forces such as van der Waals interactions, electrostatic interactions and hydrogen bonding.

Some enzymes have a *particular amino acid at the active site that is essential for their enzymatic activity*. **Trypsin** and **chymotrypsin** are digestive enzymes (proteases) that catalyse the hydrolysis of peptide bonds in proteins. The mechanism for their protease activity involves a *serine residue at the active site*, and they are called **serine proteases**.

Enzymes show a remarkable **specificity** for their substrates. Some enzymes have specificity for a *single substrate*, while others are specific for a *class of substrates*. The **enzyme specificity** can be explained by the **lock and key hypothesis**, first postulated by Emil Fischer in 1894.

The **active site** of an enzyme has a unique geometric shape that is *complementary* to the geometric shape of a substrate molecule. In this model, the **lock is the enzyme** and the **key is the substrate**. Only the correctly shaped key (substrate) fits into the key hole (active site) of the lock (enzyme). Incorrectly shaped substrate molecules do not fit into the enzyme.

A diagrammatic representation of the **lock and key model** is shown in Figure 20.7.

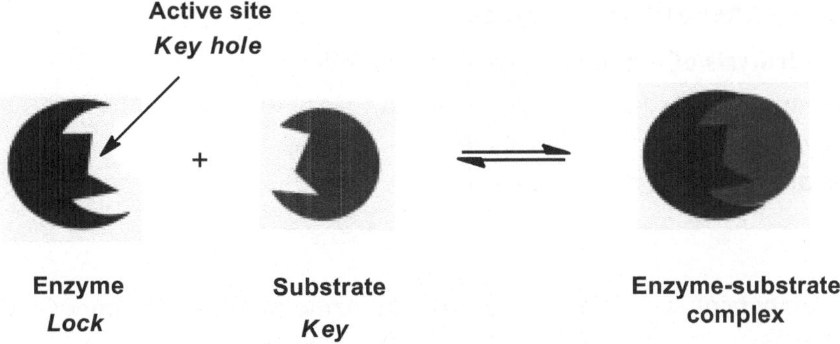

Figure 20.7: A diagrammatic representation of the lock and key model.

The lock-and-key model is rigid in nature, which cannot explain all the experimental evidences. A modified model of enzyme action is known as the **induced fit model**, proposed by Daniel Koshland in 1958.

The induced fit model assumes that the enzyme is partially flexible and the substrate plays a role in determining the final shape of the enzyme. Once the substrate binds, a *conformational change* in the enzyme is induced, which changes its shape that finally allows the enzyme to bind the substrate more effectively. This is called the induced fit, which gives a *more precise fit* of the substrate to the active site of the enzyme.

A diagrammatic representation of the **induced fit model** is shown in Figure 20.8.

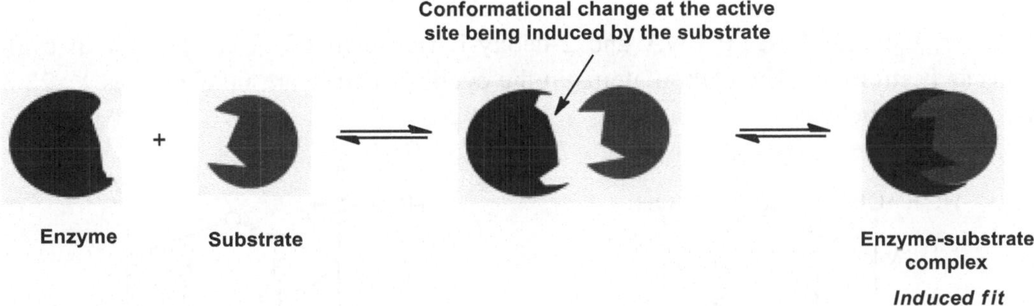

Figure 20.8: A diagrammatic representation of the induced fit model.

20.4 NUCLEIC ACIDS

The nucleus of a living cell contains *chromosomes* that are made up of nucleic acids and proteins. Nucleic acids store genetic or hereditary information, which is transmitted from one generation to succeeding generations.

Like proteins, **nucleic acids are also polymers**. The **monomeric units** in nucleic acids are compounds that are called **nucleotides**. Thus nucleic acids are **polynucleotides**.

There are two types of nucleic acids: deoxyribonucleic acid (DNA) and ribonucleic acid (RNA).

20.4.1 Nucleosides and Nucleotides

Complete **hydrolysis of a nucleic acid** gives the following:

- A five-carbon **sugar** (or monosaccharide)
- A nitrogen-containing heterocyclic **base**
- A **phosphate ion** (or phosphoric acid).

Sugars

The sugar component is an aldopentose. As the name implies, **ribonucleic acid (RNA)** contains **D-ribose** as the sugar component, whereas the sugar component in **deoxyribonucleic acid (DNA)** is **2-deoxy-D-ribose** (in which the OH at 2-position of ribose is replaced by H).

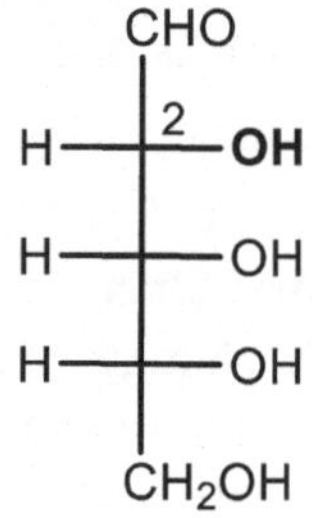

D-Ribose

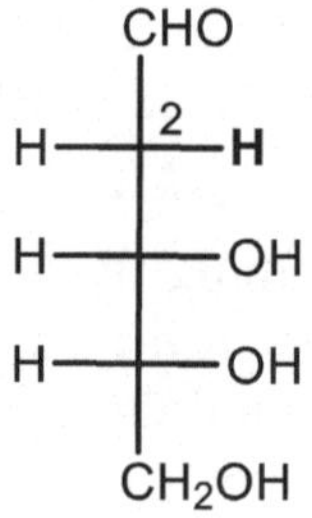

2-Deoxy-D-ribose

However, both D-ribose in RNA and 2-deoxy-D-ribose in DNA are present in cyclic forms as **β anomers**. The carbon atoms in the cyclic structure are numbered as 1′, 2′, 3′, etc., as shown below.

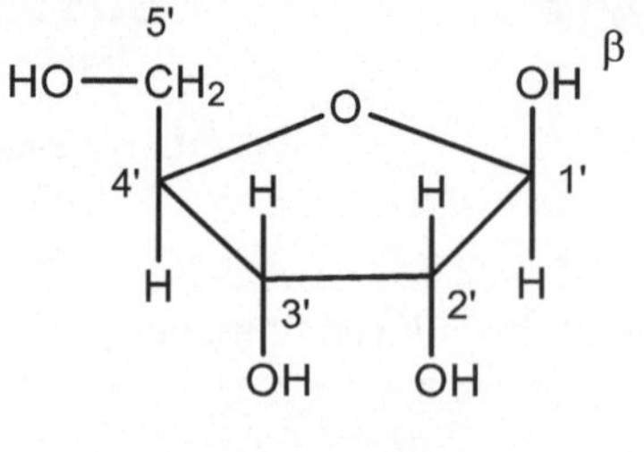

β-D-Ribose

Present in RNA

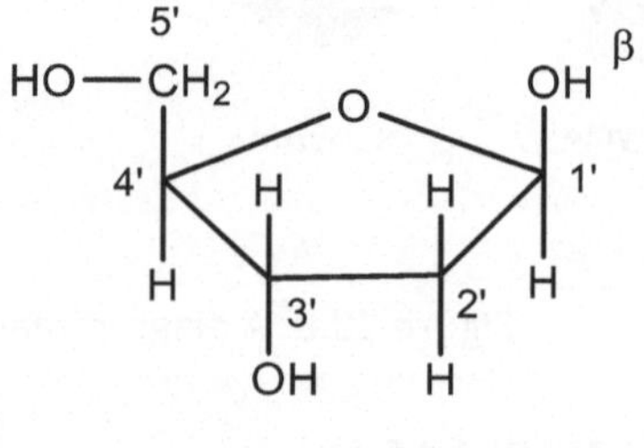

2-Deoxy-β-D-ribose

Present in DNA

Bases

DNA contains four heterocyclic bases: adenine (A), guanine (G), cytosine (C) and thymine (T).

RNA also contains four bases but in RNA, **uracil (U)** replaces thymine (T), the other three bases being the same as in DNA.

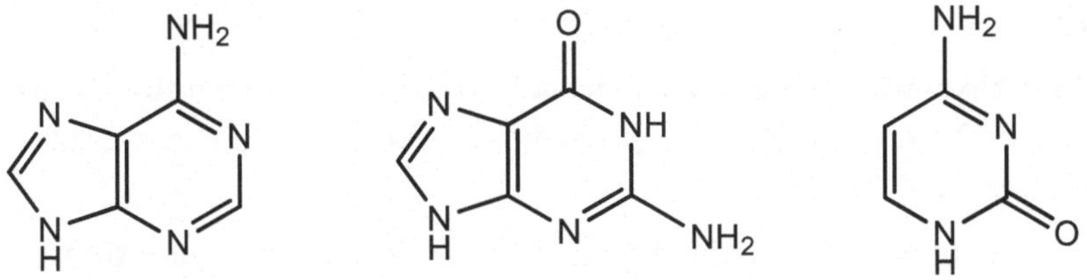

Adenine (A) **Guanine (G)** **Cytosine (C)**

Three bases present in DNA or RNA

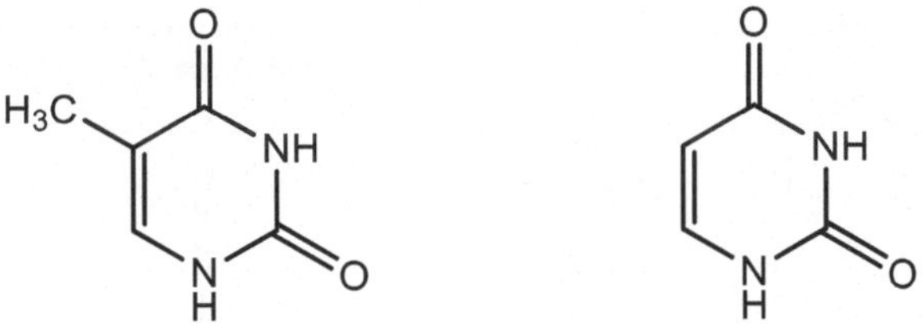

Thymine (T)
Present in DNA

Uracil (U)
Present in RNA

Nucleosides

A **nucleoside** contains either β-D-ribose or 2-deoxy-β-D-ribose and *one* of the four bases. Thus a nucleoside consists of a **sugar** and a **base**.

An example of a **nucleoside** obtained from RNA is **adenosine**, which can be formed from **β-D-ribose** (sugar) and **adenine** (base).

The structure of **adenosine** is shown in Figure 20.9.

The base is attached to C1′ (anomeric carbon) of the sugar replacing the OH group. This linkage is called a **β-N-glycosidic linkage**.

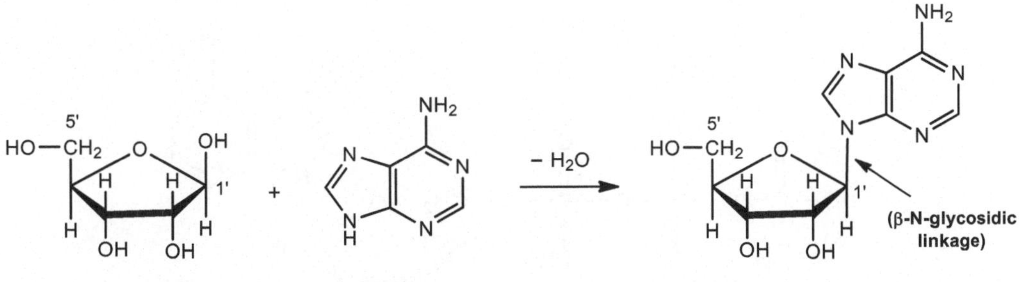

β-D-Ribose + **Adenine** $\xrightarrow{-H_2O}$ **Adenosine** *A nucleoside* (β-N-glycosidic linkage)

Figure 20.9: Structure of adenosine (a nucleoside) obtained from RNA.

Nucleotides

Addition of a phosphate group to a nucleoside results in a nucleotide. Thus a nucleotide contains either β-D-ribose or 2-deoxy-β-D-ribose, *one* of the four bases, and a phosphate group as the phosphate ester.

An example of a **nucleotide** obtained from RNA is **adenosine monophosohate (AMP),** which can be formed from **β-D-ribose** (sugar), **adenine** (base) and **phosphoric acid.**

The structure of **adenosine monophosphate** is shown in Figure 20.10.

The phosphate group is attached to C5′ of the monosaccharide ring. This linkage is called a **phosphate ester linkage.**

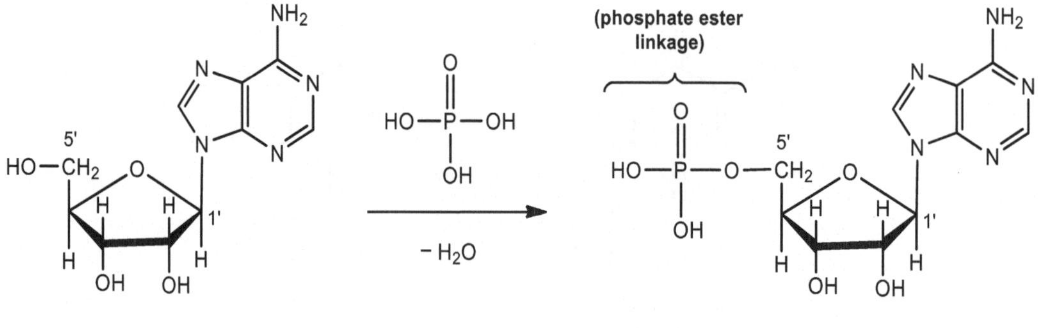

Adenosine
A nucleoside

$- H_2O$

(phosphate ester linkage)

Adenosine monophosphate (AMP)
A nucleotide

Figure 20.10: Structure of adenosine monophosphate (a nucleotide) obtained from RNA.

PROBLEM 20.8: Adenosine triphosphate (ATP) is the source of energy for storage and use in living cells. Draw the structure for ATP.

Answer

ATP is a nucleoside triphosphate, consisting of β-D-ribose, adenine and three serially connected phosphate groups.

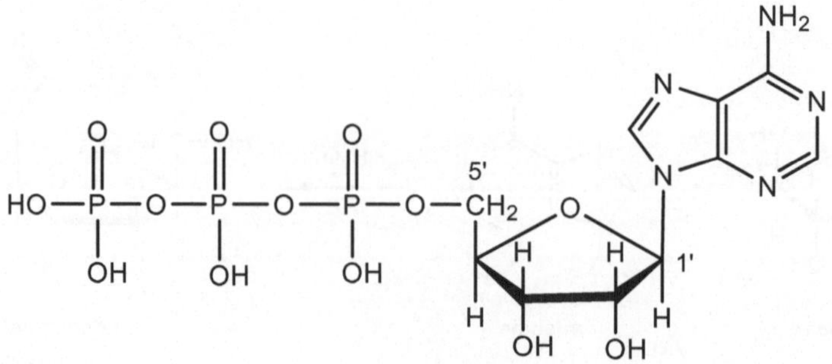

Adenosine triphosphate (ATP)

20.4.2 Polynucleotides

Nucleic acids (DNA or RNA) are polynucleotides, in which nucleotides are joined together by **phophodiester linkages** between C5′ of *one* ribose (or 2-deoxyribose) and C3′ of *another.*

A **dinucleotide segment** of a DNA chain is shown in Figure 20.11. Note that the sugar unit in a nucleotide of DNA is 2-deoxyribose.

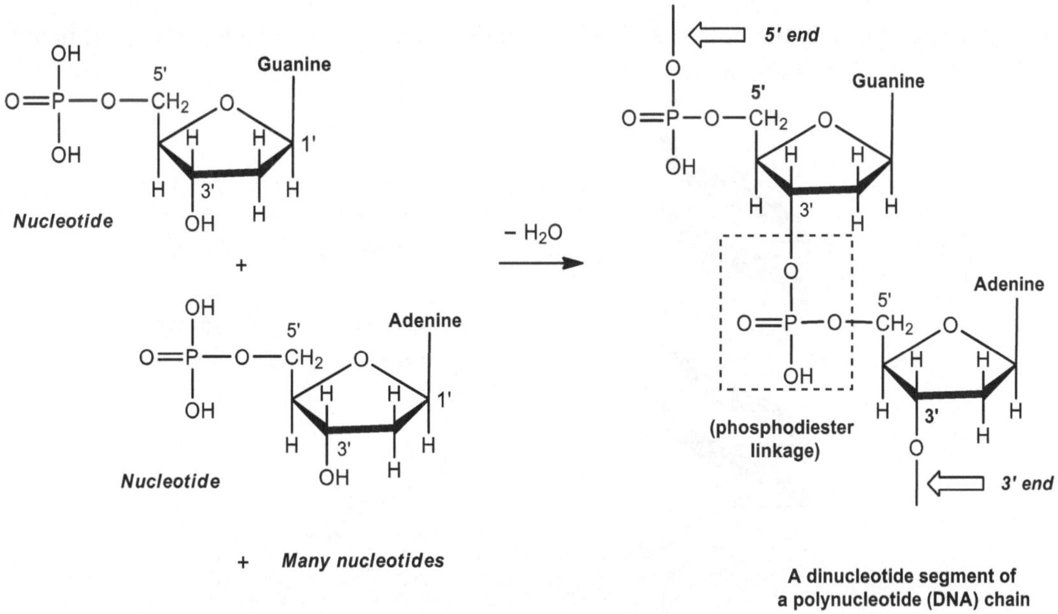

Figure 20.11: A dinucleotide segment of a DNA chain.

Note: A polynucleotide chain or a portion of it (such as a dinucleotide segment) has a **3′ end** and a **5′ end**.

20.4.3 Primary Structure of Nucleic Acids

The nucleic acid has a linear chain with a *backbone* of **sugar and phosphate units**, and the **heterocyclic bases** *protrude from the chain* at regular intervals.

The sequence of nucleotides in the nucleic acid chain is called its primary structure (cf. the primary structure of a protein).

A schematic diagram for the primary structure of a nucleic acid is shown here.

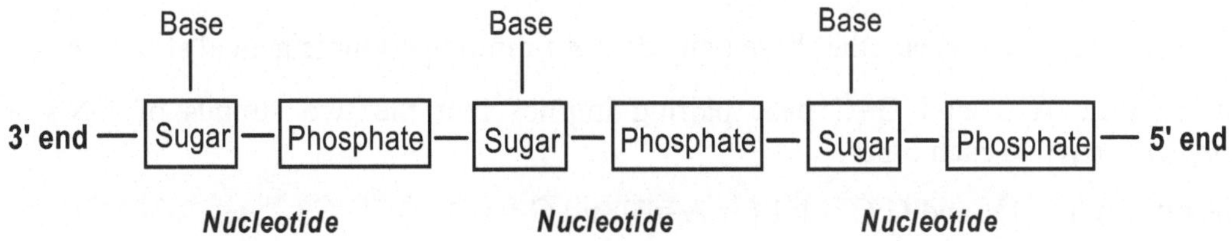

Note: It is the *sequence of bases* along the DNA chain that contains the genetic information (see Section 20.4.5).

20.4.4 Secondary Structure of Nucleic Acids: Double Helix Structure of DNA

In 1953, James D. Watson and Francis H. C. Crick proposed a **double helix** as a model for the secondary structure of DNA. The double helix structure was verified thereafter by Maurice H. F. Wilkins by X-ray analysis. Watson, Crick and Wilkins shared the Nobel Prize in physiology or medicine in 1962.

The polynucleotide backbone of DNA is a right-handed helix. Two such helices are wound about each other to form a **double helix**, as shown in Figure 20.12.

Note that the two polynucleotide chains or strands run in *opposite* directions, as indicated by 3′ → 5′ direction of the arrows.

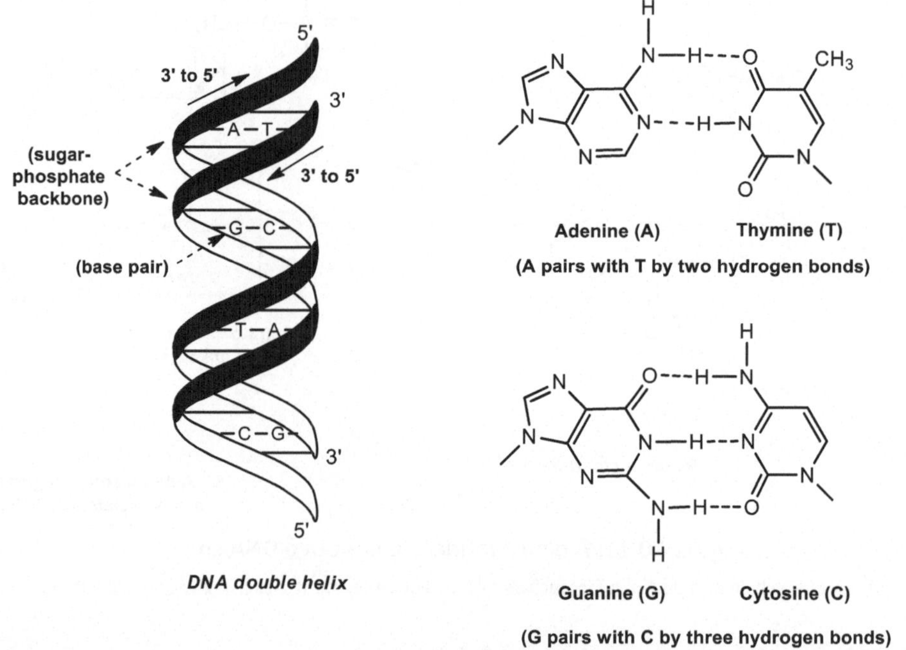

Figure 20.12: A schematic diagram of the DNA double helix and hydrogen bonding in adenine–thymine (A–T) and guanine–cytosine (G–C) base pairs. All horizontal lines on the inside of the helix indicate base pairs; however, bases are not shown on all the lines for clarity.

The two chains are held together by hydrogen bonds between adenine (A) and thymine (T) and between guanine (G) and cytosine (C) on *opposite* strands. The A–T base pairing involves two hydrogen bonds, whereas the G–C base pairing involves three hydrogen bonds (see Figure 20.12).

Hydrogen bonds between other base pairs do not fit into the double helical structure.

The specific A–T and G–C base pairing implies that the two strands of DNA are *complementary* to each other.

The ratio A:T = 1:1 and G:C = 1:1 (or A = T and G = C).

Note: The secondary structure of RNA is usually a *single stranded helix*. There are three types of RNA, called messenger RNA (mRNA), ribosomal RNA (rRNA), and transfer RNA (tRNA).

20.4.5 Genetic Code and Synthesis of Proteins

The synthesis of proteins is all important to the function of a cell. A **segment of the DNA molecule** that contains the information necessary to direct the synthesis of a protein is called the **gene**.

The genetic information flows from

DNA ⟶ RNA ⟶ Proteins

This is known as the **central dogma of molecular biology**. The first part (DNA → RNA) of the central dogma is called **transcription** and the second part (RNA → Proteins) is called **translation**.

Thus, **protein synthesis** involves two major processes, namely, transcription and translation.

Transcription

Transcription takes place in the cell nucleus. During transcription, the genetic information in a segment of DNA or gene is copied down to a type of RNA, called **messenger RNA (mRNA)**. In this process, an enzyme called *RNA polymerase* unwinds a part of the DNA helix and assembles a strand of mRNA by 'reading' the sequence of bases on one of the strands of DNA.

The assemblage occurs through **complementary base pairing**: if there is a G on the DNA, there will be a C on the RNA (**G–C** base pairing), but if there is an A on the DNA, there will be U on the RNA (**A–U** base pairing). Remember that U (not T) is present in RNA.

Translation

Translation is the process by which the base sequence of mRNA is translated into a sequence of amino acids during protein synthesis.

After transcription, the mRNA leaves the nucleus and moves to the **ribosome** in the cytoplasm. Ribosomes are organelles that are composed of **ribosomal RNA (rRNA)** and proteins. One of the functions of rRNA is to bind the mRNA to the ribosome.

At the ribosome, mRNA calls up a series of **transfer RNA (tRNA)** molecules, each of which is loaded with a particular amino acid. *A tRNA binds to a specific site of mRNA.* This site has three adjacent nucleotide bases (a triplet) called the **codon**, which codes for an amino acid.

Which **mRNA triplet** corresponds to which **amino acid** is called the **genetic code**. The Indian- American biochemist Har Gobind Khorana shared the Nobel Prize in physiology or medicine in 1968 with Robert W. Holley and Marshall W. Nirenberg for deciphering the genetic code.

An RNA has four bases (A, U, G, C); thus there are $4^3 = 64$ possible codons. Of these, **3 codons (UAA, UAG, UGA)** do not code for amino acids but indicate the **stop signal** for the protein synthesis. The remaining **61 codons specify the 20 different amino acids** that make up proteins. The **AUG** codon which codes for methionine also indicates the **start signal** for the protein synthesis.

The **genetic code** (mRNA triplets or codons) that specifies different amino acids is shown in Table 20.2. The stop codons (UAA, UAG, UGA) are not included.

Table 20.2: Genetic code: mRNA triplets (codons) specifying different amino acids

Amino acid	mRNA triplet (codon)	Amino acid	mRNA triplet (codon)	Amino acid	mRNA triplet (codon)
Gly	GGA	Ser	AGC	Asn	AAC
	GGC		AGU		AAU
	GGG		UCA	Gln	CAA
	GGU		UCG		CAG
			UCC		
Ala	GCA		UCU	Pro	CCA
	GCC				CCC
	GCG	Thr	ACA		CCG
	GCU		ACC		CCU
			ACG		
Val	GUA		ACU	Asp	GAC
	GUC				GAU
	GUG	Cys	UGC	Glu	GAA
	GUU		UGU		GAG
Leu	CUA	Met	AUG	Lys	AAA
	CUC				AAG
	CUG	Phe	UUU	Arg	AGA
	CUU		UUC		AGG
	UUA				CGA
	UUG	Tyr	UAC		CGC
			UAU		CGG
Ile	AUA				CGU
	AUC	Trp	UGG	His	CAC
	AUU				CAU

Note: The **DNA triplet** in the gene that corresponds to an mRNA triplet can be easily derived using complementary A–U and G–C base pairing. For example, an mRNA triplet (GCU) for alanine corresponds to the DNA triplet (CGA) in the gene.

20.4.6 Polymerase Chain Reaction (PCR)

The **polymerase chain reaction (PCR)** is a revolutionary method in molecular biology that is used to make millions to billions of copies of a specific segment of DNA.

The method is simple and inexpensive. With the advent of PCR technique, it is possible to amplify the target DNA sequence exponentially to obtain sufficient amounts required for molecular and genetic studies from very low quantities (even a single molecule) of source DNA. The PCR was developed by the American biochemist Kary B. Mullis in the 1980s, and he was awarded the Nobel Prize in chemistry in 1993 for his pioneering work.

PCR is based on using the ability of the enzyme *DNA polymerase* to synthesize new strands of DNA. The DNA sample is first heated when the DNA double helix denatures or separates into two pieces of single-stranded DNA. Then the enzyme DNA polymerase synthesizes or builds two new strands of DNA, using the original strands as templates. This leads to the duplication of the original DNA, with each of the new molecules having one old strand and one new strand of DNA. The cycle of denaturation and synthesis of new DNA is repeated.

Each cycle doubles the amount of target DNA; thus the amount of DNA increases exponentially. After n cycles, the amount of DNA will be increased 2^n times. This means that after 20 cycles, there is roughly 1 million times as much DNA. The entire cyclic process of PCR has been automated and can be completed in a few hours.

PCR is widely used in human genetics and in forensic sciences including DNA fingerprinting (see below). It is a valuable tool in the diagnosis of genetic disorders such as muscular dystrophy and cystic fibrosis, and in the detection of viruses that cause diseases like AIDS, hepatitis and measles.

DNA fingerprinting

About 99.9% of the DNA between two individuals is the same. It is the remaining 0.1% that makes a person unique. This amount is not quite small; this means that there are about 3 million base pairs that are different between two persons. This variation in the DNA profile of an individual characterizes his uniqueness, like his actual fingerprint.

DNA fingerprinting gives a DNA pattern unique to an individual from a tiny sample of human material such as blood, skin or hair. It is the same for every cell in all body parts and cannot be altered by any treatment, while the actual fingerprint can be altered by surgery. Thus DNA fingerprinting is much more superior for the purpose of identification.

DNA fingerprinting is used in forensics to identify a suspect in criminal investigation, to determine the paternity of an individual, and to identify a dead body that is too old or damaged to be recognizable.

Reverse transcription–polymerase chain reaction (RT-PCR)

A variation of the polymerase chain reaction (PCR) is known as **the reverse transcription–polymerase chain reaction (RT-PCR)**. This method combines *reverse transcription* (RNA → DNA) with PCR.

In RT-PCR, RNA molecules are first converted into their complementary DNA (cDNA) sequences by the enzyme *reverse transcriptase*. The cDNA then serves as a template for exponential amplification using the standard PCR procedure.

RT-PCR is used to detect an RNA target from a specific organism such as a virus. For example, the coronavirus SARS-CoV-2 that causes coronavirus disease (COVID) has RNA as the genetic material. It is an RNA virus, which distinguishes it from a DNA virus such as varicella-zoster virus (VZV) that causes chickenpox. When the COVID-19 pandemic began, RT-PCR tests were the first to be developed and widely used to diagnose the disease.

SUPPLEMENTARY PROBLEMS

20.9 Write the structure of product in each of the following reactions. Would the product in each case be optically active?

(a)

```
        CHO
         |
    H ---+--- OH
         |              HNO3
    H ---+--- OH     ---------->
         |
       CH2OH
```

D-Erythrose

(b)

```
        CHO
         |
   HO ---+--- H
         |              HNO3
    H ---+--- OH     ---------->
         |
       CH2OH
```

D-Threose

[*Ans*: (a) optically inactive tartaric acid; (b) optically active tartaric acid]

20.10 D-Ribose and D-arabinose are aldopentoses that form the same osazone on treatment with phenylhydrazine. The structure of D-ribose is given below. Predict the structure of D-arabinose.

```
        CHO
         |
    H ---+--- OH
         |
    H ---+--- OH
         |
    H ---+--- OH
         |
       CH2OH
```

D-Ribose

[*Hint*: D-arabinose is the C2 epimer of D-ribose]

20.11 When oleic acid is heated to 200°C in the presence of a small amount of selenium, it isomerises to a compound called elaidic acid. Give a structure for elaidic acid.

[*Hint*: the reaction involves a *cis-trans* isomerisation]

20.12 The sex hormone isolated from the urine of pregnant women is estrone, which is a metabolized form of estradiol. Write the structure for estrone (a ketone).

20.13 Four different dipeptides can be prepared using glycine and alanine. Write their structures and label them using three-letter abbreviations for the amino acids.

20.14 Glutathione is a tripeptide (γ-Glu-Cys-Gly) present in most living cells. It acts as an important antioxidant. But it is not a typical tripeptide because the peptide bond formed by glutamic acid involves its γ-CO_2H group instead of the normal α-CO_2H group. Give the structure of glutathione.

[*Ans*: $H_2N—\underset{\alpha}{\overset{CO_2H}{\overset{|}{C}H}}—\underset{\beta}{CH_2}—\underset{\gamma}{CH_2}—\underset{\underset{O}{\|}}{C}—NH—\overset{CH_2SH}{\overset{|}{C}H}—\underset{\underset{O}{\|}}{C}—NH—CH_2—CO_2H$]

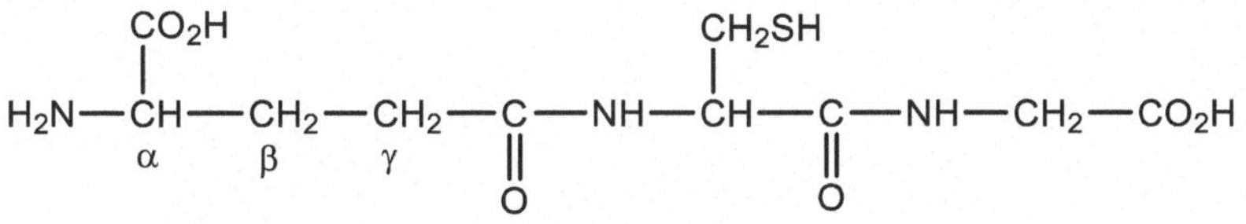

20.15 Draw a parallel β-sheet showing the network of hydrogen bonds between adjacent β-strands. How many atoms are present in the hydrogen-bonded ring structures?

[*Hint*: each ring is 12-membered]

20.16 Give the DNA triplets for the start codon (AUG) and the stop codons (UAA, UAG, UGA).
[*Ans*: TAC; ATT, ATC, ACT]

Further Reading

There are a large number of excellent organic textbooks, which can be used to consolidate the understanding of the subject. Some of the titles are given here.

GENERAL

G. M. Hornby and J. M. Peach, *Foundations of Organic Chemistry*, Oxford University Press: New York, 1992.

R. T. Morrison and R. N. Boyd, *Organic Chemistry*, 6th Edition, Prentice Hall, New Jersey, 1992; Adapted 7th Edition (co-author: S. K. Bhttacharjee), Pearson India Education, 2011.

T. W. G. Solomons, *Organic Chemistry*, 6th Edition, Wiley, New York, 1996; Global Edition (co-authors: C. B. Fryhle and S. A. Snyder), 2017.

S. H. Pine, *Organic Chemistry*, 5th Edition, McGraw Hill, New York, 1987.

J. Clayden, N. Greeves and S. Warren, *Organic Chemistry*, 2nd Edition, Oxford University Press, New York, 2012.

G. M. Loudon, *Organic Chemistry*, 4th Edition, Oxford University Press, New York, 2002.

P. Y. Bruice, *Organic Chemistry*, 6th Edition, Pearson, London, 2010.

J. McMurry, *Organic Chemistry*, 9th Edition, Brooks/Cole, California, 2015.

H. Hart, L. E. Craine and D. J. Hart, *Organic Chemistry – A short Course*, Houghton Mifflin Company, New York, 1999.

F. A. Carey and R. J. Sundberg, *Advanced Organic Chemistry*, 5th Edition, Part A, Springer, New York, 2007.

I. T. Sidhwani and R. K. Sharma, *An Introductory Text on Green Chemistry*, Wiley India, 2020.

STRUCTURE

W. G. Richards and P. R. Scott, *Energy Levels in Atoms and Molecules*, Oxford University Press, Oxford, 1994.

E. Cartmell and G. W. A. Fowles, *Valency and Molecular Structure*, 4th Edition, Butterworths, London, 1977.

M. Orchin, F. Kaplan, R. S. Macomber, R. M. Wilson and H. Zimmer, *The Vocabulary of Organic Chemistry*, Wiley, New York, 1980.

I. Fleming, *Molecular Orbitals and Organic Chemical Reactions*, Student Edition, Wiley, Chichester, 2009.

D. K. Mandal, *Pericyclic Chemistry: Orbital Mechanisms and Stereochemistry*, Chapter 1; Elsevier, Cambridge, 2018.

A. Bassendale, *The Third Dimension in Organic Chemistry*, Wiley, New York, 1984.

E. L. Eliel, S. H. Wilen and L. N. Mander, *Stereochemistry of Organic Compounds*, Wiley, New York, 1994.

M. J. T. Robinson, *Organic Stereochemistry*, Oxford University Press, Oxford, 2000.

D. K. Mandal, *Stereochemistry and Organic Reactions: Conformation, Configuration, Stereoelectronic Effects and Asymmetric Synthesis*, Chapters 1–3, Academic Press/ Elsevier, Cambridge, 2021.

W. Kemp, *Organic Spectroscopy*, 3rd Edition, ELBS, Hampshire, 1991.

S. B. Duckett and B. C. Gilbert, *Foundations of Spectroscopy*, Oxford University Press, New York, 2000.

L. M. Harwood and T. D. W. Claridge, *Introduction to Organic Spectroscopy*, Oxford University Press, New York, 1996.

REACTIONS

H. Maskill, *Mechanisms of Organic Reactions*, Oxford University Press, New York, 1996.

P. Sykes, *A Guidebook to Mechanism in Organic Chemistry*, 6th Edition, Longman, New York, 1988.

R. B. Grossman, *The Art of Writing Reasonable Organic Reaction Mechanisms*, 2nd Edition, Springer, New York, 2003.

C. J. Moody and G. H. Whitham, *Reactive Intermediates*, Oxford University Press, New York, 1992.

G. D. Meakins, *Functional Groups: Characteristics and Interconversions*, Oxford University Press, New York, 1996.

J. Jones, *Core Carbonyl Chemistry*, Oxford University Press, New York, 1997.

M. Sainsbury, *Aromatic Chemistry*, Oxford University Press, New York, 1992.

L. M. Harwood, *Polar Rearrangements*, Oxford University Press, New York, 1992.

T. J. Donohoe, *Oxidation and Reduction in Organic Synthesis*, Oxford University Press, New York, 2000.

SYNTHESIS

S. Warren, Organic Synthesis: The Disconnection Approach, Wiley, New York, 1982.

C. L. Willis and M. Wills, Organic Synthesis, Oxford University Press, New York, 1995.

R. O. C. Norman and J. M. Coxon, Principles of Organic Synthesis, 3rd Edition, ELBS, London, 1993.

W. Carruthers and I. Coldham, Modern Methods of Organic Synthesis, 4th Edition, Cambridge University Press, Cambridge, 2004.

J. Robertson, Protective Group Chemistry, Oxford University Press, New York, 2000.

J. Saunders, Top Drugs: Top Synthetic Routes, Oxford University Press, New York, 2000.

POLYMERS AND BIOMOLECULES

C. E. Carraher, Jr., *Introduction to Polymer Chemistry*, CRC Press, New York, 2006.

B. G. Davis and A. J. Fairbanks, *Carbohydrate Chemistry*, Oxford University Press, New York, 2002.

J. Jones, *Amino Acid and Peptide Synthesis*, Oxford University Press, New York, 1997.

D. Voet and J. G. Voet, *Biochemistry*, 3rd Edition, Wiley, New York, 2004.

Index

B

C

D

E

N

O

P

T

U

V

W

Z